Engineering Mechanics
STATICS
FOURTH EDITION

Anthony Bedford • Wallace Fowler
University of Texas at Austin

Upper Saddle River, New Jersey 07458

Library of Congress Cataloging-in-Publication Data
Bedford, A.
 Engineering mechanics: statics / Anthony Bedford and Wallace Fowler.—4th ed.
  p.   cm.
 Includes index.
 ISBN 0-13-146323-3
 1. Statics. I. Fowler, Wallace. II. Title.
CIP Data Available.

Vice President and Editorial Director, ECS: *Marcia J. Horton*
Executive Editor: *Eric Svendsen*
Associate Editor: *Dee Bernhard*
Vice President and Director of Production and Manufacturing, ESM: *David W. Riccardi*
Executive Managing Editor: *Vince O'Brien*
Managing Editor: *David A. George*
Marketing Manager: *Holly Stark*
Production Editor: *Craig Little*
Director of Creative Services: *Paul Belfanti*
Assistant Manager, Formatting: *Allyson Graesser*
Electronic Composition: *Clara Bartunek, Judith R. Wilkens*
Creative Director: *Carole Anson*
Art Director: *Kenny Beck*
Interior Designer: *Judith Matz-Coniglio*
Cover Designer: *Susan Anderson-Smith*
Art Editor: *Xiaohong Zhu*
Manufacturing Manager: *Trudy Pisciotti*
Manufacturing Buyer: *Lisa McDowell*

About the Cover: Photograph of City of Arts and Science Park, Valencia, Spain by Jose Fuste Rega/Corbis. Used by permission.

The author and publisher of this book have used their best efforts in preparing this book. These efforts include the development, research, and testing of the theories and programs to determine their effectiveness. The author and publisher make no warranty of any kind, expressed or implied, with regard to these programs or the documentation contained in this book. The author and publisher shall not be liable in any event for incidental or consequential damages in connection with, or arising out of, the furnishing, performance, or use of these programs.

Photo Credits: *Chapter 1, 2, 3, 4, 6, 8, and 9 Openers:* Tony Stone Images, *Chapter 1:* Figure 1.5, Couresy of NASA/JPL/Caltech, *Chapter 2:* Figure P2.25, CORBIS, Figure P2.121/2.122, NASA, *Chapter 3:* Figure P3.20, CORBIS; Figure 3.23, NASA; Figure 3.24, Photri Microstock, Inc., *Chapter 4:* Figure P4.185, CORBIS, *Chapter 5:* Opener, NASA Headquarters, *Chapter 6:* Opener, Photo Researchers, Inc.; Figure 6.3, Brownie Harris/The Stock Market; Figure 6.15 and 6.19, Marshall Henrichs; Figure 6.17, Pierre Berger/Photo Researchers, Inc., *Chapter 9:* Figure 9.13a, John Reader/Photo Researchers, Inc.; Figure 9.20a, Courtesy of SKF Industries; Figure P9.47/9.48, Rainbow; Figure P9.125/P9.126, Omni-Photo Communications, Inc., *Chapter 10:* Opener, SuperStock, Inc.; Figure 10.19, Warner Dieterich/The Image Bank; Figure 10.23, G+J Images/The Image Bank; Figure 10.23(a), Steve Niedorf/The Image Bank, *Chapter 11:* Opener, Photo Researchers Inc. All other photos were taken by the authors.

MATLAB is a registered trademark of The MathWorks, Inc., 3 Apple Hill Drive, Natick, MA 01760-2098. Mathcad is a registered trademark of MathSoft Engineering and Education, 101 Main St., Cambridge, MA 02142-1521.All other product names, brand names, and company names are trademarks or registered trademarks of their respective owners.

Printed in the United States of America

10 9 8 7 6 5 4 3 2 1

ISBN 0-13-146323-3

Pearson Education Ltd., *London*
Pearson Education Australia Pty., Limited, *Sydney*
Pearson Education Singapore, Pte. Ltd
Pearson Education North Asia Ltd., *Hong Kong*
Pearson Education Canada, Ltd., *Toronto*
Pearson Educación de Mexico, S.A. de C.V.
Pearson Education—Japan, *Tokyo*
Pearson Education Malaysia, Pte. Ltd.
Pearson Education, *Upper Saddle River, New Jersey*

CONTENTS

Preface   viii

About the Authors   xiii

1   Introduction   3

1.1   Engineering and Mechanics   4

1.2   Learning Mechanics   4
Problem Solving   4
Calculators and Computers   5
Engineering Applications   5
Subsequent Use of This Text   5

1.3   Fundamental Concepts   5
Numbers   5
Space and Time   6
Newton's Laws   6

1.4   Units   8
International System of Units   8
U.S. Customary Units   8
Angular Units   9
Conversion of Units   9

1.5   Newtonian Gravitation   14

2   Vectors   18

Vector Operations and Definitions   20

2.1   Scalars and Vectors   20

2.2   Rules for Manipulating Vectors   20
Vector Addition   21
Product of a Scalar and a Vector   22
Vector Subtraction   23
Unit Vectors   23

Cartesian Components   28

2.3   Components in Two Dimensions   28
Manipulating Vectors in Terms of Components   28
Position Vectors in Terms of Components   29

2.4   Components in Three Dimensions   42
Magnitude of a Vector in Terms of Components   43
Direction Cosines   43
Position Vectors in Terms of Components   44
Components of a Vector Parallel to a Given Line   45

Products of Vectors   59

2.5   Dot Products   59
Definition   59
Dot Products in Terms of Components   59
Vector Components Parallel and Normal to a Line   60

2.6   Cross Products   66
Definition   66
Cross Products in Terms of Components   67
Evaluating a 3 × 3 Determinant   68

2.7   Mixed Triple Products   69
Chapter Summary   *76*
Review Problems   *78*

3   Forces   80

3.1   Types of Forces   82
Terminology   82
Gravitational Forces   82
Contact Forces   83

3.2   Analysis of Force   87
Equilibrium   87
Free-Body Diagrams   87

3.3   Two-Dimensional Force Systems   89

3.4   Three-Dimensional Force Systems   105
COMPUTATIONAL MECHANICS   114
Chapter Summary   *120*
Review Problems   *121*

4   Systems of Forces and Moments   126

4.1   Two-Dimensional Description of the Moment   128

4.2   The Moment Vector   141
Magnitude of the Moment   141
Direction of the Moment   141
Relation to the Two-Dimensional Description   143
Varignon's Theorem   144

4.3   Moment of a Force About a Line   154
Definition   154
Applications   156

4.4   Couples   168

4.5   Equivalent Systems   178
Conditions for Equivalence   178
Demonstration of Equivalence   178

4.6   Representing Systems by Equivalent Systems   183
Representing a System by a Force and a Couple   183

Representing a System by a Wrench   188
COMPUTATIONAL MECHANICS   200
Chapter Summary   203
Review Problems   205

---

5   Objects in Equilibrium   210

5.1   The Equilibrium Equations   212

5.2   Two-Dimensional Applications   212
Supports   212
Free-Body Diagrams   216
The Scalar Equilibrium Equations   217

5.3   Statically Indeterminate Objects   237
Redundant Supports   238
Improper Supports   240

5.4   Three-Dimensional Applications   243
Supports   243
The Scalar Equilibrium Equations   248

5.5   Two-Force and Three-Force Members   260
Two-Force Members   260
Three-Force Members   262
COMPUTATIONAL MECHANICS   267
Chapter Summary   270
Review Problems   272

---

6   Structures in Equilibrium   276

6.1   Trusses   278

6.2   The Method of Joints   280

6.3   The Method of Sections   292

6.4   Space Trusses   298

6.5   Frames and Machines   302
Analyzing the Entire Structure   303
Analyzing the Members   303
COMPUTATIONAL MECHANICS   325
Chapter Summary   329
Review Problems   330

---

7   Centroids and Centers of Mass   334

Centroids   336

7.1   Centroids of Areas   337

7.2   Centroids of Composite Areas   344

7.3   Distributed Loads   351

Describing a Distributed Load   351
Determining Force and Moment   351
The Area Analogy   352

7.4   Centroids of Volumes and Lines   358
Definitions   358
Centroids of Composite Volumes and Lines   363

7.5   The Pappus–Guldinus Theorems   369
First Theorem   369
Second Theorem   370

Centers of Mass   374

7.6   Definition of the Center of Mass   374

7.7   Centers of Mass of Objects   375

7.8   Centers of Mass of Composite Objects   380
Chapter Summary   388
Review Problems   389

8   Moments of Inertia   394

Areas   396

8.1   Definitions   396

8.2   Parallel-Axis Theorems   402

8.3   Rotated and Principal Axes   416
Rotated Axes   417
Moment of Inertia About the x' Axis   417
Moment of Inertia About the y' Axis   418
Principal Axes   418
Mohr's Circle   422

Masses   428

8.4   Simple Objects   428
Slender Bars   428
Thin Plates   429

8.5   Parallel-Axis Theorem   432
Chapter Summary   441
Review Problems   442

9   Friction   446

9.1   Theory of Dry Friction   448
Coefficients of Friction   449
Angles of Friction   451

9.2   Applications   469
Wedges   469
Threads   471
Journal Bearings   479
Thrust Bearings and Clutches   481

Belt Friction   488
COMPUTATIONAL MECHANICS   496
Chapter Summary   *500*
Review Problems   *501*

10   Internal Forces and Moments   506

Beams   508

10.1   Axial Force, Shear Force, and Bending Moment   508

10.2   Shear Force and Bending Moment Diagrams   514

10.3   Relations Between Distributed Load, Shear Force, and Bending Moment   520
Construction of the Shear Force Diagram   522
Construction of the Bending Moment Diagram   524

Cables   532

10.4   Loads Distributed Uniformly Along Straight Lines   533
Shape of the Cable   533
Tension of the Cable   534
Length of the Cable   534

10.5   Loads Distributed Uniformly Along Cables   537
Shape of the Cable   537
Tension of the Cable   538
Length of the Cable   538

10.6   Discrete Loads   542
Determining the Configuration and Tensions   542
Comments on Continuous and Discrete Models   543
COMPUTATIONAL MECHANICS   547

Liquids and Gases   550

10.7   Pressure and the Center of Pressure   550

10.8   Pressure in a Stationary Liquid   552
Chapter Summary   *561*
Review Problems   *563*

11   Virtual Work and Potential Energy   566

11.1   Virtual Work   568
Work   568
Principle of Virtual Work   569
Application to Structures   571

11.2   Potential Energy   580
Examples of Conservative Forces   580
Principle of Virtual Work for Conservative Forces   582
Stability of Equilibrium   582

COMPUTATIONAL MECHANICS   590
Chapter Summary   592
Review Problems   593

APPENDICES

A   Review of Mathematics   597

A.1   Algebra   597
Quadratic Equations   597
Natural Logarithms   597

A.2   Trigonometry   598

A.3   Derivatives   598

A.4   Integrals   599

A.5   Taylor Series   600

B   Properties of Areas and Lines   601

B.1   Areas   601

B.2   Lines   603

C   Properties of Volumes and Homogeneous Objects   604

Answers to Even-Numbered Problems   607

Index   617

Preface

Our original objective in writing this book was to present the foundations and applications of statics as we do in the classroom. We used many sequences of figures, emulating the gradual development of a figure by a teacher explaining a concept. We stressed the importance of visual analysis in gaining understanding, especially through the use of free-body diagrams. Because inspiration is so conducive to learning, we based many of our examples and problems on a variety of modern engineering applications. With encouragement and help from many students and fellow teachers who have used the book, we continue and expand upon these themes in this edition.

Examples That Teach

Each of our examples follows a three-part framework—**Strategy/Solution/Critical Thinking**—that is designed to help students develop engineering problem-solving skills. In the Strategy sections, we demonstrate how to plan the solution to a problem. The Solution presents the detailed steps needed to arrive at the required results. Experienced instructors know that many aspects of a given problem can be explained most effectively after it has been solved. With the Critical Thinking sections, we introduce this important element of classroom teaching. We point out important features of solutions, indicate how the methods used can be extended to other types of problems, and comment on the engineering motivations of particular problems.

Further, you will find each section ends with a section of simple "Study Questions" designed to help students understand their reading.

Engineering Design

In recent years, many teachers have begun to introduce concepts from engineering design in their mechanics courses, and this trend is being further encouraged by ABET requirements. Without compromising our emphasis on fundamental mechanics, we include design considerations in many examples and problems. Optional Design Examples provide more detailed discussions of applications of statics in engineering design. Brief sections of problems called Design Experiences are based on the Design Examples and can be assigned at the discretion of the instructor. Many chapters conclude with Design Projects that offer students more extensive participation in design.

Computational Mechanics

Some instructors prefer to teach statics without emphasizing the use of a computer. Others use statics as an opportunity to introduce students to the use of computers in engineering, having them either write their own programs in a lower level language or use higher level problem-solving software. Our book is suitable for each of these approaches. We provide optional, self-contained Computational Mechanics sections with examples and problems designed for solution by a programmable calculator or computer. In addition, tutorials on using Mathcad and MATLAB in engineering mechanics are available on the web. See the Supplements section for further information.

Consistent Use of Color

To help students recognize and interpret elements of figures, we use consistent identifying colors:

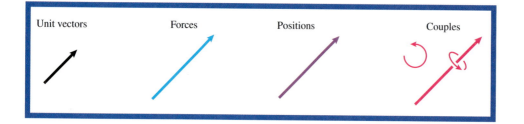

New to the Fourth Edition

Positive responses from users and reviewers have led us to retain the basic organization, content, and features of previous editions. During our preparation of this edition, we reexamined how we presented each concept, example, figure, summary statement, and problem. Where necessary, we made changes, additions, or deletions to simplify and clarify the presentation. In response to requests, we made the following notable changes:

Critical Thinking Sections—We have revised and expanded the discussion sections featured in our past editions' examples to provide a stronger emphasis on critical thinking. Strategy and Critical Thinking sections are now featured in every example. Each Critical Thinking section was reviewed by Chad M. Landis, Rice University, and Walter Gerstle, University of New Mexico, who provided us with feedback and suggestions for making them more helpful and educational for students.

New Art Program—Nearly every reviewer emphasized that their students needed help visualizing engineering situations. They also noted that students responded more favorably to situations appearing more "real." In response, we have tried to rework the key figures in this text with exceptional "Photo-realistic" clarity. We hope this will better help students visualize situations, and provide a stronger connection to the real world.

Sequences of Figures in Examples—In text material and the examples, we continue and expand upon our use of sequences of figures to achieve the clarity of classroom presentations of concepts.

In addition,
- We have added new examples where users indicated more were needed. Many of the new examples continue our emphasis on realistic and motivational applications and engineering design.
- We include 450 new and revised problems. As with the examples, many of the new problems focus on placing statics within the context of engineering practice. In this edition, problems that are relatively lengthier or more difficult have been indicated with an asterisk.
- An extensive new supplement program now includes OneKey—a web-based system that helps you better manage your courses, assess student progress, and more. See the Supplements description for complete information.

Commitment to Students and Instructors

In revising the textbook and solution manual, we have taken precautions to ensure accuracy to the best of our ability. We have each solved the new problems in an effort to be sure that their answers are correct and that they are of an appropriate level of difficulty. Karim Nohra of the University of South Florida, Scott Hendricks of VPI, and Kurt Norlin of Laurel Technical Services further verified the text, examples, problems, and solutions manual to help ensure accuracy. Any errors that remain are the responsibility of the authors. We welcome communication from students and instructors concerning errors or areas for improvement. Our mailing address is Department of Aerospace Engineering and Engineering Mechanics, University of Texas at Austin, Austin, Texas 78712. Our electronic mail address is abedford@mail.utexas.edu

Supplements

Student Supplements

The *Statics* Study Pack is designed to give students the tools to improve their skills drawing free-body diagrams, and to help them review for tests. (The *Statics* Study Pack is available bundled at no additional cost with the fourth edition). It contains a tutorial on free-body diagrams with 50 practice problems of increasing difficulty with complete solutions. Further strategies and tips help students understand how to use the diagrams in solving the accompanying problems. This supplement and accompanying chapter-by-chapter review material was prepared by Peter Schiavone of the University of Alberta

The *Statics* Study Pack is also available as a stand-alone item. Order stand-alone Study Packs with the ISBN 0-13-150287-5.

MATLAB and Mathcad Tutorials—Each tutorial discusses a basic mechanics concept, and then shows how to solve a specific problem related to this concept using MATLAB and Mathcad. There are 20 tutorials each for MATLAB and Mathcad, and are available to instructors in PDF format for distribution to students. Worksheets were developed by Ronald Larsen and Stephen Hunt of Montana State University–Bozeman.

Instructor Supplements

OneKey–OneKey is an on-line solution perfect for helping manage your class and preparing lectures, quizzes, and tests. Using OneKey, professors can quickly create an online course tailored to their specific needs. OneKey contains complete electronic solutions files of homework and makes it easy for you to post homework solutions at a protected online site for student review. Five hundred additional mechanics problems and solutions—organized by chapter, topic, and level of difficulty—are also available for student study and test prep. Further, OneKey gives you access to a Math Topic review, MATLAB and Mathcad student tutorials, animations, simulations, the Research Navigator, and PHGradeAssist, Prentice Hall's on-line algorithmic homework generator.

To learn more about OneKey, visit www.prenhall.com/onekey, or contact your local PH rep. Prentice Hall will be happy to host a OneKey site for you, or OneKey is available in course management cartridges for schools using this technology. Access and use for professors is free. Student access codes are free with their Bedford/Fowler textbooks, or available for sale as stand-alone items. For further information and ordering ISBNs, contact your local PH rep., or email engineering@prenhall.com.

Web Assessment Software—Using **PHGradeAssist**, students solve problems from the text that have randomized parameters, so that each student solves a problem with slightly different numbers. After students have submitted their answers, they receive the correct answers and, if necessary, can continue to work similar problems until they are successful. By using the optional course management system, instructors can have students' results recorded electronically. Contact your PH representative for more information. This supplement is available through the Bedford/Fowler OneKey system.

Instructor's Solutions Manual—This supplement, available to instructors, contains completely worked-out solutions. Each solution comes with the problem statement as well as associated artwork.

Instructors' Resource Center on CD—This CD contains PowerPoint slides and JPEG files of all art from the text. It also contains sets of PowerPoint slides showing each example and electronic files of solutions.

Acknowledgments

The following colleagues provided reviews based on their knowledge and teaching experience that greatly helped us in preparing the fourth editions of *Statics* and *Dynamics*.

Shaaban Abdallah
University of Cincinnati
George G. Adams
Northeastern University
Haim Baruh
Rutgers University
David M. Bayer
University of North Carolina
Glenn Beltz
University of California–Santa Barbara
Mitsunori Denda
Rutgers University
Bogdan I. Epureanu
University of Michigan
Walter Gerstle
University of New Mexico
Paul R. Heyliger
Colorado State University
Robert W. Hinks
Arizona State University
Chad M. Landis
Rice Unversity
John B. Ligon
Michigan Tech University

Mark T. Lusk
Colorado School of Mines
Nels Madsen
Auburn University
James R. Matthews
University of New Mexico
Mohammad Noori
North Carolina State University
Corrado Poli
University of Massachusetts–Amherst
Yitshak Ram
Louisiana State University
Edwin C. Rossow
Northwestern University
Kenneth Sawyers
Lehigh University
Richard A. Scott
University of Michigan
Brian Self
U.S. Air Force Academy
William Semke
University of North Dakota
Dennis VandenBrink
Western Michigan University

Reviewers for the Previous Editions

Students and instructors have made insightful comments and suggested improvements that we have incorporated into this edition. The following academic colleagues have critically reviewed the book and given us many valuable suggestions:

Edward E. Adams
Michigan Technological University
Raid S. Al-Akkad
University of Dayton
Jerry L. Anderson
Memphis State University
James G. Andrews
University of Iowa
Robert J. Asaro
University of California, San Diego
Leonard B. Baldwin
University of Wyoming
Gautam Batra
University of Nebraska
Mary Bergs
Marquette University

Spencer Brinkerhoff
Northern Arizona University
L.M. Brock
University of Kentucky
William (Randy) Burkett
Texas Tech University
Donald Carlson
University of Illinois
Major Robert M. Carpenter
U.S. Military Academy
Douglas Carroll
University of Missouri, Rolla
Paul C. Chan
New Jersey Institute of Technology
Namas Chandra
Florida State University

James Cheney
University of California, Davis

Ravinder Chona
Texas A & M University

Anthony DeLuzio
Merrimack College

Mitsunori Denda
Rutgers University

James F. Devine
University of South Florida

Craig Douglas
University of Massachusetts, Lowell

Marijan Dravinski
University of Southern California

S. Olani Durrant
Brigham Young University

Estelle Eke
California State University, Sacramento

William Ferrante
University of Rhode Island

Robert W. Fitzgerald
Worcester Polytechnic Institute

George T. Flowers
Auburn University

Mark Frisina
Wentworth Institute

Robert W. Fuessle
Bradley University

William Gurley
University of Tennessee, Chattanooga

John Hansberry
University of Massachusetts, Dartmouth

W. C. Hauser
California Polytechnic University, Pomona

Linda Hayes
University of Texas–Austin

R. Craig Henderson
Tennessee Technological University

James Hill
University of Alabama

Allen Hoffman
Worcester Polytechnic Institute

Edward E. Hornsey
University of Missouri, Rolla

Robert A. Howland
University of Notre Dame

Joe Ianelli
University of Tennessee, Knoxville

Ali Iranmanesh
Gadsden State Community College

David B. Johnson
Southern Methodist University

E. O. Jones, Jr.
Auburn University

Serope Kalpakjian
Illinois Institute of Technology

Kathleen A. Keil
California Polytechnic University, San Luis Obispo

Yohannes Ketema
University of Minnesota

Seyyed M. H. Khandani
Diablo Valley College

Charles M. Krousgrill
Purdue University

B. Kent Lall
Portland State University

Kenneth W. Lau
University of Massachusetts, Lowell

Norman Laws
University of Pittsburgh

William M. Lee
U.S. Naval Academy

Donald G. Lemke
University of Illinois, Chicago

Richard J. Leuba
North Carolina State University

Richard Lewis
Louisiana Technological University

Bertram Long
Northeastern University

V. J. Lopardo
U.S. Naval Academy

Frank K. Lu
University of Texas, Arlington

K. Madhaven
Christian Brothers College

Gary H. McDonald
University of Tennessee

James McDonald
Texas Technical University

Jim Meagher
California Polytechnic State University, San Luis Obispo

Lee Minardi
Tufts University

Norman Munroe
Florida International University

Shanti Nair
University of Massachusetts, Amherst

Saeed Niku
California Polytechnic State University, San Luis Obispo

Harinder Singh Oberoi
Western Washington University

James O'Connor
University of Texas, Austin

Samuel P. Owusu-Ofori
North Carolina A& T State University

Venkata Panchakarla
Florida State University

Assimina A Pelegri
Rutgers University
Noel C. Perkins
University of Michigan
David J. Purdy
Rose-Hulman Institute of Technology
Colin E Ratcliffe
U.S. Naval Academy
Daniel Riahi
University of Illinois
Charles Ritz
California Polytechnic State University, Pomona
George Rosborough
University of Colorado, Boulder
Robert Schmidt
University of Detroit
Robert J. Schultz
Oregon State University
Patricia M. Shamamy
Lawrence Technological University
Sorin Siegler
Drexel University
L. N. Tao
Illinois Institute of Technology
Craig Thompson
Western Wyoming Community College

John Tomko
Cleveland State University
Kevin Z. Truman
Washington University
John Valasek
Texas A &M University
Dennis VandenBrink
Western Michigan University
Thomas J. Vasko
University of Hartford
Mark R. Virkler
University of Missouri, Columbia
William H. Walston, Jr.
University of Maryland
Reynolds Watkins
Utah State University
Charles White
Northeastern University
Norman Wittels
Worcester Polytechnic Institute
Julius P. Wong
University of Louisville
T. W. Wu
University of Kentucky
Constance Ziemian
Bucknell University

We have again had the pleasure of working with friendly, professional, and energetic people at Prentice Hall. Our editor, Eric Svendsen, provided philosophical guidance in addition to organizing and managing the large collaborative effort required by a book of this kind. Brian Hoehl provided skillful liaison between Eric and us and kept the work moving smoothly. Dee Bernhard was our interface to the reviewers, and was instrumental in making that process work well. Xiaohong Zhu made important contributions to the new art program and helped with engineer/artist communications. Craig Little devoted large amounts of time and effort to shepherding the book through the production process and to helping us with its many details. Susan Anderson-Smith provided graceful cover designs. Kurt Norlin helped us with checking, made many editorial suggestions, and prepared the list of answers. Scott Hendricks and Karim Nohra joined us in checking everything and also oversaw the revision of the solution manual. We are grateful to the developers of the supplements that are so helpful to students. Peter Schiavone developed the *Statics* Study Pack that accompanies the book, and Stephen Hunt and Ronald Larsen wrote the MATLAB/Mathcad tutorials.

We thank the many users of the previous editions who have shared their experiences with us and suggested revisions. And we again thank Nancy and Marsha, who are still waiting for us to stop working weekends.

Anthony Bedford and Wallace Fowler
Austin, Texas

About the Authors

Anthony Bedford is Professor Emeritus with the Department of Aerospace Engineering and Engineering Mechanics at the University of Texas at Austin. He received the B.S. degree from the University of Texas at Austin, the M.S. degree from the California Institute of Technology, and the Ph.D. degree at Rice University in 1967. He has industrial experience at Douglas Aircraft Company, TRW, and Sandia National Laboratories. He has been on the faculty of the University of Texas at Austin since 1968.

Dr. Bedford's main professional activity has been education and research in engineering mechanics. He has written technical papers on mixture theory, wave propagation, and the mechanics of high velocity impacts, and is the author of the books *Hamilton's Principle in Continuum Mechanics, Introduction to Elastic Wave Propagation* (with D. S. Drumheller), and *Mechanics of Materials* (with K. M. Liechti).

Wallace T. Fowler holds the Paul D. and Betty Robertson Meek Professorship in Engineering with the Department of Aerospace Engineering and Engineering-Mechanics at the University of Texas at Austin. Dr. Fowler received the B.A., M.S., and Ph.D. degrees from the University of Texas at Austin, and has been on the faculty there since 1965. During the fall of 1976, he was on the staff of the United States Air Force Test Pilot School, Edwards Air Force Base, California, and during 1981–1982, he was a visiting professor at the United States Air Force Academy. He is currently the Director of the Texas Space Grant Consortium.

Dr. Fowler's areas of teaching and research are dynamics, orbital mechanics, and spacecraft mission design. He is author or coauthor of technical papers on trajectory optimization, attitude dynamics, and space mission planning and has also published papers on the theory and practice of engineering teaching. He has received numerous teaching awards including the Chancellor's Council Outstanding Teaching Award, the General Dynamics Teaching Excellence Award, the Halliburton Education Foundation Award of Excellence, the ASEE Fred Merryfield Design Award, and the AIAA-ASEE Distinguished Aerospace Educator Award. He is a member of the Academy of Distinguished Teachers at the University of Texas at Austin. He is a licensed professional engineer, a member of several technical societies, and a Fellow of both the American Institute of Aeronautics and Astronautics and the American Society for Engineering Education. During 2000–2001, he served as president of the American Society for Engineering Education.

Engineering Mechanics
STATICS

Introduction

How do engineers design and construct the devices we use, from simple objects such as chairs and pencil sharpeners to complicated ones such as dams, cars, airplanes, and spacecraft? They must have a deep understanding of the physics underlying the design of these devices, and they must be able to use mathematical models to predict their behavior. Students of engineering begin to learn how to analyze and predict the behaviors of physical systems by studying mechanics.

◀ The architects and engineers are guided by the principles of statics during each step of the design and construction of a building. Statics is one of the sciences underlying the art of structural design.

1.1 Engineering and Mechanics

How can engineers design complex systems and predict their characteristics before they are constructed? Engineers have always relied on their knowledge of previous designs, experiments, ingenuity, and creativity to develop new designs. Modern engineers add a powerful technique: They develop mathematical equations based on the physical characteristics of the devices they design. With these mathematical models, engineers predict the behavior of their designs, modify them, and test them prior to their actual construction. Aerospace engineers use mathematical models to predict the paths the space shuttle will follow in flight. Civil engineers use mathematical models to analyze the effects of loads on buildings and foundations.

At its most basic level, mechanics is the study of forces and their effects. Elementary mechanics is divided into *statics*, the study of objects in equilibrium, and *dynamics*, the study of objects in motion. The results obtained in elementary mechanics apply directly to many fields of engineering. Mechanical and civil engineers who design structures use the equilibrium equations derived in statics. Civil engineers who analyze the responses of buildings to earthquakes and aerospace engineers who determine the trajectories of satellites use the equations of motion derived in dynamics.

Mechanics was the first analytical science; consequently fundamental concepts, analytical methods, and analogies from mechanics are found in virtually every field of engineering. Students of chemical and electrical engineering gain a deeper appreciation for basic concepts in their fields, such as equilibrium, energy, and stability, by learning them in their original mechanical contexts. By studying mechanics, they retrace the historical development of these ideas.

1.2 Learning Mechanics

Mechanics consists of broad principles that govern the behavior of objects. In this book we describe these principles and provide examples that demonstrate some of their applications. Although it is essential that you practice working problems similar to these examples, and we include many problems of this kind, our objective is to help you understand the principles well enough to apply them to situations that are new to you. Each generation of engineers confronts new problems.

Problem Solving

In the study of mechanics you learn problem-solving procedures you will use in succeeding courses and throughout your career. Although different types of problems require different approaches, the following steps apply to many of them:

- Identify the information that is given and the information, or answer, you must determine. It's often helpful to restate the problem in your own words. When appropriate, make sure you understand the physical system or model involved.
- Develop a *strategy* for the problem. This means identifying the principles and equations that apply and deciding how you will use them to solve the problem. Whenever possible, draw diagrams to help visualize and solve the problem.
- Whenever you can, try to predict the answer. This will develop your intuition and will often help you recognize an incorrect answer.
- Solve the equations and, whenever possible, interpret your results and compare them with your prediction. This last step is a *reality check*. Is your answer reasonable?

Calculators and Computers

Most of the problems in this book are designed to lead to an algebraic expression with which to calculate the answer in terms of given quantities. A calculator with trigonometric and logarithmic functions is sufficient to determine the numerical value of such answers. The use of a programmable calculator or a computer with problem-solving software such as Mathcad or MATLAB is convenient, but be careful not to become too reliant on tools you will not have during tests.

Sections headed "Computational Mechanics" contain examples and problems that are suitable for solution with a programmable calculator or a computer.

Engineering Applications

Although the problems are designed primarily to help you learn mechanics, many of them illustrate uses of mechanics in engineering. Sections headed "Application to Engineering" describe how mechanics is applied in various fields of engineering.

We also include problems that emphasize two essential aspects of engineering:

- *Design.* Some problems ask you to choose values of parameters to satisfy stated design criteria.
- *Safety.* Some problems ask you to evaluate the safety of devices and choose values of parameters to satisfy stated safety requirements.

Subsequent Use of This Text

This book contains tables and information you will find useful in subsequent engineering courses and throughout your engineering career. In addition, you will often want to review fundamental engineering subjects, both during the remainder of your formal education and when you are a practicing engineer. The most efficient way to do so is by using the textbooks with which you are familiar. Your engineering textbooks will form the core of your professional library.

1.3 Fundamental Concepts

Some topics in mechanics will be familiar to you from everyday experience or from previous exposure to them in mathematics and physics courses. In this section we briefly review the foundations of elementary mechanics.

Numbers

Engineering measurements, calculations, and results are expressed in numbers. You need to know how we express numbers in the examples and problems and how to express the results of your own calculations.

Significant Digits This term refers to the number of meaningful (that is, accurate) digits in a number, counting to the right starting with the first nonzero digit. The two numbers 7.630 and 0.007630 are each stated to four significant digits. If only the first four digits in the number 7,630,000 are known to be accurate, this can be indicated by writing the number in scientific notation as 7.630×10^6.

If a number is the result of a measurement, the significant digits it contains are limited by the accuracy of the measurement. If the result of a measurement is stated to be 2.43, this means that the actual value is believed to be closer to 2.43 than to 2.42 or 2.44.

Numbers may be rounded off to a certain number of significant digits. For example, we can express the value of π to three significant digits, 3.14, or we can express it to six significant digits, 3.14159. When you use a calculator or computer, the number of significant digits is limited by the number of digits the machine is designed to carry.

Use of Numbers in This Book   You should treat numbers given in problems as exact values and not be concerned about how many significant digits they contain. If a problem states that a quantity equals 32.2, you can assume its value is 32.200 . . . .We generally express intermediate results and answers in the examples and the answers to the problems to at least three significant digits. If you use a calculator, your results should be that accurate. Be sure to avoid round-off errors that occur if you round off intermediate results when making a series of calculations. Instead, carry through your calculations with as much accuracy as you can by retaining values in your calculator.

Space and Time

Space simply refers to the three-dimensional universe in which we live. Our daily experiences give us an intuitive notion of space and the locations, or positions, of points in space. The distance between two points in space is the length of the straight line joining them.

Measuring the distance between points in space requires a unit of length. We use both the International System of units, or SI units, and U.S. Customary units. In SI units, the unit of length is the meter (m). In U.S. Customary units, the unit of length is the foot (ft).

Time is, of course, familiar—our lives are measured by it. The daily cycles of light and darkness and the hours, minutes, and seconds measured by our clocks and watches give us an intuitive notion of time. Time is measured by the intervals between repeatable events, such as the swings of a clock pendulum or the vibrations of a quartz crystal in a watch. In both SI units and U.S. Customary units, the unit of time is the second (s). The minute (min), hour (h), and day are also frequently used.

If the position of a point in space relative to some reference point changes with time, the rate of change of its position is called its *velocity*, and the rate of change of its velocity is called its *acceleration*. In SI units, the velocity is expressed in meters per second (m/s) and the acceleration is expressed in meters per second per second, or meters per second squared (m/s^2). In U.S. Customary units, the velocity is expressed in feet per second (ft/s) and the acceleration is expressed in feet per second squared (ft/s^2).

Newton's Laws

Elementary mechanics was established on a firm basis with the publication in 1687 of *Philosophiae Naturalis Principia Mathematica*, by Isaac Newton. Although highly original, it built on fundamental concepts developed by many others during a long and difficult struggle toward understanding (Fig. 1.1).

Newton stated three "laws" of motion, which we express in modern terms:

1. *When the sum of the forces acting on a particle is zero, its velocity is constant. In particular, if the particle is initially stationary, it will remain stationary.*

2. *When the sum of the forces acting on a particle is not zero, the sum of the forces is equal to the rate of change of the linear momentum of the particle. If the mass is constant, the sum of the forces is equal to the product of the mass of the particle and its acceleration.*

3. *The forces exerted by two particles on each other are equal in magnitude and opposite in direction.*

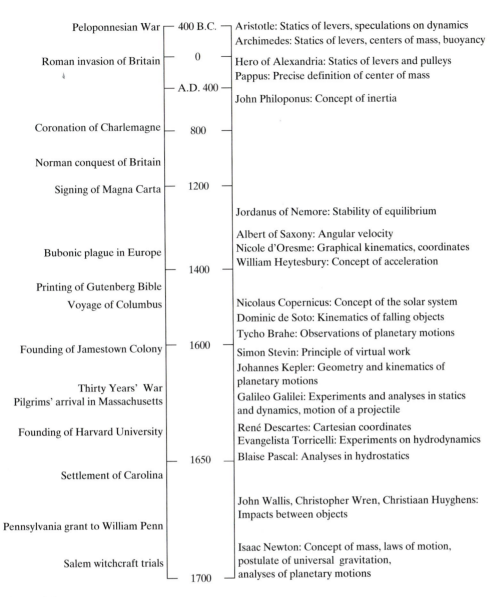

Figure 1.1
Chronology of developments in mechanics up to the publication of Newton's
Principia in relation to other events in history.

Notice that we did not define force and mass before stating Newton's laws. The
modern view is that these terms are defined by the second law. To demonstrate,
suppose that we choose an arbitrary object and define it to have unit mass. Then
we define a unit of force to be the force that gives our unit mass an acceleration
of unit magnitude. In principle, we can then determine the mass of any object:
We apply a unit force to it, measure the resulting acceleration, and use the second law to determine the mass. We can also determine the magnitude of any
force: We apply it to our unit mass, measure the resulting acceleration, and use
the second law to determine the force.

 Thus Newton's second law gives precise meanings to the terms *mass* and
force. In SI units, the unit of mass is the kilogram (kg). The unit of force is
the newton (N), which is the force required to give a mass of one kilogram an
acceleration of one meter per second squared. In U.S. Customary units, the unit

of force is the pound (lb). The unit of mass is the slug, which is the amount of mass accelerated at one foot per second squared by a force of one pound.

Although the results we discuss in this book are applicable to many of the problems met in engineering practice, there are limits to the validity of Newton's laws. For example, they don't give accurate results if a problem involves velocities that are not small compared to the velocity of light $(3 \times 10^8 \text{ m/s})$. Einstein's special theory of relativity applies to such problems. Elementary mechanics also fails in problems involving dimensions that are not large compared to atomic dimensions. Quantum mechanics must be used to describe phenomena on the atomic scale.

Study Questions

1. What is the definition of the significant digits of a number?
2. What are the units of length, mass, and force in the SI system?

1.4 Units

The SI system of units has become nearly standard throughout the world. In the United States, U.S. Customary units are also used. In this section we summarize these two systems of units and explain how to convert units from one system to another.

International System of Units

In SI units, length is measured in meters (m) and mass in kilograms (kg). Time is measured in seconds (s), although other familiar measures such as minutes (min), hours (h), and days are also used when convenient. Meters, kilograms, and seconds are called the *base units* of the SI system. Force is measured in newtons (N). Recall that these units are related by Newton's second law: One newton is the force required to give an object of one kilogram mass an acceleration of one meter per second squared:

$$1 \text{ N} = (1 \text{ kg})(1 \text{ m/s}^2) = 1 \text{ kg-m/s}^2.$$

Because the newton can be expressed in terms of the base units, it is called a *derived unit.*

To express quantities by numbers of convenient size, multiples of units are indicated by prefixes. The most common prefixes, their abbreviations, and the multiples they represent are shown in Table 1.1. For example, 1 km is 1 kilometer, which is 1000 m, and 1 Mg is 1 megagram, which is 10^6 g, or 1000 kg. We frequently use kilonewtons (kN).

Table 1.1 The common prefixes used in SI units and the multiples they represent.

Prefix	Abbreviation	Multiple
nano-	n	10^{-9}
micro-	μ	10^{-6}
milli-	m	10^{-3}
kilo-	k	10^{3}
mega-	M	10^{6}
giga-	G	10^{9}

U.S. Customary Units

In U.S. Customary units, length is measured in feet (ft) and force is measured in pounds (lb). Time is measured in seconds (s). These are the base units of the U.S. Customary system. In this system of units, mass is a derived unit. The unit of mass is the slug, which is the mass of material accelerated at one foot per second squared by a force of one pound. Newton's second law states that

$$1 \text{ lb} = (1 \text{ slug})(1 \text{ ft/s}^2).$$

From this expression we obtain

$$1 \text{ slug} = 1 \text{ lb-s}^2/\text{ft}.$$

We use other U.S. Customary units such as the mile (1 mi = 5280 ft) and the inch (1 ft = 12 in). We also use the kilopound (kip), which is 1000 lb.

Angular Units

In both SI and U.S. Customary units, angles are normally expressed in radians (rad). We show the value of an angle θ in radians in Fig. 1.2. It is defined to be the ratio of the part of the circumference subtended by θ to the radius of the circle. Angles are also expressed in degrees. Since there are 360 degrees (360°) in a complete circle, and the complete circumference of the circle is $2\pi R$, 360° equals 2π rad.

Equations containing angles are nearly always derived under the assumption that angles are expressed in radians. Therefore, when you want to substitute the value of an angle expressed in degrees into an equation, you should first convert it into radians. A notable exception to this rule is that many calculators are designed to accept angles expressed in either degrees or radians when you use them to evaluate functions such as $\sin \theta$.

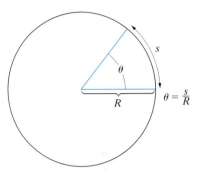

Figure 1.2
Definition of an angle in radians.

Conversion of Units

Many situations arise in engineering practice that require values expressed in one kind of unit to be converted into values in other units. For example, if some of the data to be used in an equation are given in SI units and some are given in U.S. Customary units, they must all be expressed in terms of one system of units before they are substituted into the equation. Converting units is straightforward, although it must be done with care.

Suppose that we want to express 1 mile per hour (mi/h) in terms of feet per second (ft/s). Because 1 mile equals 5280 feet and 1 hour equals 3600 seconds, we can treat the expressions

$$\left(\frac{5280 \text{ ft}}{1 \text{ mi}}\right) \text{ and } \left(\frac{1 \text{ h}}{3600 \text{ s}}\right)$$

as ratios whose values are 1. In this way, we obtain

$$1 \text{ mi/h} = (1 \text{ mi/h})\left(\frac{5280 \text{ ft}}{1 \text{ mi}}\right)\left(\frac{1 \text{ h}}{3600 \text{ s}}\right) = 1.47 \text{ ft/s}.$$

Some useful unit conversions are given in Table 1.2.

Table 1.2 Unit conversions.

Time	1 minute	=	60 seconds
	1 hour	=	60 minutes
	1 day	=	24 hours
Length	1 foot	=	12 inches
	1 mile	=	5280 feet
	1 inch	=	25.4 millimeters
	1 foot	=	0.3048 meters
Angle	2π radians	=	360 degrees
Mass	1 slug	=	14.59 kilograms
Force	1 pound	=	4.448 newtons

Study Questions

1. What are the base units of the SI and U.S. Customary systems?
2. What is the definition of an angle in radians?

Example 1.1   **Converting Units of Pressure**

The pressure exerted at a point of the hull of the deep submersible vehicle in Fig. 1.3 is 3.00×10^6 Pa (pascals). A pascal is 1 newton per square meter. Determine the pressure in pounds per square foot.

Figure 1.3
Deep Submersible Vehicle.

Strategy
From Table 1.2, 1 pound = 4.448 newtons and 1 foot = 0.3048 meters. With these unit conversions we can calculate the pressure in pounds per square foot.

Solution
The pressure (to three significant digits) is

$$3.00 \times 10^6 \text{ N/m}^2 = (3.00 \times 10^6 \text{ N/m}^2)\left(\frac{1 \text{ lb}}{4.448 \text{ N}}\right)\left(\frac{0.3048 \text{ m}}{1 \text{ ft}}\right)^2$$

$$= 62,700 \text{ lb/ft}^2.$$

Critical Thinking
How could we have obtained this result in a more direct way? Notice from the table of unit conversions in the inside front cover that 1 Pa = 0.0209 lb/ft^2. Therefore,

$$3.00 \times 10^6 \text{ N/m}^2 = (3.00 \times 10^6 \text{ N/m}^2)\left(\frac{0.0209 \text{ lb/ft}^2}{1 \text{ N/m}^2}\right)$$

$$= 62,700 \text{ lb/ft}^2.$$

Suppose that in Einstein's equation

$$E = mc^2,$$

the mass m is in kilograms and the velocity of light c is in meters per second.
(a) What are the SI units of E?
(b) If the value of E in SI units is 20, what is its value in U.S. Customary base units?

Strategy
(a) Since we know the units of the terms m and c, we can deduce the units of E from the given equation.
(b) We can use the unit conversions for mass and length from Table 1.2 to convert E from SI units to U.S. Customary units.

Solution
(a) From the equation for E,

$$E = (m \text{ kg})(c \text{ m/s})^2,$$

the SI units of E are kg-m^2/s^2.

(b) From Table 1.2, 1 slug = 14.59 kg and 1 ft = 0.3048 m. Therefore,

$$1 \text{ kg-m}^2/\text{s}^2 = (1 \text{ kg-m}^2/\text{s}^2)\left(\frac{1 \text{ slug}}{14.59 \text{ kg}}\right)\left(\frac{1 \text{ ft}}{0.3048 \text{ m}}\right)^2$$

$$= 0.738 \text{ slug-ft}^2/\text{s}^2.$$

The value of E in U.S. Customary units is

$$E = (20)(0.738) = 14.8 \text{ slug-ft}^2/\text{s}^2.$$

Critical Thinking
In part (a), how did we know that we could determine the units of E by determining the units of mc^2? The dimensions, or units, of each term in an equation must be the same. For example, in the equation $a + b = c$, the dimensions of each of the terms a, b, and c must be the same. The equation is said to be *dimensionally homogeneous*. This requirement is nicely expressed by the colloquial phrase "Don't compare apples and oranges."

Problems

1.1 Express the fractions $\frac{1}{3}$ and $\frac{2}{3}$ to three significant digits.

1.2 The base of natural logarithms is $e = 2.718281828\ldots$.
(a) Express e to five significant digits.
(b) Determine the value of e^2 to five significant digits.
(c) Use the value of e you obtained in part (a) to determine the value of e^2 to five significant digits.
[Part (c) demonstrates the hazard of using rounded-off values in calculations.]

1.3 A machinist drills a circular hole in a panel with radius $r = 5$ mm. Determine the circumference C and the area A of the hole to four significant digits.

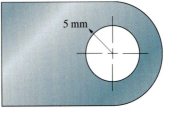

5 mm

Problem 1.3

1.4 The opening in the soccer goal is 24 ft wide and 8 ft high. Use these values to determine its dimensions in meters to three significant digits

Problem 1.4

1.5 The coordinates (in meters) of point A are $x_A = 3$, $y_A = 7$, and the coordinates of point B are $x_B = 10$, $y_B = 2$. Determine the length of the straight line from A to B to three significant digits.

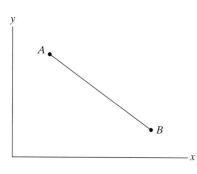

Problem 1.5

1.6 Suppose that you have just purchased a Ferrari F355 coupe and you want to know whether you can use your set of SAE (U.S. Customary unit) wrenches to work on it. You have wrenches with widths $w = 1/4$ in, $1/2$ in, $3/4$ in, and 1 in, and the car has nuts with dimensions $n = 5$ mm, 10 mm, 15 mm, 20 mm, and 25 mm. Defining a wrench to fit if w is no more than 2% larger than n, which of your wrenches can you use?

Problem 1.6

1.7 On August 20, 1974, Nolan Ryan threw the first baseball pitch measured at over 100 mi/h. The measured speed was 100.9 mi/h. Determine the speed of the pitch to four significant digits
(a) in ft/s;
(b) in km/h.

1.8 On March 18, 1999, an experimental Maglev (magnetic levitation) train in Japan reached a maximum speed of 552 km/h. What was its velocity in mi/h to three significant digits?

Problem 1.8

1.9 In May 1963, in the last flight of Project Mercury, Astronaut L. Gordon Cooper traveled a distance of 546,167 miles in 1 day, 10 hours, 19 minutes, and 49 seconds. Determine his average speed (the distance traveled divided by the time required) to three significant digits (a) in mi/h; (b) in km/h.

1.10 Engineers who study shock waves sometimes express velocity in millimeters per microsecond (mm/μs). Suppose the velocity of a wavefront is measured and determined to be 5 mm/μs. Determine its velocity (a) in m/s; (b) in mi/s.

1.11 The kinetic energy of a particle of mass m is defined to be $\frac{1}{2}mv^2$, where v is the magnitude of the particle's velocity. If the value of the kinetic energy of a particle at a given time is 200 when m is in kilograms and v is in meters per second, what is the value when m is in slugs and v is in feet per second?

1.12 The acceleration due to gravity at sea level in SI units is $g = 9.81$ m/s^2. By converting units, use this value to determine the acceleration due to gravity at sea level in U.S. Customary units.

1.13 A *furlong per fortnight* is a facetious unit of velocity, perhaps made up by a student as a satirical comment on the bewildering variety of units engineers must deal with. A furlong is 660 ft (1/8 mile). A fortnight is 2 weeks (14 nights). If you walk to class at 2 m/s, what is your speed in furlongs per fortnight to three significant digits?

1.14 The cross-sectional area of a beam is 480 in^2. What is its cross-sectional area in m^2?

1.15 The cross-sectional area of the C12×30 American Standard Channel steel beam is $A = 8.81$ in^2. What is its cross-sectional area in mm^2?

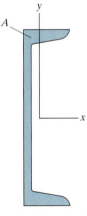

Problem 1.15

1.16 A pressure transducer measures a value of 300 lb/in^2. Determine the value of the pressure in pascals. A pascal (Pa) is one newton per meter squared.

1.17 A horsepower is 550 ft-lb/s. A watt is 1 N-m/s. Determine the number of watts generated by (a) the Wright brothers' 1903 airplane, which had a 12-horsepower engine, and (b) a modern passenger jet with a power of 100,000 horsepower at cruising speed.

Wright
Brothers' Flyer
(shown to scale)

Boeing 747

Problem 1.17

1.18 In SI units, the universal gravitational constant $G = 6.67 \times 10^{-11}$ N-m^2/kg^2. Determine the value of G in U.S. Customary base units.

1.19 The moment of inertia of the rectangular area about the x axis is given by the equation

$$I = \tfrac{1}{3}bh^3.$$

The dimensions of the area are $b = 200$ mm and $h = 100$ mm. Determine the value of I to four significant digits in terms of (a) mm^4, (b) m^4, and (c) in^4.

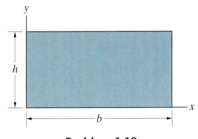

Problem 1.19

1.20 In the equation

$$T = \tfrac{1}{2}I\omega^2,$$

the term I is in kg-m^2 and ω is in s^{-1}.

(a) What are the SI units of T?

(b) If the value of T is 100 when I is in kg-m^2 and ω is in s^{-1}, what is the value of T when it is expressed in terms of U.S. Customary base units?

1.21 The equation

$$\sigma = \frac{My}{I}$$

is used in the mechanics of materials to determine normal stresses in beams.

(a) When this equation is expressed in terms of SI base units, M is in newton-meters (N-m), y is in meters (m), and I is in meters to the fourth power (m^4). What are the SI units of σ?

(b) If $M = 2000$ N-m, $y = 0.1$ m, and $I = 7 \times 10^{-5}$ m^4, what is the value of σ in U.S. Customary base units?

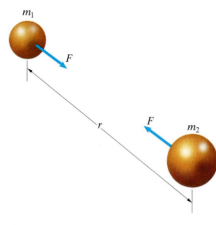

Figure 1.4
The gravitational forces between two particles are equal in magnitude and directed
along the line between them.

1.5 Newtonian Gravitation

Newton postulated that the gravitational force between two particles of mass m_1
and m_2 that are separated by a distance r (Fig. 1.4) is

$$F = \frac{Gm_1m_2}{r^2}, \tag{1.1}$$

where G is called the universal gravitational constant. Based on this postulate,
he calculated the gravitational force between a particle of mass m_1 and a homogeneous sphere of mass m_2 and found that it is also given by Eq. (1.1), with
r denoting the distance from the particle to the center of the sphere. Although
the earth is not a homogeneous sphere, we can use this result to approximate
the weight of an object of mass m due to the gravitational attraction of the
earth. We have

$$W = \frac{Gmm_E}{r^2}, \tag{1.2}$$

where m_E is the mass of the earth and r is the distance from the center of the
earth to the object. Notice that the weight of an object depends on its location
relative to the center of the earth, whereas the mass of the object is a measure
of the amount of matter it contains and doesn't depend on its position.

When an object's weight is the only force acting on it, the resulting acceleration is called the acceleration due to gravity. In this case, Newton's second law states that $W = ma$, and from Eq. (1.2) we see that the acceleration
due to gravity is

$$a = \frac{Gm_E}{r^2}. \tag{1.3}$$

The *acceleration due to gravity at sea level* is denoted by g. Denoting the
radius of the earth by R_E, we see from Eq. (1.3) that $Gm_E = gR_E^2$. Substituting
this result into Eq. (1.3), we obtain an expression for the acceleration due to
gravity at a distance r from the center of the earth in terms of the acceleration
due to gravity at sea level:

$$a = g\frac{R_E^2}{r^2}. \tag{1.4}$$

Since the weight of the object $W = ma$, the weight of an object at a distance
r from the center of the earth is

$$W = mg\frac{R_E^2}{r^2}. \tag{1.5}$$

At sea level $(r = R_E)$, the weight of an object is given in terms of its mass
by the simple relation

$$W = mg. \tag{1.6}$$

The value of g varies from location to location on the surface of the earth.
The values we use in examples and problems are $g = 9.81$ m/s^2 in SI units and
$g = 32.2$ ft/s^2 in U.S. Customary units.

Study Questions

1. Does the weight of an object depend on its location?
2. If you know an object's mass, how do you determine its weight at sea level?

Example 1.3   Determining an Object's Weight

When the Mars Exploration Rover was fully assembled, its mass was 180 kg. The acceleration due to gravity at the surface of Mars is 3.68 m/s^2 and the radius of Mars is 3390 km.

(a) What was the rover's weight when it was at sea level on Earth?

(b) What is the rover's weight on the surface of Mars?

(c) The entry phase began when the spacecraft reached the Mars atmospheric entry interface point at 3522 km from the center of Mars. What was the rover's weight at that point?

Figure 1.5
Mars Exploration Rover being assembled.

Strategy

The rover's weight at sea level on Earth is given by Eq. (1.6) with $g = 9.81$ m/s^2.

We can determine the weight on the surface of Mars by using Eq. (1.6) with the acceleration due to gravity equal to 3.68 m/s^2.

To determine the rover's weight as it began the entry phase, we can write an equation for Mars equivalent to Eq. (1.5).

Solution

(a) The weight at sea level on Earth is

$$W = mg$$

$$= (180 \text{ kg})(9.81 \text{ m/s}^2)$$

$$= 1770 \text{ N } (397 \text{ lb}).$$

(b) Let $g_M = 3.68$ m/s^2 be the acceleration due to gravity at the surface of Mars. Then the weight of the rover on the surface of Mars is

$$W = mg_M$$

$$= (180 \text{ kg})(3.68 \text{ m/s}^2)$$

$$= 662 \text{ N } (149 \text{ lb}).$$

(c) Let $R_M = 3390$ km be the radius of Mars. From Eq. (1.5), the rover's weight when it is 3522 km above the center of Mars is

$$W = mg_M \frac{R_M^2}{r^2}$$

$$= (180 \text{ kg})(3.68 \text{ m/s}^2)\frac{(3{,}390{,}000 \text{ m})^2}{(3{,}522{,}000 \text{ m})^2}$$

$$= 614 \text{ N } (138 \text{ lb}).$$

Critical Thinking

In part (c), how did we know that we could apply Eq. (1.5) to Mars? Equation (1.5) is applied to Earth based on modeling it as a homogeneous sphere. It can be applied to other celestial objects under the same assumption. The accuracy of the results depends on how aspherical and inhomogeneous the object is.

Problems

1.22 Let W be your weight at sea level in pounds. (a) What is your weight at sea level in newtons? (b) What is your mass in kilograms?

1.23 The acceleration due to gravity is 1.62 m/s^2 on the surface of the moon and 9.81 m/s^2 on the surface of the earth. A female astronaut's mass is 57 kg. What is the maximum allowable mass of her spacesuit and equipment if the engineers don't want the total weight on the moon of the woman, her spacesuit, and equipment to exceed 180 N?

1.24 A person has a mass of 50 kg.
(a) The acceleration due to gravity at sea level is $g = 9.81$ m/s^2. What is the person's weight at sea level?
(b) The acceleration due to gravity on the surface of the moon is 1.62 m/s^2. What would the person weigh on the moon?

1.25 The acceleration due to gravity at sea level is $g = 9.81$ m/s^2. The radius of the earth is 6370 km. The universal gravitational constant $G = 6.67 \times 10^{-11}$ N-m^2/kg^2. Use this information to determine the mass of the earth.

1.26 A person weighs 180 lb at sea level. The radius of the earth is 3960 mi. What force is exerted on the person by the gravitational attraction of the earth if he is in a space station in orbit 200 mi above the surface of the earth?

1.27 The acceleration due to gravity on the surface of the moon is 1.62 m/s^2. The radius of the moon is $R_M = 1738$ km. Determine the acceleration due to gravity of the moon at a point 1738 km above its surface.

Strategy: Write an equation equivalent to Eq. (1.4) for the acceleration due to gravity of the moon.

1.28 If an object is near the surface of the earth, the variation of its weight with distance from the center of the earth can often be neglected. The acceleration due to gravity at sea level is $g = 9.81$ m/s^2. The radius of the earth is 6370 km. The weight of an object at sea level is mg, where m is its mass. At what height above the surface of the earth does the weight of the object decrease to $0.99mg$?

1.29 The centers of two oranges are 1 m apart. The mass of each orange is 0.2 kg. What gravitational force do they exert on each other? (The universal gravitational constant $G = 6.67 \times 10^{-11}$ N-m^2/kg^2.)

1.30 At a point between the earth and the moon, the magnitude of the force exerted on an object by the earth's gravity equals the magnitude of the force exerted on the object by the moon's gravity. What is the distance from the center of the earth to that point to three significant digits? The distance from the center of the earth to the center of the moon is 383,000 km, and the radius of the earth is 6370 km. The radius of the moon is 1738 km, and the acceleration due to gravity at its surface is 1.62 m/s^2.

CHAPTER

2

Vectors

If an object is subjected to several forces that have different magnitudes and act in different directions, how can the magnitude and direction of the resulting total force on the object be determined? Forces are vectors and must be added according to the definition of vector addition. In engineering we deal with many quantities that have both magnitude and direction and can be expressed and analyzed as vectors. In this chapter we review vector operations, express vectors in terms of components, and present examples of engineering applications of vectors.

◄ Vectors can specify the positions of points of a structure. Vectors are used to describe and analyze quantities that have magnitude and direction, including positions, forces, moments, velocities, and accelerations.

VECTOR OPERATIONS AND DEFINITIONS

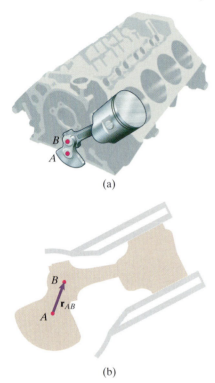

(a)

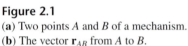

(b)

Figure 2.1
(**a**) Two points A and B of a mechanism.
(**b**) The vector $\mathbf{r}_{AB}$ from A to B.

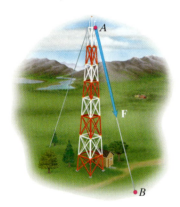

Figure 2.2
Representing the force cable AB exerts on
the tower by a vector $\mathbf{F}$.

Engineers designing a structure must analyze the positions of its members and the forces acting on them. When designing a machine, they must analyze the velocities and accelerations of its moving parts. These and many other physical quantities important in engineering can be represented by vectors and analyzed by vector operations. Here we review fundamental vector operations and definitions.

2.1 Scalars and Vectors

A physical quantity that is completely described by a real number is called a *scalar*. Time is a scalar quantity. Mass is also a scalar quantity. For example, you completely describe the mass of a car by saying that its value is 1200 kg.

In contrast, you have to specify both a nonnegative real number, or *magnitude*, and a direction to describe a vector quantity. Two vector quantities are equal only if both their magnitudes and their directions are equal.

The position of a point in space relative to another point is a vector quantity. To describe the location of a city relative to your home, it is not enough to say that it is 100 miles away. You must say that it is 100 miles west of your home. Force is also a vector quantity. When you push a piece of furniture across the floor, you apply a force of magnitude sufficient to move the furniture and you apply it in the direction you want the furniture to move.

We will represent vectors by boldfaced letters, $\mathbf{U}$, $\mathbf{V}$, $\mathbf{W}$, ..., and will denote the magnitude of a vector $\mathbf{U}$ by $|\mathbf{U}|$. A vector is represented graphically by an arrow. The direction of the arrow indicates the direction of the vector, and the length of the arrow is defined to be proportional to the magnitude. For example, consider the points A and B of the mechanism in Fig. 2.1a. We can specify the position of point B relative to point A by the vector $\mathbf{r}_{AB}$ in Fig. 2.1b. The direction of $\mathbf{r}_{AB}$ indicates the direction from point A to point B. If the distance between the two points is 200 mm, the magnitude $|\mathbf{r}_{AB}| = 200$ mm.

The cable AB in Fig. 2.2 helps support the television transmission tower. We can represent the force the cable exerts on the tower by a vector $\mathbf{F}$ as shown. If the cable exerts an 800-N force on the tower, $|\mathbf{F}| = 800$ N. (A cable suspended in this way will exhibit some sag, or curvature, and the tension will vary along its length. For now, we assume that the curvature in suspended cables and ropes and the variations in their tensions can be neglected. This assumption is approximately valid if the weight of the rope or cable is small in comparison to the tension. We discuss and analyze suspended cables and ropes in more detail in Chapter 10.)

2.2 Rules for Manipulating Vectors

Vectors are a convenient means for representing physical quantities that have magnitude and direction, but that is only the beginning of their usefulness. Just as real numbers are manipulated with the familiar rules for addition, subtraction, multiplication, and so forth, there are rules for manipulating vectors. These rules provide powerful tools for engineering analysis.

Vector Addition

When an object moves from one location in space to another, we say it under-goes a *displacement*. If we move a book (or, speaking more precisely, some point of a book) from one location on a table to another, as shown in Fig. 2.3a, we can represent the displacement by the vector **U**. The direction of **U** indicates the direction of the displacement, and $|\mathbf{U}|$ is the distance the book moves.

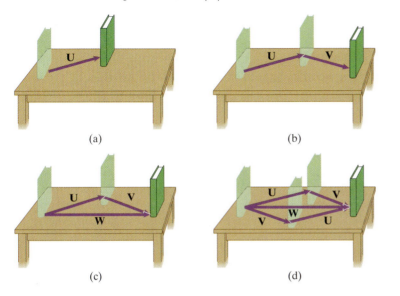

(a)

(b)

(c)

(d)

Figure 2.3
(a) A displacement represented by the vector **U**.
(b) The displacement **U** followed by the displacement **V**.
(c) The displacements **U** and **V** are equivalent to the displacement **W**.
(d) The final position of the book doesn't depend on the order of the displacements.

Suppose that we give the book a second displacement **V**, as shown in Fig. 2.3b. The two displacements **U** and **V** are equivalent to a single displace-ment of the book from its initial position to its final position, which we repre-sent by the vector **W** in Fig. 2.3c. Notice that the final position of the book is the same whether we first give it the displacement **U** and then the displacement **V** or we first give it the displacement **V** and then the displacement **U** (Fig. 2.3d). The displacement **W** is defined to be the sum of the displacements **U** and **V**:

$$\mathbf{U} + \mathbf{V} = \mathbf{W}.$$

The definition of vector addition is motivated by the addition of dis-placements. Consider the two vectors **U** and **V** shown in Fig. 2.4a. If we place them head to tail (Fig. 2.4b), their sum is defined to be the vector from the tail of **U** to the head of **V** (Fig. 2.4c). This is called the *triangle rule* for vector ad-dition. Figure 2.4d demonstrates that the sum is independent of the order in

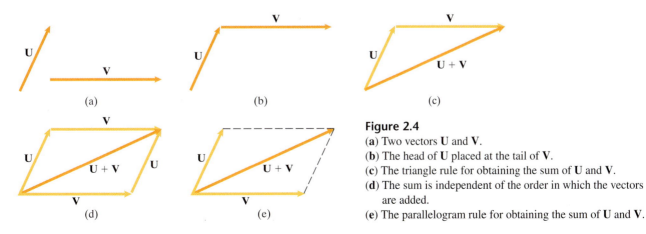

(a)

(b)

(c)

(d)

(e)

Figure 2.4
(a) Two vectors **U** and **V**.
(b) The head of **U** placed at the tail of **V**.
(c) The triangle rule for obtaining the sum of **U** and **V**.
(d) The sum is independent of the order in which the vectors are added.
(e) The parallelogram rule for obtaining the sum of **U** and **V**.

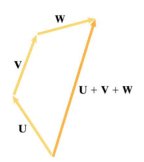

Figure 2.5
Sum of the three vectors **U**, **V**, and **W**.

Figure 2.6
Three vectors **U**, **V**, and **W** whose sum is zero.

which the vectors are placed head to tail. From this figure we obtain the *parallelogram rule* for vector addition (Fig. 2.4e).

The definition of vector addition implies that

$$\mathbf{U} + \mathbf{V} = \mathbf{V} + \mathbf{U} \qquad \text{Vector addition is commutative.} \qquad (2.1)$$

and

$$(\mathbf{U} + \mathbf{V}) + \mathbf{W} = \mathbf{U} + (\mathbf{V} + \mathbf{W}) \qquad \text{Vector addition is associative.} \qquad (2.2)$$

for any vectors **U**, **V**, and **W**. These results mean that when two or more vectors are added, the order in which they are added doesn't matter. The sum can be obtained by placing the vectors head to tail in any order, and the vector from the tail of the first vector to the head of the last one is the sum (Fig. 2.5). If the sum of two or more vectors is zero, they form a closed polygon when they are placed head to tail (Fig.2.6).

A physical quantity is called a vector if it has magnitude and direction and obeys the definition of vector addition. We have seen that displacement is a vector. The position of a point in space relative to another point is also a vector quantity. In Fig. 2.7, the vector $\mathbf{r}_{AC}$ from A to C is the sum of $\mathbf{r}_{AB}$ and $\mathbf{r}_{BC}$. A force has direction and magnitude, but do forces obey the definition of vector addition? For now we will assume that they do. When we discuss dynamics, we will show that Newton's second law implies that force is a vector.

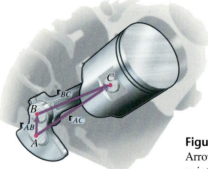

Figure 2.7
Arrows denoting the relative positions of points are vectors.

Product of a Scalar and a Vector

The product of a scalar (real number) a and a vector **U** is a vector written as $a\mathbf{U}$. Its magnitude is $|a||\mathbf{U}|$, where $|a|$ is the absolute value of the scalar a. The direction of $a\mathbf{U}$ is the same as the direction of **U** when a is positive and is opposite to the direction of **U** when a is negative.

The product $(-1)\mathbf{U}$ is written as $-\mathbf{U}$ and is called "the negative of the vector **U**." It has the same magnitude as **U** but the opposite direction. The division of a vector **U** by a scalar a is defined to be the product

$$\frac{\mathbf{U}}{a} = \left(\frac{1}{a}\right)\mathbf{U}.$$

Figure 2.8 shows a vector **U** and the products of **U** with the scalars 2, −1, and 1/2.

The definitions of vector addition and the product of a scalar and a vector imply that

$$a(b\mathbf{U}) = (ab)\mathbf{U}, \qquad \text{The product is associative with} \qquad (2.3)$$
$$\text{respect to scalar multiplication.}$$

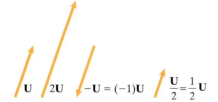

Figure 2.8
A vector **U** and some of its scalar multiples.

$$(a + b)\mathbf{U} = a\mathbf{U} + b\mathbf{U}, \quad \text{The products is distributive} \quad (2.4)$$
with respect to scalar addition.

and

$$a(\mathbf{U} + \mathbf{V}) = a\mathbf{U} + a\mathbf{V}, \quad \text{The products is distributive} \quad (2.5)$$
with respect to vector addition.

for any scalars a and b and vectors $\mathbf{U}$ and $\mathbf{V}$. We will need these results when we discuss components of vectors.

Vector Subtraction

The difference of two vectors $\mathbf{U}$ and $\mathbf{V}$ is obtained by adding $\mathbf{U}$ to the vector $(-1)\mathbf{V}$:

$$\mathbf{U} - \mathbf{V} = \mathbf{U} + (-1)\mathbf{V}. \quad (2.6)$$

Consider the two vectors $\mathbf{U}$ and $\mathbf{V}$ shown in Fig. 2.9a. The vector $(-1)\mathbf{V}$ has the same magnitude as the vector $\mathbf{V}$ but is in the opposite direction (Fig. 2.9b). In Fig. 2.9c, we add the vector $\mathbf{U}$ to the vector $(-1)\mathbf{V}$ to obtain $\mathbf{U} - \mathbf{V}$.

Unit Vectors

A *unit vector* is simply a vector whose magnitude is 1. A unit vector specifies a direction and also provides a convenient way to express a vector that has a particular direction. If a unit vector $\mathbf{e}$ and a vector $\mathbf{U}$ have the same direction, we can write $\mathbf{U}$ as the product of its magnitude $|\mathbf{U}|$ and the unit vector $\mathbf{e}$ (Fig. 2.10),

$$\mathbf{U} = |\mathbf{U}|\mathbf{e}.$$

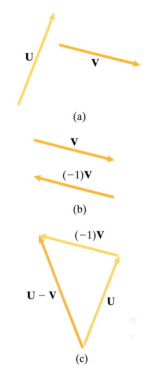

(a)

(b)

(c)

Figure 2.9
(a) Two vectors $\mathbf{U}$ and $\mathbf{V}$.
(b) The vectors $\mathbf{V}$ and $(-1)\mathbf{V}$.
(c) The sum of $\mathbf{U}$ and $(-1)\mathbf{V}$ is the vector difference $\mathbf{U} - \mathbf{V}$.

Figure 2.10
Since $\mathbf{U}$ and $\mathbf{e}$ have the same direction, the vector $\mathbf{U}$ equals the product of its magnitude with $\mathbf{e}$.

Any vector $\mathbf{U}$ can be regarded as the product of its magnitude and a unit vector that has the same direction as $\mathbf{U}$. Dividing both sides of this equation by $|\mathbf{U}|$ yields

$$\frac{\mathbf{U}}{|\mathbf{U}|} = \mathbf{e},$$

so *dividing any vector by its magnitude yields a unit vector that has the same direction*.

Study Questions
1. What is the triangle rule for vector addition?
2. Vector addition is commutative. What does that mean?
3. If you multiply a vector $\mathbf{U}$ by a number a, what do you know about the resulting vector $a\,\mathbf{U}$?
4. What is a unit vector?

Example 2.1	Adding Vectors

Figure 2.11 is an initial design sketch of part of the roof of a sports stadium that is to be supported by the cables *AB* and *AC*. The forces the cables exert on the pylon to which they are attached are represented by the vectors $\mathbf{F}_{AB}$ and $\mathbf{F}_{AC}$. The magnitudes of the forces are $|\mathbf{F}_{AB}| = 100$ kN and $|\mathbf{F}_{AC}| = 60$ kN. Determine the magnitude and direction of the sum of the forces exerted on the pylon by the cables (a) graphically and (b) by using trigonometry.

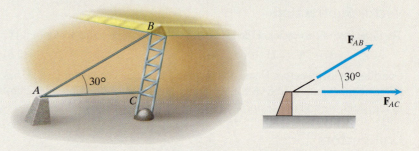

Figure 2.11

Strategy

(a) By drawing the parallelogram rule for adding the two forces *with the vectors drawn to scale*, we can measure the magnitude and direction of their sum.
(b) We will calculate the magnitude and direction of the sum of the forces by applying the laws of sines and cosines (Appendix A, Section A.2) to the triangles formed by the parallelogram rule.

Solution

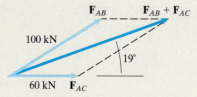

(a) Graphical solution.

(a) We graphically construct the parallelogram rule for obtaining the sum of the two forces with the lengths of $\mathbf{F}_{AB}$ and $\mathbf{F}_{AC}$ proportional to their magnitudes (Fig. a). By measuring the figure, we estimate the magnitude of the vector $\mathbf{F}_{AB} + \mathbf{F}_{AC}$ to be 155 kN and its direction to be 19° above the horizontal.

(b) Consider the parallelogram rule for obtaining the sum of the two forces (Fig. b). Since $\alpha + 30° = 180°$, the angle $\alpha = 150°$. By applying the law of cosines to the shaded triangle with the magnitudes of the vectors in kN, we obtain

$$|\mathbf{F}_{AB} + \mathbf{F}_{AC}|^2 = |\mathbf{F}_{AB}|^2 + |\mathbf{F}_{AC}|^2 - 2|\mathbf{F}_{AB}||\mathbf{F}_{AC}| \cos \alpha$$
$$= (100)^2 + (60)^2 - 2(100)(60) \cos 150°$$

and determine that the magnitude $|\mathbf{F}_{AB} + \mathbf{F}_{AC}| = 155$ kN.

To determine the angle β between the vector $\mathbf{F}_{AB} + \mathbf{F}_{AC}$ and the horizontal, we apply the law of sines to the shaded triangle:

$$\frac{\sin \beta}{|\mathbf{F}_{AB}|} = \frac{\sin \alpha}{|\mathbf{F}_{AB} + \mathbf{F}_{AC}|}.$$

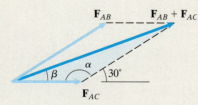

(b) Trigonometric solution.

The solution is

$$\beta = \arcsin \left(\frac{|\mathbf{F}_{AB}| \sin \alpha}{|\mathbf{F}_{AB} + \mathbf{F}_{AC}|} \right) = \arcsin \left(\frac{100 \sin 150°}{155} \right) = 18.8°$$

Critical Thinking

In engineering applications, vector operations are nearly always done analytically. So why is it worthwhile to gain experience with graphical methods? Doing so enhances your intuition about vectors and helps you understand vector operations. Also, sketching out a graphical solution can often help you formulate an analytical solution.

In this example, we determined the sum of two vectors analytically by making use of results from trigonometry with which you are familiar. But in the next section, we will show that expressing vectors in terms of perpendicular components greatly simplifies analytical vector operations.

Problems

Refer to the following diagram when solving Problems 2.1 through 2.5. The force vectors $\mathbf{F}_A$, $\mathbf{F}_B$, and $\mathbf{F}_C$ lie in the same plane.

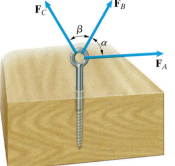

Problems 2.1–2.5

2.1 The magnitudes $|\mathbf{F}_A| = 60$ N and $|\mathbf{F}_B| = 80$ N. The angle $\alpha = 45°$. Graphically determine the magnitude of the vector $\mathbf{F}_A + \mathbf{F}_B$ and the angle between the vectors $\mathbf{F}_B$ and $\mathbf{F}_A + \mathbf{F}_B$.

Strategy: Construct the parallelogram for determining the sum of the forces, drawing the lengths of $\mathbf{F}_A$ and $\mathbf{F}_B$ proportional to their magnitudes and accurately measuring the angle α, as we did in Example 2.1. Then you can measure the magnitude of $\mathbf{F}_A + \mathbf{F}_B$ and the angle between $\mathbf{F}_B$ and $\mathbf{F}_A + \mathbf{F}_B$.

2.2 The magnitudes $|\mathbf{F}_A| = 40$ N, $|\mathbf{F}_B| = 50$ N, and $|\mathbf{F}_A + \mathbf{F}_B| = 80$ N. Assume that $0 < \alpha < 90°$. Graphically determine the angle α.

2.3 The magnitudes $|\mathbf{F}_A| = 40$ N, $|\mathbf{F}_B| = 50$ N, and $|\mathbf{F}_A + \mathbf{F}_B| = 80$ N. Assume that $0 < \alpha < 90°$. Use trigonometry to determine the angle α.

2.4 The magnitudes $|\mathbf{F}_A| = 40$ N, $|\mathbf{F}_B| = 50$ N, and $|\mathbf{F}_C| = 40$ N. The angles $\alpha = 50°$ and $\beta = 80°$. Graphically determine the magnitude of $\mathbf{F}_A + \mathbf{F}_B + \mathbf{F}_C$.

2.5 The magnitudes $|\mathbf{F}_A| = 40$ N, $|\mathbf{F}_B| = 50$ N, and $|\mathbf{F}_C| = 40$ N. The angles $\alpha = 50°$ and $\beta = 80°$. Use trigonometry to determine the magnitude of $\mathbf{F}_A + \mathbf{F}_B + \mathbf{F}_C$.

2.6 If the magnitude of the vector $\mathbf{r}_{AC}$ is 195 mm, what is the angle θ?

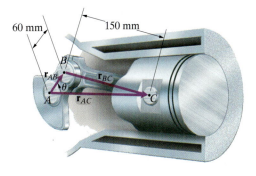

Problem 2.6

2.7 The vectors $\mathbf{F}_A$ and $\mathbf{F}_B$ represent the forces exerted on the pulley by the belt. Their magnitudes are $|\mathbf{F}_A| = 80$ N and $|\mathbf{F}_B| = 60$ N. What is the magnitude $|\mathbf{F}_A + \mathbf{F}_B|$ of the total force the belt exerts on the pulley?

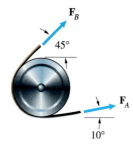

Problem 2.7

2.8 The sum of the forces $\mathbf{F}_A + \mathbf{F}_B + \mathbf{F}_C = \mathbf{0}$. The magnitude $|\mathbf{F}_A| = 100$ N and the angle $\alpha = 60°$. Determine $|\mathbf{F}_B|$ and $|\mathbf{F}_C|$.

2.9 The sum of the forces $\mathbf{F}_A + \mathbf{F}_B + \mathbf{F}_C = \mathbf{0}$. If the magnitudes $|\mathbf{F}_A| = 100$ N and $|\mathbf{F}_B| = 80$ N, what are $|\mathbf{F}_C|$ and the angle α?

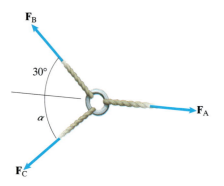

Problems 2.8/2.9

2.10 The forces acting on the sailplane are represented by three vectors. The lift $\mathbf{L}$ and drag $\mathbf{D}$ are perpendicular, the magnitude of the weight $\mathbf{W}$ is 3500 N, and $\mathbf{W} + \mathbf{L} + \mathbf{D} = \mathbf{0}$. What are the magnitudes of the lift and drag?

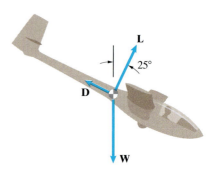

Problem 2.10

2.11 A spherical storage tank is supported by cables. The tank is subjected to three forces: the forces $\mathbf{F}_A$ and $\mathbf{F}_B$ exerted by the cables and the weight $\mathbf{W}$. The weight of the tank $|\mathbf{W}| = 600$ lb. The vector sum of the forces acting on the tank equals zero. Determine the magnitudes of $\mathbf{F}_A$ and $\mathbf{F}_B$ (a) graphically and (b) by using trigonometry.

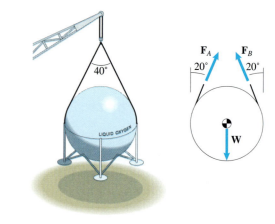

Problem 2.11

2.12 The rope ABC exerts forces $\mathbf{F}_{BA}$ and $\mathbf{F}_{BC}$ on the block at B. Their magnitudes are equal: $|\mathbf{F}_{BA}| = |\mathbf{F}_{BC}|$. The magnitude of the total force exerted on the block at B by the rope is $|\mathbf{F}_{BA} + \mathbf{F}_{BC}| = 920$ N. Determine $|\mathbf{F}_{BA}|$ (a) graphically and (b) by using trigonometry.

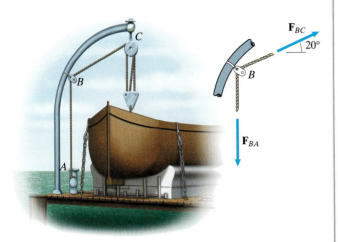

Problem 2.12

2.13 Two snowcats tow a housing unit to a new location at Mc-Murdo Base, Antarctica. (The top view is shown. The cables are horizontal.) The sum of the forces $\mathbf{F}_A$ and $\mathbf{F}_B$ exerted on the unit is parallel to the line L, and $|\mathbf{F}_A| = 1000$ lb. Determine $|\mathbf{F}_B|$ and $|\mathbf{F}_A + \mathbf{F}_B|$ (a) graphically and (b) by using trigonometry.

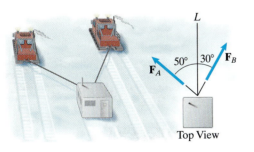

Problem 2.13

2.14 A surveyor determines that the horizontal distance from A to B is 400 m and that the horizontal distance from A to C is 600 m. Determine the magnitude of the horizontal vector $\mathbf{r}_{BC}$ from B to C and the angle α (a) graphically and (b) by using trigonometry.

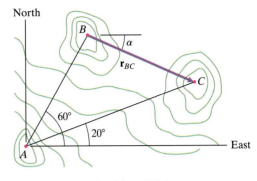

Problem 2.14

2.15 The vector $\mathbf{r}$ extends from point A to the midpoint between points B and C. Prove that

$$\mathbf{r} = \tfrac{1}{2}(\mathbf{r}_{AB} + \mathbf{r}_{AC}).$$

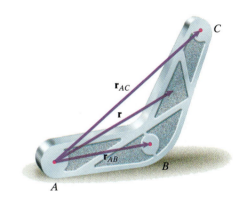

Problem 2.15

2.16 By drawing sketches of the vectors, explain why

$$\mathbf{U} + (\mathbf{V} + \mathbf{W}) = (\mathbf{U} + \mathbf{V}) + \mathbf{W}.$$

CARTESIAN COMPONENTS

Vectors are much easier to work with when they are expressed in terms of mutually perpendicular vector components. Here we explain how to express vectors in cartesian components in two and three dimensions and give examples of vector manipulations using components.

2.3 Components in Two Dimensions

Consider the vector $\mathbf{U}$ in Fig. 2.12a. By placing a cartesian coordinate system so that the vector $\mathbf{U}$ is parallel to the x–y plane, we can write it as the sum of perpendicular *vector components* $\mathbf{U}_x$ and $\mathbf{U}_y$ that are parallel to the x and y axes (Fig. 2.12b):

$$\mathbf{U} = \mathbf{U}_x + \mathbf{U}_y.$$

Figure 2.12
(a) A vector $\mathbf{U}$.
(b) The vector components $\mathbf{U}_x$ and $\mathbf{U}_y$.
(c) The vector components can be expressed in terms of $\mathbf{i}$ and $\mathbf{j}$.

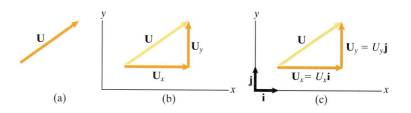

(a) (b) (c)

Then by introducing a unit vector $\mathbf{i}$ defined to point in the direction of the positive x axis and a unit vector $\mathbf{j}$ defined to point in the direction of the positive y axis (Fig. 2.12c), we can express the vector $\mathbf{U}$ in the form

$$\mathbf{U} = U_x\mathbf{i} + U_y\mathbf{j}. \tag{2.7}$$

The scalars U_x and U_y are called *scalar components of* $\mathbf{U}$. *When we refer simply to the components of a vector, we will mean its scalar components*. We will refer to U_x and U_y as the x and y components of $\mathbf{U}$.

The components of a vector specify both its direction relative to the cartesian coordinate system and its magnitude. From the right triangle formed by the vector $\mathbf{U}$ and its vector components (Fig. 2.12c), we see that the magnitude of $\mathbf{U}$ is given in terms of its components by the Pythagorean theorem:

$$|\mathbf{U}| = \sqrt{U_x^2 + U_y^2}. \tag{2.8}$$

With this equation the magnitude of a vector can be determined when its components are known.

Manipulating Vectors in Terms of Components

The sum of two vectors $\mathbf{U}$ and $\mathbf{V}$ in terms of their components is

$$\mathbf{U} + \mathbf{V} = (U_x\mathbf{i} + U_y\mathbf{j}) + (V_x\mathbf{i} + V_y\mathbf{j})$$
$$= (U_x + V_x)\mathbf{i} + (U_y + V_y)\mathbf{j}. \tag{2.9}$$

The components of $\mathbf{U} + \mathbf{V}$ are the sums of the components of the vectors $\mathbf{U}$ and $\mathbf{V}$. Notice that in obtaining this result we used Eqs. (2.2), (2.4), and (2.5).

It is instructive to derive Eq. (2.9) graphically. The summation of $\mathbf{U}$ and $\mathbf{V}$ is shown in Fig. 2.13a. In Fig. 2.13b we introduce a coordinate system and show

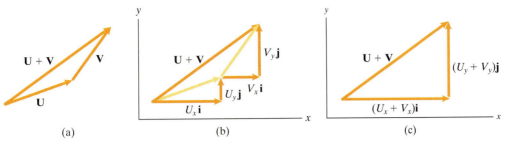

(a) (b) (c)

the components **U** and **V**. In Fig. 2.13c we add the x and y components, obtaining Eq. (2.9).

The product of a number a and a vector **U** in terms of the components of **U** is

$$a\mathbf{U} = a(U_x\mathbf{i} + U_y\mathbf{j}) = aU_x\mathbf{i} + aU_y\mathbf{j}.$$

The component of $a\mathbf{U}$ in each coordinate direction equals the product of a and the component of **U** in that direction. We used Eqs. (2.3) and (2.5) to obtain this result.

Figure 2.13
(**a**) The sum of **U** and **V**.
(**b**) The vector components of **U** and **V**.
(**c**) The sum of the components in each coordinate direction equals the component of **U** + **V** in that direction.

Position Vectors in Terms of Components

We can express the position vector of a point relative to another point in terms of the cartesian coordinates of the points. Consider point A with coordinates (x_A, y_A) and point B with coordinates (x_B, y_B). Let $\mathbf{r}_{AB}$ be the vector that specifies the position of B relative to A (Fig. 2.14a). That is, we denote the vector *from* a point A *to* a point B by $\mathbf{r}_{AB}$. We see from Fig. 2.14b that $\mathbf{r}_{AB}$ is given in terms of the coordinates of points A and B by

$$\mathbf{r}_{AB} = (x_B - x_A)\mathbf{i} + (y_B - y_A)\mathbf{j}. \tag{2.10}$$

Notice that the x component of the position vector from a point A to a point B is obtained by subtracting the x coordinate of A from the x coordinate of B, and the y component is obtained by subtracting the y coordinate of A from the y coordinate of B.

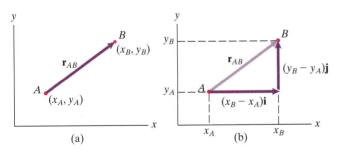

(a) (b)

Figure 2.14
(**a**) Two points A and B and the position vector $\mathbf{r}_{AB}$ from A to B.
(**b**) The components of $\mathbf{r}_{AB}$ can be determined from the coordinates of points A and B.

Study Questions

1. How are the scalar components of a vector defined in terms of a cartesian coordinate system?

2. If you know the scalar components of a vector, how can you determine its magnitude?

3. Suppose that you know the coordinates of two points A and B. How do you determine the scalar components of the position vector of point B relative to point A?

Example 2.2 **Adding Vectors in Terms of Components**

The forces acting on the sailplane in Fig. 2.15 are its weight $\mathbf{W} = -600\mathbf{j}$ (lb), the drag $\mathbf{D} = -200\mathbf{i} + 100\mathbf{j}$ (lb), and the lift $\mathbf{L}$.

(a) If the sum of the forces on the sailplane is zero, what are the components of $\mathbf{L}$?
(b) If the lift $\mathbf{L}$ has the components determined in (a) and the drag $\mathbf{D}$ increases by a factor of 2, what is the magnitude of the sum of the forces on the sailplane?

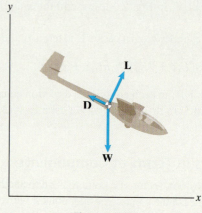

Figure 2.15

Strategy
(a) By setting the sum of the forces equal to zero, we can determine the components of $\mathbf{L}$. (b) Using the value of $\mathbf{L}$ from (a), we can determine the components of the sum of the forces and use Eq. (2.8) to determine its magnitude.

Solution
(a) We set the sum of the forces equal to zero with the forces expressed in lb:

$$\mathbf{W} + \mathbf{D} + \mathbf{L} = \mathbf{0}.$$

$$(-600\mathbf{j}) + (-200\mathbf{i} + 100\mathbf{j}) + \mathbf{L} = \mathbf{0}.$$

Solving for the lift, we obtain

$$\mathbf{L} = 200\mathbf{i} + 500\mathbf{j} \text{ (lb)}.$$

(b) If the drag increases by a factor of 2, the sum of the forces on the sailplane is

$$\mathbf{W} + 2\mathbf{D} + \mathbf{L} = (-600\mathbf{j}) + 2(-200\mathbf{i} + 100\mathbf{j}) + (200\mathbf{i} + 500\mathbf{j})$$

$$= -200\mathbf{i} + 100\mathbf{j} \text{ (lb)}.$$

From Eq. (2.8), the magnitude of the sum is

$$|\mathbf{W} + 2\mathbf{D} + \mathbf{L}| = \sqrt{(-200 \text{ lb})^2 + (100 \text{ lb})^2} = 224 \text{ lb}.$$

Critical Thinking
In part (a) of the solution, notice that we were able to determine both components of the vector $\mathbf{L}$ from one vector equation. *One vector equation in two dimensions is equivalent to two scalar equations.*

Example 2.3   Determining Components in Terms of an Angle

Hydraulic cylinders are used to exert forces in many mechanical devices. The force is exerted by pressurized liquid (hydraulic fluid) pushing against a piston within the cylinder. The hydraulic cylinder AB in Fig. 2.16 exerts a 4000-lb force **F** on the bed of the dump truck at B. Express **F** in terms of components using the coordinate system shown.

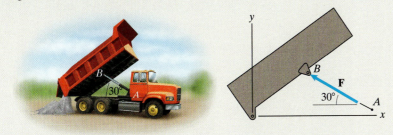

Figure 2.16

Strategy

When the direction of a vector is specified by an angle, as in this example, we can determine the values of the components from the right triangle formed by the vector and its components.

Solution

We draw the vector **F** and its vector components in Fig. a. From the resulting right triangle, we see that the magnitude of $\mathbf{F}_x$ is

$$|\mathbf{F}_x| = |\mathbf{F}| \cos 30° = (4000 \text{ lb}) \cos 30° = 3460 \text{ lb}.$$

$\mathbf{F}_x$ points in the negative x direction, so

$$\mathbf{F}_x = -3460\mathbf{i} \text{ (lb)}.$$

The magnitude of $\mathbf{F}_y$ is

$$|\mathbf{F}_y| = |\mathbf{F}| \sin 30° = (4000 \text{ lb}) \sin 30° = 2000 \text{ lb}.$$

The vector component $\mathbf{F}_y$ points in the positive y direction, so

$$\mathbf{F}_y = 2000\mathbf{j} \text{ (lb)}.$$

The vector **F**, in terms of its components, is

$$\mathbf{F} = \mathbf{F}_x + \mathbf{F}_y = -3460\mathbf{i} + 2000\mathbf{j} \text{ (lb)}.$$

The x component of **F** is -3460 lb, and the y component is 2000 lb.

Critical Thinking

When you have determined the components of a given vector, you should make sure they appear reasonable. In this example you can see from the vector's direction that the x component should be negative and the y component positive. You can also make sure that the components yield the correct magnitude. In this example,

$$|\mathbf{F}| = \sqrt{(-3460 \text{ lb})^2 + (2000 \text{ lb})^2} = 4000 \text{ lb}.$$

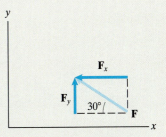

(a) The force **F** and its components form a right triangle.

Example 2.4 **Determining Components**

The cable from point A to point B exerts an 800-N force **F** on the top of the television transmission tower in Fig. 2.17. Express **F** in terms of components using the coordinate system shown.

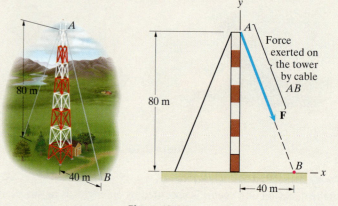

Figure 2.17

Strategy

We determine the components of **F** in three ways.

First Method   From the given dimensions we can determine the angle α between **F** and the y axis (Fig. a), then determine the components from the right triangles formed by the vector **F** and its components.

Second Method   The right triangles formed by **F** and its components are similar to the triangle OAB in Fig. a. We can determine the components of **F** by using the ratios of the sides of these similar triangles.

Third Method   From the given dimensions we can determine the components of the position vector $\mathbf{r}_{AB}$ from point A to point B (Fig. b). By dividing this vector by its magnitude, we will obtain a unit vector $\mathbf{e}_{AB}$ with the same direction as **F** (Fig. c), then obtain **F** in terms of its components by expressing it as the product of its magnitude and $\mathbf{e}_{AB}$.

Solution

First Method   Consider the force **F** and its vector components (Fig. a). The tangent of the angle α between **F** and the y axis is $\tan \alpha = 40/80 = 0.5$, so $\alpha = \arctan (0.5) = 26.6°$. From the right triangles formed by **F** and its vector components, the magnitude of $\mathbf{F}_x$ is

$$|\mathbf{F}_x| = |\mathbf{F}| \sin 26.6° = (800 \text{ N})\sin 26.6° = 358 \text{ N}$$

and the magnitude of $\mathbf{F}_y$ is

$$|\mathbf{F}_y| = |\mathbf{F}| \cos 26.6° = (800 \text{ N})\cos 26.6° = 716 \text{ N}.$$

Since $\mathbf{F}_x$ points in the positive x direction and $\mathbf{F}_y$ points in the negative y direction, the force **F** is

$$\mathbf{F} = 358\mathbf{i} - 716\mathbf{j} \text{ (N)}.$$

Second Method The length of the cable AB is $\sqrt{(80 \text{ m})^2 + (40 \text{ m})^2} = 89.4$ m. Since the triangle OAB in Fig. a is similar to the triangle formed by $\mathbf{F}$ and its vector components, it follows that

$$\frac{|\mathbf{F}_x|}{|\mathbf{F}|} = \frac{OB}{AB} = \frac{40 \text{ m}}{89.4 \text{ m}}.$$

Thus, the magnitude of $\mathbf{F}_x$ is

$$|\mathbf{F}_x| = \left(\frac{40 \text{ m}}{89.4 \text{ m}}\right)|\mathbf{F}| = \left(\frac{40 \text{ m}}{89.4 \text{ m}}\right)(800 \text{ N}) = 358 \text{ N}.$$

We can also see from the similar triangles that

$$\frac{|\mathbf{F}_y|}{|\mathbf{F}|} = \frac{OA}{AB} = \frac{80 \text{ m}}{89.4 \text{ m}},$$

so the magnitude of $\mathbf{F}_y$ is

$$|\mathbf{F}_y| = \left(\frac{80 \text{ m}}{89.4 \text{ m}}\right)|\mathbf{F}| = \left(\frac{80 \text{ m}}{89.4 \text{ m}}\right)(800 \text{ N}) = 716 \text{ N}.$$

Thus, we again obtain the result

$$\mathbf{F} = 358\mathbf{i} - 716\mathbf{j} \text{ (N)}.$$

Third Method The vector $\mathbf{r}_{AB}$ in Fig. b, with coordinates expressed in meters, is

$$\mathbf{r}_{AB} = (x_B - x_A)\mathbf{i} + (y_B - y_A)\mathbf{j} = (40 - 0)\mathbf{i} + (0 - 80)\mathbf{j}$$
$$= 40\mathbf{i} - 80\mathbf{j} \text{ (m)}.$$

We divide this vector by its magnitude to obtain a unit vector $\mathbf{e}_{AB}$ that has the same direction as the force $\mathbf{F}$ (Fig. c):

$$\mathbf{e}_{AB} = \frac{\mathbf{r}_{AB}}{|\mathbf{r}_{AB}|} = \frac{40\mathbf{i} - 80\mathbf{j} \text{ (m)}}{\sqrt{(40 \text{ m})^2 + (-80 \text{ m})^2}} = 0.447\mathbf{i} - 0.894\mathbf{j}.$$

The force $\mathbf{F}$ is equal to the product of its magnitude $|\mathbf{F}|$ and $\mathbf{e}_{AB}$:

$$\mathbf{F} = |\mathbf{F}|\mathbf{e}_{AB} = (800 \text{ N})(0.447\mathbf{i} - 0.894\mathbf{j}) = 358\mathbf{i} - 716\mathbf{j} \text{ (N)}.$$

Critical Thinking

What information is required to specify a vector? Its magnitude and direction must be prescribed. In this example, the magnitude of the vector $\mathbf{F}$ is given. Notice that the 40-m and 80-m dimensions in Fig. 2.17 define the direction of the cable from A to B, which also defines the direction of the vector. Consider how we used this information in our three methods of determining the components of the vector:

First Method We used the 40-m and 80-m dimensions to determine the angle α between the vertical and the line AB (which also specifies the direction of the vector) and then used the magnitude and the angle α to calculate the components.

Second Method We took advantage of the fact that the vector and its components form a triangle similar to the triangle OAB. Then, knowing the magnitude of the vector and the 40-m and 80-m dimensions, we were able to determine the components.

Third Method Because we knew the 40-m and 80-m dimensions, we could express the position vector from A to B in terms of its components. Dividing that position vector by its magnitude yielded a unit vector *with the same direction as* $\mathbf{F}$. Then we obtained the components by multiplying the unit vector by the magnitude of $\mathbf{F}$.

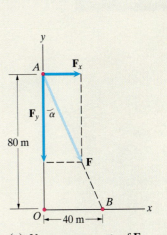

(a) Vector components of $\mathbf{F}$.

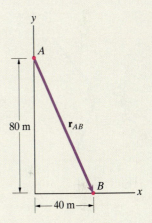

(b) The vector $\mathbf{r}_{AB}$ form A to B.

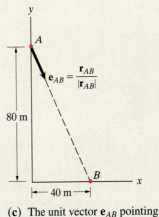

(c) The unit vector $\mathbf{e}_{AB}$ pointing from A toward B.

| Example 2.5 | Determining an Unknown Vector Magnitude |

The cables A and B in Fig. 2.18 exert forces $\mathbf{F}_A$ and $\mathbf{F}_B$ on the hook. The magnitude of $\mathbf{F}_A$ is 100 lb. The tension in cable B has been adjusted so that the total force $\mathbf{F}_A + \mathbf{F}_B$ is perpendicular to the wall to which the hook is attached.

(a) What is the magnitude of $\mathbf{F}_B$?

(b) What is the magnitude of the total force exerted on the hook by the two cables?

Strategy

The vector sum of the two forces is perpendicular to the wall, so the sum of the components parallel to the wall equals zero. From this condition we can obtain an equation for the magnitude of $\mathbf{F}_B$.

Solution

(a) In terms of the coordinate system shown in Fig. a, the components of $\mathbf{F}_A$ and $\mathbf{F}_B$ are

$$\mathbf{F}_A = |\mathbf{F}_A| \sin 40°\mathbf{i} + |\mathbf{F}_A| \cos 40°\mathbf{j},$$
$$\mathbf{F}_B = |\mathbf{F}_B| \sin 20°\mathbf{i} - |\mathbf{F}_B| \cos 20°\mathbf{j}.$$

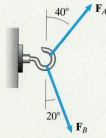

The total force is

$$\mathbf{F}_A + \mathbf{F}_B = (|\mathbf{F}_A| \sin 40° + |\mathbf{F}_B| \sin 20°)\mathbf{i}$$
$$+ (|\mathbf{F}_A| \cos 40° - |\mathbf{F}_B| \cos 20°)\mathbf{j}.$$

Now we set the component of the total force parallel to the wall (the y component) equal to zero:

$$|\mathbf{F}_A| \cos 40° - |\mathbf{F}_B| \cos 20° = 0,$$

Figure 2.18

We thus obtain an equation for the magnitude of $\mathbf{F}_B$:

$$|\mathbf{F}_B| = \frac{|\mathbf{F}_A| \cos 40°}{\cos 20°} = \frac{(100 \text{ lb})\cos 40°}{\cos 20°} = 81.5 \text{ lb}.$$

(b) Since we now know the magnitude of $\mathbf{F}_B$, we can determine the total force acting on the hook:

$$\mathbf{F}_A + \mathbf{F}_B = (|\mathbf{F}_A| \sin 40° + |\mathbf{F}_B| \sin 20°)\mathbf{i}$$
$$= [(100 \text{ lb})\sin 40° + (81.5 \text{ lb})\sin 20°]\mathbf{i} = 92.2\mathbf{i} \text{ (lb)}.$$

The magnitude of the total force is 92.2 lb.

Critical Thinking

We can obtain the solution to (a) in a less formal way. If the component of the total force parallel to the wall is zero, we see in Fig. a that the magnitude of the vertical component of $\mathbf{F}_A$ must equal the magnitude of the vertical component of $\mathbf{F}_B$:

$$|\mathbf{F}_A| \cos 40° = |\mathbf{F}_B| \cos 20°.$$

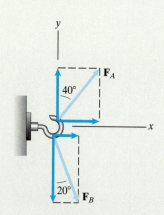

(a) Resolving $\mathbf{F}_A$ and $\mathbf{F}_B$ into components parallel and perpendicular to the wall.

Therefore the magnitude of $\mathbf{F}_B$ is

$$|\mathbf{F}_B| = \frac{|\mathbf{F}_A| \cos 40°}{\cos 20°} = \frac{(100 \text{ lb}) \cos 40°}{\cos 20°} = 81.5 \text{ lb}.$$

Problems

2.17 A force $\mathbf{F} = 40\mathbf{i} - 20\mathbf{j}$ (N). What is its magnitude $|\mathbf{F}|$?

Strategy: The magnitude of a vector in terms of its components is given by Eq. (2.8).

2.18 An engineer estimating the components of a force $\mathbf{F} = F_x\mathbf{i} + F_y\mathbf{j}$ acting on a bridge abutment has determined that $F_x = 130$ MN, $|\mathbf{F}| = 165$ MN, and F_y is negative. What is F_y?

2.19 A support is subjected to a force $\mathbf{F} = F_x\mathbf{i} + 80\mathbf{j}$ (N). If the support will safely support a force of magnitude 100 N, what is the allowable range of values of the component F_x?

2.20 If $\mathbf{F}_A = 600\mathbf{i} - 800\mathbf{j}$ (kip) and $\mathbf{F}_B = 200\mathbf{i} - 200\mathbf{j}$ (kip), what is the magnitude of the force $\mathbf{F} = \mathbf{F}_A - 2\mathbf{F}_B$?

2.21 If $\mathbf{F}_A = \mathbf{i} - 4.5\mathbf{j}$ (kN) and $\mathbf{F}_B = -2\mathbf{i} - 2\mathbf{j}$ (kN), what is the magnitude of the force $\mathbf{F} = 6\mathbf{F}_A + 4\mathbf{F}_B$?

2.22 Two perpendicular vectors $\mathbf{U}$ and $\mathbf{V}$ lie in the x–y plane. The vector $\mathbf{U} = 6\mathbf{i} - 8\mathbf{j}$ and $|\mathbf{V}| = 20$. What are the components of $\mathbf{V}$?

2.23 A fish exerts a 40-N force on the line that is represented by the vector $\mathbf{F}$. Express $\mathbf{F}$ in terms of components using the coordinate system shown.

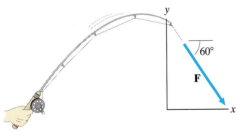

Problem 2.23

2.24 A person exerts a 60-lb force $\mathbf{F}$ to push a crate onto a truck. Express $\mathbf{F}$ in terms of components.

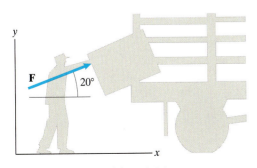

Problem 2.24

2.25 The missile's engine exerts a 260-kN force $\mathbf{F}$. Express $\mathbf{F}$ in terms of components using the coordinate system shown.

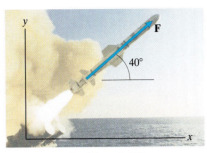

Problem 2.25

2.26 For the truss shown, express the position vector $\mathbf{r}_{AD}$ from point A to point D in terms of components. Use your result to determine the distance from point A to point D.

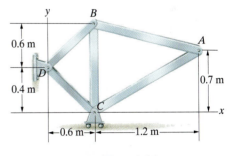

Problem 2.26

2.27 The points A, B, ... are the joints of the hexagonal structural element. Let $\mathbf{r}_{AB}$ be the position vector from joint A to joint B, $\mathbf{r}_{AC}$ the position vector from joint A to joint C, and so forth. Determine the components of the vectors $\mathbf{r}_{AC}$ and $\mathbf{r}_{AF}$.

2.28 Determine the components of the vector $\mathbf{r}_{AB} - \mathbf{r}_{BC}$.

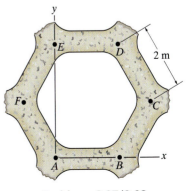

Problems 2.27/2.28

2.29 The coordinates of point A are $(1.8, 3.0)$ m. The y coordinate of point B is 0.6 m and the magnitude of the vector $\mathbf{r}_{AB}$ is 3.0 m. What are the components of $\mathbf{r}_{AB}$?

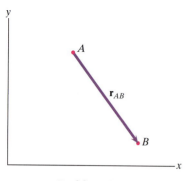

Problem 2.29

2.30 (a) Express the position vector from point A of the front-end loader to point B in terms of components.

(b) Express the position vector from point B to point C in terms of components.

(c) Use the results of (a) and (b) to determine the distance from point A to point C.

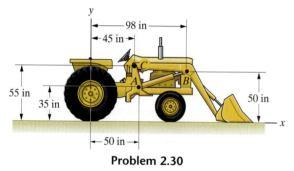

Problem 2.30

2.31 Five identical cylinders with radius $R = 0.2$ m are stacked as shown. Determine the components of the position vectors (a) from point A to point B and (b) from point B to point E.

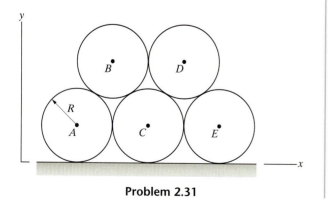

Problem 2.31

2.32 Determine the position vector $\mathbf{r}_{AB}$ in terms of its components if (a) $\theta = 30°$; (b) $\theta = 225°$.

2.33 Determine the position vector $\mathbf{r}_{BC}$ in terms of its components if (a) $\theta = 30°$; (b) $\theta = 225°$.

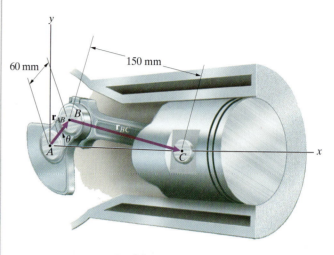

Problems 2.32/2.33

2.34 A surveyor measures the location of point A and determines that $\mathbf{r}_{OA} = 400\mathbf{i} + 800\mathbf{j}$ (m). He wants to determine the location of a point B so that $|\mathbf{r}_{AB}| = 400$ m and $|\mathbf{r}_{OA} + \mathbf{r}_{AB}| = 1200$ m. What are the cartesian coordinates of point B?

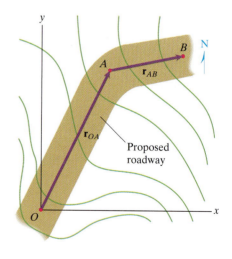

Problem 2.34

2.35 The magnitude of the position vector $\mathbf{r}_{BA}$ from point B to point A is 6 m and the magnitude of the position vector $\mathbf{r}_{CA}$ from point C to point A is 4 m. What are the components of $\mathbf{r}_{BA}$?

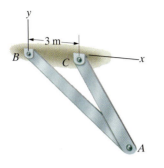

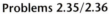

Problems 2.35/2.36

2.36 In Problem 2.35, determine the components of a unit vector $\mathbf{e}_{CA}$ that points from point C toward point A.

Strategy: Determine the components of $\mathbf{r}_{CA}$ and then divide the vector $\mathbf{r}_{CA}$ by its magnitude.

2.37 The x and y coordinates of points A, B, and C of the sailboat are shown.

(a) Determine the components of a unit vector that is parallel to the forestay AB and points from A toward B.

(b) Determine the components of a unit vector that is parallel to the backstay BC and points from C toward B.

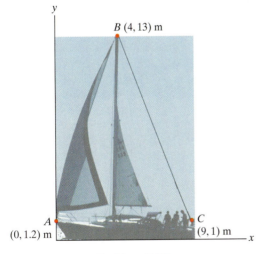

Problem 2.37

2.38 The length of the bar AB is 0.6 m. Determine the components of a unit vector $\mathbf{e}_{AB}$ that points from point A toward point B.

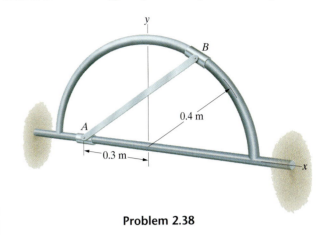

Problem 2.38

2.39 Determine the components of a unit vector that is parallel to the hydraulic actuator BC and points from B toward C.

2.40 The hydraulic actuator BC exerts a 1.2-kN force $\mathbf{F}$ on the joint at C that is parallel to the actuator and points from B toward C. Determine the components of $\mathbf{F}$.

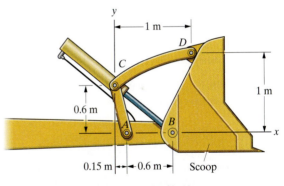

Problems 2.39/2.40

2.41 A surveyor finds that the length of the line OA is 1500 m and the length of the line OB is 2000 m.

(a) Determine the components of the position vector from point A to point B.

(b) Determine the components of a unit vector that points from point A toward point B.

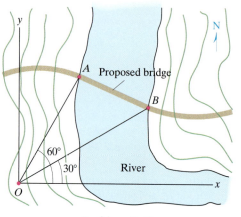

Problem 2.41

2.42 The positions at a given time of the Sun (**S**) and the planets Mercury (**M**), Venus (**V**), and Earth (**E**) are shown. The approximate distance from the Sun to Mercury is 57×10^6 km, the distance from the Sun to Venus is 108×10^6 km, and the distance from the Sun to the Earth is 150×10^6 km. Assume that the Sun and planets lie in the x–y plane. Determine the components of a unit vector that points from the Earth toward Mercury.

2.43 For the positions described in Problem 2.42, determine the components of a unit vector that points from the Earth toward Venus.

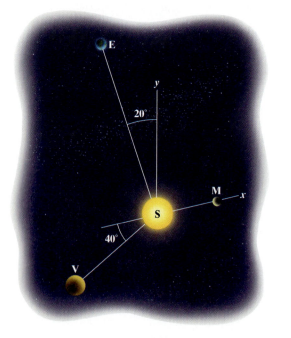

Problems 2.42/2.43

2.44 The rope ABC exerts forces $\mathbf{F}_{BA}$ and $\mathbf{F}_{BC}$ on the block at B. Their magnitudes are equal: $|\mathbf{F}_{BA}| = |\mathbf{F}_{BC}|$. The magnitude of the total force exerted on the block at B by the rope is $|\mathbf{F}_{BA} + \mathbf{F}_{BC}| = 920$ N. Determine $|\mathbf{F}_{BA}|$ by expressing the forces $\mathbf{F}_{BA}$ and $\mathbf{F}_{BC}$ in terms of components and compare your answer to the answer of Problem 2.12.

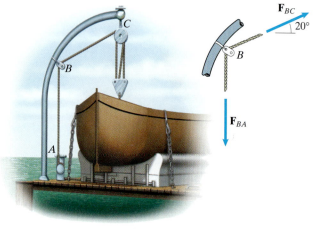

Problem 2.44

2.45 The magnitude of the horizontal force $\mathbf{F}_1$ is 5 kN and $\mathbf{F}_1 + \mathbf{F}_2 + \mathbf{F}_3 = \mathbf{0}$. What are the magnitudes of $\mathbf{F}_2$ and $\mathbf{F}_3$?

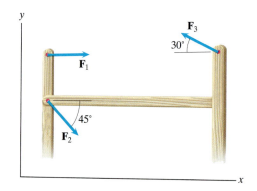

Problem 2.45

2.46 Four groups engage in a tug-of-war. The magnitudes of the forces exerted by groups *B, C,* and *D* are $|\mathbf{F}_B| = 800$ lb, $|\mathbf{F}_C| = 1000$ lb, and $|\mathbf{F}_D| = 900$ lb. If the vector sum of the four forces equals zero, what is the magnitude of $\mathbf{F}_A$ and the angle α?

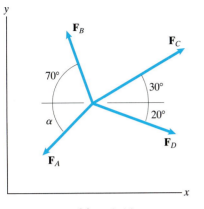

Problem 2.46

2.47 The two vernier engines of the launch vehicle exert thrusts (forces) that control the vehicle's attitude, or angular position. Each engine exerts a 5000-lb thrust. At the present instant, the thrusts are in the directions shown. (a) What is the *x* component of the force exerted on the vehicle by the vernier engines? (b) If the launch vehicle's main engines exert a 200,000-lb thrust parallel to the *y* axis, what is the *y* component of the total force on the launch vehicle?

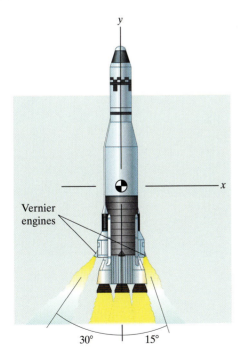

Problem 2.47

2.48 The bracket must support the two forces shown, where $|\mathbf{F}_1| = |\mathbf{F}_2| = 2$ kN. An engineer determines that the bracket will safely support a total force of magnitude 3.5 kN in any direction. Assume that $0 \leq \alpha \leq 90°$. What is the safe range of the angle α?

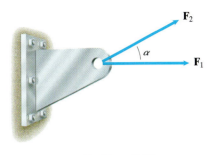

Problem 2.48

2.49 The figure shows three forces acting on a joint of a structure. The magnitude of $\mathbf{F}_C$ is 60 kN, and $\mathbf{F}_A + \mathbf{F}_B + \mathbf{F}_C = \mathbf{0}$. What are the magnitudes of $\mathbf{F}_A$ and $\mathbf{F}_B$?

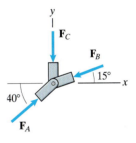

Problem 2.49

2.50 Four coplanar forces act on a beam. The forces $\mathbf{F}_B$ and $\mathbf{F}_C$ are vertical. The vector sum of the forces is zero. The magnitudes $|\mathbf{F}_B| = 10$ kN and $|\mathbf{F}_C| = 5$ kN. Determine the magnitudes of $\mathbf{F}_A$ and $\mathbf{F}_D$.

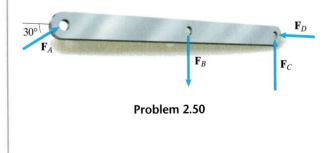

Problem 2.50

2.51 Six forces act on a beam that forms part of a building's frame. The vector sum of the forces is zero. The magnitudes $|\mathbf{F}_B| = |\mathbf{F}_E| = 20$ kN, $|\mathbf{F}_C| = 16$ kN, and $|\mathbf{F}_D| = 9$ kN. Determine the magnitudes of $\mathbf{F}_A$ and $\mathbf{F}_G$.

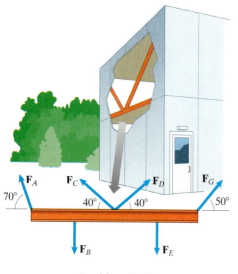

Problem 2.51

2.52 The total weight of the man and parasail is $|\mathbf{W}| = 230$ lb. The drag force $\mathbf{D}$ is perpendicular to the lift force $\mathbf{L}$. If the vector sum of the three forces is zero, what are the magnitudes of $\mathbf{L}$ and $\mathbf{D}$?

Problem 2.52

2.53 The three forces acting on the car are shown. The force $\mathbf{T}$ is parallel to the x axis and the magnitude of the force $\mathbf{W}$ is 14 kN. If $\mathbf{T} + \mathbf{W} + \mathbf{N} = \mathbf{0}$, what are the magnitudes of the forces $\mathbf{T}$ and $\mathbf{N}$?

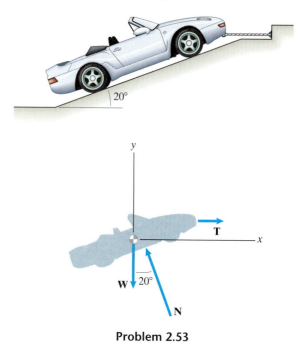

Problem 2.53

2.54 The cables A, B, and C help support a pillar that forms part of the supports of a structure. The magnitudes of the forces exerted by the cables are equal: $|\mathbf{F}_A| = |\mathbf{F}_B| = |\mathbf{F}_C|$. The magnitude of the vector sum of the three forces is 200 kN. What is $|\mathbf{F}_A|$?

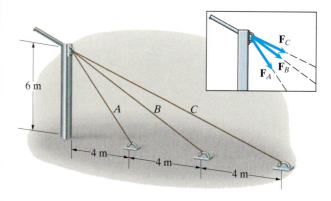

Problem 2.54

2.55 The total force exerted on the top of the mast B by the sailboat's forestay AB and backstay BC is $180\mathbf{i} - 820\mathbf{j}$ (N). What are the magnitudes of the forces exerted at B by the cables AB and BC?

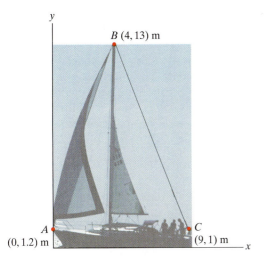

B (4, 13) m

y

A
(0, 1.2) m

C
(9, 1) m

x

Problem 2.55

2.56 The structure shown forms part of a truss designed by an architectural engineer to support the roof of an orchestra shell. The members AB, AC, and AD exert forces $\mathbf{F}_{AB}$, $\mathbf{F}_{AC}$, and $\mathbf{F}_{AD}$ on the joint A. The magnitude $|\mathbf{F}_{AB}| = 4$ kN. If the vector sum of the three forces equals zero, what are the magnitudes of $\mathbf{F}_{AC}$ and $\mathbf{F}_{AD}$?

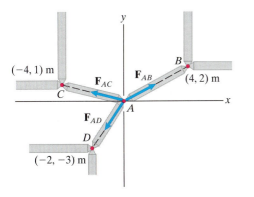

y

$(-4, 1)$ m

$\mathbf{F}_{AC}$

$\mathbf{F}_{AB}$

B

$(4, 2)$ m

C

A

x

$\mathbf{F}_{AD}$

D

$(-2, -3)$ m

Problem 2.56

2.57 The distance $s = 45$ in.
(a) Determine the unit vector $\mathbf{e}_{BA}$ that points from B toward A.
(b) Use the unit vector you obtained in (a) to determine the coordinates of the collar C.

2.58 Determine the x and y coordinates of the collar C as functions of the distance s.

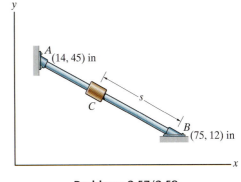

y

A
(14, 45) in

C

s

B
(75, 12) in

x

Problems 2.57/2.58

2.59 The position vector $\mathbf{r}$ goes from point A to a point on the straight line between B and C. Its magnitude is $|\mathbf{r}| = 6$ ft. Express $\mathbf{r}$ in terms of components.

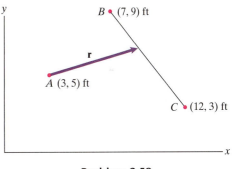

y

B   (7, 9) ft

$\mathbf{r}$

A (3, 5) ft

C   (12, 3) ft

x

Problem 2.59

2.60 Let $\mathbf{r}$ be the position vector from point C to the point that is a distance s meters from point A along the straight line between A and B. Express $\mathbf{r}$ in terms of components. (Your answer will be in terms of s.)

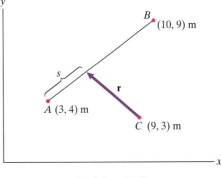

y

B
(10, 9) m

s

$\mathbf{r}$

A (3, 4) m

C (9, 3) m

x

Problem 2.60

2.4 Components in Three Dimensions

Many engineering applications require vectors to be expressed in terms of components in a three-dimensional coordinate system. In this section we explain this technique and demonstrate vector operations in three dimensions.

We first review how to draw objects in three dimensions. Consider a three-dimensional object such as a cube. If we draw the cube as it appears when the point of view is perpendicular to one of its faces, we obtain Fig. 2.19a. In this view, the cube appears two dimensional. The dimension perpendicular to the page cannot be seen. To remedy this, we move the point of view upward and to the right, obtaining Fig. 2.19b. In this *oblique* view, the third dimension is visible. The hidden edges of the cube are shown as dashed lines.

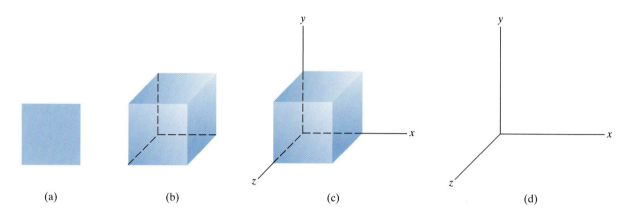

(a) (b) (c) (d)

Figure 2.19
(a) A cube viewed with the line of sight perpendicular to a face.
(b) An oblique view of the cube.
(c) A cartesian coordinate system aligned with the edges of the cube.
(d) Three-dimensional representation of the coordinate system.

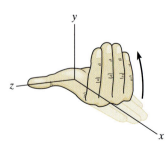

Figure 2.20
Recognizing a right-handed coordinate system.

We can use this approach to draw three-dimensional coordinate systems. In Fig. 2.19c we align the x, y, and z axes of a three-dimensional cartesian coordinate system with the edges of the cube. The three-dimensional representation of the coordinate system alone is shown in Fig. 2.19d. The coordinate system shown is said to be *right handed*. If the fingers of the right hand are pointed in the direction of the positive x axis and then bent (as in preparing to make a fist) toward the positive y axis, the thumb points in the direction of the positive z axis (Fig. 2.20). Otherwise, the coordinate system is left handed. Because some equations used in mathematics and engineering do not yield correct results when they are applied using a left-handed coordinate system, we use only right-handed coordinate systems.

We can express a vector $\mathbf{U}$ in terms of vector components $\mathbf{U}_x$, $\mathbf{U}_y$, and $\mathbf{U}_z$ parallel to the x, y, and z axes, respectively (Fig. 2.21), as

$$\mathbf{U} = \mathbf{U}_x + \mathbf{U}_y + \mathbf{U}_z. \tag{2.11}$$

(We have drawn a box around the vector to help in visualizing the directions of the vector components.) By introducing unit vectors $\mathbf{i}$, $\mathbf{j}$, and $\mathbf{k}$ that point in the positive x, y, and z directions, we can express $\mathbf{U}$ in terms of scalar components as

$$\mathbf{U} = U_x\mathbf{i} + U_y\mathbf{j} + U_z\mathbf{k}. \qquad (2.12)$$

We will refer to the scalars U_x, U_y, and U_z as the x, y, and z components of $\mathbf{U}$.

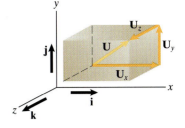

Figure 2.21
A vector $\mathbf{U}$ and its vector components.

Magnitude of a Vector in Terms of Components

Consider a vector $\mathbf{U}$ and its vector components (Fig. 2.22a). From the right triangle formed by the vectors $\mathbf{U}_y$, $\mathbf{U}_z$, and their sum $\mathbf{U}_y + \mathbf{U}_z$ (Fig. 2.22b), we can see that

$$|\mathbf{U}_y + \mathbf{U}_z|^2 = |\mathbf{U}_y|^2 + |\mathbf{U}_z|^2. \qquad (2.13)$$

The vector $\mathbf{U}$ is the sum of the vectors $\mathbf{U}_x$ and $\mathbf{U}_y + \mathbf{U}_z$. These three vectors form a right triangle (Fig. 2.22c), from which we obtain

$$|\mathbf{U}|^2 = |\mathbf{U}_x|^2 + |\mathbf{U}_y + \mathbf{U}_z|^2.$$

Substituting Eq. (2.13) into this result yields the equation

$$|\mathbf{U}|^2 = |\mathbf{U}_x|^2 + |\mathbf{U}_y|^2 + |\mathbf{U}_z|^2 = U_x^2 + U_y^2 + U_z^2.$$

Thus, the magnitude of a vector $\mathbf{U}$ is given in terms of its components in three dimensions by

$$|\mathbf{U}| = \sqrt{U_x^2 + U_y^2 + U_z^2}. \qquad (2.14)$$

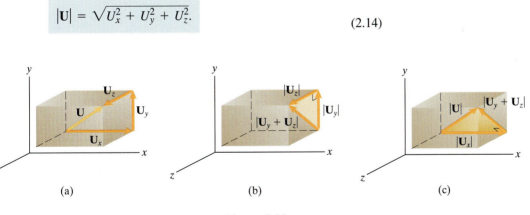

(a)   (b)   (c)

Figure 2.22
(a) A vector $\mathbf{U}$ and its vector components.
(b) The right triangle formed by the vectors $\mathbf{U}_y$, $\mathbf{U}_z$, and $\mathbf{U}_y + \mathbf{U}_z$.
(c) The right triangle formed by the vectors $\mathbf{U}$, $\mathbf{U}_x$, and $\mathbf{U}_y + \mathbf{U}_z$.

Direction Cosines

We described the direction of a vector relative to a two-dimensional cartesian coordinate system by specifying the angle between the vector and one of the coordinate axes. One of the ways we can describe the direction of a vector in three dimensions is by specifying the angles θ_x, θ_y, and θ_z between the vector and the positive coordinate axes (Fig. 2.23a).

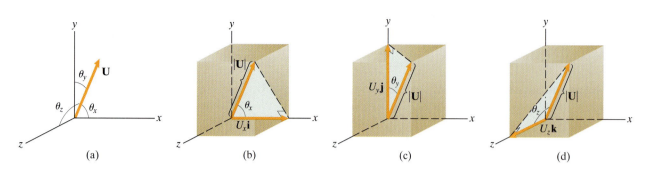

Figure 2.23
(a) A vector $\mathbf{U}$ and the angles θ_x, θ_y, and θ_z.
(b)–(d) The angles θ_x, θ_y, and θ_z and the vector components of $\mathbf{U}$.

In Figs. 2.23b–d, we demonstrate that the components of the vector $\mathbf{U}$ are respectively given in terms of the angles θ_x, θ_y, and θ_z, by

$$U_x = |\mathbf{U}| \cos \theta_x, \quad U_y = |\mathbf{U}| \cos \theta_y, \quad U_z = |\mathbf{U}| \cos \theta_z. \quad (2.15)$$

The quantities $\cos \theta_x$, $\cos \theta_y$, and $\cos \theta_z$ are called the *direction cosines* of $\mathbf{U}$. The direction cosines of a vector are not independent. If we substitute Eqs. (2.15) into Eq. (2.14), we find that the direction cosines satisfy the relation

$$\cos^2 \theta_x + \cos^2 \theta_y + \cos^2 \theta_z = 1. \quad (2.16)$$

Suppose that $\mathbf{e}$ is a unit vector with the same direction as $\mathbf{U}$, so that

$$\mathbf{U} = |\mathbf{U}|\mathbf{e}.$$

In terms of components, this equation is

$$U_x\mathbf{i} + U_y\mathbf{j} + U_z\mathbf{k} = |\mathbf{U}|(e_x\mathbf{i} + e_y\mathbf{j} + e_z\mathbf{k}).$$

Thus the relations between the components of $\mathbf{U}$ and $\mathbf{e}$ are

$$U_x = |\mathbf{U}|e_x, \quad U_y = |\mathbf{U}|e_y, \quad U_z = |\mathbf{U}|e_z.$$

By comparing these equations to Eqs. (2.15), we see that

$$\cos \theta_x = e_x, \quad \cos \theta_y = e_y, \quad \cos \theta_z = e_z.$$

The direction cosines of a vector $\mathbf{U}$ are the components of a unit vector with the same direction as $\mathbf{U}$.

Position Vectors in Terms of Components

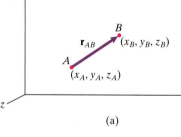

(a)

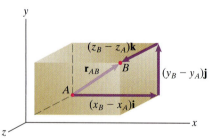

Figure 2.24
(a) The position vector from point A to point B.
(b) The components of $\mathbf{r}_{AB}$ can be determined from the coordinates of points A and B.

Generalizing the two-dimensional case, we consider a point A with coordinates (x_A, y_A, z_A) and a point B with coordinates (x_B, y_B, z_B). The position vector $\mathbf{r}_{AB}$ from A to B, shown in Fig. 2.24a, is given in terms of the coordinates of A and B by

$$\mathbf{r}_{AB} = (x_B - x_A)\mathbf{i} + (y_B - y_A)\mathbf{j} + (z_B - z_A)\mathbf{k}. \quad (2.17)$$

The components are obtained by subtracting the coordinates of point A from the coordinates of point B (Fig. 2.24b).

Components of a Vector Parallel to a Given Line

In three-dimensional applications, the direction of a vector is often defined by specifying the coordinates of two points on a line that is parallel to the vector. This information can be used to determine the components of the vector.

Suppose that we know the coordinates of two points A and B on a line parallel to a vector **U** (Fig. 2.25a). We can use Eq. (2.17) to determine the position vector $\mathbf{r}_{AB}$ from A to B (Fig. 2.25b). We can divide $\mathbf{r}_{AB}$ by its magnitude to obtain a unit vector $\mathbf{e}_{AB}$ that points from A toward B (Fig. 2.25c). Since $\mathbf{e}_{AB}$ has the same direction as **U**, we can determine **U** in terms of its scalar components by expressing it as the product of its magnitude and $\mathbf{e}_{AB}$.

More generally, suppose that we know the magnitude of a vector **U** and the components of any vector **V** that has the same direction as **U**. Then $\mathbf{V}/|\mathbf{V}|$ is a unit vector with the same direction as **U**, and we can determine the components of **U** by expressing it as $\mathbf{U} = |\mathbf{U}|(\mathbf{V}/|\mathbf{V}|)$.

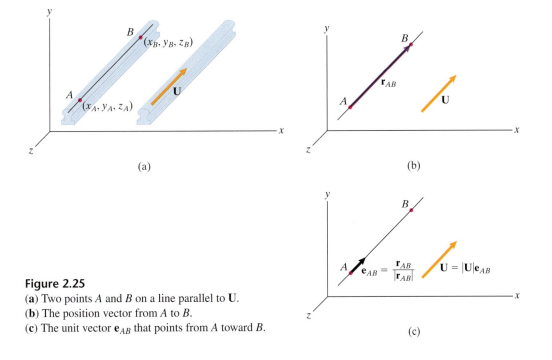

(a)

(b)

(c)

Figure 2.25
(a) Two points A and B on a line parallel to **U**.
(b) The position vector from A to B.
(c) The unit vector $\mathbf{e}_{AB}$ that points from A toward B.

Study Questions

1. How do you identify a right-handed coordinate system?
2. If you know the scalar components of a vector in three dimensions, how can you determine its magnitude?
3. What are the direction cosines of a vector? If you know them, how do you determine the components of the vector?
4. Suppose that you know the coordinates of two points A and B in three dimensions. How do you determine the components of the position vector of point B relative to point A?

Example 2.6 | **Magnitude and Direction Cosines of a Vector**

The coordinates of point C of the truss in Fig.2.26 are $x_C = 4$ m, $y_C = 0$, $z_C = 0$, and the coordinates of point D are $x_D = 2$ m, $y_D = 3$ m, $z_D = 1$ m. Let $\mathbf{r}_{CD}$ be the position vector from C to D.

(a) What is the magnitude of $\mathbf{r}_{CD}$?

(b) What are the direction cosines of $\mathbf{r}_{CD}$?

(c) Determine the components of a unit vector $\mathbf{e}_{CD}$ that points from point C toward point D.

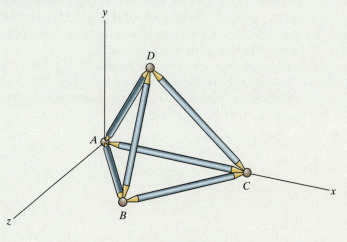

Figure 2.26

Strategy

(a) We can obtain the components of $\mathbf{r}_{CD}$ by subtracting the coordinates of C from the coordinates of D.

(b) Once the components of $\mathbf{r}_{CD}$ are known, we can determine the direction cosines from Eqs. (2.15).

(c) Dividing the vector $\mathbf{r}_{CD}$ by its magnitude yields the unit vector $\mathbf{e}_{CD}$.

Solution

(a) The vector $\mathbf{r}_{CD}$ and the coordinates of points C and D are shown in Fig. a. The components of $\mathbf{r}_{CD}$ are given by

$$\mathbf{r}_{CD} = (x_D - x_C)\mathbf{i} + (y_D - y_C)\mathbf{j} + (z_D - z_C)\mathbf{k}$$

$$= (2-4)\mathbf{i} + (3-0)\mathbf{j} + (1-0)\mathbf{k} \text{ (m)}$$

$$= -2\mathbf{i} + 3\mathbf{j} + \mathbf{k} \text{ (m)}.$$

(b) The magnitude of $\mathbf{r}_{CD}$ is

$$|\mathbf{r}_{CD}| = \sqrt{r_{CDx}{}^2 + r_{CDy}{}^2 + r_{CDz}{}^2}$$

$$= \sqrt{(-2 \text{ m})^2 + (3 \text{ m})^2 + (1 \text{ m})^2}$$

$$= 3.74 \text{ m}.$$

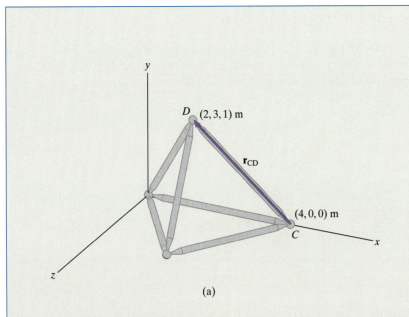

(a)

The direction cosines of $\mathbf{r}_{CD}$ are

$$\cos\theta_x = \frac{r_{CDx}}{|\mathbf{r}_{CD}|} = \frac{-2\text{ m}}{3.74\text{ m}} = -0.535,$$

$$\cos\theta_y = \frac{r_{CDy}}{|\mathbf{r}_{CD}|} = \frac{3\text{ m}}{3.74\text{ m}} = 0.802,$$

$$\cos\theta_z = \frac{r_{CDz}}{|\mathbf{r}_{CD}|} = \frac{1\text{ m}}{3.74\text{ m}} = 0.267,$$

(c) The unit vector that points from C toward D is

$$\mathbf{e}_{CD} = \frac{\mathbf{r}_{CD}}{|\mathbf{r}_{CD}|}$$

$$= \frac{-2\mathbf{i} + 3\mathbf{j} + \mathbf{k}\text{ (m)}}{3.74\text{ m}}$$

$$= -0.535\mathbf{i} + 0.802\mathbf{j} + 0.267\mathbf{k}.$$

(Notice that we already knew these components, because the direction cosines of $\mathbf{r}_{CD}$ are the components of a unit vector with the same direction as $\mathbf{r}_{CD}$.)

Critical Thinking
Why is it useful to know the components of a unit vector that points from C toward D? In Chapter 6 we will analyze the internal forces in the members of three-dimensional trusses, such as the one in this example. To express those forces in terms of their components, we will need to know the components of unit vectors parallel to the members.

Example 2.7   Determining Components in Three Dimensions

The crane in Fig. 2.27 exerts a 600-lb force $\mathbf{F}$ on the caisson. The angle between $\mathbf{F}$ and the x axis is 54°, and the angle between $\mathbf{F}$ and the y axis is 40°. The z component of $\mathbf{F}$ is positive. Express $\mathbf{F}$ in terms of components.

Figure 2.27

Strategy
Only two of the angles between the vector and the positive coordinate axes are given, but we can use Eq. (2.16) to determine the third angle. Then we can determine the components of $\mathbf{F}$ by using Eqs. (2.15).

Solution
The angles between $\mathbf{F}$ and the positive coordinate axes are related by

$$\cos^2 \theta_x + \cos^2 \theta_y + \cos^2 \theta_z = (\cos 54°)^2 + (\cos 40°)^2 + \cos^2 \theta_z = 1.$$

Solving this equation for $\cos \theta_z$, we obtain the two solutions $\cos \theta_z = 0.260$ and $\cos \theta_z = -0.260$, which tells us that $\theta_z = 74.9°$ or $\theta_z = 105.1°$. The z component of the vector $\mathbf{F}$ is positive, so the angle between $\mathbf{F}$ and the positive z axis is less than 90°. Therefore $\theta_z = 74.9°$.

The components of $\mathbf{F}$ are

$$F_x = |\mathbf{F}| \cos \theta_x = 600 \cos 54° = 353 \text{ lb},$$

$$F_y = |\mathbf{F}| \cos \theta_y = 600 \cos 40° = 460 \text{ lb},$$

$$F_z = |\mathbf{F}| \cos \theta_z = 600 \cos 74.9° = 156 \text{ lb}.$$

Critical Thinking
You are aware that knowing the square of a number does not tell you the value of the number uniquely. If $a^2 = 4$, the number a can be either 2 or −2. In this example, knowledge of the angles θ_x and θ_y allowed us to solve Eq. (2.16) for the value of $\cos^2 \theta_z$, which resulted in two possible values of the angle θ_z. There is a simple geometrical explanation for why this happened. The two angles θ_x and θ_y are sufficient to define a line parallel to the vector $\mathbf{F}$, *but not the direction of $\mathbf{F}$ along that line.* The two values of θ_z we obtained correspond to the two possible directions of $\mathbf{F}$ along the line. Additional information is needed to indicate the direction. In this example, the additional information was supplied by stating that the z component of $\mathbf{F}$ is positive.

Example 2.8	Determining Components in Three Dimensions

The tether of the balloon in Fig. 2.28 exerts an 800-N force **F** on the hook at O. The vertical line AB intersects the x–z plane at point A. The angle between the z axis and the line OA is 60°, and the angle between the line OA and **F** is 45°. Express **F** in terms of components.

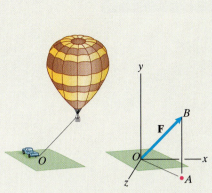

Figure 2.28

Strategy

We can determine the components of **F** from the given geometric information in two steps. First, we express **F** as the sum of two vector components parallel to the lines OA and AB. The component parallel to AB is the vector component $\mathbf{F}_y$. Then we can use the component parallel to OA to determine the vector components $\mathbf{F}_x$ and $\mathbf{F}_z$.

Solution

In Fig. a, we express **F** as the sum of its y component $\mathbf{F}_y$ and the component $\mathbf{F}_h$ parallel to OA. The magnitude of $\mathbf{F}_y$ is

$$|\mathbf{F}_y| = |\mathbf{F}| \sin 45° = (800 \text{ N}) \sin 45° = 566 \text{ N},$$

and the magnitude of $\mathbf{F}_h$ is

$$|\mathbf{F}_h| = |\mathbf{F}| \cos 45° = (800 \text{ N}) \cos 45° = 566 \text{ N},$$

In Fig. b, we express $\mathbf{F}_h$ in terms of the vector components $\mathbf{F}_x$ and $\mathbf{F}_z$. The magnitude of $\mathbf{F}_x$ is

$$|\mathbf{F}_x| = |\mathbf{F}_h| \sin 60° = (566 \text{ N}) \sin 60° = 490 \text{ N},$$

and the magnitude of $\mathbf{F}_z$ is

$$|\mathbf{F}_z| = |\mathbf{F}_h| \cos 60° = (566 \text{ N}) \cos 60° = 283 \text{ N}.$$

The vector components $\mathbf{F}_x$, $\mathbf{F}_y$, and $\mathbf{F}_z$ all point in the positive axis directions, so the scalar components of **F** are positive:

$$\mathbf{F} = 490\mathbf{i} + 566\mathbf{j} + 283\mathbf{k} \text{ (N)}.$$

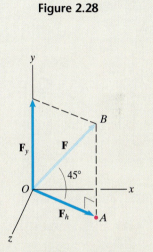

(a) Resolving **F** into vector components parallel to OA and OB.

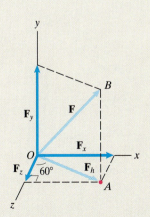

(b) Resolving $\mathbf{F}_h$ into vector components parallel to the x and z axes.

Critical Thinking

As this example demonstrates, two angles are required to specify a vector's direction relative to a three-dimensional coordinate system. The two angles used may not be defined in the same way as in the example (see Problem 2.84), but however they are defined, you can determine the components of the vector in terms of the magnitude and the two specified angles by a procedure similar to the one we used here.

Example 2.9   Vector Whose Direction Is Specified by Two Points

The bar AB in Fig. 2.29 exerts a 140-N force $\mathbf{F}$ on its support at A. The force is parallel to the bar and points toward B. Express $\mathbf{F}$ in terms of components.

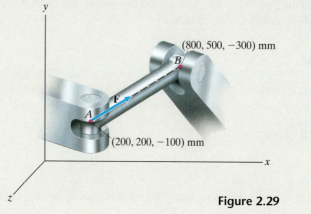

Figure 2.29

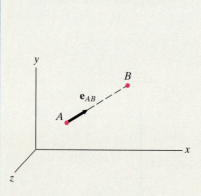

(a) The position vector $\mathbf{r}_{AB}$.

(b) The unit vector $\mathbf{e}_{AB}$ pointing from A toward B.

Strategy

Since we are given the coordinates of points A and B, we can determine the components of the position vector from A to B. By dividing the position vector by its magnitude, we can obtain a unit vector with the same direction as $\mathbf{F}$. Then by multiplying the unit vector by the magnitude of $\mathbf{F}$, we obtain $\mathbf{F}$ in terms of its components.

Solution

The position vector from A to B, with the coordinates in mm, is (Fig. a)

$$\mathbf{r}_{AB} = (x_B - x_A)\mathbf{i} + (y_B - y_A)\mathbf{j} + (z_B - z_A)\mathbf{k}$$

$$= [(800) - (200)]\mathbf{i} + [(500) - (200)]\mathbf{j} + [(-300) - (-100)]\mathbf{k}$$

$$= 600\mathbf{i} + 300\mathbf{j} - 200\mathbf{k} \text{ (mm).}$$

and its magnitude is

$$|\mathbf{r}_{AB}| = \sqrt{(600)^2 + (300)^2 + (-200)^2} = 700 \text{ mm.}$$

By dividing $\mathbf{r}_{AB}$ by its magnitude, we obtain a unit vector with the same direction as $\mathbf{F}$ (Fig. b):

$$\mathbf{e}_{AB} = \frac{\mathbf{r}_{AB}}{|\mathbf{r}_{AB}|} = \frac{6}{7}\mathbf{i} + \frac{3}{7}\mathbf{j} - \frac{2}{7}\mathbf{k}.$$

Then, in terms of its components,

$$\mathbf{F} = |\mathbf{F}|\mathbf{e}_{AB} = (140 \text{ N})\left(\frac{6}{7}\mathbf{i} + \frac{3}{7}\mathbf{j} - \frac{2}{7}\mathbf{k}\right) = 120\mathbf{i} + 60\mathbf{j} - 40\mathbf{k} \text{ (N).}$$

Critical Thinking

Prescribing the positions of two points on the line of action of a vector, as in this example, is a common method of specifying the direction of a vector in three dimensions. Notice that this example involves three distinct types of vectors: a force vector, a position vector, and a unit vector. They have the same direction relative to the coordinate system, but the magnitude of $\mathbf{F}$ is in newtons, the magnitude of $\mathbf{r}_{AB}$ is in millimeters, and $\mathbf{e}_{AB}$ is dimensionless.

Example 2.10   Determining Components in Three Dimensions

The rope in Fig. 2.30 extends from point B through a metal loop attached to the wall at A to point C. The rope exerts forces $\mathbf{F}_{AB}$ and $\mathbf{F}_{AC}$ on the loop at A with magnitudes $|\mathbf{F}_{AB}| = |\mathbf{F}_{AC}| = 200$ lb. What is the magnitude of the total force $\mathbf{F} = \mathbf{F}_{AB} + \mathbf{F}_{AC}$ exerted on the loop by the rope?

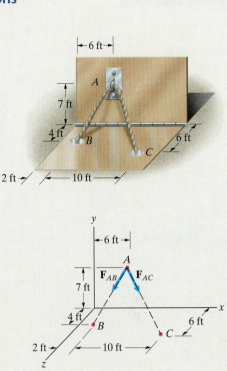

Strategy

The force $\mathbf{F}_{AB}$ is parallel to the line from A to B, and the force $\mathbf{F}_{AC}$ is parallel to the line from A to C. Since we can determine the coordinates of points A, B, and C from the given dimensions, we can determine the components of unit vectors that have the same directions as the two forces and use them to express the forces in terms of scalar components.

Solution

Let $\mathbf{r}_{AB}$ be the position vector from point A to point B and let $\mathbf{r}_{AC}$ be the position vector from point A to point C (Fig. a). From the given dimensions, the coordinates of points A, B, and C are

$$A: (6, 7, 0) \text{ ft}, \qquad B: (2, 0, 4) \text{ ft}, \qquad C: (12, 0, 6) \text{ ft}.$$

Therefore, the components of $\mathbf{r}_{AB}$ and $\mathbf{r}_{AC}$, with the coordinates in ft, are given by

$$\mathbf{r}_{AB} = (x_B - x_A)\mathbf{i} + (y_B - y_A)\mathbf{j} + (z_B - z_A)\mathbf{k}$$
$$= (2 - 6)\mathbf{i} + (0 - 7)\mathbf{j} + (4 - 0)\mathbf{k}$$
$$= -4\mathbf{i} - 7\mathbf{j} + 4\mathbf{k} \text{ (ft)}$$

and

$$\mathbf{r}_{AC} = (x_C - x_A)\mathbf{i} + (y_C - y_A)\mathbf{j} + (z_C - z_A)\mathbf{k}$$
$$= (12 - 6)\mathbf{i} + (0 - 7)\mathbf{j} + (6 - 0)\mathbf{k}$$
$$= 6\mathbf{i} - 7\mathbf{j} + 6\mathbf{k} \text{ (ft)}.$$

Figure 2.30

Their magnitudes are $|\mathbf{r}_{AB}| = 9$ ft and $|\mathbf{r}_{AC}| = 11$ ft. By dividing $\mathbf{r}_{AB}$ and $\mathbf{r}_{AC}$ by their magnitudes, we obtain unit vectors $\mathbf{e}_{AB}$ and $\mathbf{e}_{AC}$ that point in the directions of $\mathbf{F}_{AB}$ and $\mathbf{F}_{AC}$ (Fig. b):

$$\mathbf{e}_{AB} = \frac{\mathbf{r}_{AB}}{|\mathbf{r}_{AB}|} = -0.444\mathbf{i} - 0.778\mathbf{j} + 0.444\mathbf{k},$$

$$\mathbf{e}_{AC} = \frac{\mathbf{r}_{AC}}{|\mathbf{r}_{AC}|} = 0.545\mathbf{i} - 0.636\mathbf{j} + 0.545\mathbf{k}.$$

The forces $\mathbf{F}_{AB}$ and $\mathbf{F}_{AC}$ are

$$\mathbf{F}_{AB} = (200 \text{ lb})\mathbf{e}_{AB} = -88.9\mathbf{i} - 155.6\mathbf{j} + 88.9\mathbf{k} \text{ (lb)},$$
$$\mathbf{F}_{AC} = (200 \text{ lb})\mathbf{e}_{AC} = 109.1\mathbf{i} - 127.3\mathbf{j} + 109.1\mathbf{k} \text{ (lb)}.$$

The total force exerted on the loop by the rope is

$$\mathbf{F} = \mathbf{F}_{AB} + \mathbf{F}_{AC} = 20.2\mathbf{i} - 282.8\mathbf{j} + 198.0\mathbf{k} \text{ (lb)},$$

and its magnitude is

$$|\mathbf{F}| = \sqrt{(20.2)^2 + (-282.8)^2 + (198.0)^2} = 346 \text{ lb}.$$

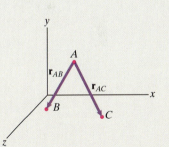

(a) The position vectors $\mathbf{r}_{AB}$ and $\mathbf{r}_{AC}$.

Critical Thinking

How do you know that the magnitude and direction of the total force exerted on the metal loop at A by the rope is given by the magnitude and direction of the vector $\mathbf{F} = \mathbf{F}_{AB} + \mathbf{F}_{AC}$? At this point in our development of mechanics, we assume that force is a vector, but have provided no proof. In the study of dynamics it is shown that Newton's second law implies that force is a vector.

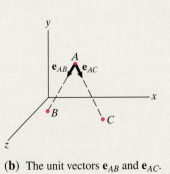

(b) The unit vectors $\mathbf{e}_{AB}$ and $\mathbf{e}_{AC}$.

Example 2.11 | **Determining Components of a Force**

The cable AB in Fig. 2.31 exerts a 50-N force $\mathbf{T}$ on the collar at A. Express $\mathbf{T}$ in terms of components.

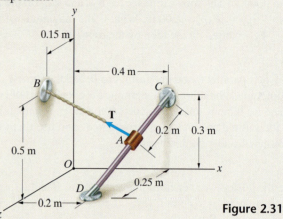

Figure 2.31

Strategy

Let $\mathbf{r}_{AB}$ be the position vector from A to B. We will divide $\mathbf{r}_{AB}$ by its magnitude to obtain a unit vector $\mathbf{e}_{AB}$ having the same direction as the force $\mathbf{T}$. Then we can obtain $\mathbf{T}$ in terms of scalar components by expressing it as the product of its magnitude and $\mathbf{e}_{AB}$. To begin this procedure, we must first determine the coordinates of the collar A. We will do so by obtaining a unit vector $\mathbf{e}_{CD}$ pointing from C toward D and multiplying it by 0.2 m to determine the position of the collar A relative to C.

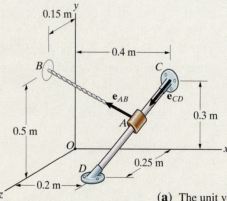

(a) The unit vectors $\mathbf{e}_{AB}$ and $\mathbf{e}_{CD}$.

Solution

Determining the Coordinates of Point A   The position vector from C to D, with the coordinates in meters, is

$$\mathbf{r}_{CD} = (0.2 - 0.4)\mathbf{i} + (0 - 0.3)\mathbf{j} + (0.25 - 0)\mathbf{k}$$

$$= -0.2\mathbf{i} - 0.3\mathbf{j} + 0.25\mathbf{k} \text{ (m)}.$$

Dividing this vector by its magnitude, we obtain the unit vector $\mathbf{e}_{CD}$ (Fig. a):

$$\mathbf{e}_{CD} = \frac{\mathbf{r}_{CD}}{|\mathbf{r}_{CD}|} = \frac{-0.2\mathbf{i} - 0.3\mathbf{j} + 0.25\mathbf{k}}{\sqrt{(-0.2)^2 + (-0.3)^2 + (0.25)^2}}$$

$$= -0.456\mathbf{i} - 0.684\mathbf{j} + 0.570\mathbf{k}.$$

Using this vector, we obtain the position vector from C to A:

$$\mathbf{r}_{CA} = (0.2\text{m})\mathbf{e}_{CD} = -0.091\mathbf{i} - 0.137\mathbf{j} + 0.114\mathbf{k}\ (\text{m}).$$

The position vector from the origin of the coordinate system to C is $\mathbf{r}_{OC} = 0.4\mathbf{i} + 0.3\mathbf{j}$ (m), so the position vector from the origin to A is

$$\mathbf{r}_{OA} = \mathbf{r}_{OC} + \mathbf{r}_{CA} = (0.4\mathbf{i} + 0.3\mathbf{j}) + (-0.091\mathbf{i} - 0.137\mathbf{j} + 0.114\mathbf{k})$$

$$= 0.309\mathbf{i} + 0.163\mathbf{j} + 0.114\mathbf{k}\ (\text{m}).$$

The coordinates of A are $(0.309, 0.163, 0.114)$ m.

Determining the Components of T Using the coordinates of point A, we find that the position vector from A to B is

$$\mathbf{r}_{AB} = (0 - 0.309)\mathbf{i} + (0.5 - 0.163)\mathbf{j} + (0.15 - 0.114)\mathbf{k}$$

$$= -0.309\mathbf{i} + 0.337\mathbf{j} + 0.036\mathbf{k}\ (\text{m}).$$

Dividing this vector by its magnitude, we obtain the unit vector $\mathbf{e}_{AB}$ (Fig. a).

$$\mathbf{e}_{AB} = \frac{\mathbf{r}_{AB}}{|\mathbf{r}_{AB}|} = \frac{-0.309\mathbf{i} + 0.337\mathbf{j} + 0.036\mathbf{k}\ (\text{m})}{\sqrt{(-0.309\ \text{m})^2 + (0.337\ \text{m})^2 + (0.036\ \text{m})^2}}$$

$$= -0.674\mathbf{i} + 0.735\mathbf{j} + 0.079\mathbf{k}.$$

The force $\mathbf{T}$ is

$$\mathbf{T} = |\mathbf{T}|\mathbf{e}_{AB} = (50\ \text{N})(-0.674\mathbf{i} + 0.735\mathbf{j} + 0.079\mathbf{k})$$

$$= -33.7\mathbf{i} + 36.7\mathbf{j} + 3.9\mathbf{k}\ (\text{N}).$$

Critical Thinking

Look at the two ways unit vectors were used in this example. The unit vector $\mathbf{e}_{CD}$ was used to obtain the components of the position vector $\mathbf{r}_{CA}$, which made it possible to determine the coordinates of point A. The coordinates of point A were then used to determine the unit vector $\mathbf{e}_{AB}$, which was used to express the force $\mathbf{T}$ in terms of its components.

Problems

2.61 A vector $\mathbf{U} = 3\mathbf{i} - 4\mathbf{j} - 12\mathbf{k}$. What is its magnitude?

Strategy: The magnitude of a vector is given in terms of its components by Eq. (2.14).

2.62 The vector $\mathbf{e} = \frac{1}{3}\mathbf{i} + \frac{2}{3}\mathbf{j} + e_z\mathbf{k}$ is a unit vector. Determine the component e_z.

2.63 An engineer determines that the attachment point will be subjected to a force $\mathbf{F} = 20\mathbf{i} + F_y\mathbf{j} - 45\mathbf{k}$ (kN). If the attachment point will safely support a force of 80-kN magnitude in any direction, what is the acceptable range of values of F_y?

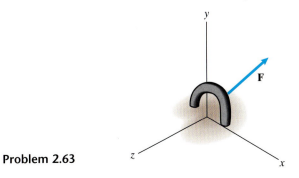

Problem 2.63

2.64 A vector $\mathbf{U} = U_x\mathbf{i} + U_y\mathbf{j} + U_z\mathbf{k}$. Its magnitude $|\mathbf{U}| = 30$. Its components are related by the equations $U_y = -2U_x$ and $U_z = 4U_y$. Determine the components.

2.65 An object is acted upon by two forces $\mathbf{F}_1 = 20\mathbf{i} + 30\mathbf{j} - 24\mathbf{k}$ (kN) and $\mathbf{F}_2 = -60\mathbf{i} + 20\mathbf{j} + 40\mathbf{k}$ (kN). What is the magnitude of the total force acting on the object?

2.66 Two vectors $\mathbf{U} = 3\mathbf{i} - 2\mathbf{j} + 6\mathbf{k}$ and $\mathbf{V} = 4\mathbf{i} + 12\mathbf{j} - 3\mathbf{k}$.
(a) Determine the magnitudes of $\mathbf{U}$ and $\mathbf{V}$.
(b) Determine the magnitude of the vector $3\mathbf{U} + 2\mathbf{V}$.

2.67 A vector $\mathbf{U} = 40\mathbf{i} - 70\mathbf{j} - 40\mathbf{k}$.
(a) What is its magnitude?
(b) What are the angles θ_x, θ_y, and θ_z between $\mathbf{U}$ and the positive coordinate axes?

   Strategy: Since you know the components of $\mathbf{U}$, you can determine the angles θ_x, θ_y, and θ_z from Eqs. (2.15).

2.68 A force vector is given in terms of its components by $\mathbf{F} = 10\mathbf{i} - 20\mathbf{j} - 20\mathbf{k}$ (N).
(a) What are the direction cosines of $\mathbf{F}$?
(b) Determine the components of a unit vector $\mathbf{e}$ that has the same direction as $\mathbf{F}$.

2.69 The cable exerts a force $\mathbf{F}$ on the hook at O whose magnitude is 200 N. The angle between the vector $\mathbf{F}$ and the x axis is 40°, and the angle between the vector $\mathbf{F}$ and the y axis is 70°.
(a) What is the angle between the vector $\mathbf{F}$ and the z axis?
(b) Express $\mathbf{F}$ in terms of components.

   Strategy: (a) Because you know the angles between the vector $\mathbf{F}$ and the x and y axes, you can use Eq. (2.16) to determine the angle between $\mathbf{F}$ and the z axis. (Observe from the figure that the angle between $\mathbf{F}$ and the z axis is clearly within the range $0 < \theta_z < 180°$.)
(b) The components of $\mathbf{F}$ can be obtained with Eqs. (2.15).

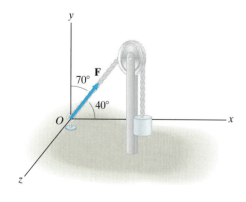

Problem 2.69

2.70 A unit vector has direction cosines $\cos\theta_x = -0.5$ and $\cos\theta_y = 0.2$. Its z component is positive. Express it in terms of components.

2.71 The airplane's engines exert a total thrust force $\mathbf{T}$ of 200-kN magnitude. The angle between $\mathbf{T}$ and the x axis is 120°, and the angle between $\mathbf{T}$ and the y axis is 130°. The z component of $\mathbf{T}$ is positive.
(a) What is the angle between $\mathbf{T}$ and the z axis?
(b) Express $\mathbf{T}$ in terms of components.

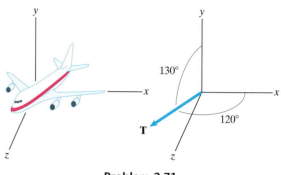

Problem 2.71

Refer to the following diagram when solving Problems 2.72 through 2.76.

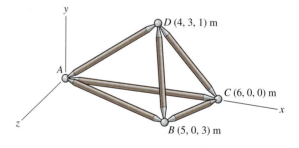

Problems 2.72–2.76

2.72 Determine the components of the position vector $\mathbf{r}_{BD}$ from point B to point D. Use your result to determine the distance from B to D.

2.73 What are the direction cosines of the position vector $\mathbf{r}_{BD}$ from point B to point D?

2.74 Determine the components of the unit vector $\mathbf{e}_{CD}$ that points from point C toward point D.

2.75 What are the direction cosines of the unit vector $\mathbf{e}_{CD}$ that points from point C toward point D?

2.76 The bar *CD* exerts a force **F** on the joint at point *D* that points from point *C* toward point *D*. Its magnitude is $|\mathbf{F}| = 40$ kN. Express **F** in terms of components.

2.77 Astronauts on the space shuttle use radar to determine the magnitudes and direction cosines of the position vectors of two satellites *A* and *B*. The vector $\mathbf{r}_A$ from the shuttle to satellite *A* has magnitude 2 km, and direction cosines $\cos\theta_x = 0.768$, $\cos\theta_y = 0.384$, $\cos\theta_z = 0.512$. The vector $\mathbf{r}_B$ from the shuttle to satellite *B* has magnitude 4 km and direction cosines $\cos\theta_x = 0.743$, $\cos\theta_y = 0.557$, $\cos\theta_z = -0.371$. What is the distance between the satellites?

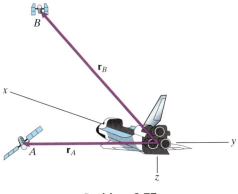

Problem 2.77

2.78 Archaeologists measure a pre-Columbian ceremonial structure and obtain the dimensions shown. Determine (a) the magnitude and (b) the direction cosines of the position vector from point *A* to point *B*.

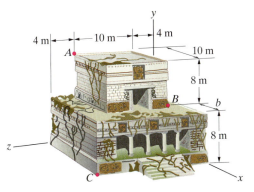

Problem 2.78

2.79 Consider the structure described in Problem 2.78. After returning to the United States, an archaeologist discovers that a graduate student has erased the only data file containing the dimension *b*. But from recorded GPS data he is able to calculate that the distance from point B to point C is 16.61 m.

(a) What is the distance *b*?

(b) Determine the direction cosines of the position vector from *B* to *C*.

2.80 Observers at *A* and *B* use theodolites to measure the direction from their positions to a rocket in flight. If the coordinates of the rocket's position at a given instant are (4, 4, 2) km, determine the direction cosines of the vectors $\mathbf{r}_{AR}$ and $\mathbf{r}_{BR}$ that the observers would measure at that instant.

2.81 * Suppose that the coordinates of the rocket's position are unknown. At a given instant, the person at *A* determines that the direction cosines of $\mathbf{r}_{AR}$ are $\cos\theta_x = 0.535$, $\cos\theta_y = 0.802$, and $\cos\theta_z = 0.267$, and the person at *B* determines that the direction cosines of $\mathbf{r}_{BR}$ are $\cos\theta_x = -0.576$, $\cos\theta_y = 0.798$, and $\cos\theta_z = -0.177$. What are the coordinates of the rocket's position at that instant?

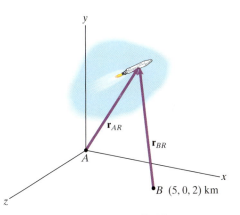

Problems 2.80/2.81

2.82 * The height of Mount Everest was originally measured by a surveyor in the following way. He first measured the altitudes of two points and the horizontal distance between them. For example, suppose that the points A and B are 3000 m above sea level and are 10,000 m apart. He then used a theodolite to measure the direction cosines of the vector $\mathbf{r}_{AP}$ from point A to the top of the mountain P and the vector $\mathbf{r}_{BP}$ from point B to P. Suppose that the direction cosines of $\mathbf{r}_{AP}$ are $\cos\theta_x = 0.5179$, $\cos\theta_y = 0.6906$, and $\cos\theta_z = 0.5048$, and the direction cosines of $\mathbf{r}_{BP}$ are $\cos\theta_x = -0.3743$, $\cos\theta_y = 0.7486$, and $\cos\theta_z = 0.5472$. Using this data, determine the height of Mount Everest above sea level.

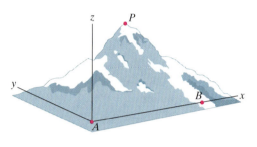

Problem 2.82

2.83 The distance from point O to point A is 20 ft. The straight line AB is parallel to the y axis, and point B is in the x–z plane. Express the vector $\mathbf{r}_{OA}$ in terms of components.

Strategy: You can express $\mathbf{r}_{OA}$ as the sum of a vector from O to B and a vector from B to A. You can then express the vector from O to B as the sum of vector components parallel to the x and z axes. See Example 2.8.

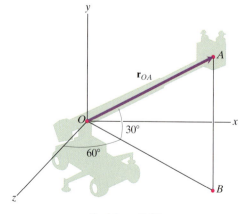

Problem 2.83

2.84 The pole supporting the sign is parallel to the x axis and is 2 m long. Point A is contained in the y–z plane. Express the position vector $\mathbf{r}$ from the origin to the end of the pole in terms of components.

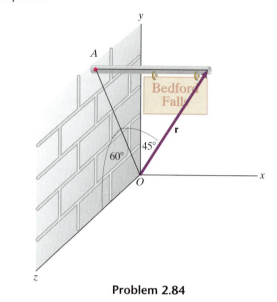

Problem 2.84

2.85 The straight line from the head of the force vector $\mathbf{F}$ to point A is parallel to the y axis, and point A is contained in the x–z plane. The x component of $\mathbf{F}$ is $F_x = 100$ N.
(a) What is the magnitude of $\mathbf{F}$?
(b) Determine the angles θ_x, θ_y, and θ_z between $\mathbf{F}$ and the positive coordinate axes.

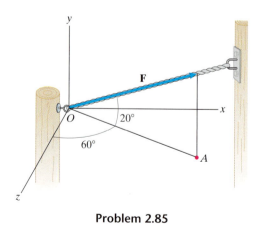

Problem 2.85

2.86 The position of a point P on the surface of the earth is specified by the longitude λ, measured from the point G on the equator directly south of Greenwich, England, and the latitude L measured from the equator. Longitude is given as west (W) longitude or east (E) longitude, indicating whether the angle is measured west or east from point G. Latitude is given as north (N) latitude or south (S) latitude, indicating whether the angle is measured north or south from the equator. Suppose that P is at longitude 30° W and latitude 45° N. Let R_E be the radius of the earth. Using the coordinate system shown, determine the components of the position vector of P relative to the center of the earth. (Your answer will be in terms of R_E.)

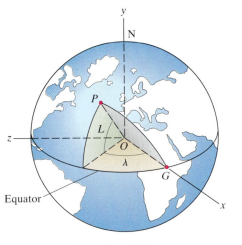

Problem 2.86

2.87 An engineer calculates that the magnitude of the axial force in one of the beams of a geodesic dome is $|\mathbf{P}| = 7.65$ kN. The cartesian coordinates of the endpoints A and B of the straight beam are $(-12.4, 22.0, -18.4)$ m and $(-9.2, 24.4, -15.6)$ m, respectively. Express the force $\mathbf{P}$ in terms of components.

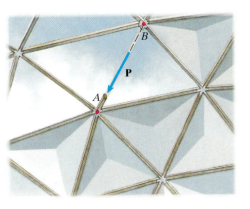

Problem 2.87

2.88 The cable BC exerts an 8-kN force $\mathbf{F}$ on the bar AB at B.
(a) Determine the components of a unit vector that points from point B toward point C.
(b) Express $\mathbf{F}$ in terms of components.

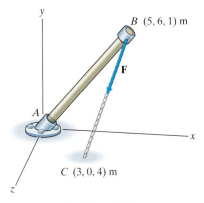

Problem 2.88

2.89 A cable extends from point C to point E. It exerts a 50-lb force $\mathbf{T}$ on the plate at C that is directed along the line from C to E. Express $\mathbf{T}$ in terms of components.

2.90 What are the direction cosines of the force $\mathbf{T}$ in Problem 2.89?

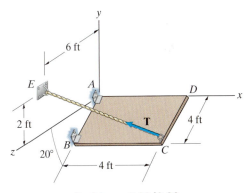

Problems 2.89/2.90

2.91 The cable AB exerts a 200-lb force $\mathbf{F}_{AB}$ at point A that is directed along the line from A to B. Express $\mathbf{F}_{AB}$ in terms of components.

2.92 Cable AB exerts a 200-lb force $\mathbf{F}_{AB}$ at point A that is directed along the line from A to B. The cable AC exerts a 100-lb force $\mathbf{F}_{AC}$ at point A that is directed along the line from A to C. Determine the magnitude of the total force exerted at point A by the two cables.

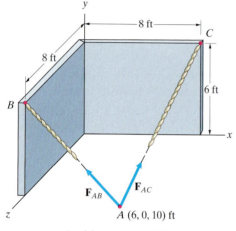

Problems 2.91/2.92

2.93 The 70-m-tall tower is supported by three cables that exert forces $\mathbf{F}_{AB}$, $\mathbf{F}_{AC}$, and $\mathbf{F}_{AD}$ on it. The magnitude of each force is 2 kN. Express the total force exerted on the tower by the three cables in terms of components.

2.94 The magnitude of the force $\mathbf{F}_{AB}$ is 2 kN. The x and z components of the vector sum of the forces exerted on the tower by the three cables are zero. What are the magnitudes of $\mathbf{F}_{AC}$ and $\mathbf{F}_{AD}$?

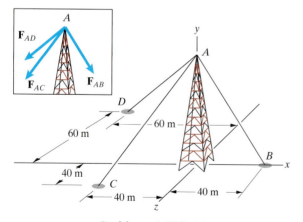

Problems 2.93/2.94

2.95 Express the position vector from point O to the collar at A in terms of components.

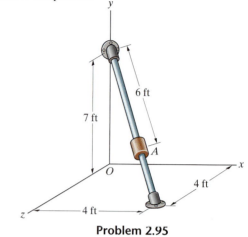

Problem 2.95

2.96 The cable AB exerts a 32-lb force $\mathbf{T}$ on the collar at A. Express $\mathbf{T}$ in terms of components.

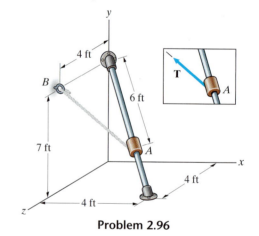

Problem 2.96

2.97 The circular bar has a 4-m radius and lies in the $x\,y$ plane. Express the position vector from point B to the collar at A in terms of components.

2.98 The cable AB exerts a 60-N force $\mathbf{T}$ on the collar at A that is directed along the line from A toward B. Express $\mathbf{T}$ in terms of components.

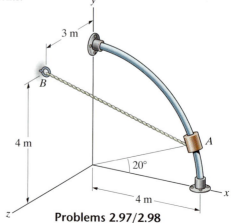

Problems 2.97/2.98

PRODUCTS OF VECTORS

Two kinds of products of vectors, the dot and cross products, have been found to have applications in science and engineering, especially in mechanics and electromagnetic field theory. We use both of these products in Chapter 4 to evaluate moments of forces about points and lines. We discuss them here so that our discussion of moments will not be interrupted by the details of vector operations.

2.5 Dot Products

The dot product of two vectors has many uses, including determining the components of a vector parallel and perpendicular to a given line and determining the angle between two lines in space.

Definition

Consider two vectors **U** and **V** (Fig. 2.32a). The *dot product* of **U** and **V**, denoted by **U · V** (hence the name "dot product"), is defined to be the product of the magnitude of **U**, the magnitude of **V**, and the cosine of the angle θ between **U** and **V** when they are placed tail to tail (Fig. 2.32b):

$$\mathbf{U} \cdot \mathbf{V} = |\mathbf{U}||\mathbf{V}| \cos \theta. \tag{2.18}$$

Because the result of the dot product is a scalar, the dot product is sometimes called the scalar product. The units of the dot product are the product of the units of the two vectors. *Notice that the dot product of two nonzero vectors is equal to zero if and only if the vectors are perpendicular.*

The dot product has the properties

$$\mathbf{U} \cdot \mathbf{V} = \mathbf{V} \cdot \mathbf{U}, \qquad \text{The dot product is commutative.} \tag{2.19}$$

$$a(\mathbf{U} \cdot \mathbf{V}) = (a\mathbf{U}) \cdot \mathbf{V} = \mathbf{U} \cdot (a\mathbf{V}), \qquad \text{The dot product is associative with respect to scalar multiplication.} \tag{2.20}$$

and

$$\mathbf{U} \cdot (\mathbf{V} + \mathbf{W}) = \mathbf{U} \cdot \mathbf{V} + \mathbf{U} \cdot \mathbf{W}, \qquad \text{The dot product is associative with respect to vector addition.} \tag{2.21}$$

for any scalar a and vectors **U**, **V**, and **W**.

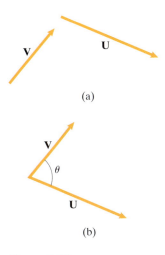

Figure 2.32
(a) The vectors **U** and **V**.
(b) The angle θ between **U** and **V** when the two vectors are placed tail to tail.

Dot Products in Terms of Components

In this section we derive an equation that allows you to determine the dot product of two vectors if you know their scalar components. The derivation also results in an equation for the angle between the vectors. The first step is to determine the dot products formed from the unit vectors **i**, **j**, and **k**. Let us evaluate the dot product **i · i**. The magnitude $|\mathbf{i}| = 1$, and the angle between two identical vectors placed tail to tail is zero, so we obtain

$$\mathbf{i} \cdot \mathbf{i} = |\mathbf{i}||\mathbf{i}| \cos (0) = (1)(1)(1) = 1.$$

The dot product of **i** and **j** is

$$\mathbf{i} \cdot \mathbf{j} = |\mathbf{i}||\mathbf{j}| \cdot \cos (90°) = (1)(1)(0) = 0.$$

Continuing in this way, we obtain

$$
\begin{aligned}
\mathbf{i}\cdot\mathbf{i} &= 1, & \mathbf{i}\cdot\mathbf{j} &= 0, & \mathbf{i}\cdot\mathbf{k} &= 0, \\
\mathbf{j}\cdot\mathbf{i} &= 0, & \mathbf{j}\cdot\mathbf{j} &= 1, & \mathbf{j}\cdot\mathbf{k} &= 0, \\
\mathbf{k}\cdot\mathbf{i} &= 0, & \mathbf{k}\cdot\mathbf{j} &= 0, & \mathbf{k}\cdot\mathbf{k} &= 1.
\end{aligned}
\tag{2.22}
$$

The dot product of two vectors $\mathbf{U}$ and $\mathbf{V}$, expressed in terms of their components, is

$$
\begin{aligned}
\mathbf{U}\cdot\mathbf{V} &= (U_x\mathbf{i} + U_y\mathbf{j} + U_z\mathbf{k})\cdot(V_x\mathbf{i} + V_y\mathbf{j} + V_z\mathbf{k}) \\
&= U_xV_x(\mathbf{i}\cdot\mathbf{i}) + U_xV_y(\mathbf{i}\cdot\mathbf{j}) + U_xV_z(\mathbf{i}\cdot\mathbf{k}) \\
&\quad + U_yV_x(\mathbf{j}\cdot\mathbf{i}) + U_yV_y(\mathbf{j}\cdot\mathbf{j}) + U_yV_z(\mathbf{j}\cdot\mathbf{k}) \\
&\quad + U_zV_x(\mathbf{k}\cdot\mathbf{i}) + U_zV_y(\mathbf{k}\cdot\mathbf{j}) + U_zV_z(\mathbf{k}\cdot\mathbf{k}).
\end{aligned}
$$

In obtaining this result, we used Eqs. (2.20) and (2.21). Substituting Eqs. (2.22) into this expression, we obtain an equation for the dot product in terms of the scalar components of the two vectors:

$$
\mathbf{U}\cdot\mathbf{V} = U_xV_x + U_yV_y + U_zV_z.
\tag{2.23}
$$

To obtain an equation for the angle θ in terms of the components of the vectors, we equate the expression for the dot product given by Eq. (2.23) to the definition of the dot product, Eq. (2.18), and solve for $\cos\theta$:

$$
\cos\theta = \frac{\mathbf{U}\cdot\mathbf{V}}{|\mathbf{U}||\mathbf{V}|} = \frac{U_xV_x + U_yV_y + U_zV_z}{|\mathbf{U}||\mathbf{V}|}.
\tag{2.24}
$$

Vector Components Parallel and Normal to a Line

In some engineering applications a vector must be expressed in terms of vector components that are parallel and normal (perpendicular) to a given line. The component of a vector parallel to a line is called the *projection* of the vector onto the line. For example, when the vector represents a force, the projection of the force onto a line is the component of the force in the direction of the line.

We can determine the components of a vector parallel and normal to a line by using the dot product. Consider a vector $\mathbf{U}$ and a straight line L (Fig. 2.33a). We can express $\mathbf{U}$ as the sum of vector components $\mathbf{U}_p$ and $\mathbf{U}_n$ that are parallel and normal to L (Fig. 2.33b).

The Parallel Component In terms of the angle θ between $\mathbf{U}$ and the vector component $\mathbf{U}_p$, the magnitude of $\mathbf{U}_p$ is

$$
|\mathbf{U}_p| = |\mathbf{U}|\cos\theta.
\tag{2.25}
$$

Let $\mathbf{e}$ be a unit vector parallel to L (Fig. 2.34). The dot product of $\mathbf{e}$ and $\mathbf{U}$ is

$$
\mathbf{e}\cdot\mathbf{U} = |\mathbf{e}||\mathbf{U}|\cos\theta = |\mathbf{U}|\cos\theta.
$$

Comparing this result with Eq. (2.25), we see that the magnitude of $\mathbf{U}_p$ is

$$
|\mathbf{U}_p| = \mathbf{e}\cdot\mathbf{U}.
$$

Therefore the parallel vector component, or projection of $\mathbf{U}$ onto L, is

$$
\mathbf{U}_p = (\mathbf{e}\cdot\mathbf{U})\mathbf{e}.
\tag{2.26}
$$

(This equation holds even if $\mathbf{e}$ doesn't point in the direction of $\mathbf{U}_p$. In that case, the angle $\theta > 90°$ and $\mathbf{e}\cdot\mathbf{U}$ is negative.) When the components of a vector and

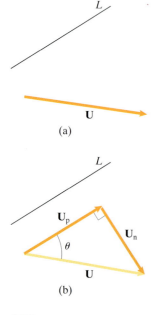

(a)

(b)

Figure 2.33
(a) A vector $\mathbf{U}$ and line L.
(b) Resolving $\mathbf{U}$ into components parallel and normal to L.

the components of a unit vector **e** parallel to a line L are known, we can use Eq. (2.26) to determine the component of the vector parallel to L.

The Normal Component   Once the parallel vector component has been determined, we can obtain the normal vector component from the relation $\mathbf{U} = \mathbf{U}_p + \mathbf{U}_n$:

$$\mathbf{U}_n = \mathbf{U} - \mathbf{U}_p. \qquad (2.27)$$

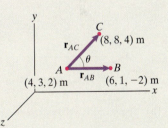

Figure 2.34
The unit vector **e** is parallel to L.

Study Questions

1. What is the definition of the dot product?

2. The dot product is commutative. What does that mean?

3. If you know the components of two vectors **U** and **V**, how can you determine their dot product?

4. How can you use the dot product to determine the vector components of a vector parallel and normal to a line?

Example 2.12	**Using the Dot Product to Determine an Angle**

What is the angle θ between the lines AB and AC in Fig. 2.35?

Strategy

We know the coordinates of the points A, B, and C, so we can determine the components of the vector $\mathbf{r}_{AB}$ from A to B and the vector $\mathbf{r}_{AC}$ from A to C (Fig. a). Then we can use Eq. (2.24) to determine θ.

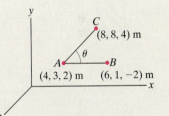

Figure 2.35

Solution

The vectors $\mathbf{r}_{AB}$ and $\mathbf{r}_{AC}$, with the coordinates in meters, are

$$\mathbf{r}_{AB} = (6 - 4)\mathbf{i} + (1 - 3)\mathbf{j} + (-2 - 2)\mathbf{k} = 2\mathbf{i} - 2\mathbf{j} - 4\mathbf{k} \text{ (m)},$$

$$\mathbf{r}_{AC} = (8 - 4)\mathbf{i} + (8 - 3)\mathbf{j} + (4 - 2)\mathbf{k} = 4\mathbf{i} + 5\mathbf{j} + 2\mathbf{k} \text{ (m)}.$$

Their magnitudes are

$$|\mathbf{r}_{AB}| = \sqrt{(2 \text{ m})^2 + (-2 \text{ m})^2 + (-4 \text{ m})^2} = 4.90 \text{ m},$$

$$|\mathbf{r}_{AC}| = \sqrt{(4 \text{ m})^2 + (5 \text{ m})^2 + (2 \text{ m})^2} = 6.71 \text{ m}.$$

The dot product of $\mathbf{r}_{AB}$ and $\mathbf{r}_{AC}$ is

$$\mathbf{r}_{AB} \cdot \mathbf{r}_{AC} = (2 \text{ m})(4 \text{ m}) + (-2 \text{ m})(5 \text{ m}) + (-4 \text{ m})(2 \text{ m}) = -10 \text{ m}^2.$$

Therefore,

$$\cos\theta = \frac{\mathbf{r}_{AB} \cdot \mathbf{r}_{AC}}{|\mathbf{r}_{AB}||\mathbf{r}_{AC}|} = \frac{-10 \text{ m}^2}{(4.90 \text{ m})(6.71 \text{ m})} = -0.304.$$

The angle $\theta = \arccos(-0.304) = 107.7°$.

(a) The position vectors $\mathbf{r}_{AB}$ and $\mathbf{r}_{AC}$.

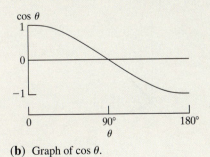

(b) Graph of $\cos\theta$.

Critical Thinking

What does it mean if the dot product of two vectors is negative? From Eq. (2.18) and the graph of the cosine (Fig. b), you can see that the dot product is negative, as it is in this example, only if the enclosed angle between the two vectors is greater than 90°.

Example 2.13 Vector Components Parallel and Normal to a Line

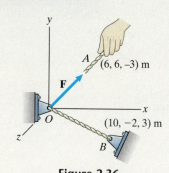

Figure 2.36

Suppose that you pull on the cable OA in Fig. 2.36, exerting a 50-N force $\mathbf{F}$ at O. What are the vector components of $\mathbf{F}$ parallel and normal to the cable OB?

Strategy

Expressing $\mathbf{F}$ as the sum of vector components parallel and normal to OB (Fig. a), we can determine the vector components by using Eqs. (2.26) and (2.27). But to apply them, we must first express $\mathbf{F}$ in terms of scalar components and determine the scalar components of a unit vector parallel to OB. We can obtain the components of $\mathbf{F}$ by determining the components of the unit vector pointing from O toward A and multiplying them by $|\mathbf{F}|$.

Solution

The position vectors from O to A and from O to B are (Fig. b)

$$\mathbf{r}_{OA} = 6\mathbf{i} + 6\mathbf{j} - 3\mathbf{k}\ (\text{m}),$$

$$\mathbf{r}_{OB} = 10\mathbf{i} - 2\mathbf{j} + 3\mathbf{k}\ (\text{m}).$$

Their magnitudes are $|\mathbf{r}_{OA}| = 9$ m and $|\mathbf{r}_{OB}| = 10.6$ m. Dividing these vectors by their magnitudes, we obtain unit vectors that point from the origin toward A and B (Fig. c):

$$\mathbf{e}_{OA} = \frac{\mathbf{r}_{OA}}{|\mathbf{r}_{OA}|} = \frac{6\mathbf{i} + 6\mathbf{j} - 3\mathbf{k}\ (\text{m})}{9\ \text{m}} = 0.667\mathbf{i} + 0.667\mathbf{j} - 0.333\mathbf{k},$$

$$\mathbf{e}_{OB} = \frac{\mathbf{r}_{OB}}{|\mathbf{r}_{OB}|} = \frac{10\mathbf{i} - 2\mathbf{j} + 3\mathbf{k}\ (\text{m})}{10.6\ \text{m}} = 0.941\mathbf{i} - 0.188\mathbf{j} + 0.282\mathbf{k}.$$

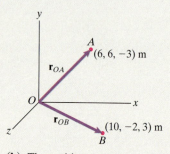

(a) The components of $\mathbf{F}$ parallel and normal to OB.

The force $\mathbf{F}$ in terms of scalar components is

$$\mathbf{F} = |\mathbf{F}|\mathbf{e}_{OA} = (50\ \text{N})(0.667\mathbf{i} + 0.667\mathbf{j} - 0.333\mathbf{k})$$
$$= 33.3\mathbf{i} + 33.3\mathbf{j} - 16.7\mathbf{k}\ (\text{N}).$$

Taking the dot product of $\mathbf{e}_{OB}$ and $\mathbf{F}$, we obtain

$$\mathbf{e}_{OB} \cdot \mathbf{F} = (0.941)(33.3\ \text{N}) + (-0.188)(33.3\ \text{N}) + (0.282)(-16.7\ \text{N})$$
$$= 20.4\ \text{N}.$$

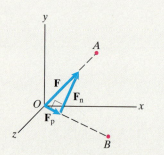

(b) The position vectors $\mathbf{r}_{OA}$ and $\mathbf{r}_{OB}$.

The parallel vector component of $\mathbf{F}$ is

$$\mathbf{F}_\text{p} = (\mathbf{e}_{OB} \cdot \mathbf{F})\mathbf{e}_{OB} = (20.4\ \text{N})(0.941\mathbf{i} - 0.188\mathbf{j} + 0.282\mathbf{k})$$
$$= 19.2\mathbf{i} - 3.83\mathbf{j} + 5.75\mathbf{k}\ (\text{N}),$$

and the normal vector component is

$$\mathbf{F}_\text{n} = \mathbf{F} - \mathbf{F}_\text{p} = 14.2\mathbf{i} + 37.2\mathbf{j} - 22.4\mathbf{k}\ (\text{N}).$$

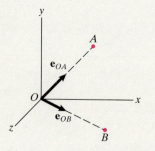

(c) The unit vectors $\mathbf{e}_{OA}$ and $\mathbf{e}_{OB}$.

Critical Thinking

How can you confirm that two vectors are perpendicular? It is clear from Eq. (2.18) that the dot product of two non-zero vectors is zero if and only if the enclosed angle between them is 90°. We can use this diagnostic test to confirm that the components of $\mathbf{F}$ determined in this example are perpendicular. Evaluating the dot product of $\mathbf{F}_\text{p}$ and $\mathbf{F}_\text{n}$ in terms of their components in newtons, we obtain

$$\mathbf{F}_\text{p} \cdot \mathbf{F}_\text{n} = (19.2)(14.2) + (-3.83)(37.2) + (5.75)(-22.4) = 0.$$

Problems

2.99 Determine the dot product $\mathbf{U} \cdot \mathbf{V}$ of the vectors $\mathbf{U} = 4\mathbf{i} + 6\mathbf{j} - 10\mathbf{k}$ and $\mathbf{V} = -8\mathbf{i} + 12\mathbf{j} + 2\mathbf{k}$.

Strategy: The vectors are expressed in terms of their components, so you can use Eq. (2.23) to determine their dot product.

2.100 Determine the dot product $\mathbf{U} \cdot \mathbf{V}$ of the vectors $\mathbf{U} = 40\mathbf{i} + 20\mathbf{j} + 60\mathbf{k}$ and $\mathbf{V} = -30\mathbf{i} + 15\mathbf{k}$.

2.101 What is the dot product of the position vector $\mathbf{r} = -10\mathbf{i} + 25\mathbf{j}$ (m) and the force vector $\mathbf{F} = 300\mathbf{i} + 250\mathbf{j} + 300\mathbf{k}$ (N)?

2.102 Suppose that the dot product of two vectors $\mathbf{U}$ and $\mathbf{V}$ is $\mathbf{U} \cdot \mathbf{V} = 0$. If $|\mathbf{U}| \neq 0$, what do you know about the vector $\mathbf{V}$?

2.103 Two *perpendicular* vectors are given in terms of their components by $\mathbf{U} = U_x\mathbf{i} - 4\mathbf{j} + 6\mathbf{k}$ and $\mathbf{V} = 3\mathbf{i} + 2\mathbf{j} - 3\mathbf{k}$. Use the dot product to determine the component U_x.

2.104 The three vectors

$$\mathbf{U} = U_x\mathbf{i} + 3\mathbf{j} + 2\mathbf{k},$$

$$\mathbf{V} = -3\mathbf{i} + V_y\mathbf{j} + 3\mathbf{k},$$

$$\mathbf{W} = -2\mathbf{i} + 4\mathbf{j} + W_z\mathbf{k}$$

are mutually perpendicular. Use the dot product to determine the components U_x, V_y, and W_z.

2.105 The magnitudes $|\mathbf{U}| = 10$ and $|\mathbf{V}| = 20$.
(a) Use Eq. (2.18) to determine $\mathbf{U} \cdot \mathbf{V}$.
(b) Use Eq. (2.23) to determine $\mathbf{U} \cdot \mathbf{V}$.

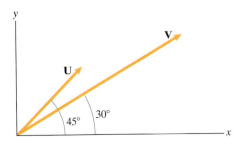

Problem 2.105

2.106 By evaluating the dot product $\mathbf{U} \cdot \mathbf{V}$, prove the identity $\cos(\theta_1 - \theta_2) = \cos\theta_1\cos\theta_2 + \sin\theta_1\sin\theta_2$.

Strategy: Evaluate the dot product both by using Eq. (2.18) and by using Eq. (2.23).

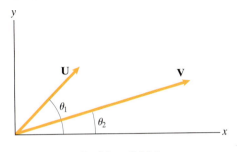

Problem 2.106

2.107 Use the dot product to determine the angle between the forestay (cable AB) and the backstay (cable BC) of the sailboat.

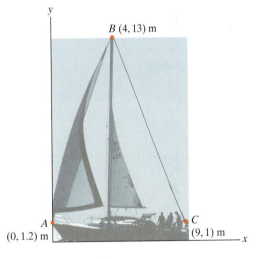

Problem 2.107

2.108 Determine the angle θ between the lines AB and AC
(a) by using the law of cosines (see Appendix A);
(b) by using the dot product.

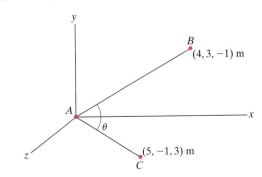

Problem 2.108

2.109 The ship O measures the positions of the ship A and the airplane B and obtains the coordinates shown. What is the angle θ between the lines of sight OA and OB?

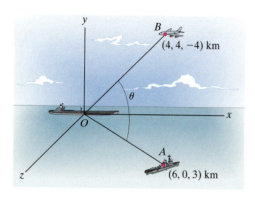

Problem 2.109

2.110 Astronauts on the space shuttle use radar to determine the magnitudes and direction cosines of the position vectors of two satellites A and B. The vector $\mathbf{r}_A$ from the shuttle to satellite A has magnitude 2 km and direction cosines $\cos \theta_x = 0.768$, $\cos \theta_y = 0.384$, $\cos \theta_z = 0.512$. The vector $\mathbf{r}_B$ from the shuttle to satellite B has magnitude 4 km and direction cosines $\cos \theta_x = 0.743$, $\cos \theta_y = 0.557$, $\cos \theta_z = -0.371$. What is the angle θ between the vectors $\mathbf{r}_A$ and $\mathbf{r}_B$?

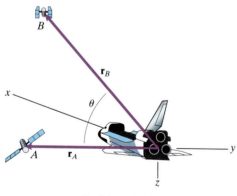

Problem 2.110

2.111 The cable BC exerts an 800-N force $\mathbf{F}$ on the bar AB at B. Use Eq. (2.26) to determine the vector component of $\mathbf{F}$ parallel to the bar.

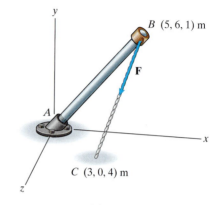

Problem 2.111

2.112 The force $\mathbf{F} = 21\mathbf{i} + 14\mathbf{j}$ (kN). Determine its vector components parallel and normal to the line OA.

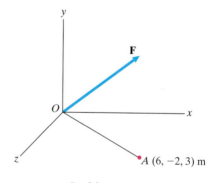

Problem 2.112

2.113 At the instant shown, the Harrier's thrust vector is $\mathbf{T} = 17{,}000\mathbf{i} + 68{,}000\mathbf{j} - 8{,}000\mathbf{k}$ (N) and its velocity vector is $\mathbf{v} = 7.3\mathbf{i} + 1.8\mathbf{j} - 0.6\mathbf{k}$ (m/s). The quantity $P = |\mathbf{T}_p||\mathbf{v}|$, where $\mathbf{T}_p$ is the vector component of $\mathbf{T}$ parallel to $\mathbf{v}$, is the power currently being transferred to the airplane by its engine. Determine the value of P.

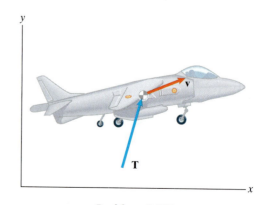

Problem 2.113

2.114 Cables extend from A to B and from A to C. The cable AC exerts a 1000-lb force **F** at A.

(a) What is the angle between the cables AB and AC?

(b) Determine the vector component of **F** parallel to the cable AB.

2.115 Let $\mathbf{r}_{AB}$ be the position vector from point A to point B. Determine the vector component of $\mathbf{r}_{AB}$ parallel to the cable AC.

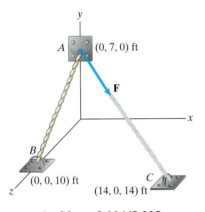

Problems 2.114/2.115

2.116 The force $\mathbf{F} = 10\mathbf{i} + 12\mathbf{j} - 6\mathbf{k}$ (N). Determine the vector components of **F** parallel and normal to the line OA.

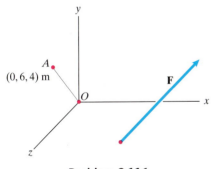

Problem 2.116

2.117 The rope AB exerts a 50-N force **T** on collar A. Determine the vector component of **T** parallel to the bar CD.

2.118 In Problem 2.117, determine the vector component of **T** normal to the bar CD.

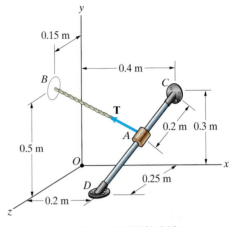

Problems 2.117/2.118

2.119 The disk A is at the midpoint of the sloped surface. The string from A to B exerts a 0.2-lb force **F** on the disk. If you express **F** in terms of vector components parallel and normal to the sloped surface, what is the component normal to the surface?

2.120 In Problem 2.119, what is the vector component of **F** parallel to the surface?

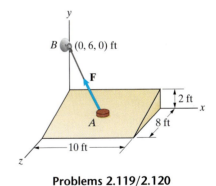

Problems 2.119/2.120

2.121 An astronaut in a maneuvering unit approaches a space station. At the present instant, the station informs him that his position relative to the origin of the station's coordinate system is $\mathbf{r}_G = 50\mathbf{i} + 80\mathbf{j} + 180\mathbf{k}$ (m) and his velocity is $\mathbf{v} = -2.2\mathbf{j} - 3.6\mathbf{k}$ (m/s). The position of an airlock is $\mathbf{r}_A = -12\mathbf{i} + 20\mathbf{k}$ (m). Determine the angle between his velocity vector and the line from his position to the airlock's position.

2.122 In Problem 2.121, determine the vector component of the astronaut's velocity parallel to the line from his position to the airlock's position.

Problems 2.121/2.122

2.123 Point P is at longitude 30°W and latitude 45°N on the Atlantic Ocean between Nova Scotia and France. (See Problem 2.86.) Point Q is at longitude 60°E and latitude 20°N in the Arabian Sea. Use the dot product to determine the shortest distance along the surface of the earth from P to Q in terms of the radius of the earth R_E.

Strategy: Use the dot product to determine the angle between the lines OP and OQ; then use the definition of an angle in radians to determine the distance along the surface of the earth from P to Q.

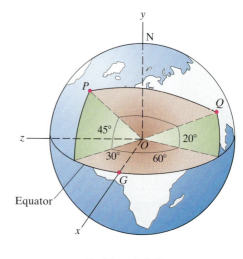

Problem 2.123

2.6 Cross Products

Like the dot product, the cross product of two vectors has many applications, including determining the rate of rotation of a fluid particle and calculating the force exerted on a charged particle by a magnetic field. Because of its usefulness for determining moments of forces, the cross product is an indispensable tool in mechanics. In this section we show you how to evaluate cross products and give examples of simple applications.

Definition

Consider two vectors $\mathbf{U}$ and $\mathbf{V}$ (Fig. 2.37a). The *cross product* of $\mathbf{U}$ and $\mathbf{V}$, denoted $\mathbf{U} \times \mathbf{V}$, is defined by

$$\mathbf{U} \times \mathbf{V} = |\mathbf{U}||\mathbf{V}| \sin \theta \, \mathbf{e}. \tag{2.28}$$

The angle θ is the angle between $\mathbf{U}$ and $\mathbf{V}$ when they are placed tail to tail (Fig. 2.37b). The vector $\mathbf{e}$ is a unit vector defined to be perpendicular to both $\mathbf{U}$ and $\mathbf{V}$. Since this leaves two possibilities for the direction of $\mathbf{e}$, the vectors $\mathbf{U}$, $\mathbf{V}$, and $\mathbf{e}$ are defined to be a right-handed system. The *right-hand rule* for determining the direction of $\mathbf{e}$ is shown in Fig. 2.37c. If the fingers of the right hand are pointed in the direction of the vector **U** (the first vector in the cross product) and then bent toward the vector **V** (the second vector in the cross product), the thumb points in the direction of **e**.

Because the result of the cross product is a vector, it is sometimes called the vector product. The units of the cross product are the product of the units of the two vectors. Notice that the cross product of two nonzero vectors is equal to zero if and only if the two vectors are parallel.

An interesting property of the cross product is that it is *not* commutative. Eq. (2.28) implies that the magnitude of the vector **U** × **V** is equal to the magnitude of the vector **V** × **U**, but the right-hand rule indicates that they are opposite in direction (Fig. 2.38). That is,

$$\mathbf{U} \times \mathbf{V} = -\mathbf{V} \times \mathbf{U}. \qquad \text{The cross product is \textit{not} commutative.} \qquad (2.29)$$

The cross product also satisfies the relations

$$a\,(\mathbf{U} \times \mathbf{V}) = (a\mathbf{U}) \times \mathbf{V} = \mathbf{U} \times (a\mathbf{V}) \qquad (2.30)$$

The cross product is associative with respect to scalar multiplication.

and

$$\mathbf{U} \times (\mathbf{V} + \mathbf{W}) = (\mathbf{U} \times \mathbf{V}) + (\mathbf{U} \times \mathbf{W}) \qquad (2.31)$$

The cross product is distributive with respect to vector addition.

for any scalar a and vectors **U**, **V**, and **W**.

Cross Products in Terms of Components

To obtain an equation for the cross product of two vectors in terms of their components, we must determine the cross products formed from the unit vectors **i**, **j**, and **k**. Since the angle between two identical vectors placed tail to tail is zero, it follows that

$$\mathbf{i} \times \mathbf{i} = |\mathbf{i}||\mathbf{i}|\,\sin(0)\mathbf{e} = \mathbf{0}.$$

The cross product **i** × **j** is

$$\mathbf{i} \times \mathbf{j} = |\mathbf{i}||\mathbf{j}|\,\sin 90°\mathbf{e} = \mathbf{e},$$

where **e** is a unit vector perpendicular to **i** and **j**. Either **e** = **k** or **e** = −**k**. Applying the right-hand rule, we find that **e** = **k** (Fig. 2.39). Therefore,

$$\mathbf{i} \times \mathbf{j} = \mathbf{k}.$$

Continuing in this way, we obtain

$$\mathbf{i} \times \mathbf{i} = \mathbf{0}, \qquad \mathbf{i} \times \mathbf{j} = \mathbf{k}, \qquad \mathbf{i} \times \mathbf{k} = -\mathbf{j},$$

$$\mathbf{j} \times \mathbf{i} = -\mathbf{k}, \qquad \mathbf{j} \times \mathbf{j} = \mathbf{0}, \qquad \mathbf{j} \times \mathbf{k} = \mathbf{i},$$

$$\mathbf{k} \times \mathbf{i} = \mathbf{j}, \qquad \mathbf{k} \times \mathbf{j} = -\mathbf{i}, \qquad \mathbf{k} \times \mathbf{k} = \mathbf{0}. \qquad (2.32)$$

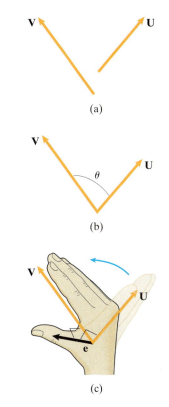

Figure 2.37
(**a**) The vectors **U** and **V**.
(**b**) The angle θ between the vectors when they are placed tail to tail.
(**c**) Determining the direction of **e** by the right-hand rule.

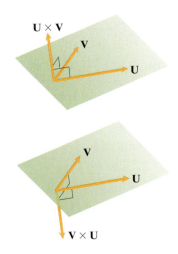

Figure 2.38
Directions of **U** × **V** and **V** × **U**.

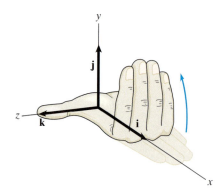

Figure 2.39
The right-hand rule indicates that $\mathbf{i} \times \mathbf{j} = \mathbf{k}$.

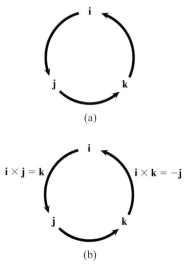

(a)

(b)

Figure 2.40
(a) Arrange the unit vectors in a circle with arrows to indicate their order.
(b) You can use the circle to determine their cross products.

These results can be remembered easily by arranging the unit vectors in a circle, as shown in Fig. 2.40a. The cross product of adjacent vectors is equal to the third vector with a positive sign if the order of the vectors in the cross product is the order indicated by the arrows and a negative sign otherwise. For example, in Fig. 2.40b we see that $\mathbf{i} \times \mathbf{j} = \mathbf{k}$, but $\mathbf{i} \times \mathbf{k} = -\mathbf{j}$.

The cross product of two vectors $\mathbf{U}$ and $\mathbf{V}$, expressed in terms of their components, is

$$
\begin{aligned}
\mathbf{U} \times \mathbf{V} &= (U_x\mathbf{i} + U_y\mathbf{j} + U_z\mathbf{k}) \times (V_x\mathbf{i} + V_y\mathbf{j} + V_z\mathbf{k}) \\
&= U_xV_x(\mathbf{i} \times \mathbf{i}) + U_xV_y(\mathbf{i} \times \mathbf{j}) + U_xV_z(\mathbf{i} \times \mathbf{k}) \\
&\quad + U_yV_x(\mathbf{j} \times \mathbf{i}) + U_yV_y(\mathbf{j} \times \mathbf{j}) + U_yV_z(\mathbf{j} \times \mathbf{k}) \\
&\quad + U_zV_x(\mathbf{k} \times \mathbf{i}) + U_zV_y(\mathbf{k} \times \mathbf{j}) + U_zV_z(\mathbf{k} \times \mathbf{k}).
\end{aligned}
$$

By substituting Eqs. (2.32) into this expression, we obtain the equation

$$
\begin{aligned}
\mathbf{U} \times \mathbf{V} = (U_yV_z - U_zV_y)\mathbf{i} &- (U_xV_z - U_zV_x)\mathbf{j} \\
&+ (U_xV_y - U_yV_x)\mathbf{k}.
\end{aligned} \tag{2.33}
$$

This result can be compactly written as the determinant

$$
\mathbf{U} \times \mathbf{V} = \begin{vmatrix} \mathbf{i} & \mathbf{j} & \mathbf{k} \\ U_x & U_y & U_z \\ V_x & V_y & V_z \end{vmatrix}. \tag{2.34}
$$

This equation is based on Eqs. (2.32), which we obtained using a right-handed coordinate system. It gives the correct result for the cross product only if a right-handed coordinate system is used to determine the components of $\mathbf{U}$ and $\mathbf{V}$.

Evaluating a 3 × 3 Determinant

A 3 × 3 determinant can be evaluated by repeating its first two columns and evaluating the products of the terms along the six diagonal lines:

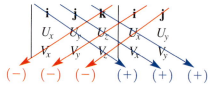

Adding the terms obtained from the diagonals that run downward to the right (blue arrows) and subtracting the terms obtained from the diagonals that run downward to the left (red arrows) gives the value of the determinant:

$$
\begin{vmatrix} \mathbf{i} & \mathbf{j} & \mathbf{k} \\ U_x & U_y & U_z \\ V_x & V_y & V_z \end{vmatrix} = \begin{aligned} &U_yV_z\mathbf{i} + U_zV_x\mathbf{j} + U_xV_y\mathbf{k} \\ -&U_yV_x\mathbf{k} - U_zV_y\mathbf{i} - U_xV_z\mathbf{j}. \end{aligned}
$$

A 3 × 3 determinant can also be evaluated by expressing it as

$$
\begin{vmatrix} \mathbf{i} & \mathbf{j} & \mathbf{k} \\ U_x & U_y & U_z \\ V_x & V_y & V_z \end{vmatrix} = \mathbf{i}\begin{vmatrix} U_y & U_z \\ V_y & V_z \end{vmatrix} - \mathbf{j}\begin{vmatrix} U_x & U_z \\ V_x & V_z \end{vmatrix} + \mathbf{k}\begin{vmatrix} U_x & U_y \\ V_x & V_y \end{vmatrix}.
$$

The terms on the right are obtained by multiplying each element of the first row of the 3×3 determinant by the 2×2 determinant obtained by crossing out that element's row and column. For example, the first element of the first row, $\mathbf{i}$, is multiplied by the 2×2 determinant

$$\begin{vmatrix} \mathbf{i} & \mathbf{j} & \mathbf{k} \\ U_x & U_y & U_z \\ V_x & V_y & V_z \end{vmatrix}.$$

Be sure to remember that the second term is subtracted. Expanding the 2×2 determinants, we obtain the value of the determinant:

$$\begin{vmatrix} \mathbf{i} & \mathbf{j} & \mathbf{k} \\ U_x & U_y & U_z \\ V_x & V_y & V_z \end{vmatrix} = (U_y V_z - U_z V_y)\mathbf{i} - (U_x V_z - U_z V_x)\mathbf{j} + (U_x V_y - U_y V_x)\mathbf{k}.$$

2.7 Mixed Triple Products

In Chapter 4, when we discuss the moment of a force about a line, we will use an operation called the *mixed triple product*, defined by

$$\mathbf{U} \cdot (\mathbf{V} \times \mathbf{W}). \tag{2.35}$$

In terms of the scalar components of the vectors,

$$
\begin{aligned}
\mathbf{U} \cdot (\mathbf{V} \times \mathbf{W}) &= (U_x\mathbf{i} + U_y\mathbf{j} + U_z\mathbf{k}) \cdot \begin{vmatrix} \mathbf{i} & \mathbf{j} & \mathbf{k} \\ V_x & V_y & V_z \\ W_x & W_y & W_z \end{vmatrix} \\
&= (U_x\mathbf{i} + U_y\mathbf{j} + U_z\mathbf{k}) \cdot [(V_yW_z - V_zW_y)\mathbf{i} \\
&\quad - (V_xW_z - V_zW_x)\mathbf{j} + (V_xW_y - V_yW_x)\mathbf{k}] \\
&= U_x(V_yW_z - V_zW_y) - U_y(V_xW_z - V_zW_x) \\
&\quad + U_z(V_xW_y - V_yW_x).
\end{aligned}
$$

This result can be expressed as the determinant

$$\mathbf{U} \cdot (\mathbf{V} \times \mathbf{W}) = \begin{vmatrix} U_x & U_y & U_z \\ V_x & V_y & V_z \\ W_x & W_y & W_z \end{vmatrix}. \tag{2.36}$$

Interchanging any two of the vectors in the mixed triple product changes the sign but not the absolute value of the result. For example,

$$\mathbf{U} \cdot (\mathbf{V} \times \mathbf{W}) = -\mathbf{W} \cdot (\mathbf{V} \times \mathbf{U}).$$

If the vectors $\mathbf{U}$, $\mathbf{V}$, and $\mathbf{W}$ in Fig. 2.41 form a right-handed system, it can be shown that the volume of the parallelepiped equals $\mathbf{U} \cdot (\mathbf{V} \times \mathbf{W})$.

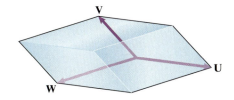

Figure 2.41
Parallelepiped defined by the vectors $\mathbf{U}$, $\mathbf{V}$, and $\mathbf{W}$.

Study Questions

1. What is the definition of the cross product?
2. If you know the components of two vectors **U** and **V**, how can you determine their cross product?
3. If the cross product of two vectors is zero, what does that mean?

Example 2.14 Calculating the Cross Product

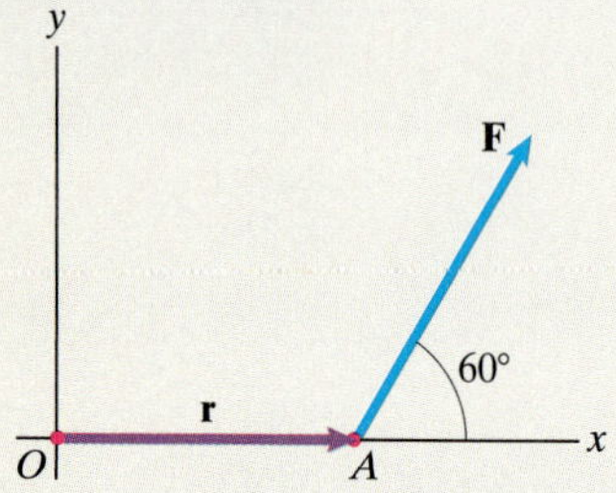

Figure 2.42

The magnitude of the force **F** in Fig. 2.42 is 100 lb. The magnitude of the vector **r** from point O to point A is 8 ft.

(a) Use the definition of the cross product to determine $\mathbf{r} \times \mathbf{F}$.

(b) Use Eq. (2.34) to determine $\mathbf{r} \times \mathbf{F}$.

Strategy

(a) We know the magnitudes of **r** and **F** and the angle between them when they are placed tail to tail. Since both vectors lie in the x–y plane, the unit vector **k** is perpendicular to both **r** and **F**. We therefore have all the information we need to determine $\mathbf{r} \times \mathbf{F}$ directly from the definition.

(b) We can determine the components of **r** and **F** and use Eq. (2.34) to determine $\mathbf{r} \times \mathbf{F}$.

Solution

(a) Using the definition of the cross product gives

$$\mathbf{r} \times \mathbf{F} = |\mathbf{r}||\mathbf{F}| \sin\theta\, \mathbf{e} = (8 \text{ ft})(100 \text{ lb}) \sin 60^\circ\, \mathbf{e} = 693\mathbf{e} \text{ (ft-lb)}.$$

Since **e** is defined to be perpendicular to **r** and **F**, either $\mathbf{e} = \mathbf{k}$ or $\mathbf{e} = -\mathbf{k}$. Pointing the fingers of the right hand in the direction of **r** and closing them toward **F**, the right-hand rule indicates that $\mathbf{e} = \mathbf{k}$. Therefore,

$$\mathbf{r} \times \mathbf{F} = 693\mathbf{k} \text{ (ft-lb)}.$$

(b) The vector $\mathbf{r} = 8\mathbf{i}$ (ft). The vector **F**, in terms of scalar components, is

$$\mathbf{F} = 100 \cos 60^\circ\, \mathbf{i} + 100 \sin 60^\circ\, \mathbf{j} \text{ (lb)}.$$

From Eq. (2.34), the cross product, with the components of **r** in feet and the components of **F** in pound, is

$$\mathbf{r} \times \mathbf{F} = \begin{vmatrix} \mathbf{i} & \mathbf{j} & \mathbf{k} \\ r_x & r_y & r_z \\ F_x & F_y & F_z \end{vmatrix} = \begin{vmatrix} \mathbf{i} & \mathbf{j} & \mathbf{k} \\ 8 & 0 & 0 \\ 100 \cos 60^\circ & 100 \sin 60^\circ & 0 \end{vmatrix}$$

$$= (8 \text{ ft})(100 \sin 60^\circ \text{ lb})\mathbf{k} = 693\mathbf{k} \text{ (ft-lb)}.$$

Critical Thinking

We designed this example so that the cross product of **r** and **F** could be evaluated both by applying the definition and by using Eq. (2.34), to demonstrate that they yield the same result. But in most applications of the cross product it is not practical to use the definition, and Eq. (2.34) must be used.

Example 2.15 Minimum Distance from a Point to a Line

Consider the straight lines OA and OB in Fig. 2.43.
(a) Determine the components of a unit vector that is perpendicular to both OA and OB.
(b) What is the minimum distance from point A to the line OB?

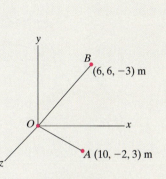

Figure 2.43

Strategy

(a) Let $\mathbf{r}_{OA}$ and $\mathbf{r}_{OB}$ be the position vectors from O to A and from O to B (Fig. a). Since the cross product $\mathbf{r}_{OA} \times \mathbf{r}_{OB}$ is perpendicular to $\mathbf{r}_{OA}$ and $\mathbf{r}_{OB}$, we will determine it and divide it by its magnitude to obtain a unit vector perpendicular to the lines OA and OB.

(b) The minimum distance from A to the line OB is the length d of the straight line from A to OB that is perpendicular to OB (Fig. b). We can see that $d = |\mathbf{r}_{OA}| \sin \theta$, where θ is the angle between $\mathbf{r}_{OA}$ and $\mathbf{r}_{OB}$. From the definition of the cross product, the magnitude of $\mathbf{r}_{OA} \times \mathbf{r}_{OB}$ is $|\mathbf{r}_{OA}||\mathbf{r}_{OB}| \sin \theta$, so we can determine d by dividing the magnitude of $\mathbf{r}_{OA} \times \mathbf{r}_{OB}$ by the magnitude of $\mathbf{r}_{OB}$.

Solution

(a) The components of $\mathbf{r}_{OA}$ and $\mathbf{r}_{OB}$ are

$$\mathbf{r}_{OA} = 10\mathbf{i} - 2\mathbf{j} + 3\mathbf{k} \text{ (m)},$$

$$\mathbf{r}_{OB} = 6\mathbf{i} + 6\mathbf{j} - 3\mathbf{k} \text{ (m)}.$$

By using Eq. (2.34), we obtain $\mathbf{r}_{OA} \times \mathbf{r}_{OB}$:

$$\mathbf{r}_{OA} \times \mathbf{r}_{OB} = \begin{vmatrix} \mathbf{i} & \mathbf{j} & \mathbf{k} \\ 10 & -2 & 3 \\ 6 & 6 & -3 \end{vmatrix} = -12\mathbf{i} + 48\mathbf{j} + 72\mathbf{k} \text{ (m}^2\text{)}.$$

This vector is perpendicular to $\mathbf{r}_{OA}$ and $\mathbf{r}_{OB}$. Dividing it by its magnitude, we obtain a unit vector $\mathbf{e}$ that is perpendicular to the lines OA and OB:

$$\mathbf{e} = \frac{\mathbf{r}_{OA} \times \mathbf{r}_{OB}}{|\mathbf{r}_{OA} \times \mathbf{r}_{OB}|} = \frac{-12\mathbf{i} + 48\mathbf{j} + 72\mathbf{k} \text{ (m}^2\text{)}}{\sqrt{(-12 \text{ m}^2)^2 + (48 \text{ m}^2)^2 + (72 \text{ m}^2)^2}}$$

$$= -0.137\mathbf{i} + 0.549\mathbf{j} + 0.824\mathbf{k}.$$

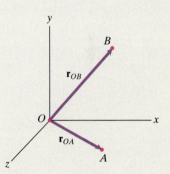

(a) The vectors $\mathbf{r}_{OA}$ and $\mathbf{r}_{OB}$.

(b) From Fig. b, the minimum distance d is

$$d = |\mathbf{r}_{OA}| \sin \theta.$$

The magnitude of $\mathbf{r}_{OA} \times \mathbf{r}_{OB}$ is

$$|\mathbf{r}_{OA} \times \mathbf{r}_{OB}| = |\mathbf{r}_{OA}||\mathbf{r}_{OB}| \sin \theta.$$

Solving this equation for $\sin \theta$, we find that the distance d is

$$d = |\mathbf{r}_{OA}| \left(\frac{|\mathbf{r}_{OA} \times \mathbf{r}_{OB}|}{|\mathbf{r}_{OA}||\mathbf{r}_{OB}|} \right) = \frac{|\mathbf{r}_{OA} \times \mathbf{r}_{OB}|}{|\mathbf{r}_{OB}|}$$

$$= \frac{\sqrt{(-12 \text{ m}^2)^2 + (48 \text{ m}^2)^2 + (72 \text{ m}^2)^2}}{\sqrt{(6 \text{ m})^2 + (6 \text{ m})^2 + (-3 \text{ m})^2}} = 9.71 \text{ m}.$$

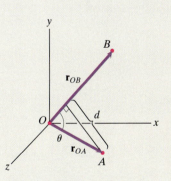

(b) The minimum distance d from A to the line OB.

Critical Thinking

This example is an illustration of the power of vector methods. Determining the minimum distance from point A to the line OB can be formulated as a minimization problem in differential calculus, but the vector solution we present is far simpler.

Example 2.16 Component of a Vector Perpendicular to a Plane

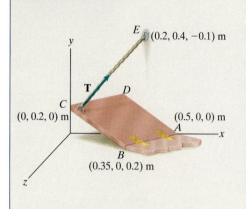

Figure 2.44

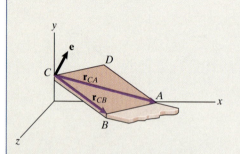

(a) Determining a unit vector perpendicular to the door.

The rope CE in Fig. 2.44 exerts a 500-N force $\mathbf{T}$ on the door $ABCD$. What is the magnitude of the component of $\mathbf{T}$ perpendicular to the door?

Strategy

We are given the coordinates of the corners A, B, and C of the door. By taking the cross product of the position vector $\mathbf{r}_{CB}$ from C to B and the position vector $\mathbf{r}_{CA}$ from C to A, we will obtain a vector that is perpendicular to the door. We can divide the resulting vector by its magnitude to obtain a unit vector perpendicular to the door and then apply Eq. (2.26) to determine the component of $\mathbf{T}$ perpendicular to the door.

Solution

The components of $\mathbf{r}_{CB}$ and $\mathbf{r}_{CA}$ are

$$\mathbf{r}_{CB} = 0.35\mathbf{i} - 0.2\mathbf{j} + 0.2\mathbf{k} \ (\text{m}),$$

$$\mathbf{r}_{CA} = 0.5\mathbf{i} - 0.2\mathbf{j} \ (\text{m}).$$

Their cross product is

$$\mathbf{r}_{CB} \times \mathbf{r}_{CA} = \begin{vmatrix} \mathbf{i} & \mathbf{j} & \mathbf{k} \\ 0.35 & -0.2 & 0.2 \\ 0.5 & -0.2 & 0 \end{vmatrix} = 0.04\mathbf{i} + 0.1\mathbf{j} + 0.03\mathbf{k} \ (\text{m}^2).$$

Dividing this vector by its magnitude, we obtain a unit vector $\mathbf{e}$ that is perpendicular to the door (Fig. a):

$$\mathbf{e} = \frac{\mathbf{r}_{CB} \times \mathbf{r}_{CA}}{|\mathbf{r}_{CB} \times \mathbf{r}_{CA}|} = \frac{0.04\mathbf{i} + 0.1\mathbf{j} + 0.03\mathbf{k} \ (\text{m}^2)}{\sqrt{(0.04 \ \text{m}^2)^2 + (0.1 \ \text{m}^2)^2 + (0.03 \ \text{m}^2)^2}}$$

$$= 0.358\mathbf{i} + 0.894\mathbf{j} + 0.268\mathbf{k}.$$

To use Eq. (2.26), we must express $\mathbf{T}$ in terms of its scalar components. The position vector from C to E is

$$\mathbf{r}_{CE} = 0.2\mathbf{i} + 0.2\mathbf{j} - 0.1\mathbf{k} \ (\text{m}),$$

so we can express the force $\mathbf{T}$ as

$$\mathbf{T} = |\mathbf{T}|\frac{\mathbf{r}_{CE}}{|\mathbf{r}_{CE}|} = (500 \ \text{N})\frac{0.2\mathbf{i} + 0.2\mathbf{j} - 0.1\mathbf{k} \ (\text{m})}{\sqrt{(0.2 \ \text{m})^2 + (0.2 \ \text{m})^2 + (-0.1 \ \text{m})^2}}$$

$$= 333\mathbf{i} + 333\mathbf{j} - 167\mathbf{k} \ (\text{N}).$$

The component of $\mathbf{T}$ parallel to the unit vector $\mathbf{e}$, which is the component of $\mathbf{T}$ perpendicular to the door, is

$$(\mathbf{e} \cdot \mathbf{T})\mathbf{e} = [(0.358)(333 \ \text{N}) + (0.894)(333 \ \text{N}) + (0.268)(-167 \ \text{N})]\mathbf{e}$$

$$= 373\mathbf{e} \ (\text{N}).$$

The magnitude of the component of $\mathbf{T}$ perpendicular to the door is 373 N.

Critical Thinking

Why is it useful to determine the component of the force $\mathbf{T}$ perpendicular to the door? If the y axis is vertical and the rope CE is the only thing preventing the hinged door from falling, you can see intuitively that it is the component of the force perpendicular to the door that holds it in place. We analyze problems of this kind in Chapter 5.

Problems

2.124 (a) Determine the cross product $\mathbf{U} \times \mathbf{V}$ of the vectors $\mathbf{U} = 4\mathbf{i} + 6\mathbf{j} - 10\mathbf{k}$ and $\mathbf{V} = -8\mathbf{i} + 12\mathbf{j} + 2\mathbf{k}$. (b) Use the dot product to prove that the vector $\mathbf{U} \times \mathbf{V}$ is perpendicular to $\mathbf{U}$ and perpendicular to $\mathbf{V}$.

   Strategy: The vectors are expressed in terms of their components, so you can use Eq. (2.34) to determine their cross product.

2.125 Two vectors $\mathbf{U} = 3\mathbf{i} + 2\mathbf{j}$ and $\mathbf{V} = 2\mathbf{i} + 4\mathbf{j}$.
(a) What is the cross product $\mathbf{U} \times \mathbf{V}$?
(b) What is the cross product $\mathbf{V} \times \mathbf{U}$?

2.126 What is the cross product $\mathbf{r} \times \mathbf{F}$ of the position vector $\mathbf{r} = 2\mathbf{i} + 2\mathbf{j} + 2\mathbf{k}$ (m) and the force $\mathbf{F} = 20\mathbf{i} - 40\mathbf{k}$ (N)?

2.127 Determine the cross product $\mathbf{r} \times \mathbf{F}$ of the position vector $\mathbf{r} = 4\mathbf{i} - 12\mathbf{j} + 3\mathbf{k}$ (m) and the force $\mathbf{F} = 16\mathbf{i} - 22\mathbf{j} - 10\mathbf{k}$ (kN).

2.128 Suppose that the cross product of two vectors $\mathbf{U}$ and $\mathbf{V}$ is $\mathbf{U} \times \mathbf{V} = \mathbf{0}$. If $|\mathbf{U}| \neq 0$, what do you know about the vector $\mathbf{V}$?

2.129 The cross product of two vectors $\mathbf{U}$ and $\mathbf{V}$ is $\mathbf{U} \times \mathbf{V} = -30\mathbf{i} + 40\mathbf{k}$. The vector $\mathbf{V} = 4\mathbf{i} - 2\mathbf{j} + 3\mathbf{k}$. Determine the components of $\mathbf{U}$.

2.130 The magnitudes $|\mathbf{U}| = 10$ and $|\mathbf{V}| = 20$.
(a) Use the definition of the cross product to determine $\mathbf{U} \times \mathbf{V}$.
(b) Use the definition of the cross product to determine $\mathbf{V} \times \mathbf{U}$.
(c) Use Eq. (2.34) to determine $\mathbf{U} \times \mathbf{V}$.
(d) Use Eq. (2.34) to determine $\mathbf{V} \times \mathbf{U}$.

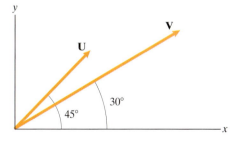

Problem 2.130

2.131 The force $\mathbf{F} = 10\mathbf{i} - 4\mathbf{j}$ (N). Determine the cross product $\mathbf{r}_{AB} \times \mathbf{F}$.

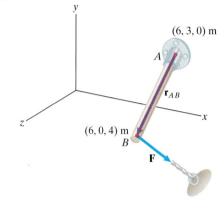

Problem 2.131

2.132 By evaluating the cross product $\mathbf{U} \times \mathbf{V}$, prove the identity $\sin(\theta_1 - \theta_2) = \sin \theta_1 \cos \theta_2 - \cos \theta_1 \sin \theta_2$.

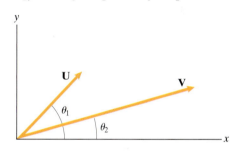

Problem 2.132

2.133 Use the cross product to determine the components of a unit vector $\mathbf{e}$ that is normal to both of the vectors $\mathbf{U} = 8\mathbf{i} - 6\mathbf{j} + 4\mathbf{k}$ and $\mathbf{V} = 3\mathbf{i} + 7\mathbf{j} + 9\mathbf{k}$.

2.134 (a) What is the cross product $\mathbf{r}_{OA} \times \mathbf{r}_{OB}$?
(b) Determine a unit vector $\mathbf{e}$ that is perpendicular to $\mathbf{r}_{OA}$ and $\mathbf{r}_{OB}$.

2.135 Use the cross product to determine the length of the shortest straight line from point B to the straight line that passes through points O and A.

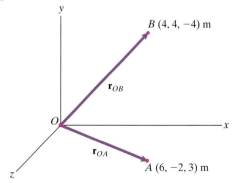

Problems 2.134/2.135

2.136 The cable BC exerts a 1000-lb force $\mathbf{F}$ on the hook at B. Determine $\mathbf{r}_{AB} \times \mathbf{F}$.

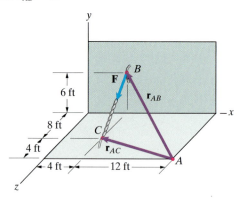

Problem 2.136

2.137 The force vector $\mathbf{F}$ points along the straight line from point A to point B. Its magnitude is $|\mathbf{F}| = 20$ N. The coordinates of points A and B are $x_A = 6$ m, $y_A = 8$ m, $z_A = 4$ m and $x_B = 8$ m, $y_B = 1$ m, $z_B = -2$ m.

(a) Express the vector $\mathbf{F}$ in terms of its components.

(b) Use Eq. (2.34) to determine the cross products $\mathbf{r}_A \times \mathbf{F}$ and $\mathbf{r}_B \times \mathbf{F}$.

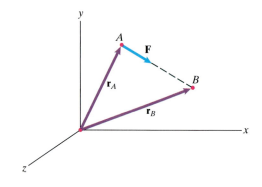

Problem 2.137

2.138 The rope AB exerts a 50-N force $\mathbf{T}$ on the collar at A. Let $\mathbf{r}_{CA}$ be the position vector from point C to point A. Determine the cross product $\mathbf{r}_{CA} \times \mathbf{T}$.

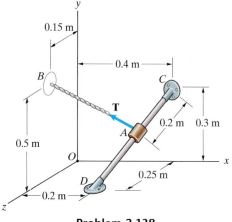

Problem 2.138

2.139 The straight line L is collinear with the force vector $\mathbf{F}$. Let D be the perpendicular distance from an arbitrary point P to L. Prove that

$$D|\mathbf{F}| = |\mathbf{r} \times \mathbf{F}|,$$

where $\mathbf{r}$ is a position vector from point P to *any* point on L.

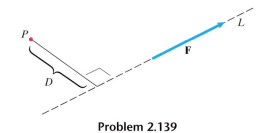

Problem 2.139

2.140 The bar AB is 6 m long and is perpendicular to the bars AC and AD. Use the cross product to determine the coordinates x_B, y_B, z_B of point B.

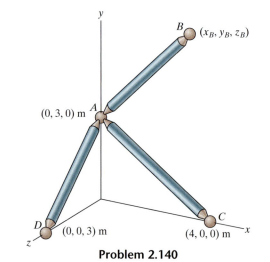

Problem 2.140

2.141 * Determine the minimum distance from point P to the plane defined by the three points A, B, and C.

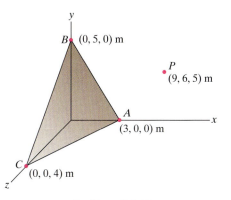

Problem 2.141

2.142 * The force vector $\mathbf{F}$ points along the straight line from point A to point B. Use Eqs. (2.28)–(2.31) to prove that

$$\mathbf{r}_B \times \mathbf{F} = \mathbf{r}_A \times \mathbf{F}.$$

Strategy: Let $\mathbf{r}_{AB}$ be the position vector from point A to point B. Express $\mathbf{r}_B$ in terms of $\mathbf{r}_A$ and $\mathbf{r}_{AB}$. Notice that the vectors $\mathbf{r}_{AB}$ and $\mathbf{F}$ are parallel.

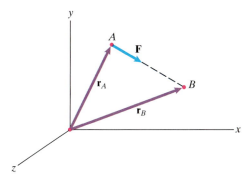

Problem 2.142

2.143 For the vectors $\mathbf{U} = 6\mathbf{i} + 2\mathbf{j} - 4\mathbf{k}$, $\mathbf{V} = 2\mathbf{i} + 7\mathbf{j}$, and $\mathbf{W} = 3\mathbf{i} + 2\mathbf{k}$, evaluate the following mixed triple products:

(a) $\mathbf{U} \cdot (\mathbf{V} \times \mathbf{W})$;
(b) $\mathbf{W} \cdot (\mathbf{V} \times \mathbf{U})$;
(c) $\mathbf{V} \cdot (\mathbf{W} \times \mathbf{U})$.

2.144 Use the mixed triple product to calculate the volume of the parallelepiped.

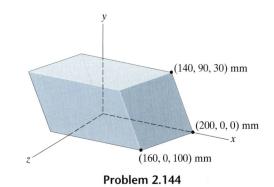

Problem 2.144

2.145 By using Eqs. (2.23) and (2.34), show that

$$\mathbf{U} \cdot (\mathbf{V} \times \mathbf{W}) = \begin{vmatrix} U_x & U_y & U_z \\ V_x & V_y & V_z \\ W_x & W_y & W_z \end{vmatrix}.$$

2.146 The vectors $\mathbf{U} = \mathbf{i} + U_y\mathbf{j} + 4\mathbf{k}$, $\mathbf{V} = 2\mathbf{i} + \mathbf{j} - 2\mathbf{k}$, and $\mathbf{W} = -3\mathbf{i} + \mathbf{j} - 2\mathbf{k}$ are coplanar (they lie in the same plane). What is the component U_y?

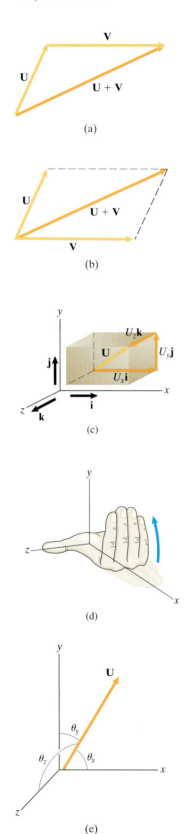

(a)

(b)

(c)

(d)

(e)

CHAPTER SUMMARY

In this chapter we have defined scalars, vectors, and vector operations. We showed how to express vectors in terms of cartesian components and carry out vector operations in terms of components. We introduced the definitions of the dot and cross products and the mixed triple product and demonstrated some applications of these operations, particularly the use of the dot product to express a vector in terms of components parallel and perpendicular to a given direction. In Chapter 3 we will use vector operations to analyze forces acting on objects in equilibrium.

A physical quantity completely described by a real number is a *scalar*. A *vector* has both *magnitude* and *direction* and satisfies a defined rule of addition. A vector is represented graphically by an arrow whose length is defined to be proportional to its magnitude.

Rules for Manipulating Vectors

The sum of two vectors is defined by the *triangle rule* (Fig. a) or the equivalent *parallelogram rule* (Fig. b).

The product of a scalar a and a vector $\mathbf{U}$ is a vector $a\mathbf{U}$ with magnitude $|a||\mathbf{U}|$. Its direction is the same as $\mathbf{U}$ when a is positive and opposite to $\mathbf{U}$ when a is negative. The product $(-1)\mathbf{U}$ is written $-\mathbf{U}$ and is called the negative of $\mathbf{U}$. The division of $\mathbf{U}$ by a is the product $(1/a)\mathbf{U}$.

A *unit vector* is a vector whose magnitude is 1. A unit vector specifies a direction. Any vector $\mathbf{U}$ can be expressed as $|\mathbf{U}|\mathbf{e}$, where $\mathbf{e}$ is a unit vector with the same direction as $\mathbf{U}$. Dividing any vector by its magnitude yields a unit vector with the same direction as the vector.

Cartesian Components

A vector $\mathbf{U}$ is expressed in terms of *scalar components* as

$$\mathbf{U} = U_x\mathbf{i} + U_y\mathbf{j} + U_z\mathbf{k} \tag{2.12}$$

(Fig. c). The coordinate system is *right-handed* (Fig. d): If the fingers of the right hand are pointed in the positive x direction and then closed toward the positive y direction, the thumb points in the z direction. The magnitude of $\mathbf{U}$ is

$$|\mathbf{U}| = \sqrt{U_x^2 + U_y^2 + U_z^2}. \tag{2.14}$$

Let θ_x, θ_y, and θ_z be the angles between $\mathbf{U}$ and the positive coordinate axes (Fig. e). Then the scalar components of $\mathbf{U}$ are

$$U_x = |\mathbf{U}|\cos\theta_x, \quad U_y = |\mathbf{U}|\cos\theta_y, \quad U_z = |\mathbf{U}|\cos\theta_z. \tag{2.15}$$

The quantities $\cos\theta_x$, $\cos\theta_y$, and $\cos\theta_z$ are the *direction cosines* of $\mathbf{U}$. They satisfy the relation

$$\cos^2\theta_x + \cos^2\theta_y + \cos^2\theta_z = 1. \tag{2.16}$$

The *position vector* $\mathbf{r}_{AB}$ from a point A with coordinates (x_A, y_A, z_A) to a point B with coordinates (x_B, y_B, z_B) is given by

$$\mathbf{r}_{AB} = (x_B - x_A)\mathbf{i} + (y_B - y_A)\mathbf{j} + (z_B - z_A)\mathbf{k}. \tag{2.17}$$

Dot Products

The dot product of two vectors **U** and **V** is

$$\mathbf{U} \cdot \mathbf{V} = |\mathbf{U}||\mathbf{V}| \cos \theta, \tag{2.18}$$

where θ is the angle between the vectors when they are placed tail to tail. The dot product of two nonzero vectors is equal to zero if and only if the two vectors are perpendicular.

In terms of scalar components,

$$\mathbf{U} \cdot \mathbf{V} = U_x V_x + U_y V_y + U_z V_z. \tag{2.23}$$

A vector **U** can be expressed as the sum of vector components $\mathbf{U}_p$ and $\mathbf{U}_n$ parallel and normal to a straight line L. In terms of a unit vector **e** that is parallel to L,

$$\mathbf{U}_p = (\mathbf{e} \cdot \mathbf{U})\mathbf{e} \tag{2.26}$$

and

$$\mathbf{U}_n = \mathbf{U} - \mathbf{U}_p. \tag{2.27}$$

Cross Products

The cross product of two vectors **U** and **V** is

$$\mathbf{U} \times \mathbf{V} = |\mathbf{U}||\mathbf{V}| \sin \theta \, \mathbf{e}, \tag{2.28}$$

where θ is the angle between the vectors **U** and **V** when they are placed tail to tail and **e** is a unit vector perpendicular to **U** and **V**. The direction of **e** is specified by the *right-hand rule*: When the fingers of the right hand are pointed in the direction of **U** (the first vector in the cross product) and closed toward **V** (the second vector in the cross product), the thumb points in the direction of **e**. The cross product of two nonzero vectors is equal to zero if and only if the two vectors are parallel.

In terms of scalar components,

$$\mathbf{U} \times \mathbf{V} = \begin{vmatrix} \mathbf{i} & \mathbf{j} & \mathbf{k} \\ U_x & U_y & U_z \\ V_x & V_y & V_z \end{vmatrix}. \tag{2.34}$$

Mixed Triple Products

The *mixed triple product* is the operation

$$\mathbf{U} \cdot (\mathbf{V} \times \mathbf{W}). \tag{2.35}$$

In terms of scalar components,

$$\mathbf{U} \cdot (\mathbf{V} \times \mathbf{W}) = \begin{vmatrix} U_x & U_y & U_z \\ V_x & V_y & V_z \\ W_x & W_y & W_z \end{vmatrix}. \tag{2.36}$$

Review Problems

2.147 The magnitude of **F** is 8 kN. Express **F** in terms of scalar components.

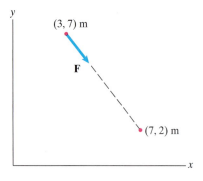

Problem 2.147

2.148 The magnitude of the vertical force **W** is 600 lb, and the magnitude of the force **B** is 1500 lb. Given that $\mathbf{A} + \mathbf{B} + \mathbf{W} = \mathbf{0}$, determine the magnitude of the force **A** and the angle α.

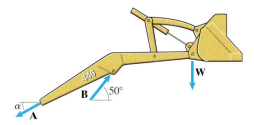

Problem 2.148

2.149 The magnitude of the vertical force vector **A** is 200 lb. If $\mathbf{A} + \mathbf{B} + \mathbf{C} = \mathbf{0}$, what are the magnitudes of the force vectors **B** and **C**?

2.150 The magnitude of the horizontal force vector **D** is 280 lb. If $\mathbf{D} + \mathbf{E} + \mathbf{F} = \mathbf{0}$, what are the magnitudes of the force vectors **E** and **F**?

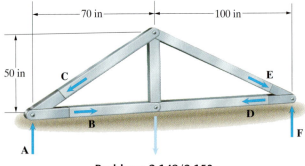

Problems 2.149/2.150

Refer to the following diagram when solving Problems 2.151 through 2.157.

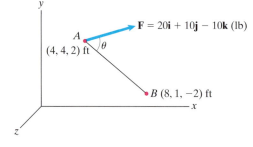

Problems 2.151–2.157

2.151 What are the direction cosines of **F**?

2.152 Determine the components of a unit vector parallel to line AB that points from A toward B.

2.153 What is the angle θ between the line AB and the force **F**?

2.154 Determine the vector component of **F** that is parallel to the line AB.

2.155 Determine the vector component of **F** that is normal to the line AB.

2.156 Determine the vector $\mathbf{r}_{BA} \times \mathbf{F}$, where $\mathbf{r}_{BA}$ is the position vector from B to A.

2.157 (a) Write the position vector $\mathbf{r}_{AB}$ from point A to point B in terms of components.

(b) A vector **R** has magnitude $|\mathbf{R}| = 200$ N and is parallel to the line from A to B. Write **R** in terms of components.

2.158 The rope exerts a force of magnitude $|\mathbf{F}| = 200$ lb on the top of the pole at B.

(a) Determine the vector $\mathbf{r}_{AB} \times \mathbf{F}$, where $\mathbf{r}_{AB}$ is the position vector from A to B.

(b) Determine the vector $\mathbf{r}_{AC} \times \mathbf{F}$, where $\mathbf{r}_{AC}$ is the position vector from A to C.

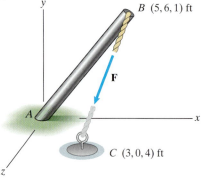

Problem 2.158

2.159 The magnitude of $\mathbf{F}_B$ is 400 N and $|\mathbf{F}_A + \mathbf{F}_B| = 900$ N. Determine the components of $\mathbf{F}_A$.

2.160 Suppose that the forces $\mathbf{F}_A$ and $\mathbf{F}_B$ have the same magnitude and $\mathbf{F}_A \cdot \mathbf{F}_B = 600$ N^2. What are $\mathbf{F}_A$ and $\mathbf{F}_B$?

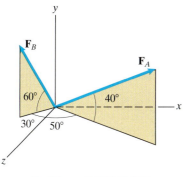

Problems 2.159/2.160

2.161 The magnitude of the force vector $\mathbf{F}_B$ is 2 kN. Express it in terms of components.

2.162 The magnitude of the vertical force vector $\mathbf{F}$ is 6 kN. Determine the vector components of $\mathbf{F}$ parallel and normal to the line from B to D.

2.163 The magnitude of the vertical force vector $\mathbf{F}$ is 6 kN. Given that $\mathbf{F} + \mathbf{F}_A + \mathbf{F}_B + \mathbf{F}_C = \mathbf{0}$, what are the magnitudes of $\mathbf{F}_A$, $\mathbf{F}_B$, and $\mathbf{F}_C$?

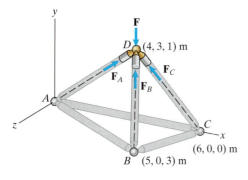

Problems 2.161–2.163

2.164 The magnitude of the vertical force $\mathbf{W}$ is 160 N. The direction cosines of the position vector from A to B are $\cos \theta_x = 0.500$, $\cos \theta_y = 0.866$, and $\cos \theta_z = 0$, and the direction cosines of the position vector from B to C are $\cos \theta_x = 0.707$, $\cos \theta_y = 0.619$, and $\cos \theta_z = -0.342$. Point G is the midpoint of the line from B to C. Determine the vector $\mathbf{r}_{AG} \times \mathbf{W}$, where $\mathbf{r}_{AG}$ is the position vector from A to G.

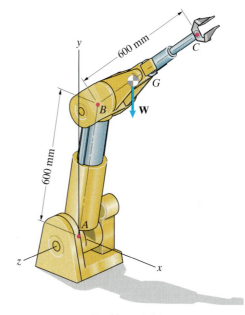

Problem 2.164

2.165 The rope CE exerts a 500-N force $\mathbf{T}$ on the hinged door.
(a) Express $\mathbf{T}$ in terms of components.
(b) Determine the vector component of $\mathbf{T}$ parallel to the line from point A to point B.

2.166 In Problem 2.165, let $\mathbf{r}_{BC}$ be the position vector from point B to point C. Determine the cross product $\mathbf{r}_{BC} \times \mathbf{T}$.

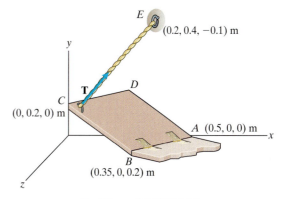

Problems 2.165/2.166

Forces

In Chapter 2 we represented forces by vectors and used vector addition to sum forces. In this chapter we discuss forces in more detail and introduce two of the most important concepts in mechanics, equilibrium and the free-body diagram. We will use free-body diagrams to identify the forces on objects and use equilibrium to determine unknown forces.

◄ The gravitational force on the climber is balanced by the forces exerted by the rope suspending him. In this chapter we use free-body diagrams to analyze forces on objects in equilibrium.

3.1 Types of Forces

Force is a familiar concept, as is evident from the words push, pull, and lift used in everyday conversation. In engineering we deal with different types of forces having a large range of magnitudes. In this section we introduce some terms used to describe forces and discuss particular forces that occur frequently in engineering applications.

Terminology

Line of Action   When a force is represented by a vector, the straight line collinear with the vector is called the *line of action* of the force (Fig. 3.1).

Systems of Forces   A *system of forces* is simply a particular set of forces. A system of forces is *coplanar*, or *two dimensional*, if the lines of action of the forces lie in a plane. Otherwise it is *three dimensional*. A system of forces is *concurrent* if the lines of action of the forces intersect at a point (Fig. 3.2a) and *parallel* if the lines of action are parallel (Fig. 3.2b).

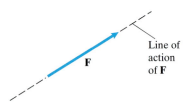

Figure 3.1
A force **F** and its line of action.

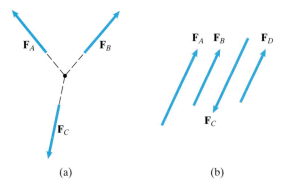

(a)   (b)

Figure 3.2
(a) Concurrent forces.
(b) Parallel forces.

External and Internal Forces   We say that a given object is subjected to an *external force* if the force is exerted by a different object. When one part of a given object is subjected to a force by another part of the same object, we say it is subjected to an *internal force*. These definitions require that you clearly define the object you are considering. For example, suppose that you are the object. When you are standing, the floor—a different object—exerts an external force on your feet. If you press your hands together, your left hand exerts an internal force on your right hand. However, if your right hand is the object you are considering, the force exerted by your left hand is an external force.

Body and Surface Forces   A force acting on an object is called a *body force* if it acts on the volume of the object and a *surface force* if it acts on its surface. The gravitational force on an object is a body force. A surface force can be exerted on an object by contact with another object. Both body and surface forces can result from electromagnetic effects.

Gravitational Forces

You are aware of the force exerted on an object by the earth's gravity whenever you pick up something heavy. We can represent the gravitational force, or weight, of an object by a vector (Fig. 3.3).

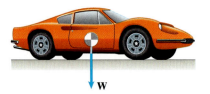

Figure 3.3
Representing an object's weight by a vector.

The magnitude of an object's weight is related to its mass m by

$$|\mathbf{W}| = mg,$$

where g is the acceleration due to gravity at sea level. We will use the values $g = 9.81 \text{ m/s}^2$ in SI units and $g = 32.2 \text{ ft/s}^2$ in U.S. Customary units.

Gravitational forces, and also electromagnetic forces, act at a distance. The objects they act on are not necessarily in contact with the objects exerting the forces. In the next section we discuss forces resulting from contacts between objects.

Contact Forces

Contact forces are the forces that result from contacts between objects. For example, you exert a contact force when you push on a wall (Fig. 3.4a). The surface of your hand exerts a force on the surface of the wall that can be represented by a vector $\mathbf{F}$ (Fig. 3.4b). The wall exerts an equal and opposite force $-\mathbf{F}$ on your hand (Fig. 3.4c). (Recall Newton's third law: The forces exerted on each other by any two particles are equal in magnitude and opposite in direction. If you have any doubt that the wall exerts a force on your hand, try pushing on the wall while standing on roller skates.)

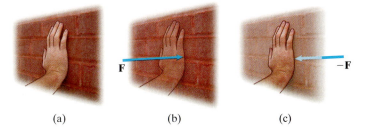

(a)   (b)   (c)

Figure 3.4
(a) Exerting a contact force on a wall by pushing on it.
(b) The vector $\mathbf{F}$ represents the force you exert on the wall.
(c) The wall exerts a force $-\mathbf{F}$ on your hand.

We will be concerned with contact forces exerted on objects by contact with the surfaces of other objects and by ropes, cables, and springs.

Surfaces Consider two plane surfaces in contact (Fig. 3.5a). We represent the force exerted on the right surface by the left surface by the vector $\mathbf{F}$ in Fig. 3.5b. We can resolve $\mathbf{F}$ into a component $\mathbf{N}$ that is normal to the surface and a component $\mathbf{f}$ that is parallel to the surface (Fig. 3.5c). The component $\mathbf{N}$ is called the *normal force*, and the component $\mathbf{f}$ is called the *friction force*. We sometimes assume that the friction force between two surfaces is negligible in comparison to the normal force, a condition we describe by saying that the surfaces are *smooth*. In this case we show only the normal force (Fig. 3.5d). When the friction force cannot be neglected, we say the surfaces are *rough*.

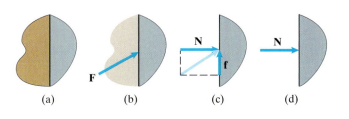

(a)   (b)   (c)   (d)

Figure 3.5
(a) Two plane surfaces in contact.
(b) The force $\mathbf{F}$ exerted on the right surface.
(c) The force $\mathbf{F}$ resolved into components normal and parallel to the surface.
(d) Only the normal force is shown when friction is neglected.

Figure 3.6
(a) Curved contacting surfaces. The dashed line indicates the plane tangent to the surfaces at their point of contact.
(b) The normal force and friction force on the right surface.

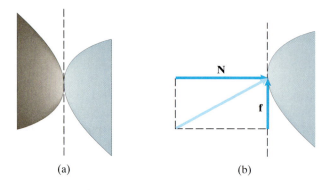

(a)   (b)

If the contacting surfaces are curved (Fig. 3.6a), the normal force and the friction force are perpendicular and parallel to the plane tangent to the surfaces at their point of contact (Fig. 3.6b).

Ropes and Cables   A contact force can be exerted on an object by attaching a rope or cable to the object and pulling on it. In Fig. 3.7a, the crane's cable is attached to a container of building materials. We can represent the force the cable exerts on the container by a vector $\mathbf{T}$ (Fig. 3.7b). The magnitude of $\mathbf{T}$ is called the *tension* in the cable, and the line of action of $\mathbf{T}$ is collinear with the cable. The cable exerts an equal and opposite force $-\mathbf{T}$ on the crane (Fig. 3.7c).

Notice that we have assumed that the cable is straight and that the tension where the cable is connected to the container equals the tension near the crane. This is approximately true if the weight of the cable is small compared to the tension. Otherwise, the cable will sag significantly and the tension will vary along its length. In Chapter 9 we will discuss ropes and cables whose weights are not small in comparison to their tensions. For now, we assume that ropes and cables are straight and that their tensions are constant along their lengths.

A *pulley* is a wheel with a grooved rim that can be used to change the direction of a rope or cable (Fig. 3.8a). For now, we assume that the tension is the

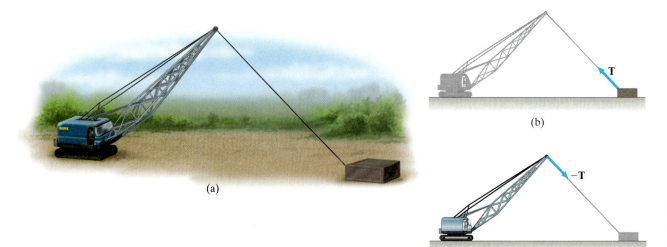

(a)

(b)

(c)

Figure 3.7
(a) A crane with its cable attached to a container.
(b) The force $\mathbf{T}$ exerted on the container by the cable.
(c) The force $-\mathbf{T}$ exerted on the crane by the cable.

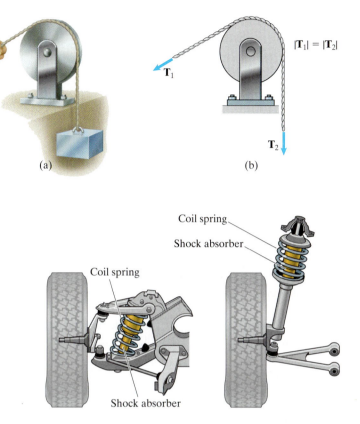

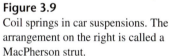

Figure 3.8
(a) A pulley changes the direction of a rope or cable.
(b) For now, you should assume that the tensions on each side of the pulley are equal.

Figure 3.9
Coil springs in car suspensions. The arrangement on the right is called a MacPherson strut.

same on both sides of a pulley (Fig. 3.8b). This is true, or at least approximately true, when the pulley can turn freely and the rope or cable either is stationary or turns the pulley at a constant rate.

Springs   Springs are used to exert contact forces in mechanical devices, for example, in the suspensions of cars (Fig. 3.9). Let's consider a coil spring whose unstretched length, the length of the spring when its ends are free, is L_0 (Fig. 3.10a). When the spring is stretched to a length L greater than L_0 (Fig. 3.10b), it pulls on the object to which it is attached with a force **F** (Fig. 3.10c). The object exerts an equal and opposite force $-$**F** on the spring (Fig. 3.10d). When the spring is compressed to a length L less than L_0 (Figs. 3.11a, b), the spring pushes on the object with a force **F** and the object exerts an equal and opposite force $-$**F** on the spring (Figs. 3.11c, d). If a spring is compressed too much, it may buckle (Fig. 3.11e). A spring designed to exert a force by being compressed is often provided with lateral support to prevent buckling, for example, by enclosing it in a cylindrical sleeve. In the car suspensions shown in Fig. 3.9, the shock absorbers within the coils prevent the springs from buckling.

The magnitude of the force exerted by a spring depends on the material it is made of, its design, and how much it is stretched or compressed relative to its unstretched length. When the change in length is not too large compared to the unstretched length, the coil springs commonly used in mechanical devices exert a force approximately proportional to the change in length:

$$|\mathbf{F}| = k|L - L_0|. \tag{3.1}$$

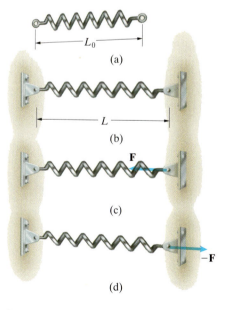

Figure 3.10
(a) A spring of unstretched length L_0.
(b) The spring stretched to a length $L > L_0$.
(c, d) The force **F** exerted by the spring and the force $-$**F** on the spring.

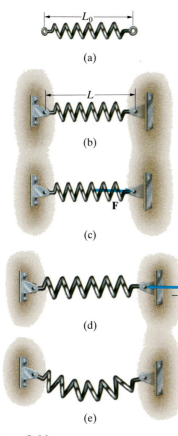

Figure 3.11
(a) A spring of length L_0.
(b) The spring compressed to a length $L < L_0$.
(c, d) The spring pushes on an object with a force **F**, and the object exerts a force $-\mathbf{F}$ on the spring.
(e) A coil spring will buckle if it is compressed too much.

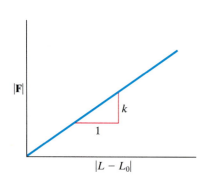

Figure 3.12
The graph of the force exerted by a linear spring as a function of its stretch or compression is a straight line with slope k.

Because the force is a linear function of the change in length (Fig. 3.12), a spring that satisfies this relation is called a *linear spring*. The value of the *spring constant* k depends on the material and design of the spring. Its dimensions are (force)/(length). Notice from Eq. (3.1) that k equals the magnitude of the force required to stretch or compress the spring a unit of length.

Suppose that the unstretched length of a spring is $L_0 = 1$ m and $k = 3000$ N/m. If the spring is stretched to a length $L = 1.2$ m, the magnitude of the pull it exerts is

$$k|L - L_0| = 3000(1.2 - 1) = 600 \text{ N.}$$

Although coil springs are commonly used in mechanical devices, we are also interested in them for a different reason. Springs can be used to *model* situations in which forces depend on displacements. For example, the force necessary to bend the steel beam in Fig. 3.13a is a linear function of the displacement δ, or

$$|\mathbf{F}| = k\delta,$$

if δ is not too large. Therefore we can model the force-deflection behavior of the beam with a linear spring (Fig. 3.13b).

Study Questions

1. What is a two-dimensional system of forces?
2. What are internal and external forces?
3. If a surface is said to be smooth, what does that mean?
4. What is the relation between the magnitude of the force exerted by a linear spring and the change in its length?

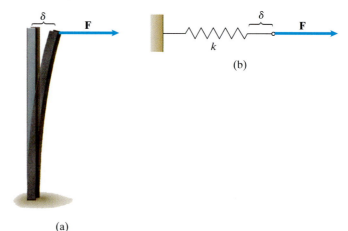

Figure 3.13
(a) A steel beam deflected by a force.
(b) Modeling the beam's behavior with a linear spring.

3.2 Analysis of Forces

Statics is concerned with the analysis of forces acting on objects in equilibrium. In this section we first explain what is meant in mechanics when an object is said to be in equilibrium. We then introduce the free-body diagram, which is the essential concept needed to identify and analyze the forces acting on an object.

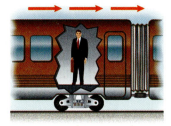

(a)

Equilibrium

In everyday conversation, equilibrium means an unchanging state—a state of balance. Before we state precisely what this term means in mechanics, let us consider some familiar examples. If you are in a building as you read this, objects you observe around you that are *at rest (stationary) relative to the building*, such as pieces of furniture, are in equilibrium. A person sitting or standing at rest relative to the building is also in equilibrium. If a train travels at constant speed on a straight track, objects within the train that are at rest relative to the train, such as the passenger seats or a passenger standing in the aisle (Fig. 3.14a), are in equilibrium. *The person at rest relative to the building and also the passenger at rest relative to the train are not accelerating.* However, if the train should begin increasing or decreasing its speed, the person standing in the aisle of the train would no longer be in equilibrium and might lose his balance (Fig. 3.14b).

(b)

Figure 3.14
(a) While the train moves at a constant speed, a person standing in the aisle is in equilibrium.
(b) If the train starts to speed up, the person is no longer in equilibrium.

We define an object to be in *equilibrium* only if each point of the object has the same constant velocity, which is referred to as *steady translation*. The velocity must be measured relative to a frame of reference in which Newton's laws are valid. Such a frame is called a *Newtonian* or *inertial reference frame*. In many engineering applications, a frame of reference that is fixed with respect to the earth can be regarded as inertial. Therefore, objects in steady translation relative to the earth can be assumed to be in equilibrium. We make this assumption throughout this book. In the examples cited in the previous paragraph, the furniture and person at rest in a building and also the passenger seats and passenger at rest within the train moving at constant speed are in steady translation relative to the earth and so are in equilibrium.

The vector sum of the external forces acting on an object in equilibrium is zero. We will use the symbol $\Sigma\, \mathbf{F}$ to denote the sum of the external forces. Thus, when an object is in equilibrium,

$$\Sigma\, \mathbf{F} = \mathbf{0}.$$

(3.2)

In some situations we can use this *equilibrium equation* to determine unknown forces acting on an object in equilibrium. The first step will be to draw a *free-body diagram* of the object to identify the external forces acting on it.

Free-Body Diagrams

A free-body diagram serves to focus attention on the object of interest and helps identify the external forces acting on it. Although in statics we are concerned only with objects in equilibrium, free-body diagrams are also used in dynamics to study the motions of objects.

Although it is one of the most important tools in mechanics, a free-body diagram is a simple concept. It is a drawing of an object and the external forces acting on it. Otherwise, nothing other than the object of interest is included. The drawing shows the object *isolated*, or *freed*, from its surroundings.

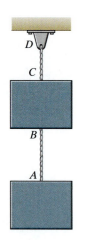

Figure 3.15
Stationary blocks suspended by cables.

Drawing a free-body diagram involves three steps:

1. *Identify the object you want to isolate*—As the following examples show, your choice is often dictated by particular forces you to determine.

2. *Draw a sketch of the object isolated from its surroundings, and show relevant dimensions and angles*—Your drawing should be reasonably accurate, but it can omit irrelevant details.

3. *Draw vectors representing all of the external forces acting on the isolated object, and label them*—Don't forget to include the gravitational force if you are not intentionally neglecting it.

A coordinate system is necessary to express the forces on the isolated object in terms of components. Often it is convenient to choose the coordinate system before drawing the free-body diagram, but in some situations the best choice of a coordinate system will not be apparent until after it has been drawn.

A simple example demonstrates how you can choose free-body diagrams to determine particular forces and also that you must distinguish carefully between external and internal forces. Two stationary blocks of equal weight W are suspended by cables in Fig. 3.15. The system is in equilibrium. Suppose that we want to determine the tensions in the two cables.

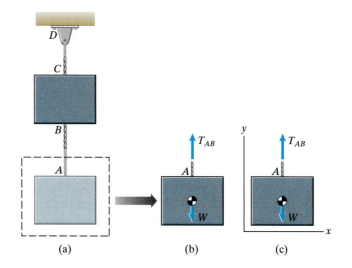

Figure 3.16
(a) Isolating the lower block and part of cable *AB*.
(b) Indicating the external forces completes the free-body diagram.
(c) Introducing a coordinate system.

(a)   (b)   (c)

To determine the tension in cable *AB*, we first isolate an "object" consisting of the lower block and part of cable *AB* (Fig. 3.16a). We then ask ourselves what forces can be exerted on our isolated object by objects not included in the diagram. The earth exerts a gravitational force of magnitude W on the block. Also, where we "cut" cable *AB*, the cable is subjected to a contact force equal to the tension in the cable (Fig. 3.16b). The arrows in this figure indicate the directions of the forces. The scalar W is the weight of the block and T_{AB} is the tension in cable *AB*. We assume that the weight of the part of cable *AB* included in the free-body diagram can be neglected in comparison to the weight of the block.

Since the free-body diagram is in equilibrium, the sum of the external forces equals zero. In terms of a coordinate system with the y axis upward (Fig. 3.16c), we obtain the equilibrium equation

$$\Sigma \mathbf{F} = T_{AB}\mathbf{j} - W\mathbf{j} = (T_{AB} - W)\mathbf{j} = \mathbf{0}.$$

Thus, the tension in cable *AB* is $T_{AB} = W$.

We can determine the tension in cable *CD* by isolating the upper block (Fig. 3.17a). The external forces are the weight of the upper block and the tensions

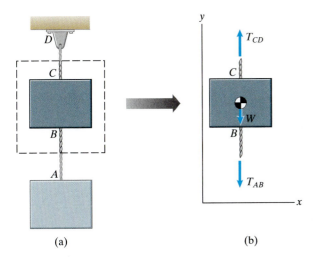

Figure 3.17
(a) Isolating the upper block to determine the tension in cable *CD*.
(b) Free-body diagram of the upper block.

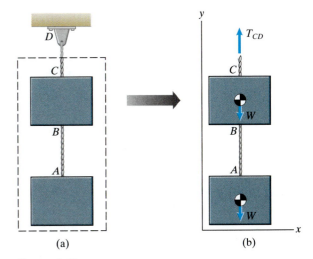

Figure 3.18
(a) An alternative choice for determining the tension in cable *CD*.
(b) Free-body diagram including both blocks and cable *AB*.

in the two cables (Fig. 3.17b). In this case we obtain the equilibrium equation

$$\Sigma \mathbf{F} = T_{CD}\mathbf{j} - T_{AB}\mathbf{j} - W\mathbf{j} = (T_{CD} - T_{AB} - W)\mathbf{j} = \mathbf{0}.$$

Since $T_{AB} = W$, we find that $T_{CD} = 2W$.

We could also have determined the tension in cable *CD* by treating the two blocks and the cable *AB* as a single object (Figs. 3.18a, b). The equilibrium equation is

$$\Sigma \mathbf{F} = T_{CD}\mathbf{j} - W\mathbf{j} - W\mathbf{j} = (T_{CD} - 2W)\mathbf{j} = \mathbf{0},$$

and we again obtain $T_{CD} = 2W$.

Why doesn't the tension in cable *AB* appear on the free-body diagram in Fig. 3.18b? Remember that only external forces are shown on free-body diagrams. Since cable *AB* is part of the free-body diagram in this case, the forces it exerts on the upper and lower blocks are internal forces.

We have described the procedure for drawing free-body diagrams. In the next section we will draw free-body diagrams of objects subjected to two-dimensional systems of forces and use them to determine unknown forces acting on objects in equilibrium.

3.3 Two-Dimensional Force Systems

Suppose that the system of external forces acting on an object in equilibrium is two dimensional (coplanar). By orienting a coordinate system so that the forces lie in the *x*–*y* plane, we can express the sum of the external forces as

$$\Sigma \mathbf{F} = (\Sigma F_x)\mathbf{i} + (\Sigma F_y)\mathbf{j} = \mathbf{0},$$

where ΣF_x and ΣF_y are the sums of the *x* and *y* components of the forces. Since a vector is zero only if each of its components is zero, we obtain two scalar equilibrium equations:

$$\Sigma F_x = 0, \qquad \Sigma F_y = 0. \tag{3.3}$$

The sums of the x and y components of the external forces acting on an object in equilibrium must each equal zero.

Example 3.1   **Using Equilibrium to Determine Forces on an Object**

For display at an automobile show, the 1440-kg car in Fig. 3.19 is held in place on the inclined surface by the horizontal cable from A to B. Determine the tension that the cable (and the fixture to which it is connected at B) must support. The car's brakes are not engaged, so the tires exert only normal forces on the inclined surface.

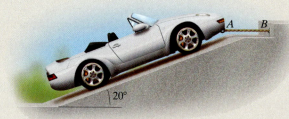

20°

Figure 3.19

(a) Isolating the car.

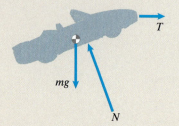

mg

N

T

(b) The completed free-body diagram shows the known and unknown external forces.

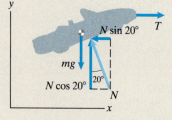

y

$N \sin 20°$ T

mg

$N \cos 20°$ 20°

N

x

(c) Introducing a coordinate system and expressing N in terms of its components.

Strategy

Since the car is in equilibrium, we can draw its free-body diagram and use Eqs. (3.3) to determine the forces exerted on the car by the cable and the inclined surface.

Solution

Draw the Free-Body Diagram   We first draw a diagram of the car isolated from its surrounding (Fig. a) and then complete the free-body diagram by showing the force exerted by the car's weight, the force T exerted by the cable, and the total normal force N exerted on the car's tires by the inclined surface (Fig. b).

Apply the Equilibrium Equations   In Fig. c, we introduce a coordinate system and resolve the normal force into x and y components. The equilibrium equations are

$$\Sigma F_x = T - N \sin 20° = 0,$$

$$\Sigma F_y = N \cos 20° - mg = 0.$$

We can solve the second equilibrium equation for N:

$$N = \frac{mg}{\cos 20°} = \frac{(1440 \text{ kg})(9.81 \text{ m/s}^2)}{\cos 20°} = 15,000 \text{ N.}$$

Then we solve the first equilibrium equation for the tension T:

$$T = N \sin 20° = 5140 \text{ N.}$$

Critical Thinking

How can you identify the external forces that act on an object? Drawing a free-body diagram helps you to do so. To isolate the car in this example, we had to remove the cable AB, which exerts the horizontal force T on the car at A that keeps the car in place on the inclined surface. We also had to remove the inclined surface, which exerts forces on the car's tires. (The example stipulated that the surface could exert only normal forces on the tires. We represented the total normal force exerted on the tires by the force N.) Finally, to isolate the car we had to remove the earth itself, which exerts the car's weight mg. *Thinking about what must be eliminated in order to isolate an object focuses your attention on those things that may exert external forces on it.*

Example 3.2 | Choosing a Free-Body Diagram

The automobile engine block in Fig. 3.20 is suspended by a system of cables. The mass of the block is 200 kg. The system is stationary. What are the tensions in cables AB and AC?

Strategy

We need a free-body diagram that is subjected to the forces we want to determine. By isolating part of the cable system near point A where the cables are joined, we can obtain a free-body diagram that is subjected to the weight of the block and the unknown tensions in cables AB and AC.

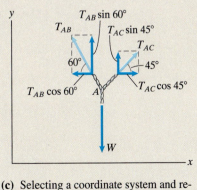

Figure 3.20

Solution

Draw the Free-Body Diagram Isolating part of the cable system near point A (Fig. a), we obtain a free-body diagram subjected to the weight of the block $W = mg = (200 \text{ kg})(9.81 \text{ m/s}^2) = 1962 \text{ N}$ and the tensions in cables AB and AC (Fig. b).

Apply the Equilibrium Equations We select the coordinate system shown in Fig. c and resolve the cable tensions into x and y components. The resulting equilibrium equations are

$$\Sigma F_x = T_{AC} \cos 45° - T_{AB} \cos 60° = 0,$$

$$\Sigma F_y = T_{AC} \sin 45° + T_{AB} \sin 60° - 1962 \text{ N} = 0.$$

Solving these equations, we find that the tensions in the cables are $T_{AB} = 1436$ N and $T_{AC} = 1016$ N.

(a) Isolating part of the cable system.
(b) The completed free-body diagram.

Critical Thinking

How can you choose a free-body diagram that permits you to determine particular unknown forces? There are no definite rules for choosing free-body diagrams. You will learn what to do in many cases from the examples we present, but you will also encounter new situations. It may be necessary to try several free-body diagrams before finding one that provides the information you need. Remember that forces you want to determine should appear as external forces on your free-body diagram, and your objective is to obtain a number of equilibrium equations equal to the number of unknown forces.

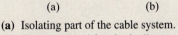

(c) Selecting a coordinate system and resolving the forces into components.

Example 3.3   Applying Equilibrium to a System of Pulleys

The mass of each pulley of the system in Fig. 3.21 is m, and the mass of the suspended object A is m_A. Determine the force T necessary for the system to be in equilibrium.

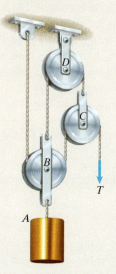

Figure 3.21

Strategy

By drawing free-body diagrams of the individual pulleys and applying equilibrium, we can relate the force T to the weights of the pulleys and the object A.

Solution

We first draw a free-body diagram of the pulley C to which the force T is applied (Fig. a). Notice that we assume the tension in the cable supported by the pulley to equal T on both sides (see Fig. 3.8). From the equilibrium equation

$$T_D - T - T - mg = 0,$$

we determine that the tension in the cable supported by pulley D is

$$T_D = 2T + mg.$$

We now know the tensions in the cables extending from pulleys C and D to pulley B in terms of T. Drawing the free-body diagram of pulley B (Fig. b), we obtain the equilibrium equation

$$T + T + 2T + mg - mg - m_A g = 0.$$

Solving, we obtain $T = m_A g/4$.

Critical Thinking

Notice that the objects we isolate in Figs. a and b include parts of the cables. The weights of those parts of cable are external forces acting on the free-body diagrams. Why didn't we include them? We tacitly assumed that the weights of those parts of cable could be neglected in comparison to the weights of the pulleys and the suspended object A. You will notice throughout the book that weights of objects are often neglected in analyzing the forces acting on them. This is a valid approximation for a given object if its weight is small compared to the other forces acting on it. But in any real engineering application, this assumption must be carefully evaluated. We discuss the weights of objects in more detail in Chapter 7.

(a) Free-body diagram of pulley C.
(b) Free-body diagram of pulley B.

T_D

C mg

T T

(a)

$T_D = 2T + mg$

T T T

B mg

$m_A g$

(b)

Design Example 3.4   Steady Flight

Figure 3.22 shows an airplane flying in the vertical plane and its free-body diagram. The forces acting on the airplane are its weight W, the thrust T exerted by its engines, and aerodynamic forces resulting from the pressure distribution on the airplane's surface. The dashed line indicates the path along which the airplane is moving. The aerodynamic forces are resolved into a component perpendicular to the path, the lift L, and a component parallel to the path, the drag D. The angle γ between the horizontal and the path is called the flight path angle, and α is the angle of attack. If the airplane remains in equilibrium for an interval of time, it is said to be in steady flight. If $\gamma = 6°$, $D = 125$ kN, $L = 680$ kN, and the mass of the airplane is 72,000 kg, what values of T and α are necessary to maintain steady flight?

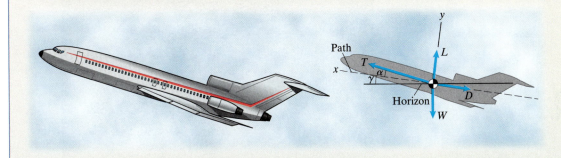

Strategy

The airplane is assumed to be in equilibrium. By applying Eqs. (3.3) to the given free-body diagram, we will obtain two equations with which to determine T and α.

Figure 3.22
External forces on an airplane in flight.

Solution

In terms of the coordinate system in Fig. 3.22, the equilibrium equations are

$$\Sigma F_x = T \cos \alpha - D - W \sin \gamma = 0, \qquad (1)$$
$$\Sigma F_y = T \sin \alpha + L - W \cos \gamma = 0, \qquad (2)$$

where the airplane's weight is $W = (72{,}000 \text{ kg})(9.81 \text{ m/s}^2) = 706{,}000$ N. We solve Eq. (2) for $\sin \alpha$, solve Eq. (1) for $\cos \alpha$, and divide to obtain an equation for $\tan \alpha$:

$$
\begin{aligned}
\tan \alpha &= \frac{W \cos \gamma - L}{W \sin \gamma + D} \\[4pt]
&= \frac{(706{,}000 \text{ N}) \cos 6° - 680{,}000 \text{ N}}{(706{,}000 \text{ N}) \sin 6° + 125{,}000 \text{ N}} \\[4pt]
&= 0.113.
\end{aligned}
$$

The angle of attack $\alpha = \arctan(0.113) = 6.44°$. Now we use Eq. (1) to determine the thrust:

$$
\begin{aligned}
T &= \frac{W \sin \gamma + D}{\cos \alpha} \\[4pt]
&= \frac{(706{,}000 \text{ N}) \sin 6° + 125{,}000 \text{ N}}{\cos 6.44°} \\[4pt]
&= 200{,}000 \text{ N}.
\end{aligned}
$$

Notice that the thrust necessary for steady flight is 28% of the airplane's weight.

Design Issues

In the examples we have considered so far, the values of certain forces acting on an object in equilibrium were given, and our goal was simply to determine the unknown forces by setting the sum of the forces equal to zero. In many situations in engineering, an object in equilibrium is subjected to forces that have different values under different conditions, and this has a profound effect on its design.

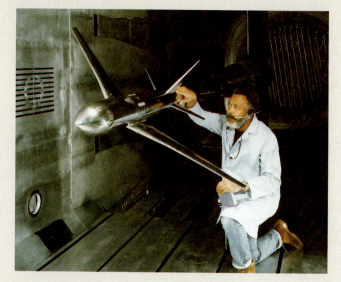

Figure 3.23
Wind tunnels are used to measure the aerodynamic forces on airplane models.

Figure 3.24
An F-15 being refueled by a KC-135 refueling plane.

When an airplane cruises at constant altitude ($\gamma = 0$), Eqs. (1) and (2) reduce to

$$T \cos \alpha = D,$$
$$T \sin \alpha + L = W.$$

The horizontal component of the thrust must equal the drag, and the sum of the vertical component of the thrust and the lift must equal the weight. For a fixed value of α, the lift and drag increase as the speed of the airplane increases. A principal design concern is to minimize D at cruising speed in order to minimize the thrust (and consequently the fuel consumption) needed to satisfy the first equilibrium equation. Much of the research on airplane design, including both theoretical analyses and model tests in wind tunnels (Fig. 3.23), is devoted to developing airplane shapes that minimize drag.

When an airplane cruises at low speed, satisfying the second equilibrium equation has the most serious implications for design. The airplane's wings must generate sufficient lift to balance its weight. This requirement is especially difficult to achieve in fast airplanes, because wings designed for low drag at high velocities do not generate as much lift at low speeds as wings that are designed for flight at lower velocities. For example, the F-15 in Fig. 3.24 must fly with a relatively large angle of attack (which increases both the lift and the vertical component of the thrust) in comparison to the refueling plane. In the case of the F-14 (Fig. 3.25), the engineers obtained both low drag at high velocities and good lift characteristics at low velocities by using variable sweep wings.

Figure 3.25
An F-14 with its wings in the takeoff and landing configuration and in the high-speed configuration.

Problems

3.1 Three forces act on a joint of a structure. The joint's weight is negligible and it is in equilibrium. The force $F_A = 4$ kN. Determine the force F_B and the vertical force F_C.

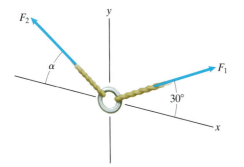

Problem 3.1

3.2 The mass of the ring is 2 kg. The y axis points upward. The angle $\alpha = 45°$.
(a) What is the ring's weight in newtons?
(b) Determine the forces F_1 and F_2.

Problem 3.2

3.3 In Problem 3.2, suppose that you want to choose the angle α so that the force F_2 is a minimum. What is the required angle α and the resulting value of F_2?

    Strategy: Draw a vector diagram of the sum of the forces acting on the ring.

3.4 The 200-kg engine block is suspended by the cables AB and AC. The angle $\alpha = 40°$. The free-body diagram obtained by isolating the part of the system within the dashed line is shown. Determine the forces T_{AB} and T_{AC}.

3.5 The 200-kg engine block is suspended by the cables AB and AC. Suppose that you don't want either of the forces T_{AB} or T_{AC} to exceed 2 kN. What is the smallest acceptable value of the angle α?

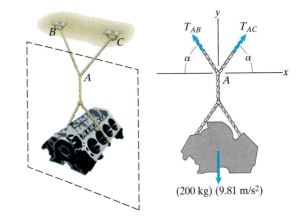

Problems 3.4/3.5

3.6 A zoologist estimates that the jaw of a predator, *Martes*, is subjected to a force P as large as 800 N. What forces T and M must be exerted by the temporalis and masseter muscles to support this value of P?

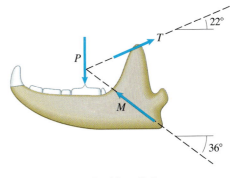

Problem 3.6

3.7 The two springs are identical, with unstretched lengths 250 mm and spring constants $k = 1200$ N/m.

(a) Draw the free-body diagram of block A.
(b) Draw the free-body diagram of block B.
(c) What are the masses of the two blocks?

3.8 The two springs are identical, with unstretched lengths of 250 mm. Suppose that their spring constant k is unknown and the sum of the masses of blocks A and B is 10 kg. Determine the value of k and the masses of the two blocks.

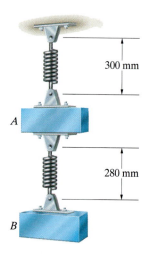

300 mm

A

280 mm

B

Problems 3.7/3.8

3.9 The inclined surface is smooth. The two springs are identical, with unstretched lengths of 250 mm and spring constants $k = 1200$ N/m. What are the masses of blocks A and B?

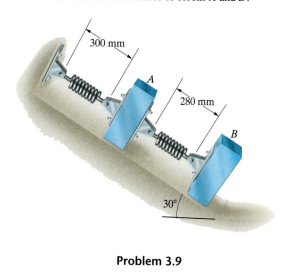

300 mm

A

280 mm

B

30°

Problem 3.9

3.10 The mass of the crane is 20,000 kg. The crane's cable is attached to a caisson whose mass is 400 kg. The tension in the cable is 1 kN.

(a) Determine the magnitudes of the normal and friction forces exerted on the crane by the level ground.
(b) Determine the magnitudes of the normal and friction forces exerted on the caisson by the level ground.

Strategy: To do part (a), draw the free-body diagram of the crane and the part of its cable within the dashed line.

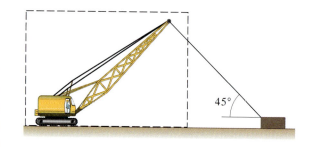

45°

Problem 3.10

3.11 The inclined surface is smooth. The 100-kg crate is held stationary by a force T applied to the cable.

(a) Draw the free-body diagram of the crate.
(b) Determine the force T.

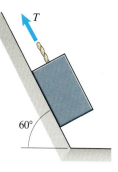

T

60°

Problem 3.11

3.12 The 1200-kg car is stationary on the sloping road.
(a) If $\alpha = 20°$, what are the magnitudes of the total normal and friction forces exerted on the car's tires by the road?
(b) The car can remain stationary only if the total friction force necessary for equilibrium is not greater than 0.6 times the total normal force. What is the largest angle α for which the car can remain stationary?

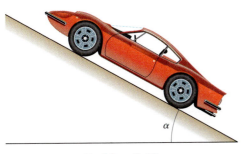

Problem 3.12

3.13 The crate is in equilibrium on the smooth surface. (Remember that "smooth" means that friction is negligible.) The spring constant is $k = 2500$ N/m and the stretch of the spring is 0.055 m. What is the mass of the crate?

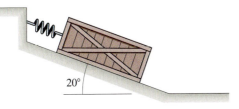

Problem 3.13

3.14 The 600-lb box is held in place on the smooth bed of the dump truck by the rope AB.
(a) If $\alpha = 25°$, what is the tension in the rope?
(b) If the rope will safely support a tension of 400 lb, what is the maximum allowable value of α?

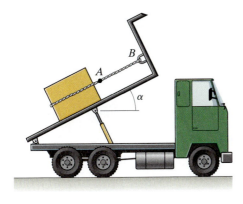

Problem 3.14

3.15 The 40-kg box is held in place on the smooth inclined surface by the horizontal cable AB. Determine the tension in the cable and the normal force exerted on the box by the inclined surface.

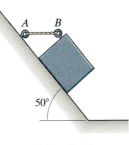

Problem 3.15

3.16 The 1360-kg car and the 2100-kg tow truck are stationary. The muddy surface on which the car's tires rest exerts negligible friction forces on them. What is the tension in the tow cable?

3.17 In Problem 3.16, determine the magnitude of the total friction force exerted on the tow truck's tires. (This is the friction force the truck's tires must exert to prevent the truck and car from sliding down the slope.)

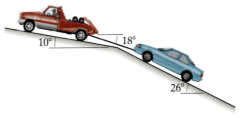

Problems 3.16/3.17

3.18 A 10-kg painting is hung with a wire supported by a nail. The length of the wire is 1.3 m.
(a) What is the tension in the wire?
(b) What is the magnitude of the force exerted on the nail by the wire?

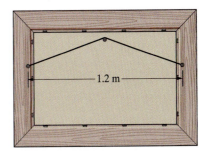

Problem 3.18

3.19 A 10-kg painting is hung with a wire supported by two nails. The length of the wire is 1.3 m.

(a) What is the tension in the wire?

(b) What is the magnitude of the force exerted on each nail by the wire? (Assume that the tension is the same in each part of the wire.)

Compare your answers to the answers to Problem 3.18.

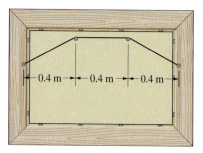

Problem 3.19

3.20 Assume that the 150-lb climber is in equilibrium. What are the tensions in the rope on the left and right sides?

3.21 If the mass of the climber shown in Problem 3.20 is 80 kg, what are the tensions in the rope on the left and right sides?

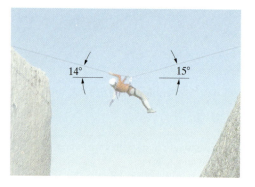

Problems 3.20/3.21

3.22 A construction worker holds a 180-kg crate in the position shown. What force must she exert on the cable?

Problem 3.22

3.23 A construction worker on the moon (acceleration due to gravity 1.62 m/s^2) holds the same crate described in Problem 3.22 in the position shown. What force must she exert on the cable?

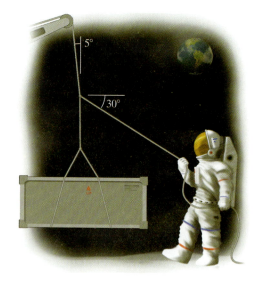

Problem 3.23

3.24 The person wants to cause the 200-lb crate to start sliding toward the right. To achieve this, the *horizontal component* of the force exerted on the crate by the rope must equal 0.35 times the normal force exerted on the crate by the floor. In Fig. a, the person pulls on the rope in the direction shown. In Fig. b, the person attaches the rope to a support as shown and pulls upward on the rope. What is the magnitude of the force he must exert on the rope in each case?

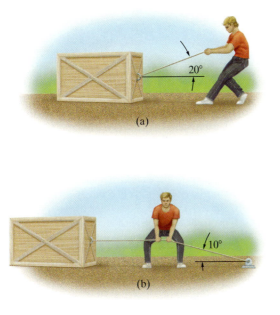

(a)

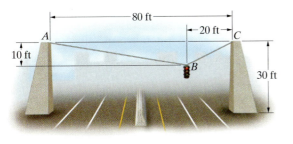

(b)

Problem 3.24

3.25 A traffic engineer wants to suspend a 200-lb traffic light above the center of the two right lanes of a four-lane thoroughfare as shown. Points A and C are at the same height. Determine the tensions in the cables AB and BC.

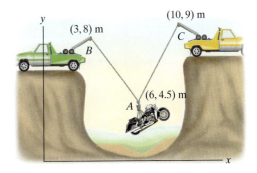

Problem 3.25

3.26 Cable AB is 3 m long and cable BC is 4 m long. Points A and C are at the same height. The mass of the suspended object is 350 kg. Determine the tensions in cables AB and BC.

3.27 The length of cable AB is adjustable. Cable BC is 4 m long. If you don't want the tension in either cable AB or cable BC to exceed 3 kN, what is the minimum acceptable length of cable AB?

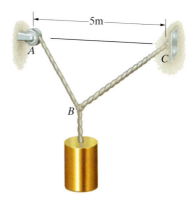

Problems 3.26/3.27

3.28 What are the tensions in the upper and lower cables? (Your answers will be in terms of W. Neglect the weight of the pulley.)

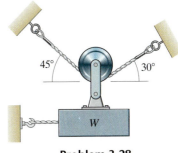

Problem 3.28

3.29 Two tow trucks lift a motorcycle out of a ravine following an accident. If the 100-kg motorcycle is in equilibrium in the position shown, what are the tensions in the cables AB and AC?

Problem 3.29

3.30 An astronaut candidate conducts experiments on an airbearing platform. While he carries out calibrations, the platform is held in place by the horizontal tethers *AB, AC,* and *AD.* The forces exerted by the tethers are the only horizontal forces acting on the platform. If the tension in tether *AC* is 2 N, what are the tensions in the other two tethers?

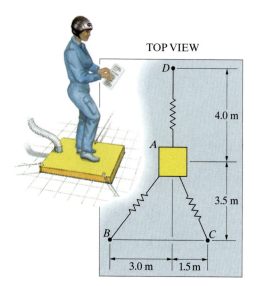

TOP VIEW

4.0 m

3.5 m

3.0 m   1.5 m

Problem 3.30

3.31 The forces exerted on the shoes and back of the 72-kg climber by the walls of the "chimney" are perpendicular to the walls exerting them. The tension in the rope is 640 N. What is the magnitude of the force exerted on his back?

10°

4°   3°

Problem 3.31

3.32 The slider *A* is in equilibrium and the bar is smooth. What is the mass of the slider?

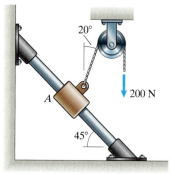

20°

200 N

A

45°

Problem 3.32

3.33 The 20-kg mass is suspended from three cables. Cable *AC* is equipped with a turnbuckle so that its tension can be adjusted and a strain gauge that allows its tension to be measured. If the tension in cable *AC* is 40 N, what are the tensions in cables *AB* and *AD*?

3.34 The 20-kg mass is suspended from three cables. Suppose that you want to adjust the tension in cable *AC* so that the tensions in cables *AC* and *AD* are equal. What is the necessary tension in cable *AC*? What is the resulting tension in cable *AB*?

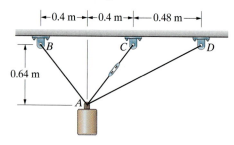

|←0.4 m→|←0.4 m→|←0.48 m →|

B   *C*   *D*

0.64 m

A

Problems 3.33/3.34

3.35 The collar *A* slides on the smooth vertical bar. The masses $m_A = 20$ kg and $m_B = 10$ kg. When $h = 0.1$ m, the spring is unstretched. When the system is in equilibrium, $h = 0.3$ m. Determine the spring constant k.

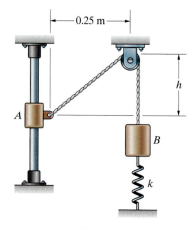

0.25 m

h

A

B

k

Problem 3.35

3.36 * Suppose that you want to design a cable system to suspend an object of weight W from the ceiling. The two wires must be identical, and the dimension b is fixed. The ratio of the tension T in each wire to its cross-sectional area A must equal a specified value $T/A = \sigma$. The "cost" of your design is the total volume of material in the two wires, $V = 2A\sqrt{b^2 + h^2}$. Determine the value of h that minimizes the cost.

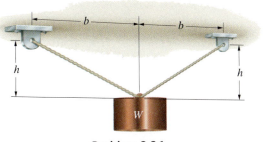

Problem 3.36

3.37 The system of cables suspends a 1000-lb bank of lights above a movie set. Determine the tensions in cables AB, CD, and CE.

3.38 A technician changes the position of the 1000-lb bank of lights by removing the cable CE. What is the tension in cable AB after the change?

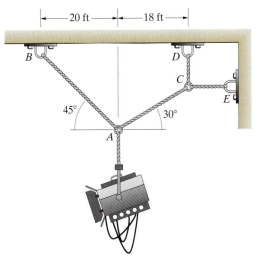

Problems 3.37/3.38

3.39 While working on another exhibit, a curator at the Smithsonian Institution pulls the suspended *Voyager* aircraft to one side by attaching three horizontal cables as shown. The mass of the aircraft is 1250 kg. Determine the tensions in the cable segments AB, BC, and CD.

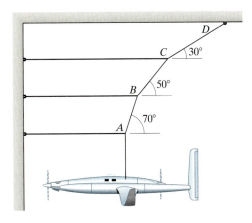

Problem 3.39

3.40 A truck dealer wants to suspend a 4000 kg truck as shown for advertising. The distance $b = 15$ m, and the sum of the lengths of the cables AB and BC is 42 m. Points A and C are at the same height. What are the tensions in the cables?

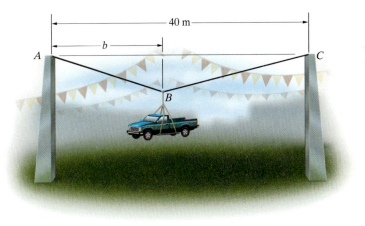

Problem 3.40

3.41 The distance $h = 12$ in, and the tension in cable AD is 200 lb. What are the tensions in cables AB and AC?

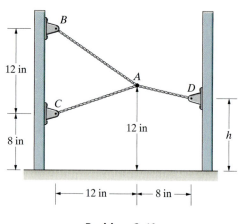

Problem 3.41

3.42 You are designing a cable system to support a suspended object of weight W. Because your design requires points A and B to be placed as shown, you have no control over the angle α, but you can choose the angle β by placing point C wherever you wish. Show that to minimize the tensions in cables AB and BC, you must choose $\beta = \alpha$ if the angle $\alpha \geq 45°$.

Strategy: Draw a diagram of the sum of the forces exerted by the three cables at A.

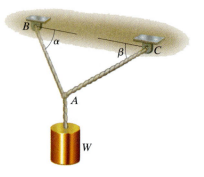

Problem 3.42

3.43 * The length of the cable ABC is 1.4 m. The 2-kN force is applied to a small pulley. The system is stationary. What is the tension in the cable?

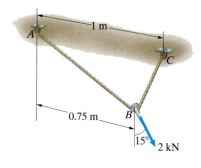

Problem 3.43

3.44 The masses $m_1 = 12$ kg and $m_2 = 6$ kg are suspended by the cable system shown. The cable BC is horizontal. Determine the angle α and the tensions in the cables AB, BC, and CD.

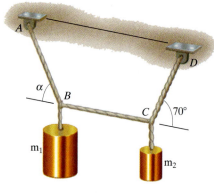

Problem 3.44

3.45 The weights $W_1 = 50$ lb and W_2 are suspended by the cable system shown. Determine the weight W_2 and the tensions in the cables AB, BC, and CD.

3.46 Assume that $W_2 = W_1/2$. If you don't want the tension anywhere in the supporting cable to exceed 200 lb, what is the largest acceptable value of W_1?

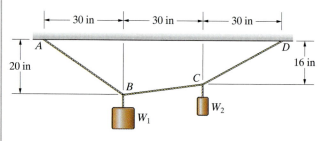

Problems 3.45/3.46

3.47 The hydraulic cylinder is subjected to three forces. An 8-kN force is exerted on the cylinder at *B* that is parallel to the cylinder and points from *B* toward *C*. The link *AC* exerts a force at *C* that is parallel to the line from *A* to *C*. The link *CD* exerts a force at *C* that is parallel to the line from *C* to *D*.

(a) Draw the free-body diagram of the cylinder. (The cylinder's weight is negligible.)

(b) Determine the magnitudes of the forces exerted by the links *AC* and *CD*.

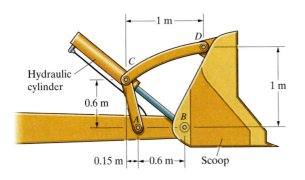

Problem 3.47

3.48 The 50-lb cylinder rests on two smooth surfaces.

(a) Draw the free-body diagram of the cylinder.

(b) If $\alpha = 30°$, what are the magnitudes of the forces exerted on the cylinder by the left and right surfaces?

3.49 Obtain an equation for the force exerted on the 50-lb cylinder by the left surface in terms of the angle α in two ways: (a) using a coordinate system with the *y* axis vertical, (b) using a coordinate system with the *y* axis parallel to the right surface.

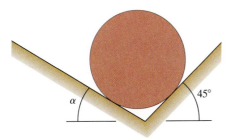

Problems 3.48/3.49

3.50 The two springs are identical, with unstretched length 0.4 m. When the 50-kg mass is suspended at *B*, the length of each spring increases to 0.6 m. What is the spring constant *k*?

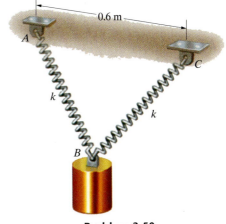

Problem 3.50

3.51 The cable *AB* is 0.5 m in length. The *unstretched* length of the spring is 0.4 m. When the 50-kg mass is suspended at *B*, the length of the spring increases to 0.45 m. What is the spring constant *k*?

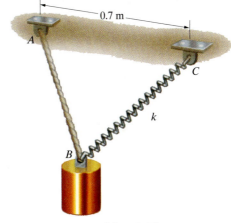

Problem 3.51

3.52 The 1440-kg car is moving at constant speed on a road with the slope shown. The aerodynamic forces on the car are the drag $D = 530$ N, which is parallel to the road, and the lift $L = 360$ N, which is perpendicular to the road. Determine the magnitudes of the total normal and friction forces exerted on the car by the road.

Problem 3.52

3.53 The inclined surface is smooth. Determine the force T that must be exerted on the cable to hold the 100-kg crate in equilibrium and compare your answer to the answer of Problem 3.11.

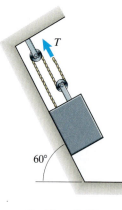

60°

Problem 3.53

3.54 The mass of each pulley of the system is m and the mass of the suspended object A is m_A. Determine the force T necessary for the system to be in equilibrium.

A

T

Problem 3.54

3.55 The mass of each pulley of the system is m and the mass of the suspended object A is m_A. Determine the force T necessary for the system to be in equilibrium.

A

T

Problem 3.55

3.56 The suspended mass $m_1 = 50$ kg. Neglecting the masses of the pulleys, determine the value of the mass m_2 necessary for the system to be in equilibrium.

3.57 The suspended mass $m_1 = 50$ kg. If the pulleys A, B, and C each have a mass of 2 kg, what mass m_2 is necessary for the system to be in equilibrium?

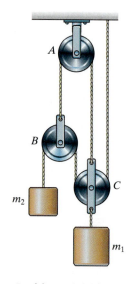

A

B

m_2

C

m_1

Problems 3.56/3.57

3.58 Pulley systems containing one, two, and three pulleys are shown. Neglecting the weights of the pulleys, determine the force T required to support the weight W in each case.

3.59 The number of pulleys in the type of system shown could obviously be extended to an arbitrary number N.

(a) Neglecting the weights of the pulleys, determine the force T required to support the weight W as a function of the number of pulleys N in the system.

(b) Using the result of part (a), determine the force T required to support the weight W for a system with 10 pulleys.

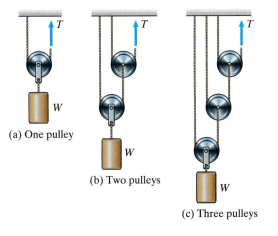

(a) One pulley

(b) Two pulleys

(c) Three pulleys

Problems 3.58/3.59

Design Experience

Problems 3.60–3.62 are related to Design Example 3.4.

3.60 A 14,000-kg airplane is in steady flight in the vertical plane. The flight path angle is $\gamma = 10°$, the angle of attack is $\alpha = 4°$, and the thrust force exerted by the engine is $T = 60$ kN. What are the magnitudes of the lift and drag forces acting on the airplane?

3.61 An airplane is in steady flight, the angle of attack $\alpha = 0$, the thrust-to-drag ratio $T/D = 2$, and the lift-to-drag ratio $L/D = 4$. What is the flight path angle γ?

3.62 An airplane glides in steady flight ($T = 0$), and its lift-to-drag ratio is $L/D = 4$.

(a) What is the flight path angle γ?

(b) If the airplane glides from an altitude of 1000 m to zero altitude, what horizontal distance does it travel?

3.4 Three-Dimensional Force Systems

The equilibrium situations we have considered so far have involved only coplanar forces. When the system of external forces acting on an object in equilibrium is three dimensional, we can express the sum of the external forces as

$$\Sigma \mathbf{F} = (\Sigma F_x)\mathbf{i} + (\Sigma F_y)\mathbf{j} + (\Sigma F_z)\mathbf{k} = \mathbf{0}.$$

Each component of this equation must equal zero, resulting in three scalar equilibrium equations:

$$\Sigma F_x = 0, \quad \Sigma F_y = 0, \quad \Sigma F_z = 0. \tag{3.4}$$

The sums of the x, y, and z components of the external forces acting on an object in equilibrium must each equal zero.

Example 3.5 | **Applying Equilibrium in Three Dimensions**

The 100-kg cylinder in Fig. 3.26 is suspended from the ceiling by cables attached at points B, C, and D. What are the tensions in cables AB, AC, and AD?

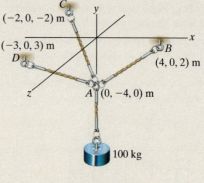

Figure 3.26

Strategy

We can determine the tensions by the same approach we used for similar two-dimensional problems. By isolating part of the cable system near point A, we can obtain a free-body diagram subjected to forces due to the tensions in the cables. Since the sums of the x, y, and z components of the external forces must each equal zero, we obtain three equations for the three unknown tensions.

Solution

Draw the Free-Body Diagram   We isolate part of the cable system near point A (Fig. a) and complete the free-body diagram by showing the forces exerted by the tensions in the cables (Fig. b). The magnitudes of the vectors $\mathbf{T}_{AB}$, $\mathbf{T}_{AC}$, and $\mathbf{T}_{AD}$ are the tensions in cables AB, AC, and AD, respectively.

Apply the Equilibrium Equations   The sum of the external forces acting on the free-body diagram is

$$\Sigma \mathbf{F} = \mathbf{T}_{AB} + \mathbf{T}_{AC} + \mathbf{T}_{AD} - (981\ \text{N})\mathbf{j} = \mathbf{0}.$$

To solve this equation for the tensions in the cables, we need to express the vectors $\mathbf{T}_{AB}$, $\mathbf{T}_{AC}$, and $\mathbf{T}_{AD}$ in terms of their components.

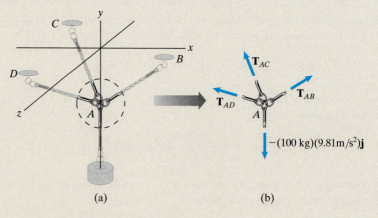

(a) (b)

(a) Isolating part of the cable system.
(b) The completed free-body diagram showing the forces exerted by the tensions in the cables.

We first determine the components of a unit vector that points in the direction of the vector $\mathbf{T}_{AB}$. Let $\mathbf{r}_{AB}$ be the position vector from point A to point B (Fig. c):

$$\mathbf{r}_{AB} = (x_B - x_A)\mathbf{i} + (y_B - y_A)\mathbf{j} + (z_B - z_A)\mathbf{k} = 4\mathbf{i} + 4\mathbf{j} + 2\mathbf{k} \text{ (m)}.$$

Dividing $\mathbf{r}_{AB}$ by its magnitude, we obtain a unit vector that has the same direction as $\mathbf{T}_{AB}$:

$$\mathbf{e}_{AB} = \frac{\mathbf{r}_{AB}}{|\mathbf{r}_{AB}|} = 0.667\mathbf{i} + 0.667\mathbf{j} + 0.333\mathbf{k}.$$

Now we can write the vector $\mathbf{T}_{AB}$ as the product of the tension T_{AB} in cable AB and $\mathbf{e}_{AB}$:

$$\mathbf{T}_{AB} = T_{AB}\mathbf{e}_{AB} = T_{AB}(0.667\mathbf{i} + 0.667\mathbf{j} + 0.333\mathbf{k}).$$

We now express the force vectors $\mathbf{T}_{AC}$ and $\mathbf{T}_{AD}$ in terms of the tensions T_{AC} and T_{AD} in cables AC and AD in the same way. The results are

$$\mathbf{T}_{AC} = T_{AC}(-0.408\mathbf{i} + 0.816\mathbf{j} - 0.408\mathbf{k}),$$

$$\mathbf{T}_{AD} = T_{AD}(-0.514\mathbf{i} + 0.686\mathbf{j} - 0.514\mathbf{k}).$$

We use these expressions to write the sum of the external forces in terms of the tensions T_{AB}, T_{AC}, and T_{AD}:

$$\Sigma \mathbf{F} = \mathbf{T}_{AB} + \mathbf{T}_{AC} + \mathbf{T}_{AD} - (981 \text{ N})\mathbf{j}$$

$$= (0.667T_{AB} - 0.408T_{AC} - 0.514T_{AD})\mathbf{i}$$

$$+ (0.667T_{AB} + 0.816T_{AC} + 0.686T_{AD} - 981 \text{ N})\mathbf{j}$$

$$+ (0.333T_{AB} - 0.408T_{AC} + 0.514T_{AD})\mathbf{k}$$

$$= \mathbf{0}.$$

The sums of the forces in the x, y, and z directions must each equal zero:

$$\Sigma F_x = 0.667T_{AB} - 0.408T_{AC} - 0.514T_{AD} = 0,$$

$$\Sigma F_y = 0.667T_{AB} + 0.816T_{AC} + 0.686T_{AD} - 981 \text{ N} = 0,$$

$$\Sigma F_z = 0.333T_{AB} - 0.408T_{AC} + 0.514T_{AD} = 0.$$

Solving these equations, we find that the tensions are $T_{AB} = 519 \text{ N}$, $T_{AC} = 636 \text{ N}$, and $T_{AD} = 168 \text{ N}$.

Critical Thinking

By using the equilibrium equation $\Sigma \mathbf{F} = \mathbf{0}$, we were able to determine the tensions in the three cables AB, AC, and AD. *One vector equation in three dimensions is equivalent to three scalar equations.* But suppose the 100-kg cylinder had been suspended from the ceiling using four cables instead of three. There would be four unknown tensions to determine, but only three scalar equilibrium equations. Such problems are said to be *statically indeterminate* and cannot generally be solved using statics alone. We discuss statically indeterminate problems in Chapter 5.

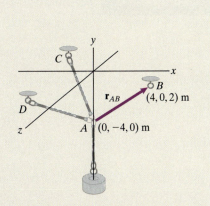

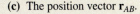

(c) The position vector $\mathbf{r}_{AB}$.

Example 3.6 | **Application of the Dot Product**

The 100-lb "slider" C in Fig. 3.27 is held in place on the smooth bar by the cable AC. Determine the tension in the cable and the force exerted on the slider by the bar.

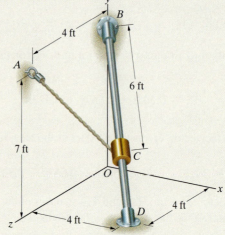

Figure 3.27

Strategy

Because we want to determine forces that act on the slider, we need to draw its free-body diagram. The external forces acting on the slider are its weight and the forces exerted on it by the cable and the bar. If we approached this example as we did the previous one, our next step would be to express the forces in terms of their components. However, we don't know the direction of the force exerted on the slider by the bar. Since the smooth bar exerts negligible friction force, we do know that the force is normal to the bar's axis. Therefore we can eliminate this force from the equation $\Sigma \mathbf{F} = \mathbf{0}$ by taking the dot product of the equation with a unit vector that is parallel to the bar.

Solution

Draw the Free-Body Diagram We isolate the slider (Fig. a) and complete the free-body diagram by showing the weight of the slider, the force $\mathbf{T}$ exerted by the tension in the cable, and the normal force $\mathbf{N}$ exerted by the bar (Fig. b).

Apply the Equilibrium Equations The sum of the external forces acting on the free-body diagram is

$$\Sigma \mathbf{F} = \mathbf{T} + \mathbf{N} - (100 \text{ lb})\mathbf{j} = \mathbf{0}. \tag{1}$$

Let $\mathbf{e}_{BD}$ be the unit vector pointing from point B toward point D. Since $\mathbf{N}$ is perpendicular to the bar, $\mathbf{e}_{BD} \cdot \mathbf{N} = 0$. Therefore,

$$\mathbf{e}_{BD} \cdot (\Sigma \mathbf{F}) = \mathbf{e}_{BD} \cdot [\mathbf{T} - (100 \text{ lb})\mathbf{j}] = 0. \tag{2}$$

Determining $\mathbf{e}_{BD}$: We determine the vector from point B to point D,

$$\mathbf{r}_{BD} = (4 - 0)\mathbf{i} + (0 - 7)\mathbf{j} + (4 - 0)\mathbf{k} = 4\mathbf{i} - 7\mathbf{j} + 4\mathbf{k} \, (\text{ft}),$$

and divide it by its magnitude to obtain the unit vector $\mathbf{e}_{BD}$:

$$\mathbf{e}_{BD} = \frac{\mathbf{r}_{BD}}{|\mathbf{r}_{BD}|} = \frac{4}{9}\mathbf{i} - \frac{7}{9}\mathbf{j} + \frac{4}{9}\mathbf{k}.$$

Expressing **T** *in terms of components*: We need to determine the coordinates of the slider C. We can write the vector from B to C in terms of the unit vector $\mathbf{e}_{BD}$,

$$\mathbf{r}_{BC} = 6\mathbf{e}_{BD} = 2.67\mathbf{i} - 4.67\mathbf{j} + 2.67\mathbf{k} \text{ (ft)},$$

and then add it to the vector from the origin O to B to obtain the vector from O to C:

$$\mathbf{r}_{OC} = \mathbf{r}_{OB} + \mathbf{r}_{BC} = 7\mathbf{j} + (2.67\mathbf{i} - 4.67\mathbf{j} + 2.67\mathbf{k})$$
$$= 2.67\mathbf{i} + 2.33\mathbf{j} + 2.67\mathbf{k} \text{ (ft)}.$$

The components of this vector are the coordinates of point C. Now we can determine a unit vector with the same direction as **T**. The vector from C to A is

$$\mathbf{r}_{CA} = (0 - 2.67)\mathbf{i} + (7 - 2.33)\mathbf{j} + (4 - 2.67)\mathbf{k}$$
$$= -2.67\mathbf{i} + 4.67\mathbf{j} + 1.33\mathbf{k} \text{ (ft)},$$

and the unit vector that points from point C toward point A is

$$\mathbf{e}_{CA} = \frac{\mathbf{r}_{CA}}{|\mathbf{r}_{CA}|} = -0.482\mathbf{i} + 0.843\mathbf{j} + 0.241\mathbf{k}.$$

Let T be the tension in the cable AC. Then we can write the vector **T** as

$$\mathbf{T} = T\mathbf{e}_{CA} = T(-0.482\mathbf{i} + 0.843\mathbf{j} + 0.241\mathbf{k}).$$

Determining **T** *and* **N**: Substituting our expressions for $\mathbf{e}_{BD}$ and **T** in terms of their components into Eq. (2) yields

$$0 = \mathbf{e}_{BD} \cdot [\mathbf{T} - (100 \text{ lb})\mathbf{j}]$$

$$= \left(\frac{4}{9}\mathbf{i} - \frac{7}{9}\mathbf{j} + \frac{4}{9}\mathbf{k}\right) \cdot [-0.482T\mathbf{i} + (0.843T - 100 \text{ lb})\mathbf{j} + 0.241T\mathbf{k}]$$

$$= -0.762T + 77.8 \text{ lb},$$

and we obtain the tension $T = 102$ lb.

Now we can determine the force exerted on the slider by the bar by using Eq. (1):

$$\mathbf{N} = -\mathbf{T} + (100 \text{ lb})\mathbf{j}$$

$$= -(102 \text{ lb})(-0.482\mathbf{i} + 0.843\mathbf{j} + 0.241\mathbf{k}) + (100 \text{ lb})\mathbf{j}$$

$$= 49.1\mathbf{i} + 14.0\mathbf{j} - 24.6\mathbf{k} \text{ (lb)}.$$

Critical Thinking

By taking the dot product of the equilibrium equation for the slider with a unit vector $\mathbf{e}_{BD}$ that is parallel to the smooth bar BD, we obtained Eq. (2), which does not contain the normal force **N**. Why does this happen? The formal answer is that $\mathbf{e}_{BD}$ is perpendicular to **N**, and so $\mathbf{e}_{BD} \cdot \mathbf{N} = 0$. But the physical interpretation of Eq. (2) provides a more compelling explanation: It states that *the component of the slider's weight parallel to the bar is balanced by the component of* **T** *parallel to the bar*. The normal force exerted on the slider by the smooth bar has no component parallel to the bar. We were therefore able to solve for the tension in the cable without knowing the normal force **N**.

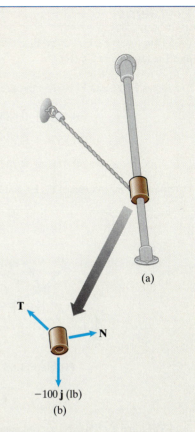

(a)

(b)

(a) Isolating the slider.
(b) Free-body diagram of the slider showing the forces exerted by its weight, the cable, and the bar.

Problems

3.63 Four forces $\mathbf{F}_1$, $\mathbf{F}_2$, $\mathbf{F}_3$, and $\mathbf{F}_4$ act on an object in equilibrium. The force $\mathbf{F}_1 = 50\mathbf{i}$ (N).

The forces $\mathbf{F}_2$, $\mathbf{F}_3$, and $\mathbf{F}_4$ point in the directions of the unit vectors

$$\mathbf{e}_2 = -0.485\mathbf{i} + 0.485\mathbf{j} - 0.728\mathbf{k},$$

$$\mathbf{e}_3 = -0.557\mathbf{i} + 0.743\mathbf{j} + 0.371\mathbf{k},$$

$$\mathbf{e}_4 = -0.371\mathbf{i} - 0.743\mathbf{j} + 0.557\mathbf{k}.$$

Determine the magnitudes of $\mathbf{F}_2$, $\mathbf{F}_3$, and $\mathbf{F}_4$.

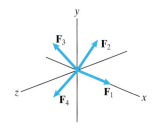

Problem 3.63

3.64 The force $\mathbf{F} = 5\mathbf{i}$ (kN) acts on point A where the cables AB, AC, and AD are joined. What are the tensions in the three cables?

Strategy: Isolate part of the cable system near point A. See Example 3.5.

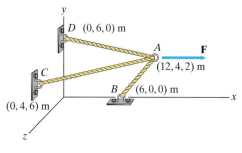

Problem 3.64

3.65 An 80-lb chandelier is suspended from three wires AB, AC, and AD of equal length. The wires are attached at points B, C, and D on the ceiling. Points B, C, and D lie on a circle of 3-ft radius and are equally spaced. (That is, they are placed at 120° intervals around the circle.) Point A is 4 ft below the ceiling. Determine the tensions in the wires.

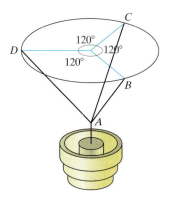

Problem 3.65

3.66 To support the tent, the tension in the rope AB must be 40 lb. What are the tensions in the ropes AC, AD, and AE?

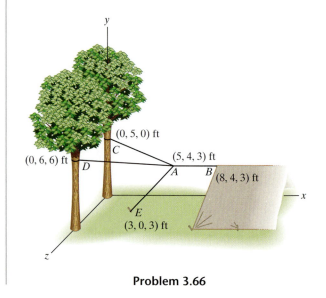

Problem 3.66

3.67 The bulldozer exerts a force $\mathbf{F} = 2\mathbf{i}$ (kip) at A. What are the tensions in cables AB, AC, and AD?

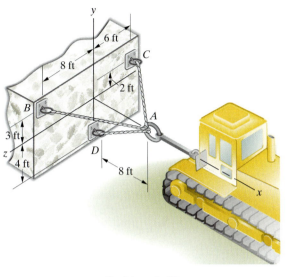

Problem 3.67

3.68 Prior to its launch, a balloon carrying a set of experiments to high altitude is held in place by groups of student volunteers holding the tethers at B, C, and D. The mass of the balloon, experiments package, and the gas it contains is 90 kg, and the buoyancy force on the balloon is 1000 N. The supervising professor conservatively estimates that each student can exert at least a 40-N tension on the tether for the necessary length of time. Based on this estimate, what minimum numbers of students are needed at B, C, and D?

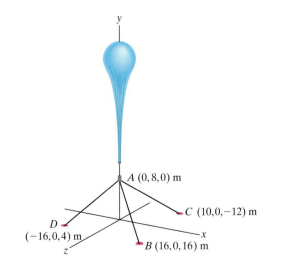

Problem 3.68

3.69 The 20-kg mass is suspended by cables attached to three vertical 2-m posts. Point A is at $(0, 1.2, 0)$ m. Determine the tensions in cables AB, AC, and AD.

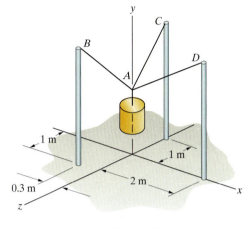

Problem 3.69

3.70 The weight of the horizontal wall section is $W = 20,000$ lb. Determine the tensions in the cables AB, AC, and AD.

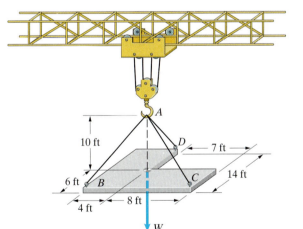

Problem 3.70

3.71 The car and the pallet supporting it in Fig. a weigh 3000 lb. They are suspended by four cables *AB, AC, AD,* and *AE.* The locations of the cable attachment points on the pallet are shown in Fig. b. The tensions in cables *AB* and *AC* are *equal.* Determine the tensions in the four cables.

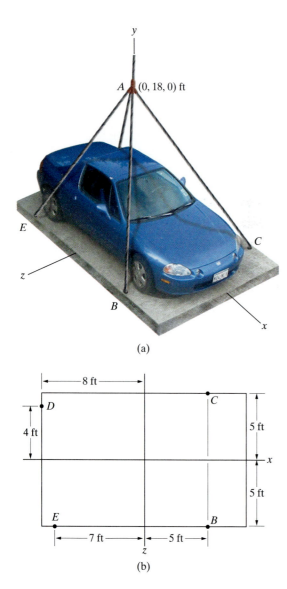

(a)

(b)

Problem 3.71

3.72 The 680-kg load suspended from the helicopter is in equilibrium. The aerodynamic drag force on the load is horizontal. The *y* axis is vertical, and cable *OA* lies in the *x–y* plane. Determine the magnitude of the drag force and the tension in cable *OA.*

3.73 The coordinates of the three cable attachment points *B, C,* and *D* are $(-3.3, -4.5, 0)$ m, $(1.1, -5.3, 1)$ m, and $(1.6, -5.4, -1)$ m, respectively. What are the tensions in cables *OB, OC,* and *OD?*

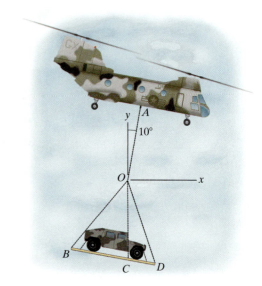

Problems 3.72/3.73

3.74 If the mass of the bar *AB* is negligible compared to the mass of the suspended object *E,* the bar exerts a force on the "ball" at *B* that points from *A* toward *B.* The mass of the object *E* is 200 kg. The *y* axis points upward. Determine the tensions in the cables *BC* and *BD.*

    Strategy: Draw a free-body diagram of the ball at *B.* (The weight of the ball is negligible.)

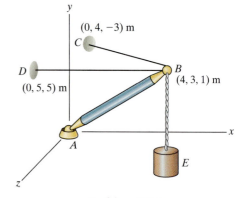

Problem 3.74

3.75 The 1350-kg car is at rest on a plane surface. The unit vector $e_n = 0.231i + 0.923j + 0.308k$ is perpendicular to the surface. The y axis points upward. Determine the magnitudes of the normal and friction forces the car's wheels exert on the surface.

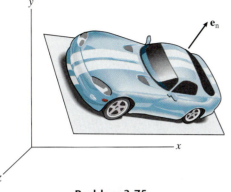

Problem 3.75

3.76 The system shown anchors a stanchion of a cable-suspended roof. If the tension in cable AB is 900 kN, what are the tensions in cables EF and EG?

3.77* The cables of the system will each safely support a tension of 1500 kN. Based on this criterion, what is the largest safe value of the tension in cable AB?

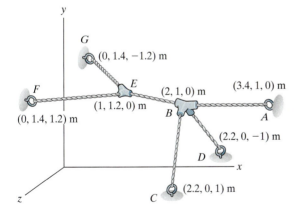

Problems 3.76/3.77

3.78 The 200-kg slider at A is held in place on the smooth vertical bar by the cable AB.
(a) Determine the tension in the cable.
(b) Determine the force exerted on the slider by the bar.

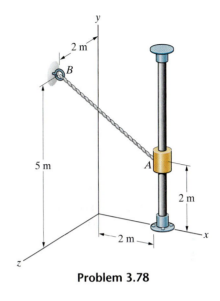

Problem 3.78

3.79 The 100-lb slider at A is held in place on the smooth circular bar by the cable AB. The circular bar is contained in the x–y plane.
(a) Determine the tension in the cable.
(b) Determine the normal force exerted on the slider by the bar.

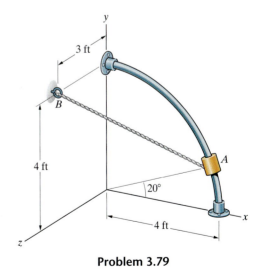

Problem 3.79

3.80 The cable AB keeps the 8-kg collar A in place on the smooth bar CD. The y axis points upward. What is the tension in the cable?

3.81* Determine the magnitude of the normal force exerted on the collar A by the smooth bar.

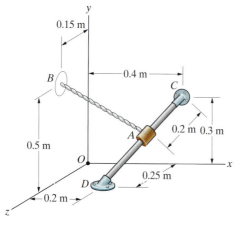

Problems 3.80/3.81

3.82* The 10-kg collar A and 20-kg collar B are held in place on the smooth bars by the 3-m cable from A to B and the force F acting on A. The force F is parallel to the bar. Determine F.

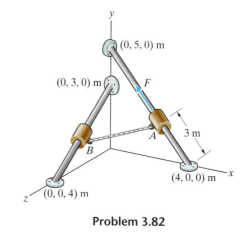

Problem 3.82

COMPUTATIONAL MECHANICS

The following examples and problems are designed for the use of a programmable calculator or computer. Example 3.7 is similar to previous examples and problems except that the solution must be calculated for a range of input quantities. Example 3.8 leads to an algebraic equation that must be solved numerically.

Computational Example 3.7 | Determining Tensions for a Range of Dimensions

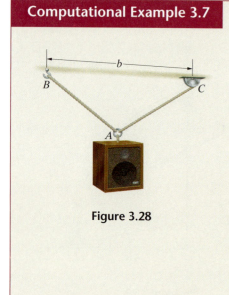

Figure 3.28

The system of cables in Fig. 3.28 is designed to suspend a load with a mass of 1000 kg from the ceiling. The dimension $b = 2$ m, and the length of cable AB is 1 m. The height of the load can be adjusted by changing the length of cable AC.

(a) Plot the tensions in cables AB and AC for values of the length of cable AC from 1.2 m to 2.2 m.

(b) Cables AB and AC can each safely support a tension equal to the weight of the load. Use the results of (a) to estimate the allowable range of the length of cable AC.

Strategy

By drawing the free-body diagram of the part of the cable system where the cables join, we can determine the tensions in the cables in terms of the length of cable AC.

Solution

(a) Let the lengths of the cables be $L_{AB} = 1$ m and L_{AC}. We can apply the law of cosines to the triangle in Fig. a to determine α in terms of L_{AC}:

$$\alpha = \arccos\left(\frac{b^2 + L_{AB}^2 - L_{AC}^2}{2bL_{AB}}\right).$$

Then we can use the law of sines to determine β:

$$\beta = \arcsin\left(\frac{L_{AB}\sin\alpha}{L_{AC}}\right).$$

Draw the Free-Body Diagram We draw the free-body diagram of the part of the cable system where the cables join in Fig. b, where T_{AB} and T_{AC} are the tensions in the cables.

Apply the Equilibrium Equations Selecting the coordinate system shown in Fig. b, the equilibrium equations are

$$\Sigma F_x = -T_{AB}\cos\alpha + T_{AC}\cos\beta = 0,$$

$$\Sigma F_y = T_{AB}\sin\alpha + T_{AC}\sin\beta - W = 0.$$

Solving these equations for the cable tensions, we obtain

$$T_{AB} = \frac{W\cos\beta}{\sin\alpha\cos\beta + \cos\alpha\sin\beta},$$

$$T_{AC} = \frac{W\cos\alpha}{\sin\alpha\cos\beta + \cos\alpha\sin\beta}.$$

To compute the results, we input a value of the length L_{AC} and calculate the angle α, then the angle β, and then the tensions T_{AB} and T_{AC}. The resulting values of T_{AB}/W and T_{AC}/W are plotted as functions of L_{AC} in Fig. 3.29. (b) The allowable range of the length of cable AC is the range over which the tensions in both cables are less than W. From Fig. 3.29 we can see that the tension T_{AB} exceeds W for values of L_{AC} less than about 1.35 m, so the safe range is $L_{AC} > 1.35$ m.

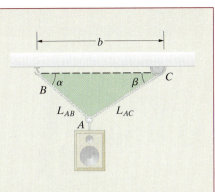

(a) Determining the angles α and β.

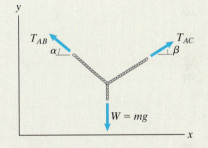

(b) Free-body diagram of part of the cable system.

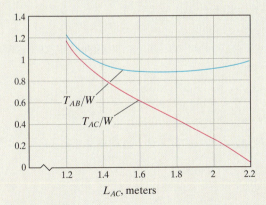

Figure 3.29
Ratios of the cable tensions to the suspended weight as functions of L_{AC}.

Crtical Thinking

As this example demonstrates, even simple problems in mechanics can require the solution of nonlinear equations. We could not obtain an analytical solution for the length L_{AC} corresponding to a given maximum tension in the two cables. But by computing the values of the tensions as a function of L_{AC}, we were able to identify the range of values for which the tensions in the cables did not exceed the load.

Computational Example 3.8

Equilibrium Position of an Object Supported by a Spring

The 12-lb collar A in Fig. 3.30 is held in equilibrium on the smooth vertical bar by the spring. The spring constant $k = 300$ 1b/ft, the unstretched length of the spring is $L_0 = b$, and the distance $b = 1$ ft. What is the distance h?

Strategy

Both the direction and the magnitude of the force exerted on the collar by the spring depend on h. By drawing the free-body diagram of the collar and applying the equilibrium equations, we can obtain an equation for h.

Solution

Draw the Free-Body Diagram   We isolate the collar (Fig. a) and complete the free-body diagram by showing its weight $W = 12$ lb, the force F exerted by the spring, and the normal force N exerted by the bar (Fig. b).

Apply the Equilibrium Equations   Selecting the coordinate system shown in Fig. b, we obtain the equilibrium equations

$$\Sigma F_x = N - \left(\frac{b}{\sqrt{h^2 + b^2}}\right)F = 0,$$

$$\Sigma F_y = \left(\frac{h}{\sqrt{h^2 + b^2}}\right)F - W = 0.$$

In terms of the length of the spring $L = \sqrt{h^2 + b^2}$, the force exerted by the spring is

$$F = k(L - L_0) = k(\sqrt{h^2 + b^2} - b).$$

Substituting this expression into the second equilibrium equation, we obtain the equation

$$\left(\frac{h}{\sqrt{h^2 + b^2}}\right)k\,(\sqrt{h^2 + b^2} - b) - W = 0.$$

Therefore, the distance h (in ft) is a root of the equation

$$f(h) = \left(\frac{h}{\sqrt{h^2 + b^2}}\right)k\,(\sqrt{h^2 + b^2} - b) - W = 0,$$

where $b = 1$ ft, $k = 300$ lb/ft, and $W = 12$ lb.

How can we solve this nonlinear algebraic equation for h? Some calculators and software are designed to obtain roots of such equations. Another approach is to calculate the value of $f(h)$ for a range of values of h and plot the results, as we have done in Fig. 3.31. From the graph we see that the solution is approximately $h = 0.45$ ft. By examining the computed results near $h = 0.45$ ft, see table.

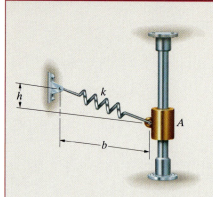

Figure 3.30

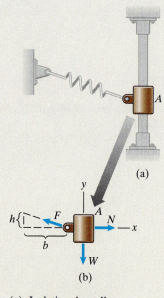

(a)

(b)

(a) Isolating the collar.
(b) The free-body diagram.

h (ft)	f (h)
0.449	−0.1818
0.450	−0.1094
0.451	−0.0368
0.452	0.0361
0.453	0.1092
0.454	0.1826

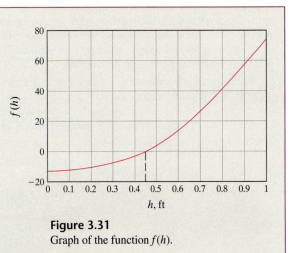

We see that the solution (to three significant digits) is $h = 0.452$ ft.

Figure 3.31
Graph of the function $f(h)$.

Critical Thinking

Just as in Example 3.7, this relatively simple mechanics problem gave rise to a nonlinear algebraic equation that we had to solve numerically. Obtaining numerical solutions to problems that cannot be solved analytically is one of the most common and important applications of computers in engineering.

Computational Problems

3.83 (a) Plot the tensions in cables AB and AC for values of d from $d = 0$ to $d = 1.8$ m.
(b) Each cable will safely support a tension of 1 kN. Use your graph to estimate the acceptable range of values of d.

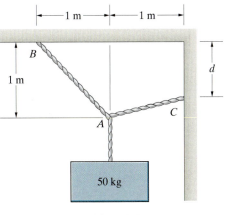

Problem 3.83

3.84 The suspended traffic light weighs 100 lb. The cables AB, BC, AD, and DE are each 11 ft long. Determine the smallest permissible length of the cable BD if the tensions in the cables must not exceed 1000 lb.

Strategy: Plot the tensions in the cables for a range of lengths of the cable BD.

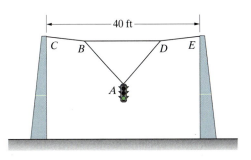

Problem 3.84

3.85 The 2000-lb scoreboard A is suspended above a sports arena by the cables AB and AC. Each cable is 160 ft long and points B and C are at the same height. Suppose you want to move the scoreboard out of the way for a tennis match by shortening cable AB while keeping the length of cable AC constant.

(a) Plot the tension in cable AB as a function of its length for values of the length from 142 ft to 160 ft.
(b) Use your graph to estimate how much you can raise the scoreboard relative to its original position if you don't want to subject cable AB to a tension greater than 6000 lb.

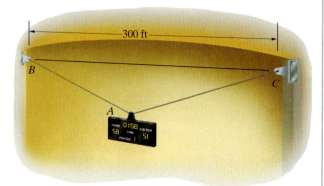

Problem 3.85

3.86 The mass of the truck is 4000 kg. The sum of the lengths of the cables AB and BC is 42 m.

(a) Draw graphs of the tensions in cables AB and BC for values of b from zero to 20 m.
(b) Each cable will safely support a tension of 60 kN. Use the results of part (a) to estimate the allowable range of the distance b.

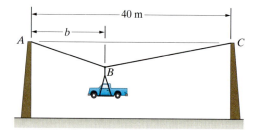

Problem 3.86

3.87 The two springs are identical, with unstretched length 0.4 m and spring constant $k = 900$ N/m. A 50-kg mass is suspended at B. What is the resulting tension in each spring?

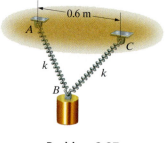

Problem 3.87

3.88 The cable AB is 0.5 m in length and the *unstretched length* of the spring BC is 0.4 m. The spring constant k is 5200 N/m. When the 50-kg mass is suspended at B, what is the resulting length of the stretched spring?

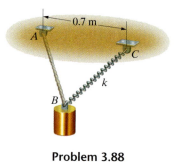

Problem 3.88

3.89 The length of the cable ABC is 1.4 m. The 2-kN force is applied to a small pulley. The system is stationary. Determine the *horizontal* distance from A to B and the tension in the cable.

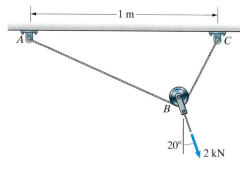

Problem 3.89

3.90 The mass of the balloon, experiments package, and the gas it contains is 90 kg, and the buoyancy force on the balloon is 1000 N. The tethers *AB, AC,* and *AD* will each safely support a tension of 500 N. The coordinates of point *A* are (0, *h*, 0). What is the minimum allowable height *h*?

3.92* The cable *AB* keeps the 8-kg collar *A* in place on the smooth bar *CD*. The *y* axis points upward. Determine the distance *s* from *C* to the collar *A* for which the tension in the cable is 150 N.

3.93* In Problem 3.92, determine the distance *s* from *C* to the collar *A* for which the magnitude of the normal force exerted on the collar *A* by the smooth bar is 50 N.

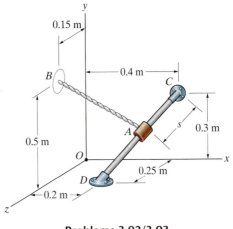

Problems 3.92/3.93

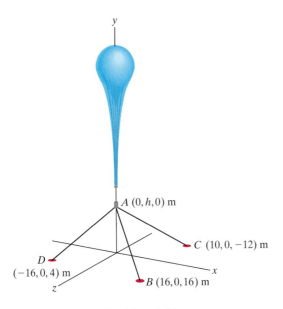

Problem 3.90

3.91 The collar *A* slides on the smooth vertical bar. The masses $m_A = 20$ kg and $m_B = 10$ kg, and the spring constant $k = 360$ N/m. When $h = 0.2$ m, the spring is unstretched. Determine the value of *h* when the system is in equilibrium.

3.94* The 10-kg collar *A* and 20-kg collar *B* slide on the smooth bars. The cable from *A* to *B* is 3 m in length. Determine the value of the distance *s* in the range $1 \le s \le 5$ m for which the system is in equilibrium.

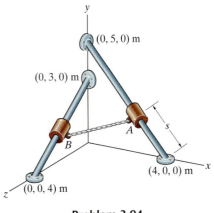

Problem 3.94

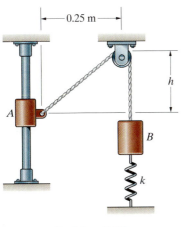

Problem 3.91

CHAPTER SUMMARY

In this chapter we discussed the forces that occur frequently in engineering applications and introduced two of the most important concepts in mechanics: the free-body diagram and equilibrium. By drawing free-body diagrams and applying the vector techniques developed in Chapter 2, we showed how unknown forces acting on objects in equilibrium can be determined from the condition that the sum of the external forces must equal zero. The sum of the moments of the external forces on an object in equilibrium must also equal zero, and this condition can be used to obtain additional information about unknown forces on objects. We will discuss moments of forces in Chapter 4. We will then apply equilibrium to individual objects in Chapter 5 and to structures in Chapter 6.

The straight line coincident with a force vector is called the *line of action* of the force. A system of forces is *coplanar*, or *two-dimensional*, if the lines of action of the forces lie in a plane. Otherwise, it is *three-dimensional*. A system of forces is *concurrent* if the lines of action of the forces intersect at a point and *parallel* if the lines of action are parallel.

An object is subjected to an *external force* if the force is exerted by a different object. When one part of an object is subjected to a force by another part of the same object, the force is *internal*.

A *body force* acts on the volume of an object, and a *surface* or *contact force* acts on its surface.

Gravitational Forces

The weight of an object is related to its mass by $W = mg$, where $g = 9.81 \ \text{m/s}^2$ in SI units and $g = 32.2 \ \text{ft/s}^2$ in U.S. Customary units.

Surfaces

Two surfaces in contact exert forces on each other that are equal in magnitude and opposite in direction. Each force can be resolved into the *normal force* and the *friction force*. If the friction force is negligible in comparison to the normal force, the surfaces are said to be *smooth*. Otherwise, they are *rough*.

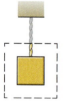

1. Choose an object to isolate.

Ropes and Cables

A rope or cable attached to an object exerts a force on the object whose magnitude is equal to the tension and whose line of action is parallel to the rope or cable at the point of attachment.

A *pulley* is a wheel with a grooved rim that can be used to change the direction of a rope or cable. When a pulley can turn freely and the rope or cable either is stationary or turns the pulley at a constant rate, the tension is approximately the same on both sides of the pulley.

2. Draw the isolated object.

Springs

The force exerted by a *linear spring* is

$$|\mathbf{F}| = k|L - L_0|, \tag{3.1}$$

where k is the *spring constant*, L is the length of the spring, and L_0 is its un-stretched length.

Free-Body Diagrams

3. Show the external forces.

A free-body diagram is a drawing of an object in which the object is isolated from its surroundings and the external forces acting on the object are shown. Drawing a free-body diagram requires the steps shown in Figs. 1–3. A coordinate system must be chosen to express the forces on the isolated object in terms of components.

Equilibrium

If an object is in equilibrium, the sum of the external forces acting on it is zero:

$$\Sigma \mathbf{F} = \mathbf{0}. \tag{3.2}$$

This implies that the sums of the external forces in the x, y, and z directions each equal zero:

$$\Sigma F_x = 0, \quad \Sigma F_y = 0, \quad \Sigma F_z = 0. \tag{3.4}$$

Review Problems

3.95 The 100-lb crate is held in place on the smooth surface by the rope AB. Determine the tension in the rope and the magnitude of the normal force exerted on the crate by the surface.

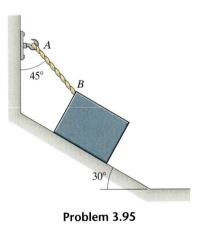

Problem 3.95

3.96 The system shown is called Russell's traction. If the sum of the downward forces exerted at A and B by the patient's leg is 32.2 lb, what is the weight W?

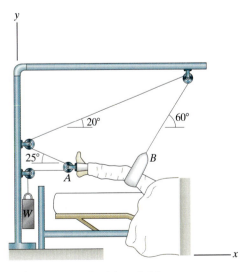
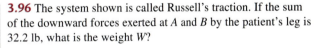

Problem 3.96

3.97 A heavy rope used as a hawser for a cruise ship sags as shown. If it weighs 200 lb, what are the tensions in the rope at A and B?

Problem 3.97

3.98 The cable AB is horizontal, and the box on the right weighs 100 lb. The surfaces are smooth.

(a) What is the tension in the cable?
(b) What is the weight of the box on the left?

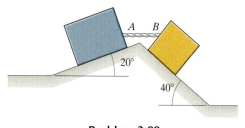

Problem 3.98

3.99 A concrete bucket used at a construction site is supported by two cranes. The 100-kg bucket contains 500 kg of concrete. Determine the tensions in the cables AB and AC.

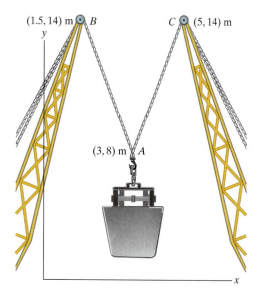

Problem 3.99

3.100 The mass of the suspended object A is m_A and the masses of the pulleys are negligible. Determine the force T necessary for the system to be in equilibrium.

Problem 3.100

3.101 The assembly A, including the pulley, weighs 60 lb. What force F is necessary for the system to be in equilibrium?

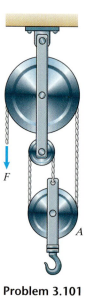

Problem 3.101

3.102 The mass of block A is 42 kg, and the mass of block B is 50 kg. The surfaces are smooth. If the blocks are in equilibrium, what is the force F?

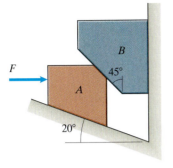

Problem 3.102

3.103 The climber A is being helped up an icy slope by two friends. His mass is 80 kg, and the direction cosines of the force exerted on him by the slope are $\cos \theta_x = -0.286$, $\cos \theta_y = 0.429$, and $\cos \theta_z = 0.857$. The y axis is vertical. If the climber is in equilibrium in the position shown, what are the tensions in the ropes AB and AC and the magnitude of the force exerted on him by the slope?

3.104 Consider the climber A being helped by his friends in Problem 3.103. To try to make the tensions in the ropes more equal, the friend at B moves to the position (4, 2, 0) m. What are the new tensions in the ropes AB and AC and the magnitude of the force exerted on the climber by the slope?

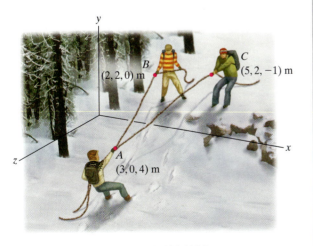

Problems 3.103/104

3.105 A climber helps his friend up an icy slope. His friend is hauling a box of supplies. If the mass of the friend is 90 kg and the mass of the supplies is 22 kg, what are the tensions in the ropes AB and CD? Assume that the slope is smooth. That is, only normal forces are exerted on the man and the box by the slope.

Problem 3.105

3.106 The small sphere of mass m is attached to a string of length L and rests on the smooth surface of a sphere of radius R. Determine the tension in the string in terms of m, L, h, and R.

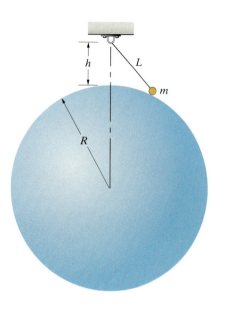

Problem 3.106

3.107 An engineer doing preliminary design studies for a new radio telescope envisions a triangular receiving platform suspended by cables from three equally spaced 40-m towers. The receiving platform has a mass of 20 Mg (megagrams) and is 10 m below the tops of the towers. What tension would the cables be subjected to?

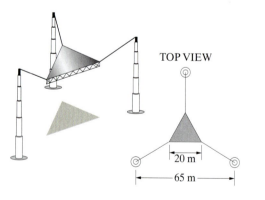

TOP VIEW

20 m

65 m

Problem 3.107

3.108 The metal disk A weighs 10 lb. It is held in place at the center of the smooth inclined surface by the strings AB and AC. What are the tensions in the strings?

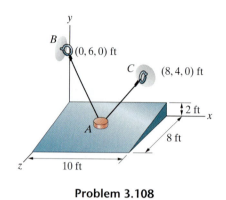

B

$(0, 6, 0)$ ft

C $(8, 4, 0)$ ft

2 ft

x

A

8 ft

z 10 ft

Problem 3.108

3.109 Cable AB is attached to the top of the vertical 3-m post, and its tension is 50 kN. What are the tensions in cables AO, AC, and AD?

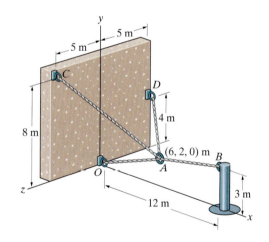

y

5 m

5 m

C

D

8 m

4 m

$(6, 2, 0)$ m B

O A

z

12 m

3 m

x

Problem 3.109

3.110* The 1350-kg car is at rest on a plane surface with its brakes locked. The unit vector $\mathbf{e}_n = 0.231\mathbf{i} + 0.923\mathbf{j} + 0.308\mathbf{k}$ is perpendicular to the surface. The y-axis points upward. The direction cosines of the cable from A to B are $\cos\theta_x = -0.816$, $\cos\theta_y = 0.408$, $\cos\theta_z = -0.408$, and the tension in the cable is 1.2 kN. Determine the magnitudes of the normal and friction forces the car's wheels exert on the surface.

3.111* The brakes of the car are released, and the car is held in place on the plane surface by the cable AB. The car's front wheels are aligned so that the tires exert no friction forces parallel to the car's longitudinal axis. The unit vector $\mathbf{e}_p = -0.941\mathbf{i} + 0.131\mathbf{j} + 0.314\mathbf{k}$ is parallel to the plane surface and aligned with the car's longitudinal axis. What is the tension in the cable?

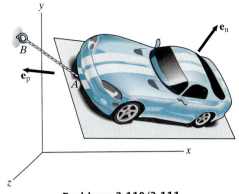

Problems 3.110/3.111

Design Project 1   A possible design for a simple scale to weigh objects is shown. The length of the string AB is 0.5 m. When an object is placed in the pan, the spring stretches and the string AB rotates. The object's weight can be determined by observing the change in the angle α.

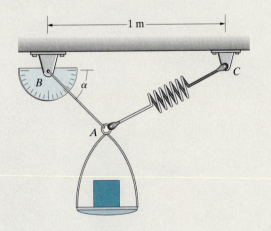

(a) Assume that objects with masses in the range 0.2–2 kg are to be weighed. Choose the unstretched length and spring constant of the spring in order to obtain accurate readings for weights in the desired range. (Neglect the weights of the pan and spring. Notice that a significant change in the angle α is needed to determine the weight accurately.)

(b) Suppose that you can use the same components—the pan, protractor, a spring, string—and also one or more pulleys. Suggest another possible configuration for the scale. Use statics to analyze your proposed configuration and compare its accuracy with that of the configuration shown for objects with masses in the range 0.2–2 kg.

Design Project 2 Suppose that the positions of points A, C, and D of the system of cables suspending the 100-kg mass are fixed, but you are free to choose the x and z coordinates of point B. Investigate the effects of different choices of the location of point B on the tensions in the cables. If the cost of cable AB is proportional to the product of the tension in the cable and its length, investigate the effect of different choices of the location of point B on the cost of the cable. Write a brief report describing the results of your investigations and recommending a location for point B.

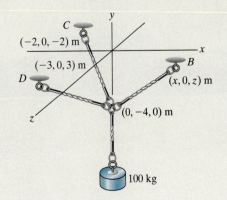

Systems of Forces and Moments

The effects of forces can depend not only on their magnitudes and directions but also on the moments, or torques, they exert. The rotations of objects such as the wheels of a vehicle, the crankshaft of an engine, and the rotor of an electric generator result from the moments of the forces exerted on them. If an object is in equilibrium, the moment about any point due to the forces acting on the object is zero. Before continuing our discussion of free-body diagrams and equilibrium, we must explain how to calculate moments and introduce the concept of equivalent systems of forces and moments.

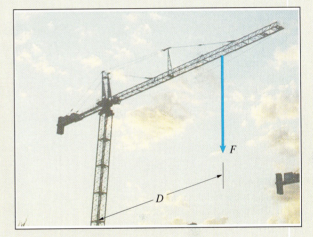

◄ Loads lifted by a building crane can exert large moments that the crane's structure must support. In this chapter we calculate moments of forces and analyze systems of forces and moments.

4.1 Two-Dimensional Description of the Moment

Consider a force of magnitude F and a point P, and let's view them in the direction perpendicular to the plane containing the force vector and the point (Fig. 4.1a). The *magnitude of the moment* of the force about P is the product DF, where D is the perpendicular distance from P to the line of action of the force (Fig. 4.1b). In this example, the force would tend to cause counterclockwise rotation about point P. That is, if we imagine that the force acts on an object that can rotate about point P, the force would cause counterclockwise rotation (Fig. 4.1c). We say that the *direction of the moment* is counterclockwise. *We define counterclockwise moments to be positive and clockwise moments to be negative.* (This is the usual convention, although we occasionally encounter situations in which it is more convenient to define clockwise moments to be positive.) Thus, the moment of the force about P is

$$M_P = DF.$$

(4.1)

Notice that if the line of action of F passes through P, the perpendicular distance $D = 0$ and the moment of F about P is zero.

Figure 4.1
(a) The force and point P.
(b) The perpendicular distance D from point P to the line of action of F.
(c) The direction of the moment is counterclockwise.

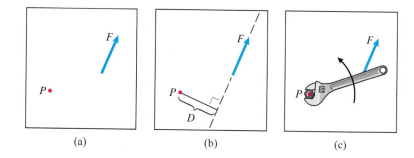

(a)   (b)   (c)

The dimensions of the moment are (distance) × (force). For example, moments can be expressed in newton-meters in SI units and in foot-pounds in U.S. Customary units.

Suppose that you want to place a television set on a shelf, and you aren't certain the attachment of the shelf to the wall is strong enough to support it. Intuitively, you place it near the wall (Fig. 4.2a), knowing that the attachment is more likely to fail if you place it away from the wall (Fig. 4.2b). What is the difference in the two cases? The magnitude and direction of the force exerted on the shelf by the weight of the television are the same in each case, but the moments exerted on the attachment are different. The moment exerted about P by its weight when it is near the wall, $M_P = -D_1 W$, is smaller in magnitude than the moment about P when it is placed away from the wall, $M_P = -D_2 W$.

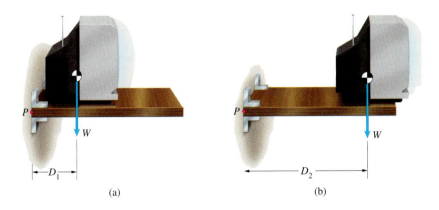

(a)

(b)

Figure 4.2
(a) Placing the television near the wall minimizes the moment exerted on the support
of the shelf at P.
(b) Placing the television far from the wall
exerts a large moment on the support at
P and could cause it to fail.

The method we describe in this section can be used to determine the
sum of the moments of a system of forces about a point if the forces are
two-dimensional (coplanar) and the point lies in the same plane. For example, consider the construction crane shown in Fig. 4.3. The sum of the moments exerted about point P by the load W_1 and the counterweight W_2 is

$$\Sigma M_P = D_1 W_1 - D_2 W_2.$$

Figure 4.3
A tower crane used in the construction of
high-rise buildings.

This moment tends to cause the top of the vertical tower to rotate and could
cause it to collapse. If the distance D_2 is adjusted so that $D_1 W_1 = D_2 W_2$, the moment about point P due to the load and the counterweight is zero.
    *If a force is expressed in terms of components, the moment of the force
about a point P is equal to the sum of the moments of its components about P.*
We prove this very useful result in the next section.

Study Questions

1. How do you determine the magnitude of the moment of a force about a point?

2. The moment of a force about a point is defined to be positive if its direction is
   counterclockwise. What does that mean?

3. If the line of action of a force passes through a point P, what do you know about
   the moment of the force about P?

Example 4.1   **Determining the Moment of a Force**

What is the moment of the 40-kN force in Fig. 4.4 about point A?

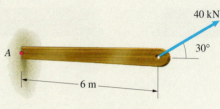

Figure 4.4

Strategy

We can calculate the moment in two ways: by determining the perpendicular distance from point A to the line of action of the force or by expressing the force in terms of components and determining the sum of the moments of the components about A.

Solution

First Method   From Fig. a, the perpendicular distance from A to the line of action of the force is

$$D = (6 \text{ m}) \sin 30° = 3 \text{ m}.$$

The magnitude of the moment of the force about A is $(3 \text{ m})(40 \text{ kN}) = 120$ kN-m, and the direction of the moment about A is counterclockwise. Therefore the moment is

$$M_A = 120 \text{ kN-m}.$$

Second Method   In Fig. b, we express the force in terms of horizontal and vertical components. The perpendicular distance from A to the line of action of the horizontal component is zero, so the horizontal component exerts no moment about A. The magnitude of the moment of the vertical component about A is $(6 \text{ m})(40 \sin 30° \text{ kN}) = 120$ kN-m, and the direction of its moment about A is counterclockwise. The moment is

$$M_A = 120 \text{ kN-m}.$$

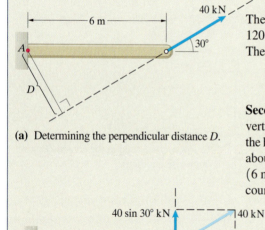

(a) Determining the perpendicular distance D.

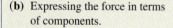

(b) Expressing the force in terms of components.

Critical Thinking

In the first method, we calculated the moment of the 40-kN force about A by determining the perpendicular distance to the line of action of the force and multiplying it by the magnitude of the force. In the second method, we first expressed the 40-kN force in terms of components and then calculated the sum of the moments of the components about A. Why do these two methods yield the same answer? This is a demonstration of a result known as Varignon's theorem, which we present in the next section. Its importance will become obvious. While in this particular example it was easy to determine the perpendicular distance to the line of action of the force, you will see in other examples and problems that this will not always be the case, and it will be much simpler to determine the moment of a force by calculating the sum of the moments of its components.

| **Example 4.2** | **Moment of a System of Forces** |

Four forces act on the machine part in Fig. 4.5. What is the sum of the moments of the forces about the origin O?

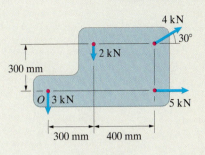

Figure 4.5

Strategy

We can determine the moments of the forces about point O directly from the given information except for the 4-kN force. We will determine its moment by expressing it in terms of components and summing the moments of the components.

Solution

Moment of the 3-kN Force   The line of action of the 3-kN force passes through O. It exerts no moment about O.

Moment of the 5-kN Force   The line of action of the 5-kN force also passes through O. It too exerts no moment about O.

Moment of the 2-kN Force   The perpendicular distance from O to the line of action of the 2-kN force is 0.3 m, and the direction of the moment about O is clockwise. The moment of the 2-kN force about O is

$$-(0.3 \text{ m})(2 \text{ kN}) = -0.600 \text{ kN-m}.$$

(Notice that we converted the perpendicular distance from millimeters into meters, obtaining the result in terms of kilonewton-meters.)

Moment of the 4-kN Force   In Fig. a, we introduce a coordinate system and express the 4-kN force in terms of x and y components. The perpendicular distance from O to the line of action of the x component is 0.3 m, and the direction of the moment about O is clockwise. The moment of the x component about O is

$$-(0.3 \text{ m})(4 \cos 30° \text{ kN}) = -1.039 \text{ kN-m}.$$

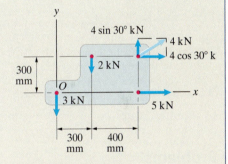

(a) Resolving the 4-kN force into components.

The perpendicular distance from point O to the line of action of the y component is 0.7 m, and the direction of the moment about O is counterclockwise. The moment of the y component about O is

$$(0.7 \text{ m})(4 \sin 30° \text{ kN}) = 1.400 \text{ kN-m}.$$

The sum of the moments of the four forces about point O is

$$\Sigma M_0 = -0.600 - 1.039 + 1.400 = -0.239 \text{ kN-m}.$$

The four forces exert a 0.239 kN-m clockwise moment about point O.

Critical Thinking

If an object is subjected to a system of known forces, why is it useful to determine the sum of the moments of the forces about a given point? As we discuss in Chapter 5, the object is in equilibrium only if the sum of the moments about *any* point is zero, so calculating the sum of the moments provides a test for equilibrium. (Notice that the object in this example is not in equilibrium.) Furthermore, in dynamics the sum of the moments of the forces acting on objects must be determined in order to analyze their angular motions.

Example 4.3 Summing Moments to Determine an Unknown Force

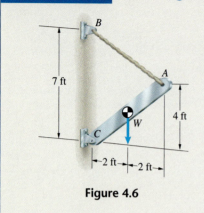

Figure 4.6

The weight $W = 300$ lb (Fig. 4.6). The sum of the moments about C due to the weight W and the force exerted on the bar CA by the cable AB is zero. What is the tension in the cable?

Strategy
Let T be the tension in cable AB. Using the given dimensions, we can express the horizontal and vertical components of the force exerted on the bar by the cable in terms of T. Then by setting the sum of the moments about C due to the weight of the bar and the force exerted by the cable equal to zero, we can obtain an equation for T.

Solution
Using similar triangles, we express the force exerted on the bar by the cable in terms of horizontal and vertical components (Fig. a). The sum of the moments about C due to the weight of the bar and the force exerted by the cable AB is

$$\Sigma M_C = 4\left(\frac{4}{5}T\right) + 4\left(\frac{3}{5}T\right) - 2W = 0.$$

Solving for T, we obtain

$$T = 0.357W = 107.1 \text{ lb}.$$

Critical Thinking
This example is a preview of the applications we consider in Chapter 5 and demonstrates why you must know how to calculate moments of forces. If the bar is in equilibrium, the sum of the moments about C is zero. Applying this condition allowed us to determine the tension in the cable. Why didn't we need to consider the force exerted on the bar by its support at C? Because we know that the moment of that force about C is zero.

(a) Resolving the force exerted by the cable into horizontal and vertical components.

Problems

4.1 The weights $W_1 = 50$ lb and $W_2 = 20$ lb. Determine the sum of the moments due to the forces exerted by the suspended weights on the bar AB (a) about point A; (b) about point B.

4.2 The masses $m_1 = 20$ kg and $m_2 = 8$ kg. Determine the sum of the moments due to the forces exerted by the suspended masses on the bar AB (a) about point A; (b) about point B.

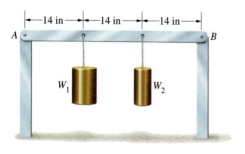

Problem 4.1

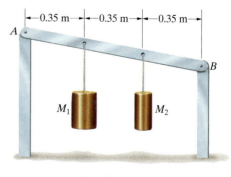

Problem 4.2

4.3 The wheels of the overhead crane exert downward forces on the horizontal I-beam at B and C. If the force at B is 40 kip and the force at C is 44 kip, determine the sum of the moments of the forces on the beam about (a) point A, (b) point D.

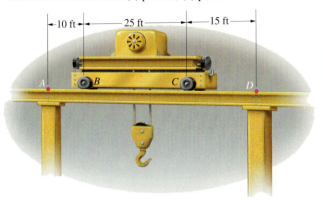

Problem 4.3

4.4 What force F applied to the pliers is required to exert a 4 N-m moment about the center of the bolt at P?

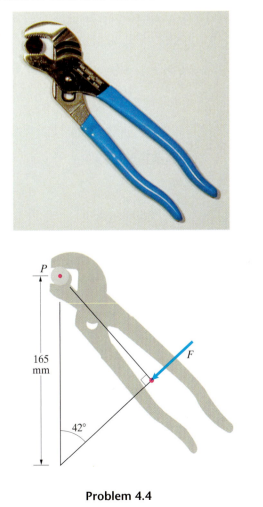

Problem 4.4

4.5 Two forces of equal magnitude F are applied to the wrench as shown. If a 50 N-m moment is required to loosen the nut, what is the necessary value of F?

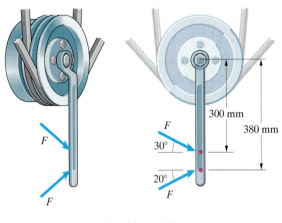

Problem 4.5

4.6 The sum of the moments of the two forces about P is zero. What is the magnitude of the force F?

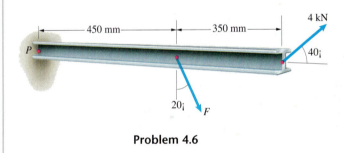

Problem 4.6

4.7 The gears exert 200-N forces on each other at their point of contact.

(a) Determine the moment about A due to the force exerted on the left gear.

(b) Determine the moment about B due to the force exerted on the right gear.

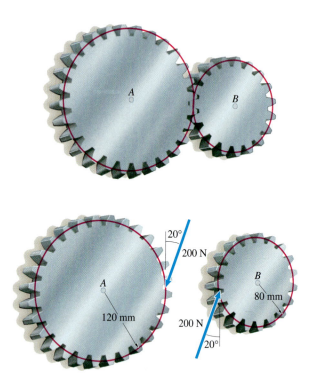

Problem 4.7

4.8 The support at the left end of the beam will fail if the moment about A of the 15-kN force F exceeds 18 kN-m. Based on this criterion, what is the largest allowable length of the beam?

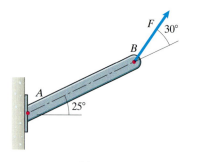

Problem 4.8

4.9 The length of the bar AP is 650 mm. The radius of the pulley is 120 mm. Equal forces $T = 50$ N are applied to the ends of the cable. What is the sum of the moments of the forces (a) about A; (b) about P?

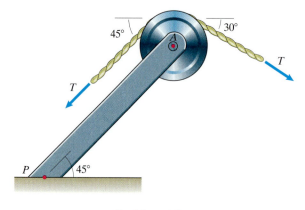

Problem 4.9

4.10 The force $F = 12$ kN. A structural engineer determines that the magnitude of the moment due to F about P should not exceed 5 kN-m. What is the acceptable range of the angle α? Assume that $0 \leq \alpha \leq 90°$.

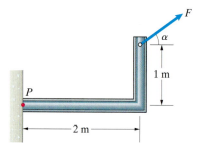

Problem 4.10

4.11 The length of bar AB is 350 mm. The moments exerted about points B and C by the vertical force F are $M_B = -1.75$ kN-m and $M_C = -4.20$ kN-m. Determine the force F and the length of bar AC.

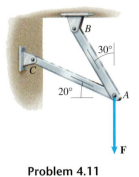

Problem 4.11

4.12 Two students attempt to loosen a lug nut with a lug wrench. One of the students exerts the two 60-lb forces; the other, having to reach around his friend, can only exert the two 30-lb forces. What torque (moment) do they exert on the nut?

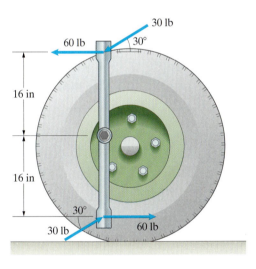

Problem 4.12

4.13 Two equal and opposite forces act on the beam. Determine the sum of the moments of the two forces (a) about point P; (b) about point Q; (c) about the point with coordinates $x = 7$ m, $y = 5$ m.

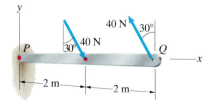

Problem 4.13

4.14 The moment exerted about point E by the weight is 299 in-lb. What moment does the weight exert about point S?

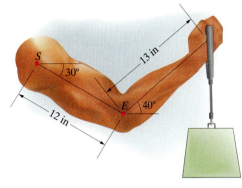

Problem 4.14

4.15 Three forces act on the square plate. Determine the sum of the moments of the forces (a) about A, (b) about B, (c) about C.

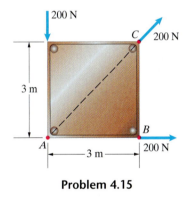

Problem 4.15

4.16 Three forces act on the piping. Determine the sum of the moments of the three forces about point P.

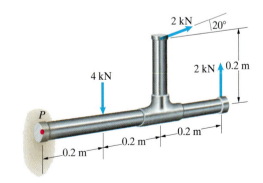

Problem 4.16

4.17 Determine the sum of the moments of the five forces acting on the Howe truss about point A.

4.18 The right support of the truss exerts an upward force of magnitude G. (Assume that the force acts at the right end of the truss.) The sum of the moments about A due to the upward force G and the five downward forces exerted on the truss is zero. What is the force G?

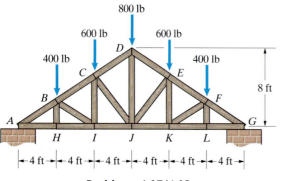

Problems 4.17/4.18

4.19 The sum of the forces F_1 and F_2 is 250 N and the sum of the moments of F_1 and F_2 about B is 700 N-m. What are F_1 and F_2?

4.20 If the two forces exert a 140 kN-m clockwise moment about A and a 20 kN-m clockwise moment about B, what are F_1 and F_2?

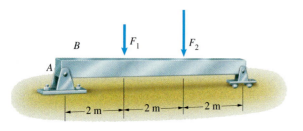

Problems 4.19/4.20

4.21 Three forces act on the car. The sum of the forces is zero and the sum of the moments of the forces about point P is zero.
(a) Determine the forces A and B.
(b) Determine the sum of the moments of the forces about point Q.

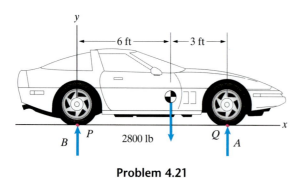

Problem 4.21

4.22 Five forces act on the piping. The vector sum of the forces is zero and the sum of the moments of the forces about point P is zero.
(a) Determine the forces A, B, and C.
(b) Determine the sum of the moments of the forces about point Q.

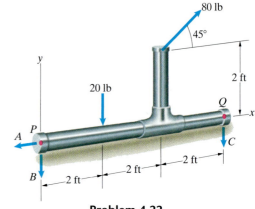

Problem 4.22

4.23 The weights (in ounces) of fish A, B, and C are 2.7, 8.1, and 2.1, respectively. The sum of the moments due to the weights of the fish about the point where the mobile is attached to the ceiling is zero. What is the weight of fish D?

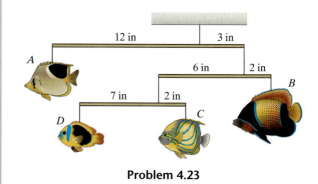

Problem 4.23

4.24 The weight $W = 1.2$ kN. The sum of the moments about A due to W and the force exerted at the end of the bar by the rope is zero. What is the tension in the rope?

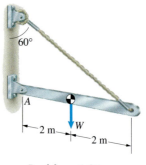

Problem 4.24

4.25 The 160-N weights of the arms AB and BC of the robotic manipulator act at their midpoints. Determine the sum of the moments of the three weights about A.

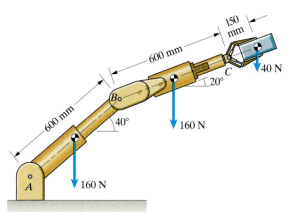

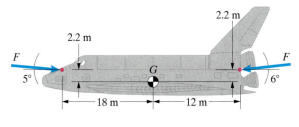

Problem 4.25

4.26 The space shuttle's attitude thrusters exert two forces of magnitude $F = 7.70$ kN. What moment do the thrusters exert about the center of mass G?

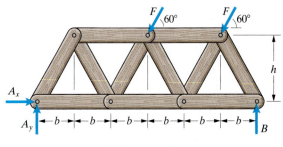

Problem 4.26

4.27 The force F exerts a 200 ft-lb counterclockwise moment about A and a 100 ft-lb clockwise moment about B. What are F and θ?

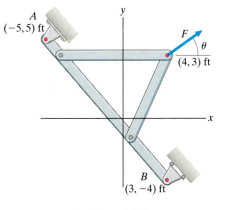

Problem 4.27

4.28 Five forces act on a link in the gear-shifting mechanism of a lawn mower. The vector sum of the five forces on the bar is zero. The sum of their moments about the point where the forces A_x and A_y act is zero.

(a) Determine the forces A_x, A_y, and B.
(b) Determine the sum of the moments of the forces about the point where the force B acts.

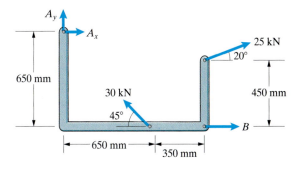

Problem 4.28

4.29 Five forces act on a model truss built by a civil engineering student as part of a design project. The dimensions are $b = 300$ mm and $h = 400$ mm and $F = 100$ N. The sum of the moments of the forces about the point where A_x and A_y act is zero. If the weight of the truss is negligible, what is the force B?

4.30 The dimensions are $b = 3$ ft and $h = 4$ ft and $F = 300$ lb. The vector sum of the forces acting on the truss is zero, and the sum of the moments of the forces about the point where A_x and A_y act is zero.

(a) Determine the forces A_x, A_y, and B.
(b) Determine the sum of the moments of the forces about the point where the force B acts.

Problems 4.29/4.30

4.31 The mass $m = 70$ kg. What is the moment about A due to the force exerted on the beam at B by the cable?

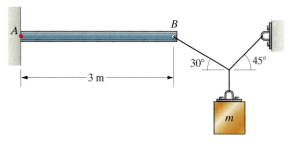

Problem 4.31

4.32 The masses $m_1 = 6$ kg and $m_2 = 12$ kg are suspended by the cable system shown. The cable BC is horizontal. Determine the angle α and the moment about point P due to the force exerted on the vertical post by the cable CD.

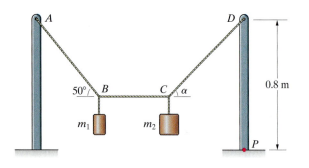

Problem 4.32

4.33 The bar AB exerts a force at B that helps support the vertical retaining wall. The force is parallel to the bar. The civil engineer wants the bar to exert a 38 kN-m moment about O. What is the magnitude of the force the bar must exert?

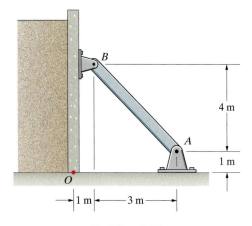

Problem 4.33

4.34 A contestant in a fly-casting contest snags his line in some grass. If the tension in the line is 5 lb, what moment does the force exerted on the rod by the line exert about point H, where he holds the rod?

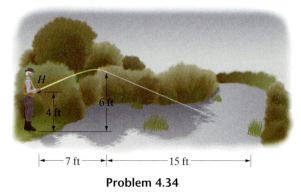

Problem 4.34

4.35 The cables AB and AC help support the tower. The tension in cable AB is 5 kN. The points A, B, C, and O are contained in the same vertical plane. (a) What is the moment about O due to the force exerted on the tower by cable AB?
(b) If the sum of the moments about O due to the forces exerted on the tower by the two cables is zero, what is the tension in cable AC?

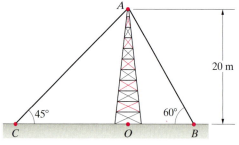

Problem 4.35

4.36 The cable from B to A (the sailboat's forestay) exerts a 230-N force at B. The cable from B to C (the backstay) exerts a 660-N force at B. The bottom of the sailboat's mast is located at $x = 4$ m, $y = 0$. What is the sum of the moments about the bottom of the mast due to the forces exerted at B by the forestay and backstay?

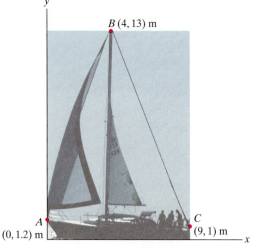

Problem 4.36

4.37 The cable AB exerts a 290-kN force on the crane's boom at B. The cable AC exerts a 148-kN force on the boom at C. Determine the sum of the moments about P due to the forces the cables AB and AC exert on the boom.

4.38 The mass of the crane's boom is 9000 kg. Its weight acts at G. The sum of the moments about P due to the boom's weight, the force exerted at B by the cable AB, and the force exerted at C by the cable AC is zero. Assume that the tensions in cables AB and AC are equal. Determine the tension in the cables.

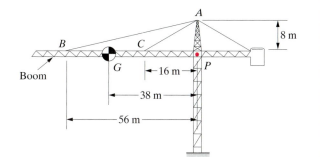

Problems 4.37/4.38

4.39 The mass of the luggage carrier and the suitcase combined is 12 kg. Their weight acts at A. The sum of the moments about the origin of the coordinate system due to the weight acting at A and the vertical force F applied to the handle of the luggage carrier is zero. Determine the force F (a) if $\alpha = 30°$; (b) if $\alpha = 50°$.

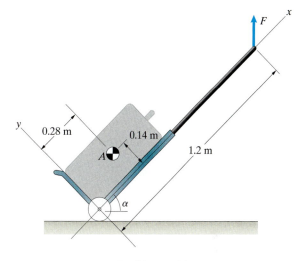

Problem 4.39

4.40 The hydraulic cylinder BC exerts a 300-kN force on the boom of the crane at C. The force is parallel to the cylinder. What is the moment of the force about A?

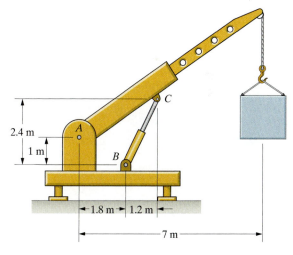

Problem 4.40

4.41 The hydraulic cylinder BC exerts a 2200-lb force on the boom of the crane at C. The force is parallel to the cylinder. The angle $\alpha = 40°$. What is the moment of the force about A?

4.42 The hydraulic cylinder exerts an 8-kN force at B that is parallel to the cylinder and points from C toward B. Determine the moments of the force about points A and D.

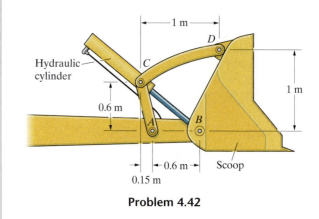

Problem 4.42

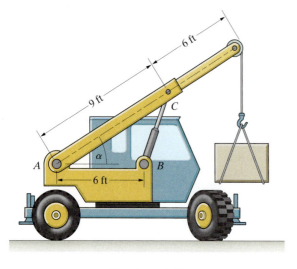

Problem 4.41

4.43 The structure shown in the diagram is one of two identical structures that support the scoop of the excavator. The bar BC exerts a 700-N force at C that points from C toward B. What is the moment of this force about K?

4.44 The bar BC exerts a force at C that points from C toward B. The hydraulic cylinder DH exerts a 1550-N force at D that points from D toward H. The sum of the moments of these two forces about K is zero. What is the magnitude of the force that bar BC exerts at C?

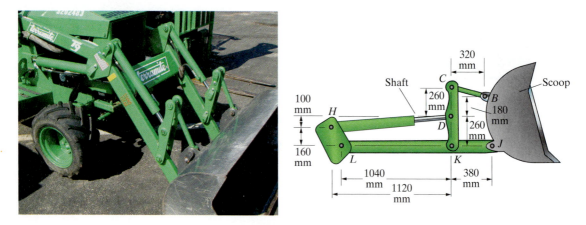

Problems 4.43/4.44

4.2 The Moment Vector

The moment of a force about a point is a vector. In this section we define this vector and explain how it is evaluated. We then show that when we use the two-dimensional description of the moment described in Section 4.1, we are specifying the magnitude and direction of the moment vector.

Consider a force vector $\mathbf{F}$ and point P (Fig. 4.7a). The *moment* of $\mathbf{F}$ about P is the vector

$$\mathbf{M}_P = \mathbf{r} \times \mathbf{F}, \tag{4.2}$$

where $\mathbf{r}$ is a position vector from P to *any* point on the line of action of $\mathbf{F}$ (Fig. 4.7b).

Magnitude of the Moment

From the definition of the cross product, the magnitude of $\mathbf{M}_P$ is

$$|\mathbf{M}_P| = |\mathbf{r}||\mathbf{F}| \sin \theta,$$

where θ is the angle between the vectors $\mathbf{r}$ and $\mathbf{F}$ when they are placed tail to tail. The perpendicular distance from P to the line of action of $\mathbf{F}$ is $D = |\mathbf{r}| \sin \theta$ (Fig. 4.7c). Therefore the magnitude of the moment $\mathbf{M}_P$ equals the product of the perpendicular distance from P to the line of action of $\mathbf{F}$ and the magnitude of $\mathbf{F}$:

$$|\mathbf{M}_P| = D|\mathbf{F}|. \tag{4.3}$$

Notice that if we know the vectors $\mathbf{M}_P$ and $\mathbf{F}$, this equation can be solved for the perpendicular distance D.

Direction of the Moment

We know from the definition of the cross product that $\mathbf{M}_P$ is perpendicular to both $\mathbf{r}$ and $\mathbf{F}$. That means that $\mathbf{M}_P$ is perpendicular to the plane containing P and $\mathbf{F}$ (Fig. 4.8a). Notice in this figure that we denote a moment by a circular arrow around the vector.

The direction of $\mathbf{M}_P$ also indicates the direction of the moment: Pointing the thumb of the right hand in the direction of $\mathbf{M}_P$, the "arc" of the fingers indicates the direction of the rotation that $\mathbf{F}$ tends to cause about P (Fig. 4.8b).

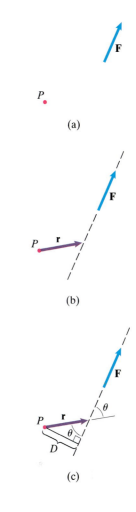

(a)

(b)

(c)

Figure 4.7
(a) The force $\mathbf{F}$ and point P.
(b) A vector $\mathbf{r}$ from P to a point on the line of action of $\mathbf{F}$.
(c) The angle θ and the perpendicular distance D.

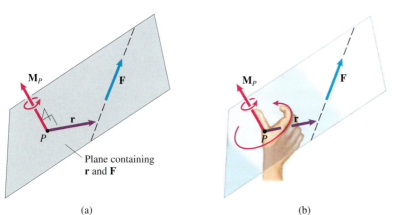

(a)

(b)

Plane containing $\mathbf{r}$ and $\mathbf{F}$

Figure 4.8
(a) $\mathbf{M}_P$ is perpendicular to the plane containing p and $\mathbf{F}$.
(b) The direction of $\mathbf{M}_P$ indicates the direction of the moment.

The result obtained from Eq. (4.2) doesn't depend on where the vector **r** intersects the line of action of **F**. Instead of using the vector **r** in Fig. 4.9a, we could use the vector **r′** in Fig. 4.9b. The vector **r** = **r′** + **u**, where **u** is parallel to **F** (Fig. 4.9c). Therefore,

$$\mathbf{r} \times \mathbf{F} = (\mathbf{r'} + \mathbf{u}) \times \mathbf{F} = \mathbf{r'} \times \mathbf{F}$$

because the cross product of the parallel vectors **u** and **F** is zero.

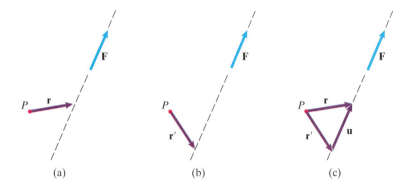

Figure 4.9
(a) A vector **r** from P to the line of action of **F**.
(b) A different vector **r′**.
(c) **r** = **r′** + **u**.

In summary, the moment of a force **F** about a point P has three properties:

1. The magnitude of $\mathbf{M}_P$ is equal to the product of the magnitude of **F** and the perpendicular distance from P to the line of action of **F**. If the line of action of **F** passes through P, $\mathbf{M}_P = \mathbf{0}$.

2. $\mathbf{M}_P$ is perpendicular to the plane containing P and **F**.

3. The direction of $\mathbf{M}_P$ indicates the direction of the moment through a right-hand rule (Fig. 4.8b). Since the cross product is not commutative, it is essential to maintain the correct sequence of the vectors in the equation $\mathbf{M}_P = \mathbf{r} \times \mathbf{F}$.

Let us determine the moment of the force **F** in Fig. 4.10a about the point P. Since the vector **r** in Eq. (4.2) can be a position vector to any point on the line of action of **F**, we can use the vector from P to the point of application of **F** (Fig. 4.10b):

$$\mathbf{r} = (12 - 3)\mathbf{i} + (6 - 4)\mathbf{j} + (-5 - 1)\mathbf{k} = 9\mathbf{i} + 2\mathbf{j} - 6\mathbf{k} \text{ (ft).}$$

The moment is

$$\mathbf{M}_P = \mathbf{r} \times \mathbf{F} = \begin{vmatrix} \mathbf{i} & \mathbf{j} & \mathbf{k} \\ 9 & 2 & -6 \\ 4 & 4 & 7 \end{vmatrix} = 38\mathbf{i} - 87\mathbf{j} + 28\mathbf{k} \text{ (ft-lb).}$$

The magnitude of $\mathbf{M}_P$,

$$|\mathbf{M}_P| = \sqrt{(38)^2 + (-87)^2 + (28)^2} = 99.0 \text{ ft-lb,}$$

equals the product of the magnitude of **F** and the perpendicular distance D from point P to the line of action of **F**. Therefore,

$$D = \frac{|\mathbf{M}_P|}{|\mathbf{F}|} = \frac{99.0 \text{ ft-lb}}{9 \text{ lb}} = 11.0 \text{ ft.}$$

The direction of $\mathbf{M}_P$ tells us both the orientation of the plane containing P and **F** and the direction of the moment (Fig. 4.10c).

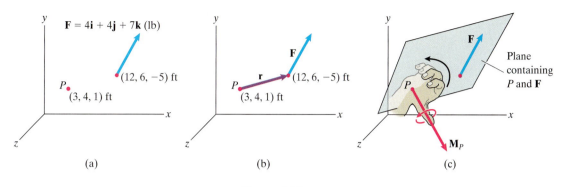

Figure 4.10
(a) A force **F** and point P.
(b) The vector **r** from P to the point of application of **F**.
(c) $\mathbf{M}_P$ is perpendicular to the plane containing P and **F**. The right-hand rule indicates the direction of the moment.

Relation to the Two-Dimensional Description

If our view is perpendicular to the plane containing the point P and the force **F**, the two-dimensional description of the moment we used in Section 4.1 specifies both the magnitude and direction of the vector $\mathbf{M}_P$. In this situation, $\mathbf{M}_P$ is perpendicular to the page, and the right-hand rule indicates whether it points out of or into the page.

For example, in Fig. 4.11a, the view is perpendicular to the x–y plane and the 10-N force is contained in the x–y plane. Suppose that we want to determine the moment of the force about the origin O. The perpendicular distance from O to the line of action of the force is 4 m. The two-dimensional description of the moment of the force about O is that its magnitude is $(4\text{ m})(10\text{ N}) = 40$ N-m and its direction is counterclockwise, or

$$M_O = 40 \text{ N-m}.$$

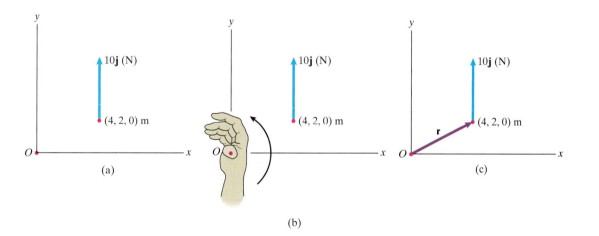

Figure 4.11
(a) The force is contained in the x–y plane.
(b) The counterclockwise direction of the moment indicates that $\mathbf{M}_O$ points out of the page.
(c) The vector **r** from O to the point of application of **F**.

That tells us that the magnitude of the vector $\mathbf{M}_O$ is 40 N-m, and the right-hand rule (Fig. 4.11b) indicates that it points out of the page. Therefore,

$$\mathbf{M}_O = 40\mathbf{k} \text{ (N-m)}.$$

We can confirm this result by using Eq. (4.2). If we let $\mathbf{r}$ be the vector from O to the point of application of the force (Fig. 4.11c),

$$\mathbf{M}_O = \mathbf{r} \times \mathbf{F} = (4\mathbf{i} + 2\mathbf{j}) \times 10\mathbf{j} = 40\mathbf{k} \text{ (N-m)}.$$

As this example illustrates, the two-dimensional description of the moment determines the moment vector. The converse is also true. The magnitude of $\mathbf{M}_O$ equals the product of the magnitude of the force and the perpendicular distance from O to the line of action of the force, 40 N-m, and the direction of the vector $\mathbf{M}_O$ indicates that the moment is counterclockwise (Fig. 4.11b).

Varignon's Theorem

Let $\mathbf{F}_1, \mathbf{F}_2, \ldots, \mathbf{F}_N$ be a concurrent system of forces whose lines of action intersect at a point Q. The moment of the system about a point P is

$$(\mathbf{r}_{PQ} \times \mathbf{F}_1) + (\mathbf{r}_{PQ} \times \mathbf{F}_2) + \cdots + (\mathbf{r}_{PQ} \times \mathbf{F}_N)$$
$$= \mathbf{r}_{PQ} \times (\mathbf{F}_1 + \mathbf{F}_2 + \cdots + \mathbf{F}_N),$$

where $\mathbf{r}_{PQ}$ is the vector from P to Q (Fig. 4.12). This result, known as *Varignon's theorem*, follows from the distributive property of the cross product, Eq. (2.31). It confirms that the moment of a force about a point P is equal to the sum of the moments of its components about P.

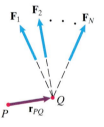

Figure 4.12
A system of concurrent forces and a point P.

Study Questions

1. When you use the equation $\mathbf{M}_P = \mathbf{r} \times \mathbf{F}$ to determine the moment of a force $\mathbf{F}$ about a point P, how do you choose the vector $\mathbf{r}$?

2. If you know the components of the vector $\mathbf{M}_P = \mathbf{r} \times \mathbf{F}$, how can you determine the product of the magnitude of $\mathbf{F}$ and the perpendicular distance from P to the line of action of $\mathbf{F}$?

3. How does the direction of the vector $\mathbf{M}_P = \mathbf{r} \times \mathbf{F}$ indicate the direction of the moment of $\mathbf{F}$ about P?

Example 4.4   Two-Dimensional Description and the Moment Vector

Determine the moment of the 400-N force in Fig. 4.13 about O.

Strategy

We will determine the moment in two ways: (a) We can use the two-dimensional description of the moment. We will express the force in terms of its components and determine the moment of each component about O by multiplying the magnitude of the component and the perpendicular distance from O to its line of action. (b) We can obtain the vector description of the moment by using Eq. (4.2).

Figure 4.13

Solution

(a) Expressing the force in terms of horizontal and vertical components (Fig. a), the two-dimensional description of the moment is

$$M_O = -(2 \text{ m})(400 \cos 30° \text{ N}) - (5 \text{ m})(400 \sin 30° \text{ N})$$
$$= -1.69 \text{ kN-m}.$$

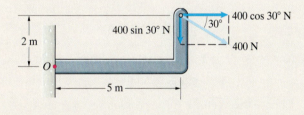

(a) Resolving the force into components.

(b) To apply Eq. (4.2), we introduce the coordinate system shown in Fig. b.
Choose the Vector r   We can let **r** be the vector from O to the point of application of the force (Fig. b):

$$\mathbf{r} = 5\mathbf{i} + 2\mathbf{j} \text{ (m)}.$$

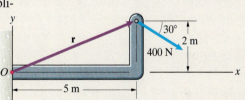

Evaluate r × F   The moment is

$$\mathbf{M}_O = \mathbf{r} \times \mathbf{F} = (5\mathbf{i} + 2\mathbf{j}) \times (400 \cos 30°\mathbf{i} - 400 \sin 30°\mathbf{j})$$
$$= -1.69\mathbf{k} \text{ (kN-m)}.$$

(b) The vector **r** from O to the point of application of the force.

Critical Thinking

In most two-dimensional situations like this one, you will find the two-dimensional description of the moment easier to use than the vector description. However, studying the relationship between the two-dimensional and vector descriptions of the moment provides insight into the vector description and demonstrates, for this special case, that the magnitude and direction of the vector specify the magnitude and direction of the moment. In three-dimensional situations like the one in the next example, the vector description of the moment is nearly always used.

Example 4.5   Determining the Moment and the Perpendicular Distance to the Line of Action

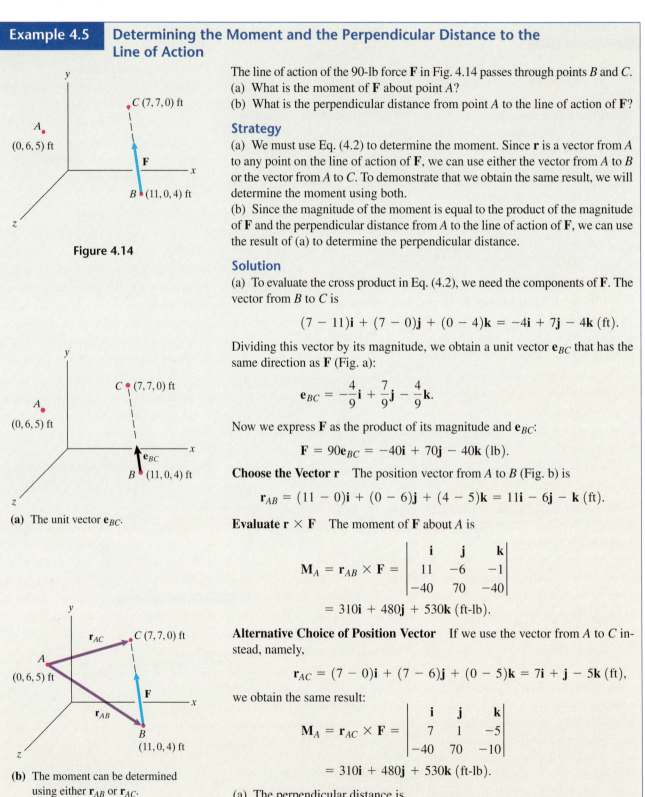

Figure 4.14

(a) The unit vector $\mathbf{e}_{BC}$.

(b) The moment can be determined using either $\mathbf{r}_{AB}$ or $\mathbf{r}_{AC}$.

The line of action of the 90-lb force $\mathbf{F}$ in Fig. 4.14 passes through points B and C.
(a) What is the moment of $\mathbf{F}$ about point A?
(b) What is the perpendicular distance from point A to the line of action of $\mathbf{F}$?

Strategy

(a) We must use Eq. (4.2) to determine the moment. Since $\mathbf{r}$ is a vector from A to any point on the line of action of $\mathbf{F}$, we can use either the vector from A to B or the vector from A to C. To demonstrate that we obtain the same result, we will determine the moment using both.
(b) Since the magnitude of the moment is equal to the product of the magnitude of $\mathbf{F}$ and the perpendicular distance from A to the line of action of $\mathbf{F}$, we can use the result of (a) to determine the perpendicular distance.

Solution

(a) To evaluate the cross product in Eq. (4.2), we need the components of $\mathbf{F}$. The vector from B to C is

$$(7 - 11)\mathbf{i} + (7 - 0)\mathbf{j} + (0 - 4)\mathbf{k} = -4\mathbf{i} + 7\mathbf{j} - 4\mathbf{k} \ (\text{ft}).$$

Dividing this vector by its magnitude, we obtain a unit vector $\mathbf{e}_{BC}$ that has the same direction as $\mathbf{F}$ (Fig. a):

$$\mathbf{e}_{BC} = -\frac{4}{9}\mathbf{i} + \frac{7}{9}\mathbf{j} - \frac{4}{9}\mathbf{k}.$$

Now we express $\mathbf{F}$ as the product of its magnitude and $\mathbf{e}_{BC}$:

$$\mathbf{F} = 90\mathbf{e}_{BC} = -40\mathbf{i} + 70\mathbf{j} - 40\mathbf{k} \ (\text{lb}).$$

Choose the Vector r   The position vector from A to B (Fig. b) is

$$\mathbf{r}_{AB} = (11 - 0)\mathbf{i} + (0 - 6)\mathbf{j} + (4 - 5)\mathbf{k} = 11\mathbf{i} - 6\mathbf{j} - \mathbf{k} \ (\text{ft}).$$

Evaluate r × F   The moment of $\mathbf{F}$ about A is

$$\mathbf{M}_A = \mathbf{r}_{AB} \times \mathbf{F} = \begin{vmatrix} \mathbf{i} & \mathbf{j} & \mathbf{k} \\ 11 & -6 & -1 \\ -40 & 70 & -40 \end{vmatrix}$$

$$= 310\mathbf{i} + 480\mathbf{j} + 530\mathbf{k} \ (\text{ft-lb}).$$

Alternative Choice of Position Vector   If we use the vector from A to C instead, namely,

$$\mathbf{r}_{AC} = (7 - 0)\mathbf{i} + (7 - 6)\mathbf{j} + (0 - 5)\mathbf{k} = 7\mathbf{i} + \mathbf{j} - 5\mathbf{k} \ (\text{ft}),$$

we obtain the same result:

$$\mathbf{M}_A = \mathbf{r}_{AC} \times \mathbf{F} = \begin{vmatrix} \mathbf{i} & \mathbf{j} & \mathbf{k} \\ 7 & 1 & -5 \\ -40 & 70 & -10 \end{vmatrix}$$

$$= 310\mathbf{i} + 480\mathbf{j} + 530\mathbf{k} \ (\text{ft-lb}).$$

(a) The perpendicular distance is

$$\frac{|\mathbf{M}_A|}{|\mathbf{F}|} = \frac{\sqrt{(310 \text{ ft-lb})^2 + (480 \text{ ft-lb})^2 + (530 \text{ ft-lb})^2}}{\sqrt{(-40 \text{ lb})^2 + (70 \text{ lb})^2 + (-40 \text{ lb})^2}} = 8.66 \text{ ft}.$$

Critical Thinking

Suppose that you were unfamiliar with the vector description of the moment and wanted to determine the moment of the force **F** about the point A in this example. You know from our discussion of the two-dimensional description that the magnitude of the moment equals the product of the perpendicular distance from A to the line of action of **F** with the magnitude of **F**. However, how would you determine the perpendicular distance in this example? In the two-dimensional description, the direction of the moment is specified by stating that it is counterclockwise or clockwise, but how can these terms be applied to Fig. 4.14? *These questions are answered by the vector description of the moment.* The magnitude of the moment vector *is* the product of the perpendicular distance from A to the line of action of **F** with the magnitude of **F**, and the direction of the moment vector defines the direction of the moment as shown in Fig. 4.10c.

Example 4.6   Applying the Moment Vector

The cables AB and AC in Fig. 4.15 extend from an attachment point A on the floor to attachment points B and C in the walls. The tension in cable AB is 10 kN, and the tension in cable AC is 20 kN. What is the sum of the moments about O due to the forces exerted on the attachment point A by the two cables?

Strategy

We must express the forces exerted on the attachment point A by the two cables in terms of their components. Then we can use Eq. (4.2) to determine the moments the forces exert about O.

Solution

Let $\mathbf{F}_{AB}$ and $\mathbf{F}_{AC}$ be the forces exerted on the attachment point A by the two cables (Fig. a). To express $\mathbf{F}_{AB}$ in terms of its components, we determine the position vector from A to B,

$$(0 - 4)\mathbf{i} + (4 - 0)\mathbf{j} + (8 - 6)\mathbf{k} = -4\mathbf{i} + 4\mathbf{j} + 2\mathbf{k} \ (\text{m}),$$

and divide it by its magnitude to obtain a unit vector $\mathbf{e}_{AB}$ with the same direction as $\mathbf{F}_{AB}$ (Fig. b):

$$\mathbf{e}_{AB} = \frac{-4\mathbf{i} + 4\mathbf{j} + 2\mathbf{k} \ (\text{m})}{\sqrt{(-4 \ \text{m})^2 + (4 \ \text{m})^2 + (2 \ \text{m})^2}} = -\frac{2}{3}\mathbf{i} + \frac{2}{3}\mathbf{j} + \frac{1}{3}\mathbf{k}.$$

Now we write $\mathbf{F}_{AB}$ as

$$\mathbf{F}_{AB} = 10\mathbf{e}_{AB} = -6.67\mathbf{i} + 6.67\mathbf{j} + 3.33\mathbf{k} \ (\text{kN}).$$

We express the force $\mathbf{F}_{AC}$ in terms of its components in the same way:

$$\mathbf{F}_{AC} = 5.71\mathbf{i} + 8.57\mathbf{j} - 17.14\mathbf{k} \ (\text{kN}).$$

Choose the Vector r   Since the lines of action of both forces pass through point A, we can use the vector from O to A to determine the moments of both forces about point O (Fig. a):

$$\mathbf{r} = 4\mathbf{i} + 6\mathbf{k} \ (\text{m}).$$

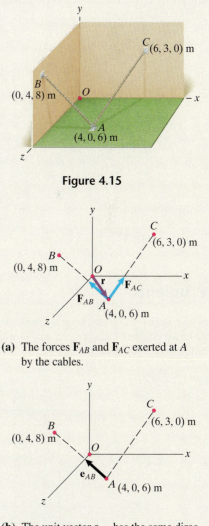

Figure 4.15

(a) The forces $\mathbf{F}_{AB}$ and $\mathbf{F}_{AC}$ exerted at A by the cables.

(b) The unit vector $\mathbf{e}_{AB}$ has the same direction as $\mathbf{F}_{AB}$.

Evaluate r × F   The sum of the moments is

$$\Sigma \, \mathbf{M}_O = (\mathbf{r} \times \mathbf{F}_{AB}) + (\mathbf{r} \times \mathbf{F}_{AC})$$

$$= \begin{vmatrix} \mathbf{i} & \mathbf{j} & \mathbf{k} \\ 4 & 0 & 6 \\ -6.67 & 6.67 & 3.33 \end{vmatrix} + \begin{vmatrix} \mathbf{i} & \mathbf{j} & \mathbf{k} \\ 4 & 0 & 6 \\ 5.71 & 8.57 & -17.14 \end{vmatrix}$$

$$= -91.4\mathbf{i} + 49.5\mathbf{j} + 61.0\mathbf{k} \text{ (kN-m).}$$

Critical Thinking

The lines of action of the forces $\mathbf{F}_{AB}$ and $\mathbf{F}_{AC}$ intersect at A. Notice that, according to Varignon's theorem, we could have summed the forces first, obtaining

$$\mathbf{F}_{AB} + \mathbf{F}_{AC} = -0.952\mathbf{i} + 15.24\mathbf{j} - 13.81\mathbf{k} \text{ (kN),}$$

and then determined the sum of the moments of the two forces about O by calculating the moment of the sum of the two forces about O:

$$\Sigma \, \mathbf{M}_O = \mathbf{r} \times (\mathbf{F}_{AB} + \mathbf{F}_{AC})$$

$$= \begin{vmatrix} \mathbf{i} & \mathbf{j} & \mathbf{k} \\ 4 & 0 & 6 \\ -0.952 & 15.24 & -13.81 \end{vmatrix}$$

$$= -91.4\mathbf{i} + 49.5\mathbf{j} + 61.0\mathbf{k} \text{ (kN-m).}$$

Problems

4.45 Use Eq. (4.2) to determine the moment of the 50-lb force about the origin O. Compare your answer with the two-dimensional description of the moment.

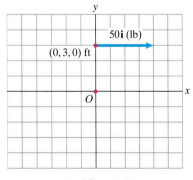

Problem 4.45

4.46 Use Eq. (4.2) to determine the moment of the 80-N force about the origin O letting $\mathbf{r}$ be the vector (a) from O to A; (b) from O to B.

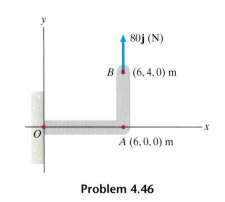

Problem 4.46

4.47 A bioengineer studying an injury sustained in throwing the javelin estimates that the magnitude of the maximum force exerted was $|\mathbf{F}| = 360$ N and the perpendicular distance from O to the line of action of $\mathbf{F}$ was 550 mm. The vector $\mathbf{F}$ and point O are contained in the x–y plane. Express the moment of $\mathbf{F}$ about the shoulder joint at O as a vector.

Problem 4.47

4.48 Use Eq.(4.2) to determine the moment of the 100-kN force (a) about A, (b) about B.

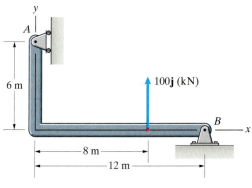

Problem 4.48

4.49 The cable AB exerts a 200-N force on the support at A that points from A toward B. Use Eq. (4.2) to determine the moment of this force about point P, (a) letting $\mathbf{r}$ be the vector from P to A; (b) letting $\mathbf{r}$ be the vector from P to B.

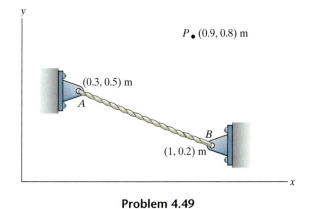

Problem 4.49

4.50 The line of action of $\mathbf{F}$ is contained in the x–y plane. The moment of $\mathbf{F}$ about O is $140\mathbf{k}$ (N-m), and the moment of $\mathbf{F}$ about A is $280\mathbf{k}$ (N-m). What are the components of $\mathbf{F}$?

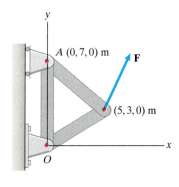

Problem 4.50

4.51 Use Eq. (4.2) to determine the sum of the moments of the three forces (a) about A, (b) about B.

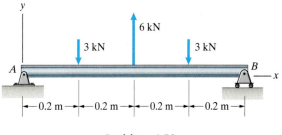

Problem 4.51

4.52 Three forces are applied to the plate. Use Eq.(4.2) to determine the sum of the moments of the three forces about the origin O.

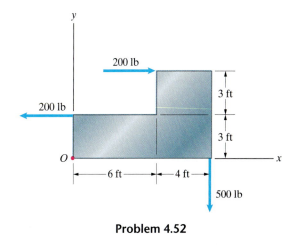

Problem 4.52

4.53 Three forces act on the plate. Use Eq. (4.2) to determine the sum of the moments of the three forces about point P.

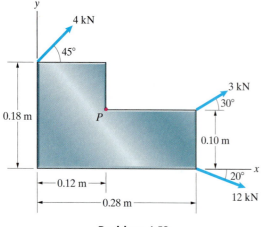

Problem 4.53

4.54 (a) Determine the magnitude of the moment of the 150-N force about A by calculating the perpendicular distance from A to the line of action of the force.
(b) Use Eq. (4.2) to determine the magnitude of the moment of the 150-N force about A.

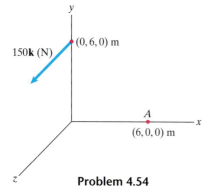

Problem 4.54

4.55 (a) Determine the magnitude of the moment of the 600-N force about A by calculating the perpendicular distance from A to the line of action of the force.
(b) Use Eq. (4.2) to determine the magnitude of the moment of the 600-N force about A.

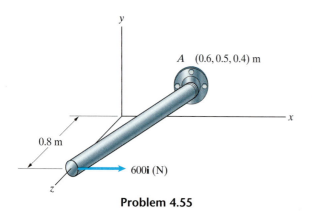

Problem 4.55

4.56 The pneumatic support AB exerts a 35-N force on the fixture at B that points from A toward B. Determine the magnitude of the moment of the force about the hinge at O.

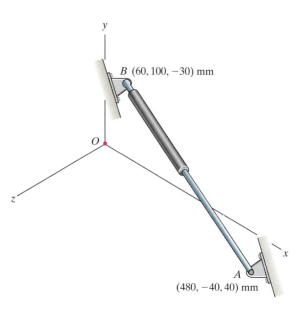

Problem 4.56

4.57 A force $\mathbf{F} = 20\mathbf{i} - 30\mathbf{j} + 60\mathbf{k}$ (lb). The moment of $\mathbf{F}$ about a point P is $\mathbf{M}_P = 450\mathbf{i} - 100\mathbf{j} - 200\mathbf{k}$ (ft-lb). What is the perpendicular distance from point P to the line of action of $\mathbf{F}$?

4.58 A force $\mathbf{F}$ is applied at the point $(8, 6, 13)$ m. Its magnitude is $|\mathbf{F}| = 90$ N, and the moment of $\mathbf{F}$ about the point $(4, 2, 6)$ is zero. What are the components of $\mathbf{F}$?

4.59 The force $\mathbf{F} = 30\mathbf{i} + 20\mathbf{j} - 10\mathbf{k}$ (N). (a) Determine the magnitude of the moment of $\mathbf{F}$ about A. (b) Suppose that you can change the direction of $\mathbf{F}$ while keeping its magnitude constant, and you want to choose a direction that maximizes the moment of $\mathbf{F}$ about A. What is the magnitude of the resulting maximum moment?

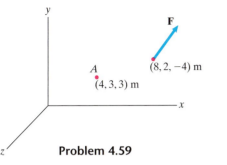

Problem 4.59

4.60 The direction cosines of the force $\mathbf{F}$ are $\cos\theta_x = 0.818$, $\cos\theta_y = 0.182$, and $\cos\theta_z = -0.545$. The support of the beam at O will fail if the magnitude of the moment of $\mathbf{F}$ about O exceeds 100 kN-m. Determine the magnitude of the largest force $\mathbf{F}$ that can safely be applied to the beam.

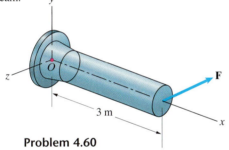

Problem 4.60

4.61 The force $\mathbf{F}$ exerted on the grip of the exercise machine points in the direction of the unit vector $\mathbf{e} = \frac{2}{3}\mathbf{i} - \frac{2}{3}\mathbf{j} + \frac{1}{3}\mathbf{k}$ and its magnitude is 120 N. Determine the magnitude of the moment of $\mathbf{F}$ about the origin O.

4.62 The force $\mathbf{F}$ points in the direction of the unit vector $\mathbf{e} = \frac{2}{3}\mathbf{i} - \frac{2}{3}\mathbf{j} + \frac{1}{3}\mathbf{k}$. The support at O will safely sup-port a moment of 560 N-m magnitude. (a) Based on this criterion, what is the largest safe magnitude of $\mathbf{F}$?
(b) If the force $\mathbf{F}$ may be exerted in any direction, what is its largest safe magnitude?

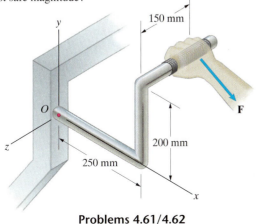

Problems 4.61/4.62

4.63 A civil engineer in Boulder, Colorado, estimates that under the severest expected Chinook winds, the total force on the highway sign will be $\mathbf{F} = 2.8\mathbf{i} - 1.8\mathbf{j}$ (kN). Let $\mathbf{M}_O$ be the moment due to $\mathbf{F}$ about the base O of the cylindrical column supporting the sign. The y component of $\mathbf{M}_O$ is called the *torsion* exerted on the cylindrical column at the base, and the component of $\mathbf{M}_O$ parallel to the x–z plane is called the *bending moment*. Determine the magnitudes of the torsion and bending moment.

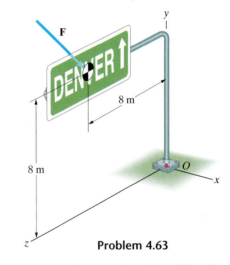

Problem 4.63

4.64 The weights of the arms OA and AB of the robotic manipulator act at their midpoints. The direction cosines of the centerline of arm OA are $\cos\theta_x = 0.500$, $\cos\theta_y = 0.866$, and $\cos\theta_z = 0$, and the direction cosines of the centerline of arm AB are $\cos\theta_x = 0.707$, $\cos\theta_y = 0.619$, and $\cos\theta_z = -0.342$. What is the sum of the moments about O due to the two forces?

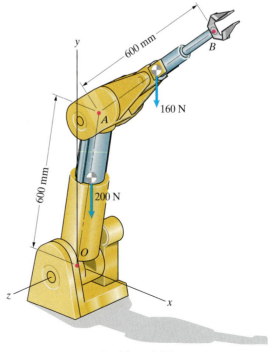

Problem 4.64

4.65 The tension in cable AB is 100 lb. If you want the magnitude of the moment due to the forces exerted on the tree by the two ropes about the base O of the tree to be 1500 ft-lb, what is the necessary tension in rope AC?

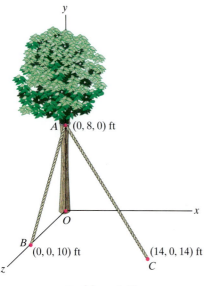

Problem 4.65

4.66 *A force $\mathbf{F}$ acts at the top end A of the pole. Its magnitude is $|\mathbf{F}| = 6$ kN and its x component is $F_x = 4$ kN. The coordinates of point A are shown. Determine the components of $\mathbf{F}$ so that the magnitude of the moment due to $\mathbf{F}$ about the base P of the pole is as large as possible.

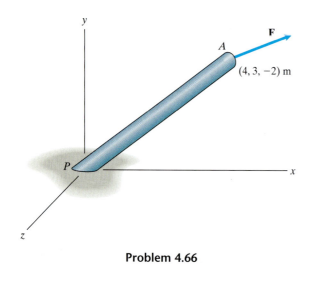

Problem 4.66

4.67 The force $\mathbf{F} = 5\mathbf{i}$ (kN) acts on the ring A where the cables AB, AC, and AD are joined. What is the sum of the moments about point D due to the force $\mathbf{F}$ and the three forces exerted on the ring by the cables?

Strategy: The ring is in equilibrium. Use what you know about the four forces acting on it.

4.68 In Problem 4.67, determine the moment about point D due to the force exerted on the ring A by the cable AB.

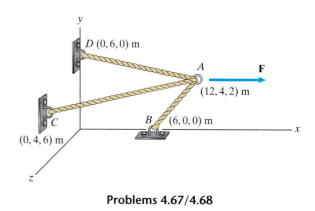

Problems 4.67/4.68

4.69 The tower is 70 m tall. The tensions in cables AB, AC, and AD are 4 kN, 2 kN, and 2 kN, respectively. Determine the sum of the moments about the origin O due to the forces exerted by the cables at point A.

4.70 Suppose that the tension in cable AB is 4 kN, and you want to adjust the tensions in cables AC and AD so that the sum of the moments about the origin O due to the forces exerted by the cables at point A is zero. Determine the tensions.

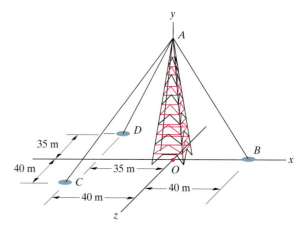

Problems 4.69/4.70

4.71 The tension in cable *AB* is 150 N. The tension in cable *AC* is 100 N. Determine the sum of the moments about *D* due to the forces exerted on the wall by the cables.

4.72 The total force exerted by the two cables in the direction perpendicular to the wall is 2 kN. The magnitude of the sum of the moments about *D* due to the forces exerted on the wall by the cables is 18 kN-m. What are the tensions in the cables?

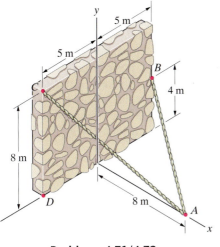

Problems 4.71/4.72

4.73 The tension in the cable *BD* is 1 kN. As a result, cable *BD* exerts a 1-kN force on the "ball" at *B* that points from *B* toward *D*. Determine the moment of this force about point *A*.

4.74 *Suppose that the mass of the suspended object *E* is 100 kg and the mass of the bar *AB* is 20 kg. Assume that the weight of the bar acts at its midpoint. If the sum of the moments about point *A* due to the weight of the bar and the forces exerted on the "ball" at *B* by the three cables *BC, BD,* and *BE* is zero, determine the tensions in the cables *BC* and *BD*.

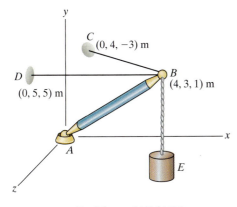

Problems 4.73/4.74

4.75 The 200-kg slider at *A* is held in place on the smooth vertical bar by the cable *AB*. Determine the moment about the bottom of the bar (point *C* with coordinates $x = 2$ m, $y = z = 0$) due to the force exerted on the slider by the cable.

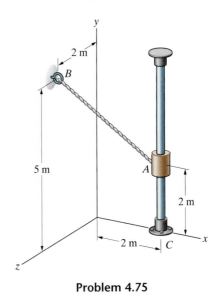

Problem 4.75

4.76 To evaluate the adequacy of the design of the vertical steel post, you must determine the moment about the bottom of the post due to the force exerted on the post at *B* by the cable *AB*. A calibrated strain gauge mounted on cable *AC* indicates that the tension in cable *AC* is 22 kN. What is the moment?

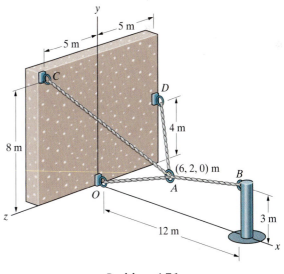

Problem 4.76

4.3 Moment of a Force About a Line

The device in Fig. 4.16, called a *capstan*, was used in the days of square-rigged sailing ships. Crewmen turned it by pushing on the handles as shown in Fig. 4.16a, providing power for such tasks as raising anchors and hoisting yards. A vertical force **F** applied to one of the handles as shown in Fig. 4.16b does not cause the capstan to turn, even though the magnitude of the moment about point P is $d|\mathbf{F}|$ in both cases.

The measure of the tendency of a force to cause rotation about a line, or axis, is called the moment of the force about the line. Suppose that a force **F** acts on an object such as a turbine that rotates about an axis L, and we resolve **F** into components in terms of the coordinate system shown in Fig. 4.17. The components F_x and F_z do not tend to rotate the turbine, just as the force parallel to the axis of the capstan did not cause it to turn. It is the component F_y that tends to cause rotation, by exerting a moment of magnitude aF_y about the turbine's axis. In this example we can determine the moment of **F** about L easily because the coordinate system is conveniently placed. We now introduce an expression that determines the moment of a force about any line.

Definition

Consider a line L and force **F** (Fig. 4.18a). Let $\mathbf{M}_P$ be the moment of **F** about an arbitrary point P on L (Fig. 4.18b). The moment of **F** about L is the component of $\mathbf{M}_P$ parallel to L, which we denote by $\mathbf{M}_L$ (Fig. 4.18c). The magnitude of the moment of **F** about L is $|\mathbf{M}_L|$, and when the thumb of the right hand is pointed in the direction of $\mathbf{M}_L$, the arc of the fingers indicates the direction of the moment about L.

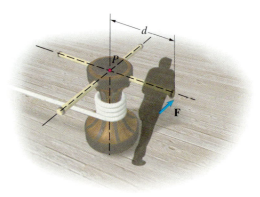

(a)

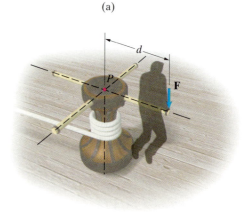

(b)

Figure 4.16
(a) Turning a capstan.
(b) A vertical force does not turn the capstan.

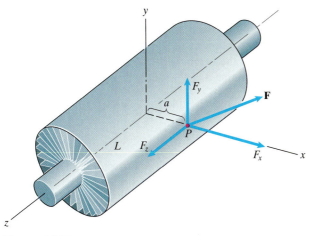

Figure 4.17
Applying a force to a turbine with axis of rotation L.

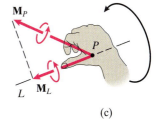

(a)

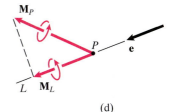

(b)

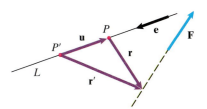

(c)

(d)

Figure 4.18
(a) The line L and force $\mathbf{F}$.
(b) $\mathbf{M}_P$ is the moment of $\mathbf{F}$ about any point P on L.
(c) The component $\mathbf{M}_L$ is the moment of $\mathbf{F}$ about L.
(d) A unit vector $\mathbf{e}$ along L.

In terms of a unit vector $\mathbf{e}$ along L (Fig. 4.18d), $\mathbf{M}_L$ is given by

$$\mathbf{M}_L = (\mathbf{e} \cdot \mathbf{M}_P)\mathbf{e}. \qquad (4.4)$$

(The unit vector $\mathbf{e}$ can point in either direction. See our discussion of vector components parallel and normal to a line in Section 2.5.) The moment $\mathbf{M}_P = \mathbf{r} \times \mathbf{F}$, so we can also express $\mathbf{M}_L$ as

$$\mathbf{M}_L = [\mathbf{e} \cdot (\mathbf{r} \times \mathbf{F})]\mathbf{e}. \qquad (4.5)$$

The mixed triple product in this expression is given in terms of the components of the three vectors by

$$\mathbf{e} \cdot (\mathbf{r} \times \mathbf{F}) = \begin{vmatrix} e_x & e_y & e_z \\ r_x & r_y & r_z \\ F_x & F_y & F_z \end{vmatrix}. \qquad (4.6)$$

Notice that the value of the scalar $\mathbf{e} \cdot \mathbf{M}_P = \mathbf{e} \cdot (\mathbf{r} \times \mathbf{F})$ determines both the magnitude and direction of $\mathbf{M}_L$. The absolute value of $\mathbf{e} \cdot \mathbf{M}_P$ is the magnitude of $\mathbf{M}_L$. If $\mathbf{e} \cdot \mathbf{M}_P$ is positive, $\mathbf{M}_L$ points in the direction of $\mathbf{e}$, and if $\mathbf{e} \cdot \mathbf{M}_P$ is negative, $\mathbf{M}_L$ points in the direction opposite to $\mathbf{e}$.

The result obtained with Eq. (4.4) or (4.5) doesn't depend on which point on L is chosen to determine $\mathbf{M}_P = \mathbf{r} \times \mathbf{F}$. If we use point P in Fig. 4.19 to determine the moment of $\mathbf{F}$ about L, we get the result given by Eq. (4.5). If we use P' instead, we obtain the same result,

$$[\mathbf{e} \cdot (\mathbf{r}' \times \mathbf{F})]\mathbf{e} = \{\mathbf{e} \cdot [(\mathbf{r} + \mathbf{u}) \times \mathbf{F}]\}\mathbf{e}$$
$$= [\mathbf{e} \cdot (\mathbf{r} \times \mathbf{F}) + \mathbf{e} \cdot (\mathbf{u} \times \mathbf{F})]\mathbf{e}$$
$$= [\mathbf{e} \cdot (\mathbf{r} \times \mathbf{F})]\mathbf{e},$$

because $\mathbf{u} \times \mathbf{F}$ is perpendicular to $\mathbf{e}$.

Figure 4.19
Using different points P and P' to determine the moment of $\mathbf{F}$ about L.

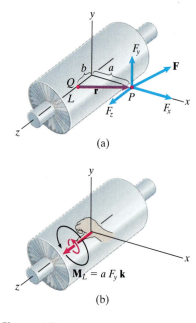

(a)

(b)

Figure 4.20

(a) An arbitrary point Q on L and the vector $\mathbf{r}$ from Q to P.

(b) $\mathbf{M}_L$ and the direction of the moment about L.

Applications

To demonstrate that $\mathbf{M}_L$ is the measure of the tendency of $\mathbf{F}$ to cause rotation about L, we return to the turbine in Fig. 4.17. Let Q be a point on L at an arbitrary distance b from the origin (Fig. 4.20a). The vector $\mathbf{r}$ from Q to P is $\mathbf{r} = a\mathbf{i} - b\mathbf{k}$, so the moment of $\mathbf{F}$ about Q is

$$\mathbf{M}_Q = \mathbf{r} \times \mathbf{F} = \begin{vmatrix} \mathbf{i} & \mathbf{j} & \mathbf{k} \\ a & 0 & -b \\ F_x & F_y & F_z \end{vmatrix} = bF_y\mathbf{i} - (aF_z + bF_x)\mathbf{j} + aF_y\mathbf{k}.$$

Since the z axis is coincident with L, the unit vector $\mathbf{k}$ is along L. Therefore the moment of $\mathbf{F}$ about L is

$$\mathbf{M}_L = (\mathbf{k} \cdot \mathbf{M}_Q)\mathbf{k} = aF_y\mathbf{k}.$$

The components F_x and F_z exert no moment about L. If we assume that F_y is positive, it exerts a moment of magnitude aF_y about the turbine's axis in the direction shown in Fig. 4.20b.

Now let's determine the moment of a force about an arbitrary line L (Fig. 4.21a). The first step is to choose a point on the line. If we choose point A (Fig. 4.21b), the vector $\mathbf{r}$ from A to the point of application of $\mathbf{F}$ is

$$\mathbf{r} = (8 - 2)\mathbf{i} + (6 - 0)\mathbf{j} + (4 - 4)\mathbf{k} = 6\mathbf{i} + 6\mathbf{j} \text{ (m)}.$$

The moment of $\mathbf{F}$ about A is

$$\mathbf{M}_A = \mathbf{r} \times \mathbf{F} = \begin{vmatrix} \mathbf{i} & \mathbf{j} & \mathbf{k} \\ 6 & 6 & 0 \\ 10 & 60 & -20 \end{vmatrix}$$

$$= -120\mathbf{i} + 120\mathbf{j} + 300\mathbf{k} \text{ (N-m)}.$$

The next step is to determine a unit vector along L. The vector from A to B is

$$(-7 - 2)\mathbf{i} + (6 - 0)\mathbf{j} + (2 - 4)\mathbf{k} = -9\mathbf{i} + 6\mathbf{j} - 2\mathbf{k} \text{ (m)}.$$

Dividing this vector by its magnitude, we obtain a unit vector $\mathbf{e}_{AB}$ that points from A toward B (Fig. 4.21c):

$$\mathbf{e}_{AB} = -\frac{9}{11}\mathbf{i} + \frac{6}{11}\mathbf{j} - \frac{2}{11}\mathbf{k}.$$

The moment of $\mathbf{F}$ about L is

$$\mathbf{M}_L = (\mathbf{e}_{AB} \cdot \mathbf{M}_A)\mathbf{e}_{AB}$$

$$= \left[\left(-\frac{9}{11}\right)(-120 \text{ N-m}) + \left(\frac{6}{11}\right)(120 \text{ N-m}) + \left(-\frac{2}{11}\right)(300 \text{ N-m})\right]\mathbf{e}_{AB}$$

$$= 109\mathbf{e}_{AB} \text{ (N-m)}.$$

The magnitude of $\mathbf{M}_L$ is 109 N-m; pointing the thumb of the right hand in the direction of $\mathbf{e}_{AB}$ indicates the direction.

If we calculate $\mathbf{M}_L$ using the unit vector $\mathbf{e}_{BA}$ that points from B toward A instead, we obtain

$$\mathbf{M}_L = -109\mathbf{e}_{BA} \text{ (N-m)}.$$

We obtain the same magnitude, and the minus sign indicates that $\mathbf{M}_L$ points in the direction opposite to $\mathbf{e}_{BA}$, so the direction of $\mathbf{M}_L$ is the same. Therefore the right-hand rule indicates the same direction (Fig. 4.21d).

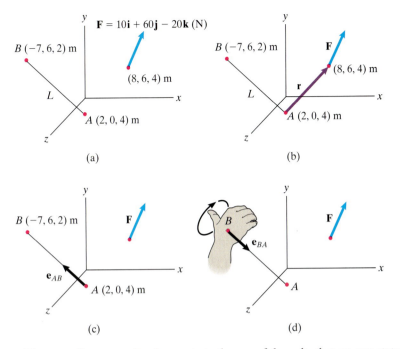

(a)

(b)

(c)

(d)

Figure 4.21
(a) A force $\mathbf{F}$ and line L.
(b) The vector $\mathbf{r}$ from A to the point of application of $\mathbf{F}$.
(c) $\mathbf{e}_{AB}$ points from A toward B.
(d) The right-hand rule indicates the direction of the moment.

The preceding examples demonstrate three useful results that we can state in more general terms:

- When the line of action of $\mathbf{F}$ is perpendicular to a plane containing L (Fig. 4.22a), the magnitude of the moment of $\mathbf{F}$ about L is equal to the product of the magnitude of $\mathbf{F}$ and the perpendicular distance D from L to the point where the line of action intersects the plane: $|\mathbf{M}_L| = |\mathbf{F}|D$.
- When the line of action of $\mathbf{F}$ is parallel to L (Fig. 4.22b), the moment of $\mathbf{F}$ about L is zero: $\mathbf{M}_L = 0$. Since $\mathbf{M}_P = \mathbf{r} \times \mathbf{F}$ is perpendicular to $\mathbf{F}$, $\mathbf{M}_P$ is perpendicular to L and the vector component of $\mathbf{M}_P$ parallel to L is zero.
- When the line of action of $\mathbf{F}$ intersects L (Fig. 4.22c), the moment of $\mathbf{F}$ about L is zero. Since we can choose any point on L to evaluate $\mathbf{M}_P$, we can use the point where the line of action of $\mathbf{F}$ intersects L. The moment $\mathbf{M}_P$ about that point is zero, so its vector component parallel to L is zero.

In summary, determining the moment of a force $\mathbf{F}$ about a point P using Eqs. (4.4)–(4.6) requires three steps:

1. Determine a vector $\mathbf{r}$—Choose any point P on L, and determine the components of a vector $\mathbf{r}$ from P to any point on the line of action of $\mathbf{F}$.
2. Determine a vector $\mathbf{e}$—Determine the components of a unit vector along L. It doesn't matter in which direction along L it points.
3. Evaluate $\mathbf{M}_L$—You can calculate $\mathbf{M}_P = \mathbf{r} \times \mathbf{F}$ and determine $\mathbf{M}_L$ by using Eq. (4.4), or you can use Eq. (4.6) to evaluate the mixed triple product and substitute the result into Eq. (4.5).

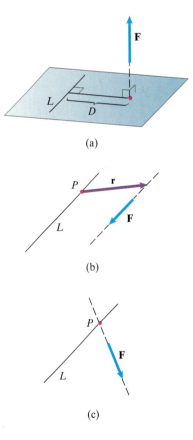

(a)

(b)

(c)

Figure 4.22
(a) $\mathbf{F}$ is perpendicular to a plane containing L.
(b) $\mathbf{F}$ is parallel to L.
(c) The line of action of $\mathbf{F}$ intersects L at P.

Study Questions

1. When you use Eq. (4.5) to determine the moment of a force $\mathbf{F}$ about a line L, how do you choose the vector $\mathbf{r}$? What is the definition of the vector $\mathbf{e}$?
2. Explain how the direction of the vector $\mathbf{M}_L$ in Eq. (4.5) indicates the direction of the moment of $\mathbf{F}$ about L.
3. What is the moment of a force $\mathbf{F}$ about a line L if the line of action of $\mathbf{F}$ passes through L? What is the moment if the line of action of $\mathbf{F}$ is parallel to L?

Example 4.7 **Moment of a Force About the *x* Axis**

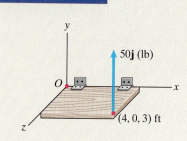

Figure 4.23

What is the moment of the 50-lb force in Fig. 4.23 about the *x* axis?

Strategy

We can determine the moment in two ways.

First Method We can use Eqs. (4.5) and (4.6). Since **r** can extend from any point on the *x* axis to the line of action of the force, we can use the vector from *O* to the point of application of the force. The vector **e** must be a unit vector along the *x* axis, so we can use either **i** or −**i**.

Second Method This example is the first of the special cases we just discussed, because the 50-lb force is perpendicular to the *x*–*z* plane. We can determine the magnitude and direction of the moment directly from the given information.

Solution

First Method *Determine a vector* **r**. The vector from *O* to the point of application of the force is (Fig. a)

$$\mathbf{r} = 4\mathbf{i} + 3\mathbf{k} \ (\text{ft}).$$

Determine a vector **e**. We can use the unit vector **i**.
Evaluate $\mathbf{M}_L$. From Eq. (4.6), the mixed triple product is

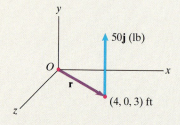

(a) The vector **r** from *O* to the point of application of the force.

$$\mathbf{i} \cdot (\mathbf{r} \times \mathbf{F}) = \begin{vmatrix} 1 & 0 & 0 \\ 4 & 0 & 3 \\ 0 & 50 & 0 \end{vmatrix} = -150 \ \text{ft-lb}.$$

Then from Eq. (4.5), the moment of the force about the *x* axis is

$$\mathbf{M}_{x \, \text{axis}} = [\mathbf{i} \cdot (\mathbf{r} \times \mathbf{F})]\mathbf{i} = -150\mathbf{i} \ (\text{ft-lb}).$$

The magnitude of the moment is 150 ft-lb, and its direction is as shown in Fig. b.

Second Method Since the 50-lb force is perpendicular to a plane (the *x*–*z* plane) containing the *x* axis, the magnitude of the moment about the *x* axis is equal to the perpendicular distance from the *x* axis to the point where the line of action of the force intersects the *x*–*z* plane (Fig. c):

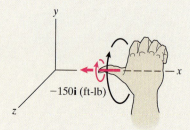

(b) The direction of the moment.

$$\left| \mathbf{M}_{x \, \text{axis}} \right| = (3 \ \text{ft})(50 \ \text{lb}) = 150 \ \text{ft-lb}.$$

Pointing the arc of the fingers in the direction of the moment about the *x* axis (Fig. c), we find that the right-hand rule indicates that $\mathbf{M}_{x \, \text{axis}}$ points in the negative *x* axis direction. Therefore,

$$\mathbf{M}_{x \, \text{axis}} = -150\mathbf{i} \ (\text{ft-lb}).$$

Critical Thinking

The hinged door in this example is designed to rotate about the *x* axis. If no other forces act on the door, you can see that the 50-lb upward force would tend to cause the door to rotate upward. It is the moment of the force about the *x* axis, and *not* the moment of the force about some point, that measures the tendency of the force to cause the door to rotate on its hinges. Furthermore, the direction of the moment of the force about the *x* axis indicates the direction in which the force tends to cause the door to rotate. (See Fig. b.)

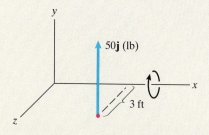

(c) The distance from the *x* axis to the point where the line of action of the force intersects the *x*–*z* plane is 3 ft. The arrow indicates the direction of the moment about the *x* axis.

| **Example 4.8** | **Moment of a Force About a Line** |

What is the moment of the force **F** in Fig. 4.24 about the bar *BC*?

Strategy
We can use Eqs. (4.5) and (4.6) to determine the moment. Since we know the coordinates of points *B* and *C*, we can determine the components of a vector **r** that extends either from *B* to the point of application of the force or from *C* to the point of application. We can also use the coordinates of points *B* and *C* to determine a unit vector along the line *BC*.

Solution
Determine a Vector r   We need a vector from any point on the line *BC* to any point on the line of action of the force. We can let **r** be the vector from *B* to the point of application of **F** (Fig. a):

$$\mathbf{r} = (4 - 0)\mathbf{i} + (2 - 0)\mathbf{j} + (2 - 3)\mathbf{k} = 4\mathbf{i} + 2\mathbf{j} - \mathbf{k} \text{ (m)}.$$

Determine a Vector e   To obtain a unit vector along the bar *BC*, we determine the vector from *B* to *C*,

$$(0 - 0)\mathbf{i} + (4 - 0)\mathbf{j} + (0 - 3)\mathbf{k} = 4\mathbf{j} - 3\mathbf{k} \text{ (m)},$$

and divide it by its magnitude (Fig. a):

$$\mathbf{e}_{BC} = \frac{4\mathbf{j} - 3\mathbf{k} \text{ (m)}}{\sqrt{(4 \text{ m})^2 + (-3 \text{ m})^2}} = 0.8\mathbf{j} - 0.6\mathbf{k}.$$

Evaluate $\mathbf{M}_L$ From Eq. (4.6), the mixed triple product is

$$\mathbf{e}_{BC} \cdot (\mathbf{r} \times \mathbf{F}) = \begin{vmatrix} 0 & 0.8 & -0.6 \\ 4 & 2 & -1 \\ -2 & 6 & 3 \end{vmatrix} = -24.8 \text{ kN-m}.$$

Substituting this result into Eq. (4.5), we obtain the moment of **F** about the bar *BC*:

$$\mathbf{M}_{BC} = [\mathbf{e}_{BC} \cdot (\mathbf{r} \times \mathbf{F})]\mathbf{e}_{BC} = -24.8\mathbf{e}_{BC} \text{ (kN-m)}.$$

The magnitude of $\mathbf{M}_{BC}$ is 24.8 kN-m, and its direction is opposite to that of $\mathbf{e}_{BC}$. The direction of the moment is shown in Fig. b.

Critical Thinking
When you use Eq. (4.4) or (4.5) to calculate the moment of a force about a line, how do you know which direction along the line the unit vector **e** should point? The answer is that it doesn't matter. In this example, instead of the unit vector $\mathbf{e}_{BC}$ that points from point *B* toward point *C*, we could have used the unit vector

$$\mathbf{e}_{CB} = -0.8\mathbf{j} + 0.6\mathbf{k}$$

that points from point *C* toward point *B*. In that case, the mixed triple product in Eq. (4.5) would be

$$\mathbf{e}_{CB} \cdot (\mathbf{r} \times \mathbf{F}) = \begin{vmatrix} 0 & -0.8 & 0.6 \\ 4 & 2 & -1 \\ -2 & 6 & 3 \end{vmatrix} = 24.8 \text{ kN-m},$$

and the moment of **F** about the bar *BC* would be

$$[\mathbf{e}_{CB} \cdot (\mathbf{r} \times \mathbf{F})]\mathbf{e}_{CB} = 24.8\mathbf{e}_{CB} \text{ (kN-m)}.$$

Because $\mathbf{e}_{CB} = -\mathbf{e}_{BC}$, we obtain the same result for the moment about the bar.

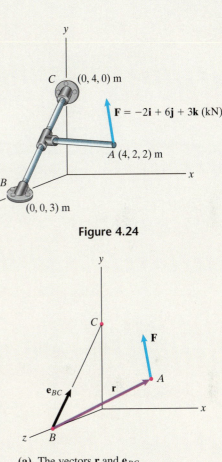

Figure 4.24

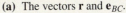

(a) The vectors **r** and $\mathbf{e}_{BC}$.

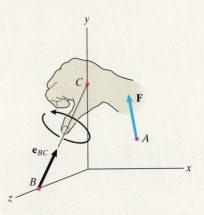

(b) The right-hand rule indicates the direction of the moment about *BC*.

Design Example 4.9 **Rotating Machines**

The crewman in Fig. 4.25 exerts the forces shown on the handles of the coffee grinder winch, where $\mathbf{F} = 4\mathbf{j} + 32\mathbf{k}$ N. Determine the total moment he exerts (a) about point O, (b) about the axis of the winch, which coincides with the x axis.

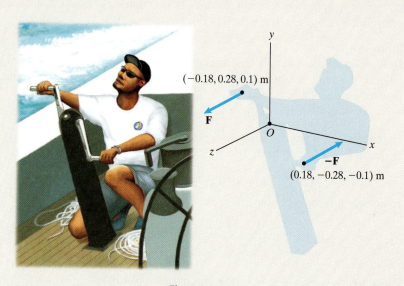

Figure 4.25

Strategy
(a) To obtain the total moment about point O, we must sum the moments of the two forces about O. Let the sum be denoted by $\Sigma\,\mathbf{M}_O$.
(b) Because point O is on the x axis, the total moment about the x axis is the component of $\Sigma\,\mathbf{M}_O$ parallel to the x axis, which is the x component of $\Sigma\,\mathbf{M}_O$.

Solution
(a) The total moment about point O is

$$\Sigma\,\mathbf{M}_O = \begin{vmatrix} \mathbf{i} & \mathbf{j} & \mathbf{k} \\ -0.18 & 0.28 & 0.1 \\ 0 & 4 & 32 \end{vmatrix} + \begin{vmatrix} \mathbf{i} & \mathbf{j} & \mathbf{k} \\ 0.18 & -0.28 & -0.1 \\ 0 & -4 & -32 \end{vmatrix}$$

$$= 17.1\mathbf{i} + 11.5\mathbf{j} - 1.4\mathbf{k} \ (\text{N-m}).$$

(b) The total moment about the x axis is the x component of $\Sigma\,\mathbf{M}_O$ (Fig. a):

$$\Sigma\,\mathbf{M}_{x\,\text{axis}} = 17.1 \ (\text{N-m}).$$

Notice that this is the result given by Eq. (4.4): Since $\mathbf{i}$ is a unit vector parallel to the x axis,

$$\Sigma\,\mathbf{M}_{x\,\text{axis}} = (\mathbf{i} \cdot \Sigma\,\mathbf{M}_O)\mathbf{i} = 17.1 \ (\text{N-m}).$$

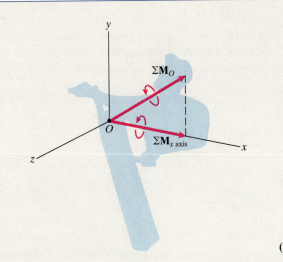

(a) The total moment about the *x* axis.

Design Issues

The winch in this example is a simple representative of a class of rotating machines that includes hydrodynamic and aerodynamic power turbines, propellers, jet engines, and electric motors and generators. The ancestors of hydrodynamic and aerodynamic power turbines—water wheels and windmills—were among the earliest machines. These devices illustrate the importance of the concept of the moment of a force about a line. Their common feature is a part designed to rotate and perform some function when it is subjected to a moment about its axis of rotation. In the case of the winch, the forces exerted on the handles by the crewman exert a moment about the axis of rotation, causing the winch to rotate and wind a rope onto a drum, trimming the boat's sails. A hydrodynamic power turbine (Fig. 4.26) has turbine blades that are subjected to forces by flowing water, exerting a moment about the axis of rotation. This moment rotates the shaft to which the blades are attached, turning an electric generator that is connected to the same shaft.

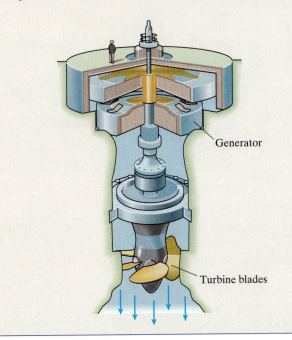

Generator

Turbine blades

Figure 4.26
A hydroelectric turbine. Water flowing through the turbine blades exerts a moment about the axis of the shaft, turning the generator.

Problems

4.77 The force $\mathbf{F} = 20\mathbf{i} + 40\mathbf{j} - 10\mathbf{k}$ (N). Use Eqs. (4.5) and (4.6) to determine the moment due to $\mathbf{F}$ about the z axis. (First see if you can write down the result without using the equations.)

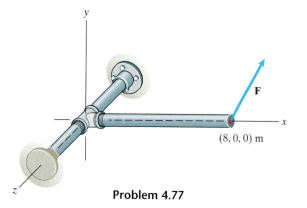

Problem 4.77

4.78 Use Eqs. (4.5) and (4.6) to determine the moment of the 20-N force about (a) the x axis, (b) the y axis, (c) the z axis. (First see if you can write down the results without using the equations.)

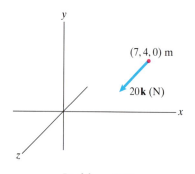

Problem 4.78

4.79 Three forces parallel to the y axis act on the rectangular plate. Use Eqs. (4.5) and (4.6) to determine the sum of the moments of the forces about the x axis. (First see if you can write down the result without using the equations.)

4.80 The three forces are parallel to the y axis. Determine the sum of the moments of the forces (a) about the y axis; (b) about the z axis.

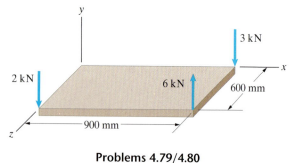

Problems 4.79/4.80

4.81 The person exerts a force $\mathbf{F} = 0.2\mathbf{i} - 0.4\mathbf{j} + 1.2\mathbf{k}$ (lb) on the gate at C. Point C lies in the x–y plane. What moment does the person exert about the gate's hinge axis, which is coincident with the y axis?

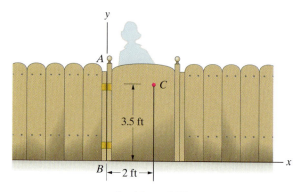

Problem 4.81

4.82 Four forces act on the plate. Their components are

$$\mathbf{F}_A = -2\mathbf{i} + 4\mathbf{j} + 2\mathbf{k} \ (\text{kN}),$$

$$\mathbf{F}_B = 3\mathbf{j} - 3\mathbf{k} \ (\text{kN}),$$

$$\mathbf{F}_C = 2\mathbf{j} + 3\mathbf{k} \ (\text{kN}),$$

$$\mathbf{F}_D = 2\mathbf{i} + 6\mathbf{j} + 4\mathbf{k} \ (\text{kN}).$$

Determine the sum of the moments of the forces (a) about the x axis; (b) about the z axis.

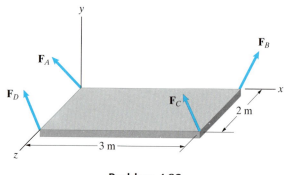

Problem 4.82

4.83 The force $\mathbf{F} = 30\mathbf{i} + 20\mathbf{j} - 10\mathbf{k}$ (lb),
(a) What is the moment of $\mathbf{F}$ about the y axis?
(b) Suppose that you keep the magnitude of $\mathbf{F}$ fixed, but you change its direction so as to make the moment of $\mathbf{F}$ about the y axis as large as possible. What is the magnitude of the resulting moment?

4.84 The moment of the force $\mathbf{F}$ about the x axis is $-80\mathbf{i}$ (ft-lb), the moment about the y axis is zero, and the moment about the z axis is $160\mathbf{k}$ (ft-lb). If $F_y = 80$ lb, what are F_x and F_z?

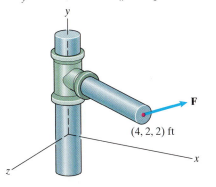

Problems 4.83/4.84

4.85 The robotic manipulator is stationary. The weights of the arms AB and BC act at their midpoints. The direction cosines of the centerline of arm AB are $\cos \theta_x = 0.500$, $\cos \theta_y = 0.866$, $\cos \theta_z = 0$, and the direction cosines of the centerline of arm BC are $\cos \theta_x = 0.707$, $\cos \theta_y = 0.619$, $\cos \theta_z = -0.342$. What total moment is exerted about the z axis by the weights of the arms?

4.86 In Problem 4.85, what total moment is exerted about the x axis by the weights of the arms?

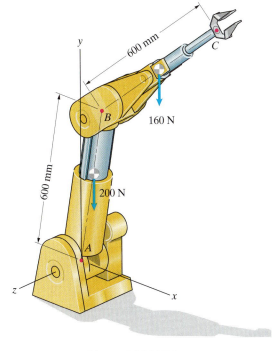

Problems 4.85/4.86

4.87 Two forces are exerted on the crankshaft by the connecting rods. The direction cosines of F_A are $\cos \theta_x = -0.182$, $\cos \theta_y = 0.818$, and $\cos \theta_z = 0.545$, and its magnitude is 4 kN. The direction cosines of F_B are $\cos \theta_x = 0.182$, $\cos \theta_y = 0.818$, and $\cos \theta_z = -0.545$, and its magnitude is 2 kN. What is the sum of the moments of the two forces about the x axis? (This is the moment that causes the crankshaft to rotate.)

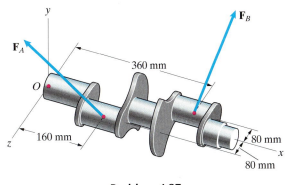
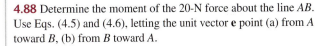

Problem 4.87

4.88 Determine the moment of the 20-N force about the line AB. Use Eqs. (4.5) and (4.6), letting the unit vector $\mathbf{e}$ point (a) from A toward B, (b) from B toward A.

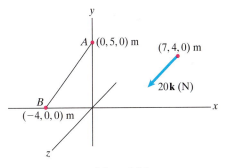

Problem 4.88

4.89 The force $\mathbf{F} = -10\mathbf{i} + 5\mathbf{j} - 5\mathbf{k}$ (kip). Determine the moment of $\mathbf{F}$ about the line AB. Draw a sketch to indicate the direction of the moment.

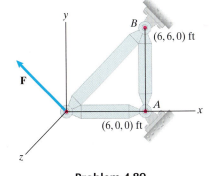

Problem 4.89

4.90 The force $\mathbf{F} = 10\mathbf{i} + 12\mathbf{j} - 6\mathbf{k}$ (N). What is the moment of $\mathbf{F}$ about the line AO? Draw a sketch to indicate the direction of the moment.

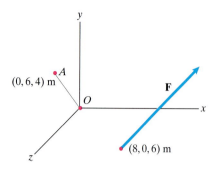

Problem 4.90

4.91 The tension in the cable AB is 1 kN. Determine the moment about the x axis due to the force exerted on the hatch by the cable at point B. Draw a sketch to indicate the direction of the moment.

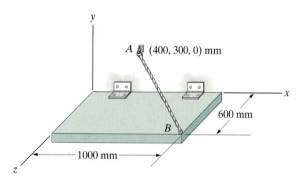

Problem 4.91

4.92 Determine the moment of the force applied at D about the straight line through the hinges A and B. (The line through A and B lies in the y–z plane.)

4.93 The tension in the cable CE is 160 lb. Determine the moment of the force exerted by the cable on the hatch at C about the straight line through the hinges A and B.

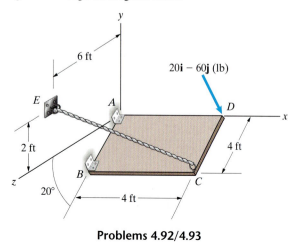

Problems 4.92/4.93

4.94 The coordinates of A are $(-2.4, 0, -0.6)$ m, and the coordinates of B are $(-2.2, 0.7, -1.2)$ m. The force exerted at B by the sailboat's main sheet AB is 130 N. Determine the moment of the force about the centerline of the mast (the y axis). Draw a sketch to indicate the direction of the moment.

Problem 4.94

4.95 The tension in cable AB is 200 lb. Determine the moments about each of the coordinate axes due to the force exerted on point B by the cable. Draw sketches to indicate the direction of the moments.

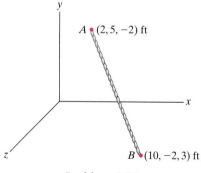

Problem 4.95

4.96 The total force exerted on the blades of the turbine by the steam nozzle is $\mathbf{F} = 20\mathbf{i} - 120\mathbf{j} + 100\mathbf{k}$ (N), and it effectively acts at the point (100, 80, 300) mm. What moment is exerted about the axis of the turbine (the *x* axis)?

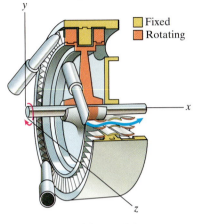

Fixed
Rotating

Problem 4.96

4.97 The tension in cable *AB* is 50 N. Determine the moment about the line *OC* due to the force exerted by the cable at *B*. Draw a sketch to indicate the direction of the moment.

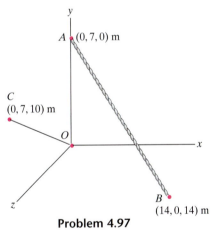

A (0, 7, 0) m
C (0, 7, 10) m
O
B (14, 0, 14) m

Problem 4.97

4.98 The tension in cable *AB* is 80 lb. What is the moment about the line *CD* due to the force exerted by the cable on the wall at *B*?

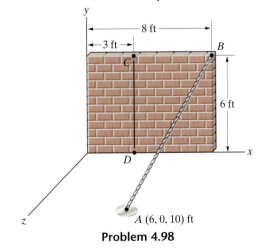

8 ft
3 ft
B
C
6 ft
D
A (6, 0, 10) ft

Problem 4.98

4.99 The magnitude of the force $\mathbf{F}$ is 0.2 N and its direction cosines are $\cos\theta_x = 0.727$, $\cos\theta_y = -0.364$, and $\cos\theta_z = 0.582$. Determine the magnitude of the moment of $\mathbf{F}$ about the axis *AB* of the spool.

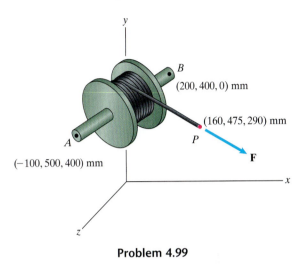

B
(200, 400, 0) mm
(160, 475, 290) mm
A
P
$\mathbf{F}$
(−100, 500, 400) mm

Problem 4.99

4.100 A motorist applies the two forces shown to loosen a lug nut. The direction cosines of $\mathbf{F}$ are $\cos\theta_x = \frac{4}{13}$, $\cos\theta_y = \frac{12}{13}$, and $\cos\theta_z = \frac{3}{13}$. If the magnitude of the moment about the *x* axis must be 32 ft-lb to loosen the nut, what is the magnitude of the forces the motorist must apply?

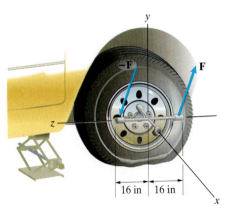

−F
F
16 in   16 in

Problem 4.100

4.101 The tension in cable AB is 2 kN. What is the magnitude of the moment about the shaft CD due to the force exerted by the cable at A? Draw a sketch to indicate the direction of the moment about the shaft.

Problem 4.101

4.102 The axis of the car's wheel passes through the origin of the coordinate system and its direction cosines are $\cos \theta_x = 0.940$, $\cos \theta_y = 0$, $\cos \theta_z = 0.342$. The force exerted on the tire by the road effectively acts at the point $x = 0$, $y = -0.36$ m, $z = 0$ and has components $\mathbf{F} = -720\mathbf{i} + 3660\mathbf{j} + 1240\mathbf{k}$ (N). What is the moment of $\mathbf{F}$ about the wheel's axis?

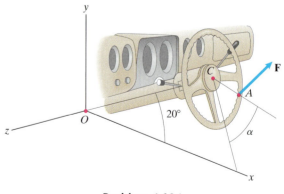

Problem 4.102

4.103 The direction cosines of the centerline OA are $\cos \theta_x = 0.500$, $\cos \theta_y = 0.866$, and $\cos \theta_z = 0$, and the direction cosines of the line AG are $\cos \theta_x = 0.707$, $\cos \theta_y = 0.619$, and $\cos \theta_z = -0.342$. What is the moment about OA due to the 250-N weight? Draw a sketch to indicate the direction of the moment about the shaft.

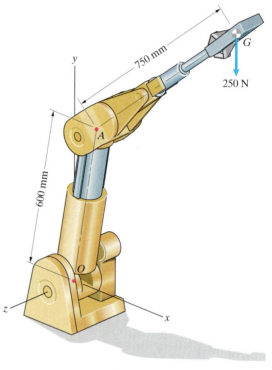

Problem 4.103

4.104 The radius of the steering wheel is 200 mm. The distance from O to C is 1 m. The center C of the steering wheel lies in the x–y plane. The driver exerts a force $\mathbf{F} = 10\mathbf{i} + 10\mathbf{j} - 5\mathbf{k}$ (N) on the wheel at A. If the angle $\alpha = 0$, what is the magnitude of the moment about the shaft OC? Draw a sketch to indicate the direction of the moment about the shaft.

Problem 4.104

4.105 *The magnitude of the force **F** is 10 N. Suppose that you want to choose the direction of the force **F** so that the magnitude of its moment about the line L is a maximum. Determine the components of **F** and the magnitude of its moment about L. (There are two solutions for **F**.)

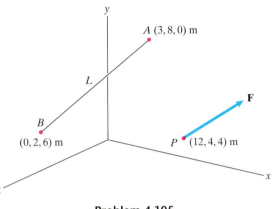

Problem 4.105

4.106 The weight W causes a tension of 100 lb in cable CD. If $d = 2$ ft, what is the moment about the z axis due to the force exerted by the cable CD at point C?

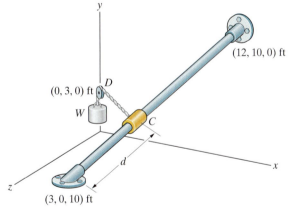

Problem 4.106

4.107 *The y axis points upward. The weight of the 4-kg rectangular plate acts at the midpoint G of the plate. The sum of the moments about the straight line through the supports A and B due to the weight of the plate and the force exerted on the plate by the cable CD is zero. What is the tension in the cable?

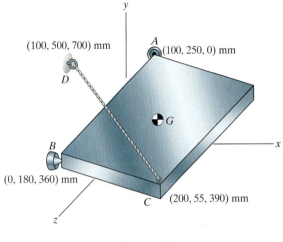

Problem 4.107

4.4 Couples

Now that we have described how to calculate the moment due to a force, consider this question: Is it possible to exert a moment on an object without subjecting it to a net force? The answer is yes, and it occurs when a compact disk begins rotating or a screw is turned by a screwdriver. Forces are exerted on these objects, but in such a way that the net force is zero while the net moment is not zero.

Two forces that have equal magnitudes, opposite directions, and different lines of action are called a *couple* (Fig. 4.27a). A couple tends to cause rotation of an object even though the vector sum of the forces is zero, and it has the remarkable property that *the moment it exerts is the same about any point*.

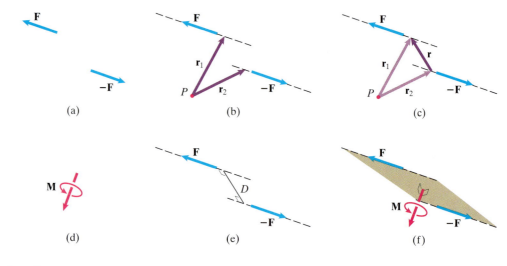

Figure 4.27
(a) A couple.
(b) Determining the moment about P.
(c) The vector $\mathbf{r} = \mathbf{r}_1 - \mathbf{r}_2$.
(d) Representing the moment of the couple.
(e) The distance D between the lines of action.
(f) $\mathbf{M}$ is perpendicular to the plane containing $\mathbf{F}$ and $-\mathbf{F}$.

The moment of a couple is simply the sum of the moments of the forces about a point P (Fig. 4.27b):

$$\mathbf{M} = [\mathbf{r}_1 \times \mathbf{F}] + [\mathbf{r}_2 \times (-\mathbf{F})] = (\mathbf{r}_1 - \mathbf{r}_2) \times \mathbf{F}.$$

The vector $\mathbf{r}_1 - \mathbf{r}_2$ is equal to the vector $\mathbf{r}$ shown in Fig. 4.27c, so we can express the moment as

$$\mathbf{M} = \mathbf{r} \times \mathbf{F}.$$

Since $\mathbf{r}$ doesn't depend on the position of P, the moment $\mathbf{M}$ is the same for *any* point P.

Because a couple exerts a moment but the sum of the forces is zero, it is often represented in diagrams simply by showing the moment (Fig. 4.27d). Like the Cheshire cat in *Alice's Adventures in Wonderland*, which vanished except for its grin, the forces don't appear; only the moment they exert is visible. But we recognize the origin of the moment by referring to it as a *moment of a couple*, or simply a *couple*.

Notice in Fig. 4.27c that $\mathbf{M} = \mathbf{r} \times \mathbf{F}$ is the moment of $\mathbf{F}$ about a point on the line of action of the force $-\mathbf{F}$. The magnitude of the moment of a force about a point equals the product of the magnitude of the force and the perpendicular distance from the point to the line of action of the force, so $|\mathbf{M}| = D|\mathbf{F}|$, where D is the perpendicular distance between the lines of action of the two forces (Fig. 4.27e). The cross product $\mathbf{r} \times \mathbf{F}$ is perpendicular to $\mathbf{r}$ and $\mathbf{F}$, which means that $\mathbf{M}$ is perpendicular to the plane containing $\mathbf{F}$ and $-\mathbf{F}$ (Fig. 4.27f). Pointing the thumb of the right hand in the direction of $\mathbf{M}$, the arc of the fingers indicates the direction of the moment.

In Fig. 4.28a, our view is perpendicular to the plane containing the two forces. The distance between the lines of action of the forces is 4 m, so the magnitude of the moment of the couple is $|\mathbf{M}| = (4 \text{ m})(2 \text{ kN}) = 8$ kN-m. The moment $\mathbf{M}$ is perpendicular to the plane containing the two forces. Pointing the arc of the fingers of the right hand counterclockwise, we find that the right-hand rule indicates that $\mathbf{M}$ points out of the page. Therefore, the moment of the couple is

$$\mathbf{M} = 8\mathbf{k} \text{ (kN-m)}.$$

We can also determine the moment of the couple by calculating the sum of the moments of the two forces about *any* point. The sum of the moments of the forces about the origin O is (Fig. 4.28b)

$$\mathbf{M} = [\mathbf{r}_1 \times (2\mathbf{j})] + [\mathbf{r}_2 \times (-2\mathbf{j})]$$
$$= [(7\mathbf{i} + 2\mathbf{j}) \times (2\mathbf{j})] + [(3\mathbf{i} + 7\mathbf{j}) \times (-2\mathbf{j})]$$
$$= 8\mathbf{k} \text{ (kN-m)}.$$

In a two-dimensional situation like this example, it isn't convenient to represent a couple by showing the moment vector, because the vector is perpendicular to the page. Instead, we represent the couple by showing its magnitude and a circular arrow that indicates its direction (Fig. 4.28c).

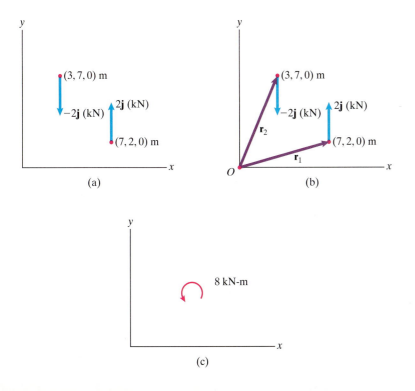

(a)

(b)

(c)

Figure 4.28
(a) A couple consisting of 2-kN forces.
(b) Determining the sum of the moments of the forces about O.
(c) Representing a couple in two dimensions.

By grasping a bar and twisting it (Fig. 4.29a), a moment can be exerted about its axis (Fig. 4.29b). Although the system of forces exerted is distributed over the surface of the bar in a complicated way, the effect is the same as if two equal and opposite forces are exerted (Fig. 4.29c). When we represent a couple as in Fig. 4.29b, or by showing the moment vector **M**, we imply that some system of forces exerts that moment. The system of forces (such as the forces exerted in twisting the bar, or the forces on the crankshaft that exert a moment on the drive shaft of a car) is nearly always more complicated than two equal and opposite forces, but the effect is the same. For this reason, we can *model* the actual system as a simple system of two forces.

Figure 4.29
(a) Twisting a bar.
(b) The moment about the axis of the bar.
(c) The same effect is obtained by applying two equal and opposite forces.

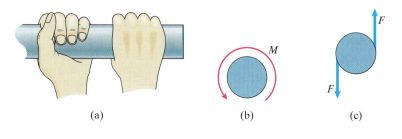

(a)   (b)   (c)

Study Questions

1. A couple consists of forces **F** and −**F** that have different lines of action. How can you determine the moment of the couple about a point P?

2. If you know the moment of a couple about a point P, what do you know about the moment of the couple about a different point P'?

3. A couple consists of forces **F** and −**F** that have different lines of action. The perpendicular distance between the lines of action is D. What is the magnitude of the moment of the couple?

Example 4.10   Determining the Moment of a Couple

The force **F** in Fig. 4.30 is $10\mathbf{i} - 4\mathbf{j}$ (N). Determine the moment of the couple and represent it as shown in Fig. 4.29b.

Strategy

We can determine the moment in two ways: We can calculate the sum of the moments of the forces about a point, or we can sum the moments of the two couples formed by the x and y components of the forces.

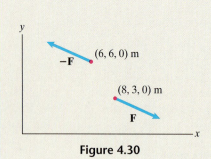

Figure 4.30

Solution

First Method   If we calculate the sum of the moments of the forces about a point on the line of action of one of the forces, the moment of that force is zero and we only need to calculate the moment of the other force. Choosing the point of application of **F** (Fig. a), we calculate the moment as

$$\mathbf{M} = \mathbf{r} \times (-\mathbf{F}) = (-2\mathbf{i} + 3\mathbf{j}) \times (-10\mathbf{i} + 4\mathbf{j}) = 22\mathbf{k} \text{ (N-m)}.$$

Second Method   The *x* and *y* components of the forces form two couples (Fig. b). We determine the moment of the original couple by summing the moments of the couples formed by the components:

Consider the 10-N couple. The magnitude of its moment is $(3\text{ m})(10\text{ N}) = 30$ N-m, and its direction is counterclockwise, indicating that the moment vector points out of the page. Therefore the moment is $30\mathbf{k}$ N-m.

The 4-N couple causes a moment of magnitude $(2\text{ m})(4\text{ N}) = 8$ N-m and its direction is clockwise, so the moment is $-8\mathbf{k}$ N-m. The moment of the original couple is

$$\mathbf{M} = 30\mathbf{k} - 8\mathbf{k} = 22\mathbf{k} \text{ (N-m)}.$$

Its magnitude is 22 N-m and its direction is counterclockwise (Fig. c).

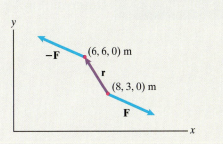

(a) Determining the moment about the point of application of **F**.

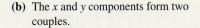

(b) The *x* and *y* components form two couples.

Critical Thinking

In the first method, how do you choose the point about which you sum the moments of the two forces? The answer is that you can use *any* point. We chose the point of application of the force **F** so that the moment due to **F** would be zero, and we would only need to calculate the moment due to the force $-\mathbf{F}$. But if we had chosen any other point we would have obtained the same result. For example, the sum of the moments about the point *P* in Fig. d is

$$\mathbf{M} = (\mathbf{r}_1 \times \mathbf{F}) + [\mathbf{r}_2 \times (-\mathbf{F})]$$

$$= \begin{vmatrix} \mathbf{i} & \mathbf{j} & \mathbf{k} \\ -2 & -4 & -3 \\ 10 & -4 & 0 \end{vmatrix} + \begin{vmatrix} \mathbf{i} & \mathbf{j} & \mathbf{k} \\ -4 & -1 & -3 \\ -10 & 4 & 0 \end{vmatrix}$$

$$= 22\mathbf{k} \text{ (N-m)}.$$

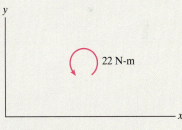

(c) Representing the moment.

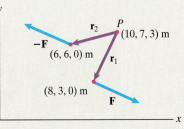

(d) Determining the moment about *P*.

Example 4.11   **Determining Unknown Forces**

Two forces A and B and a 200 ft-lb couple act on the beam in Fig. 4.31. The sum of the forces is zero, and the sum of the moments about the left end of the beam is zero. What are the forces A and B?

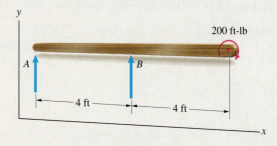

Figure 4.31

Strategy

By summing the two forces (the couple exerts no net force on the beam) and summing the moments due to the forces and the couple about the left end of the beam, we will obtain two equations in terms of the two unknown forces.

Solution

The sum of the forces is

$$\Sigma\, F_y = A + B = 0.$$

The moment of the couple (200 ft-lb clockwise) is the same about any point, so the sum of the moments about the left end of the beam is

$$\Sigma\, M_{\text{left end}} = (4\ \text{ft})\, B - 200\ \text{ft-lb} = 0.$$

The forces are $B = 50$ lb and $A = -50$ lb.

Critical Thinking

Notice that the total moment about the left end of the beam is the sum of the moment due to the force B and the moment due to the 200 ft-lb couple. As we observe in Chapter 5, if an object subjected to forces and couples is in equilibrium, the sum of the forces is zero and the sum of the moments about any point, *including moments due to couples*, is zero. In this example we needed both these conditions to determine the unknown forces A and B.

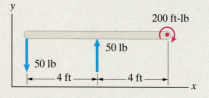

(a) The forces on the beam form a couple.

Example 4.12   Sum of the Moments Due to Two Couples

Determine the sum of the moments exerted on the pipe in Fig. 4.32 by the two couples.

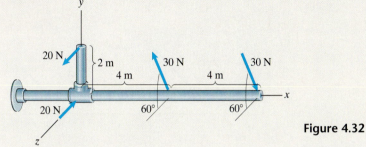

Figure 4.32

Strategy

We will express the moment exerted by each couple as a vector. To express the 30-N couple in terms of a vector, we will express the forces in terms of their components. We can then sum the moment vectors to determine the sum of the moments exerted by the couples.

Solution

Consider the 20-N couple. The magnitude of the moment of the couple is $(2 \text{ m})(20 \text{ N}) = 40$ N-m. The direction of the moment vector is perpendicular to the y–z plane, and the right-hand rule indicates that it points in the positive x axis direction. The moment of the 20-N couple is $40\mathbf{i}$ (N-m).

By resolving the 30-N forces into y and z components, we obtain the two couples in Fig. a. The moment of the couple formed by the y components is $-(30 \sin 60°)(4)\mathbf{k}$ (N-m), and the moment of the couple formed by the z components is $(30 \cos 60°)(4)\mathbf{j}$ (N-m).

The sum of the moments is therefore

$$\Sigma\, \mathbf{M} = 40\mathbf{i} + (30 \cos 60°)(4)\mathbf{j} - (30 \sin 60°)(4)\mathbf{k} \text{ (N-m)}$$

$$= 40\mathbf{i} + 60\mathbf{j} - 104\mathbf{k} \text{ (N-m)}.$$

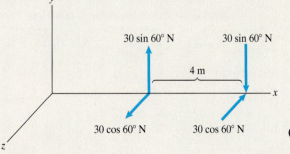

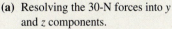

(a) Resolving the 30-N forces into y and z components.

Critical Thinking

Although the method we used in this example helps you recognize the contributions of the individual couples to the sum of the moments, it is convenient only when the orientations of the forces and their points of application relative to the coordinate system are fairly simple. When that is not the case, you can determine the sum of the moments by choosing any point and calculating the sum of the moments of the forces about that point.

Problems

4.108 Determine the moment of the couple (a) about the origin O; (b) about the point with coordinates $x = -2$ m, $y = 4$ m.

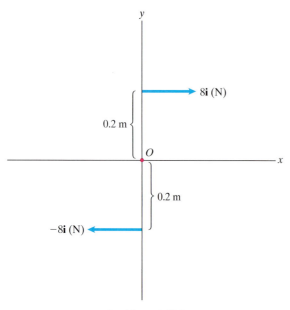

Problem 4.108

4.109 The forces are contained in the x–y plane.

(a) Determine the moment of the couple and represent it as shown in Fig. 4.28c.

(b) What is the sum of the moments of the two forces about the point $(10, -40, 20)$ ft?

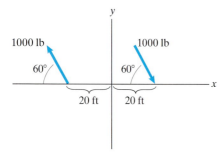

Problem 4.109

4.110 The moment of the couple is $600\mathbf{k}$ (N-m). What is the angle α?

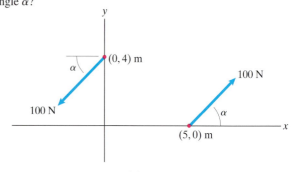

Problem 4.110

4.111 Point P is contained in the x–y plane, $|\mathbf{F}| = 100$ N, and the moment of the couple is $-500\mathbf{k}$ (N-m). What are the coordinates of P?

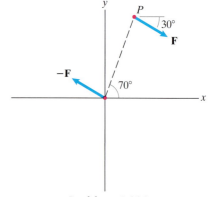

Problem 4.111

4.112 The forces are contained in the x–y plane.

(a) Determine the sum of the moments of the two couples.

(b) What is the sum of the moments of the four forces about the point $(-6, -6, 2)$ m?

(c) Represent the result of (a) as shown in Fig. 4.28c.

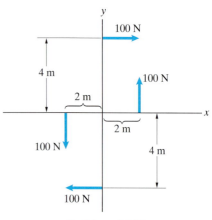

Problem 4.112

4.113 The moment of the couple is 40 kN-m counterclockwise.
(a) Express the moment of the couple as a vector.
(b) Draw a sketch showing an example of two equal and opposite forces that exert the given moment.

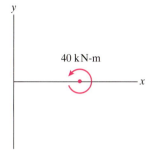

Problem 4.113

4.114 The moments of two couples are shown. What is the sum of the moments about point P?

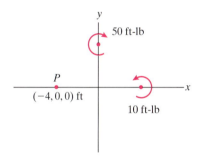

Problem 4.114

4.115 Determine the sum of the moments exerted on the plate by the two couples.

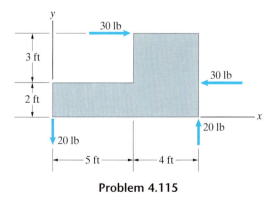

Problem 4.115

4.116 Determine the sum of the moments exerted about A by the couple and the two forces.

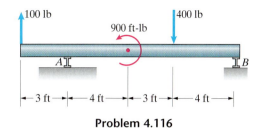

Problem 4.116

4.117 Determine the sum of the moments exerted about A by the couple and the two forces.

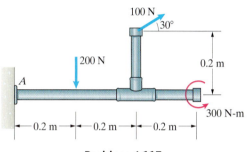

Problem 4.117

4.118 The sum of the moments about point A due to the forces and couples acting on the bar is zero.
(a) What is the magnitude of the couple C?
(b) Determine the sum of the moments about point B due to the forces and couples acting on the bar.

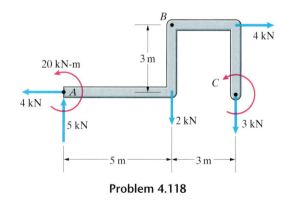

Problem 4.118

4.119 Four forces and a couple act on the beam. The vector sum of the forces is zero, and the sum of the moments about the left end of the beam is zero. What are the forces A_x, A_y, and B?

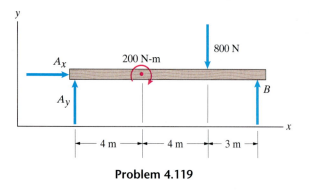

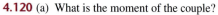

Problem 4.119

4.120 (a) What is the moment of the couple?
(b) Determine the perpendicular distance between the lines of action of the two forces.

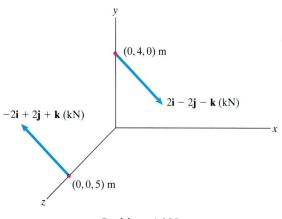

Problem 4.120

4.121 Determine the sum of the moments exerted on the plate by the three couples. (The 80-lb forces are contained in the x–z plane.)

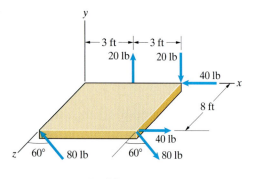

Problem 4.121

4.122 What is the magnitude of the sum of the moments exerted on the T-shaped structure by the two couples?

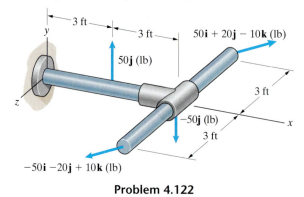

Problem 4.122

4.123 The tension in cables AB and CD is 500 N.
(a) Show that the two forces exerted by the cables on the rectangular hatch at B and C form a couple.
(b) What is the moment exerted on the plate by the cables?

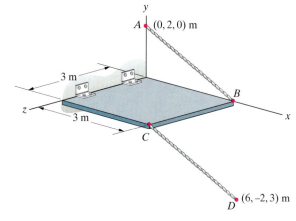

Problem 4.123

4.124 The cables *AB* and *CD* exert a couple on the vertical pipe. The tension in each cable is 8 kN. Determine the magnitude of the moment the cables exert on the pipe.

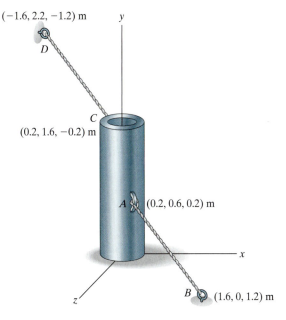

Problem 4.124

4.125 The bar is loaded by the forces

$$F_B = 2i + 6j + 3k \text{ (kN)},$$

$$F_C = i - 2j + 2k \text{ (kN)},$$

and the couple

$$M_C = 2i + j - 2k \text{ (kN-m)}.$$

Determine the sum of the moments of the two forces and the couple about *A*.

4.126 The forces

$$F_B = 2i + 6j + 3k \text{ (kN)},$$

$$F_C = i - 2j + 2k \text{ (kN)},$$

and the couple

$$M_C = M_{Cy}j + M_{Cz}k \text{ (kN-m)}.$$

Determine the values of M_{Cy} and M_{Cz} so that the sum of the moments of the two forces and the couple about *A* is zero.

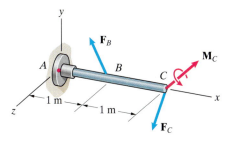

Problems 4.125/4.126

4.127 Two wrenches are used to tighten an elbow fitting. The force $F = 10k$ (lb) on the right wrench is applied at $(6, -5, -3)$ in, and the force $-F$ on the left wrench is applied at $(4, -5, 3)$ in.

(a) Determine the moment about the *x* axis due to the force exerted on the right wrench.

(b) Determine the moment of the couple formed by the forces exerted on the two wrenches.

(c) Based on the results of (a) and (b), explain why two wrenches are used.

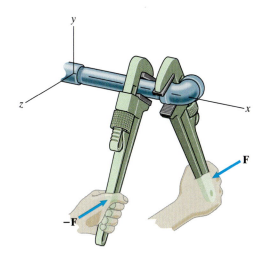

Problem 4.127

4.5 Equivalent Systems

A *system of forces and moments* is simply a particular set of forces and moments of couples. The systems of forces and moments dealt with in engineering can be complicated. This is especially true in the case of distributed forces, such as the pressure forces exerted by water on a dam. Fortunately, if we are concerned only with the total force and moment exerted, we can represent complicated systems of forces and moments by much simpler systems.

Conditions for Equivalence

We define two systems of forces and moments, designated as system 1 and system 2, to be *equivalent* if the sums of the forces are equal, or

$$(\Sigma \mathbf{F})_1 = (\Sigma \mathbf{F})_2, \tag{4.7}$$

and the sums of the moments about a point P are equal, or

$$(\Sigma \mathbf{M}_P)_1 = (\Sigma \mathbf{M}_P)_2. \tag{4.8}$$

Demonstration of Equivalence

To see what the conditions for equivalence mean, consider the systems of forces and moments in Fig. 4.33a. In system 1, an object is subjected to two forces $\mathbf{F}_A$ and $\mathbf{F}_B$ and a couple $\mathbf{M}_C$. In system 2, the object is subjected to a force $\mathbf{F}_D$ and two couples $\mathbf{M}_E$ and $\mathbf{M}_F$. The first condition for equivalence is

$$(\Sigma \mathbf{F})_1 = (\Sigma \mathbf{F})_2:$$

$$\mathbf{F}_A + \mathbf{F}_B = \mathbf{F}_D. \tag{4.9}$$

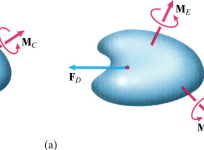

(a)

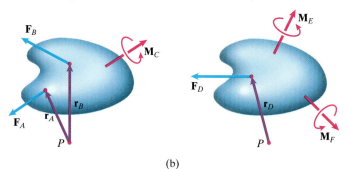

(b)

Figure 4.33
(a) Different systems of forces and moments applied to an object.
(b) Determining the sum of the moments about a point P for each system.

If we determine the sums of the moments about the point P in Fig. 4.33b, the second condition for equivalence is

$$(\Sigma \, \mathbf{M}_P)_1 = (\Sigma \, \mathbf{M}_P)_2:$$

$$(\mathbf{r}_A \times \mathbf{F}_A) + (\mathbf{r}_B \times \mathbf{F}_B) + \mathbf{M}_C = (\mathbf{r}_D \times \mathbf{F}_D) + \mathbf{M}_E + \mathbf{M}_F. \quad (4.10)$$

If these conditions are satisfied, systems 1 and 2 are equivalent.

We will use this example to demonstrate that *if the sums of the forces are equal for two systems of forces and moments and the sums of the moments about one point P are equal, then the sums of the moments about any point are equal.* Suppose that Eq. (4.9) is satisfied, and Eq. (4.10) is satisfied for the point P in Fig. 4.33b. For a different point P' (Fig. 4.34), we will show that

$$(\Sigma \, \mathbf{M}_{P'})_1 = (\Sigma \, \mathbf{M}_{P'})_2:$$

$$(\mathbf{r}_A' \times \mathbf{F}_A) + (\mathbf{r}_B' \times \mathbf{F}_B) + \mathbf{M}_C = (\mathbf{r}_D' \times \mathbf{F}_D) + \mathbf{M}_E + \mathbf{M}_F. \quad (4.11)$$

In terms of the vector $\mathbf{r}$ from P' to P, the relations between the vectors $\mathbf{r}_A'$, $\mathbf{r}_B'$, and $\mathbf{r}_D'$ in Fig. 4.34 and the vectors $\mathbf{r}_A$, $\mathbf{r}_B$, and $\mathbf{r}_D$ in Fig. 4.33b are

$$\mathbf{r}_A' = \mathbf{r} + \mathbf{r}_A, \qquad \mathbf{r}_B' = \mathbf{r} + \mathbf{r}_B, \qquad \mathbf{r}_D' = \mathbf{r} + \mathbf{r}_D.$$

Substituting these expressions into Eq. (4.11), we obtain

$$[(\mathbf{r} + \mathbf{r}_A) \times \mathbf{F}_A] + [(\mathbf{r} + \mathbf{r}_B) \times \mathbf{F}_B] + \mathbf{M}_C$$
$$= [(\mathbf{r} + \mathbf{r}_D) \times \mathbf{F}_D] + \mathbf{M}_E + \mathbf{M}_F.$$

Rearranging terms, we can write this equation as

$$[\mathbf{r} \times (\Sigma \, \mathbf{F})_1] + (\Sigma \, \mathbf{M}_P)_1 = [\mathbf{r} \times (\Sigma \, \mathbf{F})_2] + (\Sigma \, \mathbf{M}_P)_2,$$

which holds in view of Eqs. (4.9) and (4.10). The sums of the moments of the two systems about any point are equal.

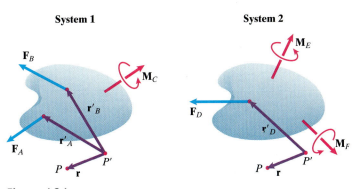

System 1   **System 2**

Figure 4.34
Determining the sums of the moments about a different point P'.

Study Questions

1. What conditions must be satisfied for two systems of forces and moments to be equivalent?

2. If the sums of the forces in two systems of forces and moments are the same, and the sums of the moments about a point P are the same, what do you know about the sums of the moments about a different point P'?

Example 4.13   **Determining Whether Systems Are Equivalent**

Three systems of forces and moments act on the beam in Fig. 4.35. Are they equivalent?

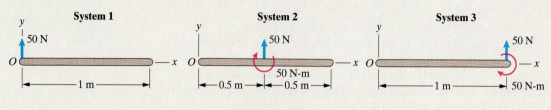

Figure 4.35

Strategy

We must check the two conditions for equivalence. Two systems are defined to be equivalent only if the sums of the forces are equal and the sums of the moments about a point are equal.

Solution

Are the Sums of the Forces Equal?   The sums of the forces are

$$(\Sigma \mathbf{F})_1 = 50\mathbf{j} \text{ (N)},$$

$$(\Sigma \mathbf{F})_2 = 50\mathbf{j} \text{ (N)},$$

$$(\Sigma \mathbf{F})_3 = 50\mathbf{j} \text{ (N)}.$$

Are the Sums of the Moments About an Arbitrary Point Equal?   The sums of the moments about the origin O are

$$(\Sigma M_O)_1 = 0,$$

$$(\Sigma M_O)_2 = (50 \text{ N})(0.5 \text{ m}) - (50 \text{ N-m}) = -25 \text{ N-m},$$

$$(\Sigma M_O)_3 = (50 \text{ N})(1 \text{ m}) - (50 \text{ N-m}) = 0.$$

Systems 1 and 3 are equivalent.

Critical Thinking

In determining whether systems of forces and moments are equivalent, you can calculate the sums of the moments about any point. In this example, the sums of the moments about the right end of the beam are

$$(\Sigma M_{\text{right end}})_1 = -(50 \text{ N})(1 \text{ m}) = -50 \text{ N-m},$$

$$(\Sigma M_{\text{right end}})_2 = -(50 \text{ N})(0.5 \text{ m}) - (50 \text{ N-m}) = -75 \text{ N-m},$$

$$(\Sigma M_{\text{right end}})_3 = -50 \text{ N-m}.$$

Example 4.14   Determining Whether Systems Are Equivalent

Two systems of forces and moments act on the rectangular plate in Fig. 4.36. Are they equivalent?

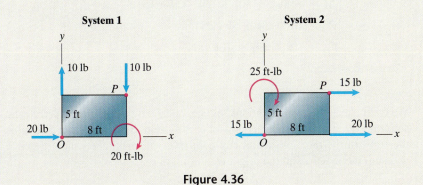

Figure 4.36

Strategy
We must check the two conditions for equivalence by summing the forces acting on the plate and summing the moments due to the forces and couple acting on the plate for each system.

Solution
Are the Sums of the Forces Equal?   The sums of the forces are

$$(\Sigma \, \mathbf{F})_1 = 20\mathbf{i} + 10\mathbf{j} - 10\mathbf{j} = 20\mathbf{i} \text{ (lb)},$$

$$(\Sigma \, \mathbf{F})_2 = 20\mathbf{i} + 15\mathbf{i} - 15\mathbf{i} = 20\mathbf{i} \text{ (lb)}.$$

Are the Sums of the Moments About an Arbitrary Point Equal?   The sums of the moments about the origin O are

$$(\Sigma \, M_O)_1 = -(8 \text{ ft})(10 \text{ lb}) - (20 \text{ ft-lb}) = -100 \text{ ft-lb},$$

$$(\Sigma \, M_O)_2 = -(5 \text{ ft})(15 \text{ lb}) - (25 \text{ ft-lb}) = -100 \text{ ft-lb}.$$

The systems are equivalent.

Critical Thinking
Why is it worth knowing that two systems of forces and moments are equivalent? From the standpoint of statics, it tells you that both systems exert the same total force and the same total moment about any point. In dynamics, it is shown that equivalent systems of forces and moments acting on an object that can be modeled as a rigid body result in the same motion of the object.

Example 4.15   Determining Whether Systems Are Equivalent

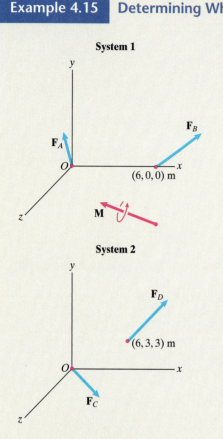

System 1

System 2

Figure 4.37

Two systems of forces and moments are shown in Fig. 4.37, where

$$\mathbf{F}_A = -10\mathbf{i} + 10\mathbf{j} - 15\mathbf{k} \text{ (kN)},$$
$$\mathbf{F}_B = 30\mathbf{i} + 5\mathbf{j} + 10\mathbf{k} \text{ (kN)},$$
$$\mathbf{M} = -90\mathbf{i} + 150\mathbf{j} + 60\mathbf{k} \text{ (kN-m)},$$
$$\mathbf{F}_C = 10\mathbf{i} - 5\mathbf{j} + 5\mathbf{k} \text{ (kN)},$$
$$\mathbf{F}_D = 10\mathbf{i} + 20\mathbf{j} - 10\mathbf{k} \text{ (kN)}.$$

Are they equivalent?

Strategy

We must check the two conditions for equivalence by summing the forces and summing the moments about a point for each system. Because we can determine the sums of the moments about any point, we will make the simplest choice and sum moments about the origin O.

Solution

Are the Sums of the Forces Equal?   The sums of the forces are

$$(\Sigma \mathbf{F})_1 = \mathbf{F}_A + \mathbf{F}_B = 20\mathbf{i} + 15\mathbf{j} - 5\mathbf{k} \text{ (kN)}.$$
$$(\Sigma \mathbf{F})_2 = \mathbf{F}_C + \mathbf{F}_D = 20\mathbf{i} + 15\mathbf{j} - 5\mathbf{k} \text{ (kN)}.$$

Are the Sums of the Moments About an Arbitrary Point Equal?   The sum of the moments about the origin O in system 1 is

$$(\Sigma \mathbf{M}_O)_1 = (6\mathbf{i} \times \mathbf{F}_B) + \mathbf{M}$$

$$= \begin{vmatrix} \mathbf{i} & \mathbf{j} & \mathbf{k} \\ 6 & 0 & 0 \\ 30 & 5 & 10 \end{vmatrix} + (-90\mathbf{i} + 150\mathbf{j} + 60\mathbf{k})$$

$$= -90\mathbf{i} + 90\mathbf{j} + 90\mathbf{k} \text{ (kN-m)}.$$

The sum of the moments about O in system 2 is

$$(\Sigma \mathbf{M}_O)_2 = (6\mathbf{i} + 3\mathbf{j} + 3\mathbf{k}) \times \mathbf{F}_D = \begin{vmatrix} \mathbf{i} & \mathbf{j} & \mathbf{k} \\ 6 & 3 & 3 \\ 10 & 20 & -10 \end{vmatrix}$$

$$= -90\mathbf{i} + 90\mathbf{j} + 90\mathbf{k} \text{ (kN-m)}.$$

The systems are equivalent.

Critical Thinking

System 1 consists of two forces and a couple. System 2, consisting of two forces, is simpler. Yet the two systems are equivalent, exerting the same total force and the same total moment about any point. If system 1 acts on an object and the total force and total moment it exerts are the only things that concern us, we could assume that the object is subjected to the simpler system 2 instead. Clearly it would be easier to deal with. This raises an interesting question: Could an even simpler system be found that is equivalent to system 2 (which means it would also be equivalent to system 1)? We discuss these ideas in the following section.

4.6 Representing Systems by Equivalent Systems

If we are concerned only with the total force and total moment exerted on an object by a given system of forces and moments, we can *represent* the system by an equivalent one. By this we mean that instead of showing the actual forces and couples acting on an object, we would show a different system that exerts the same total force and moment. In this way, we can replace a given system by a less complicated one to simplify the analysis of the forces and moments acting on an object and to gain a better intuitive understanding of their effects on the object.

Representing a System by a Force and a Couple

Let's consider an arbitrary system of forces and moments and a point P (system 1 in Fig. 4.38). We can represent this system by one consisting of a single force acting at P and a single couple (system 2). The conditions for equivalence are

$$(\Sigma \mathbf{F})_2 = (\Sigma \mathbf{F})_1:$$

$$\mathbf{F} = (\Sigma \mathbf{F})_1$$

and

$$(\Sigma \mathbf{M}_P)_2 = (\Sigma \mathbf{M}_P)_1:$$

$$\mathbf{M} = (\Sigma \mathbf{M}_P)_1.$$

These conditions are satisfied if $\mathbf{F}$ equals the sum of the forces in system 1 and $\mathbf{M}$ equals the sum of the moments about P in system 1.

Thus *no matter how complicated a system of forces and moments may be, it can be represented by a single force acting at a given point and a single couple*. Three particular cases occur frequently in practice:

Representing a Force by a Force and a Couple   A force $\mathbf{F}_P$ acting at a point P (system 1 in Fig. 4.39a) can be represented by a force $\mathbf{F}$ acting at a different point Q and a couple $\mathbf{M}$ (system 2). The moment of system 1 about point Q is $\mathbf{r} \times \mathbf{F}_P$, where $\mathbf{r}$ is the vector from Q to P (Fig. 4.39b). The conditions for equivalence are

$$(\Sigma \mathbf{F})_2 = (\Sigma \mathbf{F})_1:$$

$$\mathbf{F} = \mathbf{F}_P$$

and

$$(\Sigma \mathbf{M}_Q)_2 = (\Sigma \mathbf{M}_Q)_1:$$

$$\mathbf{M} = \mathbf{r} \times \mathbf{F}_P.$$

The systems are equivalent if the force $\mathbf{F}$ equals the force $\mathbf{F}_P$ and the couple $\mathbf{M}$ equals the moment of $\mathbf{F}_P$ about Q.

Concurrent Forces Represented by a Force

A system of concurrent forces whose lines of action intersect at a point P (system 1 in Fig. 4.40) can be represented by a single force whose line of action intersects P (system 2). The sums of the forces in the two systems are equal if

$$\mathbf{F} = \mathbf{F}_1 + \mathbf{F}_2 + \cdots + \mathbf{F}_N.$$

The sum of the moments about P equals zero for each system, so the systems are equivalent if the force $\mathbf{F}$ equals the sum of the forces in system 1.

System 1

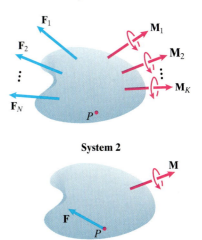

System 2

Figure 4.38
(a) An arbitrary system of forces and moments.
(b) A force acting at P and a couple.

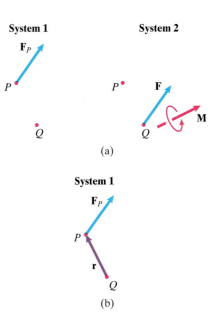

(a)

(b)

Figure 4.39
(a) System 1 is a force $\mathbf{F}_P$ acting at point P. System 2 consists of a force $\mathbf{F}$ acting at point Q and a couple $\mathbf{M}$.
(b) Determining the moment of system 1 about point Q.

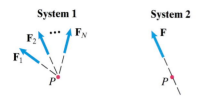

Figure 4.40
A system of concurrent forces and a
system consisting of a single force **F**.

Parallel Forces Represented by a Force   A system of parallel forces whose sum is not zero can be represented by a single force **F** (Fig. 4.41). We demonstrate this result in Example 4.19.

Study Questions

1. If you represent a system of forces and moments by a force **F** acting at a point *P* and a couple **M**, how do you determine **F** and **M**?

2. If you represent a system of concurrent forces by a single force **F**, what condition must be satisfied by the line of action of **F**?

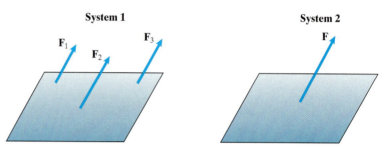

Figure 4.41
A system of parallel forces and a system
consisting of a single force **F**.

Example 4.16   Representing a Force by a Force and Couple

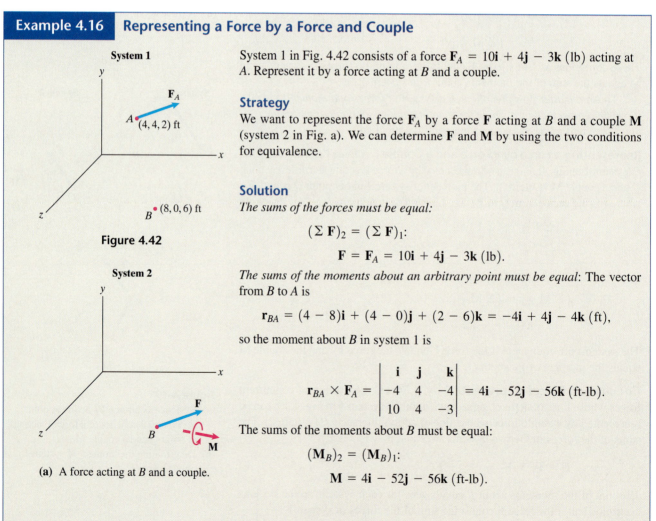

Figure 4.42

System 1 in Fig. 4.42 consists of a force $\mathbf{F}_A = 10\mathbf{i} + 4\mathbf{j} - 3\mathbf{k}$ (lb) acting at *A*. Represent it by a force acting at *B* and a couple.

Strategy

We want to represent the force $\mathbf{F}_A$ by a force **F** acting at *B* and a couple **M** (system 2 in Fig. a). We can determine **F** and **M** by using the two conditions for equivalence.

Solution

The sums of the forces must be equal:

$$(\Sigma\, \mathbf{F})_2 = (\Sigma\, \mathbf{F})_1\text{:}$$

$$\mathbf{F} = \mathbf{F}_A = 10\mathbf{i} + 4\mathbf{j} - 3\mathbf{k}\ \text{(lb)}.$$

The sums of the moments about an arbitrary point must be equal: The vector from *B* to *A* is

$$\mathbf{r}_{BA} = (4 - 8)\mathbf{i} + (4 - 0)\mathbf{j} + (2 - 6)\mathbf{k} = -4\mathbf{i} + 4\mathbf{j} - 4\mathbf{k}\ \text{(ft)},$$

so the moment about *B* in system 1 is

$$\mathbf{r}_{BA} \times \mathbf{F}_A = \begin{vmatrix} \mathbf{i} & \mathbf{j} & \mathbf{k} \\ -4 & 4 & -4 \\ 10 & 4 & -3 \end{vmatrix} = 4\mathbf{i} - 52\mathbf{j} - 56\mathbf{k}\ \text{(ft-lb)}.$$

The sums of the moments about *B* must be equal:

$$(\mathbf{M}_B)_2 = (\mathbf{M}_B)_1\text{:}$$

$$\mathbf{M} = 4\mathbf{i} - 52\mathbf{j} - 56\mathbf{k}\ \text{(ft-lb)}.$$

(a) A force acting at *B* and a couple.

Critical Thinking

In the equivalent system in Fig. a, why didn't we specify the point of application of the couple **M**? The reason is that the system is equivalent no matter where **M** acts, because the moment due to a couple is the same about any point. But there is a caveat to this freedom of choice. If the force $\mathbf{F}_A$ in Fig. 4.42 acts on an object, the system in Fig. a can be considered equivalent only if the force **F** and the couple **M** both act on the object.

Example 4.17   Representing a System by a Simpler Equivalent System

System 1 in Fig. 4.43 consists of two forces and a couple acting on a pipe. Represent system 1 by (a) a single force acting at the origin O of the coordinate system and a single couple and (b) a single force.

Strategy

(a) We can represent system 1 by a force **F** acting at the origin and a couple M (system 2 in Fig. a) and use the conditions for equivalence to determine **F** and **M**.
(b) Suppose that we place the force **F** with its point of application a distance D along the x axis (system 3 in Fig. b). The sums of the forces in systems 2 and 3 are equal. If we can choose the distance D so that the moment about O in system 3 equals **M**, system 3 will be equivalent to system 2 and therefore equivalent to system 1.

System 1

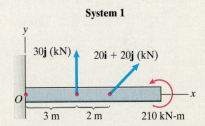

Figure 4.43

Solution

(a) The conditions for equivalence are

$$(\Sigma\,\mathbf{F})_2 = (\Sigma\,\mathbf{F})_1:$$

$$\mathbf{F} = 30\mathbf{j} + (20\mathbf{i} + 20\mathbf{j})\ (\text{kN}) = 20\mathbf{i} + 50\mathbf{j}\ (\text{kN}),$$

and

$$(\Sigma\,M_O)_2 = (\Sigma\,M_O)_1:$$

$$M = (30\ \text{kN})(3\ \text{m}) + (20\ \text{kN})(5\ \text{m}) + 210\ \text{kN-m}$$

$$= 400\ \text{kN-m}.$$

(b) The sums of the forces in systems 2 and 3 are equal. Equating the sums of the moments about O yields

$$(\Sigma\,M_O)_3 = (\Sigma\,M_O)_2:$$

$$(50\ \text{kN})D = 400\ \text{kN-m},$$

and we find that system 3 is equivalent to system 2 if $D = 8$ m.

System 2

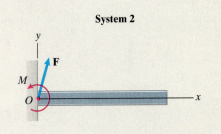

(a) A force **F** acting at O and a couple M.

System 3

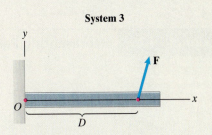

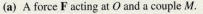

(b) A system consisting of the force **F** acting at a point on the x axis.

Critical Thinking

In part (b), why did we assume that the point of application of the force is on the x axis? In order to represent the system in Fig. a by a single force, we needed to place the line of action of the force so that the force would exert a 400 kN-m counterclockwise moment about O. Placing the point of application of the force a distance D along the x axis was simply a convenient way to accomplish that.

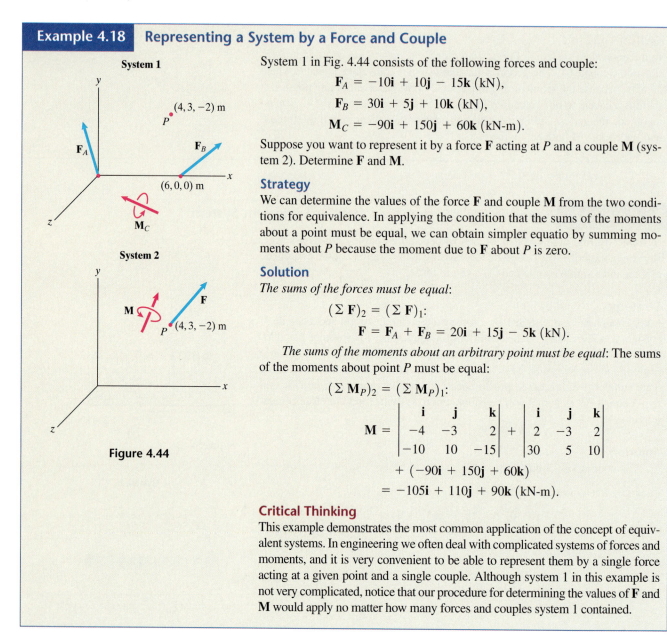

Example 4.18 Representing a System by a Force and Couple

System 1

$(4, 3, -2)$ m
P

F_A

F_B

$(6, 0, 0)$ m

M_C

System 2

F

M

P $(4, 3, -2)$ m

Figure 4.44

System 1 in Fig. 4.44 consists of the following forces and couple:

$$F_A = -10i + 10j - 15k \text{ (kN)},$$
$$F_B = 30i + 5j + 10k \text{ (kN)},$$
$$M_C = -90i + 150j + 60k \text{ (kN-m)}.$$

Suppose you want to represent it by a force **F** acting at P and a couple **M** (system 2). Determine **F** and **M**.

Strategy

We can determine the values of the force **F** and couple **M** from the two conditions for equivalence. In applying the condition that the sums of the moments about a point must be equal, we can obtain simpler equatio by summing moments about P because the moment due to **F** about P is zero.

Solution

The sums of the forces must be equal:

$$(\Sigma \, F)_2 = (\Sigma \, F)_1:$$
$$F = F_A + F_B = 20i + 15j - 5k \text{ (kN)}.$$

The sums of the moments about an arbitrary point must be equal: The sums of the moments about point P must be equal:

$$(\Sigma \, M_P)_2 = (\Sigma \, M_P)_1:$$

$$M = \begin{vmatrix} i & j & k \\ -4 & -3 & 2 \\ -10 & 10 & -15 \end{vmatrix} + \begin{vmatrix} i & j & k \\ 2 & -3 & 2 \\ 30 & 5 & 10 \end{vmatrix}$$
$$+ \, (-90i + 150j + 60k)$$
$$= -105i + 110j + 90k \text{ (kN-m)}.$$

Critical Thinking

This example demonstrates the most common application of the concept of equivalent systems. In engineering we often deal with complicated systems of forces and moments, and it is very convenient to be able to represent them by a single force acting at a given point and a single couple. Although system 1 in this example is not very complicated, notice that our procedure for determining the values of **F** and **M** would apply no matter how many forces and couples system 1 contained.

Example 4.19 Representing Parallel Forces by a Single Force

System 1 in Fig. 4.45 consists of parallel forces. Suppose you want to represent it by a force **F** (system 2). What is **F**, and where does its line of action intersect the x–z plane?

Strategy

We can determine **F** from the condition that the sums of the forces in the two systems must be equal. For the two systems to be equivalent, we must choose the point of application P so that the sums of the moments about a point are equal. This condition will tell us where the line of action intersects the x–z plane.

Solution

The sums of the forces must be equal.

$$(\Sigma \mathbf{F})_2 = (\Sigma \mathbf{F})_1:$$

$$\mathbf{F} = 30\mathbf{j} + 20\mathbf{j} - 10\mathbf{j} \text{ (lb)} = 40\mathbf{j} \text{ (lb)}.$$

The sums of the moments about an arbitrary point must be equal: Let the coordinates of point P be (x, y, z). The sums of the moments about the origin O must be equal.

$$(\Sigma \mathbf{M}_O)_2 = (\Sigma \mathbf{M}_O)_1:$$

$$\begin{vmatrix} \mathbf{i} & \mathbf{j} & \mathbf{k} \\ x & y & z \\ 0 & 40 & 0 \end{vmatrix} = \begin{vmatrix} \mathbf{i} & \mathbf{j} & \mathbf{k} \\ 6 & 0 & 2 \\ 0 & 30 & 0 \end{vmatrix} + \begin{vmatrix} \mathbf{i} & \mathbf{j} & \mathbf{k} \\ 2 & 0 & 4 \\ 0 & -10 & 0 \end{vmatrix}$$

$$+ \begin{vmatrix} \mathbf{i} & \mathbf{j} & \mathbf{k} \\ -3 & 0 & -2 \\ 0 & 20 & 0 \end{vmatrix}$$

Expanding the determinants, we obtain

$$[20 \text{ ft-lb} + (40 \text{ lb})z]\mathbf{i} + [100 \text{ ft-lb} - (40 \text{ lb})x]\mathbf{k} = \mathbf{0}.$$

The sums of the moments about the origin are equal if

$$x = 2.5 \text{ ft},$$
$$z = -0.5 \text{ ft}.$$

The systems are equivalent if $\mathbf{F} = 40\mathbf{j}$ (lb) and its line of action intersects the x–z plane at $x = 2.5$ ft and $z = -0.5$ ft. Notice that we did not obtain an equation for the y coordinate of P. The systems are equivalent if $\mathbf{F}$ is applied at any point along the line of action.

Critical Thinking

In this example we could have determined the x and z coordinates of point P in a simpler way. Since the sums of the moments about any point must be equal for the systems to be equivalent, the sums of the moments about any *line* must also be equal. Equating the sums of the moments about the x axis yields

$$(\Sigma M_{x\text{ axis}})_2 = (\Sigma M_{x\text{ axis}})_1:$$

$$-(40 \text{ lb})z = -(30 \text{ lb})(2 \text{ ft}) + (10 \text{ lb})(4 \text{ ft}) + (20 \text{ lb})(2 \text{ ft}),$$

and we obtain $z = -0.5$ ft. Also, equating the sums of the moments about the z axis gives

$$(\Sigma M_{z\text{ axis}})_2 = (\Sigma M_{z\text{ axis}})_1:$$

$$(40 \text{ lb})x = (30 \text{ lb})(6 \text{ ft}) - (10 \text{ lb})(2 \text{ ft}) - (20 \text{ lb})(3 \text{ ft}),$$

and we obtain $x = 2.5$ ft.

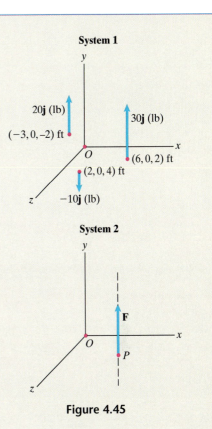

Figure 4.45

Representing a System by a Wrench

We have shown that *any* system of forces and moments can be represented by a single force acting at a given point and a single couple. This raises an interesting question: What is the simplest system that can be equivalent to any system of forces and moments?

To consider this question, let us begin with an arbitrary force $\mathbf{F}$ acting at a point P and an arbitrary couple $\mathbf{M}$ (system 1 in Fig. 4.46a) and see whether we can represent this system by a simpler one. For example, can we represent it by the force $\mathbf{F}$ acting at a different point Q and no couple (Fig 4.46b)? The sum of the forces is the same as in system 1. If we can choose the point Q so that $\mathbf{r} \times \mathbf{F} = \mathbf{M}$, where $\mathbf{r}$ is the vector from P to Q (Fig. 4.46c), the sum of the moments about P is the same as in system 1 and the systems are equivalent. But the vector $\mathbf{r} \times \mathbf{F}$ is perpendicular to $\mathbf{F}$, so it can equal $\mathbf{M}$ only if $\mathbf{M}$ is perpendicular to $\mathbf{F}$. That means that, in general, we can't represent system 1 by the force $\mathbf{F}$ alone.

However, we can represent system 1 by the force $\mathbf{F}$ acting at a point Q and the component of $\mathbf{M}$ that is parallel to $\mathbf{F}$. Figure 4.46d shows system 1 with a coordinate system placed so that $\mathbf{F}$ is along the y axis and $\mathbf{M}$ is contained in the x–y plane. In terms of this coordinate system, we can express the force and cou-

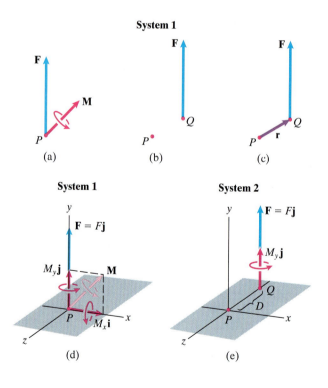

Figure 4.46
(a) System 1 is a single force and a single couple.
(b) Can system 1 be represented by a single force and no couple?
(c) The moment of $\mathbf{F}$ about P is $\mathbf{r} \times \mathbf{F}$.
(d) $\mathbf{F}$ is along the y axis, and $\mathbf{M}$ is contained in the x–y plane.
(e) System 2 is the force $\mathbf{F}$ and the component of $\mathbf{M}$ parallel to $\mathbf{F}$.

ple as $\mathbf{F} = F\mathbf{j}$ and $\mathbf{M} = M_x\mathbf{i} + M_y\mathbf{j}$. System 2 in Fig. 4.46e consists of the force $\mathbf{F}$ acting at a point on the z axis and the component of $\mathbf{M}$ parallel to $\mathbf{F}$. If we choose the distance D so that $D = M_x/F$, system 2 is equivalent to system 1. The sum of the forces in each system is $\mathbf{F}$. The sum of the moments about P in system 1 is $\mathbf{M}$, and the sum of the moments about P in system 2 is

$$(\Sigma\,\mathbf{M}_P)_2 = [(-D\mathbf{k}) \times (F\mathbf{j})] + M_y\mathbf{j} = M_x\mathbf{i} + M_y\mathbf{j} = \mathbf{M}.$$

A force $\mathbf{F}$ and a couple $\mathbf{M}_p$ that is parallel to $\mathbf{F}$ is called a *wrench. It is the simplest system that can be equivalent to an arbitrary system of forces and moments.*

How can we represent a given system of forces and moments by a wrench? If the system is a single force or a single couple or if it consists of a force $\mathbf{F}$ and a couple that is parallel to $\mathbf{F}$, it is a wrench, and we can't simplify it further. If the system is more complicated than a single force and a single couple, we can begin by choosing a convenient point P and representing the system by a force $\mathbf{F}$ acting at P and a couple $\mathbf{M}$ (Fig. 4.47a). Then representing this system by a wrench requires two steps:

1. Determine the components of $\mathbf{M}$ parallel and normal to $\mathbf{F}$ (Fig. 4.47b).
2. The wrench consists of the force $\mathbf{F}$ acting at a point Q and the parallel component $\mathbf{M}_P$ (Fig. 4.47c). To achieve equivalence, the point Q must be chosen so that the moment of $\mathbf{F}$ about P equals the normal component $\mathbf{M}_n$ (Fig. 4.47d)—that is, so that $\mathbf{r}_{PQ} \times \mathbf{F} = \mathbf{M}_n$.

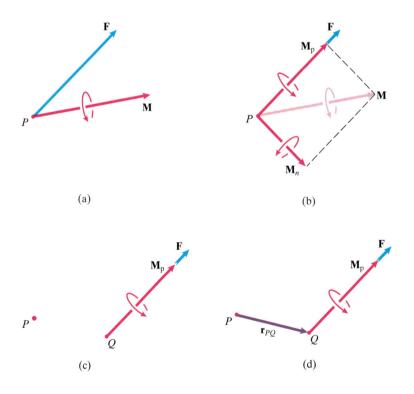

(a)

(b)

(c)

(d)

Figure 4.47
(a) If necessary, first represent the system by a single force and a single couple.
(b) The components of $\mathbf{M}$ parallel and normal to $\mathbf{F}$.
(c) The wrench.
(d) Choose Q so that the moment of $\mathbf{F}$ about P equals the normal component of $\mathbf{M}$.

Example 4.20   **Representing a Force and Couple by a Wrench**

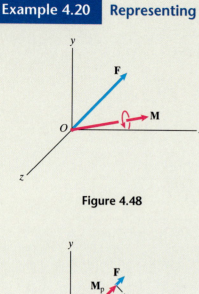

Figure 4.48

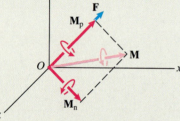

(a) Resolving $\mathbf{M}$ into components parallel and normal to $\mathbf{F}$.

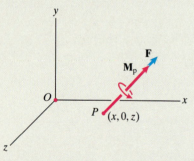

(b) The wrench acting at a point in the x–z plane.

The system in Fig. 4.48 consists of the force and couple

$$\mathbf{F} = 3\mathbf{i} + 6\mathbf{j} + 2\mathbf{k} \text{ (N)},$$

$$\mathbf{M} = 12\mathbf{i} + 4\mathbf{j} + 6\mathbf{k} \text{ (N-m)}.$$

Represent it by a wrench, and determine where the line of action of the wrench's force intersects the x–z plane.

Strategy

The wrench is the force $\mathbf{F}$ and the component of $\mathbf{M}$ parallel to $\mathbf{F}$ (Figs. a, b). We must choose the point of application P so that the moment of $\mathbf{F}$ about O equals the normal component $\mathbf{M}_n$. By letting P be an arbitrary point of the x–z plane, we can determine where the line of action of $\mathbf{F}$ intersects that plane.

Solution

Dividing $\mathbf{F}$ by its magnitude, we obtain a unit vector $\mathbf{e}$ with the same direction as $\mathbf{F}$:

$$\mathbf{e} = \frac{\mathbf{F}}{|\mathbf{F}|} = \frac{3\mathbf{i} + 6\mathbf{j} + 2\mathbf{k} \text{ (N)}}{\sqrt{(3\text{ N})^2 + (6\text{ N})^2 + (2\text{ N})^2}} = 0.429\mathbf{i} + 0.857\mathbf{j} + 0.286\mathbf{k}.$$

We can use $\mathbf{e}$ to calculate the component of $\mathbf{M}$ parallel to $\mathbf{F}$:

$$\mathbf{M}_p = (\mathbf{e} \cdot \mathbf{M})\mathbf{e} = [(0.429)(12\text{ N-m}) + (0.857)(4\text{ N-m}) + (0.286)(6\text{ N-m})]\mathbf{e}$$

$$= 4.408\mathbf{i} + 8.816\mathbf{j} + 2.939\mathbf{k} \text{ (N-m)}.$$

The component of $\mathbf{M}$ normal to $\mathbf{F}$ is

$$\mathbf{M}_n = \mathbf{M} - \mathbf{M}_p = 7.592\mathbf{i} - 4.816\mathbf{j} + 3.061\mathbf{k} \text{ (N-m)}.$$

The wrench is shown in Fig. b. Let the coordinates of P be $(x, 0, z)$. The moment of $\mathbf{F}$ about O is

$$\mathbf{r}_{OP} \times \mathbf{F} = \begin{vmatrix} \mathbf{i} & \mathbf{j} & \mathbf{k} \\ x & 0 & z \\ 3 & 6 & 2 \end{vmatrix} = -6z\mathbf{i} - (2x - 3z)\mathbf{j} + 6x\mathbf{k} \text{ (N-m)}.$$

By equating this moment to $\mathbf{M}_n$, or

$$-6z\mathbf{i} - (2x - 3z)\mathbf{j} + 6x\mathbf{k} \text{ (N-m)} = 7.592\mathbf{i} - 4.816\mathbf{j} + 3.061\mathbf{k} \text{ (N-m)},$$

we obtain the equations

$$-6z = 7.592,$$

$$-2x + 3z = -4.816,$$

$$6x = 3.061.$$

Solving these equations, we find the coordinates of point P are $x = 0.510$ m, $z = -1.265$ m.

Critical Thinking

Why did we place point P at an arbitrary point $(x, 0, z)$ in the x–z plane? Our objective was to place the line of action of the force $\mathbf{F}$ of the wrench so as to satisfy the condition that the moment of $\mathbf{F}$ about O would equal $\mathbf{M}_n$. Placing the point of application of $\mathbf{F}$ at a point $(x, 0, z)$ and then using this condition to determine x and z was a convenient way to determine the necessary location of the line of action. The point $(x, 0, z) = (0.510, 0, -1.265)$ m is the intersection of the line of action with the x–z plane.

Problems

4.128 Two systems of forces act on the beam. Are they equivalent?

Strategy: Check the two conditions for equivalence. The sums of the forces must be equal, and the sums of the moments about an arbitrary point must be equal.

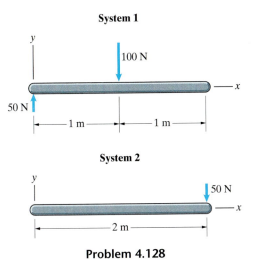

System 1

System 2

Problem 4.128

4.129 Two systems of forces and moments act on the beam. Are they equivalent?

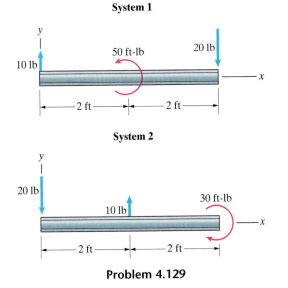

System 1

System 2

Problem 4.129

4.130 Four systems of forces and moments act on an 8-m beam. Which systems are equivalent?

4.131 The four systems can be made equivalent by adding a couple to one of the systems. Which system is it, and what couple must be added?

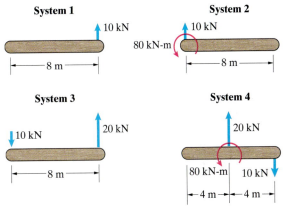

System 1

System 2

System 3

System 4

Problems 4.130/4.131

4.132 System 1 is a force **F** acting at a point O. System 2 is the force **F** acting at a different point O' along the same line of action. Explain why these systems are equivalent. (This simple result is called the *principle of transmissibility*.)

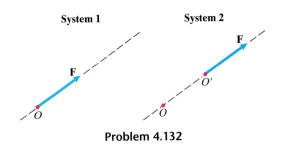

System 1

System 2

Problem 4.132

4.133 The vector sum of the forces exerted on the log by the cables is the same in the two cases. Show that the systems of forces exerted on the log are equivalent.

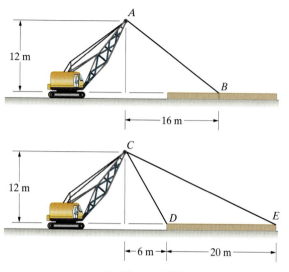

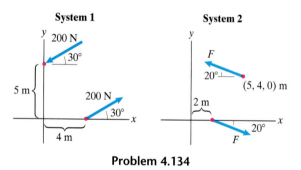

Problem 4.133

4.134 Systems 1 and 2 each consist of a couple. If they are equivalent, what is F?

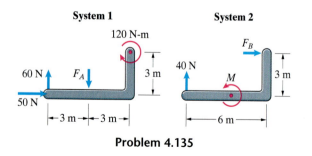

Problem 4.134

4.135 Two equivalent systems of forces and moments act on the L-shaped bar. Determine the forces F_A and F_B and the couple M.

System 1

System 2

Problem 4.135

4.136 Two equivalent systems of forces and moments act on the plate. Determine the force F and the couple M.

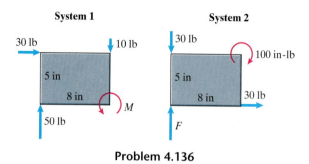

Problem 4.136

4.137 In system 1, four forces act on the rectangular flat plate. The forces are perpendicular to the plate and the 400-kN force acts at its midpoint. In system 2, no forces or couples act on the plate. Systems 1 and 2 are equivalent. What are the forces F_1, F_2, and F_3?

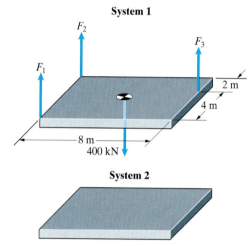

Problem 4.137

4.138 Three forces and a couple are applied to a beam (system 1).
(a) If you represent system 1 by a force applied at A and a couple (system 2), what are **F** and M?
(b) If you represent system 1 by the force **F** (system 3), what is the distance D?

System 1

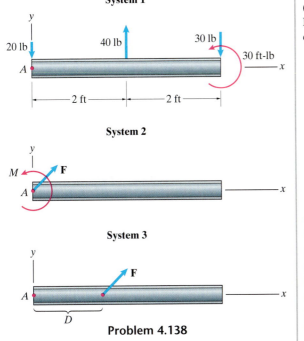

System 2

System 3

Problem 4.138

4.139 Represent the two forces and couple acting on the beam by a force **F**. Determine **F** and determine where its line of action intersects the x axis.

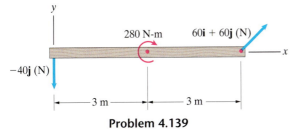

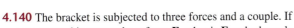

Problem 4.139

4.140 The bracket is subjected to three forces and a couple. If you represent this system by a force **F**, what is **F** and where does its line of action intersect the x axis?

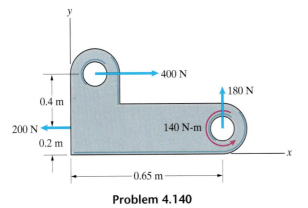

Problem 4.140

4.141 The vector sum of the forces acting on the beam is zero, and the sum of the moments about the left end of the beam is zero.
(a) Determine the forces A_x and A_y, and the couple M_A.
(b) Determine the sum of the moments about the right end of the beam.
(c) If you represent the 600-N force, the 200-N force, and the 30 N-m couple by a force **F** acting at the left end of the beam and a couple M, what are **F** and M?

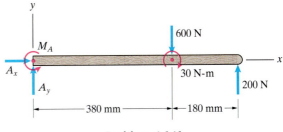

Problem 4.141

4.142 The vector sum of the forces acting on the truss is zero, and the sum of the moments about the origin O is zero.
(a) Determine the forces A_x, A_y, and B.
(b) If you represent the 2-kip, 4-kip, and 6-kip forces by a force **F**, what is **F**, and where does its line of action intersect the y axis?
(c) If you replace the 2-kip, 4-kip, and 6-kip forces by the force you determined in (b), what are the vector sum of the forces acting on the truss and the sum of the moments about O?

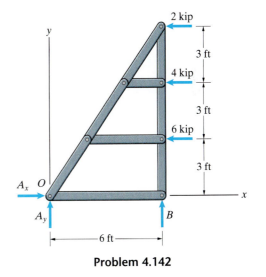

Problem 4.142

4.143 The distributed force exerted on part of a building foundation by the soil is represented by five forces. If you represent them by a force **F**, what is **F**, and where does its line of action intersect the x axis?

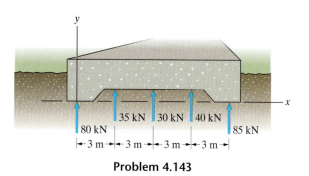

Problem 4.143

4.144 At a particular instant, aerodynamic forces distributed over the airplane's surface exert the 88-kN and 16-kN vertical forces and the 22 kN-m counterclockwise couple shown. If you represent these forces and couple by a system consisting of a force **F** acting at the center of mass G and a couple M, what are **F** and M?

4.145 If you represent the two forces and couple acting on the airplane by a force **F**, what is **F**, and where does its line of action intersect the x axis?

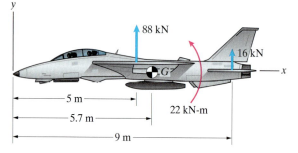

Problems 4.144/4.145

4.146 The system is in equilibrium. If you represent the forces F_{AB} and F_{AC} by a force **F** acting at A and a couple M, what are **F** and M?

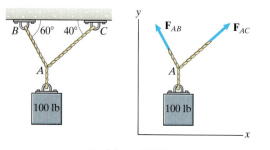

Problem 4.146

4.147 Three forces act on the beam.
(a) Represent the system by a force **F** acting at the origin O and a couple M.
(b) Represent the system by a single force. Where does the line of action of the force intersect the x axis?

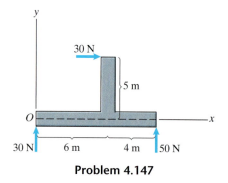

Problem 4.147

4.148 The tension in cable AB is 400 N, and the tension in cable CD is 600 N.
(a) If you represent the forces exerted on the left post by the cables by a force **F** acting at the origin O and a couple M, what are **F** and M?
(b) If you represent the forces exerted on the left post by the cables by the force **F** alone, where does its line of action intersect the y axis?

4.149 The tension in each of the cables AB and CD is 400 N. If you represent the forces exerted on the right post by the cables by a force **F**, what is **F**, and where does its line of action intersect the y axis?

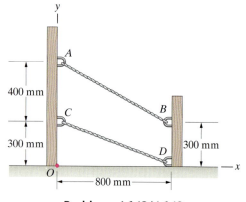

Problems 4.148/4.149

4.150 If you represent the three forces acting on the beam cross section by a force **F**, what is **F**, and where does its line of action intersect the *x* axis?

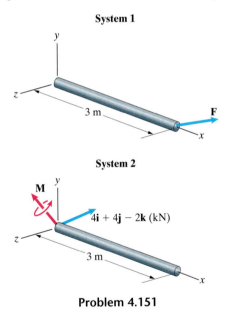

Problem 4.150

4.151 The two systems of forces and moments acting on the beam are equivalent. Determine the force **F** and the couple **M**.

System 1

System 2

Problem 4.151

4.152 The wall bracket is subjected to the force shown.
(a) Determine the moment exerted by the force about the *z* axis.
(b) Determine the moment exerted by the force about the *y* axis.
(c) If you represent the force by a force **F** acting at *O* and a couple **M**, what are **F** and **M**?

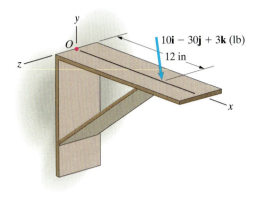

Problem 4.152

4.153 A basketball player executes a "slam dunk" shot, then hangs momentarily on the rim, exerting the two 100-lb forces shown. The dimensions are $h = 14\frac{1}{2}$ in, and $r = 9\frac{1}{2}$ in, and the angle $\alpha = 120°$.
(a) If you represent the forces he exerts by a force **F** acting at *O* and a couple **M**, what are **F** and **M**?
(b) The glass backboard will shatter if $|\mathbf{M}| > 4000$ in-lb. Does it break?

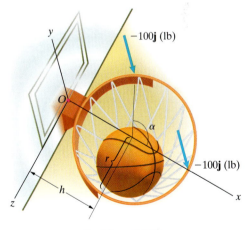

Problem 4.153

4.154 The three forces are parallel to the *x* axis.

(a) If you represent the three forces by a force **F** acting at the origin *O* and a couple **M**, what are **F** and **M**?

(b) If you represent the forces by a single force, what is the force, and where does its line of action intersect the *y–z* plane?

    Strategy: In (b), assume that the force acts at a point $(0, y, z)$ of the *y–z* plane, and use the conditions for equivalence to determine the force and the coordinates *y* and *z*. (See Example 4.20.)

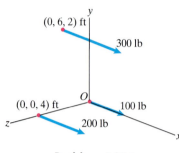

Problem 4.154

4.155 The normal forces exerted on the car's tires by the road are

$$\mathbf{N}_A = 5104\mathbf{j} \ (\text{N}),$$

$$\mathbf{N}_B = 5027\mathbf{j} \ (\text{N}),$$

$$\mathbf{N}_C = 3613\mathbf{j} \ (\text{N}),$$

$$\mathbf{N}_D = 3559\mathbf{j} \ (\text{N}).$$

If you represent these forces by a single equivalent force **N**, what is **N** and where does its line of action intersect the *x–z* plane?

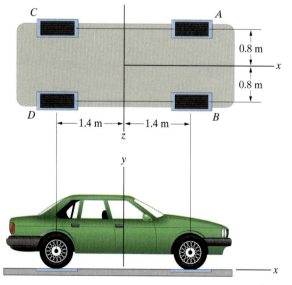

Problem 4.155

4.156 Two forces act on the beam. If you represent them by a force **F** acting at *C* and a couple **M**, what are **F** and **M**?

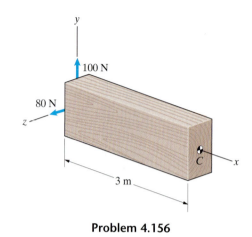

Problem 4.156

4.157 An axial force of magnitude *P* acts on the beam. If you represent it by a force **F** acting at the origin *O* and a couple **M**, what are **F** and **M**?

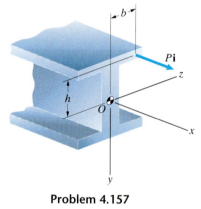

Problem 4.157

4.158 The brace is being used to remove a screw.

(a) If you represent the forces acting on the brace by a force **F** acting at the origin *O* and a couple **M**, what are **F** and **M**?

(b) If you represent the forces acting on the brace by a force **F**′ acting at a point P with coordinates (x_P, y_P, z_P) and a couple **M**′, what are **F**′ and **M**′?

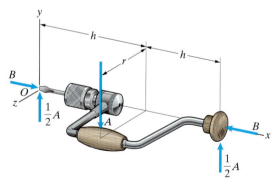

Problem 4.158

4.159 Two forces and a couple act on the cube. If you represent them by a force **F** acting at point P and a couple **M**, what are **F** and **M**?

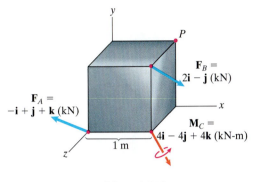

Problem 4.159

4.160 The two shafts are subjected to the torques (couples) shown.

(a) If you represent the two couples by a force **F** acting at the origin O and a couple **M**, what are **F** and **M**?

(b) What is the magnitude of the total moment exerted by the two couples?

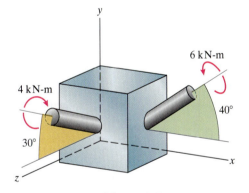

Problem 4.160

4.161 The two systems of forces and moments acting on the bar are equivalent. If

$$\mathbf{F}_A = 30\mathbf{i} + 30\mathbf{j} - 20\mathbf{k}\ (\text{kN}),$$

$$\mathbf{F}_B = 40\mathbf{i} - 20\mathbf{j} + 25\mathbf{k}\ (\text{kN}),$$

$$\mathbf{M}_B = 10\mathbf{i} + 40\mathbf{j} - 10\mathbf{k}\ (\text{kN-m}),$$

what are **F** and **M**?

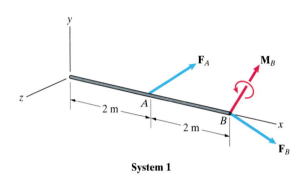

System 1

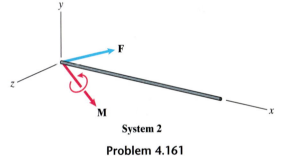

System 2

Problem 4.161

4.162 Point G is at the center of the block. The forces are

$$\mathbf{F}_A = -20\mathbf{i} + 10\mathbf{j} + 20\mathbf{k}\ (\text{lb}),$$

$$\mathbf{F}_B = 10\mathbf{j} - 10\mathbf{k}\ (\text{lb}).$$

If you represent the two forces by a force **F** acting at G and a couple **M**, what are **F** and **M**?

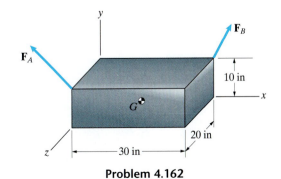

Problem 4.162

4.163 The engine above the airplane's fuselage exerts a thrust $T_0 = 16$ kip, and each of the engines under the wings exerts a thrust $T_U = 12$ kip. The dimensions are $h = 8$ ft, $c = 12$ ft, and $b = 16$ ft. If you represent the three thrust forces by a force **F** acting at the origin O and a couple **M**, what are **F** and **M**?

4.164 Consider the airplane described in Problem 4.163 and suppose that the engine under the wing to the pilot's right loses thrust.
(a) If you represent the two remaining thrust forces by a force **F** acting at the origin O and a couple **M**, what are **F** and **M**?
(b) If you represent the two remaining thrust forces by the force **F** alone, where does its line of action intersect the x–y plane?

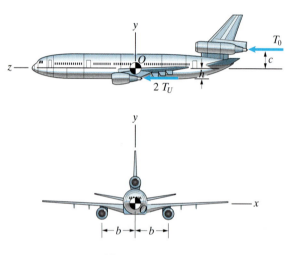

Problems 4.163/4.164

4.165 The tension in cable AB is 100 lb, and the tension in cable CD is 60 lb. Suppose that you want to replace these two cables by a single cable EF so that the force exerted on the wall at E is equivalent to the two forces exerted by cables AB and CD on the walls at A and C. What is the tension in cable EF, and what are the coordinates of points E and F?

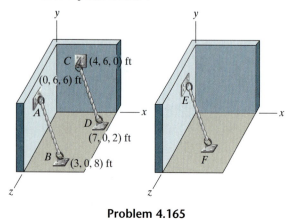

Problem 4.165

4.166 The distance $s = 4$ m. If you represent the force and the 200-N-m couple by a force **F** acting at the origin O and a couple **M**, what are **F** and **M**?

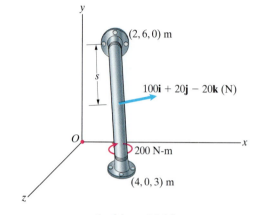

Problem 4.166

4.167 The force **F** and couple **M** in system 1 are

$$F = 12i + 4j - 3k \text{ (lb)},$$

$$M = 4i + 7j + 4k \text{ (ft-lb)}.$$

Suppose you want to represent system 1 by a wrench (system 2). Determine the couple M_p and the coordinates x and z where the line of action of the force intersects the x–z plane.

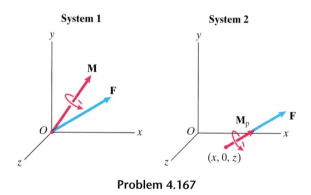

Problem 4.167

4.168 A system consists of a force **F** acting at the origin *O* and a couple **M**, where

$$\mathbf{F} = 10\mathbf{i} \ (\text{lb}), \qquad \mathbf{M} = 20\mathbf{j} \ (\text{ft-lb}).$$

If you represent the system by a wrench consisting of the force **F** and a parallel couple $\mathbf{M}_\text{p}$, what is $\mathbf{M}_\text{p}$, and where does the line of action of **F** intersect the *y–z* plane?

4.169 A system consists of a force **F** acting at the origin *O* and a couple **M**, where

$$\mathbf{F} = \mathbf{i} + 2\mathbf{j} + 5\mathbf{k} \ (\text{N}), \qquad \mathbf{M} = 10\mathbf{i} + 8\mathbf{j} - 4\mathbf{k} \ (\text{N-m}).$$

If you represent it by a wrench consisting of the force **F** and a parallel couple $\mathbf{M}_\text{p}$, (a) determine $\mathbf{M}_\text{p}$, and determine where the line of action of **F** intersects (b) the *x–z* plane, (c) the *y–z* plane.

4.170 Consider the force **F** acting at the origin *O* and the couple **M** given in Example 4.20. If you represent this system by a wrench, where does the line of action of the force intersect the *x–y* plane?

4.171 Consider the force **F** acting at the origin *O* and the couple **M** given in Example 4.20. If you represent this system by a wrench, where does the line of action of the force intersect the plane *y* = 3 m?

4.172 A wrench consists of a force of magnitude 100 N acting at the origin *O* and a couple of magnitude 60 N-m. The force and couple point in the direction from *O* to the point (1, 1, 2) m. If you represent the wrench by a force **F** acting at the point (5, 3, 1) m and a couple **M**, what are **F** and **M**?

4.173 System 1 consists of two forces and a couple. Suppose that you want to represent it by a wrench (system 2). Determine the force **F**, the couple $\mathbf{M}_\text{p}$, and the coordinates *x* and *z* where the line of action of **F** intersects the *x–z* plane.

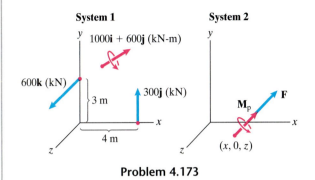

Problem 4.173

4.174 A plumber exerts the two forces shown to loosen a pipe.
(a) What total moment does he exert about the axis of the pipe?
(b) If you represent the two forces by a force **F** acting at *O* and a couple **M**, what are **F** and **M**?
(c) If you represent the two forces by a wrench consisting of the force **F** and a parallel couple $\mathbf{M}_\text{p}$, what is $\mathbf{M}_\text{p}$, and where does the line of action of **F** intersect the *x–y* plane?

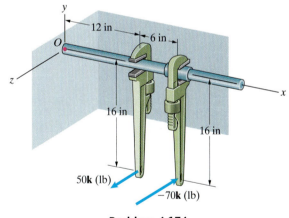

Problem 4.174

COMPUTATIONAL MECHANICS

The following example and problems are designed for the use of a programmable calculator or computer.

Computational Example 4.21

The radius R of the steering wheel in Fig. 4.49 is 200 mm. The distance from O to C is 1 m. The center C of the steering wheel lies in the x–y plane. The force $\mathbf{F} = \sin \alpha (10\mathbf{i} + 10\mathbf{j} - 5\mathbf{k})$ N. Determine the value of α at which the magnitude of the moment of $\mathbf{F}$ about the shaft OC of the steering wheel is a maximum. What is the maximum magnitude?

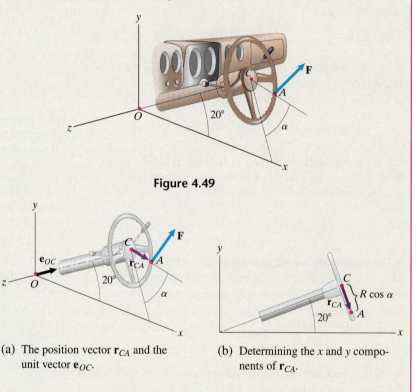

Figure 4.49

(a) The position vector $\mathbf{r}_{CA}$ and the unit vector $\mathbf{e}_{OC}$.

(b) Determining the x and y components of $\mathbf{r}_{CA}$.

Strategy

We will determine the moment of $\mathbf{F}$ about OC in terms of the angle α and obtain a graph of the moment as a function of α.

Solution

In terms of the vector $\mathbf{r}_{CA}$ from point C on the shaft to the point of application of the force, and the unit vector $\mathbf{e}_{OC}$ that points along the shaft from point O toward point C (Fig. a), the moment of $\mathbf{F}$ about the shaft is

$$\mathbf{M}_{OC} = [\mathbf{e}_{OC} \cdot (\mathbf{r}_{CA} \times \mathbf{F})]\mathbf{e}_{OC}.$$

From Fig. a, the unit vector $\mathbf{e}_{OC}$ is

$$\mathbf{e}_{OC} = \cos 20° \, \mathbf{i} + \sin 20° \, \mathbf{j},$$

and the z component of $\mathbf{r}_{CA}$ is $-R \sin \alpha$. By viewing the steering wheel with the z axis perpendicular to the page (Fig. b), we can see that the x component of $\mathbf{r}_{CA}$ is $R \cos \alpha \sin 20°$ and the y component is $-R \cos \alpha \cos 20°$, so

$$\mathbf{r}_{CA} = R(\cos \alpha \sin 20° \, \mathbf{i} - \cos \alpha \cos 20° \mathbf{j} - \sin \alpha \, \mathbf{k}).$$

The magnitude of $\mathbf{M}_{OC}$ is the absolute value of the scalar

$$\mathbf{e}_{OC} \cdot (\mathbf{r}_{CA} \times \mathbf{F}) = \begin{vmatrix} \cos 20° & \sin 20° & 0 \\ R \cos \alpha \sin 20° & -R \cos \alpha \cos 20° & -R \sin \alpha \\ 10 \sin \alpha & 10 \sin \alpha & -5 \sin \alpha \end{vmatrix}$$

$$= R[5 \sin \alpha \cos \alpha + 10(\cos 20° - \sin 20°) \sin^2 \alpha].$$

Computing the absolute value of this expression as a function of α, we obtain the graph shown in Fig. 4.50. The magnitude of the moment is an extremum at values of α of approximately 70° and 250°. The table gives the computed values of α near 70°. We can see that the maximum value is approximately 1.38 N-m. The value of the moment at $\alpha = 250°$ is also 1.38 N-m.

Critical Thinking
In addition to making it possible for us to estimate the maximum magnitude of the moment exerted on the steering wheel by the force, notice how much information is provided by the graph of the magnitude of the moment as a function of the angle α in Fig. 4.50. By using a computer to carry out large numbers of computations and present them graphically, you can gain insight into how solutions of problems in mechanics are affected by their parameters.

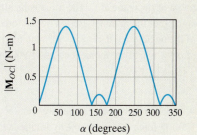

Figure 4.50
Magnitude of the moment as a function of α.

α	$|\mathbf{M}_{OC}|$ (N-m)
67°	1.3725
68°	1.3749
69°	1.3764
70°	1.3769
71°	1.3765
72°	1.3751
73°	1.3728

Computational Problems

4.175 The unstretched length of the spring is 1 m, and the spring constant is $k = 20$ N/m..

(a) Draw a graph of the moment about A due to the force exerted by the spring on the circular bar at B for values of the angle α from zero to 90°.

(b) Use the graph to estimate the value α at which the maximum moment occurs and the corresponding value of the maximum moment.

4.176 The exercise equipment shown is used by resting the elbow on the fixed pad and rotating the forearm to stretch the elastic cord AB. The cord behaves like a linear spring, and its unstretched length is 1 ft. Suppose you want to design the equipment so that the maximum moment that will be exerted about the elbow joint E as the forearm is rotated will be 60 ft-lb. What should the spring constant k of the elastic cord be?

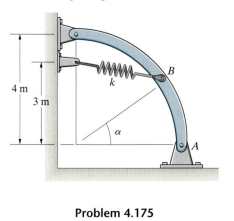

Problem 4.175

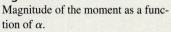

Problem 4.176

4.177 The hydraulic cylinder BC exerts a 2200-lb force on the boom of the crane at C. The force is parallel to the cylinder. Draw a graph of the moment exerted by the force about A as a function of the angle α for $0 \leq \alpha \leq 90°$, and use it to estimate the values of α for which the moment equals 12,000 ft-lb.

4.178 In Problem 4.177, the moment about A exerted by the 2200-lb force exerted by the hydraulic cylinder BC depends on the angle α. Estimate the maximum value of the moment and the angle α at which it occurs.

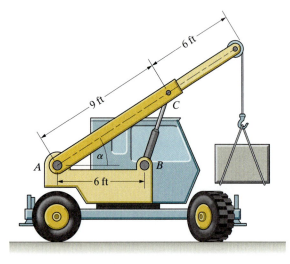

Problems 4.177/4.178

4.179 The support cable extends from the top of the 3-m column at A to a point B on the line L. The tension in the cable is 2 kN. The line L intersects the ground at the point (3, 0, 1) m and is parallel to the unit vector $\mathbf{e} = \frac{2}{7}\mathbf{i} + \frac{6}{7}\mathbf{j} - \frac{3}{7}\mathbf{k}$. The distance along L from the ground to point B is denoted s. What is the range of values of s for which the magnitude of the moment about O due to the force exerted by the cable at A exceeds 5.6 kN-m?

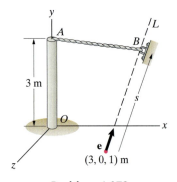

Problem 4.179

4.180 Consider Problem 4.106. Determine the distance d that causes the moment about the z axis due to the force exerted by the cable CD at point C to be a maximum. What is the maximum moment?

4.181 The rod AB supports the open hood of the car. The force exerted by the rod on the hood at B is parallel to the rod. The rod must exert a moment of 100 ft-lb magnitude about the x axis to support the hood. Draw a graph of the magnitude of the force the rod must exert as a function of the distance d from $d = 1$ ft to $d = 4$ ft. If you were designing the support, what value of d would you choose, and what is the corresponding magnitude of the force the rod must exert on the hood?

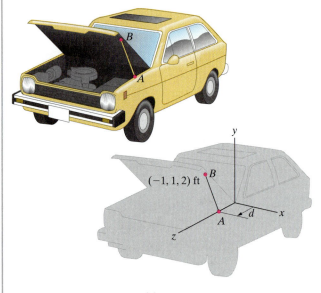

Problem 4.181

4.182 Consider the system shown in Problem 4.148. The forces exerted on the left post by cables AB and CD can be represented by a single force $\mathbf{F}$. Determine the tensions in the cables so that $|\mathbf{F}| = 600$ N and the line of action of $\mathbf{F}$ intersects the y axis at $y = 400$ mm.

4.183 Suppose you want to represent the force and the 200-N-m couple in Problem 4.166 by a force $\mathbf{F}$ and a couple $\mathbf{M}$, and choose the distance s so that the magnitude of $\mathbf{M}$ is a minimum. Determine s, $\mathbf{F}$, and $\mathbf{M}$.

CHAPTER SUMMARY

In this chapter we have defined the moment of a force about a point and about a line and explained how to evaluate them. We introduced the concept of a couple and defined equivalent systems of forces and moments. We can now apply two consequences of equilibrium: The sum of the forces equals zero, and the sum of the moments about any point equals zero. We will consider individual objects in Chapter 5 and structures in Chapter 6.

Moment of a Force About a Point

The moment of a force about a point is the measure of the tendency of the force to cause rotation about the point. The *moment* of a force $\mathbf{F}$ about a point P is the vector

$$\mathbf{M}_P = \mathbf{r} \times \mathbf{F}, \tag{4.2}$$

where $\mathbf{r}$ is a position vector from P to *any* point on the line of action of $\mathbf{F}$. The magnitude of $\mathbf{M}_P$ is equal to the product of the perpendicular distance D from P to the line of action of $\mathbf{F}$ and the magnitude of $\mathbf{F}$ (Fig. a):

$$|\mathbf{M}_P| = D|\mathbf{F}|. \tag{4.3}$$

The vector $\mathbf{M}_P$ is perpendicular to the plane containing P and $\mathbf{F}$. When the thumb of the right hand points in the direction of $\mathbf{M}_P$, the arc of the fingers indicates the sense of the rotation that $\mathbf{F}$ tends to cause about P. The dimensions of the moment are (distance) $\times$ (force).

If a force is resolved into components, the moment of the force about a point P is equal to the sum of the moments of its components about P. If the line of action of a force passes through a point P, the moment of the force about P is zero.

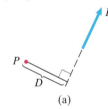

(a)

Moment of a Force About a Line

The moment of a force about a line is the measure of the tendency of the force to cause rotation about the line. Let P be any point on a line L and let $\mathbf{M}_P$ be the moment about P of a force $\mathbf{F}$ (Fig. b). The moment $\mathbf{M}_L$ of $\mathbf{F}$ about L is the vector component of $\mathbf{M}_P$ parallel to L. If $\mathbf{e}$ is a unit vector along L, then

$$\mathbf{M}_L = (\mathbf{e} \cdot \mathbf{M}_P)\mathbf{e} = [\mathbf{e} \cdot (\mathbf{r} \times \mathbf{F})]\mathbf{e}. \tag{4.4}$$

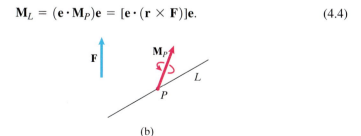

(b)

When the line of action of $\mathbf{F}$ is perpendicular to a plane containing L, $|\mathbf{M}_L|$ is equal to the product of the magnitude of $\mathbf{F}$ and the perpendicular distance D from L to the point where the line of action intersects the plane. When the line of action of $\mathbf{F}$ is parallel to L or intersects L, $\mathbf{M}_L = 0$.

Couples

Two forces that have equal magnitudes, opposite directions, and do not have the same line of action are called a *couple*. The moment **M** of a couple is the same about any point. The magnitude of **M** is equal to the product of the magnitude of one of the forces and the perpendicular distance between the lines of action, and its direction is perpendicular to the plane containing the lines of action.

Because a couple exerts a moment but no net force, it can be represented by showing the moment vector (Fig. c), or it can be represented in two dimensions by showing the magnitude of the moment and a circular arrow to indicate the direction (Fig. d). The moment represented in this way is called the *moment of a couple*, or simply a *couple*.

(c)   (d)

Equivalent Systems

Two systems of forces and moments are defined to be *equivalent* if the sums of the forces are equal, or

$$(\Sigma \mathbf{F})_1 = (\Sigma \mathbf{F})_2, \tag{4.7}$$

and the sums of the moments about a point P are equal, or

$$(\Sigma \mathbf{M}_P)_1 = (\Sigma \mathbf{M}_P)_2. \tag{4.8}$$

If the sums of the forces are equal and the sums of the moments about one point are equal, the sums of the moments about any point are equal.

Representing Systems by Equivalent Systems

If the system of forces and moments acting on an object is represented by an equivalent system, the equivalent system exerts the same total force and total moment on the object.

Any system can be represented by an equivalent system consisting of a force **F** acting at a given point P and a couple **M** (Fig. e). The simplest system that can be equivalent to any system of forces and moments is the *wrench*, which is a force **F** and a couple $\mathbf{M}_p$ that is parallel to **F** (Fig. f)).

A system of concurrent forces can be represented by a single force. A system of parallel forces whose sum is not zero can be represented by a single force.

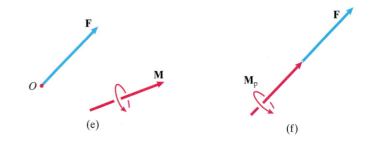

(e)   (f)

Review Problems

4.184 Determine the moment of the 200-N force about A.

(a) What is the two-dimensional description of the moment?

(b) Express the moment as a vector.

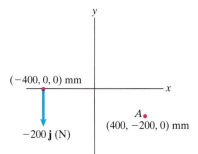

Problem 4.184

4.185 The Leaning Tower of Pisa is approximately 55 m tall and 7 m in diameter. The horizontal displacement of the top of the tower from the vertical is approximately 5 m. Its mass is approximately 3.2×10^6 kg. If you model the tower as a cylinder and assume that its weight acts at the center, what is the magnitude of the moment exerted by the weight about the point at the center of the tower's base?

Problem 4.185

4.186 The cable AB exerts a 300-N force on the support A that points from A toward B. Determine the magnitude of the moment the force exerts about point P.

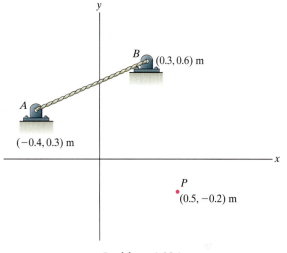

Problem 4.186

4.187 Three forces act on the structure. The sum of the moments due to the forces about A is zero. Determine the magnitude of the force F.

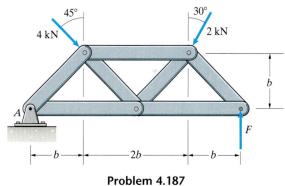

Problem 4.187

4.188 Determine the moment of the 400-N force (a) about A, (b) about B.

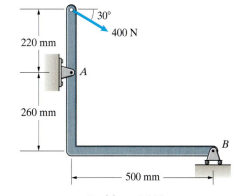

Problem 4.188

4.189 Determine the sum of the moments exerted about A by the three forces and the couple.

4.190 If you represent the three forces and the couple by an equivalent system consisting of a force $\mathbf{F}$ acting at A and a couple $\mathbf{M}$, what are the magnitudes of $\mathbf{F}$ and $\mathbf{M}$?

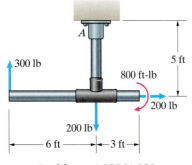

Problems 4.189/4.190

4.191 The vector sum of the forces acting on the beam is zero, and the sum of the moments about A is zero.
(a) What are the forces A_x, A_y, and B?
(b) What is the sum of the moments about B?

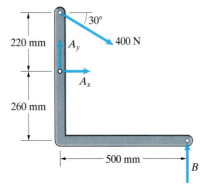

Problem 4.191

4.192 To support the ladder, the force exerted at B by the hydraulic piston AB must exert a moment about C equal in magnitude to the moment about C due to the ladder's 450-lb weight. What is the magnitude of the force exerted at B?

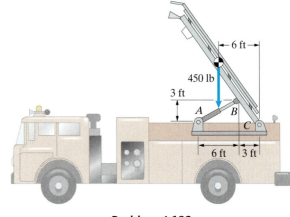

Problem 4.192

4.193 The force $\mathbf{F} = -60\mathbf{i} + 60\mathbf{j}$ (lb).
(a) Determine the moment of $\mathbf{F}$ about point A.
(b) What is the perpendicular distance from point A to the line of action of $\mathbf{F}$?

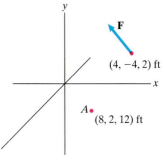

Problem 4.193

4.194 The 20-kg mass is suspended by cables attached to three vertical 2-m posts. Point A is at $(0, 1.2, 0)$ m. Determine the moment about the base E due to the force exerted on the post BE by the cable AB.

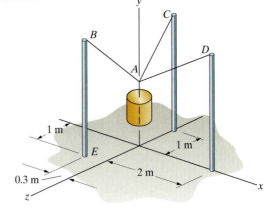

Problem 4.194

4.195 Three forces of equal magnitude are applied parallel to the sides of an equilateral triangle.

(a) Show that the sum of the moments of the forces is the same about any point.

(b) Determine the magnitude of the moment.

    Strategy: To do (a), resolve one of the forces into vector components parallel to the other two forces.

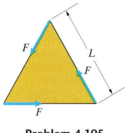

Problem 4.195

4.196 The bar *AB* supporting the lid of the grand piano exerts a force $\mathbf{F} = -6\mathbf{i} + 35\mathbf{j} - 12\mathbf{k}$ (lb) at *B*. The coordinates of *B* are (3, 4, 3) ft. What is the moment of the force about the hinge line of the lid (the *x* axis)?

Problem 4.196

4.197 Determine the moment of the vertical 800-lb force about point *C*.

4.198 Determine the moment of the vertical 800-lb force about the straight line through points *C* and *D*.

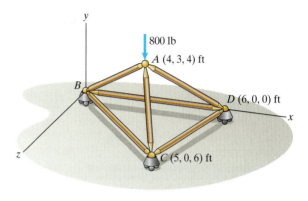

Problems 4.197/4.198

4.199 The system of cables and pulleys supports the 300-lb weight of the work platform. If you represent the upward force exerted at *E* by cable *EF* and the upward force exerted at *G* by cable *GH* by a single equivalent force $\mathbf{F}$, what is $\mathbf{F}$, and where does its line of action intersect the *x* axis?

4.200 Consider the system in Problem 4.199.

(a) What are the tensions in cables *AB* and *CD*?

(b) If you represent the forces exerted by the cables at *A* and *C* by a single equivalent force $\mathbf{F}$, what is $\mathbf{F}$, and where does its line of action intersect the *x* axis?

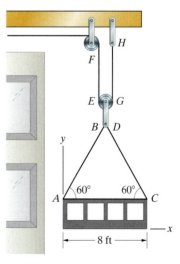

Problems 4.199/4.200

4.201 The two systems are equivalent. Determine the forces A_x and A_y, and the couple M_A.

4.202 If you represent the equivalent systems in Problem 4.201 by a force **F** acting at the origin and a couple M, what are **F** and M?

4.203 If you represent the equivalent systems in Problem 4.201 by a force **F**, what is **F**, and where does its line of action intersect the x-axis?

System 1

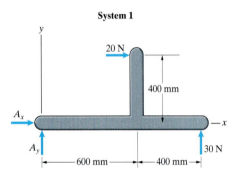

System 2

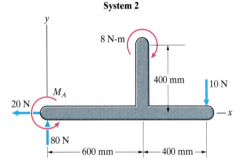

Problems 4.201–4.203

4.204 The two systems are equivalent. If

$$\mathbf{F} = -100\mathbf{i} + 40\mathbf{j} + 30\mathbf{k} \text{ (lb)},$$

$$\mathbf{M}' = -80\mathbf{i} + 120\mathbf{j} + 40\mathbf{k} \text{ (in-lb)},$$

determine **F**′ and M.

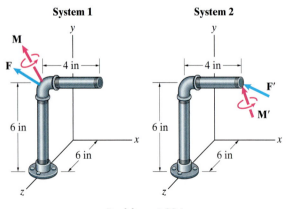

Problem 4.204

4.205 The tugboats A and B exert forces $F_A = 1$ kN and $F_B = 1.2$ kN on the ship. The angle $\theta = 30°$. If you represent the two forces by a force **F** acting at the origin O and a couple M, what are **F** and M?

4.206 The tugboats A and B exert forces $F_A = 600$ N and $F_B = 800$ N on the ship. The angle $\theta = 45°$. If you represent the two forces by a force **F**, what is **F**, and where does its line of action intersect the y-axis?

4.207 The tugboats A and B want to exert two forces on the ship that are equivalent to a force **F** acting at the origin O of 2-kN magnitude. If $F_A = 800$ N, determine the necessary values of F_B and θ.

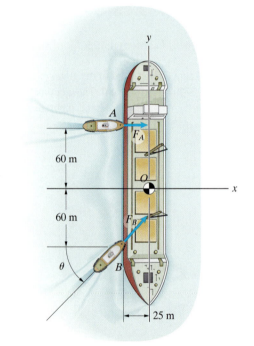

Problems 4.205–4.207

4.208 If you represent the forces exerted by the floor on the table legs by a force **F** acting at the origin *O* and a couple **M**, what are **F** and **M**?

4.209 If you represent the forces exerted by the floor on the table legs by a force **F**, what is **F**, and where does its line of action intersect the *x*–*z* plane?

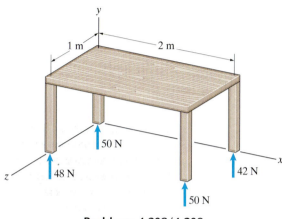

Problems 4.208/4.209

4.210 Two forces are exerted on the crankshaft by the connecting rods. The direction cosines of $\mathbf{F}_A$ are $\cos\theta_x = -0.182$, $\cos\theta_y = 0.818$, and $\cos\theta_z = 0.545$, and its magnitude is 4 kN. The direction cosines of $\mathbf{F}_B$ are $\cos\theta_x = 0.182$, $\cos\theta_y = 0.818$, and $\cos\theta_z = -0.545$, and its magnitude is 2 kN. If you represent the two forces by a force **F** acting at the origin *O* and a couple **M**, what are **F** and **M**?

4.211 If you represent the two forces exerted on the crankshaft in Problem 4.210 by a wrench consisting of a force **F** and a parallel couple $\mathbf{M}_p$, what are **F** and $\mathbf{M}_p$, and where does the line of action of **F** intersect the *x*–*z* plane?

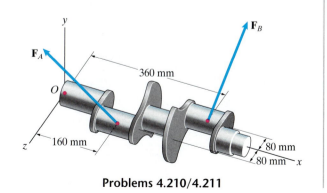

Problems 4.210/4.211

Design Project A relatively primitive device for exercising the biceps muscle is shown. Suggest an improved configuration for the device. You can use elastic cords (which behave like linear springs), weights, and pulleys. Seek a design such that the variation of the moment about the elbow joint as the device is used is small in comparison to the design shown. Give consideration to the safety of your device, its reliability, and the requirement to accommodate users having a range of dimensions and strengths. Choosing specific dimensions, determine the range of the magnitude of the moment exerted about the elbow joint as your device is used.

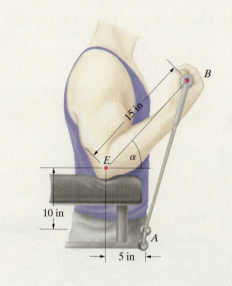

CHAPTER
5

Objects in Equilibrium

Building on concepts developed in Chapters 3 and 4, we first state the general equilibrium equations. We describe various ways that structural members can be supported, or held in place. Using free-body diagrams and equilibrium equations, we then show how to determine unknown forces and couples exerted on structural members by their supports. The principal motivation for this procedure is that it is the initial step in answering an essential question in structural analysis: How do engineers design structural elements so that they will support the loads to which they are subjected?

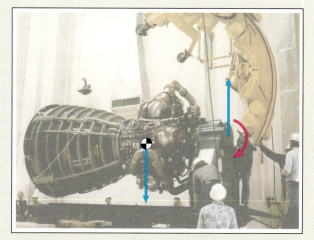

◀ A Space Shuttle main engine being held in equilibrium by a support. In this chapter we use the equilibrium equations to determine forces and couples exerted on objects by their supports.

5.1 The Equilibrium Equations

In Chapter 3 we defined an object to be in equilibrium when it is stationary or in steady translation relative to an inertial reference frame. When an object acted upon by a system of forces and moments is in equilibrium, the following conditions are satisfied:

1. The sum of the forces is zero:

$$\Sigma \mathbf{F} = \mathbf{0}. \tag{5.1}$$

2. The sum of the moments about any point is zero:

$$\Sigma \mathbf{M}_{\text{any point}} = \mathbf{0}. \tag{5.2}$$

Before we consider specific applications, some general observations about these equations are in order.

From our discussion of equivalent systems of forces and moments in Chapter 4, Eqs. (5.1) and (5.2) imply that the system of forces and moments acting on an object in equilibrium is equivalent to a system consisting of no forces and no couples. This provides insight into the nature of equilibrium. From the standpoint of the total force and total moment exerted on an object in equilibrium, the effects are the same as if no forces or couples acted on the object. This observation also makes it clear that if the sum of the forces on an object is zero and the sum of the moments about one point is zero, then the sum of the moments about every point is zero.

Figure 5.1 shows an object subjected to concurrent forces $\mathbf{F}_1, \mathbf{F}_2, \ldots, \mathbf{F}_N$ and no couples. If the sum of these forces is zero, or

$$\mathbf{F}_1 + \mathbf{F}_2 + \cdots + \mathbf{F}_N = \mathbf{0}, \tag{5.3}$$

the conditions for equilibrium are satisfied, because the moment about point P is zero. The only condition imposed by equilibrium on a set of concurrent forces is that their sum is zero.

To determine the sum of the moments about a line L due to a system of forces and moments acting on an object, we choose any point P on the line and determine the sum of the moments $\Sigma \mathbf{M}_P$ about P (Fig. 5.2). Then the sum of the moments about the line is the component of $\Sigma \mathbf{M}_P$ parallel to the line. If the object is in equilibrium, $\Sigma \mathbf{M}_P = \mathbf{0}$. We see that the sum of the moments about any line due to the forces and couples acting on an object in equilibrium is zero. This result is useful in certain types of problems.

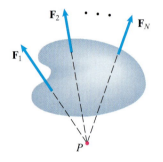

Figure 5.1
An object subjected to concurrent forces.

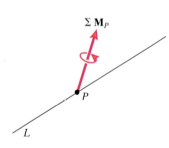

Figure 5.2
The sum of the moments $\Sigma \mathbf{M}_P$ about a point P on the line L.

5.2 Two-Dimensional Applications

Many engineering applications involve two-dimensional systems of forces and moments. These include the forces and moments exerted on many beams and planar structures, pliers, some cranes and other machines, and some types of bridges and dams. In this section we discuss supports, free-body diagrams, and the equilibrium equations for two-dimensional applications.

Supports

When you are standing, the floor supports you. When you sit in a chair with your feet on the floor, the chair and floor support you. In this section we are concerned with the ways objects can be supported, or held in place. Forces and couples exerted on an object by its supports are called *reactions*, expressing the fact that the supports "react" to the other forces and couples, or *loads*, acting on the object. For example, a bridge is held up by the reactions exerted by its supports, and the loads are the forces exerted by the weight of the bridge itself, the traffic crossing it, and the wind.

Some very common kinds of supports are represented by stylized models called *support conventions*. Actual supports often closely resemble the support conventions, but even when they don't, we represent them by these conventions if the actual supports exert the same (or approximately the same) reactions as the models.

The Pin Support   Figure 5.3a shows a *pin support*. The diagram represents a bracket to which an object (such as a beam) is attached by a smooth pin that passes through the bracket and the object. The side view is shown in Fig. 5.3b.

To understand the reactions that a pin support can exert, it's helpful to imagine holding a bar attached to a pin support (Fig. 5.3c). If you try to move the bar without rotating it (that is, translate the bar), the support exerts a reactive force that prevents this movement. However, you can rotate the bar about the axis of the pin. The support cannot exert a couple about the pin axis to prevent rotation. Thus a pin support can't exert a couple about the pin axis, but it can exert a force on an object in any direction, which is usually expressed by representing the force in terms of components (Fig. 5.3d). The arrows indicate the directions of the reactions if A_x and A_y are positive. If you determine A_x or A_y to be negative, the reaction is in the direction opposite to that of the arrow.

The pin support is used to represent any real support capable of exerting a force in any direction but not exerting a couple. Pin supports are used in many common devices, particularly those designed to allow connected parts to rotate relative to each other (Fig. 5.4).

The Roller Support   The convention called a *roller support* (Fig. 5.5a) represents a pin support mounted on wheels. Like the pin support, it cannot exert a couple about the axis of the pin. Since it can move freely in the direction parallel to the surface on which it rolls, it can't exert a force parallel to the surface but can only exert a force normal (perpendicular) to this surface (Fig. 5.5b). Figures 5.5c–e are other commonly used conventions equivalent to the roller support. The wheels of vehicles and wheels supporting parts of machines are roller supports if the friction forces exerted on them are negligible in comparison to

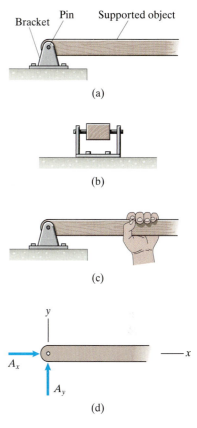

Figure 5.3
(a) A pin support.
(b) Side view showing the pin passing through the beam.
(c) Holding a supported bar.
(d) The pin support is capable of exerting two components of force.

Figure 5.4
Pin supports in a pair of scissors and a stapler.

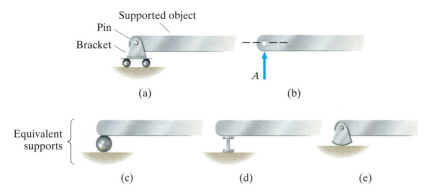

Figure 5.5
(a) A roller support.
(b) The reaction consists of a force normal to the surface.
(c)–(e) Supports equivalent to the roller support.

Figure 5.6
Supporting an object with a plane smooth surface.

the normal forces. A plane smooth surface can also be modeled by a roller support (Fig. 5.6). Beams and bridges are sometimes supported in this way so that they will be free to undergo thermal expansion and contraction.

The supports shown in Fig. 5.7 are similar to the roller support in that they cannot exert a couple and can only exert a force normal to a particular direction. (Friction is neglected.) In these supports, the supported object is attached to a pin or slider that can move freely in one direction but is constrained in the perpendicular direction. Unlike the roller support, these supports can exert a normal force in either direction.

Figure 5.7
Supports similar to the roller support except that the normal force can be exerted in either direction.

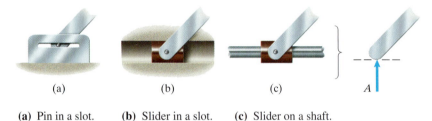

(a)   (b)   (c)   A

(a) Pin in a slot.   (b) Slider in a slot.   (c) Slider on a shaft.

The Fixed Support   The *fixed support* shows the supported object literally built into a wall (Fig. 5.8a). This convention is also called a *built-in*. To understand the reactions, imagine holding a bar attached to a fixed support (Fig. 5.8b). If you try to translate the bar, the support exerts a reactive force that prevents translation, and if you try to rotate the bar, the support exerts a reactive couple that prevents rotation. A fixed support can exert two components of force and a couple (Fig. 5.8c). The term M_A is the couple exerted by the support, and the curved arrow indicates its direction. Fence posts and lampposts have fixed supports. The attachments of parts connected so that they cannot move or rotate relative to each other, such as the head of a hammer and its handle, can be modeled as fixed supports.

Table 5.1 summarizes the support conventions commonly used in two-dimensional applications, including those we discussed in Chapter 3. Although the number of conventions may appear daunting, the examples and problems will help you become familiar with them. You should also observe how various objects you see in your everyday experience are supported and think about whether each support could be represented by one of the conventions.

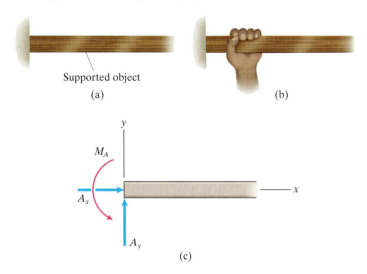

Supported object
(a)

(b)

Figure 5.8
(a) Fixed support.
(b) Holding a supported bar.
(c) The reactions a fixed support is capable
   of exerting.

y

M_A

A_x

x

A_y

(c)

Table 5.1 Supports used in two-dimensional applications.

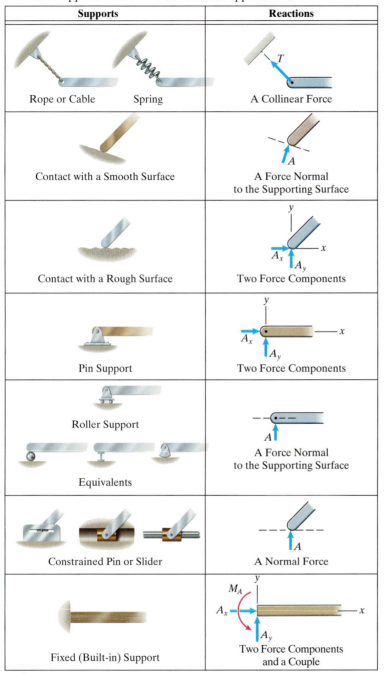

Supports	Reactions
Rope or Cable Spring	A Collinear Force
Contact with a Smooth Surface	A Force Normal to the Supporting Surface
Contact with a Rough Surface	Two Force Components
Pin Support	Two Force Components
Roller Support / Equivalents	A Force Normal to the Supporting Surface
Constrained Pin or Slider	A Normal Force
Fixed (Built-in) Support	Two Force Components and a Couple

Free-Body Diagrams

We introduced free-body diagrams in Chapter 3 and used them to determine forces acting on simple objects in equilibrium. By using the support conventions, we can model more elaborate objects and construct their free-body diagrams in a systematic way.

For example, the beam in Fig. 5.9a has a pin support at the left end and a roller support at the right end and is loaded by a force F. The roller support rests on a surface inclined at 30° to the horizontal. To obtain the free-body diagram of the beam, we first isolate it from its supports (Fig. 5.9b), since the free-body diagram must contain no object other than the beam. We complete the free-body diagram by showing the reactions that may be exerted on the beam by the supports (Fig. 5.9c). Notice that the reaction B exerted by the roller support is normal to the surface on which the support rests.

The object in Fig. 5.10a has a fixed support at the left end. A cable passing over a pulley is attached to the object at two points. We isolate it from its supports

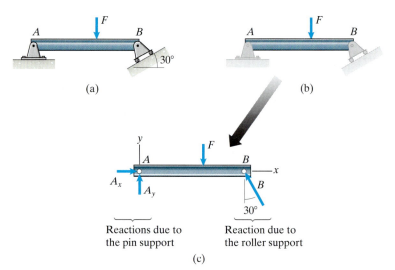

(a)   (b)

Reactions due to the pin support   Reaction due to the roller support

(c)

Figure 5.9
(a) A beam with pin and roller supports.
(b) Isolating the beam from its supports.
(c) The completed free-body diagram.

(Fig. 5.10b) and complete the free-body diagram by showing the reactions at the fixed support and the forces exerted by the cable (Fig. 5.10c). *Don't forget the couple at a fixed support.* Since we assume the tension in the cable is the same on both sides of the pulley, the two forces exerted by the cable have the same magnitude T.

Once you have obtained the free-body diagram of an object in equilibrium to identify the loads and reactions acting on it, you can apply the equilibrium equations.

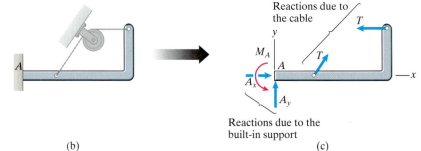

Reactions due to the cable

Reactions due to the built-in support

(a)   (b)   (c)

Figure 5.10
(a) An object with a fixed support.
(b) Isolating the object.
(c) The completed free-body diagram.

The Scalar Equilibrium Equations

When the loads and reactions on an object in equilibrium form a two-dimensional system of forces and moments, they are related by three scalar equilibrium equations:

$$\Sigma F_x = 0, \tag{5.4}$$

$$\Sigma F_y = 0, \tag{5.5}$$

$$\Sigma M_{\text{any point}} = 0. \tag{5.6}$$

A natural question is whether more than one equation can be obtained from Eq. (5.6) by evaluating the sum of the moments about more than one point. The answer is yes, and in some cases it is convenient to do so. But there is a catch—the additional equations will not be independent of Eqs. (5.4)–(5.6). In other words, *more than three independent equilibrium equations cannot be obtained from a two-dimensional free-body diagram, which means we can solve for at most three unknown forces or couples*. We discuss this point further in Section 5.3.

The seesaw found on playgrounds, consisting of a board with a pin support at the center that allows it to rotate, is a simple and familiar example that illustrates the role of Eq. (5.6). If two people of unequal weight sit at the seesaw's ends, the heavier person sinks to the ground (Fig. 5.11a). To obtain equilibrium, that person must move closer to the center (Fig. 5.11b).

We draw the free-body diagram of the seesaw in Fig. 5.11c, showing the weights of the people W_1 and W_2 and the reactions at the pin support. Evaluating the sum of the moments about A, we find that the equilibrium equations are

$$\Sigma F_x = A_x = 0, \tag{5.7}$$

$$\Sigma F_y = A_y - W_1 - W_2 = 0, \tag{5.8}$$

$$\Sigma M_{\text{point } A} = D_1 W_1 - D_2 W_2 = 0. \tag{5.9}$$

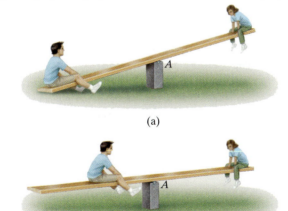

(a)

(b)

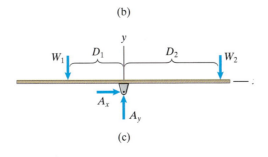

(c)

Figure 5.11
(a) If both people sit at the ends of the seesaw, the heavier one sinks.
(b) The seesaw and people in equilibrium.
(c) The free-body diagram of the seesaw, showing the weights of the people and the reactions at the pin support.

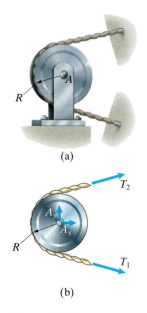

(a)

(b)

Figure 5.12
(a) A pulley of radius R.
(b) Free-body diagram of the pulley and part of the cable.

Thus, $A_x = 0, A_y = W_1 + W_2$, and $D_1 W_1 = D_2 W_2$. The last condition indicates the relation between the positions of the two persons necessary for equilibrium.

To demonstrate that an additional independent equation is not obtained by evaluating the sum of the moments about a different point, we can sum the moments about the right end of the seesaw:

$$\Sigma M_{\text{right end}} = (D_1 + D_2)W_1 - D_2 A_y = 0.$$

This equation is a linear combination of Eqs. (5.8) and (5.9):

$$(D_1 + D_2)W_1 - D_2 A_y = \underbrace{-D_2(A_y - W_1 - W_2)}_{\text{Eq. (5.8)}}$$

$$\underbrace{+(D_1 W_1 - D_2 W_2)}_{\text{Eq. (5.9)}} = 0.$$

Until now we have assumed in examples and problems that the tension in a rope or cable is the same on both sides of a pulley. Consider the pulley in Fig. 5.12a. In its free-body diagram in Fig. 5.12b, we do not assume that the tensions are equal. Summing the moments about the center of the pulley, we obtain the equilibrium equation

$$\Sigma M_{\text{point } A} = RT_1 - RT_2 = 0.$$

The tensions must be equal if the pulley is in equilibrium. However, notice that we have assumed that the pulley's support behaves like a pin support and cannot exert a couple on the pulley. When that is not true—for example, due to friction between the pulley and the support—the tensions are not necessarily equal.

Study Questions

1. What is a pin support? What reactions can it exert on an object subjected to a two-dimensional system of forces and moments?
2. What is a roller support? What reactions can it exert on an object subjected to a two-dimensional system of forces and moments?
3. How many independent equilibrium equations can you obtain from a two-dimensional free-body diagram?

| **Example 5.1** | **Reactions at Pin and Roller Supports** |

The beam in Fig. 5.13 has a pin at A and roller supports at B and is subjected to a 2-kN force. What are the reactions at the supports?

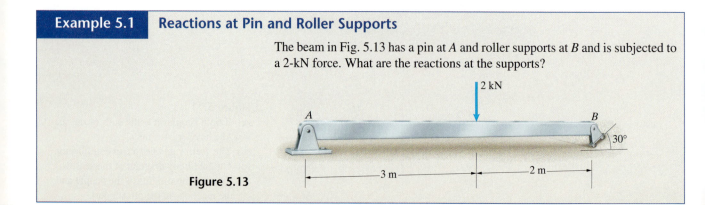

Figure 5.13

Strategy

To determine the reactions exerted on the beam by its supports, we must draw a free-body diagram of the beam *isolated from the supports*. The free-body diagram must show all external forces and couples acting on the beam, *including the reactions exerted by the supports*. Then by applying the equilibrium equations, we can determine the unknown reactions.

Solution

Draw the Free-Body Diagram   We isolate the beam from its supports and show the loads and the reactions that may be exerted by the pin and roller supports (Fig. a). There are three unknown reactions: two components of force A_x and A_y at the pin support and a force B at the roller support.

Apply the Equilibrium Equations   Summing the moments about point A, the equilibrium equations are

$$\Sigma F_x = A_x - B \sin 30° = 0,$$

$$\Sigma F_y = A_y + B \cos 30° - 2 \text{ kN} = 0,$$

$$\Sigma M_{\text{point } A} = (5 \text{ m})(B \cos 30°) - (3 \text{ m})(2 \text{ kN}) = 0$$

Solving these equations, the reactions are $A_x = 0.69$ kN, $A_y = 0.80$ kN, and $B = 1.39$ kN. The load and reactions are shown in Fig. b. It is good practice to show your answers in this way and confirm that the equilibrium equations are satisfied:

$$\Sigma F_x = 0.69 \text{ kN} - (1.39 \text{ kN}) \sin 30° = 0,$$

$$\Sigma F_y = 0.80 \text{ kN} + (1.39 \text{ kN}) \cos 30° - 2 \text{ kN} = 0,$$

$$\Sigma M_{\text{point } A} = (5 \text{ m})(1.39 \text{ kN}) \cos 30° - (3 \text{ m})(2 \text{ kN}) = 0.$$

Critical Thinking

In drawing the free-body diagram of the beam, how did we choose the directions of the reactions A_x, A_y, and B? Whenever possible, you should try to choose the correct directions of reactions when you draw free-body diagrams, as we did in this example, because it helps develop your physical intuition. But if you choose an incorrect direction for a reaction in drawing the free-body diagram of a single object, it doesn't matter. The value you obtain from the equilibrium equations for that reaction will be negative, which indicates that its actual direction is opposite the direction you chose. (This is fortunate, because you will encounter problems in which it will not be practical to try to predict the directions beforehand.) For example, in Fig. c we draw the free-body diagram of the beam with the component A_y pointed downward. From this free-body diagram, we obtain the equilibrium equations

$$\Sigma F_x = A_x - B \sin 30° = 0,$$

$$\Sigma F_y = -A_y + B \cos 30° - 2 \text{ kN} = 0,$$

$$\Sigma M_{\text{point } A} = (5 \text{ m})(B \cos 30°) - (3 \text{ m})(2 \text{ kN}) = 0.$$

Solving, we obtain $A_x = 0.69$ kN, $A_y = -0.80$ kN, and $B = 1.39$ kN. The negative value of A_y indicates that the vertical force exerted on the beam by the pin support at A is in the direction opposite to the arrow in Fig. c. That is, the force is 0.80 kN upward. Thus we again obtain the reactions shown in Fig. b.

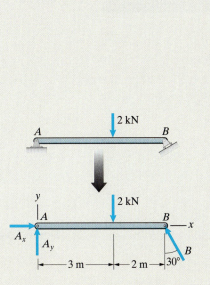

(a) Drawing the free-body diagram of the beam.

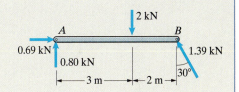

(b) The load and reaction.

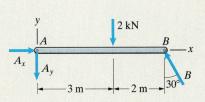

(c) An alternative free-body diagram.

Example 5.2	Reactions at a Fixed Support

The object in Fig. 5.14 has a fixed support at A and is subjected to two forces and a couple. What are the reactions at the support?

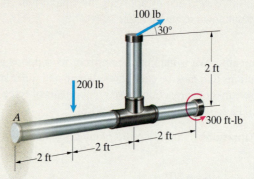

Figure 5.14

Strategy

We will obtain a free-body diagram by isolating the object from the fixed support at A and showing the reactions exerted at A, *including the couple that may be exerted by a fixed support.* Then we can determine the unknown reactions by applying the equilibrium equations.

Solution

Draw the Free-Body Diagram We isolate the object from its support and show the reactions at the fixed support (Fig. a). There are three unknown reactions: two force components A_x and A_y and a couple M_A. (Remember that we can choose the directions of these arrows arbitrarily.) We also resolve the 100-lb force into its components.

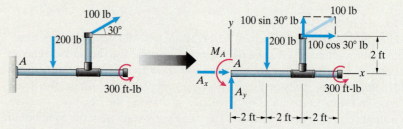

(a) Drawing the free-body diagram.

Apply the Equilibrium Equations Summing the moments about point A, the equilibrium equations are

$$\Sigma F_x = A_x + 100 \cos 30° \text{ lb} = 0,$$

$$\Sigma F_y = A_y - 200 \text{ lb} + 100 \sin 30° \text{ lb} = 0,$$

$$\Sigma M_{\text{point } A} = M_A + 300 \text{ ft-lb} - (2\text{ft})(200 \text{ lb}) - (2 \text{ ft})(100 \cos 30° \text{ lb})$$
$$+ (4 \text{ ft})(100 \sin 30° \text{ lb}) = 0.$$

Solving these equations, we obtain the reactions $A_x = -86.6$ lb, $A_y = 150.0$ lb, and $M_A = 73.2$ ft-lb.

Critical Thinking

Why don't the 300 ft-lb couple and the couple M_A exerted by the fixed support appear in the first two equilibrium equations? Remember that a couple exerts no net force. Also, because the moment due to a couple is the same about any point, the moment about A due to the 300 ft-lb counterclockwise couple is 300 ft-lb counterclockwise.

Example 5.3	Reactions on a Car's Tires

The 2800-lb car in Fig. 5.15 is stationary. Determine the normal forces exerted on the front and rear tires by the road.

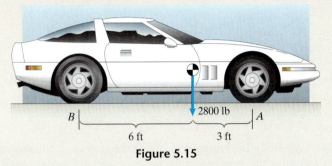

B | 2800 lb | A

6 ft   3 ft

Figure 5.15

Strategy
We can draw the free-body diagram of the car, showing the forces exerted on its tires by the road at A and B, and apply the equilibrium equations to determine the forces on the front and rear tires.

Solution
Draw the Free-Body Diagram In Fig. a we isolate the car and show its weight and the reactions exerted by the road. There are two unknown reactions: the forces A and B exerted on the front and rear tires.

Apply the Equilibrium Equations The forces have no x components. Summing the moments about point B, the equilibrium equations are

$$\Sigma F_y = A + B - 2800 \text{ lb} = 0,$$

$$\Sigma M_{\text{point } B} = -(6 \text{ ft})(2800 \text{ lb}) + (9 \text{ ft})A = 0.$$

Solving these equations, the reactions are $A = 1867$ lb and $B = 933$ lb.

Critical Thinking
This example doesn't fall within our definition of a two-dimensional system of forces and moments because the forces acting on the car are not coplanar. Let's examine why you can analyze problems of this kind as if they were two dimensional.

In Fig. b we show an oblique view of the free-body diagram of the car. In this view you can see the forces acting on the individual tires. The total normal force on the front tires is $A_L + A_R = A$, and the total normal force on the rear tires is $B_L + B_R = B$. The sum of the forces in the y direction is

$$\Sigma F_y = A_L + A_R + B_L + B_R - 2800 \text{ lb}$$

$$= A + B - 2800 \text{ lb} = 0.$$

Since the sum of the moments about any line due to the forces and couples acting on an object in equilibrium is zero, the sum of the moments about the z axis due to the forces acting on the car is zero:

$$\Sigma M_{z \text{ axis}} = (9 \text{ ft})(A_L + A_R) - (6 \text{ ft})(2800 \text{ lb})$$

$$= (9 \text{ ft})A - (6 \text{ ft})(2800 \text{ lb}) = 0.$$

Thus we obtain the same equilibrium equations we did when we solved the problem using a two-dimensional analysis.

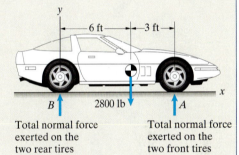

Total normal force exerted on the two rear tires

Total normal force exerted on the two front tires

(a) The free-body diagram.

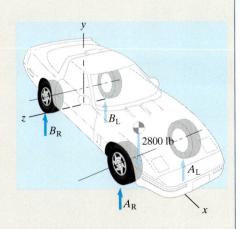

(b) An oblique view showing the forces on the individual tires.

Example 5.4   Choosing the Point About Which to Evaluate Moments

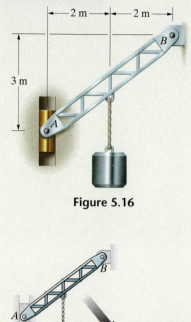

Figure 5.16

The structure AB in Fig. 5.16 supports a suspended 2-Mg (megagram) mass. The structure is attached to a slider in a vertical slot at A and has a pin support at B. What are the reactions at A and B?

Strategy
We will draw the free-body diagram of the structure and the suspended mass by removing the supports at A and B. Notice that the support at A can exert only a horizontal reaction. Then we can use the equilibrium equations to determine the reactions at A and B.

Solution

Draw the Free-Body Diagram   We isolate the structure and mass from the supports and show the reactions at the supports and the force exerted by the weight of the 2000-kg mass (Fig. a). The slot at A can exert only a horizontal force on the slider.

Apply the Equilibrium Equations   Summing moments about point B, we find that the equilibrium equations are

$$\Sigma F_x = A + B_x = 0,$$

$$\Sigma F_y = B_y - (2000)(9.81) \text{ N} = 0,$$

$$\Sigma M_{\text{point } B} = (3 \text{ m})A + (2 \text{ m})[(2000)(9.81) \text{ N}] = 0.$$

The reactions are $A = -13.1$ kN, $B_x = 13.1$ kN, and $B_y = 19.6$ kN.

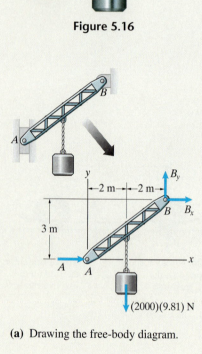

(a) Drawing the free-body diagram.

Critical Thinking
Although the point about which moments are evaluated in writing equilibrium equations can be chosen arbitrarily, a careful choice can often simplify your solution. In this example, point B lies on the lines of action of the two unknown reactions B_x and B_y. By evaluating moments about B, we obtained an equation containing only one unknown, the reaction at A.

Design Example 5.5   Design for Human Factors

Figure 5.17 shows an airport luggage carrier and its free-body diagram when it is held in equilibrium in the tilted position. If the luggage carrier supports a weight $W = 50$ lb, the angle $\alpha = 30°$, $a = 8$ in, $b = 16$ in, and $d = 48$ in, what force F must the user exert?

Strategy
The unknown reactions on the free-body diagram are the force F and the normal force N exerted by the floor. If we sum moments about the center of the wheel C, we obtain an equation in which F is the only unknown reaction.

Solution

Summing moments about C,

$$\Sigma M_{(\text{point } C)} = d(F \cos \alpha) + a(W \sin \alpha) - b(W \cos \alpha) = 0,$$

and solving for F, we obtain

$$F = \frac{(b - a \tan \alpha)W}{d}. \qquad (1)$$

Substituting the values of W, α, a, b, and d yields the solution $F = 11.9$ lb.

Design Issues

Design that accounts for human physical dimensions, capabilities, and characteristics is a special challenge. This art is called design for human factors. Here we consider a simple device, the airport luggage carrier in Fig. 5.17, and show how consideration of its potential users *and the constraints imposed by the equilibrium equations* affect its design.

The user moves the carrier by grasping the bar at the top, tilting it, and walking while pulling the carrier. The height of the handle (the dimension h) needs to be comfortable. Since $h = R + d \sin \alpha$, if we choose values of h and the wheel radius R, we obtain a relation between the length of the carrier's handle d and the tilt angle α:

$$d = \frac{h - R}{\sin \alpha}. \qquad (2)$$

Substituting this expression for d into Eq. (1), we obtain

$$F = \frac{\sin \alpha(b - a \tan \alpha)W}{h - R}. \qquad (3)$$

Suppose that based on statistical data on human dimensions, we decide to design the carrier for convenient use by persons up to 6 ft 2 in tall, which corresponds to a dimension h of approximately 36 in. Let $R = 3$ in, $a = 6$ in, and $b = 12$ in. The resulting value of F/W as a function of α is shown in Fig. 5.18. At $\alpha = 63°$, the force the user must exert is zero, which means the weight of the luggage acts at a point directly above the wheels. This would be the optimum solution if the user could maintain exactly that value of α. However, α inevitably varies, and the resulting changes in F make it difficult to control the carrier. In addition, the relatively steep angle would make the carrier awkward to pull. From this point of view, it is desirable to choose a design within the range of values of α in which F varies slowly, say, $30° \leq \alpha \leq 45°$. (Even though the force the user must exert is large in this range of α in comparison with larger values of α, it is only about 13% of the weight.) Over this range of α, the dimension d varies from 5.5 ft to 3.9 ft. A smaller carrier is desirable for lightness and ease of storage, so we choose $d = 4$ ft for our preliminary design.

We have chosen the dimension d based on particular values of the dimensions R, a, and b. In an actual design study, we would carry out the analysis for expected ranges of values of these parameters. Our final design would also reflect decisions based on safety (for example, there must be adequate means to secure the luggage and no sharp projections), reliability (the frame must be sufficiently strong and the wheels must have adequate and reliable bearings), and the cost of manufacture.

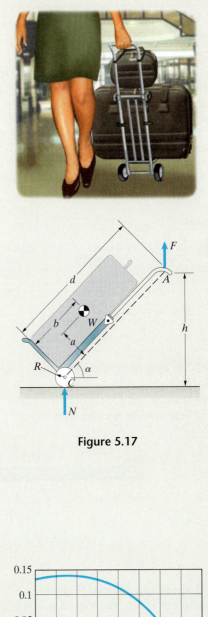

Figure 5.17

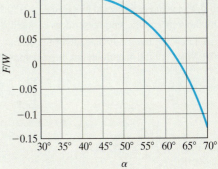

Figure 5.18
Graph of the ratio F/W as a function of α.

Problems

Assume that objects are in equilibrium. In the statements of the answers, x components are positive to the right and y components are positive upward.

5.1 The beam has pin and roller supports and is subjected to a 4-kN load.

(a) Draw the free-body diagram of the beam.
(b) Determine the reactions at the supports.

    Strategy: (a) Draw a diagram of the beam isolated from its supports. Complete the free-body diagram of the beam by adding the 4-kN load and the reactions due to the pin and roller supports (see Table 5.1). (b) Use the scalar equilibrium equations (5.4)–(5.6) to determine the reactions.

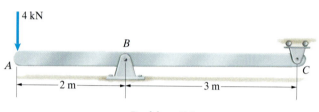

Problem 5.1

5.2 The beam has a fixed support and is loaded by a 2-kN force and a 6 kN-m couple.

(a) Draw the free-body diagram of the beam.
(b) Determine the reactions at the support.

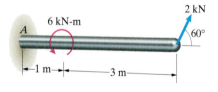

Problem 5.2

5.3 The beam is subjected to a load $F = 400$ N and is supported by the rope and the smooth surfaces at A and B.

(a) Draw the free-body diagram of the beam.
(b) What are the magnitudes of the reactions at A and B?

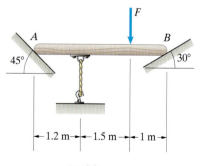

Problem 5.3

5.4 (a) Draw the free-body diagram of the beam.
(b) Determine the reactions at the supports.

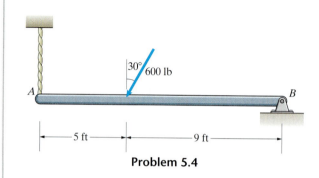

Problem 5.4

5.5 (a) Draw the free-body diagram of the 60-lb drill press, assuming that the surfaces at A and B are smooth.
(b) Determine the reactions at A and B.

Problem 5.5

5.6 The masses of the person and the diving board are 54 kg and 36 kg, respectively. Assume that they are in equilibrium.

(a) Draw the free-body diagram of the diving board.
(b) Determine the reactions at the supports A and B.

5.8 The distance $x = 9$ m.

(a) Draw the free-body diagram of the beam.
(b) Determine the reactions at the supports.

5.9 An engineer analyzes the beam and determines that each support will safely withstand a force of magnitude 20 kN. Based on this criterion, what is the range of values of the distance x at which the 10-kN force can safely be applied to the beam? Assume that $0 \le x \le 16$ m.

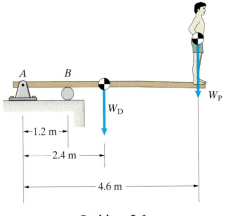

Problem 5.6

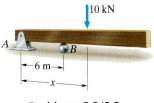

Problems 5.8/5.9

5.7 The ironing board has supports at A and B that can be modeled as roller supports.

(a) Draw the free-body diagram of the ironing board.
(b) Determine the reactions at A and B.

5.10 (a) Draw the free-body diagram of the beam.
(b) Determine the reactions at the supports.

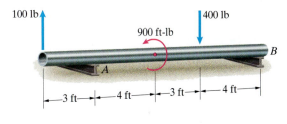

Problem 5.10

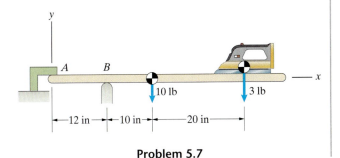

Problem 5.7

5.11 The person exerts 20-N forces on the pliers. The free-body diagram of one part of the pliers is shown. Notice that the pin at C connecting the two parts of the pliers behaves like a pin support. Determine the reactions at C and the force B exerted on the pliers by the bolt.

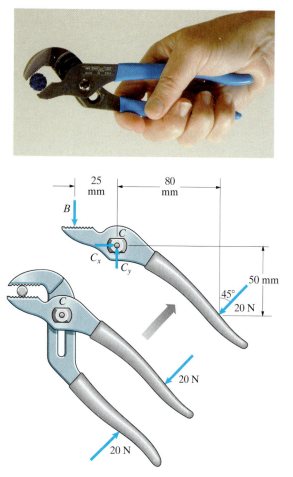

Problem 5.11

5.12 (a) Draw the free-body diagram of the beam.
(b) Determine the reactions at the pin support A.

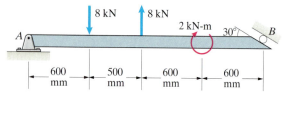

Problem 5.12

5.13 (a) Draw the free-body diagram of the beam.
(b) Determine the reactions at the supports.

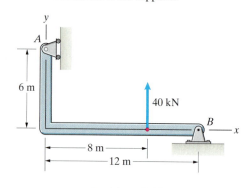

Problem 5.13

5.14 (a) Draw the free-body diagram of the beam.
(b) If $F = 4$ kN, what are the reactions at A and B?

5.15 A structural engineer determines that the support at A can safely be subjected to a force of 12 kN magnitude and the support at B can safely be subjected to a force of 15 kN magnitude. Based on this criterion, what is the largest acceptable magnitude of the force F?

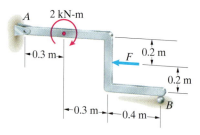

Problems 5.14/5.15

5.16 The person doing push-ups pauses in the position shown. His mass is 80 kg. Assume that his weight W acts at the point shown. The dimensions shown are $a = 250$ mm, $b = 740$ mm, and $c = 300$ mm. Determine the normal force exerted by the floor (a) on each hand, (b) on each foot.

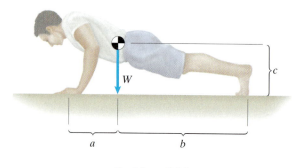

Problem 5.16

5.17 With each of the devices shown you can support a load R by applying a force F. They are called levers of the first, second, and third class.

(a) The ratio R/F is called the *mechanical advantage*. Determine the mechanical advantage of each lever.
(b) Determine the magnitude of the reaction at A for each lever. (Express your answers in terms of F.)

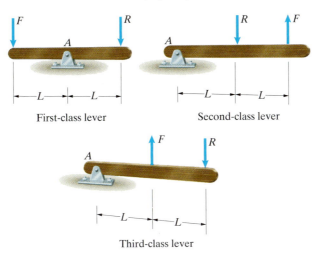

First-class lever

Second-class lever

Third-class lever

Problem 5.17

5.18 A portion of one of the decks of Frank Lloyd Wright's Fallingwater is isolated by passing an imaginary plane A through the deck. The mass of the isolated part of the deck is 14,700 kg. Treating the plane A as a fixed support, determine the reactions at A. (These are internal reactions that the deck's material, reinforced concrete, must support at the plane A.)

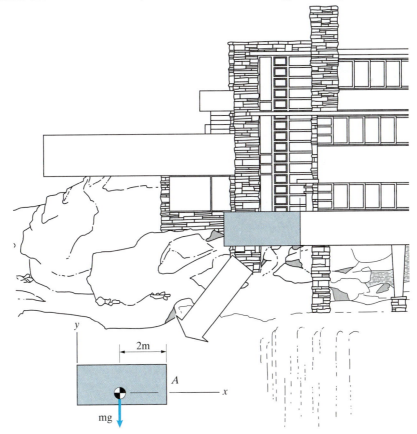

Problem 5.18

5.19 (a) Draw the free-body diagram of the beam.
(b) Determine the tension in the cable and the reactions at A.

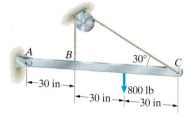

Problem 5.19

5.20 The unstretched length of the spring CD is 350 mm. Suppose that you want the lever ABC to exert a 120-N normal force on the smooth surface at A. Determine the necessary value of the spring constant k and the resulting reactions at B.

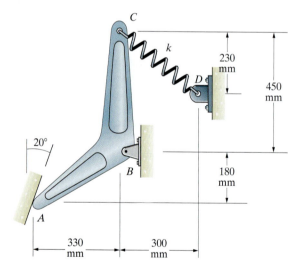

Problem 5.20

5.21 The mobile is in equilibrium. The fish B weighs 27 oz. Determine the weights of the fish A, C, and D. (The weights of the crossbars are negligible.)

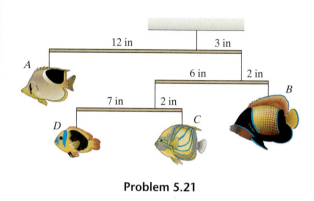

Problem 5.21

5.22 The car's wheelbase (the distance between the wheels) is 2.82 m. The mass of the car is 1760 kg and its weight acts at the point $x = 2.00$ m, $y = 0.68$ m. If the angle $\alpha = 15°$, what is the total normal force exerted on the two rear tires by the sloped ramp?

5.23 The car can remain in equilibrium on the sloped ramp only if the total friction force exerted on its tires does not exceed 0.8 times the total normal force exerted on the two rear tires. What is the largest angle α for which it can remain in equilibrium?

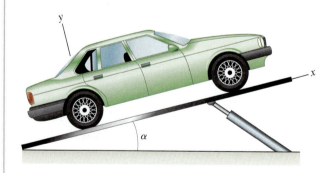

Problems 5.22/5.23

5.24 The 14.5-lb chain saw is subjected to the loads at A by the log it cuts. Determine the reactions R, B_x, and B_y that must be applied by the person using the saw to hold it in equilibrium.

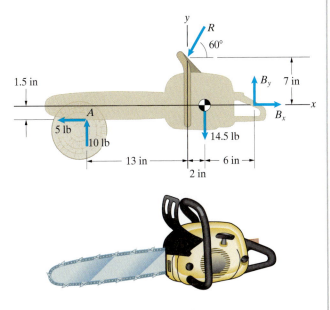

Problem 5.24

5.25 The mass of the trailer is 2.2 Mg (megagrams). The distances $a = 2.5$ m and $b = 5.5$ m. The truck is stationary, and the wheels of the trailer can turn freely, which means the road exerts no horizontal force on them. The hitch at B can be modeled as a pin support.

(a) Draw the free-body diagram of the trailer.
(b) Determine the total normal force exerted on the rear tires at A and the reactions exerted on the trailer at the pin support B.

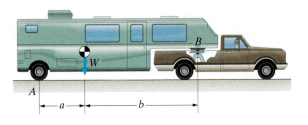

Problem 5.25

5.26 The total weight of the wheelbarrow and its load is $W = 100$ lb.
(a) If $F = 0$, what are the vertical reactions at A and B?
(b) What force F is necessary to lift the support at A off the ground?

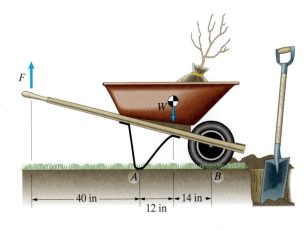

Problem 5.26

5.27 The airplane's weight is $W = 2400$ lb. Its brakes keep the rear wheels locked. The front (nose) wheel can turn freely, and so the ground exerts no horizontal force on it. The force T exerted by the airplane's propeller is horizontal.

(a) Draw the free-body diagram of the airplane. Determine the reaction exerted on the nose wheel and the total normal reaction exerted on the rear wheels
(b) when $T = 0$;
(c) when $T = 250$ lb.

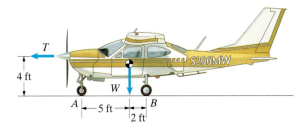

Problem 5.27

5.28 The forklift is stationary. The front wheels are free to turn, and the rear wheels are locked. The distances are $a = 1.25$ m, $b = 0.50$ m, and $c = 1.40$ m. The weight of the load is $W_L = 2$ kN, and the weight of the truck and operator is $W_F = 8$ kN. What are the reactions at A and B?

Problem 5.28

5.29 Paleontologists speculate that the stegosaur could stand on its hind limbs for short periods to feed. Based on the free-body diagram shown and assuming that $m = 2000$ kg, determine the magnitudes of the forces B and C exerted by the ligament–muscle brace and vertebral column, and determine the angle α.

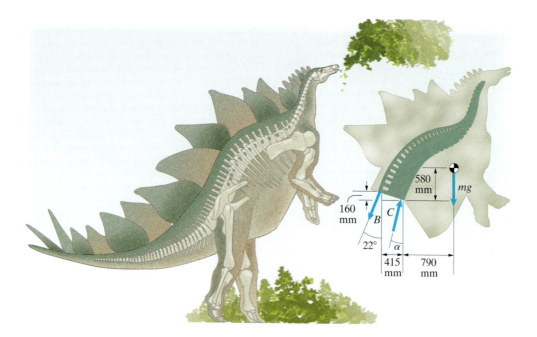

Problem 5.29

5.30 The weight of the fan is $W = 20$ lb. Its base has four equally spaced legs of length $b = 12$ in and $h = 36$ in. What is the largest thrust T exerted by the fan's propeller for which the fan will remain in equilibrium?

5.31 Consider the fan described in Problem 5.30. As a safety criterion, an engineer decides that the vertical reaction on any of the fan's legs should not be less than 20% of the fan's weight. If the thrust T is 1 lb when the fan is set on its highest speed, what is the maximum safe value of h?

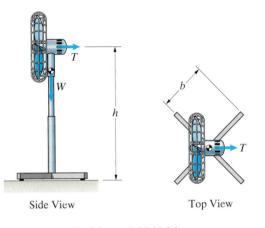

Side View   Top View

Problems 5.30/5.31

5.32 To decrease costs, an engineer considers supporting a fan with three equally spaced legs instead of the four-leg configuration shown in Problem 5.30. For the same values of b, h, and W, show that the largest thrust T for which the fan will remain in equilibrium with three legs is related to the value with four legs by

$$T_{\text{three legs}} = (1/\sqrt{2})T_{\text{four legs}}.$$

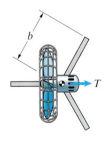

Problem 5.32

5.33 A force $F = 400$ N acts on the bracket. What are the reactions at A and B?

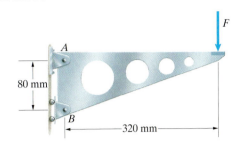

Problem 5.33

5.34 The sign's weight $W_s = 32$ lb acts at the point shown. The 10-lb weight of the bar AD acts at the midpoint of the bar. Determine the tension in the cable AE and the reactions at D.

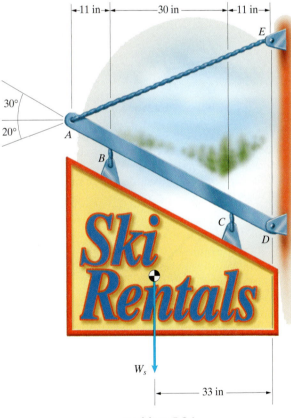

Problem 5.34

5.35 The device shown, called a *swape* or *shadoof*, helps a person lift a heavy load. (Devices of this kind were used in Egypt at least as early as 1550 B.C. and are still in use in various parts of the world.) The dimensions $a = 3.6$ m and $b = 1.2$ m. The mass of the bar and counterweight is 90 kg, and their weight W acts at the point shown. The mass of the load being lifted is 45 kg. Determine the vertical force the person must exert to support the stationary load (a) when the load is just above the ground (the position shown); (b) when the load is 1 m above the ground. Assume that the rope remains vertical.

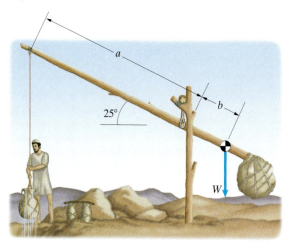

Problem 5.35

5.36 This structure, called a *truss*, has a pin support at A and a roller support at B and is loaded by two forces. Determine the reactions at the supports.

   Strategy: Draw a free-body diagram, treating the entire truss as a single object.

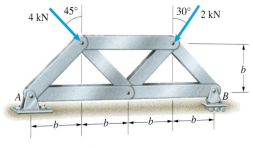

Problem 5.36

5.37 An Olympic gymnast is stationary in the "iron cross" position. The weight of his left arm and the weight of his body *not including his arms* are shown. The distances are $a = b = 9$ in and $c = 13$ in. Treat his shoulder S as a fixed support, and determine the magnitudes of the reactions at his shoulder. That is, determine the force and couple his shoulder must support.

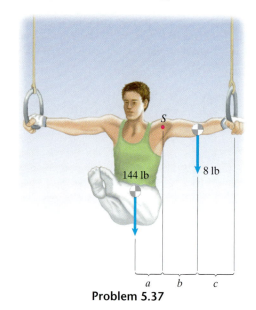

Problem 5.37

5.38 Determine the reactions at A.

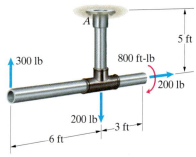

Problem 5.38

5.39 The car's brakes keep the rear wheels locked, and the front wheels are free to turn. Determine the forces exerted on the front and rear wheels by the road when the car is parked (a) on an upslope with $\alpha = 15°$; (b) on a downslope with $\alpha = -15°$.

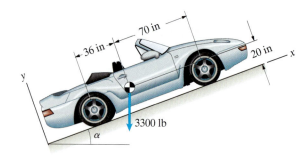

Problem 5.39

5.40 The weight W of the bar acts at its center. The surfaces are smooth. What is the tension in the horizontal string?

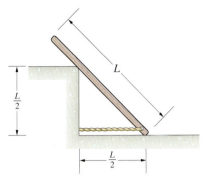

Problem 5.40

5.41 The mass of the bar is 36 kg and its weight acts at its midpoint. The spring is unstretched when $\alpha = 0$. The bar is in equilibrium when $\alpha = 30°$. Determine the spring constant k.

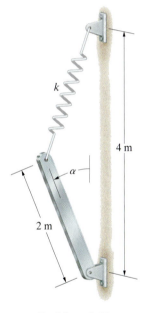

Problem 5.41

5.42 The plate is supported by a pin in a smooth slot at B. What are the reactions at the supports?

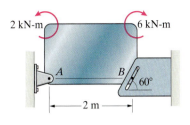

Problem 5.42

5.43 Determine the reactions at the supports.

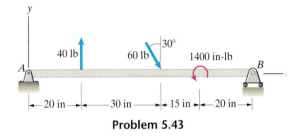

Problem 5.43

5.44 Consider the beam in Problem 5.43.
(a) If you represent the system of forces and moments consisting of the 40-lb and 60-lb forces and the 1400 in-lb couple by an equivalent system consisting of a single force $\mathbf{F}$ as shown, what is $\mathbf{F}$, and where does its line of action cross the x axis?
(b) Assume that the beam is subjected only to the force you determined in part (a) and determine the reactions at the supports. Compare your answers to the answers to Problem 5.43.

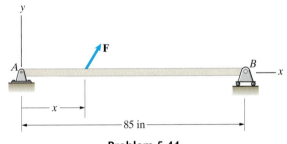

Problem 5.44

5.45 The bicycle brake on the right is pinned to the bicycle's frame at *A*. Determine the force exerted by the brake pad on the wheel rim at *B* in terms of the cable tension *T*.

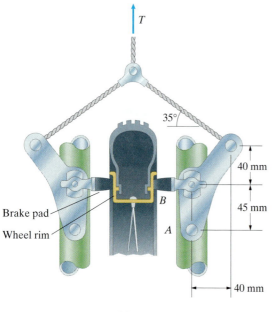

Problem 5.45

5.46 The mass of each of the suspended weights is 80 kg. Determine the reactions at the supports at *A* and *E*.

5.47 The suspended weights are each of mass *m*. The supports at *A* and *E* will each safely support a force of 6 kN magnitude. Based on this criterion, what is the largest safe value of *m*?

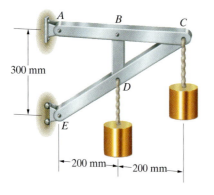

Problems 5.46/5.47

5.48 The tension in cable *BC* is 100 lb. Determine the reactions at the fixed support.

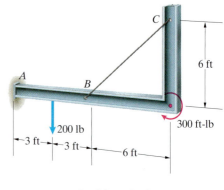

Problem 5.48

5.49 The tension in cable *AB* is 2 kN. What are the reactions at *C* in the two cases?

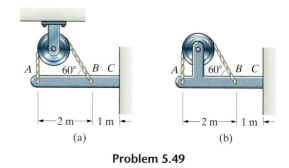

(a)   (b)

Problem 5.49

5.50 Determine the reactions at the supports.

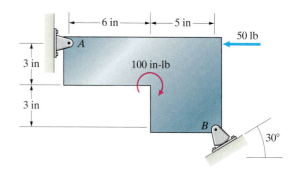

Problem 5.50

5.51 The weight $W = 2$ kN. Determine the tension in the cable and the reactions at A.

5.52 The cable will safely support a tension of 6 kN. Based on this criterion, what is the largest safe value of the weight W?

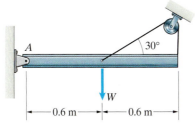

Problems 5.51/5.52

5.53 The blocks being compressed by the clamp exert a 200-N force on the pin at D that points from A toward D. The threaded shaft BE exerts a force on the pin at E that points from B toward E.
(a) Draw a free-body diagram of the arm DCE of the clamp, assuming that the pin at C behaves like a pin support.
(b) Determine the reactions at C.

5.54 The blocks being compressed by the clamp exert a 200-N force on the pin at A that points from D toward A. The threaded shaft BE exerts a force on the pin at B that points from E toward B.
(a) Draw a free-body diagram of the arm ABC of the clamp, assuming that the pin at C behaves like a pin support.
(b) Determine the reactions at C.

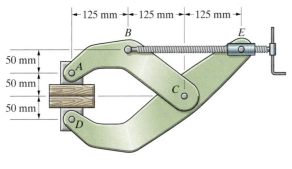

Problems 5.53/5.54

5.55 Suppose that you want to design the safety valve to open when the difference between the pressure p in the circular pipe (diameter $= 150$ mm) and atmospheric pressure is 10 MPa (megapascals; a pascal is 1 N/m^2). The spring is compressed 20 mm when the valve is closed. What should the value of the spring constant be?

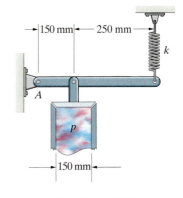

Problem 5.55

5.56 The 10-lb weight of the bar AB acts at the midpoint of the bar. The length of the bar is 3 ft. Determine the tension in the string BC and the reactions at A.

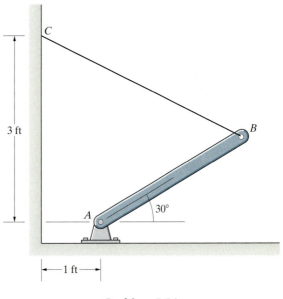

Problem 5.56

5.57 The crane's arm has a pin support at A. The hydraulic cylinder BC exerts a force on the arm at C in the direction parallel to BC. The crane's arm has a mass of 200 kg, and its weight can be assumed to act at a point 2 m to the right of A. If the mass of the suspended box is 800 kg and the system is in equilibrium, what is the magnitude of the force exerted by the hydraulic cylinder?

5.58 In Problem 5.57, what is the magnitude of the force exerted on the crane's arm by the pin support at A?

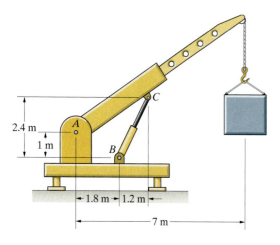
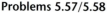

Problems 5.57/5.58

5.59 A speaker system is suspended by the cables attached at D and E. The mass of the speaker system is 130 kg, and its weight acts at G. Determine the tensions in the cables and the reactions at A and C.

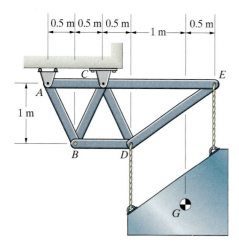

Problem 5.59

5.60 The weight $W_1 = 1000$ lb. Neglect the weight of the bar AB. The cable goes over a pulley at C. Determine the weight W_2 and the reactions at the pin support A.

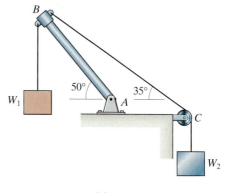

Problem 5.60

5.61 The dimensions $a = 2$ m and $b = 1$ m. The couple $M = 2400$ N-m. The spring constant is $k = 6000$ N/m, and the spring would be unstretched if $h = 0$. The system is in equilibrium when $h = 2$ m and the beam is horizontal. Determine the force F and the reactions at A.

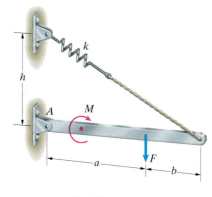

Problem 5.61

5.62 The bar is 1 m long, and its weight W acts at its midpoint. The distance $b = 0.75$ m, and the angle $\alpha = 30°$. The spring constant is $k = 100$ N/m, and the spring is unstretched when the bar is vertical. Determine W and the reactions at A.

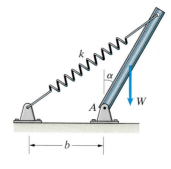

Problem 5.62

5.63 The boom derrick supports a suspended 15-kip load. The booms *BC* and *DE* are each 20 ft long. The distances are $a = 15$ ft and $b = 2$ ft, and the angle $\theta = 30°$. Determine the tension in cable *AB* and the reactions at the pin supports *C* and *D*.

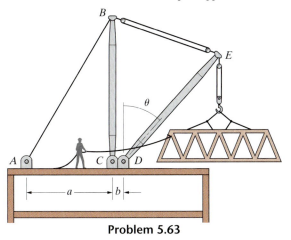

Problem 5.63

5.64 The arrangement shown controls the elevators of an airplane. (The elevators are the horizontal control surfaces in the airplane's tail.) The elevators are attached to member *EDG*. Aerodynamic pressures on the elevators exert a clockwise couple of 120 in-lb. Cable *BG* is slack, and its tension can be neglected. Determine the force *F* and the reactions at the pin support *A*.

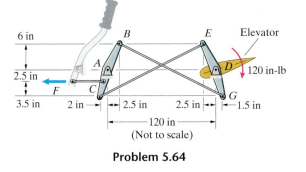

Problem 5.64

Problems 5.65–5.68 are related to Design Example 5.5.

5.65 In Fig. 5.17, suppose that $\alpha = 40°$, $d = 1$ m, $a = 200$ mm, $b = 500$ mm, $R = 75$ mm, and the mass of the luggage is 40 kg. Determine *F* and *N*.

5.66 In Fig. 5.17, suppose that $\alpha = 35°$, $d = 46$ in, $a = 10$ in, $b = 14$ in, $R = 3$ in, and you don't want the user to have to exert a force *F* larger than 20 lb. What is the largest luggage weight that can be placed on the carrier?

5.67 One of the difficulties in making design decisions is that you don't know how the user will place the luggage on the carrier in Example 5.5. Suppose you assume that the point where the weight acts may be anywhere within the "envelope" $R \le a \le 0.75c$ and $0 \le b \le 0.75d$. If $\alpha = 30°$, $c = 14$ in, $d = 48$ in, $R = 3$ in, and $W = 80$ lb, what is the largest force *F* the user will have to exert for any luggage placement?

5.68 In our design of the luggage carrier in Example 5.5, we assumed a user that would hold the carrier's handle at $h = 36$ in above the floor. We assumed that $R = 3$ in, $a = 6$ in, and $b = 12$ in, and we chose the dimension $d = 4$ ft. The resulting ratio of the force the user must exert to the weight of the luggage is $F/W = 0.132$. Suppose that people with a range of heights use this carrier. Obtain a graph of F/W as a function of *h* for $24 \le h \le 36$ in.

5.3 Statically Indeterminate Objects

In Section 5.2 we discussed examples in which we were able to use the equilibrium equations to determine unknown forces and couples acting on objects in equilibrium. It is important to be aware of two common situations in which this procedure doesn't lead to a solution. First, the free-body diagram of an object can have more unknown forces or couples than the number of independent equilibrium equations that can be obtained. For example, because no more than three independent equilibrium equations can be obtained from a given free-body diagram in a two-dimensional problem, if there are more than three unknowns they can't all be determined from the equilibrium equations alone. This occurs, for example, when an object has more supports than the minimum number necessary to maintain it in equilibrium. Such an object is said to have *redundant supports*. The second situation is when the supports of an object are improperly designed such that they cannot maintain equilibrium under the loads acting on it. The object is said to have *improper supports*. In either situation, the object is said to be *statically indeterminate*.

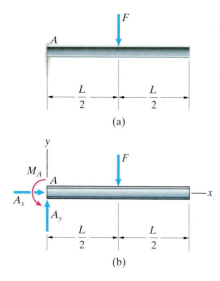

Figure 5.19
(a) A beam with a fixed support.
(b) The free-body diagram has three unknown reactions.

Engineers use redundant supports whenever possible for strength and safety. Some designs, however, require that the object be incompletely supported so that it is free to undergo certain motions. These two situations—more supports than necessary for equilibrium or not enough—are so common that we consider them in detail.

Redundant Supports

Consider a beam with a fixed support (Fig. 5.19a). From its free-body diagram (Fig. 5.19b), we obtain the equilibrium equations

$$\Sigma F_x = A_x = 0,$$

$$\Sigma F_y = A_y - F = 0,$$

$$\Sigma M_{\text{point } A} = M_A - \left(\frac{L}{2}\right)F = 0.$$

Assuming we know the load F, we have three equations and three unknown reactions, for which we obtain the solutions $A_x = 0$, $A_y = F$, and $M_A = FL/2$.

Now suppose we add a roller support at the right end of the beam (Fig. 5.20a). From the new free-body diagram (Fig. 5.20b), we obtain the equilibrium equations

$$\Sigma F_x = A_x = 0, \tag{5.10}$$

$$\Sigma F_y = A_y - F + B = 0, \tag{5.11}$$

$$\Sigma M_{\text{point } A} = M_A - \left(\frac{L}{2}\right)F + LB = 0. \tag{5.12}$$

Now we have three equations and four unknown reactions. Although the first equation tells us that $A_x = 0$, we can't solve the two equations (5.11) and (5.12) for the three reactions A_y, B, and M_A.

When faced with this situation, students often attempt to sum the moments about another point, such as point B, to obtain an additional equation:

$$\Sigma M_{\text{point } B} = M_A + \left(\frac{L}{2}\right)F - LA_y = 0.$$

Unfortunately, this doesn't help. This is not an independent equation, but is a linear combination of Eqs. (5.11) and (5.12):

$$\Sigma M_{\text{point } B} = M_A + \left(\frac{L}{2}\right)F - LA_y$$

$$= \underbrace{M_A - \left(\frac{L}{2}\right)F + LB}_{\text{Eq. (5.12)}} - \underbrace{L(A_y - F + B)}_{\text{Eq. (5.11)}}.$$

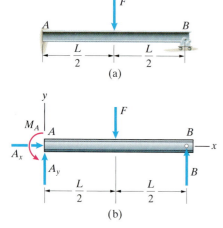

Figure 5.20
(a) A beam with fixed and roller supports.
(b) The free-body diagram has four unknown reactions.

As this example demonstrates, each support added to an object results in additional reactions. The difference between the number of reactions and the number of independent equilibrium equations is called the *degree of redundancy*. Even if an object is statically indeterminate due to redundant supports, it may be possible to determine some of the reactions from the equilibrium equations. Notice that in our previous example we were able to determine the reaction A_x even though we could not determine the other reactions.

Since redundant supports are so ubiquitous, you may wonder why we devote so much effort to teaching you how to analyze objects whose reactions can be determined with the equilibrium equations. We want to develop your understanding of equilibrium and give you practice writing equilibrium equations. The reactions on an object with redundant supports *can* be determined by supplementing the equilibrium equations with additional equations that relate the forces and couples acting on the object to its deformation, or change in shape. Thus obtaining the equilibrium equations is the first step of the solution.

Example 5.6   Recognizing a Statically Indeterminate Object

The beam in Fig. 5.21 has two pin supports and is loaded by a 2-kN force.
(a) Show that the beam is statically indeterminate.
(b) Determine as many reactions as possible.

Strategy
The beam is statically indeterminate if its free-body diagram has more unknown reactions than the number of independent equilibrium equations we can obtain. But even if this is the case, we may be able to solve the equilibrium equations for some of the reactions.

Figure 5.21

Solution

Draw the Free-Body Diagram   We draw the free-body diagram of the beam in Fig. a. There are four unknown reactions—A_x, A_y, B_x, and B_y—and we can write only three independent equilibrium equations. Therefore the beam is statically indeterminate.

Apply the Equilibrium Equations   Summing the moments about point A, the equilibrium equations are

$$\Sigma F_x = A_x + B_x = 0,$$

$$\Sigma F_y = A_y + B_y - 2 \text{ kN} = 0,$$

$$\Sigma M_{\text{point } A} = (5 \text{ m})B_y - (3 \text{ m})(2 \text{ kN}) = 0.$$

We can solve the third equation for B_y and then solve the second equation for A_y. The results are $A_y = 0.8$ kN and $B_y = 1.2$ kN. The first equation tells us that $B_x = -A_x$, but we can't solve for their values.

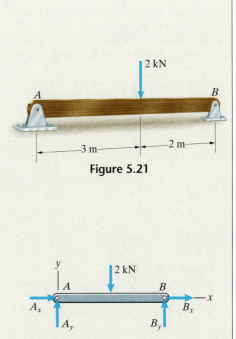

(a) The free-body diagram of the beam.

Critical Thinking
Why can't the reactions on objects with redundant supports be determined from the equilibrium equations alone? This example illustrates why. The two pin supports exert horizontal forces on the beam. The equilibrium equations tell us that these forces must be equal and opposite (Fig. b), but do not determine their magnitude or direction. Notice that the free-body diagram in Fig. b is in equilibrium for any value of T, and the forces may point outward from the beam as shown or inward toward the beam. We must know more than the fact that the beam is in equilibrium to determine the horizontal forces. Methods for solving statically indeterminate problems are developed in mechanics of materials.

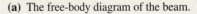

(b) The supports alone can exert reactions on the beam.

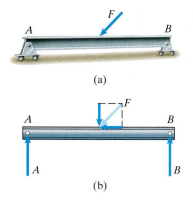

(a)

(b)

Figure 5.22
(a) A beam with two roller supports is not in equilibrium when subjected to the load shown.
(b) The sum of the forces in the horizontal direction is not zero.

Improper Supports

We say that an object has improper supports if it will not remain in equilibrium under the action of the loads exerted on it. Thus an object with improper supports will move when the loads are applied. In two-dimensional problems, this can occur in two ways:

1. *The supports can exert only parallel forces.* This leaves the object free to move in the direction perpendicular to the support forces. If the loads exert a component of force in that direction, the object is not in equilibrium. Figure 5.22a shows an example of this situation. The two roller supports can exert only vertical forces, while the force F has a horizontal component. The beam will move horizontally when F is applied. This is particularly apparent from the free-body diagram (Fig. 5.22b). The sum of the forces in the horizontal direction cannot be zero because the roller supports can exert only vertical reactions.

2. *The supports can exert only concurrent forces.* If the loads exert a moment about the point where the lines of action of the support forces intersect, the object is not in equilibrium. For example, consider the beam in Fig. 5.23a. From its free-body diagram (Fig. 5.23b) we see that the reactions A and B exert no moment about the point P, where their lines of action intersect, but the load F does. The sum of the moments about point P is not zero, and the beam will rotate when the load is applied.

Except for problems that deal explicitly with improper supports, objects in our examples and problems have proper supports. You should develop the habit of examining objects in equilibrium and thinking about why they are properly supported for the loads acting on them.

Figure 5.23
(a) A beam with roller supports on sloped surfaces.
(b) The sum of the moments about point P is not zero.

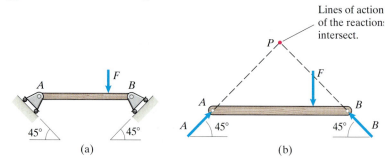

(a)   (b)

Example 5.7　Proper and Improper Supports

State whether each L-shaped bar in Fig. 5.24 is properly or improperly supported. If a bar is properly supported, determine the reactions at its supports.

Strategy

By drawing the free-body diagram of each bar, we can determine whether the reactions of the supports can exert only parallel or concurrent forces on it. If so, we can then recognize whether the applied load results in the bar not being in equilibrium.

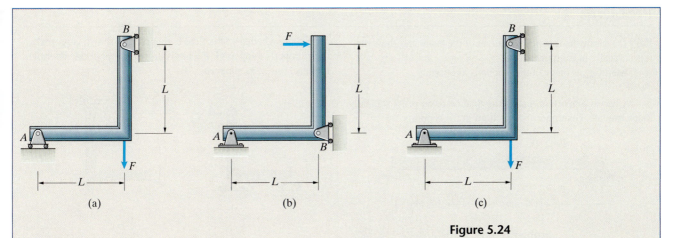

Figure 5.24

Solution

We draw the free-body diagrams of the bars in Fig. 5.25.

Bar (a)   The lines of action of the reactions due to the two roller supports intersect at P, and the load F exerts a moment about P. This bar is improperly supported.

Bar (b)   The lines of action of the reactions intersect at A, and the load F exerts a moment about A. This bar is also improperly supported.

Bar (c)   The three support forces are neither parallel nor concurrent. This bar is properly supported. The equilibrium equations are

$$\Sigma F_x = A_x - B = 0,$$

$$\Sigma F_y = A_y - F = 0,$$

$$\Sigma M_{\text{point } A} = BL - FL = 0.$$

Solving these equations, the reactions are $A_x = F, A_y = F,$ and $B = F$.

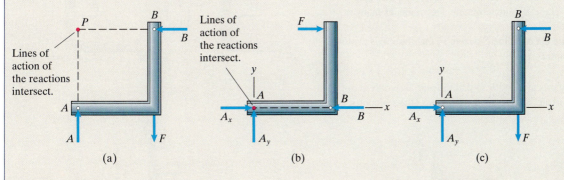

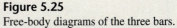

Figure 5.25
Free-body diagrams of the three bars.

Critical Thinking

An essential part of learning mechanics is developing your intuition about the behaviors of the physical systems we study. In this example, think about the effects of the loads on the three systems in Fig. 5.24, and see if you can predict whether they are properly supported. Will the loads cause the bars to move or not? Then see if your judgment is confirmed by the analyses given in the example.

Problems

5.69 (a) Draw the free-body diagram of the beam and show that it is statically indeterminate.
(b) Determine as many of the reactions as possible.

5.70 Choose supports at A and B so that the beam is not statically indeterminate. Determine the reactions at the supports.

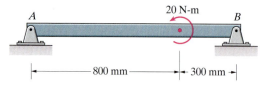

Problems 5.69/5.70

5.71 (a) Draw the free-body diagram of the beam and show that it is statically indeterminate. (The external couple M_0 is known.)
(b) By an analysis of the beam's deflection, it is determined that the vertical reaction B exerted by the roller support is related to the couple M_0 by $B = 2M_0/L$. What are the reactions at A?

5.72 Choose supports at A and B so that the beam is not statically indeterminate. Determine the reactions at the supports.

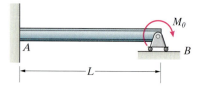

Problems 5.71/5.72

5.73 Draw the free-body diagram of the L-shaped pipe assembly and show that it is statically indeterminate. Determine as many of the reactions as possible.
Strategy: Place the coordinate system so that the x axis passes through points A and B.

5.74 Choose supports at A and B so that the pipe assembly is not statically indeterminate. Determine the reactions at the supports.

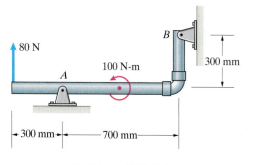

Problems 5.73/5.74

5.75 State whether each of the L-shaped bars shown is properly or improperly supported. If a bar is properly supported, determine the reactions at its supports.

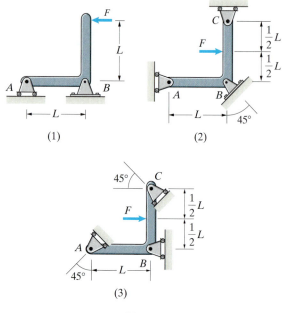

Problem 5.75

5.76 State whether each of the L-shaped bars shown is properly or improperly supported. If a bar is properly supported, determine the reactions at its supports.

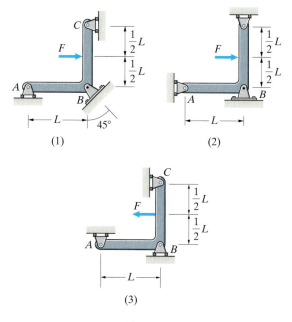

Problem 5.76

5.4 Three-Dimensional Applications

We have seen that when an object in equilibrium is subjected to a two-dimensional system of forces and moments, no more than three independent equilibrium equations can be obtained. In the case of a three-dimensional system of forces and moments, up to six independent equilibrium equations can be obtained. The three components of the sum of the forces must equal zero and the three components of the sum of the moments about a point must equal zero. The procedure for determining the reactions on an object subjected to a three-dimensional system of forces and moments—drawing a free-body diagram and applying the equilibrium equations—is the same as in two dimensions. But it is necessary to become familiar with the support conventions used in three-dimensional applications.

Supports

We present five conventions frequently used in three-dimensional problems. Again, even when actual supports do not physically resemble these models, we represent them by the models if they exert the same (or approximately the same) reactions.

The Ball and Socket Support   In the *ball and socket support*, the supported object is attached to a ball enclosed within a spherical socket (Fig. 5.26a). The socket permits the ball to rotate freely (friction is neglected) but prevents it from translating in any direction.

Imagine holding a bar attached to a ball and socket support (Fig. 5.26b). If you try to translate the bar (move it without rotating it) in any direction, the support exerts a reactive force to prevent the motion. However, you can rotate the bar about the support. The support cannot exert a couple to prevent rotation. Thus a ball and socket support can't exert a couple but can exert three components of force (Fig. 5.26c). It is the three-dimensional analog of the two-dimensional pin support.

The human hip joint is an example of a ball and socket support (Fig. 5.27). The support of the gear shift lever of a car can be modeled as a ball and socket support within the lever's range of motion.

The Roller Support   The *roller support* (Fig. 5.28a) is a ball and socket support that can roll freely on a supporting surface. A roller support can exert only a force normal to the supporting surface (Fig. 5.28b). The rolling "casters" sometimes used to support furniture legs are supports of this type.

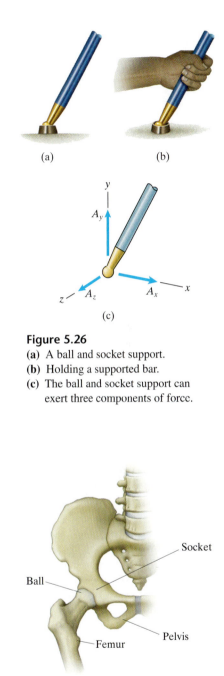

Figure 5.26
(a) A ball and socket support.
(b) Holding a supported bar.
(c) The ball and socket support can exert three components of force.

Figure 5.27
The human femur is attached to the pelvis by a ball and socket support.

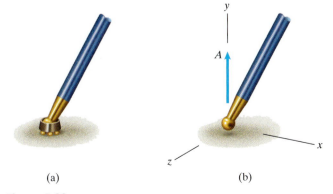

(a)   (b)

Figure 5.28
(a) A roller support.
(b) The reaction is normal to the supporting surface.

The Hinge   The hinge support is the familiar device used to support doors. It permits the supported object to rotate freely about a line, the *hinge axis*. An object is attached to a hinge in Fig. 5.29a. The z axis of the coordinate system is aligned with the hinge axis.

   If you imagine holding a bar attached to a hinge (Fig. 5.29b), notice that you can rotate the bar about the hinge axis. The hinge cannot exert a couple about the hinge axis (the z axis) to prevent rotation. However, you can't rotate the bar about the x or y axis because the hinge can exert couples about those axes to resist the motion. In addition, you can't translate the bar in any direction. The reactions a hinge can exert on an object are shown in Fig. 5.29c. There are three components of force, A_x, A_y, and A_z, and couples about the x and y axes, M_{Ax} and M_{Ay}.

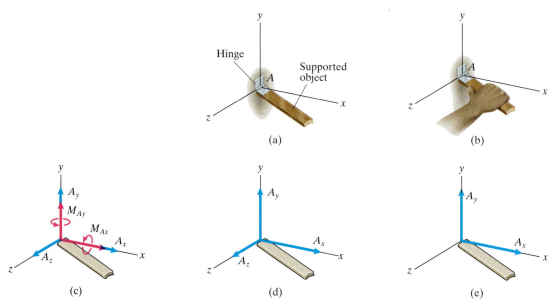

(a)

(b)

(c)

(d)

(e)

Figure 5.29
(a) A hinge. The z axis is aligned with the hinge axis.
(b) Holding a supported bar.
(c) In general, a hinge can exert five reactions: three force components and two couple components.
(d) The reactions when the hinge exerts no couples.
(e) The reactions when the hinge exerts neither couples nor a force parallel to the hinge axis.

   In some situations, either a hinge exerts no couples on the object it supports, or they are sufficiently small to neglect. An example of the latter case is when the axes of the hinges supporting a door are properly aligned (the axes of the individual hinges coincide). In these situations the hinge exerts only forces on an object (Fig. 5.29d). Situations also arise in which a hinge exerts no couples on an object and exerts no force in the direction of the hinge axis. (The hinge may actually be designed so that it cannot support a force parallel to the hinge axis.) Then the hinge exerts forces only in the directions perpendicular to the hinge axis (Fig. 5.29e). In examples and problems, we indicate when a hinge does not exert all five of the reactions in Fig. 5.29c.

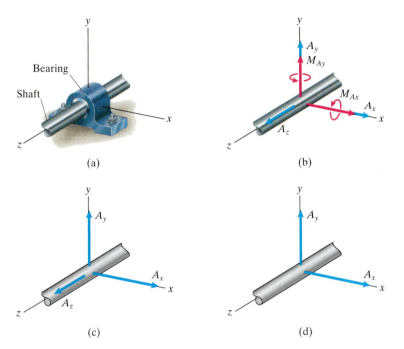

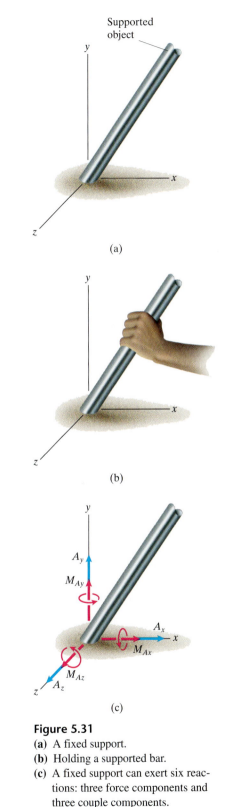

(a)

Figure 5.30
(a) A bearing. The z axis is aligned with the axis of the shaft.
(b) In general, a bearing can exert five reactions: three force components and two couple components.
(c) The reactions when the bearing exerts no couples.
(d) The reactions when the bearing exerts neither couples nor a force parallel to the axis of the shaft.

The Bearing

The type of bearing shown in Fig. 5.30a supports a circular shaft while permitting it to rotate about its axis. The reactions are identical to those exerted by a hinge. In the most general case (Fig. 5.30b), the bearing can exert a force on the supported shaft in each coordinate direction and can exert couples about axes perpendicular to the shaft but cannot exert a couple about the axis of the shaft.

As in the case of the hinge, situations can occur in which the bearing exerts no couples (Fig. 5.30c) or exerts no couples and no force parallel to the shaft axis (Fig. 5.30d). Some bearings are designed in this way for specific applications. In examples and problems, we indicate when a bearing does not exert all of the reactions in Fig. 5.30b.

The Fixed Support

You are already familiar with the fixed, or built-in, support (Fig. 5.31a). Imagine holding a bar with a fixed support (Fig. 5.31b). You cannot translate it in any direction, and you cannot rotate it about any axis. The support is capable of exerting forces A_x, A_y, and A_z in each coordinate direction and couples M_{Ax}, M_{Ay}, and M_{Az} about each coordinate axis (Fig. 5.31c).

Table 5.2 summarizes the support conventions commonly used in three-dimensional applications.

Figure 5.31
(a) A fixed support.
(b) Holding a supported bar.
(c) A fixed support can exert six reactions: three force components and three couple components.

Table 5.2 Supports used in three-dimensional applications.

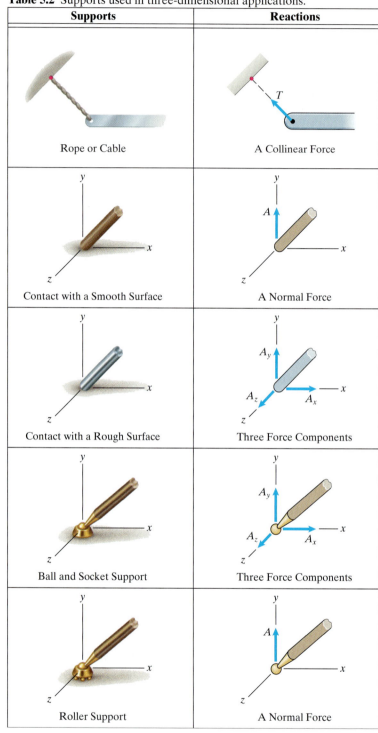

Supports	Reactions
Rope or Cable	A Collinear Force
Contact with a Smooth Surface	A Normal Force
Contact with a Rough Surface	Three Force Components
Ball and Socket Support	Three Force Components
Roller Support	A Normal Force

Table 5.2 *continued*

Supports	Reactions

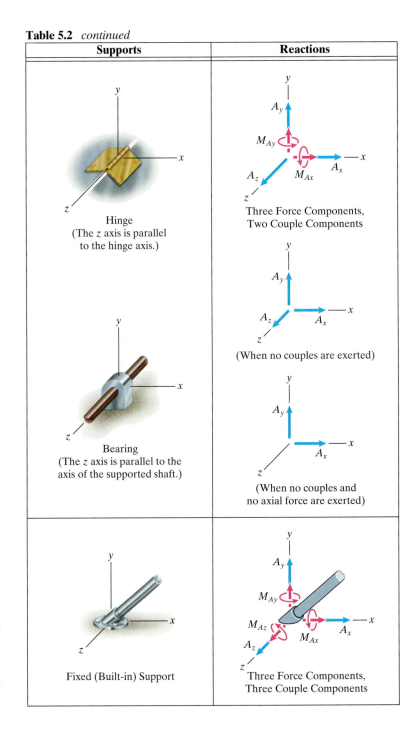

Hinge
(The z axis is parallel
to the hinge axis.)

Three Force Components,
Two Couple Components

(When no couples are exerted)

Bearing
(The z axis is parallel to the
axis of the supported shaft.)

(When no couples and
no axial force are exerted)

Fixed (Built-in) Support

Three Force Components,
Three Couple Components

The Scalar Equilibrium Equations

When an object is in equilibrium, the system of forces and couples acting on it satisfy Eqs. (5.1) and (5.2). The sum of the forces is zero and the sum of the moments about any point is zero. Expressing these equations in terms of cartesian components in three dimensions yields the six scalar equilibrium equations.

$$\Sigma F_x = 0, \tag{5.13}$$

$$\Sigma F_y = 0, \tag{5.14}$$

$$\Sigma F_z = 0, \tag{5.15}$$

$$\Sigma M_x = 0, \tag{5.16}$$

$$\Sigma M_y = 0, \tag{5.17}$$

$$\Sigma M_z = 0. \tag{5.18}$$

The sums of the moments can be evaluated about any point. Although more equations can be obtained by summing moments about other points, they will not be independent of these equations. *More than six independent equilibrium equations cannot be obtained from a given free-body diagram, so at most six unknown forces or couples can be determined.*

The steps required to determine reactions in three dimensions are familiar from the two-dimensional applications we have discussed. First obtain a free-body diagram by isolating an object and showing the loads and reactions acting on it, then use Eqs. (5.13)-(5.18) to determine the reactions.

Study Questions

1. What is a ball and socket support? What reactions can it exert on an object?
2. In general, a hinge support can exert five reactions on an object. What are they?
3. If an object has a fixed support and any additional supports, it is statically indeterminate. Why is this true?

Example 5.8 **Determining Reactions in Three Dimensions**

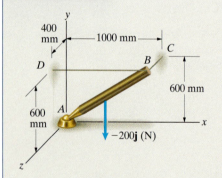

Figure 5.32

The bar *AB* in Fig. 5.32 is supported by the cables *BC* and *BD* and a ball and socket support at *A*. Cable *BC* is parallel to the *z* axis, and cable *BD* is parallel to the *x* axis. The 200-N weight of the bar acts at its midpoint. What are the tensions in the cables and the reactions at *A*?

Strategy

We must obtain the free-body diagram of the bar *AB* by isolating it from the support at *A* and the two cables. Then we can use the equilibrium equations to determine the reactions at *A* and the tensions in the cables.

Solution

Draw the Free-Body Diagram In Fig. a we isolate the bar and show the reactions that may be exerted on it. The ball and socket support can exert three components of force, A_x, A_y, and A_z. The terms T_{BC} and T_{BD} represent the tensions in the cables.

Apply the Equilibrium Equations The sums of the forces in each coordinate direction equal zero:

$$\Sigma F_x = A_x - T_{BD} = 0,$$
$$\Sigma F_y = A_y - 200 \text{ N} = 0, \tag{1}$$
$$\Sigma F_z = A_z - T_{BC} = 0.$$

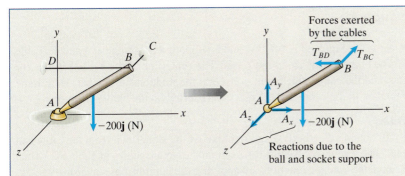

(a) Obtaining the free-body diagram of the bar.

Let $\mathbf{r}_{AB}$ be the position vector from A to B. The sum of the moments about A, with forces in N and distances in m, is

$$\Sigma \mathbf{M}_{\text{point } A} = [\mathbf{r}_{AB} \times (-T_{BC}\mathbf{k})] + [\mathbf{r}_{AB} \times (-T_{BD}\mathbf{i})]$$

$$+ [\tfrac{1}{2}\mathbf{r}_{AB} \times (-200\mathbf{j})]$$

$$= \begin{vmatrix} \mathbf{i} & \mathbf{j} & \mathbf{k} \\ 1 & 0.6 & 0.4 \\ 0 & 0 & -T_{BC} \end{vmatrix} + \begin{vmatrix} \mathbf{i} & \mathbf{j} & \mathbf{k} \\ 1 & 0.6 & 0.4 \\ -T_{BD} & 0 & 0 \end{vmatrix}$$

$$+ \begin{vmatrix} \mathbf{i} & \mathbf{j} & \mathbf{k} \\ 0.5 & 0.3 & 0.2 \\ 0 & -200 & 0 \end{vmatrix}$$

$$= (-0.6T_{BC} + 40)\mathbf{i} + (T_{BC} - 0.4T_{BD})\mathbf{j} + (0.6T_{BD} - 100)\mathbf{k}.$$

The components of this vector (the sums of the moments about the three coordinate axes) each equal zero:

$$\Sigma M_x = -(0.6 \text{ m})T_{BC} + 40 \text{ N-m} = 0,$$

$$\Sigma M_y = (1 \text{ m})T_{BC} - (0.4 \text{ m})T_{BD} = 0,$$

$$\Sigma M_z = (0.6 \text{ m})T_{BD} - 100 \text{ N-m} = 0.$$

Solving these equations, we obtain the tensions in the cables:

$$T_{BC} = 66.7 \text{ N}, \qquad T_{BD} = 166.7 \text{ N}.$$

(Notice that we needed only two of the three equations to obtain the two tensions. The third equation is redundant.)

Then from Eqs. (1) we obtain the reactions at the ball and socket support:

$$A_x = 166.7 \text{ N}, \qquad A_y = 200 \text{ N}, \qquad A_z = 66.7 \text{ N}.$$

Critical Thinking

How does a structural engineer find out whether this bar will support the loads acting on it without failing or *design the bar so that it will support those forces*? You can see that doing so would require consideration of the properties of the bar's material and the geometry of its cross section. But *the initial step in answering this question is determining all of the forces that act on the bar*, as we have done in this example. Also, the tensions in the cables BC and BD must be determined to ascertain whether they will support those tensions. If they do not, then cables must be chosen that will support them.

Example 5.9 | Reactions at a Hinge Support

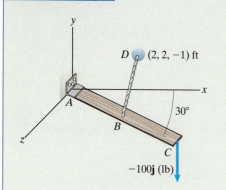

Figure 5.33

The bar AC in Fig. 5.33 is 4 ft long and is supported by a hinge at A and the cable BD. The hinge axis is along the z axis. The centerline of the bar lies in the x–y plane, and the cable attachment point B is the midpoint of the bar. Determine the tension in the cable and the reactions exerted on the bar by the hinge.

Strategy

We will obtain a free-body diagram of bar AC by isolating it from the cable and hinge. (The reactions the hinge can exert on the bar are shown in Table 5.2.) Then we can determine the reactions by applying the equilibrium equations.

Solution

Draw the Free-Body Diagram We isolate the bar from the hinge support and the cable and show the reactions they exert (Fig. a). The terms A_x, A_y, and A_z are the components of force exerted by the hinge, and the terms M_{Ax} and M_{Ay} are the couples exerted by the hinge about the x and y axes. (Remember that the hinge cannot exert a couple on the bar about the hinge axis.) The term T is the tension in the cable.

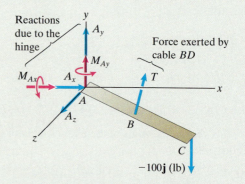

(a) The free-body diagram of the bar.

Apply the Equilibrium Equations To write the equilibrium equations, we must first express the cable force in terms of its components. The coordinates of point B are $(2 \cos 30°, -2 \sin 30°, 0)$ ft, so the position vector from B to D is

$$\mathbf{r}_{BD} = (2 - 2 \cos 30°)\mathbf{i} + [2 - (-2 \sin 30°)]\mathbf{j} + (-1 - 0)\mathbf{k}$$

$$= 0.268\mathbf{i} + 3\mathbf{j} - \mathbf{k} \text{ (ft).}$$

We divide this vector by its magnitude to obtain a unit vector $\mathbf{e}_{BD}$ that points from point B toward point D:

$$\mathbf{e}_{BD} = \frac{\mathbf{r}_{BD}}{|\mathbf{r}_{BD}|} = 0.084\mathbf{i} + 0.945\mathbf{j} - 0.315\mathbf{k}.$$

Now we can write the cable force as the product of its magnitude and $\mathbf{e}_{BD}$:

$$T\mathbf{e}_{BD} = T(0.084\mathbf{i} + 0.945\mathbf{j} - 0.315\mathbf{k}).$$

The sums of the forces in each coordinate direction must equal zero:

$$\Sigma F_x = A_x + 0.084T = 0,$$

$$\Sigma F_y = A_y + 0.945T - 100 \text{ lb} = 0, \tag{1}$$

$$\Sigma F_z = A_z - 0.315T = 0.$$

If we sum moments about A, the resulting equations do not contain the unknown reactions A_x, A_y, and A_z. The position vectors from A to B and from A to C are

$$\mathbf{r}_{AB} = 2 \cos 30°\mathbf{i} - 2 \sin 30°\mathbf{j} \text{ (ft)},$$

$$\mathbf{r}_{AC} = 4 \cos 30°\mathbf{i} - 4 \sin 30°\mathbf{j} \text{ (ft)}.$$

The sum of the moments about A, with forces in lb and distances in ft, is

$$\Sigma \mathbf{M}_{\text{point } A} = M_{Ax}\mathbf{i} + M_{Ay}\mathbf{j} + [\mathbf{r}_{AB} \times (T\mathbf{e}_{BD})] + [\mathbf{r}_{AC} \times (-100\mathbf{j})]$$

$$= M_{Ax}\mathbf{i} + M_{Ay}\mathbf{j} + \begin{vmatrix} \mathbf{i} & \mathbf{j} & \mathbf{k} \\ 1.732 & -1 & 0 \\ 0.084T & 0.945T & -0.315T \end{vmatrix}$$

$$+ \begin{vmatrix} \mathbf{i} & \mathbf{j} & \mathbf{k} \\ 3.464 & -2 & 0 \\ 0 & -100 & 0 \end{vmatrix}$$

$$= (M_{Ax} + 0.315T)\mathbf{i} + (M_{Ay} + 0.546T)\mathbf{j}$$

$$+ (1.72T - 346)\mathbf{k} = 0.$$

From this vector equation, we obtain the scalar equations

$$\Sigma M_x = M_{Ax} + (0.315 \text{ ft})T = 0,$$

$$\Sigma M_y = M_{Ay} + (0.546 \text{ ft})T = 0,$$

$$\Sigma M_z = (1.72 \text{ ft})T_{BD} - 346 \text{ ft-lb} = 0.$$

Solving these equations yields the reactions

$$T = 201 \text{ lb}, \qquad M_{Ax} = -63.4 \text{ ft-lb}, \qquad M_{Ay} = -109.8 \text{ ft-lb}.$$

Then from Eqs. (1) we obtain the forces exerted on the bar by the hinge:

$$A_x = -17.0 \text{ lb}, \qquad A_y = -90.2 \text{ lb}, \qquad A_z = 63.4 \text{ lb}.$$

Critical Thinking

Notice in Table 5.2 that there are three possibilities for the reactions exerted by a hinge or bearing. How do you know which one to choose? Under certain circumstances, a hinge may not exert significant couples on the object to which it is connected, and it also may not exert a significant force in the direction of the hinge axis. For example, when an object has two hinge supports and their axes are aligned (see Example 5.10), you can often assume that each individual hinge does not exert couples on the object. But in general, it requires experience to make such judgments. In upcoming examples and problems, we will indicate the reactions that you can assume are exerted by a hinge. Whenever you are in doubt, you should assume that a hinge may exert the most general set of reactions shown in Table 5.2 (three force components and two couple components).

Example 5.10   Reactions at Properly Aligned Hinges

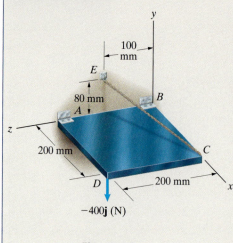

Figure 5.34

The plate in Fig. 5.34 is supported by hinges at A and B and the cable CE. The properly aligned hinges do not exert couples on the plate, and the hinge at A does not exert a force on the plate in the direction of the hinge axis. Determine the reactions at the hinges and the tension in the cable.

Strategy

We will draw the free-body diagram of the plate, using the given information about the reactions exerted by the hinges at A and B. Before the equilibrium equations can be applied, we must express the force exerted on the plate by the cable in terms of its components.

Solution

Draw the Free-Body Diagram   We isolate the plate and show the reactions at the hinges and the force exerted by the cable (Fig. a). The term T is the force exerted on the plate by cable CE.

Apply the Equilibrium Equations   Since we know the coordinates of points C and E, we can express the cable force as the product of its magnitude T and a unit vector directed from C toward E. The result is

$$T(-0.842\mathbf{i} + 0.337\mathbf{j} + 0.421\mathbf{k}).$$

The sums of the forces in each coordinate direction equal zero:

$$\Sigma F_x = A_x + B_x - 0.842T = 0,$$
$$\Sigma F_y = A_y + B_y + 0.337T - 400 = 0, \quad (1)$$
$$\Sigma F_z = B_z + 0.421T = 0.$$

If we sum the moments about B, the resulting equations will not contain the three unknown reactions at B. The sum of the moments about B, with forces in N and distances in m, is

$$\Sigma \mathbf{M}_{\text{point } B} = \begin{vmatrix} \mathbf{i} & \mathbf{j} & \mathbf{k} \\ 0.2 & 0 & 0 \\ -0.842T & 0.337T & 0.421T \end{vmatrix} + \begin{vmatrix} \mathbf{i} & \mathbf{j} & \mathbf{k} \\ 0 & 0 & 0.2 \\ A_x & A_y & 0 \end{vmatrix}$$

$$+ \begin{vmatrix} \mathbf{i} & \mathbf{j} & \mathbf{k} \\ 0.2 & 0 & 0.2 \\ 0 & -400 & 0 \end{vmatrix}$$

$$= (-0.2A_y + 80)\mathbf{i} + (-0.0842T + 0.2A_x)\mathbf{j}$$
$$+ (0.0674T - 80)\mathbf{k} = 0.$$

The scalar equations are

$$\Sigma M_x = -(0.2 \text{ m})A_y + 80 \text{ N-m} = 0,$$
$$\Sigma M_y = -(0.0842 \text{ m})T + (0.2 \text{ m})A_x = 0,$$
$$\Sigma M_z = (0.0674 \text{ m})T - 80 \text{ N-m} = 0.$$

Solving these equations, we obtain the reactions

$$T = 1187 \text{ N}, \qquad A_x = 500 \text{ N}, \qquad A_y = 400 \text{ N}.$$

Then from Eqs. (1), the reactions at B are

$$B_x = 500 \text{ N}, \qquad B_y = -400 \text{ N}, \qquad B_z = -500 \text{ N}.$$

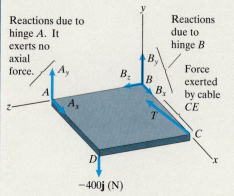

Reactions due to hinge A. It exerts no axial force.

Reactions due to hinge B

Force exerted by cable CE

(a) The free-body diagram of the plate.

Critical Thinking

"Properly aligned hinges" means hinges that are mounted on an object so that their axes are aligned. When this is the case, as in this example, it can usually be assumed that each individual hinge does not exert couples on the object. Notice that it is also assumed in this example that the hinge at A exerts no reaction parallel to the hinge axis but the hinge at B does. The hinges can be intentionally designed so that this is the case, or it can result from the way they are installed.

If our only objective in this example had been to determine the tension T, we could have done so easily by evaluating the sum of the moments about the line AB (the z-axis). Because the reactions at the hinges exert no moment about the z-axis, we obtain the equation

$$(0.2 \text{ m})(0.337T) - (0.2 \text{ m})(400 \text{ N}) = 0,$$

which yields $T = 1187$ N.

Problems

5.77 The bar AB has a fixed support at A and is loaded by the forces

$$\mathbf{F}_B = 2\mathbf{i} + 6\mathbf{j} + 3\mathbf{k} \text{ (kN)},$$

$$\mathbf{F}_C = \mathbf{i} - 2\mathbf{j} + 2\mathbf{k} \text{ (kN)}.$$

(a) Draw the free-body diagram of the bar.
(b) Determine the reactions at A.

Strategy: (a) Draw a diagram of the bar isolated from its supports. Complete the free-body diagram of the bar by adding the two external forces and the reactions due to the fixed support (see Table 5.2). (b) Use the scalar equilibrium equations (5.13)–(5.18) to determine the reactions.

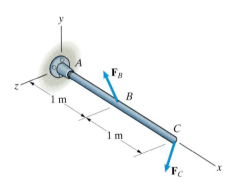

Problem 5.77

5.78 The bar AB has a fixed support at A. The tension in cable BC is 8 kN. Determine the reactions at A.

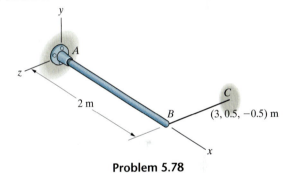

Problem 5.78

5.79 The bar AB has a fixed support at A. The collar at B is fixed to the bar. The tension in the cable BC is 10 kN.

(a) Draw the free-body diagram of the bar.
(b) Determine the reactions at A.

5.80 The magnitude of the couple exerted on the bar by the fixed support is 100 kN-m. What is the tension in the cable?

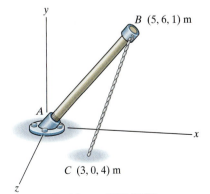

Problems 5.79/5.80

5.81 The total force exerted on the highway sign by its weight and the most severe anticipated winds is $\mathbf{F} = 2.8\mathbf{i} - 1.8\mathbf{j}$ (kN). Determine the reactions at the fixed support.

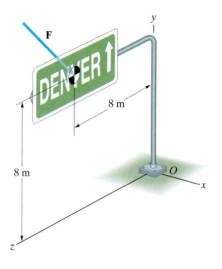

Problem 5.81

5.82 The tension in cable *AB* is 800 lb. Determine the reactions at the fixed support *C*.

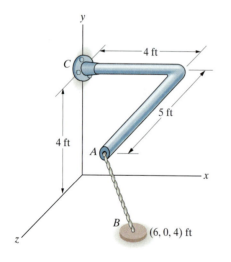

Problem 5.82

5.83 The tension in cable *AB* is 24 kN. Determine the reactions at the fixed support *D*.

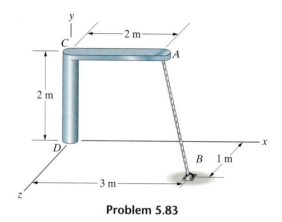

Problem 5.83

5.84 The robotic manipulator is stationary and the *y* axis is vertical. The weights of the arms *AB* and *BC* act at their mid-points. The direction cosines of the centerline of arm *AB* are $\cos \theta_x = 0.174$, $\cos \theta_y = 0.985$, $\cos \theta_z = 0$, and the direction cosines of the centerline of arm *BC* are $\cos \theta_x = 0.743$, $\cos \theta_y = 0.557$, $\cos \theta_z = -0.371$. The support at *A* behaves like a fixed support.
(a) What is the sum of the moments about *A* due to the weights of the two arms?
(b) What are the reactions at *A*?

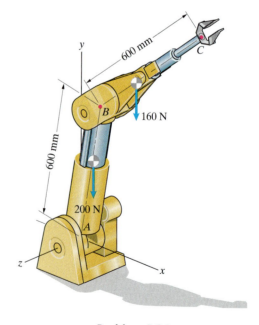

Problem 5.84

5.85 The force exerted on the grip of the exercise machine is $\mathbf{F} = 260\mathbf{i} - 130\mathbf{j}$ (N). What are the reactions at the fixed support at O?

5.86 The designer of the exercise machine assumes that the force $\mathbf{F}$ exerted on the grip will be parallel to the x–y plane and that its magnitude will not exceed 900 N. Based on these criteria, what reactions must the fixed support at O be designed to withstand?

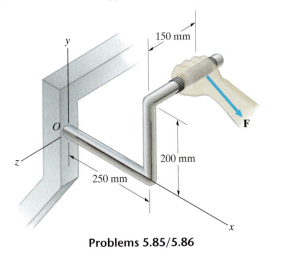

Problems 5.85/5.86

5.87 The force $\mathbf{F}$ acting on the boom ABC at C points in the direction of the unit vector $0.512\mathbf{i} - 0.384\mathbf{j} + 0.768\mathbf{k}$ and its magnitude is 8 kN. The boom is supported by a ball and socket at A and the cables BD and BE. The collar at B is fixed to the boom.
(a) Draw the free-body diagram of the boom.
(b) Determine the tensions in the cables and the reactions at A.

5.88 The cables BD and BE in Problem 5.87 will each safely support a tension of 25 kN. Based on this criterion, what is the largest acceptable magnitude of the force $\mathbf{F}$?

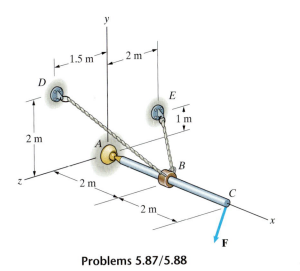

Problems 5.87/5.88

5.89 The suspended load exerts a force $F = 600$ lb at A, and the weight of the bar OA is negligible. Determine the tensions in the cables and the reactions at the ball and socket support O.

5.90 The suspended load exerts a force $F = 600$ lb at A and bar OA weighs 200 lb. Assume that the bar's weight acts at its midpoint. Determine the tensions in the cables and the reactions at the ball and socket support O.

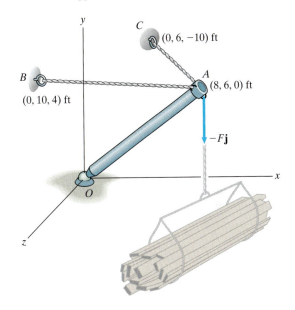

Problems 5.89/5.90

5.91 The 158,000-kg airplane is at rest on the ground ($z = 0$ is ground level). The landing gear carriages are at A, B, and C. The coordinates of the point G at which the weight of the plane acts are (3, 0.5, 5) m. What are the magnitudes of the normal reactions exerted on the landing gear by the ground?

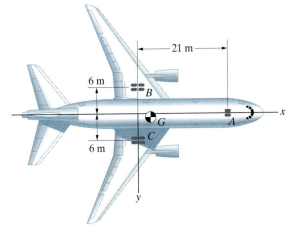

Problem 5.91

5.92 The horizontal triangular plate is suspended by the three vertical cables A, B, and C. The tension in each cable is 80 N. Determine the x and z coordinates of the point where the plate's weight effectively acts.

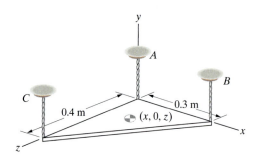

Problem 5.92

5.93 The 800-kg horizontal wall section is supported by the three vertical cables A, B, and C. What are the tensions in the cables?

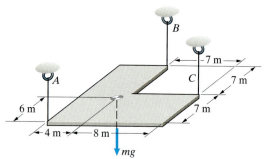

Problem 5.93

5.94 The bar AC is supported by the cable BD and a bearing at A that can rotate about the z axis. The person exerts a force $\mathbf{F} = 10\mathbf{j}$ (lb) at C. Determine the tension in the cable and the reactions at A.

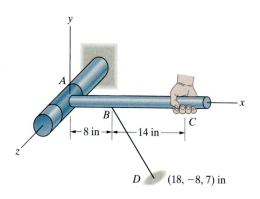

Problem 5.94

5.95 The L-shaped bar is supported by a bearing at A and rests on a smooth horizontal surface at B. The vertical force $F = 4$ kN and the distance $b = 0.15$ m. Determine the reactions at A and B.

5.96 The vertical force $F = 4$ kN and the distance $b = 0.15$ m. If you represent the reactions at A and B by an equivalent system consisting of a single force, what is the force and where does its line of action intersect the x–z plane?

5.97 The vertical force $F = 4$ kN. The bearing at A will safely support a force of 2.5-kN magnitude and a couple of 0.5 kN-m magnitude. Based on these criteria, what is the allow-able range of the distance b?

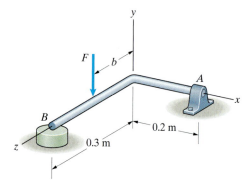

Problems 5.95–5.97

5.98 The 1.1-m bar is supported by a ball and socket support at A and the two smooth walls. The tension in the vertical cable CD is 1 kN.

(a) Draw the free-body diagram of the bar.
(b) Determine the reactions at A and B.

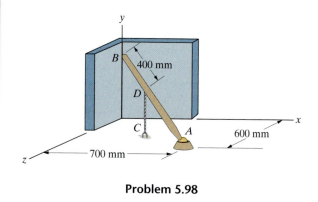

Problem 5.98

5.99 The 8-ft bar is supported by a ball and socket support at *A*, the cable *BD*, and a roller support at *C*. The collar at *B* is fixed to the bar at its midpoint. The force $\mathbf{F} = -50\mathbf{k}$ (lb). Determine the tension in cable *BD* and the reactions at *A* and *C*.

5.100 The bar is 8 ft in length. The force $\mathbf{F} = F_y\mathbf{j} - 50\mathbf{k}$ (lb). What is the largest value of F_y for which the roller support at *C* will remain on the floor?

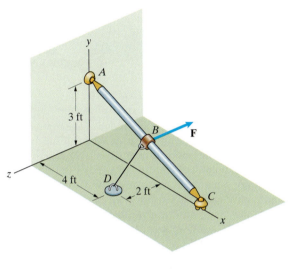

Problems 5.99/5.100

5.101 The tower is 70 m tall. The tension in each cable is 2 kN. Treat the base of the tower *A* as a fixed support. What are the reactions at *A*?

5.102 The tower is 70 m tall If the tension in cable *BC* is 2 kN, what must the tensions in cables *BD* and *BE* be if you want the couple exerted on the tower by the fixed support at *A* to be zero? What are the resulting reactions at *A*?

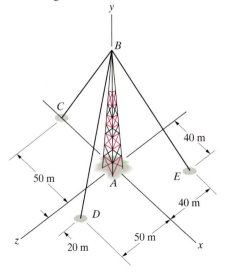

Problems 5.101/5.102

5.103 The space truss has roller supports at *B*, *C*, and *D* and is subjected to a vertical force *F* = 20 kN at *A*. What are the reactions at the roller supports?

5.104 Suppose that you don't want the reaction at any of the roller supports to exceed 15 kN. What is the largest vertical force *F* the truss can support?

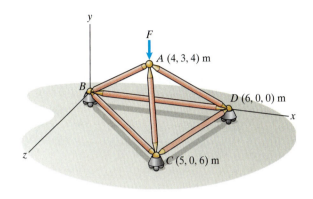

Problems 5.103/5.104

5.105 The 40-lb door is supported by hinges at *A* and *B*. The *y* axis is vertical. The hinges do not exert couples on the door, and the hinge at *B* does not exert a force parallel to the hinge axis. The weight of the door acts at its midpoint. What are the reactions at *A* and *B*?

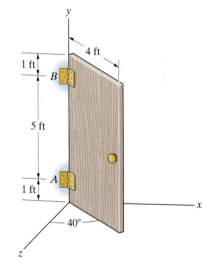

Problem 5.105

5.106 The vertical cable is attached at A. Determine the tension in the cable and the reactions at the bearing B due to the force $\mathbf{F} = 10\mathbf{i} - 30\mathbf{j} - 10\mathbf{k}$ (N).

5.107 Suppose that the z component of the force $\mathbf{F}$ is zero, but otherwise $\mathbf{F}$ is unknown. If the couple exerted on the shaft by the bearing at B is $\mathbf{M}_B = 6\mathbf{j} - 6\mathbf{k}$ N-m, what are the force $\mathbf{F}$ and the tension in the cable?

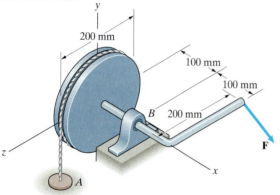

Problems 5.106/5.107

5.108 The device in Problem 5.106 is badly designed because of the couples that must be supported by the bearing at B, which would cause the bearing to "bind." (Imagine trying to open a door supported by only one hinge.) In this improved design, the bearings at B and C support no couples, and the bearing at C does not exert a force in the x direction. If the force $\mathbf{F} = 10\mathbf{i} - 30\mathbf{j} - 10\mathbf{k}$ (N), what are the tension in the vertical cable and the reactions at the bearings B and C?

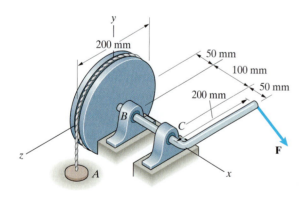

Problem 5.108

5.109 The rocket launcher is supported by the hydraulic jack DE and the bearings A and B. The bearings lie on the x axis and support shafts parallel to the x axis. The hydraulic cylinder DE exerts a force on the launcher that points along the line from D to E. The coordinates of D are (7, 0, 7) ft, and the coordinates of E are (9, 6, 4) ft. The weight $W = 30$ kip acts at (4.5, 5, 2) ft. What is the magnitude of the reaction on the launcher at E?

5.110 Consider the rocket launcher described in Problem 5.109. The bearings at A and B do not exert couples, and the bearing B does not exert a force in the x direction. Determine the reactions at A and B.

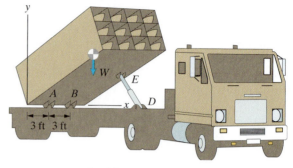

Problems 5.109/5.110

5.111 The crane's cable CD is attached to a stationary object at D. The crane is supported by the bearings E and F and the horizontal cable AB. The tension in cable AB is 8 kN. Determine the tension in the cable CD.

    Strategy: Since the reactions exerted on the crane by the bearings do not exert moments about the z axis, the sum of the moments about the z axis due to the forces exerted on the crane by the cables AB and CD equals zero. (See the discussion at the end of Example 5.10.)

5.112 The crane is supported by the horizontal cable AB and the bearings at E and F. The bearings do not exert couples, and the bearing at F does not exert a force in the z direction. The tension in cable AB is 8 kN. Determine the reactions at E and F.

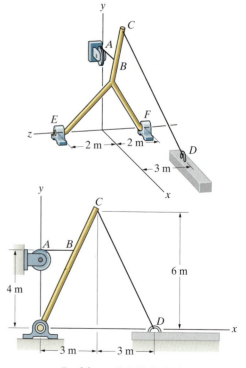

Problems 5.111/5.112

5.113 The plate is supported by hinges at A and B and the cable CE, and it is loaded by the force at D. The edge of the plate to which the hinges are attached lies in the y–z plane, and the axes of the hinges are parallel to the line through points A and B. The hinges do not exert couples on the plate. What is the tension in cable CE?

5.114 In Problem 5.113, the hinge at B does not exert a force on the plate in the direction of the hinge axis. What are the magnitudes of the forces exerted on the plate by the hinges at A and B?

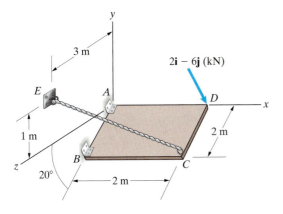

Problems 5.113/5.114

5.115 The bar ABC is supported by ball and socket supports at A and C and the cable BD. The suspended mass is 1800 kg. Determine the tension in the cable.

5.116 * In Problem 5.115, assume that the ball and socket support at A is designed so that it exerts no force parallel to the straight line from A to C. Determine the reactions at A and C.

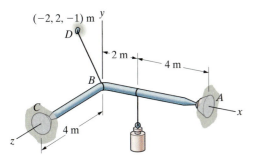

Problems 5.115/5.116

5.117 The bearings at A, B, and C do not exert couples on the bar and do not exert forces in the direction of the axis of the bar. Determine the reactions at the bearings due to the two forces on the bar.

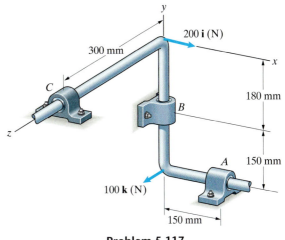

Problem 5.117

5.118 The support that attaches the sailboat's mast to the deck behaves like a ball and socket support. The line that attaches the spinnaker (the sail) to the top of the mast exerts a 200-lb force on the mast. The force is in the horizontal plane at 15° from the centerline of the boat. (See the top view.) The spinnaker pole exerts a 50-lb force on the mast at P. The force is in the horizontal plane at 45° from the centerline. (See the top view.) The mast is supported by two cables, the back stay AB and the port shroud ACD. (The fore stay AE and the starboard shroud AFG are slack, and their tensions can be neglected.) Determine the tensions in the cables AB and CD and the reactions at the bottom of the mast.

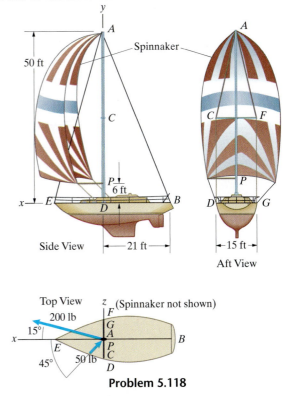

Problem 5.118

5.119 * The bar AC is supported by the cable BD and a bearing at A that can rotate about the axis AE. The person exerts a force $\mathbf{F} = 50\mathbf{j}$ (N) at C. Determine the tension in the cable.

Strategy: Use the fact that the sum of the moments about the axis AE due to the forces acting on the free-body diagram of the bar must equal zero.

5.120 * In Problem 5.119, determine the reactions at the bearing A.

Strategy: Write the couple exerted on the free-body diagram of the bar by the bearing as $\mathbf{M}_A = M_{Ax}\mathbf{i} + M_{Ay}\mathbf{j} + M_{Az}\mathbf{k}$. Then, in addition to the equilibrium equations, obtain an equation by requiring the component of $\mathbf{M}_A$ parallel to the axis AE to equal zero.

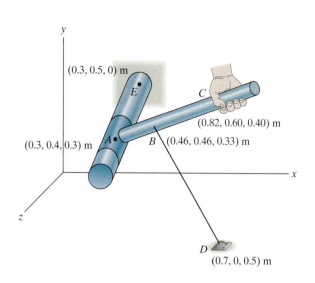

Problems 5.119/5.120

5.5 Two-Force and Three-Force Members

We have shown how the equilibrium equations are used to analyze objects that are supported and loaded in different ways. Here we discuss two particular types of loading that occur so frequently they deserve particular attention. The first type, the two-force member, is especially important and plays an important role in our analysis of structures in Chapter 6.

Two-Force Members

If the system of forces and moments acting on an object is equivalent to two forces acting at different points, we refer to the object as a *two-force member*. For example, the object in Fig. 5.35a is subjected to two sets of concurrent forces whose lines of action intersect at A and B. Since we can represent them by single forces acting at A and B (Fig. 5.35b), where $\mathbf{F} = \mathbf{F}_1 + \mathbf{F}_2 + \cdots + \mathbf{F}_N$ and $\mathbf{F}' = \mathbf{F}'_1 + \mathbf{F}'_2 + \cdots + \mathbf{F}'_M$, this object is a two-force member.

If the object is in equilibrium, what can we infer about the forces $\mathbf{F}$ and $\mathbf{F}'$? The sum of the forces equals zero only if $\mathbf{F}' = -\mathbf{F}$ (Fig. 5.35c). Furthermore, the forces $\mathbf{F}$ and $-\mathbf{F}$ form a couple, so the sum of the moments is not zero unless the lines of action of the forces lie along the line through the points A and B (Fig. 5.35d). Thus equilibrium tells us that *the two forces are equal in magnitude, are opposite in direction, and have the same line of action.* However, without additional information, we cannot determine their magnitude.

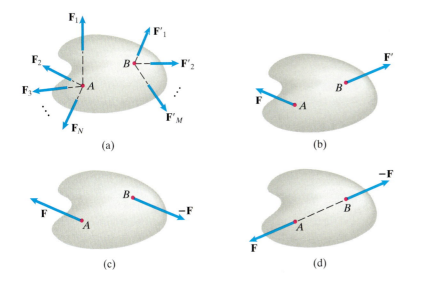

(a)   (b)

(c)   (d)

Figure 5.35
(a) An object subjected to two sets of concurrent forces.
(b) Representing the concurrent forces by two forces F and **F**'.
(c) If the object is in equilibrium, the forces must be equal and opposite.
(d) The forces form a couple unless they have the same line of action.

A cable attached at two points (Fig. 5.36a) is a familiar example of a two-force member (Fig. 5.36b). The cable exerts forces on the attachment points that are directed along the line between them (Fig. 5.36c).

A bar that has two supports that exert only forces on it (no couples) and is not subjected to any loads is a two-force member (Fig. 5.37a). Such bars are often used as supports for other objects. Because the bar is a two-force member, the lines of action of the forces exerted on the bar must lie along the line between the supports (Fig. 5.37b). Notice that, unlike the cable, the bar can exert forces at A and B either in the directions shown in Fig. 5.37c or in the opposite directions. (In other words, the cable can only pull on its supports, while the bar can either pull or push.)

In these examples we assumed that the weights of the cable and the bar could be neglected in comparison with the forces exerted on them by their supports. When that is not the case, they are clearly not two-force members.

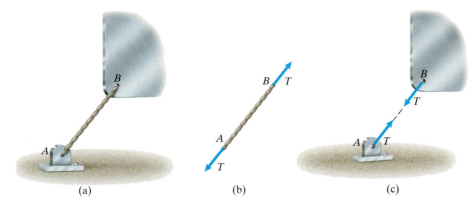

(a)   (b)   (c)

Figure 5.36
(a) A cable attached at *A* and *B*.
(b) The cable is a two-force member.
(c) The forces exerted by the cable.

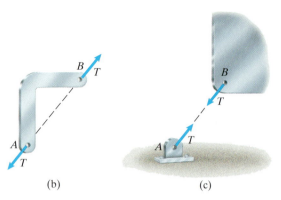

Figure 5.37

(a)   (b)   (c)

(a) The bar *AB* attaches the object to the pin support.

(b) The bar *AB* is a two-force member.

(c) The force exerted on the supported object by the bar *AB*.

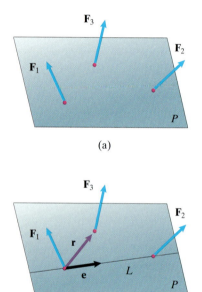

(a)

(b)

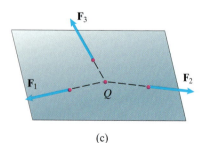

(c)

Figure 5.38
(a) The three forces and the plane *P*.
(b) Determining the moment due to force $\mathbf{F}_3$ about L.
(c) If the forces are not parallel, they must be concurrent.

Three-Force Members

If the system of forces and moments acting on an object is equivalent to three forces acting at different points, we call it a *three-force member*. We can show that if a three-force member is in equilibrium, the three forces are coplanar and are either parallel or concurrent.

We first prove that the forces are coplanar. Let them be called $\mathbf{F}_1$, $\mathbf{F}_2$, and $\mathbf{F}_3$, and let *P* be the plane containing the three points of application (Fig. 5.38a). Let *L* be the line through the points of application of $\mathbf{F}_1$ and $\mathbf{F}_2$. Since the moments due to $\mathbf{F}_1$ and $\mathbf{F}_2$ about *L* are zero, the moment due to $\mathbf{F}_3$ about *L* must equal zero (Fig. 5.38b):

$$[\mathbf{e}\cdot(\mathbf{r}\times\mathbf{F}_3)]\mathbf{e} = [\mathbf{F}_3\cdot(\mathbf{e}\times\mathbf{r})]\mathbf{e} = \mathbf{0}.$$

This equation requires that $\mathbf{F}_3$ be perpendicular to $\mathbf{e}\times\mathbf{r}$, which means that $\mathbf{F}_3$ is contained in *P*. The same procedure can be used to show that $\mathbf{F}_1$ and $\mathbf{F}_2$ are contained in *P*, so the forces are coplanar. (A different proof is required if the points of application lie on a straight line, but the result is the same.)

If the three coplanar forces are not parallel, there will be points where their lines of action intersect. Suppose that the lines of action of two of the forces intersect at a point *Q*. Then the moments of those two forces about *Q* are zero, and the sum of the moments about *Q* is zero only if the line of action of the third force also passes through *Q*. Therefore, either the forces are parallel or they are concurrent (Fig. 5.38c).

The analysis of an object in equilibrium can often be simplified by recognizing that it is a two-force or three-force member. However, in doing so we are not getting something for nothing. Once the free-body diagram of a two-force member is drawn, as shown in Figs. 5.36b and 5.37b, no further information can be obtained from the equilibrium equations. And when we require that the lines of action of nonparallel forces acting on a three-force member be coincident, we have used the fact that the sum of the moments about a point must be zero and cannot obtain further information from that condition.

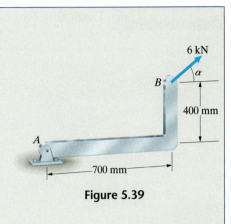

Figure 5.39

Example 5.11   **A Two-Force Member**

The L-shaped bar in Fig. 5.39 has a pin support at A and is loaded by a 6-kN force at B. Neglect the weight of the bar. Determine the angle α and the reactions at A.

Strategy

The bar is a two-force member because it is subjected only to the 6-kN force at B and the force exerted by the pin support. (If we could not neglect the weight of the bar, it would not be a two-force member.) We will determine the angle α and the reactions at A in two ways, first by applying the equilibrium equations in the usual way and then by using the fact that the bar is a two-force member.

Solution

Applying the Equilibrium Equations   We draw the free-body diagram of the bar in Fig. a, showing the reactions at the pin support. Summing moments about point A, the equilibrium equations are

$$\Sigma F_x = A_x + 6 \cos \alpha \text{ kN} = 0,$$

$$\Sigma F_y = A_y + 6 \sin \alpha \text{ kN} = 0,$$

$$\Sigma M_{\text{point } A} = (0.7 \text{ m})(6 \sin \alpha \text{ kN}) - (0.4 \text{ m})(6 \cos \alpha \text{ kN}) = 0.$$

From the third equation we see that $\alpha = \arctan(0.4/0.7)$. In the range $0 \leq \alpha \leq 360°$, this equation has the two solutions $\alpha = 29.7°$ and $\alpha = 209.7°$. Knowing α, we can determine A_x and A_y from the first two equilibrium equations. The solutions for the two values of α are

$$\alpha = 29.7°, \qquad A_x = -5.21 \text{ kN}, \qquad A_y = -2.98 \text{ kN},$$

and

$$\alpha = 209.7°, \qquad A_x = 5.21 \text{ kN}, \qquad A_y = 2.98 \text{ kN}.$$

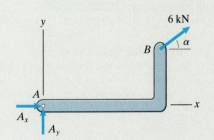

(a) The free-body diagram of the bar.

Treating the Bar as a Two-Force Member   We know that the 6-kN force at B and the force exerted by the pin support must be equal in magnitude, opposite in direction, and directed along the line between points A and B. The two possibilities are shown in Figs. b and c. Thus by recognizing that the bar is a two-force member, we immediately know the possible directions of the forces and the magnitude of the reaction at A.

In Fig. b we can see that $\tan \alpha = 0.4/0.7$, so $\alpha = 29.7°$ and the components of the reaction at A are

$$A_x = -6 \cos 29.7° \text{ kN} = -5.21 \text{ kN},$$

$$A_y = -6 \sin 29.7° \text{ kN} = -2.98 \text{ kN}.$$

In Fig. c, $\alpha = 180° + 29.7° = 209.7°$, and the components of the reaction at A are

$$A_x = 6 \cos 29.7° \text{ kN} = 5.21 \text{ kN}$$

$$A_y = 6 \sin 29.7° \text{ kN} = 2.98 \text{ kN}.$$

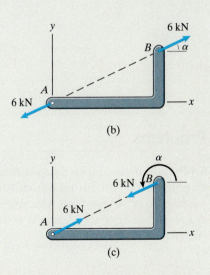

(b), (c) The possible directions of the forces.

Critical Thinking

Why is it worthwhile to recognize that an object is a two-force member? Doing so tells you the directions of the forces acting on the object and also that the forces are equal and opposite. As this example demonstrates, such information frequently simplifies the solution of a problem.

Example 5.12 Two- and Three-Force Members

The 100-lb weight of the rectangular plate in Fig. 5.40 acts at its midpoint. Determine the reactions exerted on the plate at B and C.

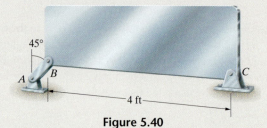

Figure 5.40

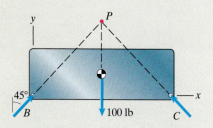

(a) The free-body diagram of the plate. The three forces must be concurrent.

Strategy

The plate is subjected to its weight and the reactions exerted by the pin supports at B and C, so it is a three-force member. Furthermore, the bar AB is a two-force member, so we know that the line of action of the reaction it exerts on the plate at B is directed along the line between A and B. We can use this information to simplify the free-body diagram of the plate.

Solution

The reaction exerted on the plate by the two-force member AB must be directed along the line between A and B, and the line of action of the weight is vertical. Since the three forces on the plate must be either parallel or concurrent, their lines of action must intersect at the point P shown in Fig. a. From the equilibrium equations

$$\Sigma F_x = B \sin 45° - C \sin 45° = 0,$$

$$\Sigma F_y = B \cos 45° + C \cos 45° - 100 \text{ lb} = 0,$$

we obtain the reactions $B = C = 70.7$ lb.

Critical Thinking

Notice how recognizing the bar AB as a two-force member allowed us to simplify the free-body diagram. Knowing the direction of the force on the plate at B allowed us to draw a single reaction instead of two component reactions. Recognizing two-force members will become even more important when we analyze structures with multiple members in Chapter 6.

Problems

5.121 The horizontal bar has a mass of 10 kg. Its weight acts at the midpoint of the bar, and it is supported by a roller support at A and the cable BC. Use the fact that the bar is a three-force member to determine the angle α, the tension in the cable BC, and the magnitude of the reaction at A.

5.122 The horizontal bar is of negligible weight. Use the fact that the bar is a three-force member to determine the angle α necessary for equilibrium.

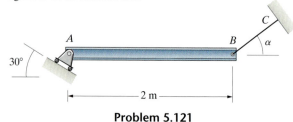

Problem 5.121

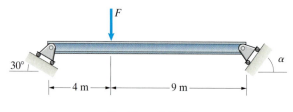

Problem 5.122

5.123 The suspended load weighs 1000 lb. The structure is a three-force member if its weight is neglected. Use this fact to determine the magnitudes of the reactions at *A* and *B*.

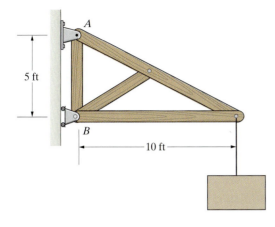

Problem 5.123

5.124 The weight *W* = 50 lb acts at the center of the disk. Use the fact that the disk is a three-force member to determine the tension in the cable and the magnitude of the reaction at the pin support.

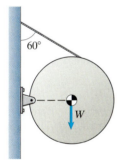

Problem 5.124

5.125 The weight *W* = 40 N acts at the center of the disk. The surfaces are rough. What force *F* is necessary to lift the disk off the floor?

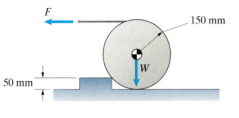

Problem 5.125

5.126 Use the fact that the horizontal bar is a three-force member to determine the angle α and the magnitudes of the reactions at *A* and *B*. Assume that $0 \leq \alpha \leq 90°$.

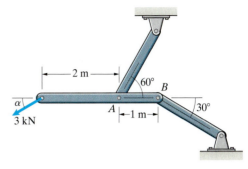

Problem 5.126

5.127 The suspended load weighs 600 lb. Use the fact that *ABC* is a three-force member to determine the magnitudes of the reactions at *A* and *B*.

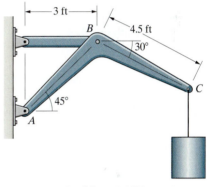

Problem 5.127

5.128 (a) Is the L-shaped bar a three-force member?
(b) Determine the magnitudes of the reactions at *A* and *B*.
(c) Are the three forces acting on the L-shaped bar concurrent?

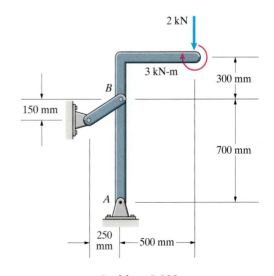

Problem 5.128

5.129 The bucket of the excavator is supported by the two-force member AB and the pin support at C. Its weight is $W = 1500$ lb. What are the reactions at C?

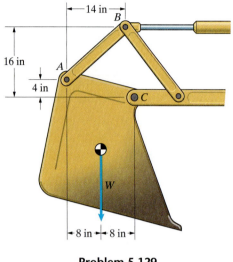

Problem 5.129

5.130 The member ACG of the front-end loader is subjected to a load $W = 2$ kN and is supported by a pin support at A and the hydraulic cylinder BC. Treat the hydraulic cylinder as a two-force member.

(a) Draw the free-body diagrams of the hydraulic cylinder and the member ACG.

(b) Determine the reactions on the member ACG.

5.131 In Problem 5.130, determine the reactions on the member ACG by using the fact that it is a three-force member.

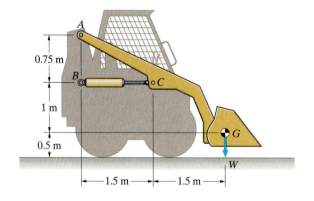

Problems 5.130/5.131

5.132 A rectangular plate is subjected to two forces A and B (Fig. a). In Fig. b, the two forces are resolved into components. By writing equilibrium equations in terms of the components A_x, A_y, B_x, and B_y, show that the two forces A and B are equal in magnitude, opposite in direction, and directed along the line between their points of application.

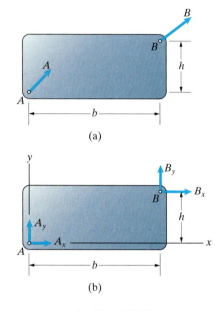

Problem 5.132

5.133 An object in equilibrium is subjected to three forces whose points of application lie on a straight line. Prove that the forces are coplanar.

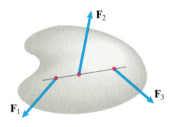

Problem 5.133

COMPUTATIONAL MECHANICS

The following example and problems are designed for the use of a programmable calculator or computer.

Computational Example 5.13

The beam in Fig. 5.41 weighs 200 lb and is supported by a pin support at A and the wire BC. The wire behaves like a linear spring with spring constant $k = 60$ lb/ft and is unstretched when the beam is in the position shown. Determine the reactions at A and the tension in the wire when the beam is in equilibrium.

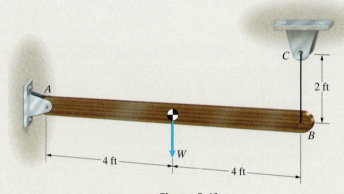

Figure 5.41

Strategy

When the beam is in equilibrium, the sum of the moments about A due to the beam's weight and the force exerted by the wire equals zero. We will obtain a graph of the sum of the moments as a function of the angle of rotation of the beam relative to the horizontal to determine the position of the beam when it is in equilibrium. Once we know the position, we can determine the tension in the wire and the reactions at A.

Solution

Let α be the angle from the horizontal to the centerline of the beam (Fig. a). The distances b and h in ft are

$$b = 8(1 - \cos \alpha),$$
$$h = 2 + 8 \sin \alpha,$$

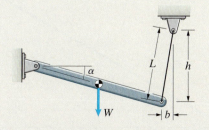

(a) Rotating the beam through an angle α.

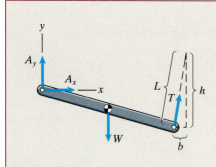

(b) The free-body diagram of the beam.

α	ΣM_A (ft-lb)
11.87°	−1.2600
11.88°	−0.5925
11.89°	0.0750
11.90°	0.7424
11.91°	1.4099

and the length of the stretched wire is

$$L = \sqrt{b^2 + h^2}.$$

The tension in the wire is

$$T = k(L - 2).$$

We draw the free-body diagram of the beam in Fig. b. In terms of the components of the force exerted by the wire,

$$T_x = \frac{b}{L}T, \qquad T_y = \frac{h}{L}T,$$

the sum of the moments about A is

$$\Sigma M_A = (8 \sin \alpha)T_x + (8 \cos \alpha)T_y - (4 \cos \alpha)W.$$

If we choose a value of α, we can sequentially evaluate these quantities. Computing ΣM_A as a function of α, we obtain the graph shown in Fig. 5.42. From the graph we estimate that $\Sigma M_A = 0$ when $\alpha = 12°$. By examining computed results near 12°, we find that the beam is in equilibrium when $\alpha = 11.89°$. The corresponding value of the tension in the wire is $T = 99.1$ lb.

To determine the reactions at A, we use the equilibrium equations

$$\Sigma F_x = A_x + T_x = 0,$$
$$\Sigma F_y = A_y + T_y - W = 0,$$

obtaining $A_x = -4.7$ lb and $A_y = 101.0$ lb.

Critical Thinking

This example is an idealized case of an important problem. Notice that the wire BC in Fig. 5.41 is one of the supports of the beam AB. The weight of the beam causes the wire to stretch, or deform, and by using a numerical solution we were able to determine how much it stretches. The point is that all supports deform to some extent when loads are applied to the objects they support. In many cases their deformations are sufficiently small that they can be ignored, as we ignored the deformation of the support at A in this example. Applications in which deformations of supports must be accounted for are studied in mechanics of materials.

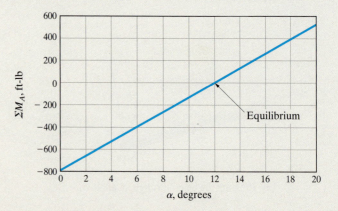

Figure 5.42
The sum of the moments as a function of α.

Computational Problems

5.134 The 10-lb weight of the bar *AB* acts at the midpoint of the bar. The length of the bar is 3 ft. Determine the value of the angle α for which the tension in the string *BC* is 6 lb. What are the resulting reactions at *A*?

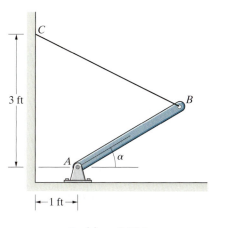

Problem 5.134

5.135 The mass of the bar is 36 kg and its weight acts at its midpoint. The spring is unstretched when $\alpha = 0$, and the spring constant is $k = 200$ N/m. Determine the values of α in the range $0 \le \alpha \le 90°$ at which the bar is in equilibrium.

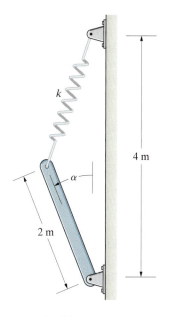

Problem 5.135

5.136 The unstretched length of the spring *CD* is 350 mm and the spring constant is $k = 3400$ N/m. Suppose that you want to choose the dimension *h* so that the lever *ABC* exerts a 280-N normal force on the smooth surface at *A*. Determine the dimension *h* and the resulting reactions at *B*.

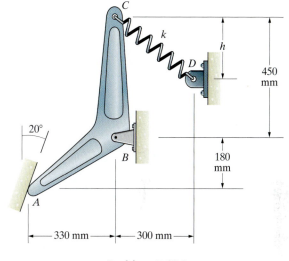

Problem 5.136

5.137 The bar is 1 m long, and its weight $W = 35$ N acts at its midpoint. The distance $b = 0.75$ m. The spring constant is $k = 100$ N/m, and the spring is unstretched when the bar is vertical. Determine the angle α and the reactions at *A*.

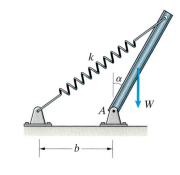

Problem 5.137

5.138 The hydraulic actuator BC exerts a force at C that points along the line from B to C. Treat A as a pin support. The mass of the suspended load is 4000 kg. If the actuator BC can exert a maximum force of 90 kN, what is the smallest permissible value of α?

5.139 The beam is in equilibrium in the position shown. Each spring has an unstretched length of 1 m. Determine the distance b and the reactions at A.

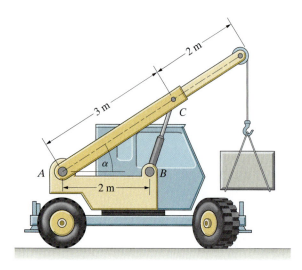

Problem 5.138

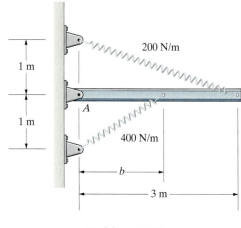

Problem 5.139

CHAPTER SUMMARY

Building on our discussions of forces in Chapter 3 and moments in Chapter 4, in this chapter we have used the equilibrium equations to analyze the forces and couples acting on many types of objects. We defined the support conventions commonly used in engineering and presented examples of their use. We discussed situations that can result in an object's being statically indeterminate. Finally, we defined two-force and three-force members. In Chapter 6 we will use the concepts and methods developed in this chapter to analyze the individual members of structures, beginning with structures consisting entirely of two-force members.

When an object is in equilibrium, the following conditions are satisfied:

1. The sum of the forces is zero:

$$\Sigma \mathbf{F} = \mathbf{0}. \tag{5.1}$$

2. The sum of the moments about any point is zero:

$$\Sigma \mathbf{M}_{\text{any point}} = \mathbf{0}. \tag{5.2}$$

Forces and couples exerted on an object by its supports are called *reactions*. The other forces and couples on the object are the *loads*. Common supports are represented by models called *support conventions*.

Two-Dimensional Applications

When the loads and reactions on an object in equilibrium form a two-dimensional system of forces and moments, they are related by three scalar equilibrium equations:

$$\Sigma F_x = 0 \tag{5.4}$$

$$\Sigma F_y = 0, \tag{5.5}$$

$$\Sigma M_{\text{any point}} = 0. \tag{5.6}$$

No more than three independent equilibrium equations can be obtained from a given two-dimensional free-body diagram.

Support conventions commonly used in two-dimensional applications are summarized in Table 5.1.

Three-Dimensional Applications

The loads and reactions on an object in equilibrium satisfy the six scalar equilibrium equations

$$\Sigma F_x = 0, \qquad \Sigma F_y = 0, \qquad \Sigma F_z = 0,$$

$$\Sigma M_x = 0, \qquad \Sigma M_y = 0, \qquad \Sigma M_z = 0. \tag{5.13)-(5.18}$$

No more than six independent equilibrium equations can be obtained from a given free-body diagram.

Support conventions commonly used in three-dimensional applications are summarized in Table 5.2.

Statically Indeterminate Objects

An object has *redundant supports* when it has more supports than the minimum number necessary to maintain it in equilibrium and *improper supports* when its supports are improperly designed to maintain equilibrium under the applied loads. In either situation, the object is *statically indeterminate*. The difference between the number of reactions and the number of independent equilibrium equations is called the *degree of redundancy*. Even if an object is statically indeterminate due to redundant supports, it may be possible to determine some of the reactions from the equilibrium equations.

Two-Force and Three-Force Members

If the system of forces and moments acting on an object is equivalent to two forces acting at different points, the object is a *two-force member*. If the object is in equilibrium, the two forces are equal in magnitude, opposite in direction, and directed along the line through their points of application. If the system of forces and moments acting on an object is equivalent to three forces acting at different points, it is a *three-force member*. If the object is in equilibrium, the three forces are coplanar and either parallel or concurrent.

Review Problems

5.140 The suspended cable weighs 12 lb.

(a) Draw the free-body diagram of the cable. (The tensions in the cable at A and B are *not* equal.)

(b) Determine the tensions in the cable at A and B.

(c) What is the tension in the cable at its lowest point?

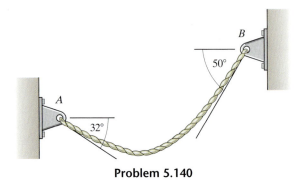

Problem 5.140

5.141 Determine the reactions at the fixed support.

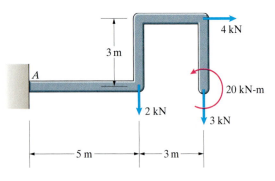

Problem 5.141

5.142 (a) Draw the free-body diagram of the 50-lb plate, and explain why it is statically indeterminate.

(b) Determine as many of the reactions at A and B as possible.

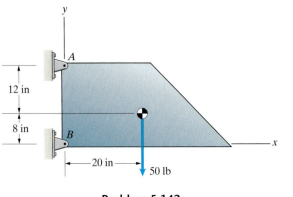

Problem 5.142

5.143 The mass of the truck is 4000 kg. Its wheels are locked, and the tension in its cable is $T = 10$ kN.

(a) Draw the free-body diagram of the truck.

(b) Determine the normal forces exerted on the truck's wheels at A and B by the road.

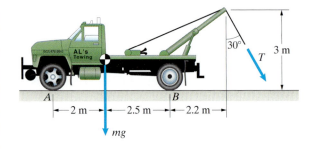

Problem 5.143

5.144 Assume that the force exerted on the head of the nail by the hammer is vertical, and neglect the hammer's weight.

(a) Draw the free-body diagram of the hammer.

(b) If $F = 10$ lb, what are the magnitudes of the force exerted on the nail by the hammer and the normal and friction forces exerted on the floor by the hammer?

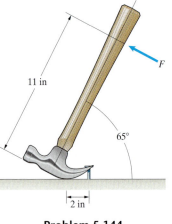

Problem 5.144

5.145 The spring constant is $k = 9600$ N/m and the unstretched length of the spring is 30 mm. Treat the bolt at A as a pin support and assume that the surface at C is smooth. Determine the reactions at A and the normal force at C.

5.146 The engineer designing the release mechanism wants the normal force exerted at C to be 120 N. If the unstretched length of the spring is 30 mm, what is the necessary value of the spring constant k?

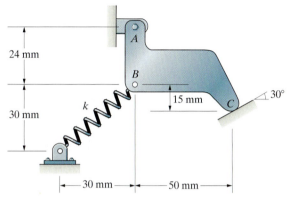

Problems 5.145/5.146

5.147 The truss supports a 90-kg suspended object. What are the reactions at the supports A and B?

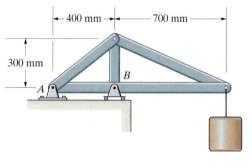

Problem 5.147

5.148 The trailer is parked on a 15° slope. Its wheels are free to turn. The hitch H behaves like a pin support. Determine the reactions at A and H.

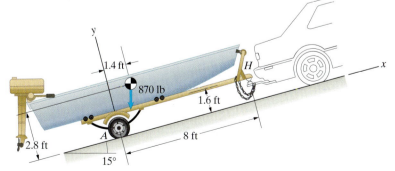

Problem 5.148

5.149 To determine the location of the point where the weight of a car acts (the *center of mass*), an engineer places the car on scales and measures the normal reactions at the wheels for two values of α, obtaining the following results.

α	A_y (kN)	B (kN).
10°	10.134	4.357
20°	10.150	3.677

What are the distances b and h?

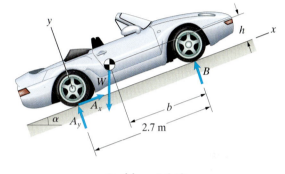

Problem 5.149

5.150 The bar is attached by pin supports to collars that slide on the two fixed bars. Its mass is 10 kg, it is 1 m in length, and its weight acts at its midpoint. Neglect friction and the masses of the collars. The spring is unstretched when the bar is vertical ($\alpha = 0$), and the spring constant is $k = 100$ N/m. Determine the values of α in the range $0 \leq \alpha \leq 60°$ at which the bar is in equilibrium.

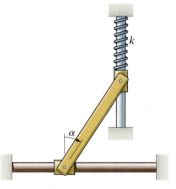

Problem 5.150

5.151 The 450-lb ladder is supported by the hydraulic cylinder *AB* and the pin support at *C*. The reaction at *B* is parallel to the hydraulic cylinder. Determine the reactions on the ladder.

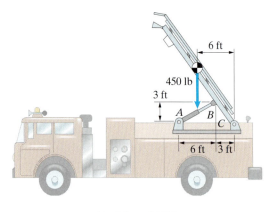

Problem 5.151

5.152 Consider the crane shown in Problem 5.138. The hydraulic actuator *BC* exerts a force at *C* that points along the line from *B* to *C*. Treat *A* as a pin support. The mass of the suspended load is 6000 kg. If the angle $\alpha = 35°$, what are the reactions at *A*?

5.153 The force exerted by the weight of the horizontal rectangular plate is 800 N. The plate is suspended by three vertical cables. The weight acts at its midpoint. What are the tensions in the cables?

5.154 The force exerted by the weight of the horizontal rectangular plate is 800 N. The weight of the rectangular plate acts at its midpoint. If you represent the reactions exerted on the plate by the three cables by a single equivalent force, what is the force, and where does its line of action intersect the plate?

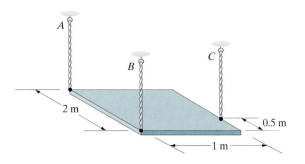

Problems 5.153/5.154

5.155 The 20-kg mass is suspended by cables attached to three vertical 2-m posts. Point *A* is at (0, 1.2, 0) m. Determine the reactions at the fixed support at *E*.

5.156 In Problem 5.155, the fixed support of each vertical post will safely support a couple of 800 N-m magnitude. Based on this criterion, what is the maximum safe value of the suspended mass?

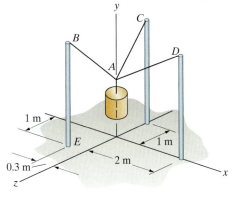

Problems 5.155/5.156

5.157 The 80-lb bar is supported by a ball and socket support at *A*, the smooth wall it leans against, and the cable *BC*. The weight of the bar acts at its midpoint.

(a) Draw the free-body diagram of the bar.
(b) Determine the tension in cable *BC* and the reactions at *A*.

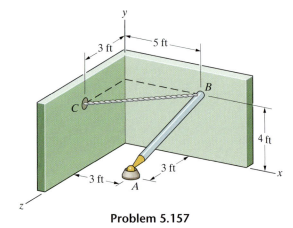

Problem 5.157

5.158 The horizontal bar of weight *W* is supported by a roller support at *A* and the cable *BC*. Use the fact that the bar is a three-force member to determine the angle α, the tension in the cable, and the magnitude of the reaction at *A*.

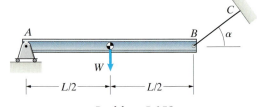

Problem 5.158

Design Project 1 The traditional wheelbarrow shown is de-signed to transport a load W while being supported by an up-ward force F applied to the handles by the user. (a) Use statics to analyze the effects of a range of choices of the dimensions a and b on the size of load that could be carried. Also consider the im-plications of these dimensions on the wheelbarrow's ease and practicality of use. (b) Suggest a different design for this classic device that achieves the same function. Use statics to compare your design to the wheelbarrow with respect to load-carrying ability and ease of use.

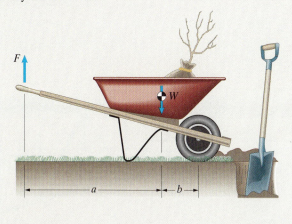

Design Project 2 The figure shows an example of the popular devices called "mobiles," which were introduced as an art form by American artist Alexander Calder (1898–1976). Suppose that you want to design a mobile representing the solar system, and have chosen colored spheres to represent the planets. The mass-es of the spheres that represent Mercury, Venus, Earth, Mars, Jupiter, Saturn, Uranus, Neptune, and Pluto are 10 g, 25 g, 25 g, 10 g, 50 g, 40 g, 40 g, 40 g, and 10 g. Assume that the cross bars and string you use are of negligible mass. Design your mobile so that the planets are in their correct order relative to the sun. Write a brief report including a drawing of your design and the analysis proving that your mobile is balanced.

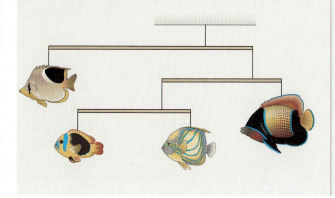

Design Project 3 The bed of the dump truck (Fig. a) is raised by two tandem hydraulic cylinders AB (Fig. b). The mass of the truck's bed and load is 16,000 kg and its weight acts at point G. (Assume that the position of point G *relative to the bed* does not change when the bed is raised.)

(a) Draw a graph of the magnitude of the total force the hy-draulic cylinders must exert to support the stationary bed for val-ues of the angle α from zero to 30°.

(b) Consider other choices for the locations of the attachment points A and B that appear to be feasible and investigate how your choices affect the magnitude of the total force the hydraulic cylin-ders must exert as α varies from zero to 30°. Also compare the costs of your choices of the attachment points to the choices shown in Fig. a, assuming that the cost of the hydraulic cylinders is proportional to the product of the maximum force they must exert as α varies from zero to 30° and their length when $\alpha = 30°$.

(c) Write a brief report presenting your investigations and mak-ing a recommendation for the locations of points A and B.

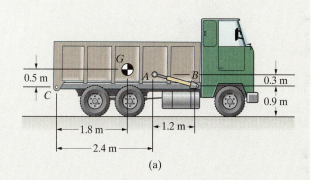

(a)

(b)

CHAPTER

6

Structures in Equilibrium

In engineering, the term *structure* can refer to any object that has the capacity to support and exert loads. In this chapter we consider structures composed of interconnected parts, or *members*. To design such a structure, or to determine whether an existing one is adequate, it is necessary to determine the forces and couples acting on the structure as a whole as well as on its individual members. We first demonstrate how this is done for the structures called trusses, which are composed entirely of two-force members. The familiar frameworks of steel members that support some highway bridges are trusses. We then consider other structures, called *frames* if they are designed to remain stationary and support loads and *machines* if they are designed to move and exert loads.

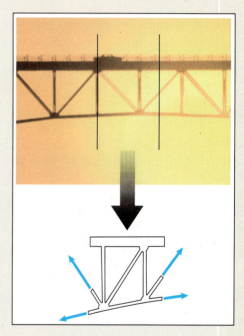

◀ The highway bridge is supported by a truss structure. In this chapter we describe techniques for determining the forces and couples acting on the individual members of structures.

Figure 6.1
A typical house is supported by trusses made of wood beams.

6.1 Trusses

We can explain the nature of truss structures such as the beams supporting a house (Fig. 6.1) by starting with very simple examples. Suppose we pin three bars together at their ends to form a triangle. If we add supports as shown in Fig. 6.2a, we obtain a structure that will support a load F. We can construct more elaborate structures by adding more triangles (Figs. 6.2b and c). The bars are the members of these structures, and the places where the bars are pinned together are called the *joints*. Even though these examples are quite simple, you can see that Fig. 6.2c, which is called a Warren truss, begins to resemble the structures used to support bridges and the roofs of houses (Fig. 6.3). If these structures are supported and loaded at their joints and we neglect the weights of the bars, each bar is a two-force member. We call such a structure a *truss*.

We draw the free-body diagram of a member of a truss in Fig. 6.4a. Because it is a two-force member, the forces at the ends, which are the sums of the forces exerted on the member at its joints, must be equal in magnitude, opposite in direction, and directed along the line between the joints. We call the force T the *axial force* in the member. When T is positive in the direction shown (that is, when the forces are directed away from each other), the member is in *tension*. When the forces are directed toward each other, the member is in *compression*.

In Fig. 6.4b, we "cut" the member by a plane and draw the free-body diagram of the part of the member on one side of the plane. We represent the system of internal forces and moments exerted by the part not included in the free-body diagram by a force **F** acting at the point P where the plane intersects the axis of the member and a couple **M**. The sum of the moments about P must equal zero, so **M** = **0**. Therefore we have a two-force member, which means

(a)

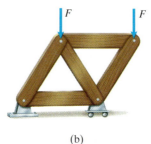

(b)

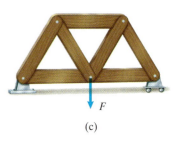

(c)

Figure 6.2
Making structures by pinning bars together to form triangles.

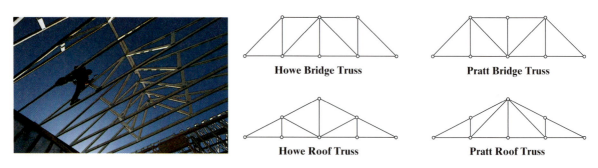

Howe Bridge Truss   **Pratt Bridge Truss**

Howe Roof Truss   **Pratt Roof Truss**

Figure 6.3
Simple examples of bridge and roof structures. (The lines represent members, and the circles represent joints.)

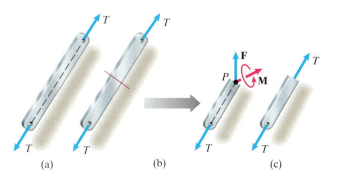

(a)   (b)   (c)

Figure 6.4
(a) Each member of a truss is a two-force member.
(b) Obtaining the free-body diagram of part of the member.
(c) The internal force is equal and opposite to the force acting at the joint, and the internal couple is zero.

that **F** must be equal in magnitude and opposite in direction to the force T acting at the joint (Fig. 6.4c). The internal force is a tension or compression equal to the tension or compression exerted at the joint. Notice the similarity to a rope or cable, in which the internal force is a tension equal to the tension applied at the ends.

Although many actual structures, including "roof trusses" and "bridge trusses," consist of bars connected at the ends, very few have pinned joints. For example, a joint of a bridge truss is shown in Fig. 6.5. The ends of the members are welded at the joint and are not free to rotate. It is obvious that such a joint can exert couples on the members. Why are these structures called trusses?

The reason is that they are designed to function as trusses, meaning that they support loads primarily by subjecting their members to axial forces. They can usually be *modeled* as trusses, treating the joints as pinned connections under the assumption that couples they exert on the members are small in comparison to axial forces. When we refer to structures with riveted joints as trusses in problems, we mean that you can model them as trusses.

In the following sections we describe two methods for determining the axial forces in the members of trusses. The method of joints is usually the preferred approach when you need to determine the axial forces in all members of a truss. When you only need to determine the axial forces in a few members, the method of sections often results in a faster solution than the method of joints.

Figure 6.5
A joint of a bridge truss.

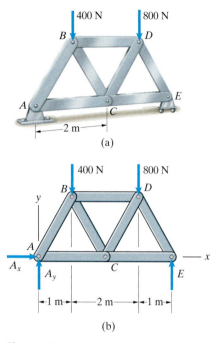

400 N   800 N

B   D

E

A

C

—2 m—

(a)

400 N   800 N

B   D

y

A

A_x

A_y C   E

—1 m—   —2 m—   —1 m—

(b)

Figure 6.6
(a) A Warren truss supporting two loads.
(b) Free-body diagram of the truss.

6.2 The Method of Joints

The method of joints involves drawing free-body diagrams of the joints of a truss one by one and using the equilibrium equations to determine the axial forces in the members. Before beginning, it is usually necessary to draw a free-body diagram of the entire truss (that is, treat the truss as a single object) and determine the reactions at its supports. For example, let's consider the Warren truss in Fig. 6.6a, which has members 2 m in length and supports loads at B and D. We draw its free-body diagram in Fig. 6.6b. From the equilibrium equations,

$$\Sigma F_x = A_x = 0,$$

$$\Sigma F_y = A_y + E - 400 \text{ N} - 800 \text{ N} = 0,$$

$$\Sigma M_{\text{point } A} = -(1 \text{ m})(400 \text{ N}) - (3 \text{ m})(800 \text{ N}) + (4 \text{ m})E = 0,$$

we obtain the reactions $A_x = 0$, $A_y = 500$ N, and $E = 700$ N.

Our next step is to choose a joint and draw its free-body diagram. In Fig. 6.7a, we isolate joint A by cutting members AB and AC. The terms T_{AB} and T_{AC} are the axial forces in members AB and AC, respectively. Although the directions of the arrows representing the unknown axial forces can be chosen arbitrarily, notice that we have chosen them so that a member is in tension if we obtain a positive value for the axial force. Consistently choosing the directions in this way helps avoid errors.

The equilibrium equations for joint A are

$$\Sigma F_x = T_{AC} + T_{AB} \cos 60° = 0,$$

$$\Sigma F_y = T_{AB} \sin 60° + 500 \text{ N} = 0.$$

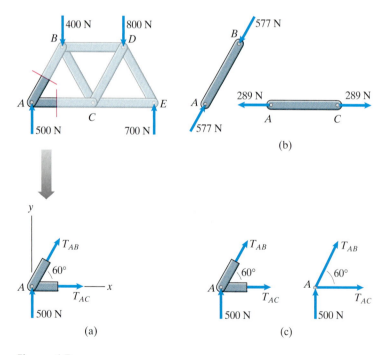

Figure 6.7
(a) Obtaining the free-body diagram of joint A.
(b) The axial forces on members AB and AC.
(c) Realistic and simple free-body diagrams of joint A.

Solving these equations, we obtain the axial forces $T_{AB} = -577$ N and $T_{AC} = 289$ N. Member AB is in compression, and member AC is in tension (Fig. 6.7b).

Although we use a realistic figure for the joint in Fig. 6.7a to help you understand the free-body diagram, in your own work you can use a simple figure showing only the forces acting on the joint (Fig. 6.7c).

We next obtain a free-body diagram of joint B by cutting members AB, BC, and BD (Fig. 6.8a). From the equilibrium equations for joint B,

$$\Sigma F_x = T_{BD} + T_{BC} \cos 60° + 577 \cos 60° \text{ N} = 0,$$

$$\Sigma F_y = -400 \text{ N} + 577 \sin 60° \text{ N} - T_{BC} \sin 60° = 0,$$

we obtain $T_{BC} = 115$ N and $T_{BD} = -346$ N. Member BC is in tension, and member BD is in compression (Fig. 6.8b). By continuing to draw free-body diagrams of the joints, we can determine the axial forces in all of the members.

In two dimensions, you can obtain only two independent equilibrium equations from the free-body diagram of a joint. Summing the moments about a point does not result in an additional independent equation because the forces are concurrent. Therefore when applying the method of joints, you should choose joints to analyze that are subjected to no more than two unknown forces. In our example, we analyzed joint A first because it was subjected to the known reaction exerted by the pin support and two unknown forces, the axial forces T_{AB} and T_{AC} (Fig. 6.7a). We could then analyze joint B because it was subjected to two known forces and two unknown forces, T_{BC} and T_{BD} (Fig. 6.8a). If we had attempted to analyze joint B first, there would have been three unknown forces.

When you determine the axial forces in the members of a truss, your task will often be simpler if you are familiar with three particular types of joints.

- **Truss joints with two collinear members and no load** (Fig. 6.9). The sum of the forces must equal zero, $T_1 = T_2$. The axial forces are equal.
- **Truss joints with two noncollinear members and no load** (Fig. 6.10). Because the sum of the forces in the x direction must equal zero, $T_2 = 0$. Therefore T_1 must also equal zero. The axial forces are zero.
- **Truss joints with three members, two of which are collinear, and no load** (Fig. 6.11). Because the sum of the forces in the x direction must equal zero, $T_3 = 0$. The sum of the forces in the y direction must equal zero, so $T_1 = T_2$. The axial forces in the collinear members are equal, and the axial force in the third member is zero.

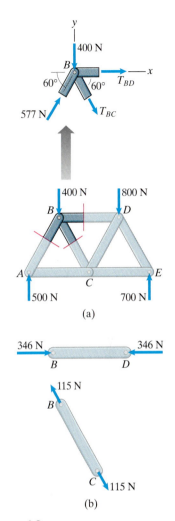

(a)

(b)

Figure 6.8
(a) Obtaining the free-body diagram of joint B.
(b) Axial forces in members BD and BC.

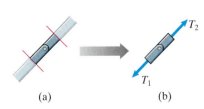

(a)   (b)

Figure 6.9
(a) A joint with two collinear members and no load.
(b) Free-body diagram of the joint.

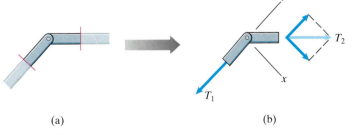

(a)   (b)

Figure 6.10
(a) A joint with two noncollinear members and no load.
(b) Free-body diagram of the joint.

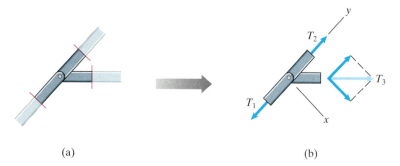

(a)   (b)

Figure 6.11
(a) A joint with three members, two of which are collinear, and no load.
(b) Free-body diagram of the joint.

Study Questions

1. What is a truss?
2. What is the method of joints?
3. How many independent equilibrium equations can you obtain from the free-body diagram of a joint?

Example 6.1   Applying the Method of Joints

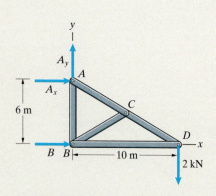

Figure 6.12

Determine the axial forces in the members of the truss in Fig. 6.12.

Strategy
We will first draw a free-body diagram of the entire truss, treating it as a single object, and determine the reactions at the supports. We can then apply the method of joints, simplifying our task by identifying any special joints of the types shown in Figs. 6.9–6.11.

Solution
Determine the Reactions at the Supports   We first draw the free-body diagram of the entire truss (Fig. a). From the equilibrium equations,

$$\Sigma F_x = A_x + B = 0,$$

$$\Sigma F_y = A_y - 2 \text{ kN} = 0,$$

$$\Sigma M_{\text{point } B} = -(6 \text{ m})A_x - (10 \text{ m})(2 \text{ kN}) = 0,$$

we obtain the reactions $A_x = -3.33$ kN, $A_y = 2$ kN, and $B = 3.33$ kN.

Identify Special Joints   Because joint C has three members, two of which are collinear, and no load, the axial force in member BC is zero, $T_{BC} = 0$, and the axial forces in the collinear members AC and CD are equal, $T_{AC} = T_{CD}$.

Draw Free-Body Diagrams of the Joints   We know the reaction exerted on joint A by the support, and joint A is subjected to only two unknown forces, the axial forces in members AB and AC. We draw its free-body diagram in Fig. b. The angle $\alpha = \arctan(5/3) = 59.0°$. The equilibrium equations for joint A are

$$\Sigma F_x = T_{AC} \sin \alpha - 3.33 \text{ kN} = 0,$$

$$\Sigma F_y = 2 \text{ kN} - T_{AB} - T_{AC} \cos \alpha = 0.$$

(a) Free-body diagram of the entire truss.

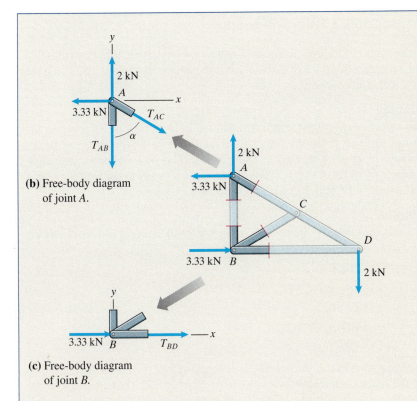

(b) Free-body diagram of joint A.

(c) Free-body diagram of joint B.

Solving these equations, we obtain $T_{AB} = 0$ and $T_{AC} = 3.89$ kN. Because the axial forces in members AC and CD are equal, $T_{CD} = 3.89$ kN.

Now we draw the free-body diagram of joint B in Fig. c. (We already know that the axial forces in members AB and BC are zero.) From the equilibrium equation

$$\Sigma\, F_x = T_{BD} + 3.33 \text{ kN} = 0,$$

we obtain $T_{BD} = -3.33$ kN. The negative sign indicates that member BD is in compression.

The axial forces in the members are

   AB: 0,

   AC: 3.89 kN in tension (T),

   BC: 0,

   BD: 3.33 kN in compression (C),

   CD: 3.89 kN in tension (T).

Critical Thinking

Observe how our solution was simplified by recognizing that joint C is the type shown in Fig. 6.11. This allowed us to determine the axial forces in all the members of the truss by analyzing only two joints.

Example 6.2   Determining the Largest Force a Truss Will Support

Figure 6.13

Each member of the truss in Fig. 6.13 will safely support a tensile force of 10 kN and a compressive force of 2 kN. What is the largest downward load F that the truss will safely support?

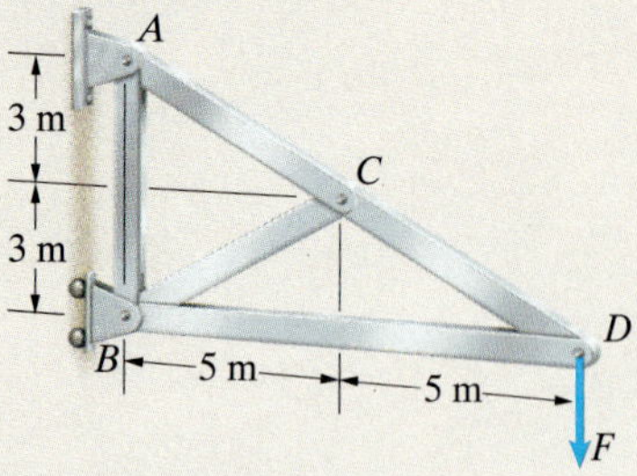

Strategy
This truss is identical to the one we analyzed in Example 6.1. By applying the method of joints in the same way, the axial forces in the members can be determined in terms of the load F. The smallest value of F that will cause a tensile force of 10 kN or a compressive force of 2 kN in any of the members is the largest value of F that the truss will support.

Solution
By using the method of joints in the same way as in Example 6.1, we obtain the axial forces:

$$AB: 0,$$
$$AC: 1.94F \text{ (T)},$$
$$BC: 0,$$
$$BD: 1.67F \text{ (C)},$$
$$CD: 1.94F \text{ (T)}.$$

For a given load F, the largest tensile force is $1.94F$ (in members AC and CD) and the largest compressive force is $1.67F$ (in member BD). The largest safe tensile force would occur when $1.94F = 10$ kN or when $F = 5.14$ kN. The largest safe compressive force would occur when $1.67F = 2$ kN or when $F = 1.20$ kN. Therefore the largest load F that the truss will safely support is 1.20 kN.

Critical Thinking
This example demonstrates why engineers analyze structures. By doing so, they can determine the loads that an existing structure will support or design a structure to support given loads. In this example, the tensile and compressive loads the members of the truss will support are given. Information of that kind must be obtained by applying the methods of mechanics of materials to the individual members. Then statics can be used, as we have done in this example, to determine the axial loads in the members in terms of the external loads on the structure.

Design Example 6.3   Bridge Design

The loads a bridge structure must support and pin supports where the structure is to be attached are shown in Fig. 6.14(1). Assigned to design the structure, a civil engineering student proposes the structure shown in Fig. 6.14(2). What are the axial forces in the members?

Strategy
The vertical members AG, BH, CI, DJ, and EK are subjected to compressive forces of magnitude F. Because of the symmetry of the structure, we can determine the axial loads in the remaining members by analyzing joints C and B.

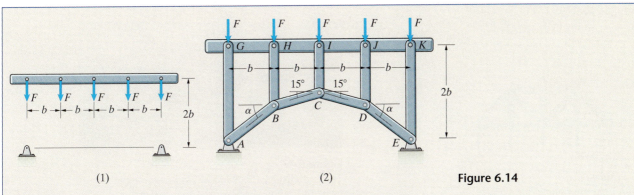

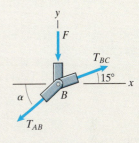

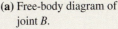

(1)   (2)   **Figure 6.14**

Solution

We will leave it as an exercise to show by drawing the free-body diagram of joint C that members BC and CD are subjected to equal compressive loads of magnitude $1.93F$. We draw the free-body diagram of joint B in Fig. a, where $T_{BC} = -1.93F$.

From the equilibrium equations

$$\Sigma F_x = -T_{AB} \cos \alpha + T_{BC} \cos 15° = 0,$$

$$\Sigma F_y = -T_{AB} \sin \alpha + T_{BC} \sin 15° - F = 0,$$

we obtain $T_{AB} = -2.39F$ and $\alpha = 38.8°$. By symmetry, $T_{DE} = T_{AB}$. The axial forces in the members are shown in Table 6.1.

(a) Free-body diagram of joint B.

Design Issues

The bridge was an early application of engineering. Although initially the solution was as primitive as laying a log between the banks, engineers constructed surprisingly elaborate bridges in the remote past. For example, archaeologists have identified foundations of the seven piers of a 120-m (400-ft) highway bridge over the Euphrates that existed in Babylon at the time of Nebuchadnezzar II (reigned 605–562 B.C.).

The basic difficulty in bridge design is that a single beam extended between the banks will fail if the distance between banks, or *span*, is too large. To meet the need for bridges of increasing strength and span, civil engineers created ingenious and aesthetic designs in antiquity and continue to do so today.

The bridge structure proposed by the student in this example, called an *arch*, is an ancient design. Notice in Table 6.1 that all the members of the structure are in compression. Because masonry (stone, brick, or concrete) is weak in tension but very strong in compression, many bridges made of these materials were designed with arched spans in the past. For the same reason, modern concrete bridges are often built with arched spans (Fig. 6.15).

TABLE 6.1 Axial forces in the members of the bridge structure.

Members	Axial Force
AG, BH, CI, DJ, EK	F (C)
AB, DE	$2.39F$ (C)
BC, CD	$1.93F$ (C)

Figure 6.15
A bridge along Highway 1 in California that is supported by concrete arches anchored in rock.

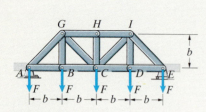

Figure 6.16
A Pratt truss supporting a bridge.

TABLE 6.2 Axial forces in the members of the Pratt truss.

Members	Axial Force
AB, BC, CD, DE	1.5F (T)
AG, EI	2.12F (C)
CG, CI	0.71F (T)
GH, HI	2F (C)
BG, DI	F (T)
CH	0

TABLE 6.3 Axial forces in the members of the suspension structure.

Members	Axial Force
BH, CI, DJ	F (T)
AB, DE	2.39F (T)
BC, CD	1.93F (T)

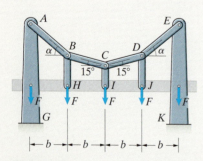

Figure 6.18
A suspension structure supporting a bridge.

Figure 6.17
The Forth Bridge (Scotland, 1890) is an example of a large truss bridge. Each main span is 520 m long.

Unlike masonry, wood and steel can support substantial forces in both compression and tension. Beginning with the wooden truss bridges designed by the architect Andrea Palladio (1518–1580), both of these materials have been used to construct a large variety of trusses to support bridges. For example, the forces in Fig. 6.14(1) can be supported by the Pratt truss shown in Fig. 6.16. Its members are subjected to both tension and compression (Table 6.2). The Forth Bridge (Fig. 6.17) has a truss structure.

Truss structures are too heavy for the largest bridges. (The Forth Bridge contains 58,000 tons of steel.) By taking advantage of the ability of relatively light cables to support large tensile forces, civil engineers use suspension structures to bridge very large spans. The system of five forces we are using as an example can be supported by the simple suspension structure in Fig. 6.18. In effect, the compression arch is inverted. (Compare Figs. 6.14(2) and 6.18.) The loads in Fig. 6.18 are "suspended" from members AB, BC, CD, and DE. Every member of this structure except the towers AG and EK is in tension (Table 6.3). The largest existing bridges, such as the Golden Gate Bridge (Fig. 6.19), consist of cable-suspended spans supported by towers.

Figure 6.19
The Golden Gate Bridge (California) has a central suspended span 1280 m (4200 ft) in length.

Problems

6.1 Determine the axial forces in the members of the truss and indicate whether they are in tension (T) or compression (C).

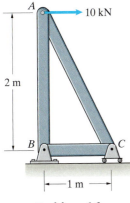

Problem 6.1

6.2 Determine the axial forces in the members of the truss and indicate whether they are in tension (T) or compression (C).

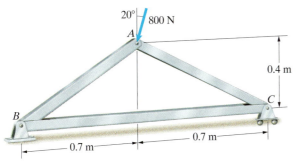

Problem 6.2

6.3 Member AB of the truss is subjected to a 1000-lb tensile force. Determine the weight W and the axial force in member AC.

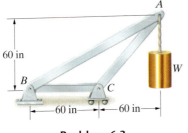

Problem 6.3

6.4 The members of the truss are all of length L. Determine the axial forces in the members and indicate whether they are in tension (T) or compression (C).

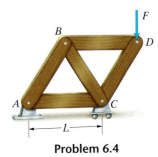

Problem 6.4

6.5 Each suspended weight has mass $m = 20$ kg. Determine the axial forces in the members of the truss and indicate whether they are in tension (T) or compression (C).

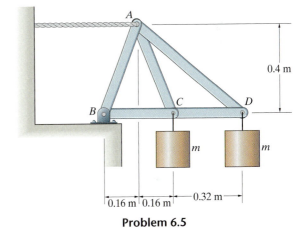

Problem 6.5

6.6 Determine the largest tensile and compressive forces that occur in the members of the truss, and indicate the members in which they occur if
(a) the dimension $h = 0.1$ m;
(b) the dimension $h = 0.5$ m.
Observe how a simple change in design affects the maximum axial loads.

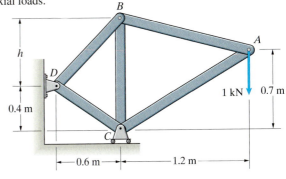

Problem 6.6

6.7 This steel truss bridge is in the Gallatin National Forest south of Bozeman, Montana. Suppose that one of the tandem trusses supporting the bridge is loaded as shown. Determine the axial forces in members AB, BC, BD, and BE.

6.8 Determine the largest tensile and compressive forces that occur in the members of the bridge truss, and indicate the members in which they occur.

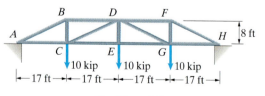

Problems 6.7/6.8

6.9 The trusses supporting the bridge in Problems 6.7 and 6.8 are called Pratt trusses. Suppose that the bridge designers had decided to use the truss shown instead, which is called a Howe truss. Determine the largest tensile and compressive forces that occur in the members, and indicate the members in which they occur. Compare your answers to the answers to Problem 6.8.

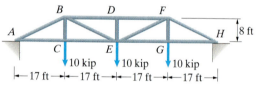

Problem 6.9

6.10 The truss shown is part of an airplane's internal structure. Determine the axial forces in members BC, BD, and BE.

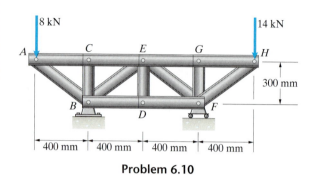

Problem 6.10

6.11 The loads $F_1 = F_2 = 8$ kN. Determine the axial forces in members BD, BE, and BG.

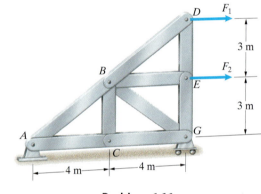

Problem 6.11

6.12 Determine the largest tensile and compressive forces that occur in the members of the truss, and indicate the members in which they occur if

(a) the dimension $h = 5$ in;
(b) the dimension $h = 10$ in.

Observe how a simple change in design affects the maximum axial loads.

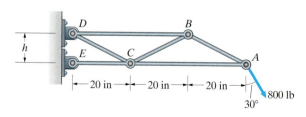

Problem 6.12

6.13 The truss supports loads at C and E. If $F = 3$ kN, what are the axial forces in members BC and BE?

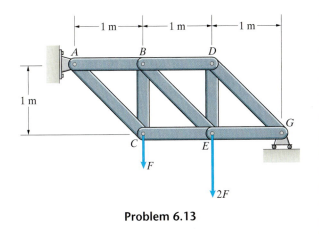

Problem 6.13

6.14 If you don't want the members of the truss to be subjected to an axial load (tension or compression) greater than 20 kN, what is the largest acceptable magnitude of the downward force F?

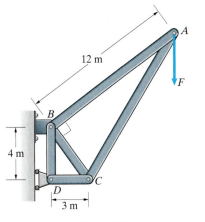

Problem 6.14

6.15 The truss is a preliminary design for a structure to attach one end of a stretcher to a rescue helicopter. Based on dynamic simulations, the design engineer estimates that the downward forces the stretcher will exert will be no greater than 1.6 kN at A and at B. What are the resulting axial forces in members CF, DF, and FG?

6.16 Upon learning of an upgrade in the helicopter's engine, the engineer designing the truss does new simulations and concludes that the downward forces the stretcher will exert at A and at B may be as large as 1.8 kN. What are the resulting axial forces in members DE, DF, and DG?

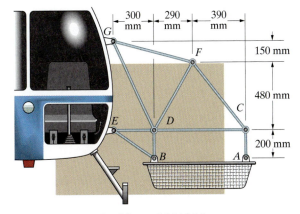
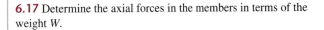

Problems 6.15/6.16

6.17 Determine the axial forces in the members in terms of the weight W.

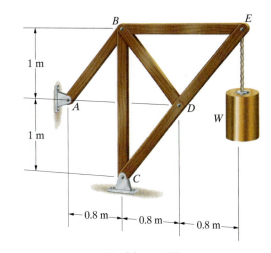

Problem 6.17

6.18 The lengths of the members of the truss are shown. The mass of the suspended crate is 900 kg. Determine the axial forces in the members.

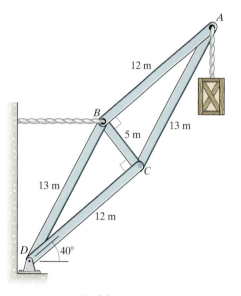

Problem 6.18

6.19 The loads $F_1 = 600$ lb and $F_2 = 300$ lb. Determine the axial forces in members AE, BD, and CD.

6.20 The loads $F_1 = 450$ lb and $F_2 = 150$ lb. Determine the axial forces in members AB, AC, and BC.

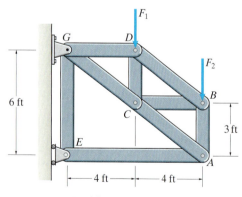

Problems 6.19/6.20

6.21 Each member of the truss will safely support a tensile force of 4 kN and a compressive force of 1 kN. Determine the largest mass m that can safely be suspended.

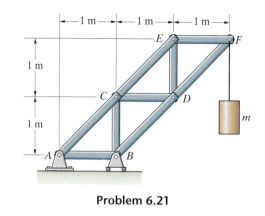

Problem 6.21

6.22 The Warren truss supporting the walkway is designed to support vertical 50-kN loads at B, D, F, and H. If the truss is subjected to these loads, what are the resulting axial forces in members BC, CD, and CE?

6.23 For the Warren truss in Problem 6.22, determine the axial forces in members DF, EF, and FG.

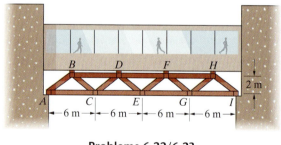

Problems 6.22/6.23

6.24 The Pratt bridge truss supports five forces ($F = 300$ kN). The dimension $L = 8$ m. Determine the axial forces in members BC, BI, and BJ.

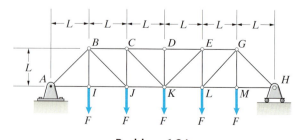

Problem 6.24

6.25 For the roof truss shown, determine the axial forces in members *AD*, *BD*, *DE*, and *DG*. Model the supports at *A* and *I* as roller supports.

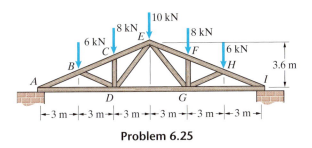

Problem 6.25

6.26 The Howe truss helps support a roof. Model the supports at *A* and *G* as roller supports. Determine the axial forces in members *AB*, *BC*, and *CD*.

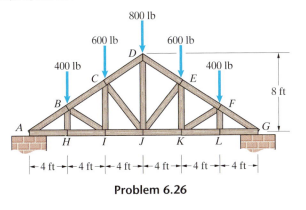

Problem 6.26

6.27 The plane truss forms part of the supports of a crane on an offshore oil platform. The crane exerts vertical 75-kN forces on the truss at *B*, *C*, and *D*. You can model the support at *A* as a pin support and model the support at *E* as a roller support that can exert a force normal to the dashed line but cannot exert a force parallel to it. The angle $\alpha = 45°$. Determine the axial forces in the members of the truss.

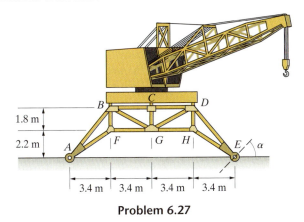

Problem 6.27

6.28 (a) Design a truss attached to the supports *A* and *B* that supports the loads applied at points *C* and *D*.
(b) Determine the axial forces in the members of the truss you designed in (a).

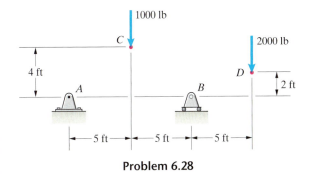

Problem 6.28

6.29 (a) Design a truss attached to the supports *A* and *B* that goes over the obstacle and supports the load applied at *C*.
(b) Determine the axial forces in the members of the truss you designed in (a).

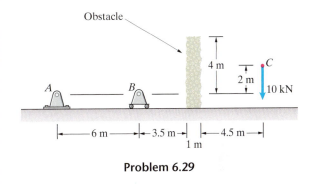

Problem 6.29

6.30 Suppose that you want to design a truss supported at *A* and *B* (Fig. a) to support a 3-kN downward load at *C*. The simplest design (Fig. b) subjects member AC to a 5-kN tensile force. Redesign the truss so that the largest tensile force is less than 3 kN.

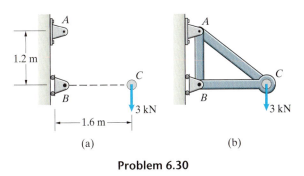

Problem 6.30

Design Experience

Problems 6.31–6.33 are related to Design Example 6.3.

6.31 The bridge structure shown in Fig. 6.14 can be given a higher arch by increasing the 15° angles to 20°. If this is done, what are the axial forces in members *AB, BC, CD,* and *DE*? Compare your answers to the values in Table 6.1.

6.32 Determine the axial forces in the Pratt truss in Fig. 6.16 and confirm the values in Table 6.2.

6.33 Determine the axial forces in the suspension bridge structure in Fig. 6.18, including the reactions exerted on the towers, and confirm the values in Table 6.3.

6.3 The Method of Sections

When we need to know the axial forces only in certain members of a truss, we often can determine them more quickly using the method of sections than using the method of joints. For example, let's reconsider the Warren truss we used to introduce the method of joints (Fig. 6.20a). It supports loads at *B* and *D*, and each member is 2 m in length. Suppose that we need to determine only the axial force in member *BC*.

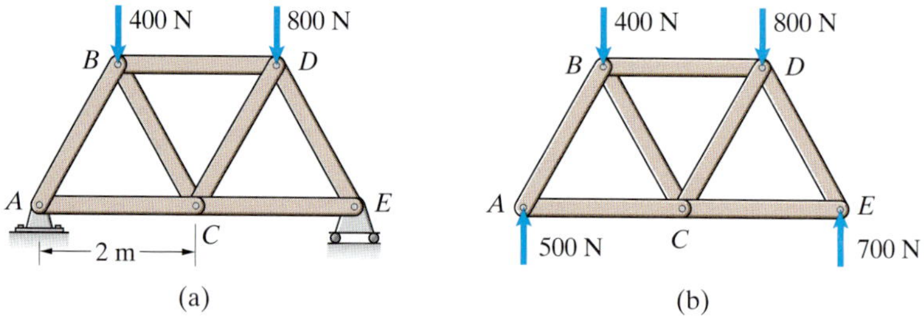

Figure 6.20
(a) A Warren truss supporting two loads.
(b) Free-body diagram of the truss, showing the reactions at the supports.

Just as in the method of joints, we begin by drawing a free-body diagram of the entire truss and determining the reactions at the supports. The results of this step are shown in Fig. 6.20b. Our next step is to cut the members *AC, BC,* and *BD* to obtain a free-body diagram of a part, or *section*, of the truss (Fig. 6.21). Summing moments about point *B*, the equilibrium equations for the section are

$$\Sigma F_x = T_{AC} + T_{BD} + T_{BC} \cos 60^\circ = 0,$$

$$\Sigma F_y = 500 \text{ N} - 400 \text{ N} - T_{BC} \sin 60^\circ = 0,$$

$$\Sigma M_{\text{point } B} = (2 \sin 60^\circ \text{ m})T_{AC} - (2 \cos 60^\circ \text{ m})(500 \text{ N}) = 0.$$

Solving them, we obtain $T_{AC} = 289$ N, $T_{BC} = 115$ N, and $T_{BD} = -346$ N.

Notice how similar this method is to the method of joints. Both methods involve cutting members to obtain free-body diagrams of parts of a truss. In the method of joints, we move from joint to joint, drawing free-body diagrams of the joints and determining the axial forces in the members as we go. In the method of sections, we try to obtain a single free-body diagram that allows us to determine the axial forces in specific members. In our example, we obtained a free-body diagram by cutting three members, including the one (member *BC*) whose axial force we wanted to determine.

In contrast to the free-body diagrams of joints, the forces on the free-body diagrams used in the method of sections are not usually concurrent, and as in our example, we can obtain three independent equilibrium equations. Although there are exceptions, it is usually necessary to choose a section that requires cutting no more than three members, or there will be more unknown axial forces than equilibrium equations.

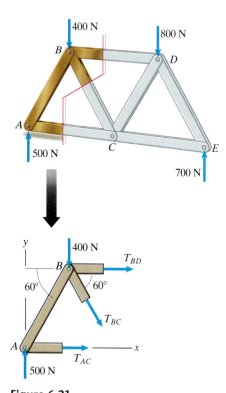

Figure 6.21
Obtaining a free-body diagram of a section of the truss.

Example 6.4   Applying the Method of Sections

The truss in Fig. 6.22 supports a 100-kN load. The horizontal members are each 1 m in length. Determine the axial force in member CJ, and state whether it is in tension or compression.

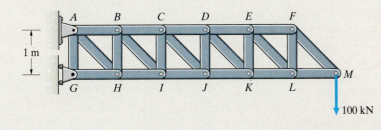

1 m

100 kN

Figure 6.22

Strategy
We need to obtain a section by cutting members that include member CJ. By cutting members CD, CJ, and IJ, we will obtain a free-body diagram with three unknown axial forces.

Solution
To obtain a section (Fig. a), we cut members CD, CJ, and IJ and draw the free-body diagram of the part of the truss on the right side of the cuts. From the equilibrium equation

$$\Sigma F_y = T_{CJ} \sin 45° - 100 \text{ kN} = 0,$$

we obtain $T_{CJ} = 141.4$ kN. The axial force in member CJ is 141.4 kN (T).

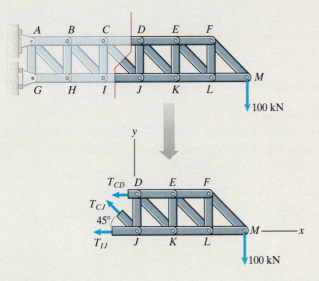

(a) Obtaining the section.

Critical Thinking
We designed this example to demonstrate that the method of sections can be very advantageous when you only need to determine the axial forces in particular members of a truss. Imagine calculating the axial force in member CJ using the method of joints. But in engineering applications it is usually necessary to know the axial forces in all the members of a truss, and in that case the two methods are comparable.

Example 6.5   Choosing an Appropriate Section

Determine the axial forces in members *DG* and *BE* of the truss in Fig. 6.23.

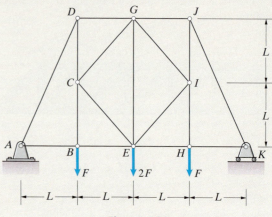

Figure 6.23

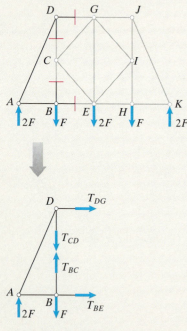

(a) Free-body diagram of the entire truss.

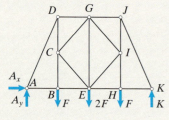

(b) A section of the truss obtained by passing planes through members *DG, CD, BC,* and *BE*.

Strategy

We can't obtain a section that involves cutting members *DG* and *BE* without cutting more than three members. However, cutting members *DG, BE, CD,* and *BC* results in a section with which we can determine the axial forces in members *DG* and *BE*.

Solution

Determine the Reactions at the Supports   We draw the free-body diagram of the entire truss in Fig. a. From the equilibrium equations,

$$\Sigma F_x = A_x = 0,$$

$$\Sigma F_y = A_y + K - F - 2F - F = 0,$$

$$\Sigma M_{\text{point } A} = -LF - (2L)(2F) - (3L)F + (4L)K = 0,$$

we obtain the reactions $A_x = 0, A_y = 2F$, and $K = 2F$.

Choose a Section

In Fig. b, we obtain a section by cutting members *DG, CD, BC,* and *BE*. Because the lines of action of T_{BE}, T_{BC}, and T_{CD} pass through point *B*, we can determine T_{DG} by summing moments about *B*:

$$\Sigma M_{\text{point } B} = -L(2F) - (2L)T_{DG} = 0.$$

The axial force $T_{DG} = -F$. Then, from the equilibrium equation

$$\Sigma F_x = T_{DG} + T_{BE} = 0,$$

we see that $T_{BE} = -T_{DG} = F$. Member *DG* is in compression, and member *BE* is in tension.

Critical Thinking

This is a clever example, but not one that is typical of problems faced in practice. The section used to solve it might not be obvious even to a person with experience analyzing structures. Notice that the free-body diagram in Fig. b is statically indeterminate, although it can be used to determine the axial forces in members *DG* and *BE*.

Problems

6.34 The truss supports a 100-kN load at *J*. The horizontal members are each 1 m in length.

(a) Use the method of joints to determine the axial force in member *DG*.
(b) Use the method of sections to determine the axial force in member *DG*.

6.35 The horizontal members are each 1 m in length. Use the method of sections to determine the axial forces in members *BC*, *CF*, and *FG*.

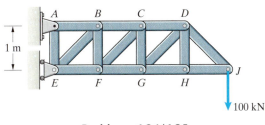

Problems 6.34/6.35

6.36 Use the method of sections to determine the axial forces in members *AB*, *BC*, and *CE*.

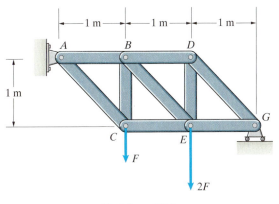

Problem 6.36

6.37 The truss supports loads at *A* and *H*. Use the method of sections to determine the axial forces in members *CE*, *BE*, and *BD*.

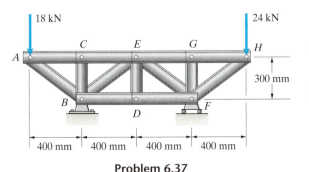

Problem 6.37

6.38 The Pratt bridge truss is loaded as shown. Use the method of sections to determine the axial forces in members *BD*, *BE*, and *CE*.

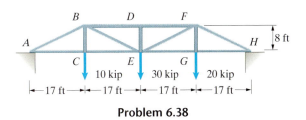

Problem 6.38

6.39 The Howe bridge truss is loaded as shown. Use the method of sections to determine the axial forces in members *BD*, *CD*, and *CE*.

6.40 For the Howe bridge truss, use the method of sections to determine the axial forces in members *DF*, *DG*, and *EG*.

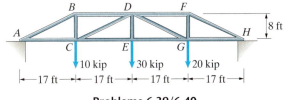

Problems 6.39/6.40

6.41 The Pratt bridge truss supports five forces $F = 340$ kN. The dimension $L = 8$ m. Use the method of sections to determine the axial force in member *JK*.

6.42 For the Pratt bridge truss in Problem 6.41, use the method of sections to determine the axial force in member *EK*.

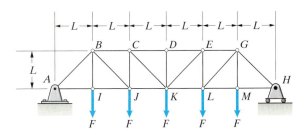

Problems 6.41/6.42

6.43 The walkway exerts vertical 50-kN loads on the Warren truss at *B, D, F,* and *H.* Use the method of sections to determine the axial force in member *CE.*

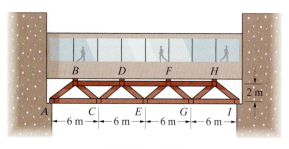

Problem 6.43

6.44 The mass *m* = 120 kg. Use the method of sections to determine the axial forces in members *BD, CD,* and *CE.*

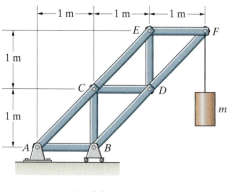

Problem 6.44

6.45 For the roof truss shown, use the method of sections to determine the axial forces in members *AD, BC,* and *BD.* Model the supports at *A* and *I* as roller supports.

6.46 For the roof truss shown, use the method of sections to determine the axial forces in members *CE, DE,* and *DG.*

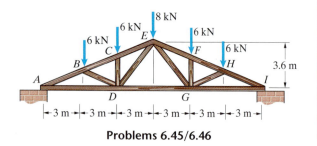

Problems 6.45/6.46

6.47 The Howe truss helps support a roof. Model the supports at *A* and *G* as roller supports.

(a) Use the method of joints to determine the axial force in member *BI.*

(b) Use the method of sections to determine the axial force in member *BI.*

6.48 Use the method of sections to determine the axial force in member *EJ.*

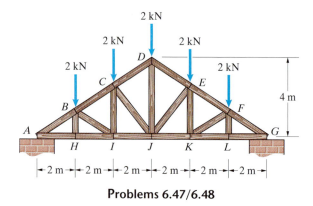

Problems 6.47/6.48

6.49 Use the method of sections to determine the axial force in member *EF.*

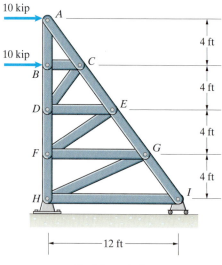

Problem 6.49

6.50 For the bridge truss shown, use the method of sections to determine the axial forces in members *CE*, *CF*, and *DF*.

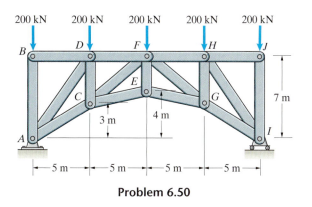

Problem 6.50

6.51 The load *F* = 20 kN and the dimension *L* = 2 m. Use the method of sections to determine the axial force in member *HK*.

Strategy: Obtain a section by cutting members *HK, HI, IJ,* and *JM*. You can determine the axial forces in members *HK* and *JM* even though the resulting free-body diagram is statically indeterminate.

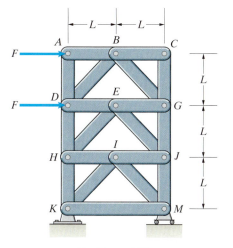

Problem 6.51

6.52 The weight of the bucket is *W* = 1000 lb. The cable passes over pulleys at *A* and *D*.

(a) Determine the axial forces in members *FG* and *HI*.

(b) By drawing free-body diagrams of sections, explain why the axial forces in members *FG* and *HI* are equal.

6.53 The weight of the bucket is *W* = 1000 lb. The cable passes over pulleys at *A* and *D*. Determine the axial forces in members *IK* and *JL*.

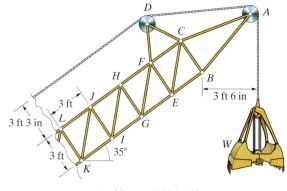

Problems 6.52/6.53

6.54 The truss supports loads at *N, P,* and *R*. Determine the axial forces in members *IL* and *KM*.

6.55 Determine the axial forces in members *HJ* and *GI*.

6.56 By drawing free-body diagrams of sections, explain why the axial forces in members *DE, FG,* and *HI* are zero.

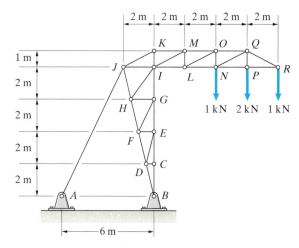

Problems 6.54–6.56

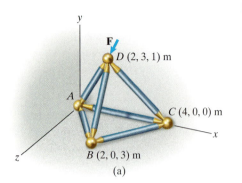

(a)

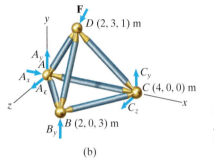

(b)

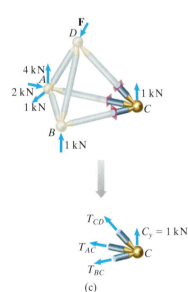

Figure 6.25
(a) A space truss supporting a load **F**.
(b) Free-body diagram of the entire truss.
(c) Obtaining the free-body diagram of joint C.

6.4 Space Trusses

We can form a simple three-dimensional structure by connecting six bars at their ends to obtain a tetrahedron, as shown in Fig. 6.24a. By adding members, we can obtain more elaborate structures (Figs. 6.24b and c). Three-dimensional structures such as these are called *space trusses* if they have joints that do not exert couples on the members (that is, the joints behave like ball and socket supports) and they are loaded and supported at their joints. Space trusses are analyzed by the same methods we described for two-dimensional trusses. The only difference is the need to cope with the more complicated geometry.

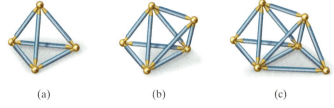

(a)   (b)   (c)

Figure 6.24
Space trusses with 6, 9, and 12 members.

Consider the space truss in Fig. 6.25a. Suppose that the load $\mathbf{F} = -2\mathbf{i} - 6\mathbf{j} - \mathbf{k}$ (kN). The joints A, B, and C rest on the smooth floor. Joint A is supported by the corner where the smooth walls meet, and joint C rests against the back wall. We can apply the method of joints to this truss.

First we must determine the reactions exerted by the supports (the floor and walls). We draw the free-body diagram of the entire truss in Fig. 6.25b. The corner can exert three components of force at A, the floor and wall can exert two components of force at C, and the floor can exert a normal force at B. Summing moments about A, we find that the equilibrium equations, with forces in kN and distances in m, are

$$\Sigma F_x = A_x - 2 = 0,$$

$$\Sigma F_y = A_y + B_y + C_y - 6 = 0,$$

$$\Sigma F_z = A_z + C_z - 1 = 0,$$

$$\Sigma M_{\text{point } A} = (\mathbf{r}_{AB} \times B_y\mathbf{j}) + [\mathbf{r}_{AC} \times (C_y\mathbf{j} + C_z\mathbf{k})] + (\mathbf{r}_{AD} \times \mathbf{F})$$

$$= \begin{vmatrix} \mathbf{i} & \mathbf{j} & \mathbf{k} \\ 2 & 0 & 3 \\ 0 & B_y & 0 \end{vmatrix} + \begin{vmatrix} \mathbf{i} & \mathbf{j} & \mathbf{k} \\ 4 & 0 & 0 \\ 0 & C_y & C_z \end{vmatrix}$$

$$+ \begin{vmatrix} \mathbf{i} & \mathbf{j} & \mathbf{k} \\ 2 & 3 & 1 \\ -2 & -6 & -1 \end{vmatrix}$$

$$= (-3B_y + 3)\mathbf{i} + (-4C_z)\mathbf{j}$$

$$+ (2B_y + 4C_y - 6)\mathbf{k} = 0.$$

Solving these equations, we obtain the reactions $A_x = 2$ kN, $A_y = 4$ kN, $A_z = 1$ kN, $B_y = 1$ kN, $C_y = 1$ kN, and $C_z = 0$.

In this example, we can determine the axial forces in members *AC, BC*, and *CD* from the free-body diagram of joint *C* (Fig. 6.25c). To write the equilibrium equations for the joint, we must express the three axial forces in terms of their components. Because member *AC* lies along the *x* axis, we express the force exerted on joint *C* by the axial force T_{AC} as the vector $-T_{AC}\mathbf{i}$. Let $\mathbf{r}_{CB}$ be the position vector from *C* to *B*:

$$\mathbf{r}_{CB} = (2 - 4)\mathbf{i} + (0 - 0)\mathbf{j} + (3 - 0)\mathbf{k} = -2\mathbf{i} + 3\mathbf{k} \text{ (m)}.$$

Dividing this vector by its magnitude to obtain a unit vector that points from *C* toward *B* yields

$$\mathbf{e}_{CB} = \frac{\mathbf{r}_{CB}}{|\mathbf{r}_{CB}|} = -0.555\mathbf{i} + 0.832\mathbf{k},$$

and we express the force exerted on joint *C* by the axial force T_{BC} as the vector

$$T_{BC}\,\mathbf{e}_{CB} = T_{BC}(-0.555\mathbf{i} + 0.832\mathbf{k}).$$

In the same way, we express the force exerted on joint *C* by the axial force T_{CD} as the vector

$$T_{CD}(-0.535\mathbf{i} + 0.802\mathbf{j} + 0.267\mathbf{k}).$$

Setting the sum of the forces on the joint equal to zero, we obtain

$$-T_{AC}\mathbf{i} + T_{BC}(-0.555\mathbf{i} + 0.832\mathbf{k})$$

$$+T_{CD}(-0.535\mathbf{i} + 0.802\mathbf{j} + 0.267\mathbf{k}) + (1 \text{ kN})\mathbf{j} = 0,$$

and then get the three equilibrium equations

$$\Sigma F_x = -T_{AC} - 0.555T_{BC} - 0.535T_{CD} = 0,$$

$$\Sigma F_y = 0.802T_{CD} + 1 \text{ kN} = 0,$$

$$\Sigma F_z = 0.832T_{BC} + 0.267T_{CD} = 0.$$

Solving these equations, the axial forces are $T_{AC} = 0.444$ kN, $T_{BC} = 0.401$ kN, and $T_{CD} = -1.247$ kN. Members *AC* and *BC* are in tension, and member *CD* is in compression. By continuing to draw free-body diagrams of the joints, we can determine the axial forces in all the members.

As our example demonstrates, three equilibrium equations can be obtained from the free-body diagram of a joint in three dimensions, so it is usually necessary to choose joints to analyze that are subjected to known forces and no more than three unknown forces.

Problems

6.57 The mass of the suspended weight is 2000 kg. Determine the axial forces in the bars AB and AC.

Strategy: Draw the free-body diagram of joint A.

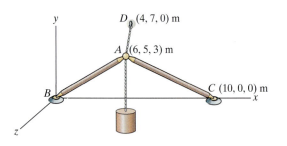

Problem 6.57

6.58 The space truss supports a vertical 10-kN load at D. The reactions at the supports at joints A, B, and C are shown. What are the axial forces in members AD, BD, and CD?

6.59 The reactions at the supports at joints A, B, and C are shown. What are the axial forces in members AB, AC, and AD?

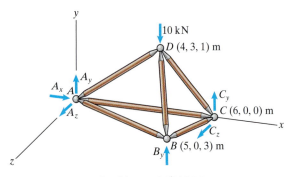

Problems 6.58/6.59

6.60 The space truss supports a vertical load F at A. Each member is of length L, and the truss rests on the horizontal surface on roller supports at B, C, and D. Determine the axial forces in members AB, AC, and AD.

6.61 For the truss in Problem 6.60, determine the axial forces in members AB, BC, and BD.

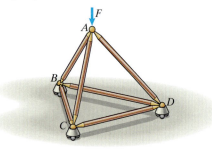

Problems 6.60/6.61

6.62 The space truss has roller supports at B, C, and D and supports a vertical 800-lb load at A. What are the axial forces in members AB, AC, and AD?

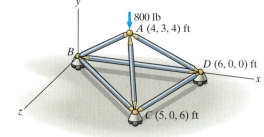

Problem 6.62

6.63 The space truss shown models an airplane's landing gear. It has ball and socket supports at C, D, and E. If the force exerted at A by the wheel is $\mathbf{F} = 40\mathbf{j}$ (kN), what are the axial forces in members AB, AC, and AD?

6.64 If the force exerted at point A of the truss in Problem 6.63 is $\mathbf{F} = 10\mathbf{i} + 60\mathbf{j} + 20\mathbf{k}$ (kN), what are the axial forces in members BC, BD, and BE?

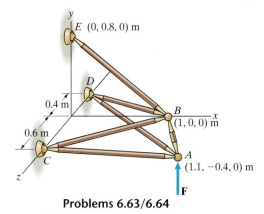

Problems 6.63/6.64

6.65 The space truss is supported by roller supports on the horizontal surface at C and D and a ball and socket support at E. The y axis points upward. The mass of the suspended object is 120 kg. The coordinates of the joints of the truss are A: (1.6, 0.4, 0) m, B: (1.0, 1.0, −0.2) m, C: (0.9, 0, 0.9) m, D: (0.9, 0, −0.6) m, and E: (0, 0.8, 0) m. Determine the axial forces in members AB, AC, and AD.

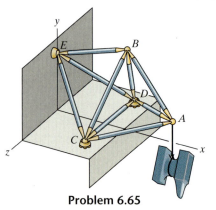

Problem 6.65

6.66 The free-body diagram of the part of the construction crane to the left of the plane is shown. The coordinates (in meters) of the joints *A*, *B*, and *C* are (1.5, 1.5, 0), (0, 0, 1), and (0, 0, −1), respectively. The axial forces P_1, P_2, and P_3 are parallel to the *x* axis. The axial forces P_4, P_5, and P_6 point in the directions of the unit vectors

$e_4 = 0.640i - 0.640j - 0.426k,$

$e_5 = 0.640i - 0.640j + 0.426k,$

$e_6 = 0.832i - 0.555k.$

The total force exerted on the free-body diagram by the weight of the crane and the load it supports is $-Fj = -44j$ (kN) acting at the point (−20, 0, 0) m. What is the axial force P_3?

Strategy: Use the fact that the moment about the line that passes through joints *A* and *B* equals zero.

6.67 In Problem 6.66, what are the axial forces P_1, P_4, and P_5?

Strategy: Write the equilibrium equations for the entire free-body diagram.

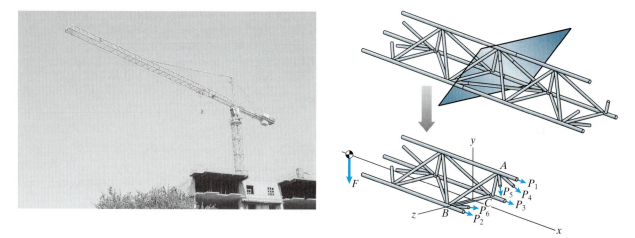

Problems 6.66/6.67

6.68 The mirror housing of the telescope is supported by a 6-bar space truss. The mass of the housing is 3 Mg (megagrams), and its weight acts at *G*. The distance from the axis of the telescope to points *A*, *B*, and *C* is 1 m, and the distance from the axis to points *D*, *E*, and *F* is 2.5 m. If the telescope axis is vertical ($\alpha = 90°$), what are the axial forces in the members of the truss?

6.69 Consider the telescope described in Problem 6.68. Determine the axial forces in the members of the truss if the angle α between the horizontal and the telescope axis is 20°.

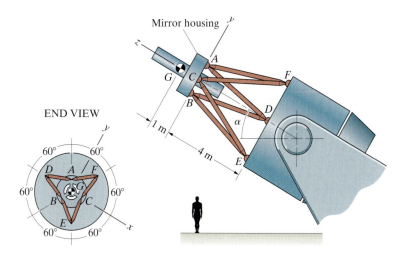

Problems 6.68/6.69

6.5 Frames and Machines

Many structures, such as the frame of a car and the human structure of bones, tendons, and muscles (Fig. 6.26), are not composed entirely of two-force members and thus cannot be modeled as trusses. In this section we consider structures of interconnected members that do not satisfy the definition of a truss. Such structures are called *frames* if they are designed to remain stationary and support loads and *machines* if they are designed to move and apply loads.

When trusses are analyzed by cutting members to obtain free-body diagrams of joints or sections, the internal forces acting at the "cuts" are simple axial forces (see Fig. 6.4). This is not generally true for frames or machines, and a different method of analysis is necessary. Instead of cutting members, you isolate entire members, or in some cases combinations of members, from the structure.

To begin analyzing a frame or machine, we draw a free-body diagram of the entire structure (that is, treat the structure as a single object) and determine the reactions at its supports. In some cases the entire structure will be statically indeterminate, but it is helpful to determine as many of the reactions as possible. We then draw free-body diagrams of individual members, or selected

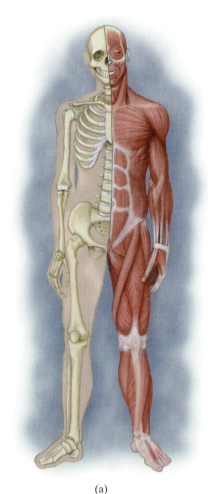

(a)   (b)

Figure 6.26
The internal structure of a person (**a**) and a car's frame (**b**) are not trusses.

combinations of members, and apply the equilibrium equations to determine the forces and couples acting on them. For example, let's consider the stationary structure in Fig. 6.27. Member BE is a two-force member, but the other three members—ABC, CD, and DEG—are not. This structure is a frame. Our objective is to determine the forces on its members.

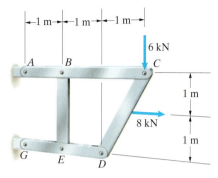

Figure 6.27
A frame supporting two loads.

Analyzing the Entire Structure

We draw the free-body diagram of the entire frame in Fig. 6.28. It is statically indeterminate: There are four unknown reactions, A_x, A_y, G_x, and G_y, whereas we can write only three independent equilibrium equations. However, notice that the lines of action of three of the unknown reactions intersect at A. Summing moments about A yields

$$\Sigma M_{\text{point }A} = (2\text{ m})G_x + (1\text{ m})(8\text{ kN}) - (3\text{ m})(6\text{ kN}) = 0,$$

and we obtain the reaction $G_x = 5$ kN. Then, from the equilibrium equation

$$\Sigma F_x = A_x + G_x + 8\text{ kN} = 0,$$

we obtain the reaction $A_x = -13$ kN. Although we cannot determine A_y or G_y from the free-body diagram of the entire structure, we can do so by analyzing the individual members.

Analyzing the Members

Our next step is to draw free-body diagrams of the members. To do so, we treat the attachment of a member to another member just as if it were a support. Looked at in this way, we can think of each member as a supported object of the kind analyzed in Chapter 5. Furthermore, the forces and couples the members exert on one another are *equal in magnitude and opposite in direction*. A simple demonstration is instructive. If you clasp your hands as shown in Fig. 6.29a and exert a force on your left hand with your right hand, your left hand exerts an equal and opposite force on your right hand (Fig. 6.29b). Similarly, if you exert a couple on your left hand, your left hand exerts an equal and opposite couple on your right hand.

In Fig. 6.30 we "disassemble" the frame and draw free-body diagrams of its members. Observe that the forces exerted on one another by the members are

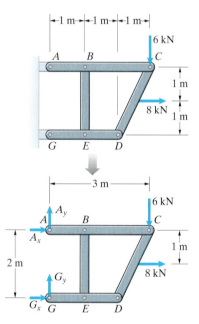

Figure 6.28
Obtaining the free-body diagram of the entire frame.

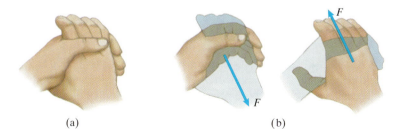

(a)   (b)

Figure 6.29
Demonstrating Newton's third law:
(a) Clasp your hands and pull on your left hand.
(b) Your hands exert equal and opposite forces.

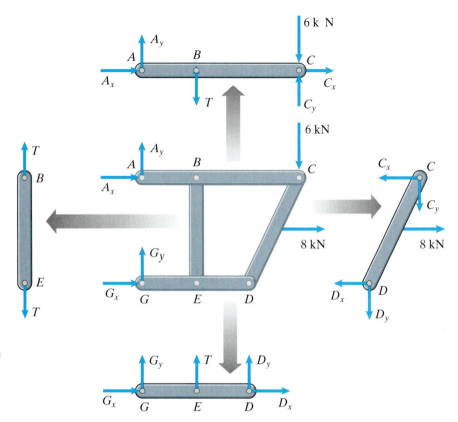

Figure 6.30
Obtaining the free-body diagrams of the members.

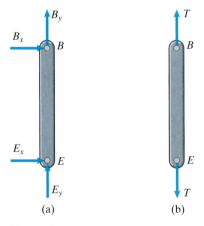

Figure 6.31
Free-body diagram of member BE:
(a) Not treating it as a two-force member.
(b) Treating it as a two-force member.

equal and opposite. For example, at point C on the free-body diagram of member ABC, the force exerted by member CD is denoted by the components C_x and C_y. The forces exerted by member ABC on member CD at point C must be equal and opposite, as shown.

We need to discuss two important aspects of these free-body diagrams before completing the analysis.

Two-Force Members   Member BE is a two-force member, and we have taken this into account in drawing its free-body diagram in Fig. 6.30. The force T is the axial force in member BE, and an equal and opposite force is subjected on member ABC at B and on member GED at E.

Recognizing two-force members in frames and machines and drawing their free-body diagrams as we have done will reduce the number of unknowns and will greatly simplify the analysis. In our example, if we did not treat member BE as a two-force member, its free-body diagram would have four unknown forces (Fig. 6.31a). By treating it as a two-force member (Fig. 6.31b), we reduce the number of unknown forces by three.

Loads Applied at Joints   A question arises when a load is applied at a joint: Where does the load appear on the free-body diagrams of the individual members? The answer is that you can place the load on *any one* of the members attached at the joint. For example, in Fig. 6.27, the 6-kN load acts at the joint where members ABC and CD are connected. In drawing the free-body diagrams of the individual members (Fig. 6.30), we assumed that the 6-kN load acted on member ABC. The force components C_x and C_y on the free-body diagram of member ABC are the forces exerted by the member CD.

To explain why we can draw the free-body diagrams in this way, let us assume that the 6-kN force acts on the pin connecting members *ABC* and *CD*, and draw separate free-body diagrams of the pin and the two members (Fig. 6.32a). The force components C'_x and C'_y are the forces exerted by the pin on member *ABC*, and C_x and C_y are the forces exerted by the pin on member *CD*. If we superimpose the free-body diagrams of the pin and member *ABC*, we obtain the two free-body diagrams in Fig. 6.32b, which is the way we drew them in Fig. 6.30. Alternatively, by superimposing the free-body diagrams of the pin and member *CD*, we obtain the two free-body diagrams in Fig. 6.32c.

Thus if a load acts at a joint, it can be placed on any one of the members attached at the joint when drawing the free-body diagrams of the individual members. Just make sure not to place it on more than one member.

To detect errors in the free-body diagrams of the members, it is helpful to "reassemble" them (Fig. 6.33a). The forces at the connections between the members cancel (they are internal forces once the members are reassembled), and the free-body diagram of the entire structure is recovered (Fig. 6.33b).

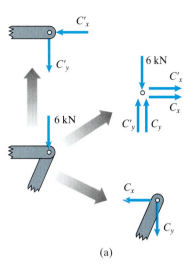

(a)

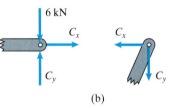

(b)

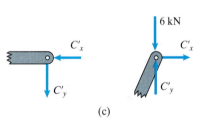

(c)

Figure 6.32
(a) Drawing free-body diagrams of the pin and the two members.
(b) Superimposing the pin on member *ABC*.
(c) Superimposing the pin on member *CD*.

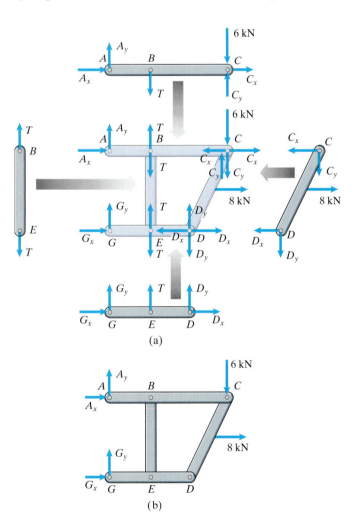

(a)

(b)

Figure 6.33
(a) "Reassembling" the free-body diagrams of the individual members.
(b) The free-body diagram of the entire frame is recovered.

Our final step is to apply the equilibrium equations to the free-body diagrams of the members (Fig. 6.34). In two dimensions, we can obtain three independent equilibrium equations from the free-body diagram of each member of a structure that we do not treat as a two-force member. (By assuming that the forces on a two-force member are equal and opposite axial forces, we have already used the three equilibrium equations for that member.) In this example, there are three members in addition to the two-force member, so we can write $(3)(3) = 9$ independent equilibrium equations, and there are nine unknown forces: A_x, A_y, C_x, C_y, D_x, D_y, G_x, G_y, and T.

Recall that we determined that $A_x = -13$ kN and $G_x = 5$ kN from our analysis of the entire structure. The equilibrium equations we obtained from the free-body diagram of the entire structure are not independent of the equilibrium equations obtained from the free-body diagrams of the members, but by using them to determine A_x and G_x, we get a head start on solving the equations for the members. Consider the free-body diagram of member ABC (Fig. 6.34a). Because we know A_x, we can determine C_x from the equation

$$\Sigma F_x = A_x + C_x = 0,$$

obtaining $C_x = -A_x = 13$ kN. Now consider the free-body diagram of GED (Fig. 6.34b). We can determine D_x from the equation

$$\Sigma F_x = G_x + D_x = 0,$$

obtaining $D_x = -G_x = -5$ kN. Now consider the free-body diagram of member CD (Fig. 6.34c). Because we know C_x, we can determine C_y by summing moments about D:

$$\Sigma M_{\text{point } D} = (2 \text{ m})C_x - (1 \text{ m})C_y - (1 \text{ m})(8 \text{ kN}) = 0.$$

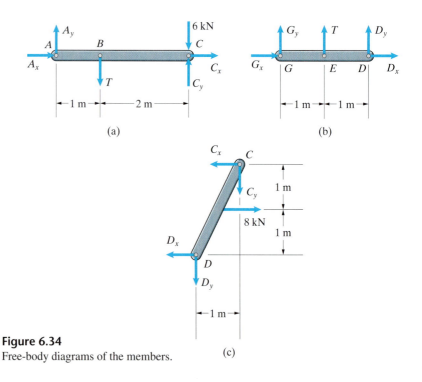

Figure 6.34
Free-body diagrams of the members.

We obtain $C_y = 18$ kN. Then, from the equation

$$\Sigma F_y = -C_y - D_y = 0,$$

we find that $D_y = -C_y = -18$ kN. Now we can return to the free-body diagrams of members ABC and GED to determine A_y and G_y. Summing moments about point B of member ABC yields

$$\Sigma M_{\text{point } B} = -(1 \text{ m})A_y + (2 \text{ m})C_y - (2 \text{ m})(6 \text{ kN}) = 0,$$

and we obtain $A_y = 2C_y - 12$ kN $= 24$ kN. Then, summing moments about point E of member GED, we have

$$\Sigma M_{\text{point } E} = (1 \text{ m})D_y - (1 \text{ m})G_y = 0,$$

from which we obtain $G_y = D_y = -18$ kN. Finally, from the free-body diagram of member GED, we use the equilibrium equation

$$\Sigma F_y = D_y + G_y + T = 0,$$

which gives us the result $T = -D_y - G_y = 36$ kN. The forces on the members are shown in Fig. 6.35. As this example demonstrates, determination of the forces on the members can often be simplified by carefully choosing the order in which the equations are solved.

We see that determining the forces and couples on the members of frames and machines involves two steps:

1. Determine the reactions at the supports—Draw the free-body diagram of the entire structure, and determine the reactions at its supports. Although this step is not essential, it can greatly simplify your analysis of the members. If the free-body diagram is statically indeterminant, determine as many of the reactions as possible.

2. Analyze the members—Draw free-body diagrams of the members, and apply the equilibrium equations to determine the forces acting on them. You can simplify this step by identifying two-force members. If a load acts at a joint of the structure, you can place the load on the free-body diagram of any one of the members attached at that joint.

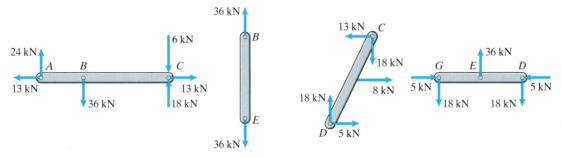

Figure 6.35
Forces on the members of the frame.

Example 6.6 | Analyzing a Frame

The frame in Fig. 6.36 is subjected to a 200-N-m couple. Determine the forces and couples on its members.

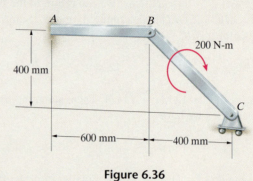

Figure 6.36

Strategy
We will first draw a free-body diagram of the entire frame, treating it as a single object, and attempt to determine the reactions at the supports. We will then draw free-body diagrams of the individual members and use the equilibrium equations to determine the forces and couples acting on them.

Solution
Determine the Reactions at the Supports We draw the free-body diagram of the entire frame in Fig. a. The term M_A is the couple exerted by the fixed support. From the equilibrium equations

$$\Sigma F_x = A_x = 0,$$

$$\Sigma F_y = A_y + C = 0,$$

$$\Sigma M_{\text{point } A} = M_A - 200 \text{ N-m} + (1 \text{ m})C = 0,$$

we obtain the reaction $A_x = 0$. We can't determine A_y, M_A, or C from this free-body diagram.

Analyze the Members We "disassemble" the frame to obtain the free-body diagrams of the members in Fig. b. The equilibrium equations for member BC are

$$\Sigma F_x = -B_x = 0,$$

$$\Sigma F_y = -B_y + C = 0,$$

$$\Sigma M_{\text{point } B} = -200 \text{ N-m} + (0.4 \text{ m})C = 0.$$

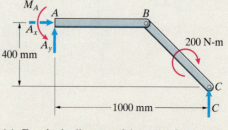

(a) Free-body diagram of the entire frame.

Solving these equations, we obtain $B_x = 0$, $B_y = 500$ N, and $C = 500$ N. The equilibrium equations for member AB are

$$\Sigma F_x = A_x + B_x = 0,$$

$$\Sigma F_y = A_y + B_y = 0,$$

$$\Sigma M_{\text{point } A} = M_A + (0.6 \text{ m})B_y = 0.$$

Because we already know A_x, B_x, and B_y, we can solve these equations for A_y and M_A. The results are $A_y = -500$ N and $M_A = -300$ N-m. This completes the solution (Fig. c).

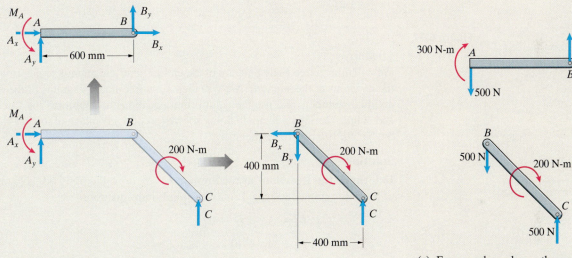

(b) Obtaining the free-body diagrams of the members.

(c) Forces and couples on the members.

Critical Thinking

We were able to solve the equilibrium equations for member BC without having to consider the free-body diagram of member AB. We were then able to solve the equilibrium equations for member AB. By choosing the members with the fewest unknowns to analyze first, you will often be able to solve them sequentially, but in some cases you will have to solve the equilibrium equations for the members simultaneously.

Even though we were unable to determine the four reactions A_x, A_y, M_A, and C with the three equilibrium equations obtained from the free-body diagram of the entire frame, we were able to determine them from the free-body diagrams of the individual members. By drawing free-body diagrams of the members, we gained three equations because we obtained three equilibrium equations from each member but only two new unknowns, B_x and B_y.

Example 6.7   **Determining Forces on Members of a Frame**

The frame in Fig. 6.37 supports a suspended weight $W = 40$ lb. Determine the forces on members *ABCD* and *CEG*.

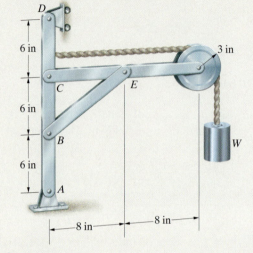

Figure 6.37

Strategy

We will draw a free-body diagram of the entire frame and attempt to determine the reactions at the supports. We will then draw free-body diagrams of the individual members and use the equilibrium equations to determine the forces and couples acting on them. In doing so, we can take advantage of the fact that the bar *BE* is a two-force member.

Solution

Determine the Reactions at the Supports   We draw the free-body diagram of the entire frame in Fig. a. From the equilibrium equations

$$\Sigma F_x = A_x - D = 0,$$

$$\Sigma F_y = A_y - 40 \text{ lb} = 0,$$

$$\Sigma M_{\text{point } A} = (18 \text{ in})D - (19 \text{ in})(40 \text{ lb}) = 0,$$

we obtain the reactions $A_x = 42.2$ lb, $A_y = 40$ lb, and $D = 42.2$ lb.

Analyze the Members   We obtain the free-body diagrams of the members in Fig. b. Notice that *BE* is a two-force member. The angle $\alpha = \arctan(6/8) = 36.9°$.

The free-body diagram of the pulley has only two unknown forces. From the equilibrium equations

$$\Sigma F_x = G_x - 40 \text{ lb} = 0,$$

$$\Sigma F_y = G_y - 40 \text{ lb} = 0,$$

we obtain $G_x = 40$ lb and $G_y = 40$ lb. There are now only three unknown forces on the free-body diagram of member *CEG*. From the equilibrium equations

$$\Sigma F_x = -C_x - R \cos \alpha - 40 \text{ lb} = 0,$$

$$\Sigma F_y = -C_y - R \sin \alpha - 40 \text{ lb} = 0,$$

$$\Sigma M_{\text{point } C} = -(8 \text{ in})R \sin \alpha - (16 \text{ in})(40 \text{ lb}) = 0,$$

we obtain $C_x = 66.7$ lb, $C_y = 40$ lb, and $R = -133.3$ lb, completing the solution (Fig. c).

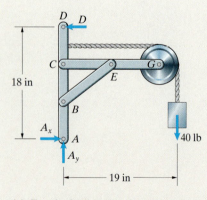

(a) Free-body diagram of the entire frame.

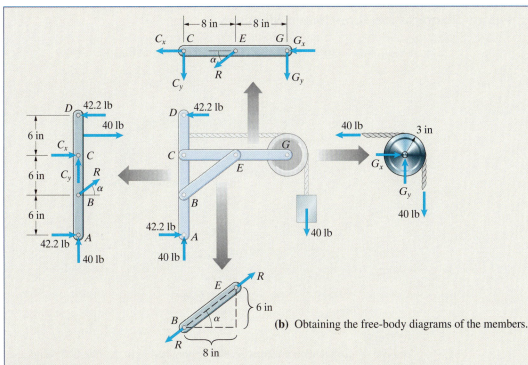

(b) Obtaining the free-body diagrams of the members.

Critical Thinking

In problems of this kind, the reactions on the individual members of the frame can be determined from the free-body diagrams of the members. Why did we draw the free-body diagram of the entire frame and solve the associated equilibrium equations? The reason is that it gave us a head start on solving the equilibrium equations for the members. In this example, when we drew the free-body diagrams of the members we already knew the reactions at A and D, which simplified the remaining analysis. Analyzing the entire frame can also provide a check on your work. Notice that we did not use the equilibrium equations for member $ABCD$. We can check our analysis by confirming that this member is in equilibrium (Fig. c):

$$\Sigma F_x = 42.2 \text{ lb} - 133.3 \cos 36.9° \text{ lb} + 66.7 \text{ lb} + 40 \text{ lb} - 42.2 \text{ lb} = 0,$$

$$\Sigma F_y = 40 \text{ lb} - 133.3 \sin 36.9° \text{ lb} + 40 \text{ lb} = 0,$$

$$\Sigma M_{\text{point } A} = (6 \text{ in})(133.3 \cos 36.9° \text{ lb}) -$$

$$(12 \text{ in})(66.7 \text{ lb}) - (15 \text{ in})(40 \text{ lb}) + (18 \text{ in})(42.2 \text{ lb}) = 0.$$

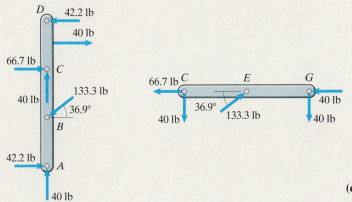

(c) Forces on members $ABCD$ and CEG.

| Example 6.8 | Free-Body Diagrams for Three Joined Members |

Determine the forces on the members of the frame in Fig. 6.38.

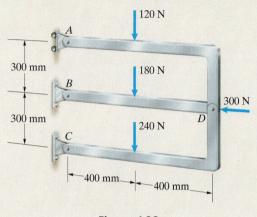

Figure 6.38

Strategy

You can confirm that no information can be obtained from the free-body diagram of the entire frame. To analyze the members, we must deal with an interesting challenge at joint D, where a load acts and three members are connected. We will obtain the free-body diagrams of the members by first isolating member AD, then separating members BD and CD.

Solution

We first isolate member AD from the rest of the structure, introducing the reactions D_x and D_y (Fig. a). We then separate members BD and CD, introducing equal and opposite forces E_x and E_y (Fig. b). In this step we could have placed the 300-N load and the forces D_x and D_y on either free-body diagram.

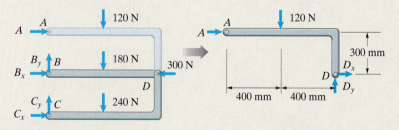

(a) Isolating member AD.

Only three unknown forces act on member AD. From the equilibrium equations

$$\Sigma F_x = A + D_x = 0,$$

$$\Sigma F_y = D_y - 120 \text{ N} = 0,$$

$$\Sigma M_{\text{point } D} = -(0.3 \text{ m})A + (0.4 \text{ m})(120 \text{ N}) = 0,$$

we obtain $A = 160$ N, $D_x = -160$ N, and $D_y = 120$ N. Now we consider the free-body diagram of member BD. From the equation

$$\Sigma M_{\text{point } D} = -(0.8 \text{ m})B_y + (0.4 \text{ m})(180 \text{ N}) = 0,$$

we obtain $B_y = 90$ N. Now we use the equation

$$\Sigma F_y = B_y - D_y + E_y - 180 \text{ N} = 90 \text{ N} - 120 \text{ N} + E_y - 180 \text{ N} = 0,$$

obtaining $E_y = 210$ N. Now that we know E_y, there are only three unknown forces on the free-body diagram of member CD. From the equilibrium equations

$$\Sigma F_x = C_x - E_x = 0,$$

$$\Sigma F_y = C_y - E_y - 240 \text{ N} = C_y - 210 \text{ N} - 240 \text{ N} = 0,$$

$$\Sigma M_{\text{point } C} = (0.3 \text{ m})E_x - (0.8 \text{ m})E_y - (0.4 \text{ m})(240 \text{ N})$$

$$= (0.3 \text{ m})E_x - (0.8 \text{ m})(210 \text{ N}) - (0.4 \text{ m})(240 \text{ N}) = 0,$$

we obtain $C_x = 880$ N, $C_y = 450$ N, and $E_x = 880$ N. Finally, we return to the free-body diagram of member BD and use the equation

$$\Sigma F_x = B_x + E_x - D_x - 300 \text{ N} = B_x + 880 \text{ N} + 160 \text{ N} - 300 \text{ N} = 0$$

to obtain $B_x = -740$ N, completing the solution (Fig. c).

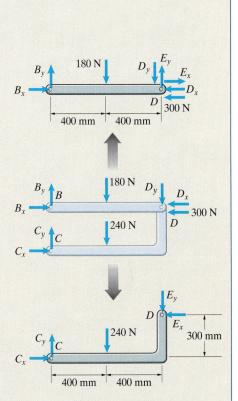

(b) Separating members BD and CD.

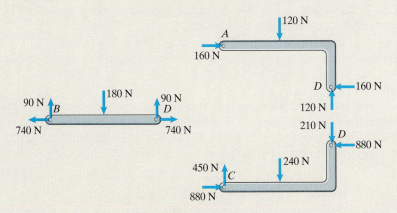

(c) Solutions for the forces on the members.

Critical Thinking

Why did we obtain free-body diagrams of the individual members by isolating one member at a time? The reason is that it simplified the bookkeeping of the reactions at joint D. This is the easiest procedure whenever three or more members are joined at a single point. The members should be isolated one at a time, and they can be removed in any order. To convince yourself that this is the preferred procedure, try drawing the individual free-body diagrams of the members in this example by isolating the three members from each other in a single step.

Example 6.9   **Analyzing a Machine**

What forces are exerted on the bolt at E in Fig. 6.39 as a result of the 150-N forces on the pliers?

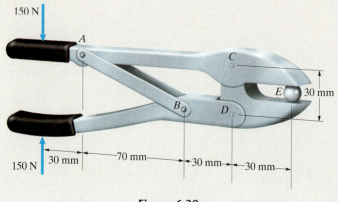

Figure 6.39

Strategy

A pair of pliers is a simple example of a machine, a structure designed to move and exert forces. The interconnections of the members are designed to create a mechanical advantage, subjecting an object to forces greater than the forces exerted by the user.

In this case there is no information to be gained from the free-body diagram of the entire structure. We must determine the forces exerted on the bolt by drawing free-body diagrams of the members.

Solution

We "disassemble" the pliers in Fig. a to obtain the free-body diagrams of the members, labeled (1), (2), and (3). The force R on free-body diagrams (1) and (3) is exerted by the two-force member AB. The angle $\alpha = \arctan(30/70) = 23.2°$. Our objective is to determine the force E exerted by the bolt.

The free-body diagram of member (3) has only three unknown forces and the 150-N load, so we can determine R, D_x, and D_y from this free-body diagram alone. The equilibrium equations are

$$\Sigma F_x = D_x + R \cos \alpha = 0,$$

$$\Sigma F_y = D_y - R \sin \alpha + 150 \text{ N} = 0,$$

$$\Sigma M_{\text{point } B} = (30 \text{ mm})D_y - (100 \text{ mm})(150 \text{ N}) = 0.$$

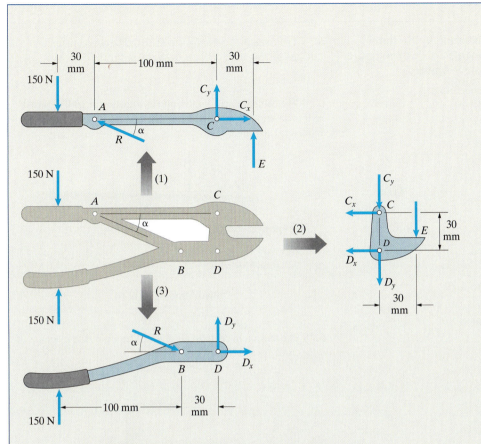

(a) Obtaining the free-body diagrams of the members.

Solving these equations, we obtain $D_x = -1517$ N, $D_y = 500$ N, and $R = 1650$ N. Knowing D_x, we can determine E from the free-body diagram of member (2) by summing moments about C:

$$\Sigma M_{\text{point } C} = -(30\text{ mm})E - (30\text{ mm})D_x = 0.$$

The force exerted on the bolt by the pliers is $E = -D_x = 1517$ N. The mechanical advantage of the pliers is $(1517\text{ N})/(150\text{ N}) = 10.1$.

Critical Thinking

What is the motivation for determining the reactions on the members of the pliers? This process is essential for machine and tool design. To design the configuration of the pliers and choose the materials and dimensions of its members, it is necessary to determine all the forces acting on the members, as we have done in this example. Once the forces are known, the methods of mechanics of materials can be used to assess the adequacy of the members to support them.

Problems

Assume that objects are in equilibrium. In the statements of the answers, *x* components are positive to the right and *y* components are positive upward.

6.70 Determine the reactions on member *AB* at *A*. (Notice that *BC* is a two-force member.)

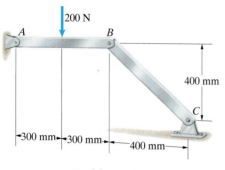

Problem 6.70

6.71 Determine the reactions on member *ABC*.

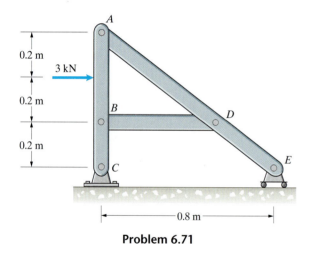

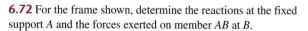

Problem 6.71

6.72 For the frame shown, determine the reactions at the fixed support *A* and the forces exerted on member *AB* at *B*.

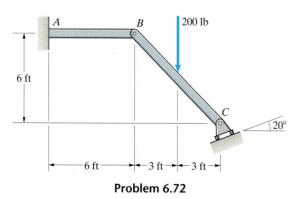

Problem 6.72

6.73 The force $F = 10$ kN. Determine the forces on member *ABC*, presenting your answers as shown in Fig. 6.35.

6.74 The cable *CE* will safely support a tension of 10 kN. Based on this criterion, what is the largest downward force *F* that can be applied to the frame?

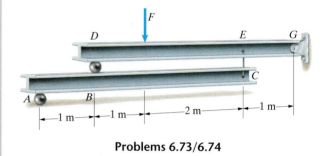

Problems 6.73/6.74

6.75 The tension in cable *BD* is 500 lb. Determine the reactions at *A* for cases (1) and (2).

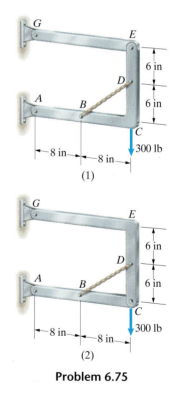

Problem 6.75

6.76 Determine the reactions on member *ABCD* at *A, C,* and *D*.

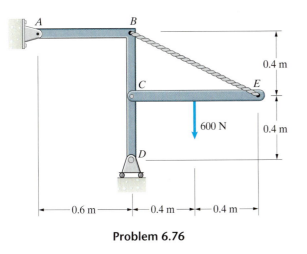

Problem 6.76

6.77 Determine the reactions on member *ABC* at *B* and *C*.

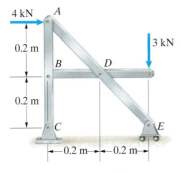

Problem 6.77

6.78 An athlete works out with a squat thrust machine. To rotate the bar *ABD*, she must exert a vertical force at *A* that causes the magnitude of the axial force in the two-force member *BC* to be 1800 N. When the bar *ABD* is on the verge of rotating, what are the reactions on the vertical bar *CDE* at *D* and *E*?

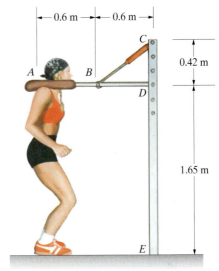

Problem 6.78

6.79 The frame supports a 6-kN vertical load at *C*. The bars *ABC* and *DEF* are horizontal. Determine the reactions on the frame at *A* and *D*.

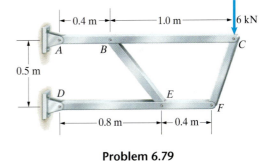

Problem 6.79

6.80 The mass *m* = 120 kg. Determine the forces on member *ABC*, presenting your answers as shown in Fig. 6.35.

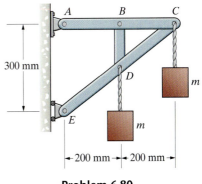

Problem 6.80

6.81 The mass of the suspended weight is 12 kg. Determine the reactions on member *ABC*.

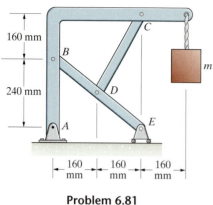

Problem 6.81

6.82 The weight of the suspended object is $W = 50$ lb. Determine the tension in the spring and the reactions at *F*. (The slotted member *DE* is vertical.)

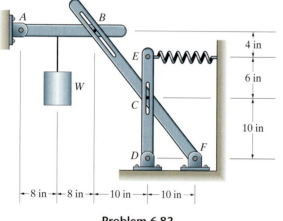

Problem 6.82

6.83 The mass $m = 50$ kg. Bar *DE* is horizontal. Determine the forces on member *ABCD*, presenting your answers as shown in Fig. 6.35.

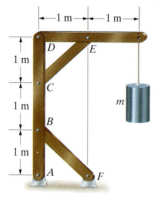

Problem 6.83

6.84 Determine the forces on member *BCD*.

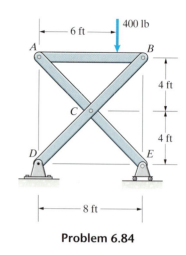

Problem 6.84

6.85 Determine the forces on member *ABC*.

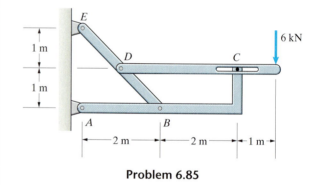

Problem 6.85

6.86 Determine the forces on member *ABD*.

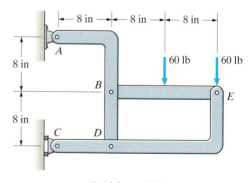

Problem 6.86

6.87 The mass $m = 12$ kg. Determine the forces on member CDE.

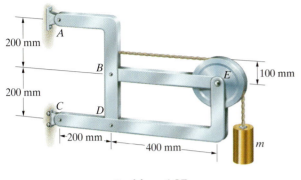

Problem 6.87

6.88 The weight $W = 80$ lb. Determine the forces on member $ABCD$.

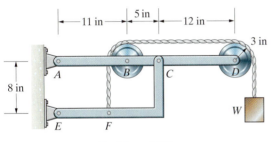

Problem 6.88

6.89 The woman using the exercise machine is holding the 80-lb weight stationary in the position shown. What are the reactions at the fixed support E and the pin support F? (A and C are pinned connections.)

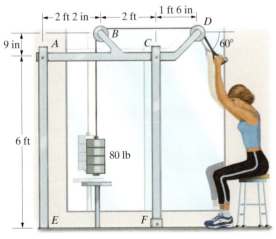

Problem 6.89

6.90 Determine the reactions on member ABC at A and B.

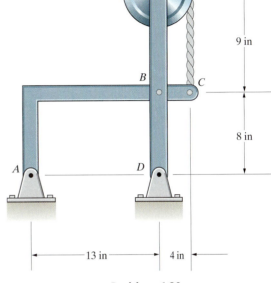

Problem 6.90

6.91 The mass of the suspended object is $m = 50$ kg. Determine the reactions on member ABC.

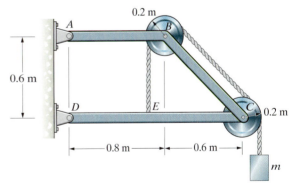

Problem 6.91

6.92 The unstretched length of the spring is L_0. Show that when the system is in equilibrium the angle α satisfies the relation $\sin \alpha = 2(L_0 - 2F/k)/L$.

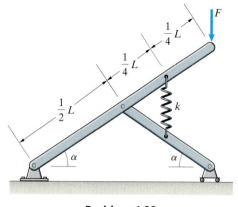

Problem 6.92

6.93 The pin support B will safely support a force of 24-kN magnitude. Based on this criterion, what is the largest mass m that the frame will safely support?

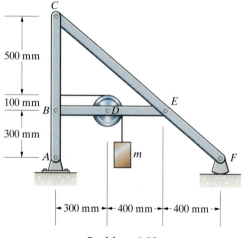

Problem 6.93

6.94 Determine the reactions at A and C.

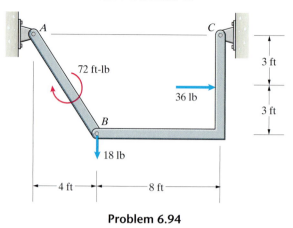

Problem 6.94

6.95 Determine the forces on member AD.

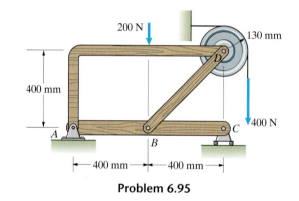

Problem 6.95

6.96 The frame shown is used to support high-tension wires. If $b = 3$ ft, $\alpha = 30°$, and $W = 200$ lb, what is the axial force in member HJ?

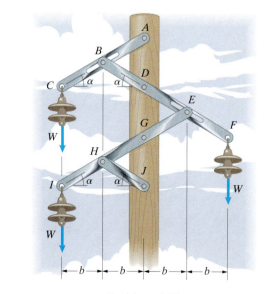

Problem 6.96

6.97 The truck and trailer are parked on a 10° slope. The 14,000-lb weight of the truck and the 8000-lb weight of the trailer act at the points shown. The truck's brakes prevent its rear wheels at *B* from turning. The truck's front wheels at *C* and the trailer's wheels at *A* can turn freely, which means they do not exert friction forces on the road. The trailer hitch at *D* behaves like a pin support. Determine the forces exerted on the truck at *B, C,* and *D*.

Strategy: Draw the individual free-body diagrams of the truck and trailer.

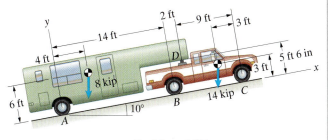

Problem 6.97

6.98 The woman exerts 20-N forces to the pliers as shown.
(a) What is the magnitude of the forces the pliers exert on the bolt at *B*?
(b) Determine the magnitude of the force the members of the pliers exert on each other at the pinned connection *C*.

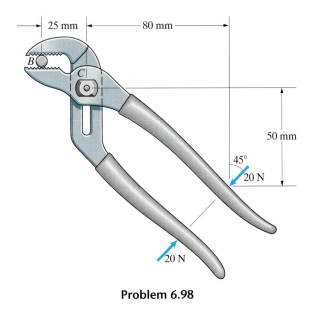

Problem 6.98

6.99 Figure a is a diagram of the bones and biceps muscle of a person's arm supporting a mass. Tension in the biceps muscle holds the forearm in the horizontal position, as illustrated in the simple mechanical model in Fig. b. The weight of the forearm is 9 N, and the mass *m* = 2 kg.
(a) Determine the tension in the biceps muscle *AB*.
(b) Determine the magnitude of the force exerted on the upper arm by the forearm at the elbow joint *C*.

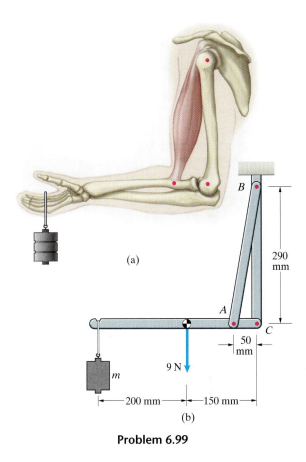

Problem 6.99

6.100 The clamp presses two blocks of wood together. Determine the magnitude of the force the members exert on each other at *C* if the blocks are pressed together with a force of 200 N.

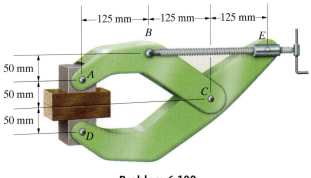

Problem 6.100

6.101 The pressure force exerted on the piston is 2 kN toward the left. Determine the couple M necessary to keep the system in equilibrium.

6.102 In Problem 6.101, determine the forces on member AB at A and B.

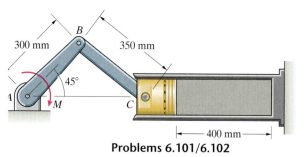

Problems 6.101/6.102

6.103 This mechanism is used to weigh mail. A package placed at A causes the weighted pointer to rotate through an angle α. Neglect the weights of the members except for the counterweight at B, which has a mass of 4 kg. If $\alpha = 20°$, what is the mass of the package at A?

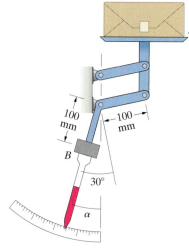

Problem 6.103

6.104 The scoop C of the front-end loader is supported by two identical arms, one on each side of the loader. One of the two arms (ABC) is visible in the figure. It is supported by a pin support at A and the hydraulic actuator BD. The sum of the other loads exerted on the arm, including its own weight, is $F = 1.6$ kN. Determine the axial force in the actuator BD and the magnitude of the reaction at A.

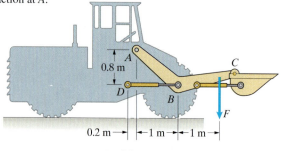

Problem 6.104

6.105 The mass of the scoop is 220 kg, and its weight acts at G. Both the scoop and the hydraulic actuator BC are pinned to the horizontal member at B. The hydraulic actuator can be treated as a two-force member. Determine the forces exerted on the scoop at B and D.

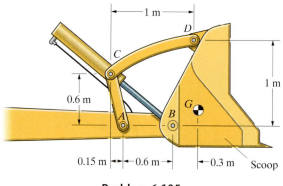

Problem 6.105

6.106 The woman exerts 20-N forces on the handles of the shears. Determine the magnitude of the forces exerted on the branch at A.

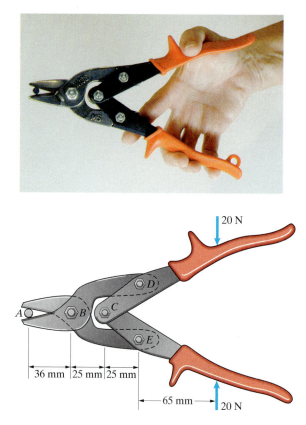

Problem 6.106

6.107 The person exerts 40-N forces on the handles of the locking wrench. Determine the magnitude of the forces the wrench exerts on the bolt at *A*.

6.108 Determine the magnitude of the force the members of the wrench exert on each other at *B* and the axial force in the two-force member *DE*.

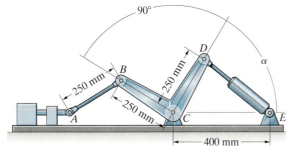

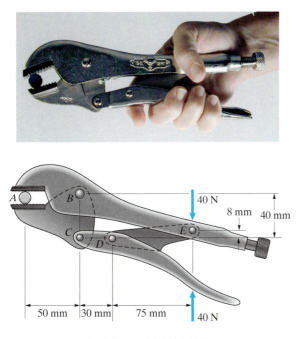

Problems 6.107/6.108

6.109 This device is designed to exert a large force on the horizontal bar at *A* for a stamping operation. If the hydraulic cylinder *DE* exerts an axial force of 800 N and $\alpha = 80°$, what horizontal force is exerted on the horizontal bar at *A*?

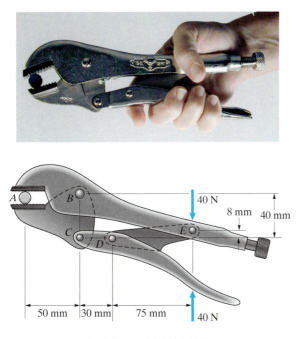

Problem 6.109

6.110 This device raises a load *W* by extending the hydraulic actuator *DE*. The bars *AD* and *BC* are 4 ft long, and the distances $b = 2.5$ ft and $h = 1.5$ ft. If $W = 300$ lb, what force must the actuator exert to hold the load in equilibrium?

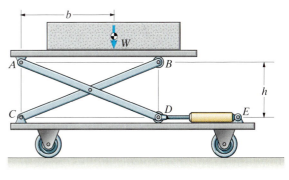

Problem 6.110

6.111 The four-bar linkage operates the forks of a fork lift truck. The force supported by the forks is $W = 8$ kN. Determine the reactions on member *CDE*.

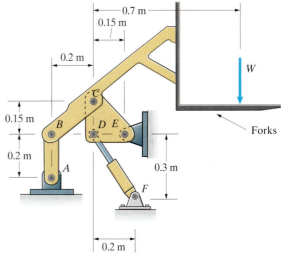

Problem 6.111

6.112 A load $W = 2$ kN is supported by the member ACG and the hydraulic actuator BC. Determine the reactions at A and the compressive axial force in the actuator BC.

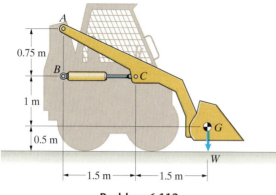

Problem 6.112

6.113 The dimensions are $a = 260$ mm, $b = 300$ mm, $c = 200$ mm, $d = 150$ mm, $e = 300$ mm, and $f = 520$ mm. The ground exerts a vertical force $F = 7000$ N on the shovel. The mass of the shovel is 90 kg and its weight acts at G. The weights of the links AB and AD are negligible. Determine the horizontal force P exerted at A by the hydraulic piston and the reactions on the shovel at C.

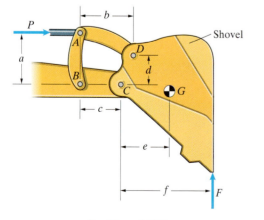

Problem 6.113

6.114 The structure shown in the diagram (one of the two identical structures that support the scoop of the excavator) supports a downward force $F = 1800$ N at G. Members BC and DH can be treated as two-force members. Determine the reactions on member CDK at K.

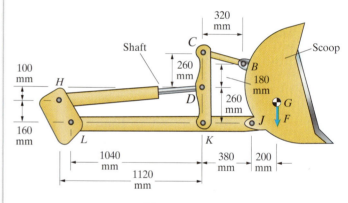

Problem 6.114

COMPUTATIONAL MECHANICS

The following example and problems are designed for the use of a programmable calculator or computer.

Computational Example 6.10

The device in Fig. 6.40 is used to compress air in a cylinder by applying a couple M to the arm AB. The pressure p in the cylinder and the net force F exerted on the piston by pressure are

$$p = p_{\text{atm}}\left(\frac{V_0}{V}\right),$$

$$F = Ap_{\text{atm}}\left(\frac{V_0}{V} - 1\right),$$

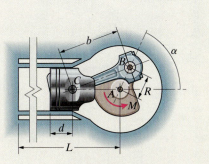

Figure 6.40

where $A = 0.02$ m^2 is the cross-sectional area of the piston, $p_{\text{atm}} = 10^5$ Pa (Pascals, or N/m^2) is atmospheric pressure, V is the volume of air in the cylinder, and V_0 is the value of V when $\alpha = 0$. The dimensions $R = 150$ mm, $b = 350$ mm, $d = 150$ mm, and $L = 1050$ mm. If M and α are initially zero and M is slowly increased until its value is 40 N-m, what are the resulting values of α and p?

Strategy

By expressing the volume of air in the cylinder in terms of α, we will determine the force exerted on the cylinder by pressure in terms of α. From the free-body diagram of the piston we will determine the axial force in the two-force member BC in terms of the pressure force on the cylinder. Then from the free-body diagram of the arm AB we will obtain a relation between M and α.

Solution

From the geometry of the arms AB and BC (Fig. a), the volume of air in the cylinder is

$$V = A(L - d - \sqrt{b^2 - R^2 \sin^2 \alpha} + R \cos \alpha).$$

When $\alpha = 0$, the volume is

$$V_0 = A(L - d - b + R).$$

Therefore, the force exerted on the piston by pressure is

$$F = Ap_{\text{atm}}\left(\frac{V_0}{V} - 1\right)$$

$$= Ap_{\text{atm}}\left(\frac{L - d - b + R}{L - d - \sqrt{b^2 - R^2 \sin^2 \alpha} + R \cos \alpha} - 1\right).$$

We draw the free-body diagrams of the piston and the arm AB in Figs. b and c, where N is the force exerted on the piston by the cylinder (friction is neglected), Q is the axial force in the two-force member BC, and A_x and A_y are

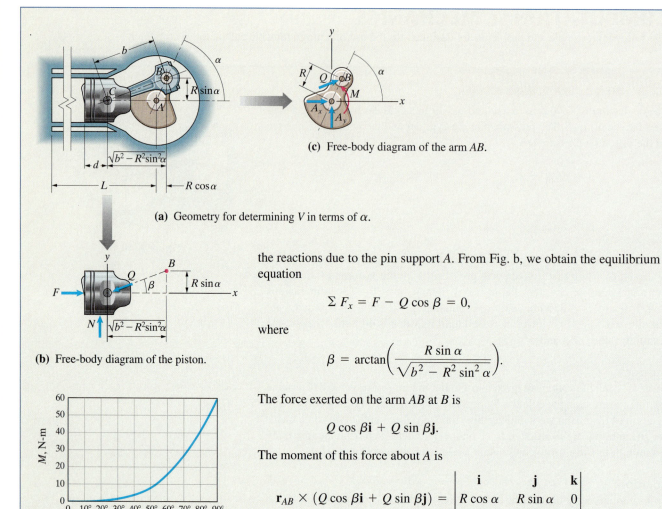

(c) Free-body diagram of the arm AB.

(a) Geometry for determining V in terms of α.

(b) Free-body diagram of the piston.

Figure 6.41
The moment M as a function of α.

α	M (N-m)
79.59°	39.9601
79.60°	39.9769
79.61°	39.9937
79.62°	40.0105
79.63°	40.0272

the reactions due to the pin support A. From Fig. b, we obtain the equilibrium equation

$$\Sigma F_x = F - Q \cos \beta = 0,$$

where

$$\beta = \arctan\left(\frac{R \sin \alpha}{\sqrt{b^2 - R^2 \sin^2 \alpha}}\right).$$

The force exerted on the arm AB at B is

$$Q \cos \beta \mathbf{i} + Q \sin \beta \mathbf{j}.$$

The moment of this force about A is

$$\mathbf{r}_{AB} \times (Q \cos \beta \mathbf{i} + Q \sin \beta \mathbf{j}) = \begin{vmatrix} \mathbf{i} & \mathbf{j} & \mathbf{k} \\ R \cos \alpha & R \sin \alpha & 0 \\ Q \cos \beta & Q \sin \beta & 0 \end{vmatrix}$$

$$= QR(\cos \alpha \sin \beta - \sin \alpha \cos \beta)\mathbf{k}.$$

Using this result, we find that the sum of the moments about A is

$$\Sigma M_{\text{point } A} = M + QR(\cos \alpha \sin \beta - \sin \alpha \cos \beta) = 0.$$

If we choose a value of α, we can sequentially calculate V, F, β, Q, and M. Computing M as a function of α, we obtain the graph shown in Fig. 6.41. The moment $M = 40$ N-m at approximately $\alpha = 80°$. By examining computed results near 80° (see table), we estimate that $\alpha = 79.61°$ when $M = 40$ N-m. Once we know α, we can calculate V and then p, obtaining

$$p = 1.148 p_{\text{atm}} = 1.148 \times 10^5 \text{ Pa}.$$

Critical Thinking

The nonlinearity in this problem, which required us to use a numerical solution, arises from two sources. One is the geometry of the crank AB and the connecting rod BC. The other is the nonlinear dependence of the pressure of the air on the displacement of the piston.

Computational Problems

6.115 (a) For each member of the truss, obtain a graph of (axial force)/F as a function of x for $0 \leq x \leq 2$ m.
(b) If you were designing this truss, what value of x would you choose based on your results in (a)?

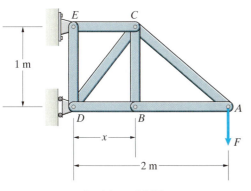

Problem 6.115

6.116 The mechanism shown is used to weigh mail. A package placed at A causes the weighted pointer to rotate through an angle α. Neglect the weights of the members except for the counterweight at B, which has a mass of 4 kg.
(a) Obtain a graph of the angle α as a function of the mass of the mail for values of the mass from 0 to 2 kg.
(b) Use the results of (a) to estimate the value of α when the mass is 1 kg.

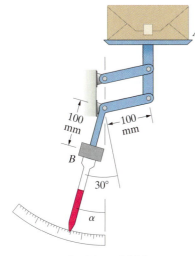

Problem 6.116

6.117 A preliminary design for a bridge structure is shown. The forces F are the loads the structure must support at $G, H, I, J,$ and K. Plot the axial forces in members AB and BC as a function of the angle β. Use your graphs to estimate the value of β for which the maximum compressive load in any member of the bridge does not exceed $2F$. Draw a sketch of the resulting design.

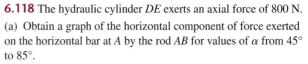
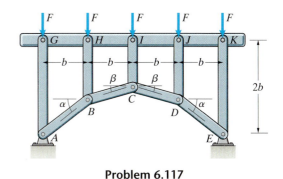

Problem 6.117

6.118 The hydraulic cylinder DE exerts an axial force of 800 N.
(a) Obtain a graph of the horizontal component of force exerted on the horizontal bar at A by the rod AB for values of α from 45° to 85°.
(b) Use the results of (a) to estimate the value of α for which the horizontal force is 2 kN.

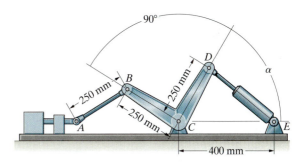

Problem 6.118

6.119 The weight of the suspended object is 10 kN. The two members have equal cross-sectional areas A, and each will safely support an axial force of $40A$ MN, where A is in square meters. Determine the value of h that minimizes the total volume of material in the two members.

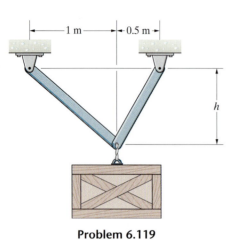

Problem 6.119

6.120 The bars AD and BC are 4 ft long, the distance $b = 2.5$ ft, and $W = 300$ lb. If the largest force the hydraulic actuator DE can exert is 1000 lb, what is the smallest height h at which the load can be supported?

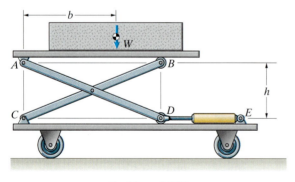

Problem 6.120

6.121 The linkage is in equilibrium under the action of the couples M_A and M_B. When $\alpha_A = 60°$, $\alpha_B = 70°$. For the range $0 \leq \alpha_A \leq 180°$, estimate the maximum positive and negative values of M_A/M_B and the values of α_A at which they occur.

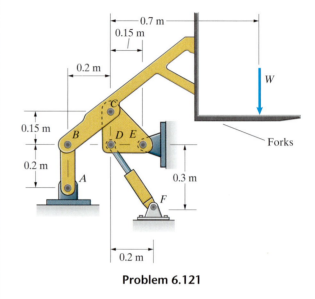

Problem 6.121

6.122 A load $W = 2$ kN is supported by the member ACG and the hydraulic actuator BC. If the actuator BC can exert a maximum axial force of 12 kN, what is the largest height above the ground at which the center of mass G can be supported?

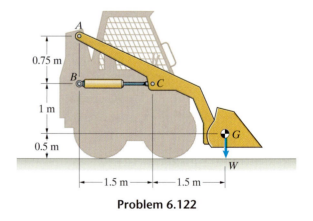

Problem 6.122

6.123 The crane exerts vertical 75-kN forces on the truss at B, C, and D. You can model the support at A as a pin support and model the support at E as a roller support that can exert a force normal to the dashed line but cannot exert a force parallel to it. Determine the value of the angle α for which the largest compressive force in any of the members is as small as possible. What are the resulting axial forces in the members?

6.124 Draw graphs of the magnitudes of the axial forces in the members BC and BD as functions of the dimension h for $0.5 \leq h \leq 1.5$ m.

6.125 Determine the value of the dimension h in the range $0.5 \leq h \leq 1.5$ m so that the magnitude of the largest axial force in any of the members, tensile or compressive, is a minimum. What are the resulting axial forces in the members?

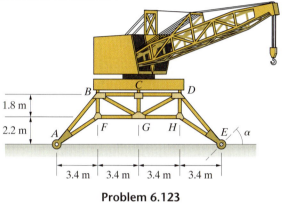

Problem 6.123

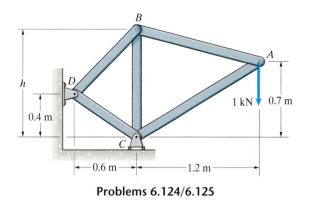

Problems 6.124/6.125

CHAPTER SUMMARY

A structure of *members* interconnected at *joints* is a *truss* if it is composed entirely of two-force members. Otherwise, it is a *frame* if it is designed to remain stationary and support loads and a *machine* if it is designed to move and exert loads.

Trusses

A member of a truss is in *tension* if the *axial forces* at the ends are directed away from each other and is in *compression* if the axial forces are directed toward each other. Before beginning to determine the axial forces in the members of a truss, it is usually necessary to draw a free-body diagram of the entire truss and determine the reactions at its supports. The axial forces in the members can be determined by two methods. The *method of joints* involves drawing free-body diagrams of the joints of a truss one by one and using the equilibrium equations to determine the axial forces in the members. In two dimensions, choose joints to analyze that are subjected to known forces and no more than two unknown forces. The *method of sections* involves drawing free-body diagrams of parts, or *sections*, of a truss and using the equilibrium equations to determine the axial forces in selected members.

A *space truss* is a three-dimensional truss. Space trusses are analyzed by the same methods used for two-dimensional trusses. Choose joints to analyze that are subjected to known forces and no more than three unknown forces.

Frames and Machines

Begin analyzing a frame or machine by drawing a free-body diagram of the entire structure and determining the reactions at its supports. If the entire structure is statically indeterminate, determine as many reactions as possible. Then draw free-body diagrams of individual members, or selected combinations of members, and apply the equilibrium equations to determine the forces and couples acting on them. Recognizing two-force members will reduce the number of unknown forces that must be determined. If a load is applied at a joint, it can be placed on the free-body diagram of *any one* of the members attached at the joint.

Review Problems

6.126 The loads $F_1 = 60$ N and $F_2 = 40$ N. (a) Draw the free-body diagram of the entire truss, and determine the reactions at its supports. (b) Determine the axial forces in the members. Indicate whether they are in tension (T) or compression (C).

6.127 The loads $F_1 = 440$ N and $F_2 = 160$ N. Determine the axial forces in the members. Indicate whether they are in tension (T) or compression (C).

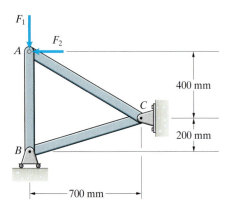

Problems 6.126/6.127

6.128 The truss supports a load $F = 10$ kN. Determine the axial forces in members AB, AC, and BC.

6.129 Each member of the truss will safely support a tensile force of 40 kN and a compressive force of 32 kN. Based on this criterion, what is the largest downward load F that can safely be applied at C?

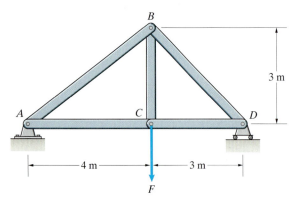

Problems 6.128/6.129

6.130 The Pratt bridge truss supports loads at F, G, and H. Determine the axial forces in members BC, BG, and FG.

6.131 Determine the axial forces in members CD, GD, and GH.

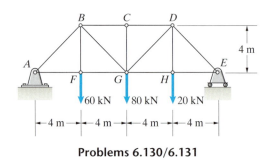

Problems 6.130/6.131

6.132 The truss supports loads at F and H. Determine the axial forces in members AB, AC, BC, BD, CD, and CE.

6.133 Determine the axial forces in members EH and FH.

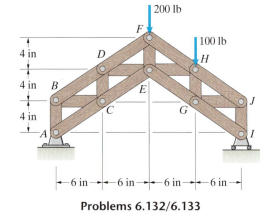

Problems 6.132/6.133

6.134 Determine the axial forces in members BD, CD, and CE.

6.135 Determine the axial forces in members DF, EF, and EG.

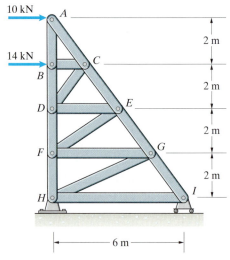

Problems 6.134/6.135

6.136 The truss supports a 400-N load at *G*. Determine the axial forces in members *AC, CD*, and *CF*.

6.137 Determine the axial forces in members *CE, EF*, and *EH*.

6.138 Which members have the largest tensile and compressive forces, and what are their values?

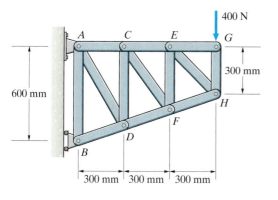

Problems 6.136–6.138

6.139 The Howe truss helps support a roof. Model the supports at *A* and *G* as roller supports. Use the method of joints to determine the axial forces in members *BC, CD, CI*, and *CJ*.

6.140 For the roof truss in Problem 6.139, use the method of sections to determine the axial forces in members *CD, CJ*, and *IJ*.

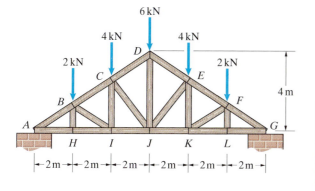

Problems 6.139/6.140

6.141 A speaker system is suspended from the truss by cables attached at *D* and *E*. The mass of the speaker system is 130 kg, and its weight acts at *G*. Determine the axial forces in members *BC* and *CD*.

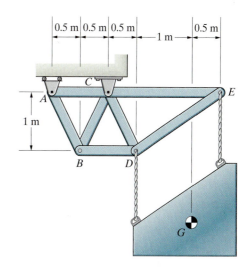

Problem 6.141

6.142 The mass of the suspended object is 900 kg. Determine the axial forces in the bars *AB* and *AC*.

Strategy: Draw the free-body diagram of joint *A*.

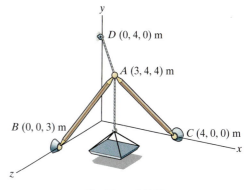

Problem 6.142

6.143 Determine the forces on member *ABC*, presenting your answers as shown in Fig. 6.35. Obtain the answers in two ways:
(a) When you draw the free-body diagrams of the individual members, place the 400-lb load on the free-body diagram of member *ABC*.
(b) When you draw the free-body diagrams of the individual members, place the 400-lb load on the free-body diagram of member *CD*.

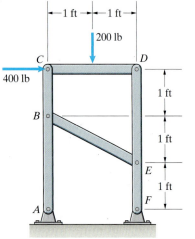

Problem 6.143

6.144 The mass *m* = 120 kg. Determine the forces on member *ABC*.

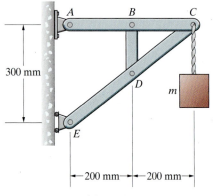
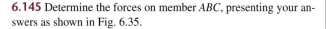

Problem 6.144

6.145 Determine the forces on member *ABC*, presenting your answers as shown in Fig. 6.35.

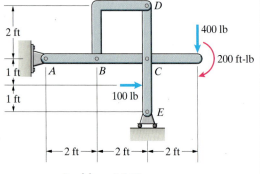

Problem 6.145

6.146 Determine the force exerted on the bolt by the bolt cutters and the magnitude of the force the members exert on each other at the pin connection *A*.

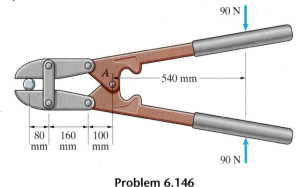

Problem 6.146

6.147 The 600-lb weight of the scoop acts at a point 1 ft 6 in to the right of the vertical line *CE*. The line *ADE* is horizontal. The hydraulic actuator *AB* can be treated as a two-force member. Determine the axial force in the hydraulic actuator *AB* and the forces exerted on the scoop at *C* and *E*.

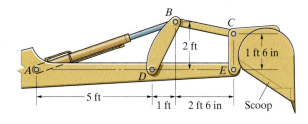

Problem 6.147

6.148 Determine the force exerted on the bolt by the bolt cutters.

6.149 Determine the magnitude of the force the members of the bolt cutters exert on each other at the pin connection *B* and the axial force in the two-force member *CD*.

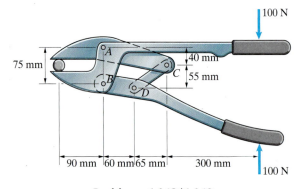

Problems 6.148/6.149

6.150 The weights $W_1 = 4$ kN and $W_2 = 10$ kN. Determine the forces on member *ACDE* at points *A* and *E*.

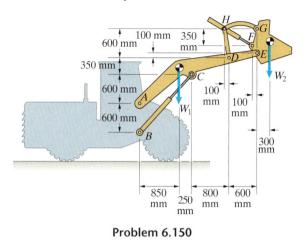

Problem 6.150

Design Project 1 Design a truss structure to support a foot bridge with an unsupported span (width) of 8 m. Make conservative estimates of the loads the structure will need to support if the pathway supported by the truss is made of wood. Consider two options: (1) Your client wants the bridge to be supported by a truss below the bridge so that the upper surface will be unencumbered by structure. (2) The client wants the truss to be above the bridge and designed so that it can serve as handrails. For each option, use statics to estimate the maximum axial forces to which the members of the structure will be subjected. Investigate alternative designs and compare the resulting axial loads.

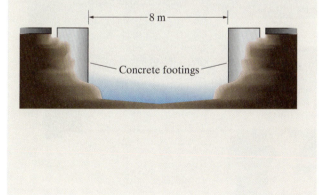

Design Project 2 The truss shown connects one end of a stretcher to a rescue helicopter. Consider alternative truss designs that support the stretcher at *A* and *B* and are supported at *E* and *G*. Compare the maximum tensile and compressive loads in the members of your designs to those in the truss shown. Assuming that the cost of a truss is proportional to the sum of the lengths of its members, compare the costs of your designs to that of the truss shown. Write a brief report describing your analysis and recommending the design you would choose.

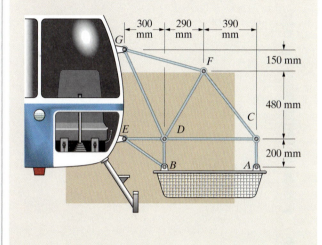

Design Project 3 Go to a fitness center and choose an exercise device that seems mechanically interesting. (For example, it may employ weights, pulleys, and levers.) By measuring dimensions (while the device is not in use), drawing sketches, and perhaps taking photographs, gather the information necessary to analyze the device. Use statics to determine the range of forces a person must exert in using the device.

Suggest changes to the design of the device (other than simply increasing weights) that will increase the maximum force the user must exert.

Prepare a brief report that (1) describes the original device; (2) presents your model and analysis of the device; (3) describes your proposed changes and any analyses supporting them; and (4) recommends the design change you would choose to increase the maximum force the user must employ.

Centroids and Centers of Mass

An object's weight does not act at a single point—it is distributed over the entire volume of the object. But the weight can be represented by a single equivalent force acting at a point called the center of mass. When the equilibrium equations are used to determine the reactions exerted on an object by its supports, the location of the center of mass must be known if the weight of the object is to be included in the analysis. The dynamic behaviors of objects also depend on the location of their centers of mass. In this chapter we define the center of mass and show how it is determined for various kinds of objects. We also introduce definitions that can be interpreted as the average positions of areas, volumes, and lines. These average positions are called centroids. Centroids coincide with the centers of mass of particular classes of objects, and they also arise in many other engineering applications.

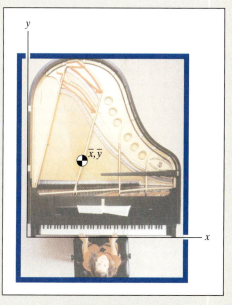

◄ The loads that the legs of the piano must be designed to support depend not only on the piano's weight but also on the position of its center of mass—the point at which the weight effectively acts.

CENTROIDS

Suppose that we want to determine the average position of a group of students sitting in a room. First, we introduce a coordinate system so that we can specify the position of each student. For example, we can align the axes with the walls of the room (Fig. 7.1a). We number the students from 1 to N and denote the position of student 1 by (x_1, y_1), the position of student 2 by (x_2, y_2), and so on. The average x coordinate, which we denote by $\bar{x}$, is the sum of their x coordinates divided by N; that is,

$$\bar{x} = \frac{x_1 + x_2 + \cdots + x_N}{N} = \frac{\sum\limits_i x_i}{N}, \tag{7.1}$$

where the symbol $\sum\limits_i$ means "sum over the range of i." The average y coordinate is

$$\bar{y} = \frac{\sum\limits_i y_i}{N}. \tag{7.2}$$

We indicate the average position by the symbol shown in Fig. 7.1b.

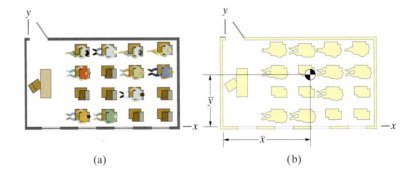

Figure 7.1
(a) A group of students in a classroom.
(b) Their average position.

(a)   (b)

Now suppose that we pass out some pennies to the students. Let the number of coins given to student 1 be c_1, the number given to student 2 be c_2, and so on. What is the average position of the coins in the room? Clearly, the average position of the coins may not be the same as the average position of the students. For example, if the students in the front of the room have more coins, the average position of the coins will be closer to the front of the room than the average position of the students.

To determine the x coordinate of the average position of the coins, we need to sum the x coordinates of the coins and divide by the number of coins. We can obtain the sum of the x coordinates of the coins by multiplying the number of coins each student has by his or her x coordinate and summing.

We can obtain the number of coins by summing the numbers $c_1, c_2, \ldots$. Thus, the average x coordinate of the coins is

$$\bar{x} = \frac{\sum_i x_i c_i}{\sum_i c_i}. \tag{7.3}$$

We can determine the average y coordinate of the coins in the same way:

$$\bar{y} = \frac{\sum_i y_i c_i}{\sum_i c_i}. \tag{7.4}$$

By assigning other meanings to $c_1, c_2, \ldots$, we can determine the average positions of other measures associated with the students. For example, we could determine the average position of their age or the average position of their height.

More generally, we can use Eqs. (7.3) and (7.4) to determine the average position of any set of quantities with which we can associate positions. An average position obtained from these equations is called a *weighted average position*, or *centroid*. The "weight" associated with position (x_1, y_1) is c_1, the weight associated with position (x_2, y_2) is c_2, and so on. In Eqs. (7.1) and (7.2), the weight associated with the position of each student is 1. When the census is taken, the centroid of the population of the United States—the average position of the population—is determined in this way. In the next section we use Eqs. (7.3) and (7.4) to determine centroids of areas.

7.1 Centroids of Areas

Consider an arbitrary area A in the x–y plane (Fig. 7.2a). Let us divide the area into parts $A_1, A_2, \ldots, A_N$ (Fig. 7.2b) and denote the positions of the parts by $(x_1, y_1), (x_2, y_2), \ldots, (x_N, y_N)$. We can obtain the centroid, or average position of the area, by using Eqs. (7.3) and (7.4) with the areas of the parts as the weights:

$$\bar{x} = \frac{\sum_i x_i A_i}{\sum_i A_i}, \qquad \bar{y} = \frac{\sum_i y_i A_i}{\sum_i A_i}. \tag{7.5}$$

A question arises if we try to carry out this procedure: What are the exact positions of the areas $A_1, A_2, \ldots, A_N$? We could reduce the uncertainty in their positions by dividing A into smaller parts, but we would still obtain only approximate values for $\bar{x}$ and $\bar{y}$. To determine the exact location of the centroid, we must take the limit as the sizes of the parts approach zero. We obtain this limit

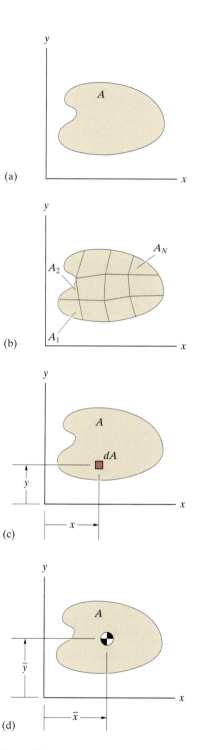

Figure 7.2
(a) The area A.
(b) Dividing A into N parts.
(c) A differential element of area dA with coordinates (x, y).
(d) The centroid of the area.

by replacing Eqs. (7.5) by the integrals

$$\bar{x} = \frac{\int_A x \, dA}{\int_A dA},$$

(7.6)

$$\bar{y} = \frac{\int_A y \, dA}{\int_A dA},$$

(7.7)

where x and y are the coordinates of the differential element of area dA (Fig. 7.2c). The subscript A on the integral signs means the integration is carried out over the entire area. The centroid of the area is shown in Fig. 7.2d.

Keeping in mind that the centroid of an area is its average position will often help you locate it. For example, the centroid of a circular area or a rectangular area obviously lies at the center of the area. If an area has "mirror image" symmetry about an axis, the centroid lies on the axis (Fig. 7.3a), and if an area is symmetric about two axes, the centroid lies at their intersection (Fig. 7.3b).

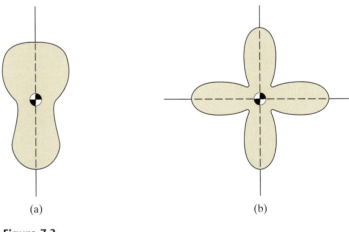

(a)   (b)

Figure 7.3
(a) An area that is symmetric about an axis.
(b) An area with two axes of symmetry.

Study Questions

1. How is a weighted average position defined?
2. How is the concept of a weighted average used to define the centroid of a plane area?
3. Why is integration generally needed to determine the exact position of the centroid of an area?

Example 7.1   Centroid of an Area by Integration

Determine the centroid of the triangular area in Fig. 7.4.

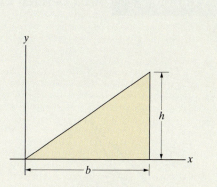

Figure 7.4

Strategy
We will determine the coordinates of the centroid by using an element of area dA in the form of a "strip" of width dx.

Solution
Let dA be the vertical strip in Fig. a. The height of the strip is $(h/b)x$, so $dA = (h/b)x\,dx$. To integrate over the entire area, we must integrate with respect to x from $x = 0$ to $x = b$. The x coordinate of the centroid is

$$\bar{x} = \frac{\int_A x\,dA}{\int_A dA} = \frac{\int_0^b x\left(\frac{h}{b}x\,dx\right)}{\int_0^b \frac{h}{b}x\,dx} = \frac{\frac{h}{b}\left[\frac{x^3}{3}\right]_0^b}{\frac{h}{b}\left[\frac{x^2}{2}\right]_0^b} = \frac{2}{3}b.$$

To determine $\bar{y}$, we let y in Eq. (7.7) be the y coordinate of the midpoint of the strip (Fig. b):

$$\bar{y} = \frac{\int_A y\,dA}{\int_A dA} = \frac{\int_0^b \frac{1}{2}\left(\frac{h}{b}x\right)\left(\frac{h}{b}x\,dx\right)}{\int_0^b \frac{h}{b}x\,dx} = \frac{\frac{1}{2}\left(\frac{h}{b}\right)^2\left[\frac{x^3}{3}\right]_0^b}{\frac{h}{b}\left[\frac{x^2}{2}\right]_0^b} = \frac{1}{3}h.$$

The centroid is shown in Fig. c.

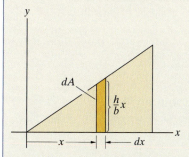

(a) An element dA in the form of a strip.

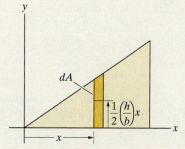

(b) The y coordinate of the midpoint of the strip is $\frac{1}{2}(h/b)x$.

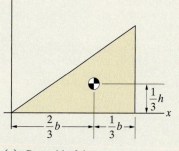

(c) Centroid of the area.

Critical Thinking
Always be alert for opportunities to check your results. In this example, we should make sure that our integration procedure gives the correct result for the area of the triangle:

$$\int_A dA = \int_0^b \frac{h}{b}x\,dx = \frac{h}{b}\left[\frac{x^2}{2}\right]_0^b = \frac{1}{2}bh.$$

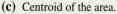

Example 7.2 | **Area Defined by Two Equations**

Determine the centroid of the area in Fig. 7.5.

Strategy

We can determine the coordinates of the centroid using an element of area in the form of a vertical strip, just as we did in Example 7.1. In this case the strip must be defined so that it extends from the lower curve ($y = x^2$) to the upper curve ($y = x$).

Solution

Let dA be the vertical strip in Fig. a. The height of the strip is $x - x^2$, so $dA = (x - x^2)\, dx$. The x coordinate of the centroid is

$$\bar{x} = \frac{\displaystyle\int_A x\, dA}{\displaystyle\int_A dA} = \frac{\displaystyle\int_0^1 x(x - x^2)\, dx}{\displaystyle\int_0^1 (x - x^2)\, dx} = \frac{\left[\dfrac{x^3}{3} - \dfrac{x^4}{4}\right]_0^1}{\left[\dfrac{x^2}{2} - \dfrac{x^3}{3}\right]_0^1} = \frac{1}{2}.$$

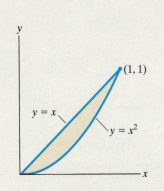

Figure 7.5

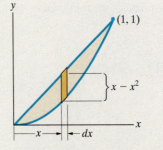

(a) A vertical strip of width dx. The height of the strip is equal to the difference in the two functions.

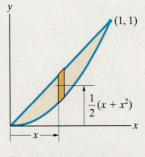

(b) The y coordinate of the midpoint of the strip.

The y coordinate of the midpoint of the strip is $x^2 + \frac{1}{2}(x - x^2) = \frac{1}{2}(x + x^2)$ (Fig. b). Substituting this expression for y in Eq. (7.7), we obtain the y coordinate of the centroid:

$$\bar{y} = \frac{\displaystyle\int_A y\, dA}{\displaystyle\int_A dA} = \frac{\displaystyle\int_0^1 \left[\frac{1}{2}(x + x^2)\right](x - x^2)\, dx}{\displaystyle\int_0^1 (x - x^2)\, dx} = \frac{\dfrac{1}{2}\left[\dfrac{x^3}{3} - \dfrac{x^5}{5}\right]_0^1}{\left[\dfrac{x^2}{2} - \dfrac{x^3}{3}\right]_0^1} = \frac{2}{5}.$$

Critical Thinking

Notice the generality of the approach we use in this example. It can be used to determine the x and y coordinates of the centroid of any area whose upper and lower boundaries are defined by two functions.

Problems

7.1 If $a = 2$, what is the x coordinate of the centroid of the area?
   Strategy: The x coordinate of the centroid is given by Eq. (7.6). For the element of area dA, use a vertical strip of width dx. (See Example 7.1.)

7.2 Determine the y coordinate of the centroid of the area if $a = 3$.

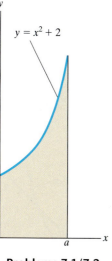

Problems 7.1/7.2

7.3 What is the x coordinate of the centroid of the area?

7.4 What is the y coordinate of the centroid of the area?

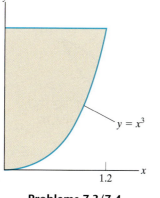

Problems 7.3/7.4

7.5 Determine the coordinates of the centroid of the area.

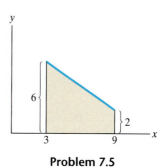

Problem 7.5

7.6 Determine x coordinate of the centroid of the area and compare your answer to the value given in Appendix B.

7.7 Determine the y coordinate of the centroid of the area and compare your answer to the value given in Appendix B.

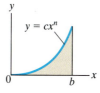

Problems 7.6/7.7

7.8 Suppose that an art student wants to paint a panel of wood as shown, with the horizontal and vertical lines passing through the centroid of the painted area, and asks you to determine the coordinates of the centroid. What are they?

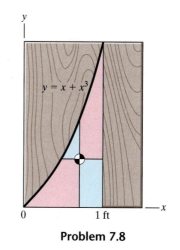

Problem 7.8

7.9 The y coordinate of the centroid of the area is $\bar{y} = 1.063$. Determine the value of the constant c and the x coordinate of the centroid.

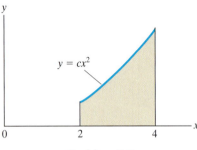

Problem 7.9

7.10 Determine the coordinates of the centroid of the metal plate's cross-sectional area.

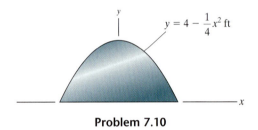

$y = 4 - \frac{1}{4}x^2$ ft

Problem 7.10

7.11 An architect wants to build a wall with the profile shown. To estimate the effects of wind loads, he must determine the wall's area and the coordinates of its centroid. What are they?

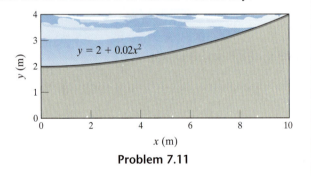

$y = 2 + 0.02x^2$

x (m)

Problem 7.11

7.12 Determine the coordinates of the centroid of the area.

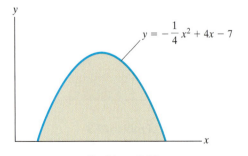

$y = -\frac{1}{4}x^2 + 4x - 7$

Problem 7.12

7.13 Determine the coordinates of the centroid of the area.

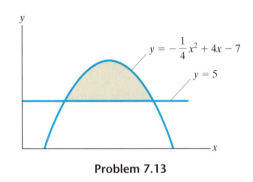

$y = -\frac{1}{4}x^2 + 4x - 7$

$y = 5$

Problem 7.13

7.14 Determine the x coordinate of the centroid of the area.

7.15 Determine the y coordinate of the centroid of the area.

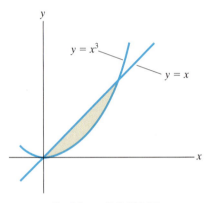

$y = x^3$

$y = x$

Problems 7.14/7.15

7.16 Determine the coordinates of the centroid of the area.

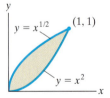

$y = x^{1/2}$ $(1, 1)$

$y = x^2$

Problem 7.16

7.17 Determine the x coordinate of the centroid of the area.

7.18 Determine the y coordinate of the centroid of the area.

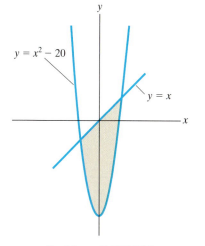

$y = x^2 - 20$

$y = x$

Problems 7.17/7.18

7.19 What is the x coordinate of the centroid of the area?

7.20 What is the y coordinate of the centroid of the area?

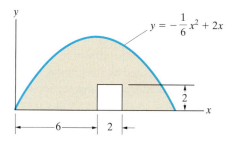

$y = -\dfrac{1}{6}x^2 + 2x$

6   2

2

Problems 7.19/7.20

7.21 An agronomist wants to measure the rainfall at the centroid of a plowed field between two roads. What are the coordinates of the point where the rain gauge should be placed?

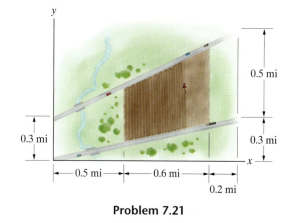

0.5 mi

0.3 mi

0.3 mi

0.5 mi   0.6 mi

0.2 mi

Problem 7.21

7.22 The cross section of an earth-fill dam is shown. Determine the coefficients a and b so that the y coordinate of the centroid of the cross section is 10 m.

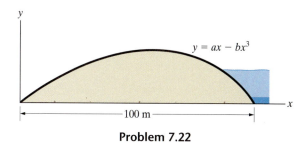

$y = ax - bx^3$

100 m

Problem 7.22

7.23 The Supermarine Spitfire used by Great Britain in World War II had a wing with an elliptical profile. Determine the coordinates of its centroid.

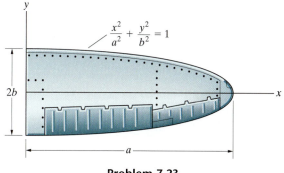

$\dfrac{x^2}{a^2} + \dfrac{y^2}{b^2} = 1$

2b

a

Problem 7.23

7.24 Determine the coordinates of the centroid of the area.

    Strategy: Write the equation for the circular boundary in the form $y = (R^2 - x^2)^{1/2}$ and use a vertical "strip" of width dx as the element of area dA.

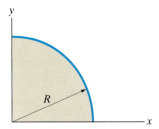

Problem 7.24

7.25 *If $R = 6$ and $b = 3$, what is the y coordinate of the centroid of the area?

7.26 *What is the x coordinate of the centroid of the area in Problem 7.25?

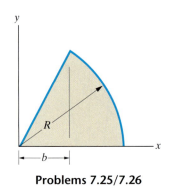

Problems 7.25/7.26

7.2 Centroids of Composite Areas

Although centroids of areas can be determined by integration, the process becomes difficult and tedious for complicated areas. In this section we describe a much easier approach that can be used if an area consists of a combination of simple areas, which we call a *composite area*. We can determine the centroid of a composite area without integration if the centroids of its parts are known.

    The area in Fig.7.6a consists of a triangle, a rectangle, and a semicircle, which we call parts 1, 2, and 3. The x-coordinate of the centroid of the composite area is

$$\bar{x} = \frac{\displaystyle\int_A x \, dA}{\displaystyle\int_A dA} = \frac{\displaystyle\int_{A_1} x \, dA + \int_{A_2} x \, dA + \int_{A_3} x \, dA}{\displaystyle\int_{A_1} dA + \int_{A_2} dA + \int_{A_3} dA}. \tag{7.8}$$

The x coordinates of the centroids of the parts are shown in Fig. 7.6b. From the equation for the x coordinate of the centroid of part 1,

$$\bar{x}_1 = \frac{\displaystyle\int_{A_1} x \, dA}{\displaystyle\int_{A_1} dA},$$

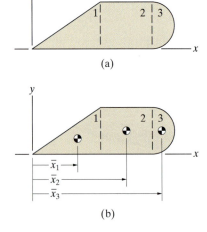

(a)

(b)

Figure 7.6
(a) A composite area composed of three simple areas.
(b) The centroids of the parts.

we obtain

$$\int_{A_1} x \, dA = \bar{x}_1 A_1.$$

Using this equation and equivalent equations for parts 2 and 3, we can write Eq. (7.8) as

$$\bar{x} = \frac{\bar{x}_1 A_1 + \bar{x}_2 A_2 + \bar{x}_3 A_3}{A_1 + A_2 + A_3}.$$

We have obtained an equation for the x coordinate of the composite area in terms of those of its parts. The coordinates of the centroid of a composite area with an arbitrary number of parts are

$$\bar{x} = \frac{\sum_i \bar{x}_i A_i}{\sum_i A_i}, \qquad \bar{y} = \frac{\sum_i \bar{y}_i A_i}{\sum_i A_i}. \qquad (7.9)$$

When we can divide an area into parts whose centroids are known, we can use these expressions to determine its centroid. The centroids of some simple areas are tabulated in Appendix B.

   We began our discussion of the centroid of an area by dividing an area into finite parts and writing equations for its weighted average position. The results, Eqs. (7.5), are approximate because of the uncertainty in the positions of the parts of the area. The exact Eqs. (7.9) are identical except that the positions of the parts are their centroids.

   The area in Fig. 7.7a consists of a triangular area with a circular hole, or cutout. Designating the triangular area (without the cutout) as part 1 of the composite area (Fig. 7.7b) and the area of the cutout as part 2 (Fig. 7.7c), we obtain the x coordinate of the centroid of the composite area:

$$\bar{x} = \frac{\displaystyle\int_{A_1} x \, dA - \int_{A_2} x \, dA}{\displaystyle\int_{A_1} dA - \int_{A_2} dA} = \frac{\bar{x}_1 A_1 - \bar{x}_2 A_2}{A_1 - A_2}.$$

This equation is identical in form to the first of Eqs. (7.9) except that the terms corresponding to the cutout are negative. As this example demonstrates, we can use Eqs. (7.9) to determine the centroids of composite areas containing cutouts by treating the cutouts as negative areas.

   We see that determining the centroid of a composite area requires three steps:

1. Choose the parts—Try to divide the composite area into parts whose centroids you know or can easily determine.

2. Determine the values for the parts—Determine the centroid and the area of each part. Watch for instances of symmetry that can simplify your task.

3. Calculate the centroid—Use Eqs. (7.9) to determine the centroid of the composite area.

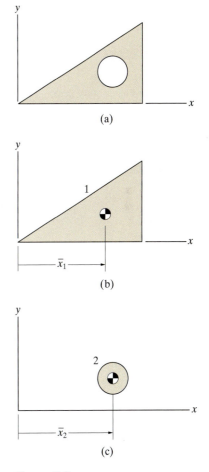

Figure 7.7
(a) An area with a cutout.
(b) The triangular area.
(c) The area of the cutout.

Example 7.3   Centroid of a Composite Area

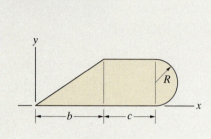

Figure 7.8

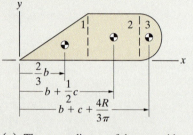

(a) The x coordinates of the centroids of the parts.

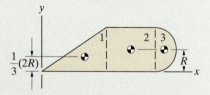

(b) The y coordinates of the centroids of the parts.

Determine the centroid of the area in Fig. 7.8.

Strategy

We must divide the area into parts (the parts are obvious in this example), determine the centroids of the parts, and apply Eqs. (7.9).

Solution

Choose the Parts   We can divide the area into a triangle, a rectangle, and a semicircle, which we call parts 1, 2, and 3, respectively.

Determine the Values for the Parts   The x coordinates of the centroids of the parts are shown in Fig. a. The x coordinates, the areas of the parts, and their products are summarized in Table 7.1.

TABLE 7.1   Information for determining the x coordinate of the centroid

	$\bar{x}_i$	A_i	$\bar{x}_i A_i$
Part 1 (triangle)	$\frac{2}{3}b$	$\frac{1}{2}b(2R)$	$(\frac{2}{3}b)\left[\frac{1}{2}b(2R)\right]$
Part 2 (rectangle)	$b + \frac{1}{2}c$	$c(2R)$	$(b + \frac{1}{2}c)[c(2R)]$
Part 3 (semicircle)	$b + c + \dfrac{4R}{3\pi}$	$\frac{1}{2}\pi R^2$	$\left(b + c + \dfrac{4R}{3\pi}\right)(\frac{1}{2}\pi R^2)$

Calculate the Centroid   The x coordinate of the centroid of the composite area is

$$\bar{x} = \frac{\bar{x}_1 A_1 + \bar{x}_2 A_2 + \bar{x}_3 A_3}{A_1 + A_2 + A_3}$$

$$= \frac{(\frac{2}{3}b)\left[\frac{1}{2}b(2R)\right] + (b + \frac{1}{2}c)[c(2R)] + \left(b + c + \dfrac{4R}{3\pi}\right)(\frac{1}{2}\pi R^2)}{\frac{1}{2}b(2R) + c(2R) + \frac{1}{2}\pi R^2}.$$

We repeat the last two steps to determine the y coordinate of the centroid. The y coordinates of the centroids of the parts are shown in Fig. b. Using the information summarized in Table 7.2, we obtain

$$\bar{y} = \frac{\bar{y}_1 A_1 + \bar{y}_2 A_2 + \bar{y}_3 A_3}{A_1 + A_2 + A_3}$$

$$= \frac{\left[\frac{1}{3}(2R)\right]\left[\frac{1}{2}b(2R)\right] + R[c(2R)] + R\left(\frac{1}{2}\pi R^2\right)}{\frac{1}{2}b(2R) + c(2R) + \frac{1}{2}\pi R^2}.$$

TABLE 7.2   Information for determining the y coordinate of the centroid

	$\bar{y}_i$	A_i	$\bar{y}_i A_i$
Part 1 (triangle)	$\frac{1}{3}(2R)$	$\frac{1}{2}b(2R)$	$\left[\frac{1}{3}(2R)\right]\left[\frac{1}{2}b(2R)\right]$
Part 2 (rectangle)	R	$c(2R)$	$R[c(2R)]$
Part 3 (semicircle)	R	$\frac{1}{2}\pi R^2$	$R(\frac{1}{2}\pi R^2)$

Example 7.4   Centroid of an Area with a Cutout

Determine the centroid of the area in Fig. 7.9.

Strategy

Instead of attempting to divide the area into parts, a simpler approach is to treat
it as a composite of a rectangular area with a semicircular cutout. Then we can
apply Eq. (7.9) by treating the cutout as a negative area.

Solution

Choose the Parts   We call the rectangle without the semicircular cutout and
the area of the cutout parts 1 and 2, respectively (Fig. a).

Determine the Values for the Parts   From Appendix B, the x coordinate of
the centroid of the cutout is

$$\bar{x}_2 = \frac{4R}{3\pi} = \frac{4(100)}{3\pi}\ \text{mm}.$$

The information for determining the x coordinate of the centroid is summarized
in Table 7.3. Notice that we treat the cutout as a negative area.

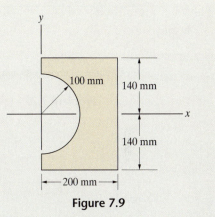

Figure 7.9

TABLE 7.3   Information for determining $\bar{x}$

	$\bar{x}_i$ (mm)	A_i (mm²)	$\bar{x}_i A_i$ (mm³)
Part 1 (rectangle)	100	(200)(280)	(100)[(200)(280)]
Part 2 (cutout)	$\dfrac{4(100)}{3\pi}$	$-\dfrac{1}{2}\pi(100)^2$	$-\dfrac{4(100)}{3\pi}\left[\dfrac{1}{2}\pi(100)^2\right]$

Calculate the Centroid   The x coordinate of the centroid is

$$\bar{x} = \frac{\bar{x}_1 A_1 + \bar{x}_2 A_2}{A_1 + A_2} = \frac{(100)[(200)(280)] - \dfrac{4(100)}{3\pi}\left[\dfrac{1}{2}\pi(100)^2\right]}{(200)(280) - \dfrac{1}{2}\pi(100)^2} = 122\ \text{mm}$$

Because of the symmetry of the area, $\bar{y} = 0$.

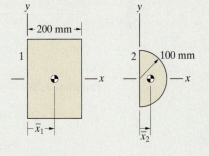

(a)   The rectangle and the semicircular
cutout.

Problems

For Problems 7.27–7.36, determine the coordinates of the centroids.

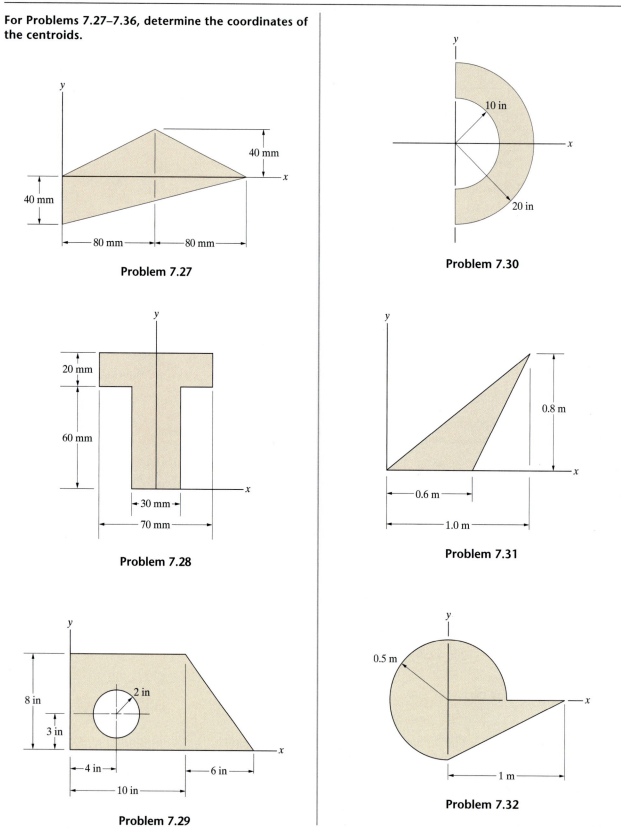

Problem 7.27

Problem 7.28

Problem 7.29

Problem 7.30

Problem 7.31

Problem 7.32

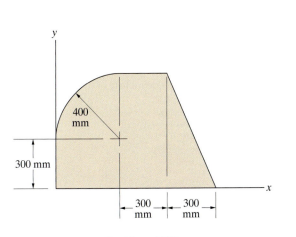

Problem 7.33

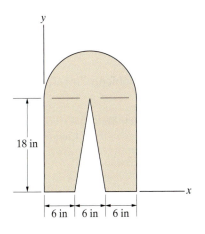

Problem 7.34

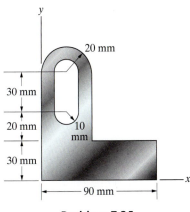

Problem 7.35

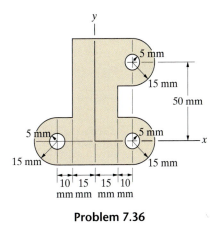

Problem 7.36

7.37 The dimensions $b = 42$ mm and $h = 22$ mm. Determine the y coordinate of the centroid of the beam's cross section.

7.38 If the cross-sectional area of the beam is 8400 mm^2 and the y coordinate of the centroid of the area is $\bar{y} = 90$ mm, what are the dimensions b and h?

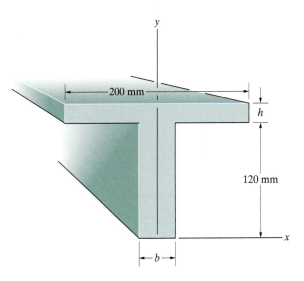

Problems 7.37/7.38

7.39 Determine the y coordinate of the centroid of the beam's cross section.

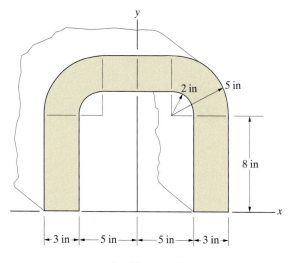

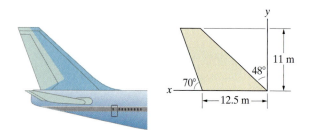

Problem 7.39

7.40 Determine the coordinates of the centroid of the airplane's vertical stabilizer.

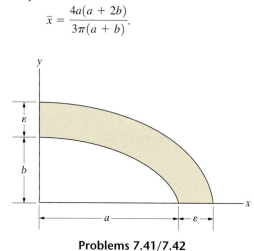

Problem 7.40

7.41 The area has elliptical boundaries. If $a = 30$ mm, $b = 15$ mm, and $\varepsilon = 6$ mm, what is the x coordinate of the centroid of the area?

7.42 By determining the x coordinate of the centroid of the area shown in Problem 7.41 in terms of a, b, and ε, and evaluating its limit as $\varepsilon \rightarrow 0$, show that the x coordinate of the centroid of a quarter-elliptical line is

$$\bar{x} = \frac{4a(a + 2b)}{3\pi(a + b)}.$$

Problems 7.41/7.42

7.43 Three sails of a New York pilot schooner are shown. The coordinates of the points are in feet. Determine the centroid of sail 1.

7.44 Determine the centroid of sail 2.

7.45 Determine the centroid of sail 3.

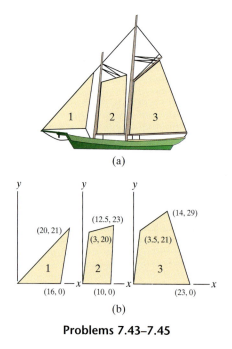

Problems 7.43–7.45

7.3 Distributed Loads

The load exerted on a beam (stringer) supporting a floor of a building is distributed over the beam's length (Fig. 7.10a). The load exerted by wind on a television transmission tower is distributed along the tower's height (Fig. 7.10b). In many engineering applications, loads are continuously distributed along lines. We will show that the concept of the centroid of an area can be useful in the analysis of objects subjected to such loads.

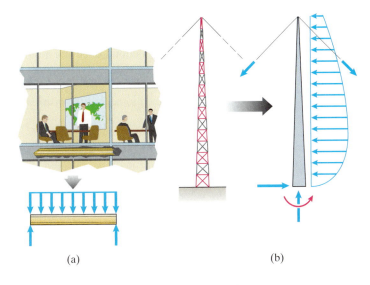

(a)   (b)

Figure 7.10
Examples of distributed forces:
(a) Uniformly distributed load exerted on a beam of a building's frame by the floor.
(b) Wind load distributed along the height of a tower.

Describing a Distributed Load

We can use a simple example to demonstrate how such loads are expressed analytically. Suppose that we pile bags of sand on a beam, as shown in Fig. 7.11a. It is clear that the load exerted by the bags is distributed over the length of the beam and that its magnitude at a given position x depends on how high the bags are piled at that position. To describe the load, we define a function w such that the *downward* force exerted on an infinitesimal element dx of the beam is $w \, dx$. With this function we can model the varying magnitude of the load exerted by the sand bags (Fig. 7.11b). The arrows in the figure indicate that the load acts in the downward direction. Loads distributed along lines, from simple examples such as a beam's own weight to complicated ones such as the lift distributed along the length of an airplane's wing, are modeled by the function w. Since the product of w and dx is a force, the dimensions of w are (force)/(length). For example, w can be expressed in newtons per meter in SI units or in pounds per foot in U.S. Customary units.

Determining Force and Moment

Let's assume that the function w describing a particular distributed load is known (Fig. 7.12a). The graph of w, is called the *loading curve*. Since the force acting on an element dx of the line is $w \, dx$, we can determine the total force F

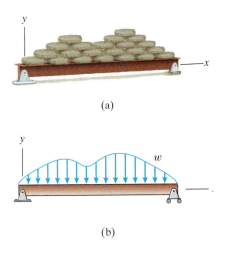

(a)

(b)

Figure 7.11
(a) Loading a beam with bags of sand.
(b) The distributed load w models the load exerted by the bags.

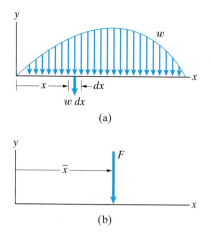

(a)

(b)

Figure 7.12
(a) A distributed load and the force exerted on a differential element dx.
(b) The equivalent force.

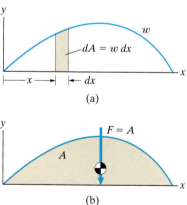

(a)

(b)

Figure 7.13
(a) Determining the "area" between the function w and the x axis.
(b) The equivalent force is equal to the "area," and the line of action passes through its centroid.

exerted by the distributed load by integrating the loading curve with respect to x:

$$F = \int_L w\, dx. \tag{7.10}$$

We can also integrate to determine the moment about a point exerted by the distributed load. For example, the moment about the origin due to the force exerted on the element dx is $xw\, dx$, so the total moment about the origin due to the distributed load is

$$M = \int_L xw\, dx. \tag{7.11}$$

When you are concerned only with the total force and moment exerted by a distributed load, you can represent it by a single equivalent force F (Fig. 7.12b). For equivalence, the force must act at a position $\bar{x}$ on the x axis such that the moment of F about the origin is equal to the moment of the distributed load about the origin:

$$\bar{x}F = \int_L xw\, dx.$$

Therefore the force F is equivalent to the distributed load if we place it at the position

$$\bar{x} = \frac{\displaystyle\int_L xw\, dx}{\displaystyle\int_L w\, dx}. \tag{7.12}$$

The Area Analogy

Notice that the term $w\, dx$ is equal to an element of "area" dA between the loading curve and the x axis (Fig. 7.13a). (We use quotation marks because $w\, dx$ is actually a force and not an area.) Interpreted in this way, Eq. (7.10) states that the total force exerted by the distributed load is equal to the "area" A between the loading curve and the x axis:

$$F = \int_L w\, dx = \int_A dA = A. \tag{7.13}$$

Substituting $w\, dx = dA$ into Eq. (7.12), we obtain

$$\bar{x} = \frac{\displaystyle\int_L xw\, dx}{\displaystyle\int_L w\, dx} = \frac{\displaystyle\int_A x\, dA}{\displaystyle\int_A dA}. \tag{7.14}$$

The force F is equivalent to the distributed load if it acts at the centroid of the "area" between the loading curve and the x axis (Fig. 7.13b). Using this analogy to represent a distributed load by an equivalent force can be very useful when the loading curve is relatively simple (see Example 7.5).

> **Study Questions**
> 1. What is the definition of the function w?
> 2. How is the force exerted by a distributed load determined from the loading curve?
> 3. How is the moment exerted by a distributed load determined from the loading curve?

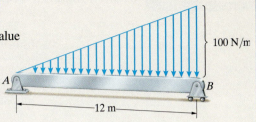

Example 7.5   Beam with a Triangular Distributed Load

The beam in Fig. 7.14 is subjected to a "triangular" distributed load whose value at B is 100 N/m.
(a) Represent the distributed load by a single equivalent force.
(b) Determine the reactions at A and B.

Figure 7.14

Strategy
(a) The magnitude of the force is equal to the "area" under the triangular loading curve, and the equivalent force acts at the centroid of the triangular "area."
(b) Once the distributed load is represented by a single equivalent force, we can apply the equilibrium equations to determine the reactions.

Solution
(a) The "area" of the triangular distributed load is one-half its base times its height, or $\frac{1}{2}(12 \text{ m}) \times (100 \text{ N/m}) = 600$ N. The centroid of the triangular "area" is located at $\bar{x} = \frac{2}{3}(12 \text{ m}) = 8$ m. We can therefore represent the distributed load by an equivalent downward force of 600-N magnitude acting at $x = 8$ m (Fig. a).

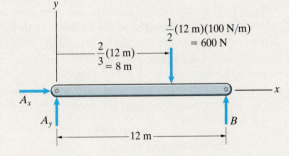

(a) Representing the distributed load by an equivalent force.

(b) From the equilibrium equations

$$\Sigma F_x = A_x = 0,$$

$$\Sigma F_y = A_y + B - 600 \text{ N} = 0,$$

$$\Sigma M_{\text{point } A} = (12 \text{ m})B - (8 \text{ m})(600 \text{ N}) = 0,$$

we obtain $A_x = 0$, $A_y = 200$ N, and $B = 400$ N.

Critical Thinking
The area analogy made it very straightforward to represent the triangular distributed load in this example by an equivalent force. For comparison, you should determine the reactions at A and B by using Eqs. (7.10) and (7.11) to calculate the force and moment exerted by the distributed load.

Example 7.6	Beam Subjected to Distributed Loads

The beam in Fig. 7.15 is subjected to two distributed loads. Determine the reactions at A and B.

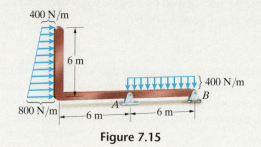

Figure 7.15

Strategy
We can easily apply the area analogy to the uniformly distributed load between A and B. We will treat the distributed load on the vertical section of the beam as the sum of uniform and triangular distributed loads and use the area analogy to represent each distributed load by an equivalent force.

Solution
We draw the free-body diagram of the beam in Fig. a, expressing the left distributed load as the sum of uniform and triangular loads. In Fig. b, we represent the three distributed loads by equivalent forces. The "area" of the uniform distributed load on the right is $(6 \text{ m}) \times (400 \text{ N/m}) = 2400 \text{ N}$, and its centroid is 3 m from B. The area of the uniform distributed load on the vertical part of the beam is $(6 \text{ m}) \times (400 \text{ N/m}) = 2400 \text{ N}$, and its centroid is located at $y = 3$ m. The area of the triangular distributed load is $\frac{1}{2}(6 \text{ m}) \times (400 \text{ N/m}) = 1200 \text{ N}$, and its centroid is located at $y = \frac{1}{3}(6 \text{ m}) = 2$ m.

From the equilibrium equations

$$\Sigma F_x = A_x + 1200 \text{ N} + 2400 \text{ N} = 0,$$

$$\Sigma F_y = A_y + B - 2400 \text{ N} = 0,$$

$$\Sigma M_{\text{point } A} = (6 \text{ m})B - (3 \text{ m})(2400 \text{ N}) - (2 \text{ m})(1200 \text{ N}) - (3 \text{ m})(2400 \text{ N}) = 0,$$

we obtain $A_x = -3600$ N, $A_y = -400$ N, and $B = 2800$ N.

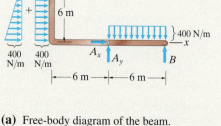

(a) Free-body diagram of the beam.

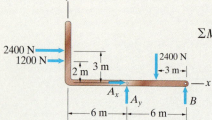

(b) Representing the distributed loads by equivalent forces.

Critical Thinking
When you analyze a problem involving distributed loads, should you always use the area analogy to represent them by equivalent forces, as we did in Example 7.5 and this example? The area analogy is useful when a loading curve is sufficiently simple that its area and the location of its centroid are easy to determine. When that is not the case, you can use Eqs. (7.10) and (7.11) to determine the force and moment exerted by a distributed load. We illustrate this approach in Example 7.7.

Example 7.7   Beam with a Distributed Load

The beam in Fig. 7.16 is subjected to a distributed load, a force, and a couple. The distributed load is $w = 300x - 50x^2 + 0.3x^4$ lb/ft. Determine the reactions at the fixed support A.

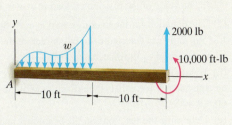

Strategy
Since we know the function w, we can use Eqs. (7.10) and (7.11) to determine the force and moment exerted on the beam by the distributed load. We can then use the equilibrium equations to determine the reactions at A.

Figure 7.16

Solution
We isolate the beam and show the reactions at the fixed support in Fig. a. The *downward* force exerted by the distributed load is

$$\int_L w \, dx = \int_0^{10} (300x - 50x^2 + 0.3x^4)dx = 4330 \text{ lb}.$$

The *clockwise* moment about A exerted by the distributed load is

$$\int_L xw \, dx = \int_0^{10} x(300x - 50x^2 + 0.3x^4) \, dx = 25,000 \text{ ft-lb}.$$

From the equilibrium equations

$$\Sigma F_x = A_x = 0,$$

$$\Sigma F_y = A_y - 4330 \text{ lb} + 2000 \text{ lb} = 0,$$

$$\Sigma M_{\text{point } A} = M_A - 25,000 \text{ ft-lb} + (20 \text{ ft})(2000 \text{ lb}) + 10,000 \text{ ft-lb} = 0,$$

we obtain $A_x = 0$, $A_y = 2330$ lb, and $M_A = -25,000$ ft-lb.

Critical Thinking
When you use Eq. (7.11), it is important to be aware that you are calculating the *clockwise* moment due to the distributed load w *about the origin* $x = 0$.

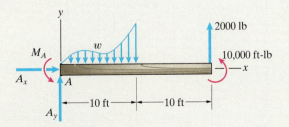

(a) Free-body diagram of the beam.

Problems

7.46 The value of the distributed load w at $x = 6$ m is 240 N/m.

(a) The equation for the loading curve is $w = 80x - 240$ N/m. Use Eq. (7.10) to determine the magnitude of the total force exerted on the beam by the distributed load.

(b) If you use the area analogy to represent the distributed load by an equivalent force, what is the magnitude of the force and where does it act?

(c) Determine the reactions on the beam at A and B.

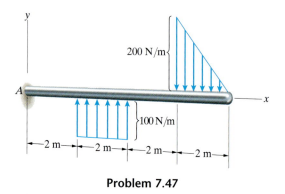

240 N/m

Problem 7.46

7.47 Determine the reactions at the fixed support A.

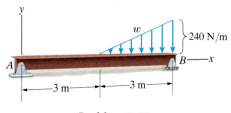

200 N/m

100 N/m

Problem 7.47

7.48 Determine the reactions at the fixed support A.

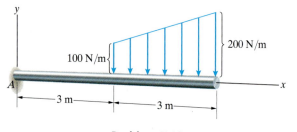

200 N/m

100 N/m

Problem 7.48

7.49 Determine the reactions at A and B.

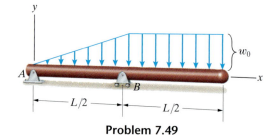

w_0

Problem 7.49

7.50 Determine the reactions at the fixed support A.

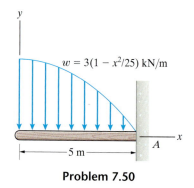

$w = 3(1 - x^2/25)$ kN/m

Problem 7.50

7.51 An engineer measures the forces exerted by the soil on a 10-m section of a building foundation and finds that they are described by the distributed load $w = -10x - x^2 + 0.2x^3$ kN/m.

(a) Determine the magnitude of the total force exerted on the foundation by the distributed load.

(b) Determine the magnitude of the moment about A due to the distributed load.

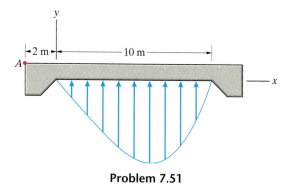

Problem 7.51

7.52 Determine the reactions on the beam at A and B.

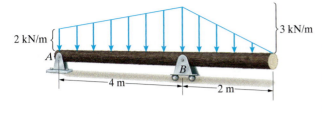

Problem 7.52

7.53 The aerodynamic lift of the wing is described by the distributed load $w = -300 \sqrt{1 - 0.04x^2}$ N/m. The mass of the wing is 27 kg, and its center of mass is located 2 m from the wing root R.
(a) Determine the magnitudes of the force and the moment about R exerted by the lift of the wing.
(b) Determine the reactions on the wing at R.

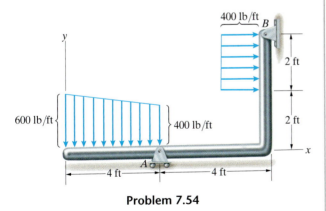

Problem 7.53

7.54 Determine the reactions on the bar at A and B.

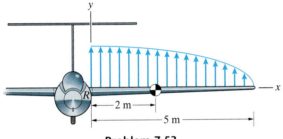

Problem 7.54

7.55 Determine the reactions at A and B.

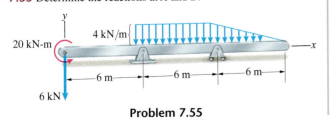

Problem 7.55

7.56 Determine the axial forces in members BD, CD, and CE of the truss and indicate whether they are in tension (T) or compression (C).

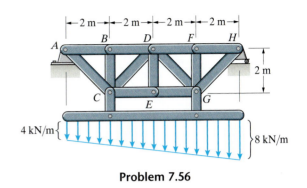

Problem 7.56

7.57 Determine the reactions on member ABC at A and B.

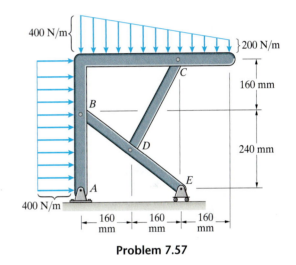

Problem 7.57

7.58 Determine the forces on member ABC of the frame.

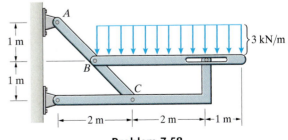

Problem 7.58

7.4 Centroids of Volumes and Lines

Here we define the centroids, or average positions, of volumes and lines, and show how to determine the centroids of composite volumes and lines. We will show in Section 7.7 that knowing the centroids of volumes and lines allows you to determine the centers of mass of certain types of objects, which tells you where their weights effectively act.

Definitions

Volumes Consider a volume V, and let dV be a differential element of V with coordinates x, y, and z (Fig. 7.17). By analogy with Eqs. (7.6) and (7.7), the coordinates of the centroid of V are

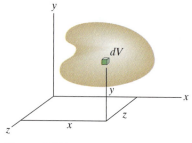

Figure 7.17
A volume V and differential element dV.

$$\bar{x} = \frac{\displaystyle\int_V x\,dV}{\displaystyle\int_V dV}, \qquad \bar{y} = \frac{\displaystyle\int_V y\,dV}{\displaystyle\int_V dV}, \qquad \bar{z} = \frac{\displaystyle\int_V z\,dV}{\displaystyle\int_V dV}. \tag{7.15}$$

The subscript V on the integral signs means that the integration is carried out over the entire volume.

If a volume has the form of a plate with uniform thickness and cross-sectional area A (Fig. 7.18a), its centroid coincides with the centroid of A and lies at the midpoint between the two faces. To show that this is true, we obtain a volume element dV by projecting an element dA of the cross-sectional area through the thickness T of the volume, so that $dV = T\,dA$ (Fig. 7.18b). Then the x and y coordinates of the centroid of the volume are

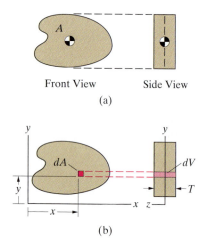

Front View   Side View
(a)

(b)

Figure 7.18
(a) A volume of uniform thickness.
(b) Obtaining dV by projecting dA through the volume.

$$\bar{x} = \frac{\displaystyle\int_V x\,dV}{\displaystyle\int_V dV} = \frac{\displaystyle\int_A xT\,dA}{\displaystyle\int_A T\,dA} = \frac{\displaystyle\int_A x\,dA}{\displaystyle\int_A dA},$$

$$\bar{y} = \frac{\displaystyle\int_V y\,dV}{\displaystyle\int_V dV} = \frac{\displaystyle\int_A yT\,dA}{\displaystyle\int_A T\,dA} = \frac{\displaystyle\int_A y\,dA}{\displaystyle\int_A dA}.$$

The coordinate $\bar{z} = 0$ by symmetry. Thus you know the centroid of this type of volume if you know (or can determine) the centroid of its cross-sectional area.

Lines The coordinates of the centroid of a line L are

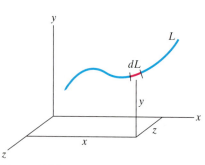

Figure 7.19
A line L and differential element dL.

$$\bar{x} = \frac{\displaystyle\int_L x\,dL}{\displaystyle\int_L dL}, \qquad \bar{y} = \frac{\displaystyle\int_L y\,dL}{\displaystyle\int_L dL}, \qquad \bar{z} = \frac{\displaystyle\int_L z\,dL}{\displaystyle\int_L dL}, \tag{7.16}$$

where dL is a differential length of the line with coordinates x, y, and z. (Fig. 7.19).

Example 7.8   Centroid of a Cone by Integration

Determine the centroid of the cone in Fig. 7.20.

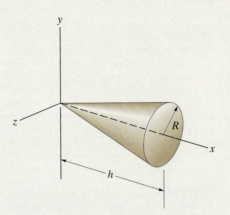

Figure 7.20

Strategy

The centroid must lie on the x axis because of symmetry. We will determine its x coordinate by using an element of volume dV in the form of a "disk" of width dx.

Solution

Let dV be the disk in Fig. a. The radius of the disk is $(R/h)x$ (Fig. b), and its volume equals the product of the area of the disk and its thickness, $dV = \pi[(R/h)x]^2 \, dx$. To integrate over the entire volume, we must integrate with respect to x from $x = 0$ to $x = h$. The x coordinate of the centroid is

$$\bar{x} = \frac{\displaystyle\int_V x \, dV}{\displaystyle\int_V dV} = \frac{\displaystyle\int_0^h x\pi \frac{R^2}{h^2}x^2 \, dx}{\displaystyle\int_0^h \pi \frac{R^2}{h^2}x^2 \, dx} = \frac{3}{4}h.$$

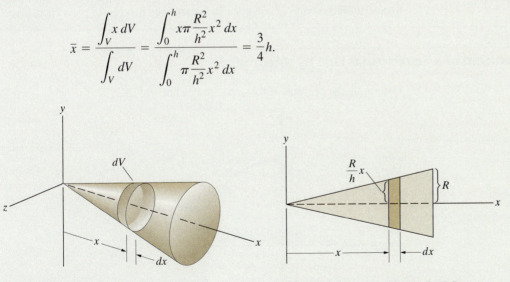

(a) An element dV in the form of a disk.   **(b)** The radius of the element is $(R/h)x$.

Critical Thinking

Notice that we determined the centroid of the cone by using an element of volume of the form $dV = A \, dx$, where A is the cone's cross-sectional area. You can use this approach to determine the centroids of other volumes of revolution and pyramidal volumes (see Problems 7.59–7.66).

Example 7.9 **Centroid of a Line by Integration**

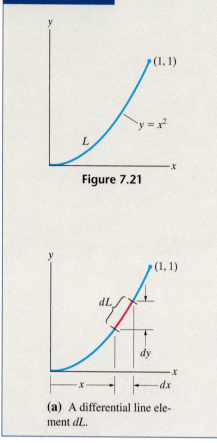

Figure 7.21

(a) A differential line element dL.

The line L in Fig. 7.21 is defined by the function $y = x^2$. Determine the x coordinate of its centroid.

Strategy

We can express a differential element dL of a line (Fig. a) in terms of dx and dy:

$$dL = \sqrt{dx^2 + dy^2} = \sqrt{1 + \left(\frac{dy}{dx}\right)^2}\ dx.$$

From the equation describing the line, the derivative $dy/dx = 2x$, so we obtain an expression for dL in terms of x:

$$dL = \sqrt{1 + 4x^2}\ dx.$$

Solution

To integrate over the entire line, we must integrate from $x = 0$ to $x = 1$. The x coordinate of the centroid is

$$\bar{x} = \frac{\displaystyle\int_L x\ dL}{\displaystyle\int_L dL} = \frac{\displaystyle\int_0^1 x\sqrt{1 + 4x^2}\ dx}{\displaystyle\int_0^1 \sqrt{1 + 4x^2}\ dx} = 0.574.$$

Critical Thinking

Our approach in this example is appropriate to determine the centroid of a line that is described by a function of the form $y = f(x)$. In Example 7.10 we show how to determine the centroid of a line that is described in terms of polar coordinates.

Example 7.10 **Centroid of a Semicircular Line by Integration**

Determine the centroid of the semicircular line in Fig. 7.22.

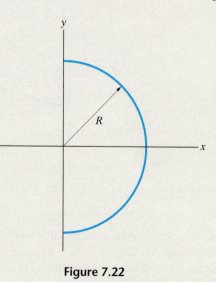

Figure 7.22

Strategy

Because of the symmetry of the line, the centroid lies on the x axis. To determine $\bar{x}$, we will integrate in terms of polar coordinates. By letting θ change by an amount $d\theta$, we obtain a differential line element of length $dL = R\,d\theta$ (Fig. a). The x coordinate of dL is $x = R\cos\theta$.

Solution

To integrate over the entire line, we must integrate with respect to θ from $\theta = -\pi/2$ to $\theta = +\pi/2$:

$$\bar{x} = \frac{\displaystyle\int_L x\,dL}{\displaystyle\int_L dL} = \frac{\displaystyle\int_{-\pi/2}^{\pi/2}(R\cos\theta)R\,d\theta}{\displaystyle\int_{-\pi/2}^{\pi/2}R\,d\theta} = \frac{R^2\big[\sin\theta\big]_{-\pi/2}^{\pi/2}}{R\big[\theta\big]_{-\pi/2}^{\pi/2}} = \frac{2R}{\pi}.$$

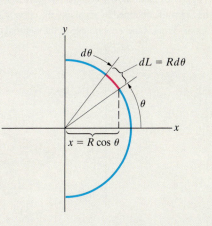

(a) A differential line element $dL = R\,d\theta$.

Critical Thinking

Notice that our integration procedure gives the correct length of the line:

$$\int_L dL = \int_{-\pi/2}^{\pi/2}R\,d\theta = R\big[\theta\big]_{-\pi/2}^{\pi/2} = \pi R.$$

Problems

7.59 Determine the coordinates of the centroid of the truncated conical volume.

Strategy: Use the method described in Example 7.8.

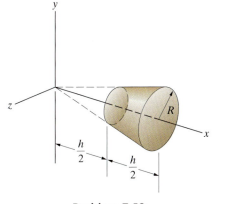

Problem 7.59

7.60 A grain storage tank has the form of a surface of revolution with the profile shown. The height of the tank is 7 m and its diameter at ground level is 10 m. Determine the volume of the tank and the height *above ground level* of the centroid of its volume.

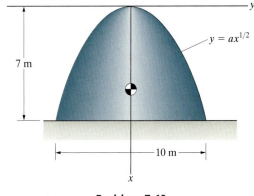

Problem 7.60

7.61 The object shown, designed to serve as a pedestal for a speaker, has a profile obtained by revolving the curve $y = 0.167x^2$ about the x axis. What is the x coordinate of the centroid of the object?

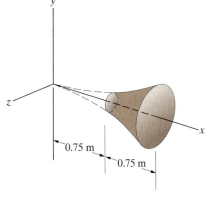

Problem 7.61

7.62 The volume of a nose cone is generated by rotating the function $y = x - 0.2x^2$ about the x axis.

(a) What is the volume of the nose cone?

(b) What is the x coordinate of the centroid of the volume?

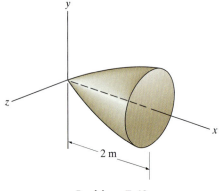

Problem 7.62

7.63 Determine the centroid of the hemispherical volume.

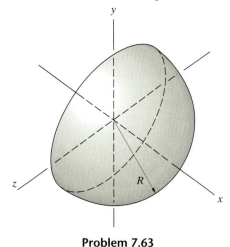

Problem 7.63

7.64 The volume consists of a segment of a sphere of radius R. Determine its centroid.

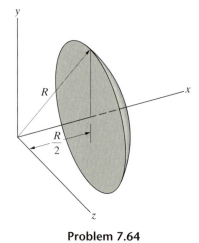

Problem 7.64

7.65 A volume of revolution is obtained by revolving the curve $x^2/a^2 + y^2/b^2 = 1$ about the x axis. Determine its centroid.

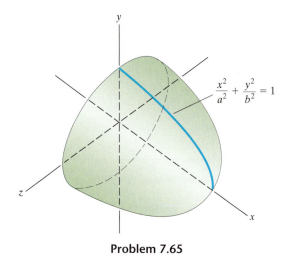

Problem 7.65

7.66 The volume of revolution has a cylindrical hole of radius R. Determine its centroid.

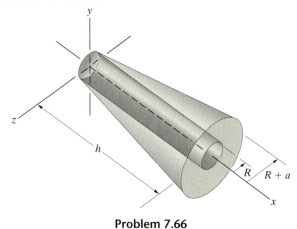

Problem 7.66

7.67 Determine the coordinates of the centroid of the line. (See Example 7.9.)

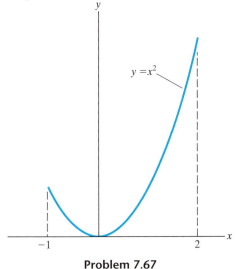

Problem 7.67

7.68 Determine the x coordinate of the centroid of the line.

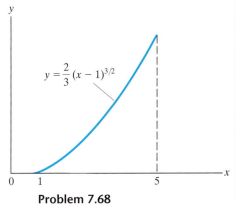

Problem 7.68

7.69 Determine the x coordinate of the centroid of the line.

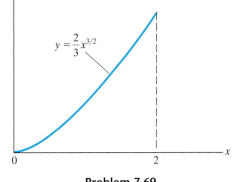

Problem 7.69

7.70 Determine the centroid of the circular arc.

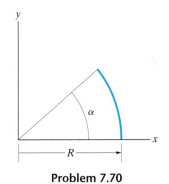

Problem 7.70

Centroids of Composite Volumes and Lines

The centroids of composite volumes and lines can be derived using the same approach we applied to areas. The coordinates of the centroid of a composite volume are

$$\bar{x} = \frac{\sum_i \bar{x}_i V_i}{\sum_i V_i}, \qquad \bar{y} = \frac{\sum_i \bar{y}_i V_i}{\sum_i V_i}, \qquad \bar{z} = \frac{\sum_i \bar{z}_i V_i}{\sum_i V_i}, \qquad (7.17)$$

and the coordinates of the centroid of a composite line are

$$\bar{x} = \frac{\sum_i \bar{x}_i L_i}{\sum_i L_i}, \qquad \bar{y} = \frac{\sum_i \bar{y}_i L_i}{\sum_i L_i}, \qquad \bar{z} = \frac{\sum_i \bar{z}_i L_i}{\sum_i L_i}. \qquad (7.18)$$

The centroids of some simple volumes and lines are tabulated in Appendices B and C.

Determining the centroid of a composite volume or line requires three steps:

1. **Choose the parts**—Try to divide the composite into parts whose centroids you know or can easily determine.
2. **Determine the values for the parts**—Determine the centroid and the volume or length of each part. Watch for instances of symmetry that can simplify your task.
3. **Calculate the centroid**—Use Eqs. (7.17) or (7.18) to determine the centroid of the composite volume or line.

Example 7.11   Centroid of a Composite Volume

Determine the centroid of the volume in Fig. 7.23.

Strategy
We must divide the volume into parts (the parts are obvious in this example), determine the centroids of the parts, and apply Eqs. (7.17).

Solution
Choose the Parts   The volume consists of a cone and a cylinder, which we call parts 1 and 2, respectively.

Determine the Values for the Parts   The centroid and volume of the cone are given in Appendix C. The x coordinates of the centroids of the parts are shown in Fig. a, and the information for determining the x coordinate of the centroid is summarized in Table 7.4.

Figure 7.23

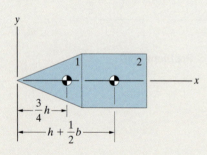

(a) The x coordinates of the centroids of the cone and cylinder.

TABLE 7.4   Information for determining $\bar{x}$

	$\bar{x}_i$	V_i	$\bar{x}_i V_i$
Part 1 (cone)	$\frac{3}{4}h$	$\frac{1}{3}\pi R^2 h$	$(\frac{3}{4}h)(\frac{1}{3}\pi R^2 h)$
Part 2 (cylinder)	$h + \frac{1}{2}b$	$\pi R^2 b$	$(h + \frac{1}{2}b)(\pi R^2 b)$

Calculate the Centroid   The x coordinate of the centroid of the composite volume is

$$\bar{x} = \frac{\bar{x}_1 V_1 + \bar{x}_2 V_2}{V_1 + V_2} = \frac{(\frac{3}{4}h)(\frac{1}{3}\pi R^2 h) + (h + \frac{1}{2}b)(\pi R^2 b)}{\frac{1}{3}\pi R^2 h + \pi R^2 b}.$$

Because of symmetry, $\bar{y} = 0$ and $\bar{z} = 0$.

Critical Thinking
We will show in Section 7.7 that the center of mass of a homogeneous object coincides with the centroid of its volume. Therefore our solution in this example determines the location of the center of mass of a homogeneous object that occupies this volume. This is one of our principal motivations for introducing the concept of a centroid in statics.

Example 7.12   Centroid of a Volume Containing a Cutout

Determine the centroid of the volume in Fig. 7.24.

Strategy
We can divide this volume into the five simple parts shown in Fig. a. Notice that parts 2 and 3 *do not* have the cutout. It is assumed to be "filled in," which simplifies the geometries of those parts. Part 5, which is the volume of the 20-mm-diameter hole, will be treated as a negative volume in Eqs. (7.17).

Solution

Choose the Parts We can divide the volume into the five simple parts shown in Fig. a. Part 5 is the volume of the 20-mm-diameter hole.

Determine the Values for the Parts The centroids of parts 1 and 3 are located at the centroids of their semicircular cross sections (Fig. b). The information for determining the x-coordinate of the centroid is summarized in Table 7.5. Part 5 is a negative volume.

TABLE 7.5 Information for determining $\bar{x}$

	$\bar{x}_i$ (mm)	V_i (mm³)	$\bar{x}_i V_i$ (mm⁴)
Part 1	$-\dfrac{4(25)}{3\pi}$	$\dfrac{\pi(25)^2}{2}(20)$	$\left[-\dfrac{4(25)}{3\pi}\right]\left[\dfrac{\pi(25)^2}{2}(20)\right]$
Part 2	100	$(200)(50)(20)$	$(100)[(200)(50)(20)]$
Part 3	$200 + \dfrac{4(25)}{3\pi}$	$\dfrac{\pi(25)^2}{2}(20)$	$\left[200 + \dfrac{4(25)}{3\pi}\right]\left[\dfrac{\pi(25)^2}{2}(20)\right]$
Part 4	0	$\pi(25)^2(40)$	0
Part 5	200	$-\pi(10)^2(20)$	$-(200[\pi(10)^2(20)])$

Calculate the Centroid The x coordinate of the centroid of the composite volume is

$$\bar{x} = \frac{\bar{x}_1 V_1 + \bar{x}_2 V_2 + \bar{x}_3 V_3 + \bar{x}_4 V_4 + \bar{x}_5 V_5}{V_1 + V_2 + V_3 + V_4 + V_5}$$

$$= \frac{\left[-\dfrac{4(25)}{3\pi}\right]\left[\dfrac{\pi(25)^2}{2}(20)\right] + (100)[(200)(50)(20)] + \left[200 + \dfrac{4(25)}{3\pi}\right]\left[\dfrac{\pi(25)^2}{2}(20)\right] + 0 - (200)[\pi(10)^2(20)]}{\dfrac{\pi(25)^2}{2}(20) + (200)(50)(20) + \dfrac{\pi(25)^2}{2}(20) + \pi(25)^2(40) - \pi(10)^2(20)}$$

$$= 72.77 \text{ mm}.$$

The z coordinates of the centroids of the parts are zero except $\bar{z}_4 = 30$ mm. Therefore the z coordinate of the centroid of the composite volume is

$$\bar{z} = \frac{\bar{z}_4 V_4}{V_1 + V_2 + V_3 + V_4 + V_5}$$

$$= \frac{30[\pi(25)^2(40)]}{\dfrac{\pi(25)^2}{2}(20) + (200)(50)(20) + \dfrac{\pi(25)^2}{2}(20) + \pi(25)^2(40) - \pi(10)^2(20)}$$

$$= 7.56 \text{ mm}.$$

Because of symmetry, $\bar{y} = 0$.

Critical Thinking

You can recognize that the volume in this example could be part of a mechanical device. Many manufactured parts have volumes that are composites of simple volumes, and the method used in this example can be used to determine their centroids and, if they are homogeneous, their centers of mass.

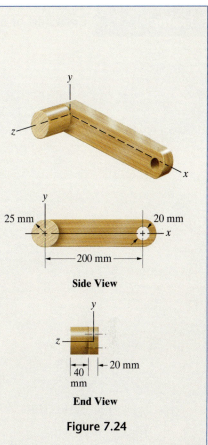

Side View

End View

Figure 7.24

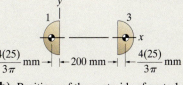

(a) Dividing the volume into five parts.

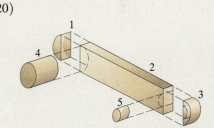

(b) Positions of the centroids of parts 1 and 3.

Example 7.13   **Centroid of a Composite Line**

Determine the centroid of the line in Fig. 7.25. The quarter-circular arc lies in the y–z plane.

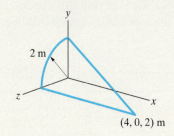

Figure 7.25

$(4, 0, 2)$ m

Strategy
We must divide the line into parts (in this case the quarter-circular arc and the two straight segments), determine the centroids of the parts, and apply Eqs. (7.18).

Solution

Choose the Parts   The line consists of a quarter-circular arc and two straight segments, which we call parts 1, 2, and 3 (Fig. a).

Determine the Values for the Parts   From Appendix B, the coordinates of the centroid of the quarter-circular arc are $\bar{x}_1 = 0$, $\bar{y}_1 = \bar{z}_1 = 2(2)/\pi$ m. The centroids of the straight segments lie at their midpoints. For segment 2, $\bar{x}_2 = 2$ m, $\bar{y}_2 = 0$, and $\bar{z}_2 = 2$ m, and for segment 3, $\bar{x}_3 = 2$ m, $\bar{y}_3 = 1$ m, and $\bar{z}_3 = 1$ m. The length of segment 3 is $L_3 = \sqrt{(4)^2 + (2)^2 + (2)^2} = 4.90$ m. This information is summarized in Table 7.6.

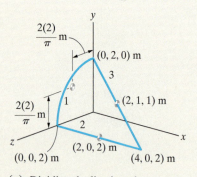

(a) Dividing the line into three parts.

TABLE 7.6   Information for determining the centroid.

	$\bar{x}_i$ (m)	$\bar{y}_i$ (m)	$\bar{z}_i$ (m)	L_i (m)
Part 1	0	$2(2)/\pi$	$2(2)/\pi$	$\pi(2)/2$
Part 2	2	0	2	4
Part 3	2	1	1	4.90

Calculate the Centroid   The coordinates of the centroid of the composite line are

$$\bar{x} = \frac{\bar{x}_1 L_1 + \bar{x}_2 L_2 + \bar{x}_3 L_3}{L_1 + L_2 + L_3} = \frac{0 + (2)(4) + (2)(4.90)}{\pi + 4 + 4.90} = 1.478 \text{ m},$$

$$\bar{y} = \frac{\bar{y}_1 L_1 + \bar{y}_2 L_2 + \bar{y}_3 L_3}{L_1 + L_2 + L_3} = \frac{[2(2)/\pi][\pi(2)/2] + 0 + (1)(4.90)}{\pi + 4 + 4.90} = 0.739 \text{ m},$$

$$\bar{z} = \frac{\bar{z}_1 L_1 + \bar{z}_2 L_2 + \bar{z}_3 L_3}{L_1 + L_2 + L_3} = \frac{[2(2)/\pi][\pi(2)/2] + (2)(4) + (1)(4.90)}{\pi + 4 + 4.90} = 1.404 \text{ m}.$$

Critical Thinking
What possible reason could you have for wanting to know the centroid (average position) of a line? In Section 7.7 we show that the center of mass of a slender homogeneous bar, which is the point at which the weight of the bar can be represented by an equivalent force, lies approximately at the centroid of the bar's axis.

Problems

For Problems 7.71–7.78, determine the centroids of the volumes.

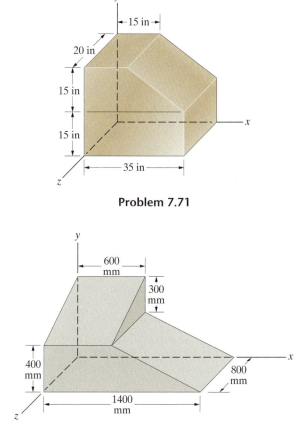

Problem 7.71

Problem 7.72

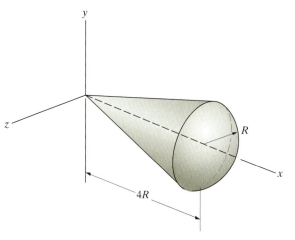

Problem 7.73

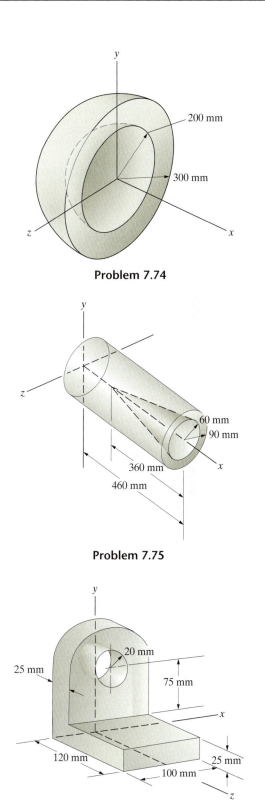

Problem 7.74

Problem 7.75

Problem 7.76

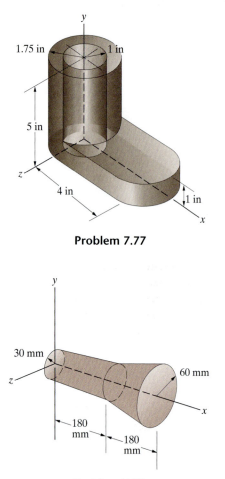

Problem 7.77

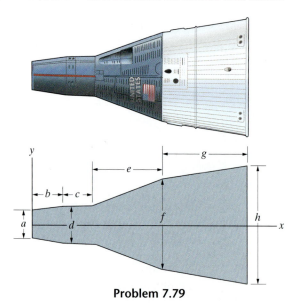

Problem 7.78

7.79 The dimensions of the *Gemini* spacecraft (in meters) are $a = 0.70$, $b = 0.88$, $c = 0.74$, $d = 0.98$, $e = 1.82$, $f = 2.20$, $g = 2.24$, and $h = 2.98$. Determine the centroid of its volume.

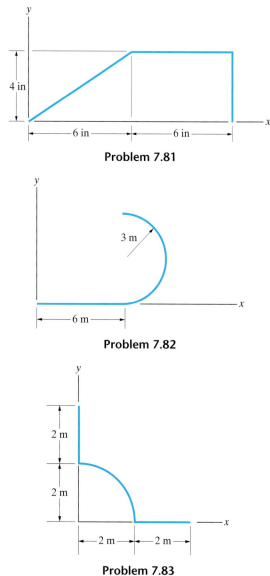

Problem 7.79

7.80 Two views of a machine element are shown. Determine the centroid of its volume.

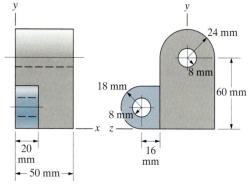

Problem 7.80

For Problems 7.81–7.83, determine the centroids of the lines.

Problem 7.81

Problem 7.82

Problem 7.83

7.84 The semicircular part of the line lies in the x–z plane. Determine the centroid of the line.

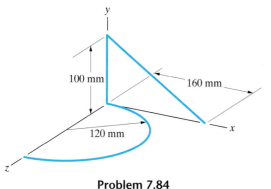

Problem 7.84

7.85 Determine the centroid of the line.

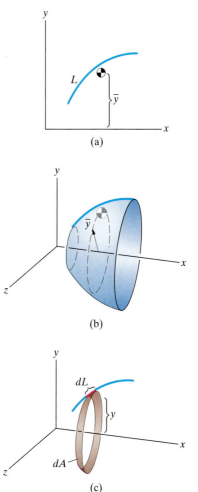

Problem 7.85

7.5 The Pappus–Guldinus Theorems

In this section we discuss two simple and useful theorems relating surfaces and volumes of revolution to the centroids of the lines and areas that generate them.

First Theorem

Consider a line L in the x–y plane that does not intersect the x axis (Fig. 7.26a). Let the coordinates of the centroid of the line be $(\bar{x}, \bar{y})$. We can generate a surface by revolving the line about the x axis (Fig. 7.26b). As the line revolves about the x axis, the centroid of the line moves in a circular path of radius $\bar{y}$.

The first Pappus–Guldinus theorem states that the area of the surface of revolution is equal to the product of the distance through which the centroid of the line moves and the length of the line:

$$A = 2\pi\bar{y}\,L. \tag{7.19}$$

To prove this result, we observe that as the line revolves about the x axis, the area dA generated by an element dL of the line is $dA = 2\pi y\,dL$, where y is the y coordinate of the element dL (Fig. 7.26c). Therefore, the total area of the surface of revolution is

$$A = 2\pi \int_L y\,dL. \tag{7.20}$$

From the definition of the y coordinate of the centroid of the line,

$$\bar{y} = \frac{\displaystyle\int_L y\,dL}{\displaystyle\int_L dL},$$

we obtain

$$\int_L y\,dL = \bar{y}L.$$

Substituting this result into Eq. (7.20), we obtain Eq. (7.19).

Figure 7.26
(a) A line L and the y coordinate of its centroid.
(b) The surface generated by revolving the line L about the x axis and the path followed by the centroid of the line.
(c) An element dL of the line and the element of area dA it generates.

Second Theorem

Consider an area A in the x–y plane that does not intersect the x axis (Fig. 7.27a). Let the coordinates of the centroid of the area be $(\bar{x}, \bar{y})$. We can generate a volume by revolving the area about the x axis (Fig. 7.27b). As the area revolves about the x axis, the centroid of the area moves in a circular path of length $2\pi\bar{y}$.

The second Pappus–Guldinus theorem states that the volume V of the volume of revolution is equal to the product of the distance through which the centroid of the area moves and the area:

$$V = 2\pi\bar{y}A. \tag{7.21}$$

As the area revolves about the x axis, the volume dV generated by an element dA of the area is $dV = 2\pi y\,dA$, where y is the y coordinate of the element dA (Fig. 7.27c). Therefore, the total volume is

$$V = 2\pi \int_A y\,dA. \tag{7.22}$$

From the definition of the y coordinate of the centroid of the area,

$$\bar{y} = \frac{\displaystyle\int_A y\,dA}{\displaystyle\int_A dA},$$

we obtain

$$\int_A y\,dA = \bar{y}A.$$

Substituting this result into Eq. (7.22), we obtain Eq. (7.21).

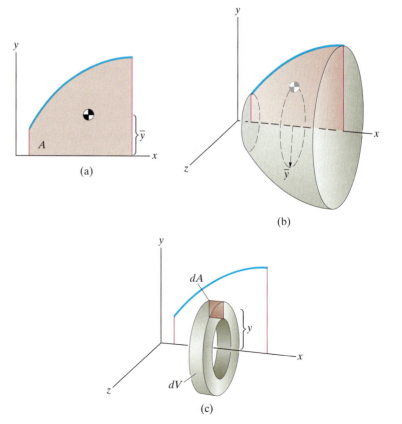

Figure 7.27

(a) An area A and the y coordinate of its centroid.

(b) The volume generated by revolving the area A about the x axis and the path followed by the centroid of the area.

(c) An element dA of the area and the element of volume dV it generates.

Example 7.14

Use the Pappus–Guldinus theorems to determine the surface area A and volume V of the cone in Fig. 7.28.

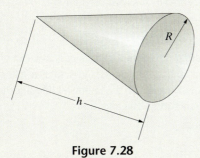

Figure 7.28

Strategy

We can generate the curved surface of the cone by revolving a straight line about an axis, and we can generate its volume by revolving a right triangular area about the axis. Since the centroids of the straight line and the triangular area are known, we can use the Pappus–Guldinus theorems to determine the area and volume of the cone.

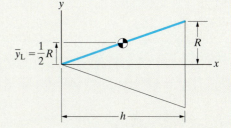

(a) The straight line that generates the curved surface of the cone.

Solution

Revolving the straight line in Fig. a about the x axis generates the curved surface of the cone. The y coordinate of the centroid of the line is $\bar{y}_L = \frac{1}{2}R$, and its length is $L = \sqrt{h^2 + R^2}$. The centroid of the line moves a distance $2\pi\bar{y}_L$ as the line revolves about the x axis, so the area of the curved surface is

$$(2\pi\bar{y})L = \pi R\sqrt{h^2 + R^2}.$$

We obtain the total surface area A of the cone by adding the area of the base,

$$A = \pi R\sqrt{h^2 + R^2} + \pi R^2.$$

Revolving the triangular area in Fig. b about the x axis generates the volume V. The y coordinate of its centroid is $\bar{y}_T = \frac{1}{3}R$, and its area is $A = \frac{1}{2}hR$, so the volume of the cone is

$$V = (2\pi\bar{y}_T)A = \frac{1}{3}\pi hR^2.$$

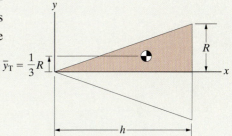

(b) The area that generates the volume of the cone.

Critical Thinking

We chose this example to demonstrate that the Pappus–Guldinus theorems yield the known surface area and volume of a cone, but the same procedure can obviously be used to determine the surface areas and volumes of other volumes of revolution. Or, as we show in the next example, the theorems can be used in the reverse way, determining centroids of lines and areas when the areas and volumes of revolution that they generate are known.

Example 7.15   **Determining Centroids with Pappus–Guldinus Theorems**

The circumference of a sphere of radius R is $2\pi R$, its surface area is $4\pi R^2$, and its volume is $\frac{4}{3}\pi R^3$. Use this information to determine (a) the centroid of a semi-circular line; (b) the centroid of a semicircular area.

Strategy

Revolving a semicircular line about an axis generates a spherical area, and re-volving a semicircular area around an axis generates a spherical volume. Know-ing the area and volume, we can use the Pappus–Guldinus theorems to determine the centroids of the generating line and area.

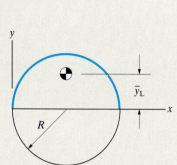

(a) Revolving a semicircular line about the x axis.

Solution

(a) Revolving the semicircular line in Fig. a about the x axis generates the sur-face area of a sphere. The length of the line is $L = \pi R$, and $\bar{y}_L$ is the y coordi-nate of its centroid. The centroid of the line moves a distance $2\pi\bar{y}_L$, so the surface area of the sphere is

$$(2\pi\bar{y}_L)L = 2\pi^2 R\bar{y}_L.$$

By equating this expression to the surface area $4\pi R^2$, we determine $\bar{y}_L$:

$$\bar{y}_L = \frac{2R}{\pi}.$$

(b) Revolving the semicircular area in Fig. b generates the sphere's volume. The area of the semicircle is $A = \frac{1}{2}\pi R^2$, and $\bar{y}_S$ is the y coordinate of its cen-troid. The centroid moves a distance $2\pi\bar{y}_S$, so the volume of the sphere is

$$(2\pi\bar{y}_S)A = \pi^2 R^2\bar{y}_S.$$

Equating this expression to the volume $\frac{4}{3}\pi R^3$, we obtain

$$\bar{y}_S = \frac{4R}{3\pi}.$$

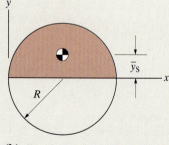

(b) Revolving a semicircular area about the x axis.

Critical Thinking

If you can obtain a result by using the Pappus–Guldinus theorems, you will often save time and effort in comparison with other approaches. Compare this example with Example 7.10, in which we use integration to determine the cen-troid of a semicircular line.

Problems

7.86 Use the first Pappus–Guldinus theorem to determine the area of the curved part of the surface of the truncated cone.

7.87 Use the second Pappus–Guldinus theorem to determine the volume of the truncated cone.

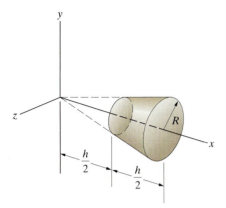

Problems 7.86/7.87

7.88 Use the second Pappus–Guldinus theorem to determine the volume generated by revolving the curve about the x axis.

7.89 Use the second Pappus–Guldinus theorem to determine the volume generated by revolving the curve about the y axis.

7.90 The length of the curve is $L = 1.479$, and the area generated by rotating it about the x axis is $A = 3.810$. Use the first Pappus–Guldinus theorem to determine the y coordinate of the centroid of the curve.

7.91 Use the first Pappus–Guldinus theorem to determine the area of the surface generated by revolving the curve about the y axis.

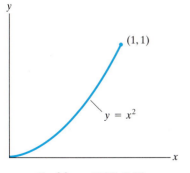

Problems 7.88–7.91

7.92 A nozzle for a large rocket engine is designed by revolving the function $y = \frac{2}{3}(x - 1)^{3/2}$ about the y axis. Use the first Pappus–Guldinus theorem to determine the surface area of the nozzle.

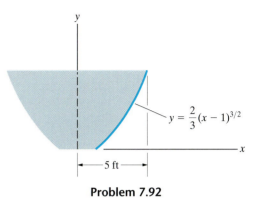

Problem 7.92

7.93 The coordinates of the centroid of the line are $\bar{x} = 332$ mm and $\bar{y} = 118$ mm. Use the first Pappus–Guldinus theorem to determine the area of the surface of revolution obtained by revolving the line about the x axis.

7.94 The coordinates of the centroid of the area between the x axis and the line are $\bar{x} = 355$ mm and $\bar{y} = 78.4$ mm. Use the second Pappus–Guldinus theorem to determine the volume obtained by revolving the area about the x axis.

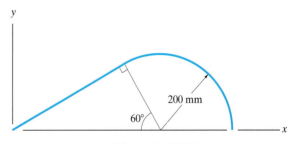

Problems 7.93/7.94

7.95 The volume of revolution contains a hole of radius R.
(a) Use integration to determine its volume.
(b) Use the second Pappus–Guldinus theorem to determine its volume.

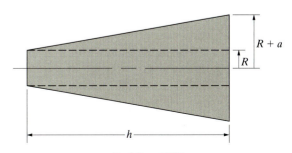

Problem 7.95

7.96 Determine the volume of the volume of revolution.

7.97 Determine the surface area of the volume of revolution.

7.98 The volume of revolution has an elliptical cross section. Determine its volume.

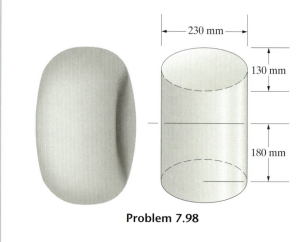

Problem **7.98**

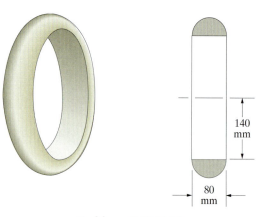

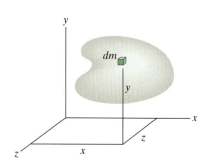

Problems **7.96/7.97**

CENTERS OF MASS

The *center of mass* of an object is the centroid, or average position, of its mass. In the following section we give the analytical definition of the center of mass and demonstrate one of its most important properties: *An object's weight can be represented by a single equivalent force acting at its center of mass.* We then discuss how to locate centers of mass and show that for particular classes of objects, the center of mass coincides with the centroid of a volume, area, or line. Finally, we show how to locate centers of mass of composite objects.

7.6 Definition of the Center of Mass

The center of mass of an object is defined by

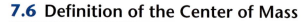

$$\bar{x} = \frac{\int_m x \, dm}{\int_m dm}, \qquad \bar{y} = \frac{\int_m y \, dm}{\int_m dm}, \qquad \bar{z} = \frac{\int_m z \, dm}{\int_m dm}, \qquad (7.23)$$

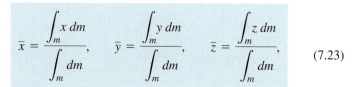

Figure 7.29
An object and differential element of mass *dm*.

where *x*, *y*, and *z* are the coordinates of the differential element of mass *dm* (Fig. 7.29). The subscripts *m* indicate that the integration must be carried out over the entire mass of the object.

Before considering how to determine the center of mass of an object, we will demonstrate that the weight of an object can be represented by a single equivalent force acting at its center of mass. Consider an element of mass dm of an object (Fig. 7.30a). If the y axis of the coordinate system points upward, the weight of dm is $-dmg\,\mathbf{j}$. Integrating this expression over the mass m, we obtain the total weight of the object,

$$\int_m - g\mathbf{j}\,dm = -mg\mathbf{j} = -W\mathbf{j}.$$

The moment of the weight of the element dm about the origin is

$$(x\mathbf{i} + y\mathbf{j} + z\mathbf{k}) \times (-dmg\,\mathbf{j}) = gz\mathbf{i}\,dm - gx\mathbf{k}\,dm.$$

Integrating this expression over m, we obtain the total moment about the origin due to the weight of the object:

$$\int_m (gz\mathbf{i}\,dm - gx\mathbf{k}\,dm) = mg\bar{z}\mathbf{i} - mg\bar{x}\,\mathbf{k} = W\bar{z}\mathbf{i} - W\bar{x}\,\mathbf{k}.$$

If we represent the weight of the object by the force $-W\mathbf{j}$ acting at the center of mass (Fig. 7.30b), the moment of this force about the origin is equal to the total moment due to the weight:

$$(\bar{x}\mathbf{i} + \bar{y}\mathbf{j} + \bar{z}\mathbf{k}) \times (-W\mathbf{j}) = W\bar{z}\mathbf{i} - W\bar{x}\,\mathbf{k}.$$

This result shows that when we are concerned only with the total force and total moment exerted by the weight of an object, we can assume that its weight acts at the center of mass.

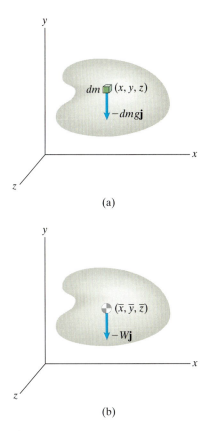

(a)

(b)

Figure 7.30
(a) Weight of the element dm.
(b) Representing the weight by a single force at the center of mass.

7.7 Centers of Mass of Objects

To apply Eqs. (7.23) to specific objects, we will change the variable of integration from mass to volume by introducing the density.

The *density* ρ of an object is defined such that the mass of a differential element dV of the volume of the object is $dm = \rho\,dV$. The dimensions of ρ are therefore (mass/volume). For example, it can be expressed in $\mathrm{kg/m^3}$ in SI units or in $\mathrm{slug/ft^3}$ in U.S. Customary units. The total mass of an object is

$$m = \int_m dm = \int_V \rho\,dV. \tag{7.24}$$

An object whose density is uniform throughout its volume is said to be *homogeneous*. In this case, the total mass equals the product of the density and the volume:

$$m = \rho \int_V dV = \rho V. \qquad \text{Homogeneous object} \tag{7.25}$$

The *weight density* is defined by $\gamma = g\rho$. It can be expressed in $\mathrm{N/m^3}$ in SI units or in $\mathrm{lb/ft^3}$ in U.S. Customary units. The weight of an element of volume dV of an object is $dW = \gamma\,dV$, and the total weight of a homogeneous object equals γV.

By substituting $dm = \rho \, dV$ into Eq. (7.23), we can express the coordinates of the center of mass in terms of volume integrals:

$$\bar{x} = \frac{\int_V \rho x \, dV}{\int_V \rho \, dV}, \qquad \bar{y} = \frac{\int_V \rho y \, dV}{\int_V \rho \, dV}, \qquad \bar{z} = \frac{\int_y \rho z \, dV}{\int_V \rho \, dV}. \qquad (7.26)$$

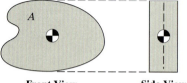

Figure 7.31
A plate of uniform thickness.

If ρ is known as a function of position in an object, these expressions determine its center of mass. Furthermore, we can use these expressions to show that the centers of mass of particular classes of objects coincide with centroids of volumes, areas, and lines:

- **The center of mass of a homogeneous object coincides with the centroid of its volume.** If an object is homogeneous, ρ = constant and Eqs. (7.26) become the equations for the centroid of the volume,

$$\bar{x} = \frac{\int_V x \, dV}{\int_V dV}, \qquad \bar{y} = \frac{\int_V y \, dV}{\int_V dV}, \qquad \bar{z} = \frac{\int_V z \, dV}{\int_V dV}.$$

- **The center of mass of a homogeneous plate of uniform thickness coincides with the centroid of its cross-sectional area** (Fig. 7.31). The center of mass of the plate coincides with the centroid of its volume, and we showed in Section 7.4 that the centroid of the volume of a plate of uniform thickness coincides with the centroid of its cross-sectional area.

- **The center of mass of a homogeneous slender bar of uniform cross-sectional area coincides approximately with the centroid of the axis of the bar** (Fig. 7.32a). The axis of the bar is defined to be the line through the centroid of its cross section. Let $dm = \rho A \, dL$, where A is the cross-sectional area of the bar and dL is a differential element of length of its axis (Fig. 7.32b). If we substitute this expression into Eqs. (7.26), they become the equations for the centroid of the axis:

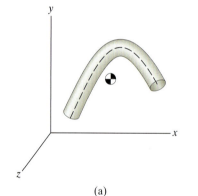

(a)

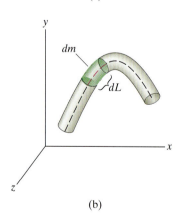

(b)

Figure 7.32
(a) A slender bar and the centroid of its axis.
(b) The element dm.

$$\bar{x} = \frac{\int_L x \, dL}{\int_L dL}, \qquad \bar{y} = \frac{\int_L y \, dL}{\int_L dL}, \qquad \bar{z} = \frac{\int_L z \, dL}{\int_L dL}.$$

This result is approximate because the center of mass of the element dm does not coincide with the centroid of the cross section in regions where the bar is curved.

Study Questions

1. If you want to represent the weight of an object as a single equivalent force, at what point must the force act?

2. How is the density of an object defined?

3. What is the relationship between the density ρ and the weight density γ?

4. If an object is homogeneous, what do you know about the position of its center of mass?

Example 7.16   Representing the Weight of an L-Shaped Bar

The mass of the homogeneous slender bar in Fig. 7.33 is 80 kg. What are the reactions at A and B?

Strategy

We can determine the reactions in two ways.

First Method   We represent the weight of each straight segment of the bar by a force acting at the center of mass of the segment.

Second Method   We determine the center of mass of the bar by determining the centroid of its axis and represent the weight of the bar by a single force acting at the center of mass.

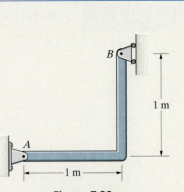

Figure 7.33

Solution

First Method   In the free-body diagram in Fig. a, we place half of the weight of the bar at the center of mass of each straight segment. From the equilibrium equations

$$\Sigma F_x = A_x - B = 0,$$

$$\Sigma F_y = A_y - (40)(9.81)\text{ N} - (40)(9.81)\text{ N} = 0,$$

$$\Sigma M_{\text{point }A} = (1\text{ m})B - (1\text{ m})[(40)(9.81)\text{ N}] - (0.5\text{ m})[(40)(9.81)\text{ N}] = 0,$$

we obtain $A_x = 589$ N, $A_y = 785$ N, and $B = 589$ N.

Second Method   We can treat the centerline of the bar as a composite line composed of two straight segments (Fig. b). The coordinates of the centroid of the composite line are

$$\bar{x} = \frac{\bar{x}_1 L_1 + \bar{x}_2 L_2}{L_1 + L_2} = \frac{(0.5)(1) + (1)(1)}{1 + 1} = 0.75\text{ m},$$

$$\bar{y} = \frac{\bar{y}_1 L_1 + \bar{y}_2 L_2}{L_1 + L_2} = \frac{(0)(1) + (0.5)(1)}{1 + 1} = 0.25\text{ m}.$$

In the free-body diagram in Fig. c, we place the weight of the bar at its center of mass. From the equilibrium equations

$$\Sigma F_x = A_x - B = 0,$$

$$\Sigma F_y = A_y - (80)(9.81)\text{ N} = 0,$$

$$\Sigma M_{\text{point }A} = (1\text{ m})B - (0.75\text{ m})[(80)(9.81)\text{ N}] = 0,$$

we again obtain $A_x = 589$ N, $A_y = 785$ N, and $B = 589$ N.

Critical Thinking

This example demonstrates the importance of this chapter's subject matter. The weight of the bar is distributed over its volume. How could we account for it in determining the reactions at A and B? Our first method was intuitively compelling. The weight of each straight segment of the bar was assumed to act at the midpoint of that segment. In our second method, we assumed that the total weight of the bar acted at its center of mass, which we determined by calculating the position of the centroid of the bar's axis. You should confirm that the total weight in Fig. c is equivalent to the individual weights of the segments in Fig. a. But the essential point is that *both these systems are equivalent to the distributed weight of the bar.*

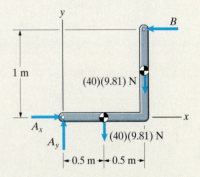

(a) Placing the weights of the straight segments at their centers of mass.

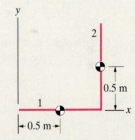

(b) Centroids of the straight segments of the axis.

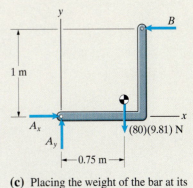

(c) Placing the weight of the bar at its center of mass.

Example 7.17   Cylinder with Nonuniform Density

Determine the mass of the cylinder in Fig. 7.34 and the position of its center of mass if (a) it is homogeneous with density ρ_0; (b) its density is given by the equation $\rho = \rho_0(1 + x/L)$.

Strategy
In (a), the mass of the cylinder is simply the product of its density and its volume and the center of mass is located at the centroid of its volume. In (b), the cylinder is inhomogeneous and we must use Eqs. (7.24) and (7.26) to determine its mass and center of mass.

Solution
(a) The volume of the cylinder is LA, so its mass is $\rho_0 LA$. Since the center of mass is coincident with the centroid of the volume of the cylinder, the coordinates of the center of mass are $\bar{x} = \frac{1}{2}L, \bar{y} = 0, \bar{z} = 0$.

(b) We can determine the mass of the cylinder by using an element of volume dV in the form of a disk of thickness dx (Fig. a). The volume $dV = A\, dx$. The mass of the cylinder is

$$m = \int_v \rho\, dV = \int_0^L \rho_0\left(1 + \frac{x}{L}\right)A\, dx = \frac{3}{2}\rho_0 AL.$$

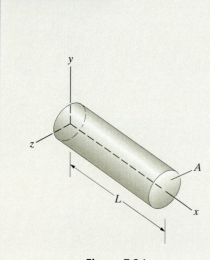

Figure 7.34

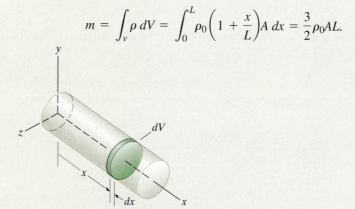

(a) An element of volume dV in the form of a disk.

The x coordinate of the center of mass is

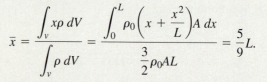

$$\bar{x} = \frac{\displaystyle\int_v x\rho\, dV}{\displaystyle\int_v \rho\, dV} = \frac{\displaystyle\int_0^L \rho_0\left(x + \frac{x^2}{L}\right)A\, dx}{\dfrac{3}{2}\rho_0 AL} = \frac{5}{9}L.$$

Because the density does not depend on y or z, we know from symmetry that $\bar{y} = 0$ and $\bar{z} = 0$.

Critical Thinking
Notice that the center of mass of the inhomogeneous cylinder is *not* located at the centroid of its volume. Its density increases from left to right, so the center of mass is located to the right of the midpoint of the cylinder. Many of the objects we deal with in engineering are not homogeneous, but it is not common for an object's density to vary continuously through its volume as in this example. More often, objects consist of assemblies of parts (composites) that have different densities because they consist of different materials. Frequently the individual parts are approximately homogeneous. We discuss the determination of the centers of mass of such composite objects in the next section.

Problems

7.99 The mass of the homogeneous flat plate is 80 kg. What are the reactions at *A* and *B*?

  Strategy: The center of mass of the plate is coincident with the centroid of its area. Determine the horizontal coordinate of the centroid and assume that the plate's weight acts there.

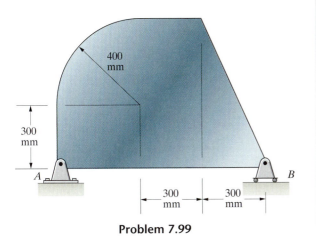

Problem 7.99

7.100 The mass of the homogeneous flat plate is 50 kg. Determine the reactions at the supports *A* and *B*.

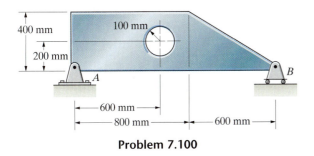

Problem 7.100

7.101 The suspended sign is a homogeneous flat plate that has a mass of 130 kg. Determine the axial forces in members *AD* and *CE*. (Notice that the *y* axis is positive downward.)

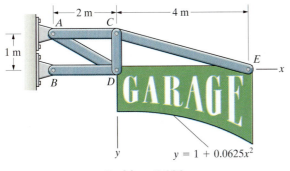

Problem 7.101

7.102 The bar has a mass of 80 kg. What are the reactions at *A* and *B*?

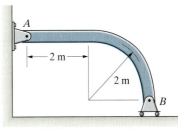

Problem 7.102

7.103 The mass of the bar per unit length is 2 kg/m. Choose the dimension *b* so that part *BC* of the suspended bar is horizontal. What is the dimension *b*, and what are the resulting reactions on the bar at *A*?

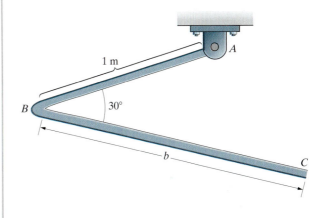

Problem 7.103

7.104 The semicircular part of the homogeneous slender bar lies in the *x*–*z* plane. Determine the center of mass of the bar.

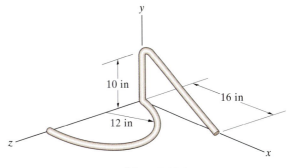

Problem 7.104

7.105 The 10-ft horizontal cylinder with 1-ft radius is supported at A and B. Its weight density is $\gamma = 100(1 - 0.002x^2)$ lb/ft^3. What are the reactions at A and B?

7.106 A horizontal cone with 800-mm length and 200-mm radius has a fixed support at A. Its density is $\rho = 6000(1+0.4x^2)$ kg/m^3, where x is in meters. What are the reactions at A?

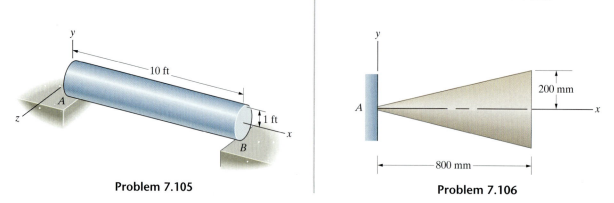

Problem 7.105

Problem 7.106

7.8 Centers of Mass of Composite Objects

The center of mass of an object consisting of a combination of parts can be determined if the centers of mass of its parts are known. The coordinates of the center of mass of a composite object composed of parts with masses $m_1, m_2, \ldots$, are

$$\bar{x} = \frac{\sum\limits_i \bar{x}_i m_i}{\sum\limits_i m_i}, \qquad \bar{y} = \frac{\sum\limits_i \bar{y}_i m_i}{\sum\limits_i m_i}, \qquad \bar{z} = \frac{\sum\limits_i \bar{z}_i m_i}{\sum\limits_i m_i}, \qquad (7.27)$$

where $\bar{x}_i, \bar{y}_i, \bar{z}_i$ are the coordinates of the centers of mass of the parts. Because the weights of the parts are related to their masses by $W_i = g m_i$, Eqs. (7.27) can also be expressed as

$$\bar{x} = \frac{\sum\limits_i \bar{x}_i W_i}{\sum\limits_i W_i}, \qquad \bar{y} = \frac{\sum\limits_i \bar{y}_i W_i}{\sum\limits_i W_i}, \qquad \bar{z} = \frac{\sum\limits_i \bar{z}_i W_i}{\sum\limits_i W_i}. \qquad (7.28)$$

When the masses or weights and the centers of mass of the parts of a composite object are known, these equations determine its center of mass.

Determining the center of mass of a composite object requires three steps:

1. Choose the parts—Try to divide the object into parts whose centers of mass you know or can easily determine.

2. Determine the values for the parts—Determine the center of mass and the mass or weight of each part. Watch for instances of symmetry that can simplify your task.

3. Calculate the center of mass—Use Eqs. (7.27) or (7.28) to determine the center of mass of the composite object.

Example 7.18 Center of Mass of a Composite Object

The L-shaped machine part in Fig. 7.35 is composed of two homogeneous bars. Bar 1 is tungsten alloy with density 14,000 kg/m³, and bar 2 is steel with density 7800 kg/m³. Determine the center of mass of the machine part.

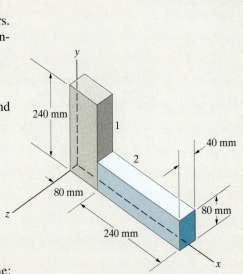

Figure 7.35

Strategy
We can determine the mass and center of mass of each homogeneous bar and use Eqs. (7.27).

Solution
The volume of bar 1 is

$$(80 \text{ mm})(240 \text{ mm})(40 \text{ mm}) = 7.68 \times 10^5 \text{ mm}^3 = 7.68 \times 10^{-4} \text{ m}^3,$$

so its mass is

$$m_1 = (7.68 \times 10^{-4} \text{ m}^3)(1.4 \times 10^4 \text{ kg/m}^3) = 10.75 \text{ kg}.$$

The center of mass of bar 1 coincides with the centroid of its volume: $\bar{x}_1 = 40$ mm, $\bar{y}_1 = 120$ mm, $\bar{z}_1 = 0$. Bar 2 has the same volume as bar 1, so its mass is

$$m_2 = (7.68 \times 10^{-4} \text{ m}^3)(7.8 \times 10^3 \text{ kg/m}^3) = 5.99 \text{ kg}.$$

The coordinates of its center of mass are $\bar{x}_2 = 200$ mm, $\bar{y}_2 = 40$ mm, $\bar{z}_2 = 0$.

Using the information summarized in Table 7.7, we obtain the x coordinate of the center of mass,

$$\bar{x} = \frac{\bar{x}_1 m_1 + \bar{x}_2 m_2}{m_1 + m_2} = \frac{(40 \text{ mm})(10.75 \text{ kg}) + (200 \text{ mm})(5.99 \text{ kg})}{10.75 \text{ kg} + 5.99 \text{ kg}} = 97.2 \text{ mm},$$

and the y coordinate,

$$\bar{y} = \frac{\bar{y}_1 m_1 + \bar{y}_2 m_2}{m_1 + m_2} = \frac{(120 \text{ mm})(10.75 \text{ kg}) + (40 \text{ mm})(5.99 \text{ kg})}{10.75 \text{ kg} + 5.99 \text{ kg}} = 91.4 \text{ mm}.$$

Because of the symmetry of the object, $\bar{z} = 0$.

TABLE 7.7 Information for determining the center of mass

	m_1 (kg)	$\bar{x}_i$ (mm)	$\bar{x}_i m_i$ (mm-kg)	$\bar{y}_i$ (mm)	$\bar{y}_i m_i$ (mm-kg)
Bar 1	10.75	40	(40)(10.75)	120	(120)(10.75)
Bar 2	5.99	200	(200)(5.99)	40	(40)(5.99)

Critical Thinking
This example illustrates the most common procedure for determining centers of mass of objects in engineering. Objects frequently consist of assemblies of parts, each of which is homogeneous. In other cases, individual parts are not homogeneous, but their masses and centers of mass are known (from data supplied by the manufacturers of the parts, for example), and so Eqs. (7.27) can be used.

Example 7.19   Center of Mass of a Composite Object

The composite object in Fig. 7.36 consists of a bar welded to a cylinder. The homogeneous bar is aluminum (weight density 168 lb/ft^3), and the homogeneous cylinder is bronze (weight density 530 lb/ft^3). Determine the center of mass of the object.

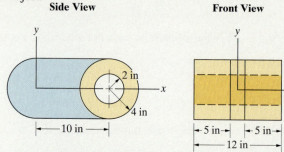

Side View   **Front View**

Figure 7.36

Strategy

We can determine the weight of each homogeneous part by multiplying its volume by its weight density. We also know that the center of mass of each part coincides with the centroid of its volume. The centroid of the cylinder is located at its center, but we must determine the location of the centroid of the bar by treating it as a composite volume.

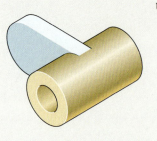

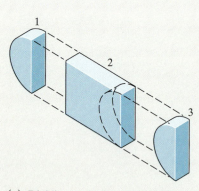

(a) Dividing the bar into three parts.

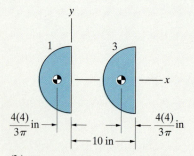

(b) The centroids of the two semicircular parts.

Solution

The volume of the cylinder is

$$V_{\text{cylinder}} = (12 \text{ in})[\pi(4 \text{ in})^2 - \pi(2 \text{ in})^2]$$

$$= 452 \text{ in}^3 = 0.262 \text{ ft}^3,$$

so its weight is

$$W_{\text{cylinder}} = (0.262 \text{ ft}^3)(530 \text{ lb/ft}^3) = 138.8 \text{ lb}.$$

The x coordinate of its center of mass is $\overline{x}_{\text{cylinder}} = 10$ in. The volume of the bar is

$$V_{\text{bar}} = (10 \text{ in})(8 \text{ in})(2 \text{ in}) + \tfrac{1}{2}\pi(4 \text{ in})^2(2 \text{ in}) - \tfrac{1}{2}\pi(4 \text{ in})^2(2 \text{ in})$$

$$= 160 \text{ in}^3 = 0.0926 \text{ ft}^3,$$

and its weight is

$$W_{\text{bar}} = (0.0926 \text{ ft}^3)(168 \text{ lb/ft}^3) = 15.6 \text{ lb}.$$

We can determine the centroid of the volume of the bar by treating it as a composite volume consisting of three parts (Fig. a). Part 3 is a semicircular "cutout." The centroids of part 1 and the semicircular cutout 3 are located at the centroids of their semicircular cross sections (Fig b). Using the information summarized in Table 7.8, we have

$$\overline{x}_{\text{bar}} = \frac{\overline{x}_1 V_1 + \overline{x}_2 V_2 + \overline{x}_3 V_3}{V_1 + V_2 + V_3}$$

$$= \frac{-\dfrac{4(4)}{3\pi}[\tfrac{1}{2}\pi(4)^2(2)] + 5[(10)(8)(2)] - \left[10 - \dfrac{4(4)}{3\pi}\right][\tfrac{1}{2}\pi(4)^2(2)]}{\tfrac{1}{2}\pi(4)^2(2) + (10)(8)(2) - \tfrac{1}{2}\pi(4)^2(2)}$$

$$= 1.86 \text{ in}.$$

TABLE 7.8 Information for determining the x coordinate of the centroid of the bar

	$\bar{x}_i$ (in)	V_i (in^3)	$\bar{x}_i V_i$ (in^4)
Part 1	$-\dfrac{4(4)}{3\pi}$	$\frac{1}{2}\pi(4)^2(2)$	$-\dfrac{4(4)}{3\pi}[\frac{1}{2}\pi(4)^2(2)]$
Part 2	5	$(10)(8)(2)$	$5[(10)(8)(2)]$
Part 3	$10 - \dfrac{4(4)}{3\pi}$	$-\frac{1}{2}\pi(4)^2(2)$	$-\left[10 - \dfrac{4(4)}{3\pi}\right][\frac{1}{2}\pi(4)^2(2)]$

Therefore, the x coordinate of the center of mass of the composite object is

$$\bar{x} = \frac{\bar{x}_{\text{bar}} W_{\text{bar}} + \bar{x}_{\text{cylinder}} W_{\text{cylinder}}}{W_{\text{bar}} + W_{\text{cylinder}}}$$

$$= \frac{(1.86 \text{ in})(15.6 \text{ lb}) + (10 \text{ in})(138.8 \text{ lb})}{15.6 \text{ lb} + 138.8 \text{ lb}}$$

$$= 9.18 \text{ in}.$$

Because of the symmetry of the bar, the y and z coordinates of its center of mass are $\bar{y} = 0$ and $\bar{z} = 0$.

Critical Thinking

The composite object in this example is not homogeneous, which means we could not assume that its center of mass coincides with the centroid of its volume. But the bar and the cylinder are each homogeneous, so we *could* determine their individual centers of mass by finding the centroids of their volumes. The primary challenge in this example was determining the centroid of the volume of the bar with its semicircular end and semicircular cutout.

Design Example 7.20	Centers of Mass of Vehicles

A car is placed on a platform that measures the normal force exerted by each tire independently (Fig. 7.37). Measurements made with the platform horizontal and with the platform tilted at $\alpha = 15°$ are shown in Table 7.9. Determine the position of the car's center of mass.

TABLE 7.9 Measurements of the normal forces exerted by the tires

Wheelbase = 2.82 m		
Track = 1.55 m	**Measured Loads (N)**	
	$\alpha = 0$	$\alpha = 15°$
Left front wheel, N_{LF}	5104	4463
Right front wheel, N_{RF}	5027	4396
Left rear wheel, N_{LR}	3613	3956
Right rear wheel, N_{RR}	3559	3898

Figure 7.37

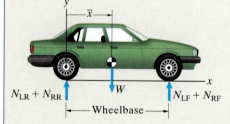

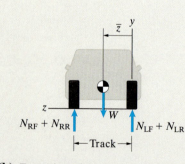

(a) Side view of the free-body diagram with the platform horizontal.

(b) Front view of the free-body diagram with the platform horizontal.

Strategy

The given measurements tell us the normal reactions exerted on the car's tires by the platform. By drawing free-body diagrams of the car in the two positions and applying equilibrium equations, we will obtain equations that can be solved for the unknown coordinates of the car's center of mass.

Solution

We draw the free-body diagram of the car when the platform is in the horizontal position in Figs. a and b. The car's weight is

$$W = N_{LF} + N_{RF} + N_{LR} + N_{RR}$$

$$= 5104 + 5027 + 3613 + 3559$$

$$= 17{,}303 \text{ N}.$$

From Fig. a, we obtain the equilibrium equation

$$\Sigma M_{z \text{ axis}} = (\text{wheelbase})(N_{LF} + N_{RF}) - \bar{x}W = 0,$$

which we can solve for $\bar{x}$:

$$\bar{x} = \frac{(\text{wheelbase})(N_{LF} + N_{RF})}{W}$$

$$= \frac{(2.82 \text{ m})(5104 \text{ N} + 5027 \text{ N})}{17{,}303 \text{ N}}$$

$$= 1.651 \text{ m}.$$

From Fig. b,

$$\Sigma M_{x \text{ axis}} = \bar{z}W - (\text{track})(N_{RF} + N_{RR}) = 0,$$

which we can solve for $\bar{z}$:

$$\bar{z} = \frac{(\text{track})(N_{RF} + N_{RR})}{W}$$

$$= \frac{(1.55 \text{ m})(5027 \text{ N} + 3559 \text{ N})}{17{,}303 \text{ N}}$$

$$= 0.769 \text{ m}.$$

Now that we know $\bar{x}$, we can determine $\bar{y}$ from the free-body diagram of the car when the platform is in the tilted position (Fig. c). From the equilibrium equation

$$\Sigma M_{z\,\text{axis}} = (\text{wheelbase})(N_{\text{LF}} + N_{\text{RF}}) + \bar{y}W \sin 15° - \bar{x}W \cos 15°$$

$$= 0,$$

we obtain

$$\bar{y} = \frac{\bar{x}W \cos 15° - (\text{wheelbase})(N_{\text{LF}} + N_{\text{RF}})}{W \sin 15°}$$

$$= \frac{(1.651 \text{ m})(17{,}303 \text{ N}) \cos 15° - (2.82 \text{ m})(4463 \text{ N} + 4396 \text{ N})}{(17{,}303 \text{ N}) \sin 15°}$$

$$= 0.584 \text{ m}.$$

Notice that we could not have determined $\bar{y}$ without the measurements made with the car in the tilted position.

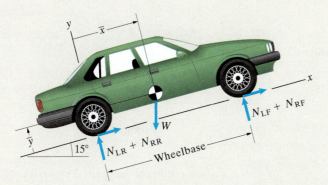

(**c**) Side view of the free-body diagram with the platform tilted.

Design Issues

The location of the center of mass of a vehicle affects its operation and performance. The forces exerted on the suspensions and wheels of cars and train coaches, the tractions their wheels create, and their dynamic behaviors are affected by the locations of their centers of mass. Not only are the performances of airplanes affected by the locations of their centers of mass, they cannot fly unless their centers of mass lie within prescribed bounds. For engineers who design vehicles, the position of the center of mass is one of the principal parameters governing decisions about the configuration of the vehicle and the layout of its contents. In testing new designs of both land vehicles and airplanes, the position of the center of mass is affected by the configuration of the particular vehicle and the weights and locations of stowage and passengers. It is often necessary to locate the center of mass experimentally by a technique such as the one we have described. Such experimental measurements are also used to confirm center of mass locations predicted by calculations made during design.

Problems

7.107 The circular cylinder is made of aluminum (Al) with density 2700 kg/m³ and iron (Fe) with density 7860 kg/m³.

(a) Determine the centroid of the volume of the cylinder.

(b) Determine the center of mass of the cylinder.

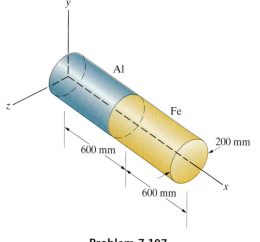

Problem 7.107

7.108 The cylindrical tube is made of aluminum with density 2700 kg/m³. The cylindrical plug is made of steel with density 7800 kg/m³. Determine the coordinates of the center of mass of the composite object.

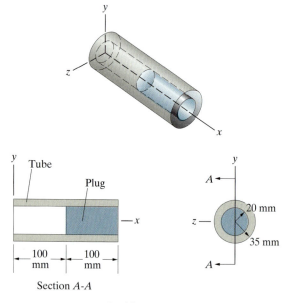

Problem 7.108

7.109 The truncated conical bar is made of bronze with weight density 0.28 lb/in³ and titanium with weight density 0.16 lb/in³. Determine the coordinates of the center of mass of the bar.

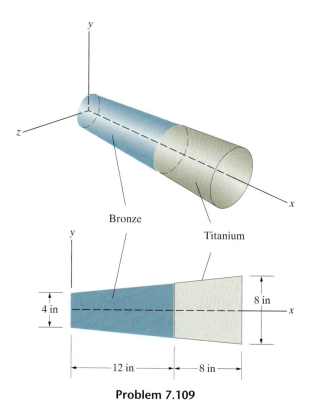

Problem 7.109

7.110 A machine consists of three parts. The masses and the locations of the centers of mass of two of the parts are

Part	Mass (kg)	$\bar{x}$ (mm)	$\bar{y}$ (mm)	$\bar{z}$ (mm)
1	2.0	100	50	−20
2	4.5	150	70	0

The mass of part 3 is 2.5 kg. The design engineer wants to position part 3 so that the center of mass location of the machine is $\bar{x} = 120$ mm, $\bar{y} = 80$ mm, $\bar{z} = 0$. Determine the necessary position of the center of mass of part 3.

7.111 Two views of a machine element are shown. Part 1 is aluminum alloy with density 2800 kg/m³, and part 2 is steel with density 7800 kg/m³. Determine the coordinates of its center of mass.

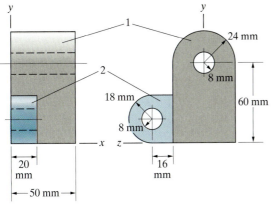

Problem 7.111

7.112 The airplane's total weight is $W = 2400$ lb. The weight of the engine and propeller is 525 lb and their combined center of mass is 6.5 ft to the left of point B. If these items are to be removed for maintenance, will it be necessary to place a support under the airplane's tail to prevent it from falling?

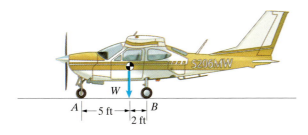

Problem 7.112

7.113 With its engine removed, the mass of the car is 1100 kg and its center of mass is at C. The mass of the engine is 220 kg.

(a) Suppose that you want to place the center of mass E of the engine so that the center of mass of the car is midway between the front wheels A and the rear wheels B. What is the distance b?

(b) If the car is parked on a 15° slope facing up the slope, what total normal force is exerted by the road on the rear wheels B?

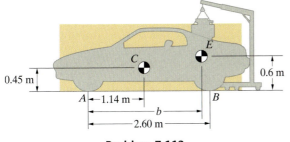

Problem 7.113

7.114 The airplane is parked with its landing gear resting on scales. The weights measured at A, B, and C are 30 kN, 140 kN, and 146 kN, respectively. After a crate is loaded onto the plane, the weights measured at A, B, and C are 31 kN, 142 kN, and 147 kN, respectively. Determine the mass and the x and y coordinates of the center of mass of the crate.

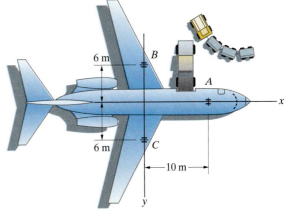

Problem 7.114

Design Experience

Problems 7.115 and 7.116 are related to Design Example 7.20.

7.115 A suitcase with a mass of 90 kg is placed in the trunk of the car described in Example 7.20. The position of the center of mass of the suitcase is $\bar{x}_s = -0.533$ m, $\bar{y}_s = 0.762$ m, $\bar{z}_s = -0.305$ m. If the suitcase is regarded as part of the car, what is the new position of the car's center of mass?

7.116 A group of engineering students constructs a miniature device of the kind described in Example 7.20 and uses it to determine the center of mass of a miniature vehicle. The data they obtain are shown in the following table:

Wheelbase = 36 in		
Track = 30 in	**Measured Loads (lb)**	
	$\alpha = 0$	$\alpha = 10°$
Left front wheel, N_{LF}	35	32
Right front wheel, N_{RF}	36	33
Left rear wheel, N_{LR}	27	34
Right rear wheel, N_{RR}	29	30

Determine the center of mass of the vehicle. Use the same coordinate system as in Design Example 7.20.

CHAPTER SUMMARY

Centroids

A *centroid* is a weighted average position. The coordinates of the centroid of an area A in the x–y plane are

$$\bar{x} = \frac{\int_A x \, dA}{\int_A dA}, \qquad \bar{y} = \frac{\int_A y \, dA}{\int_A dA}. \tag{7.6), (7.7}$$

The coordinates of the centroid of a *composite area* composed of parts $A_1, A_2, \ldots$, are

$$\bar{x} = \frac{\sum_i \bar{x}_i A_i}{\sum_i A_i}, \qquad \bar{y} = \frac{\sum_i \bar{y}_i A_i}{\sum_i A_i}. \tag{7.9}$$

Similar equations define the centroids of volumes [Eqs. (7.15) and (7.17)] and lines [(Eqs. (7.16) and (7.18)].

Distributed Forces

A force distributed along a line is described by a function w, defined such that the force on a differential element dx of the line is $w \, dx$. The force exerted by a distributed load is

$$F = \int_L w \, dx, \tag{7.10}$$

and the moment about the origin is

$$M = \int_L xw \, dx. \tag{7.11}$$

The force F is equal to the "area" between the function w, and the x axis and is equivalent to the distributed load if it is placed at the centroid of the "area."

The Pappus–Guldinus Theorems

First Theorem   Consider a line of length L in the x–y plane with centroid $\bar{x}, \bar{y}$. The area of the surface generated by revolving the line about the x axis is

$$A = 2\pi\bar{y}L. \tag{7.19}$$

Second Theorem   Let A be an area in the x–y plane with centroid $\bar{x}, \bar{y}$. The volume generated by revolving A about the x axis is

$$V = 2\pi\bar{y}A. \tag{7.21}$$

Centers of Mass

The *center of mass* of an object is the centroid of its mass. The weight of an object can be represented by a single equivalent force acting at its center of mass.

The *density* ρ is defined such that the mass of a differential element of volume is $dm = \rho \, dV$. An object whose density is uniform throughout its volume is said to be *homogeneous*. The *weight density* $\gamma = g\rho$.

The coordinates of the center of mass of an object are

$$\bar{x} = \frac{\int_V \rho x \, dV}{\int_V \rho \, dV}, \quad \bar{y} = \frac{\int_V \rho y \, dV}{\int_V \rho \, dV}, \quad \bar{z} = \frac{\int_V \rho z \, dV}{\int_V \rho \, dV}. \qquad (7.26)$$

The center of mass of a homogeneous object coincides with the centroid of its volume. The center of mass of a homogeneous plate of uniform thickness coincides with the centroid of its cross-sectional area. The center of mass of a homogeneous slender bar of uniform cross-sectional area coincides approximately with the centroid of the axis of the bar.

Review Problems

7.117 Determine the centroid of the area by letting dA be a vertical strip of width dx.

7.118 Determine the centroid of the area by letting dA be a horizontal strip of height dy.

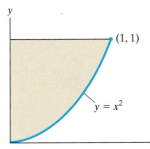

Problems 7.117/7.118

7.119 Determine the centroid of the area.

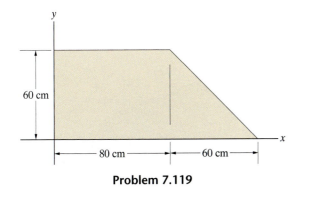

Problem 7.119

7.120 Determine the centroid of the area.

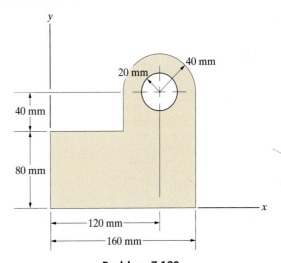

Problem 7.120

7.121 The cantilever beam is subjected to a triangular distributed load. What are the reactions at A?

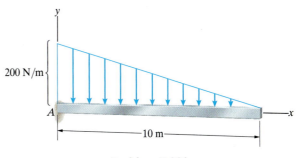

Problem 7.121

7.122 What is the axial load in member *BD* of the frame?

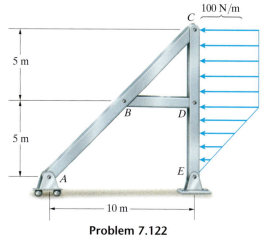

Problem 7.122

7.123 An engineer estimates that the maximum wind load on the 40-m tower in Fig. a is described by the distributed load in Fig. b. The tower is supported by three cables, *A*, *B*, and *C*, from the top of the tower to equally spaced points 15 m from the bottom of the tower (Fig. c). If the wind blows from the west and cables *B* and *C* are slack, what is the tension in cable *A*? (Model the base of the tower as a ball and socket support.)

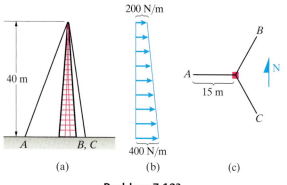

(a) (b) (c)

Problem 7.123

7.124 Determine the reactions on member *ABCD* at *A* and *D*.

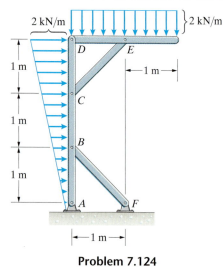

Problem 7.124

7.125 Estimate the centroid of the volume of the *Apollo* lunar return configuration (not including its rocket nozzle) by treating it as a cone and a cylinder.

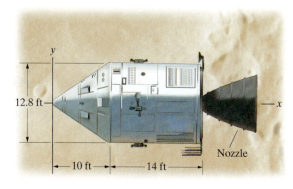

Problem 7.125

7.126 The shape of the rocket nozzle of the *Apollo* lunar return configuration is approximated by revolving the curve shown around the *x* axis. In terms of the coordinate system shown, determine the centroid of the volume of the nozzle.

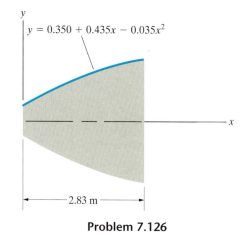

$$y = 0.350 + 0.435x - 0.035x^2$$

Problem 7.126

7.127 Determine the coordinates of the centroid of the volume.

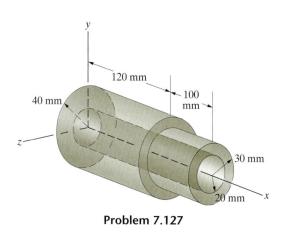

Problem 7.127

7.128 Determine the surface area of the volume of revolution.

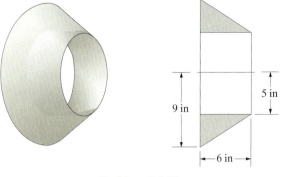

5 in

9 in

6 in

Problem 7.128

7.129 Determine the *y* coordinate of the center of mass of the homogeneous steel plate.

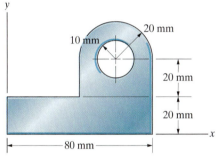

20 mm

10 mm

20 mm

20 mm

80 mm

Problem 7.129

7.130 Determine the *x* coordinate of the center of mass of the homogeneous steel plate.

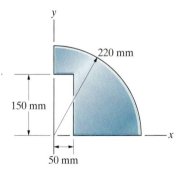

220 mm

150 mm

50 mm

Problem 7.130

7.131 The area of the homogeneous plate is 10 ft². The vertical reactions on the plate at *A* and *B* are 80 lb and 84 lb, respectively. Suppose that you want to equalize the reactions at *A* and *B* by drilling a 1-ft-diameter hole in the plate. What horizontal distance from *A* should the center of the hole be? What are the resulting reactions at *A* and *B*?

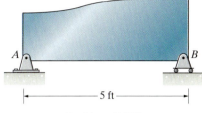

A *B*

5 ft

Problem 7.131

7.132 The plate is of uniform thickness and is made of homogeneous material whose mass per unit area of the plate is 2 kg/m². The vertical reactions at *A* and *B* are 6 N and 10 N, respectively. What is the *x* coordinate of the centroid of the hole?

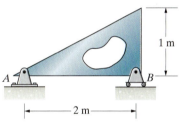

1 m

A *B*

2 m

Problem 7.132

7.133 Determine the center of mass of the homogeneous sheet of metal.

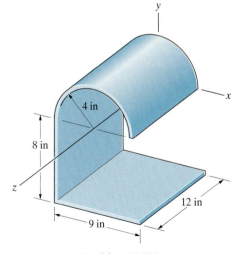

4 in

8 in

9 in

12 in

Problem 7.133

7.134 Determine the center of mass of the homogeneous object.

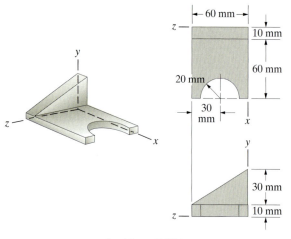

Problem 7.134

7.135 Determine the center of mass of the homogeneous object.

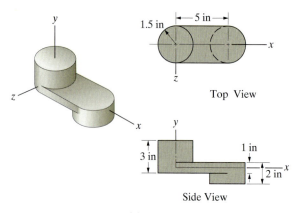

Problem 7.135

7.136 The arrangement shown can be used to determine the location of the center of mass of a person. A horizontal board has a pin support at A and rests on a scale that measures weight at B. The distance from A to B is 2.3 m. When the person is not on the board, the scale at B measures 90 N.

(a) When a 63-kg person is in position (1), the scale at B measures 496 N. What is the x coordinate of the person's center of mass?

(b) When the same person is in position (2), the scale measures 523 N. What is the x coordinate of his center of mass?

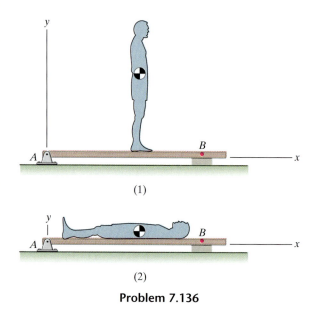

(1)

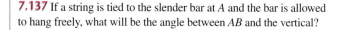

(2)

Problem 7.136

7.137 If a string is tied to the slender bar at A and the bar is allowed to hang freely, what will be the angle between AB and the vertical?

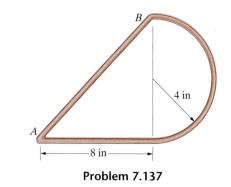

Problem 7.137

7.138 When the truck is unloaded, the total reactions at the front and rear wheels are $A = 54$ kN and $B = 36$ kN. The density of the load of gravel is $\rho = 1600$ kg/m^3. The dimension of the load in the z direction is 3 m, and its surface profile, given by the function shown, does not depend on z. What are the total reactions at the front and rear wheels of the loaded truck?

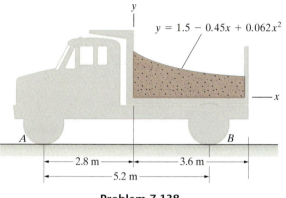

$$y = 1.5 - 0.45x + 0.062x^2$$

Problem 7.138

7.139 The mass of the moon is 0.0123 times the mass of the earth. If the moon's center of mass is 383,000 km from the center of mass of the earth, what is the distance from the center of mass of the earth to the center of mass of the earth–moon system?

Design Project

7.140 Construct a homogeneous thin flat plate with the shape shown in Fig. a. (Use the cardboard back of a pad of paper to construct the plate. Choose your dimensions so that the plate is as large as possible.) Calculate the location of the center of mass of the plate. Measuring as carefully as possible, mark the center of mass clearly on both sides of the plate. Then carry out the following experiments.

(a) Balance the plate on your finger (Fig. b) and observe that it balances at its center of mass. Explain the result of this experiment by drawing a free-body diagram of the plate.

(b) This experiment requires a needle or slender nail, a length of string, and a small weight. Tie the weight to one end of the string and make a small loop at the other end. Stick the needle through the plate at any point other than its center of mass. Hold the needle horizontal so that the plate hangs freely from it (Fig. c). Use the loop to hang the weight from the needle, and let the weight hang freely so that the string lies along the face of the plate. Observe that the string passes through the center of mass of the plate. Repeat this experiment several times, sticking the needle through various points on the plate. Explain the results of this experiment by drawing a free-body diagram of the plate.

(c) Hold the plate so that the plane of the plate is vertical, and throw the plate upward, spinning it like a Frisbee. Observe that the plate spins about its center of mass.

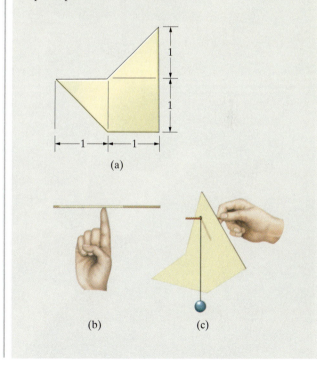

Moments of Inertia

Quantities called moments of inertia arise repeatedly in analyses of engineering problems. Moments of inertia of areas are used in the study of distributed forces and in calculating deflections of beams. The moment exerted by the pressure on a submerged flat plate can be expressed in terms of the moment of inertia of the plate's area. In dynamics, mass moments of inertia are used in calculating the rotational motions of objects. We show how to calculate the moments of inertia of simple areas and objects and then use results called parallel-axis theorems to calculate moments of inertia of more complex areas and objects.

◀ A beam's resistance to bending and ability to support loads depend on a property of its cross section called the moment of inertia. In this chapter we define and calculate moments of inertia of areas.

AREAS

8.1 Definitions

Consider an area A in the x–y plane (Fig. 8.1a). Four moments of inertia of A are defined:

1. Moment of inertia about the x axis:

$$I_x = \int_A y^2 \, dA, \qquad (8.1)$$

where y is the y coordinate of the differential element of area dA (Fig. 8.1b). This moment of inertia is sometimes expressed in terms of the *radius of gyration* about the x axis, k_x, which is defined by

$$I_x = k_x^2 A. \qquad (8.2)$$

2. Moment of inertia about the y axis:

$$I_y = \int_A x^2 \, dA, \qquad (8.3)$$

where x is the x coordinate of the element dA (Fig. 8.1b). The radius of gyration about the y axis, k_y, is defined by

$$I_y = k_y^2 A. \qquad (8.4)$$

3. Product of inertia:

$$I_{xy} = \int_A xy \, dA. \qquad (8.5)$$

4. Polar moment of inertia:

$$J_O = \int_A r^2 \, dA, \qquad (8.6)$$

where r is the radial distance from the origin of the coordinate system to dA (Fig. 8.1b). The radius of gyration about the origin, k_O, is defined by

$$J_O = k_O^2 A. \qquad (8.7)$$

The polar moment of inertia is equal to the sum of the moments of inertia about the x and y axes:

$$J_O = \int_A r^2 \, dA = \int_A (y^2 + x^2) \, dA = I_x + I_y.$$

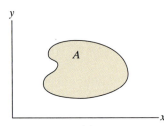

(a)

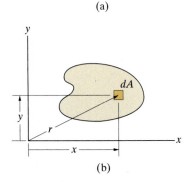

(b)

Figure 8.1
(a) An area A in the x–y plane.
(b) A differential element of A.

Substituting the expressions for the moments of inertia in terms of the radii of gyration into this equation, we obtain

$$k_O^2 = k_x^2 + k_y^2.$$

The dimensions of the moments of inertia of an area are $(\text{length})^4$, and the radii of gyration have dimensions of length. Notice that the definitions of the moments of inertia I_x, I_y, and J_O and the radii of gyration imply that they have positive values for any area. They cannot be negative or zero.

If an area A is symmetric about the x axis, for each element dA with coordinates (x, y), there is a corresponding element dA with coordinates $(x, -y)$, as shown in Fig. 8.2. The contributions of these two elements to the product of inertia I_{xy} of the area cancel: $xy\, dA + (-xy)\, dA = 0$. This means that the product of inertia of the area is zero. The same kind of argument can be used for an area that is symmetric about the y axis. *If an area is symmetric about either the x axis or the y axis, its product of inertia is zero.*

In the examples that follow, we demonstrate that the moments and products of inertia of areas can be evaluated using the same types of integrals with which we determined centroids of areas in Chapter 7.

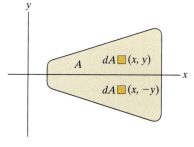

Figure 8.2

Example 8.1	Moments of Inertia of a Triangular Area

Determine I_x, I_y, and I_{xy} for the triangular area in Fig. 8.3.

Strategy
Equation (8.3) for the moment of inertia about the y axis is very similar in form to the equation for the x coordinate of the centroid of an area, and we can evaluate it for this triangular area in exactly the same way: by using a differential element of area dA in the form of a vertical strip of width dx. We then show that I_x and I_{xy} can be evaluated by using the same element of area.

Solution
Let dA be the vertical strip in Fig. a. The equation describing the triangular area's upper boundary is $f(x) = (h/b)x$, so $dA = f(x)\, dx = (h/b)x\, dx$. To integrate over the entire area, we must integrate with respect to x from $x = 0$ to $x = b$.

Moment of Inertia About the y Axis

$$I_y = \int_A x^2\, dA = \int_0^b x^2 f(x)\, dx$$

$$= \int_0^b x^2\left(\frac{h}{b}x\right) dx = \frac{h}{b}\left[\frac{x^4}{4}\right]_0^b = \frac{1}{4}hb^3.$$

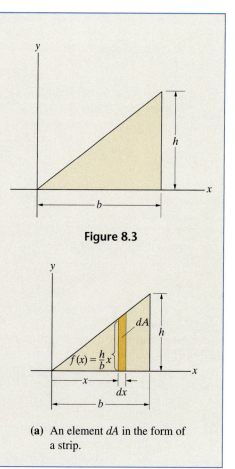

Figure 8.3

(a) An element dA in the form of a strip.

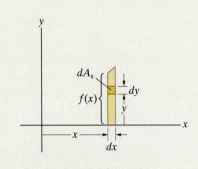

(b) An element of the strip element dA.

Moment of Inertia About the x Axis   We will first determine the moment of inertia of the strip dA about the x axis while holding x and dx fixed. In terms of the element of area $dA_s = dx\,dy$ shown in Fig. b,

$$(I_x)_{\text{strip}} = \int_{\text{strip}} y^2\,dA_s = \int_0^{f(x)} (y^2\,dx)\,dy$$

$$= \left[\frac{y^3}{3}\right]_0^{f(x)} dx \quad \frac{1}{3} = [f(x)]^3\,dx.$$

Integrating this expression with respect to x from $x = 0$ to $x = b$, we obtain the value of I_x for the entire area:

$$I_x = \int_0^b \frac{1}{3}[f(x)]^3\,dx = \int_0^b \frac{1}{3}\left(\frac{h}{b}x\right)^3 dx$$

$$= \frac{h^3}{3b^3}\left[\frac{x^4}{4}\right]_0^b = \frac{1}{12}bh^3.$$

Product of Inertia   We first evaluate the product of inertia of the strip dA, holding x and dx fixed (Fig. b):

$$(I_{xy})_{\text{strip}} = \int_{\text{strip}} xy\,dA_s = \int_0^{f(x)} (xy\,dx)\,dy$$

$$= \left[\frac{y^2}{2}\right]_0^{f(x)} x\,dx = \frac{1}{2}[f(x)]^2\,x\,dx.$$

We integrate this expression with respect to x from $x = 0$ to $x = b$ to obtain the value of I_{xy} for the entire area:

$$I_{xy} = \int_0^b \frac{1}{2}[f(x)]^2 x\,dx = \int_0^b \frac{1}{2}\left(\frac{h}{b}x\right)^2 x\,dx$$

$$= \frac{h^2}{2b^2}\left[\frac{x^4}{4}\right]_0^b = \frac{1}{8}b^2h^2.$$

Critical Thinking

We chose this example so that you can confirm that we obtain the results tabulated for a triangular area in Appendix B. But notice that you can use the same procedure to obtain the moments of inertia of other areas whose boundaries are described by functions of the form $y = f(x)$.

Example 8.2 Moments of Inertia of a Circular Area

Determine the moments of inertia and radii of gyration of the circular area in Fig. 8.4.

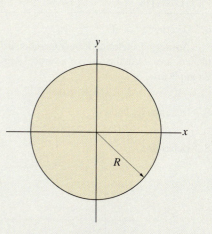

Strategy

We will first determine the polar moment of inertia J_O by integrating in terms of polar coordinates. We know from the symmetry of the area that $I_x = I_y$, and since $I_x + I_y = J_O$, the moments of inertia I_x and I_y are each equal to $\frac{1}{2}J_O$. We also know from the symmetry of the area that $I_{xy} = 0$.

Figure 8.4

Solution

By letting r change by an amount dr, we obtain an annular element of area $dA = 2\pi r\, dr$ (Fig. a). The polar moment of inertia is

$$J_O = \int_A r^2\, dA = \int_0^R 2\pi r^3\, dr = 2\pi \left[\frac{r^4}{4}\right]_0^R = \frac{1}{2}\pi R^4,$$

and the radius of gyration about O is

$$k_O = \sqrt{\frac{J_O}{A}} = \sqrt{\frac{(1/2)\pi R^4}{\pi R^2}} = \frac{1}{\sqrt{2}}R.$$

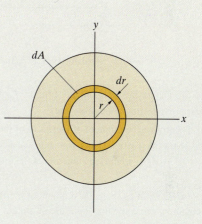

The moments of inertia about the x and y axes are

$$I_x = I_y = \frac{1}{2}J_O = \frac{1}{4}\pi R^4,$$

and the radii of gyration about the x and y axes are

$$k_x = k_y = \sqrt{\frac{I_x}{A}} = \sqrt{\frac{(1/4)\pi R^4}{\pi R^2}} = \frac{1}{2}R.$$

(a) An annular element dA.

The product of inertia is zero:

$$I_{xy} = 0.$$

Critical Thinking

The symmetry of this example saved us from having to integrate to determine I_x, I_y, and I_{xy}. Be alert for symmetry that can shorten your work. In particular, remember that $I_{xy} = 0$ if the area is symmetric about either the x or the y axis.

Problems

8.1 Determine I_y and k_y.

    Strategy: Determine I_y by integration using a differential element of area dA in the form of a vertical strip of width dx. See Example 8.1.

8.2 Determine I_x and k_x.

8.3 Determine I_{xy}.

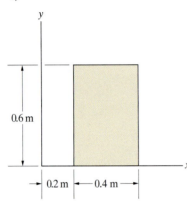

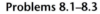

Problems 8.1–8.3

8.4 (a) Determine the moment of inertia I_y of the beam's rectangular cross section about the y axis.

(b) Determine the moment of inertia $I_{y'}$ of the beam's cross section about the y' axes. Using your numerical values, show that $I_y = I_{y'} + d_x^2 A$, where A is the area of the cross section.

8.5 (a) Determine the polar moment of inertia J_O of the beam's rectangular cross section about the origin O.

(b) Determine the polar moment of inertia $J_{O'}$ of the beam's cross section about the origin O'. Using your numerical values, show that $J_O = J_{O'} + (d_x^2 + d_y^2)A$, where A is the area of the cross section.

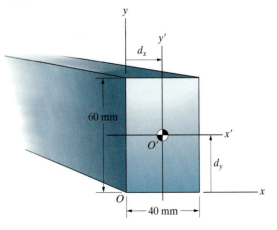

Problems 8.4/8.5

8.6 Determine I_y and k_y.

8.7 Determine J_O and k_O.

8.8 Determine I_{xy}.

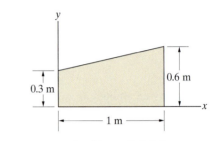

Problems 8.6–8.8

8.9 Determine I_y.

8.10 Determine I_x.

8.11 Determine J_O.

8.12 Determine I_{xy}.

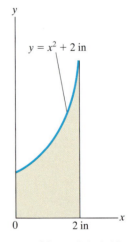

Problems 8.9–8.12

8.13 Determine I_y and k_y.

8.14 Determine I_x and k_x.

8.15 Determine J_O and k_O.

8.16 Determine I_{xy}.

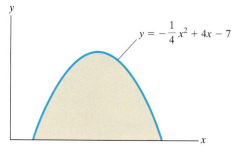

$$y = -\frac{1}{4}x^2 + 4x - 7$$

Problems 8.13–8.16

8.17 Determine I_y and k_y.

8.18 Determine I_x and k_x.

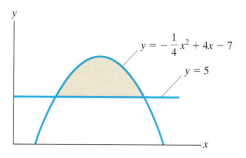

$$y = -\frac{1}{4}x^2 + 4x - 7$$

$y = 5$

Problems 8.17/8.18

8.19 (a) Determine I_y and k_y by letting dA be a vertical strip of width dx.

(b) The polar moment of inertia of a circular area with its center at the origin is $J_O = \frac{1}{2}\pi R^4$. Explain how you can use this information to confirm your answer to (a).

8.20 (a) Determine I_x and k_x by letting dA be a horizontal strip of height dy.

(b) The polar moment of inertia of a circular area with its center at the origin is $J_O = \frac{1}{2}\pi R^4$. Explain how you can use this information to confirm your answer to (a).

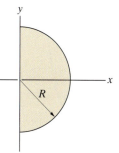

R

Problems 8.19/8.20

8.21 Determine the moments of inertia I_x and I_y.

Strategy: Use the procedure described in Example 8.2 to determine J_O, then use the symmetry of the area to determine I_x and I_y.

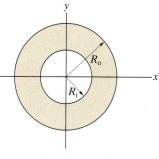

R_o

R_i

Problem 8.21

8.22 What are the values of I_y and k_y for the elliptical area of the airplane's wing?

8.23 What are the values of I_x and k_x for the elliptical area of the airplane's wing?

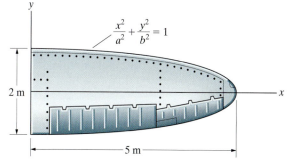

$$\frac{x^2}{a^2} + \frac{y^2}{b^2} = 1$$

2 m

5 m

Problems 8.22/8.23

8.24 Determine I_y and k_y.

8.25 Determine I_x and k_x.

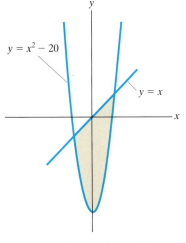

$y = x^2 - 20$

$y = x$

Problems 8.24/8.25

8.26 A vertical plate of area A is beneath the surface of a stationary body of water. The pressure of the water subjects each element dA of the surface of the plate to a force $(p_O + \gamma y)\, dA$, where p_O is the pressure at the surface of the water and γ is the weight density of the water. Show that the magnitude of the moment about the x axis due to the pressure on the front face of the plate is

$$M_{x\text{-axis}} = p_O \bar{y} A + \gamma I_x,$$

where $\bar{y}$ is the y coordinate of the centroid of A and I_x is the moment of inertia of A about the x axis.

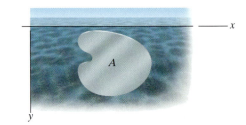

A

Problem 8.26

8.2 Parallel-Axis Theorems

The values of the moments of inertia of an area depend on the position of the coordinate system relative to the area.

In some situations the moments of inertia of an area are known in terms of a particular coordinate system but we need their values in terms of a different coordinate system. When the coordinate systems are parallel, the desired moments of inertia can be obtained by using the theorems we describe in this section. Furthermore, these theorems make it possible for us to determine the moments of inertia of a composite area when the moments of inertia of its parts are known.

Suppose that we know the moments of inertia of an area A in terms of a coordinate system $x'y'$ with its origin at the centroid of the area, and we wish to determine the moments of inertia in terms of a parallel coordinate system xy (Fig. 8.5a). We denote the coordinates of the centroid of A in the xy coordinate system by (d_x, d_y), and $d = \sqrt{d_x^2 + d_y^2}$ is the distance from the origin of the xy coordinate system to the centroid (Fig. 8.5b).

Figure 8.5
(a) The area A and the coordinate systems $x'y'$ and xy.
(b) The differential element dA.

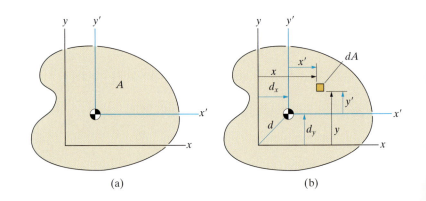

(a)   (b)

We need two preliminary results before deriving the parallel-axis theorems. In terms of the $x'y'$ coordinate system, the coordinates of the centroid of A are

$$\bar{x}' = \frac{\displaystyle\int_A x'\,dA}{\displaystyle\int_A dA}, \qquad \bar{y}' = \frac{\displaystyle\int_A y'\,dA}{\displaystyle\int_A dA}.$$

But the origin of the $x'y'$ coordinate system is located at the centroid of A, so $\bar{x}' = 0$ and $\bar{y}' = 0$. Therefore,

$$\int_A x'\,dA = 0, \qquad \int_A y'\,dA = 0. \qquad (8.8)$$

Moment of Inertia About the x Axis   In terms of the xy coordinate system, the moment of inertia of A about the x axis is

$$I_x = \int_A y^2\,dA, \qquad (8.9)$$

where y is the coordinate of the element of area dA relative to the xy coordinate system. From Fig. 8.5b, we see that $y = y' + d_y$, where y' is the coordinate of dA relative to the $x'y'$ coordinate system. Substituting this expression into Eq. (8.9), we obtain

$$I_x = \int_A (y' + d_y)^2\,dA = \int_A (y')^2\,dA + 2d_y\int_A y'\,dA + d_y^2\int_A dA.$$

The first integral on the right is the moment of inertia of A about the x'axis. From Eq. (8.8), the second integral on the right equals zero. Therefore, we obtain

$$I_x = I_{x'} + d_y^2 A. \qquad (8.10)$$

This is a *parallel-axis theorem*. It relates the moment of inertia of A about the x'axis through the centroid to the moment of inertia about the parallel axis x (Fig. 8.6).

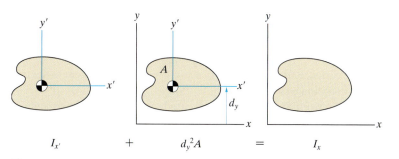

$$I_{x'} \qquad + \qquad d_y^2 A \qquad = \qquad I_x$$

Figure 8.6
The parallel-axis theorem for the moment of inertia about the x axis.

Moment of Inertia About the y Axis   In terms of the xy coordinate system, the moment of inertia of A about the y axis is

$$I_y = \int_A x^2\, dA = \int_A (x' + d_x)^2\, dA$$

$$= \int_A (x')^2\, dA + 2d_x \int_A x'\, dA + d_x^2 \int_A dA.$$

From Eq. (8.8), the second integral on the right equals zero. Therefore, the parallel-axis theorem that relates the moment of inertia of A about the y'axis through the centroid to the moment of inertia about the parallel axis y is

$$I_y = I_{y'} + d_x^2 A. \tag{8.11}$$

Product of Inertia   In terms of the xy coordinate system, the product of inertia is

$$I_{xy} = \int_A xy\, dA = \int_A (x' + d_x)(y' + d_y)\, dA$$

$$= \int_A x'y'\, dA + d_y \int_A x'\, dA + d_x \int_A y'\, dA + d_x d_y \int_A dA.$$

The second and third integrals equal zero from Eq. (8.8). We see that the parallel-axis theorem for the product of inertia is

$$I_{xy} = I_{x'y'} + d_x d_y A. \tag{8.12}$$

Polar Moment of Inertia   The polar moment of inertia $J_O = I_x + I_y$. Summing Eqs. (8.10) and (8.11), the parallel axis theorem for the polar moment of inertia is

$$J_O = J_O' + (d_x^2 + d_y^2)A = J_O' + d^2 A, \tag{8.13}$$

where d is the distance from the origin of the $x'y'$ coordinate system to the origin of the xy coordinate system.

How can the parallel-axis theorems be used to determine the moments of inertia of a composite area? Suppose that we want to determine the moment of inertia about the y axis of the area in Fig. 8.7a. We can divide it into a triangle, a semicircle, and a circular cutout, denoted as parts 1, 2, and 3 (Fig. 8.7b). By using the parallel-axis theorem for I_y, we can determine the moment of inertia of each part about the y axis. For example, the moment of inertia of part 2 (the semicircle) about the y axis is (Fig. 8.7c)

$$(I_y)_2 = (I_{y'})_2 + (d_x)_2^2 A_2.$$

We must determine the values of $(I_{y'})_2$ and $(d_x)_2$. Moments of inertia and centroid locations for some simple areas are tabulated in Appendix B. Once this procedure is carried out for each part, the moment of inertia of the composite area is

$$I_y = (I_y)_1 + (I_y)_2 - (I_y)_3.$$

Notice that the moment of inertia of the circular cutout is subtracted.

We see that determining a moment of inertia of a composite area in terms of a given coordinate system involves three steps:

1. **Choose the parts**—Try to divide the composite area into parts whose moments of inertia you know or can easily determine.

2. **Determine the moments of inertia of the parts**—Determine the moment of inertia of each part in terms of a parallel coordinate system with its origin at the centroid of the part, and then use the parallel-axis theorem to determine the moment of inertia in terms of the given coordinate system.

3. **Sum the results**—Sum the moments of inertia of the parts (or subtract in the case of a cutout) to obtain the moment of inertia of the composite area.

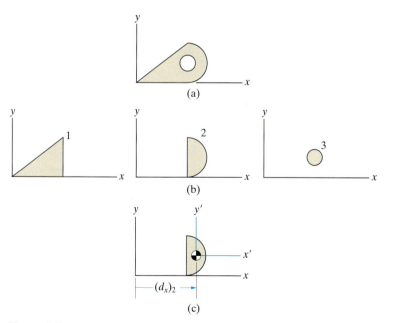

Figure 8.7
(a) A composite area.
(b) The three parts of the area.
(c) Determining $(I_y)_2$.

Example 8.3   Demonstration of the Parallel-Axis Theorems

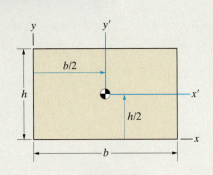

Figure 8.8

The moments of inertia of the rectangular area in Fig. 8.8 in terms of the $x'y'$ coordinate system are $I_{x'} = \frac{1}{12}bh^3$, $I_{y'} = \frac{1}{12}hb^3$, $I_{x'y'} = 0$, and $J'_O = \frac{1}{12}(bh^3 + hb^3)$. (See Appendix B.) Determine its moments of inertia in terms of the xy coordinate system.

Strategy
The $x'y'$ coordinate system has its origin at the centroid of the area and is parallel to the xy coordinate system. We can use the parallel-axis theorems to determine the moments of inertia of A in terms of the xy coordinate system.

Solution
The coordinates of the centroid in terms of the xy coordinate system are $d_x = b/2$, $d_y = h/2$. The moment of inertia about the x axis is

$$I_x = I_{x'} + d_y^2 A = \frac{1}{12}bh^3 + \left(\frac{1}{2}h\right)^2 bh = \frac{1}{3}bh^3.$$

The moment of inertia about the y axis is

$$I_y = I_{y'} + d_x^2 A = \frac{1}{12}hb^3 + \left(\frac{1}{2}b\right)^2 bh = \frac{1}{3}hb^3.$$

The product of inertia is

$$I_{xy} = I_{x'y'} + d_x d_y A = 0 + \left(\frac{1}{2}b\right)\left(\frac{1}{2}h\right)bh = \frac{1}{4}b^2h^2.$$

The polar moment of inertia is

$$J_O = J'_O + d^2 A = \frac{1}{12}(bh^3 + hb^3) + \left[\left(\frac{1}{2}b\right)^2 + \left(\frac{1}{2}h\right)^2\right]bh$$

$$= \frac{1}{3}(bh^3 + hb^3).$$

Critical Thinking
Notice that we could also have determined J_O by using the relation

$$J_O = I_x + I_y = \frac{1}{3}bh^3 + \frac{1}{3}hb^3.$$

We designed this example so that you can confirm that the parallel-axis theorems yield the results in Appendix B for a rectangular area. But the same procedure can be used to obtain the moments of inertia of the area in terms of any coordinate system that is parallel to the $x'y'$ system.

Example 8.4   Moments of Inertia of a Composite Area

Determine I_x, k_x, and I_{xy} for the composite area in Fig. 8.9.

Strategy

We can divide this area into two rectangles. We must use the parallel-axis theorem to determine I_x and I_{xy} for each rectangle in terms of the xy coordinate system and sum the results for the rectangles to determine I_x and I_{xy} for the composite area. Then we can use Eq. (8.2) to determine the radius of gyration k_x for the composite area.

Solution

Choose the Parts   We can determine the moments of inertia by dividing the area into the rectangular parts 1 and 2 shown in Fig. a.

Determine the Moments of Inertia of the Parts   For each part, we introduce a coordinate system $x'y'$ with its origin at the centroid of the part (Fig. b). The moments of inertia of the rectangular parts in terms of these coordinate systems are given in Appendix B. We then use the parallel-axis theorem to determine the moment of inertia of each part about the x axis (Table 8.1).

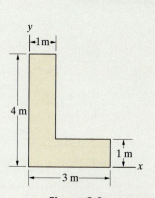

Figure 8.9

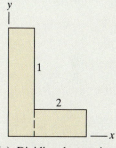

(a) Dividing the area into rectangles 1 and 2.

TABLE 8.1 Determining the moments of inertia of the parts about the x axis.

	d_y (m)	A (m²)	$I_{x'}$ (m⁴)	$I_x = I_{x'} + d_y^2 A$ (m⁴)
Part 1	2	(1)(4)	$\frac{1}{12}(1)(4)^3$	21.33
Part 2	0.5	(2)(1)	$\frac{1}{12}(2)(1)^3$	0.67

Sum the Results   The moment of inertia of the composite area about the x axis is

$$I_x = (I_x)_1 + (I_x)_2 = 21.33 \text{ m}^4 + 0.67 \text{ m}^4 = 22.00 \text{ m}^4.$$

The sum of the areas is $A = A_1 + A_2 = 6 \text{ m}^2$, so the radius of gyration about the x axis is

$$k = \sqrt{\frac{I_x}{A}} = \sqrt{\frac{22 \text{ m}^4}{6 \text{ m}^2}} = 1.91 \text{ m}.$$

Repeating this procedure, we determine I_{xy} for each part in Table 8.2. The product of inertia of the composite area is

$$I_{xy} = (I_{xy})_1 + (I_{xy})_2 = 4 \text{ m}^4 + 2 \text{ m}^4 = 6 \text{ m}^4.$$

TABLE 8.2 Determining the products of inertia of the parts in terms of the xy coordinate system

	d_x (m)	d_y (m)	A (m²)	$I_{x'y'}$	$I_{xy} = I_{x'y'} + d_x d_y A$ (m⁴)
Part 1	0.5	2	(1)(4)	0	4
Part 2	2	0.5	(2)(1)	0	2

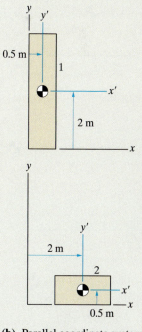

(b) Parallel coordinate systems $x'y'$ with origins at the centroids of the parts.

Critical Thinking

The moments of inertia you obtain do not depend on how you divide a composite area into parts, and you will often have a choice of convenient ways to divide a given area. See Problem 8.28, in which we divide the composite area in this example in a different way.

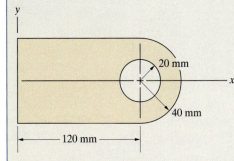

y

20 mm

40 mm

x

120 mm

Figure 8.10

| **Example 8.5** | **Moments of Inertia of a Composite Area** |

Determine I_y and k_y for the composite area in Fig. 8.10.

Strategy
We can divide this area into a rectangle *without the semicircular cutout*, a semicircle *without the semicircular cutout*, and a circular cutout. We can use a parallel-axis theorem to determine I_y for each part in terms of the xy coordinate system. Then, by adding the values for the rectangle and semicircle and subtracting the value for the circular cutout, we can determine I_y for the composite area. Then we can use Eq. (8.4) to determine the radius of gyration k_y for the composite area.

Solution

Choose the Parts   We divide the area into a rectangle, a semicircle, and the circular cutout, calling them parts 1, 2, and 3, respectively (Fig. a).

Determine the Moments of Inertia of the Parts   The moments of inertia of the parts in terms of the $x'y'$ coordinate systems and the location of the centroid of the semicircular part are given in Appendix B. In Table 8.3 we use the parallel-axis theorem to determine the moment of inertia of each part about the y axis.

TABLE 8.3 Determining the moments of inertia of the parts.

	d_x (mm)	A (mm^2)	$I_{y'}$ (mm^4)	$I_y = I_{y'} + d_x^2 A$ (mm^4)
Part 1	60	$(120)(80)$	$\frac{1}{12}(80)(120)^3$	4.608×10^7
Part 2	$120 + \dfrac{4(40)}{3\pi}$	$\frac{1}{2}\pi(40)^2$	$\left(\dfrac{\pi}{8} - \dfrac{8}{9\pi}\right)(40)^4$	4.744×10^7
Part 3	120	$\pi(20)^2$	$\frac{1}{4}\pi(20)^4$	1.822×10^7

Sum the Results   The moment of inertia of the composite area about the y axis is

$$I_y = (I_y)_1 + (I_y)_2 - (I_y)_3 = (4.608 + 4.744 - 1.822) \times 10^7 \text{ mm}^4$$
$$= 7.530 \times 10^7 \text{ mm}^4.$$

The total area is

$$A = A_1 + A_2 - A_3 = (120\text{mm})(80\text{mm}) + \frac{1}{2}\pi(40\text{mm})^2 - \pi(20\text{mm})^2$$
$$= 1.086 \times 10^4 \text{ mm}^2,$$

so the radius of gyration about the y axis is

$$k_y = \sqrt{\frac{I_y}{A}} = \sqrt{\frac{7.530 \times 10^7 \text{ mm}^4}{1.086 \times 10^4 \text{ mm}^2}} = 83.3 \text{ mm}.$$

Critical Thinking
Integration is an additive process, which is why the moments of inertia of composite areas can be determined by adding (or, in the case of a cutout, subtracting) the moments of inertia of the parts. But you can't determine the radii of gyration of composite areas by adding or subtracting the radii of gyration of the parts. This can be seen from the equations relating the moments of inertia, radii of gyration, and area. For this example, we can demonstrate it numerically. The operation

$$(k_y)_1 + (k_y)_2 - (k_y)_3 = \sqrt{\frac{(I_y)_1}{A_1}} + \sqrt{\frac{(I_y)_2}{A_2}} - \sqrt{\frac{(I_y)_3}{A_3}} = 86.3 \text{ mm}$$

does not yield the correct radius of gyration of the composite area.

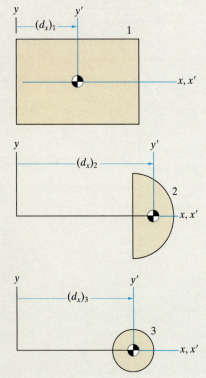

y   y'

$(d_x)_1$ 1

x, x'

y   y'

$(d_x)_2$

2

x, x'

y   y'

$(d_x)_3$

3

x, x'

(a) Parts 1, 2, and 3.

Design Example 8.6   Beam Design

The equal areas in Fig. 8.11 are candidates for the cross section of a beam. (A beam with the second cross section shown is called an I-beam.) Compare their moments of inertia about the x axis.

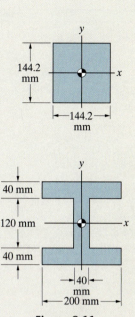

Strategy
We can obtain the moment of inertia of the square cross section from Appendix B. We will divide the I-beam cross section into three rectangles and use the parallel-axis theorem to determine its moment of inertia by the same procedure used in Examples 8.4 and 8.5.

Solution

Square Cross Section   From Appendix B, the moment of inertia of the square cross section about the x axis is

$$I_x = \frac{1}{12}(144.2 \text{ mm})(144.2 \text{ mm})^3 = 3.60 \times 10^7 \text{ mm}^4.$$

I-Beam Cross Section   We can divide the area into the rectangular parts shown in Fig. a. Introducing coordinate systems $x'y'$ with their origins at the centroids of the parts (Fig. b), we use the parallel-axis theorem to determine the moments of inertia about the x axis (Table 8.4). Their sum is

$$I_x = (I_x)_1 + (I_x)_2 + (I_x)_3 = (5.23 + 0.58 + 5.23) \times 10^7 \text{ mm}^4$$

$$= 11.03 \times 10^7 \text{ mm}^4.$$

Figure 8.11

The moment of inertia of the I-beam about the x axis is 3.06 times that of the square cross section of equal area.

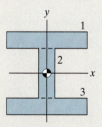

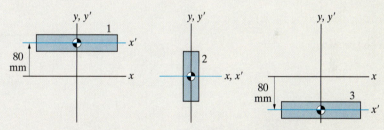

(a) Dividing the I-beam cross section into parts.

(b) Parallel coordinate systems $x'y'$ with origins at the centroids of the parts.

TABLE 8.4 Determining the moments of inertia of the parts about the x axis.

	d_y (mm)	A (mm^2)	$I_{x'}$ (mm^4)	$I_x = I_{x'} + d_y^2 A$ (mm^4)
Part 1	80	(200)(40)	$\frac{1}{12}(200)(40)^3$	5.23×10^7
Part 2	0	(40)(120)	$\frac{1}{12}(40)(120)^3$	0.58×10^7
Part 3	−80	(200)(40)	$\frac{1}{12}(200)(40)^3$	5.23×10^7

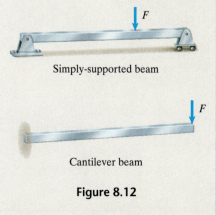

Simply-supported beam

Cantilever beam

Figure 8.12

Design Issues
A *beam* is a bar of material that supports lateral loads, meaning loads perpendicular to the axis of the bar. Two common types of beams are shown in Fig. 8.12 supporting a lateral load F. A beam with pinned ends is called a *simply supported* beam, and a beam with a single, built-in support is called a *cantilever beam*.

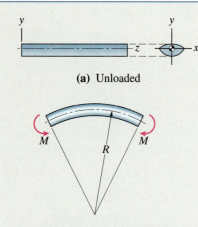

(a) Unloaded

(b) Subjected to couples at the ends.

Figure 8.13
A beam with symmetrical cross section.

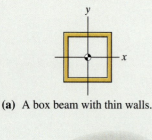

(a) A box beam with thin walls.

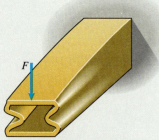

(b) Failure by buckling.

(c) Stabilizing the walls with a filler.

Figure 8.15

The lateral loads on a beam cause it to bend, and it must be stiff, or resistant to bending, to support them. It is shown in mechanics of materials that a beam's resistance to bending depends directly on the moment of inertia of its cross-sectional area. Consider the beam in Fig. 8.13a. The cross section is symmetric about the y axis and the origin of the coordinate system is placed at its centroid. If the beam consists of a homogeneous structural material such as steel and it is subjected to couples at the ends, as shown in Fig. 8.13b, it bends into a circular arc of radius

$$R = \frac{EI_x}{M},$$

where I_x is the moment of inertia of the beam cross section about the x axis. The "modulus of elasticity" E has different values for different materials. (This equation holds only if M is small enough so that the beam returns to its original shape when the couples are removed. The bending in Fig. 8.13b is exaggerated.) Thus, the amount the beam bends for a given value of M depends on the material and the moment of inertia of its cross section. Increasing I_x increases the value of R, which means the resistance of the beam to bending is increased.

This explains in large part the cross sections of many of the beams you see in use—for example, in highway overpasses and in the frames of buildings. They are configured to increase their moments of inertia. The cross sections in Fig. 8.14 all have the same area. The numbers are the ratios of the moment of inertia I_x to the value of I_x for the solid square cross section.

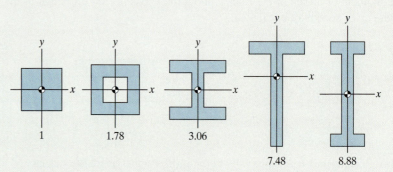

Figure 8.14
Typical beam cross sections and the ratio of I_x to the value for a solid square beam of equal cross-sectional area.

However, configuring the cross section of a beam to increase its moment of inertia can be carried too far. The "box" beam in Fig. 8.15a has a value of I_x that is four times as large as a solid square beam of the same cross-sectional area, but its walls are so thin they may "buckle," as shown in Fig. 8.15b. The stiffness implied by the beam's large moment of inertia is not realized because it becomes geometrically unstable. One solution used by engineers to achieve a large moment of inertia in a relatively light beam while avoiding failure due to buckling is to stabilize its walls by filling the beam with a light material such as honeycombed metal or foamed plastic (Fig. 8.15c).

Problems

8.27 Determine I_x and k_x for the composite area by dividing it into rectangles 1 and 2 as shown, and compare your results to those of Example 8.4.

8.28 Determine I_y and k_y for the composite area.

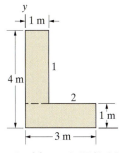
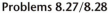

Problems 8.27/8.28

8.29 Determine I_x and k_x.

8.30 Determine I_y and k_y.

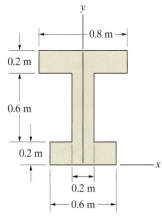

Problems 8.29/8.30

8.31 Determine I_x and k_x.

8.32 Determine I_y and k_y.

8.33 Determine J_O and k_O.

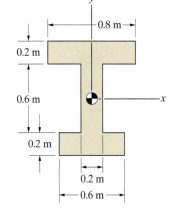

Problems 8.31–8.33

8.34 If you design the beam cross section so that $I_x = 6.4 \times 10^5$ mm^4, what are the resulting values of I_y and J_O?

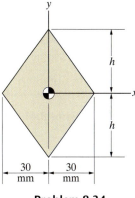

Problem 8.34

8.35 Determine I_y and k_y.

8.36 Determine I_x and k_x.

8.37 Determine I_{xy}.

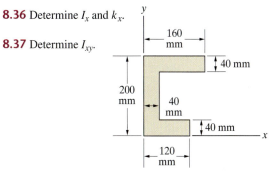

Problems 8.35–8.37

8.38 Determine I_x and k_x.

8.39 Determine I_y and k_y.

8.40 Determine I_{xy}.

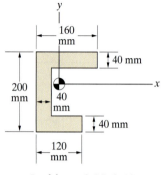

Problems 8.38–8.40

8.41 Determine I_x and k_x.

8.42 Determine J_O and k_O.

8.43 Determine I_{xy}.

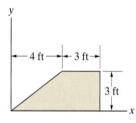

Problems 8.41–8.43

8.44 Determine I_x and k_x.

8.45 Determine J_O and k_O.

8.46 Determine I_{xy}.

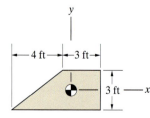

Problems 8.44–8.46

8.47 Determine I_x and k_x.

8.48 Determine J_O and k_O.

8.49 Determine I_{xy}.

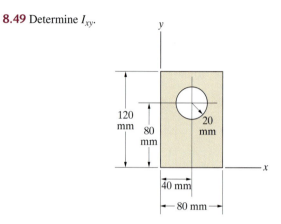

Problems 8.47–8.49

8.50 Determine I_x and k_x.

8.51 Determine I_y and k_y.

8.52 Determine J_O and k_O.

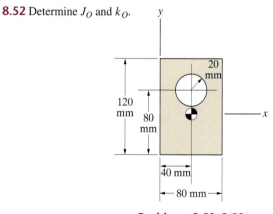

Problems 8.50–8.52

8.53 Determine I_y and k_y.

8.54 Determine J_O and k_O.

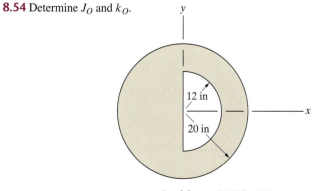

Problems 8.53/8.54

8.55 Determine I_y and k_y if $h = 3$ m.

8.56 Determine I_x and k_x if $h = 3$ m.

8.57 If $I_y = 5$ m^4, what is the dimension h?

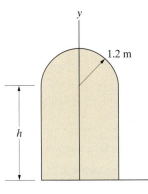

1.2 m

y

h

x

Problems 8.55–8.57

8.58 Determine I_y and k_y.

8.59 Determine I_x and k_x.

8.60 Determine I_{xy}.

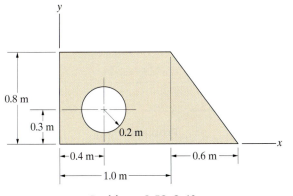

y

0.8 m

0.3 m

0.2 m

←0.4 m→

←0.6 m→

1.0 m

x

Problems 8.58–8.60

8.61 Determine I_y and k_y.

8.62 Determine I_x and k_x.

8.63 Determine I_{xy}.

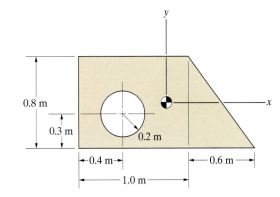

y

0.8 m

0.3 m

0.2 m

←0.4 m→

←0.6 m→

1.0 m

x

Problems 8.61–8.63

8.64 Determine I_y and k_y.

8.65 Determine I_x and k_x.

8.66 Determine I_{xy}.

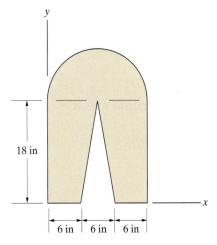

y

18 in

6 in 6 in 6 in

x

Problems 8.64–8.66

8.67 Determine I_y and k_y.

8.68 Determine J_O and k_O.

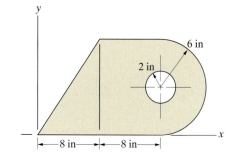
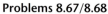

Problems 8.67/8.68

8.69 Determine I_y and k_y.

8.70 Determine I_x and k_x.

8.71 Determine I_{xy}.

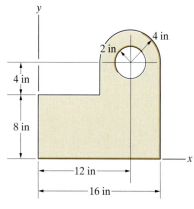

Problems 8.69–8.71

8.72 Determine I_y and k_y.

8.73 Determine I_x and k_x.

8.74 Determine I_{xy}.

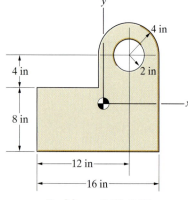

Problems 8.72–8.74

8.75 Determine I_y and k_y.

8.76 Determine J_O and k_O.

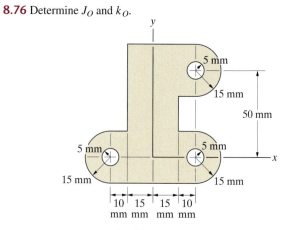

Problems 8.75/8.76

8.77 Determine I_x and I_y for the beam's cross section.

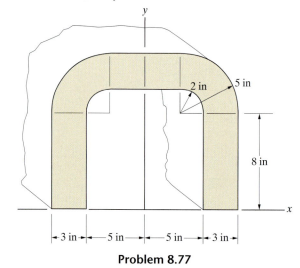
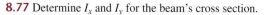

Problem 8.77

8.78 Determine I_x and I_y for the beam's cross section.

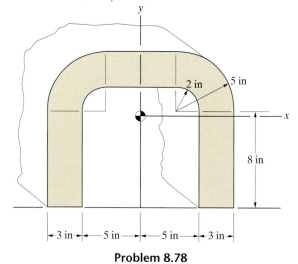
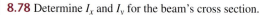

Problem 8.78

8.79 The area $A = 2 \times 10^4$ mm^2. Its moment of inertia about the y axis is $I_y = 3.2 \times 10^8$ mm^4. Determine its moment of inertia about the $\hat{y}$ axis.

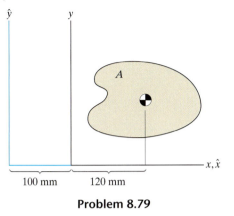

100 mm 120 mm

Problem 8.79

8.80 The area $A = 100$ in^2 and it is *symmetric* about the x' axis. The moments of inertia $I_{x'} = 420$ in^4, $I_{y'} = 580$ in^4, $J_O = 11,000$ in^4, and $I_{xy} = 4800$ in^4. What are I_x and I_y?

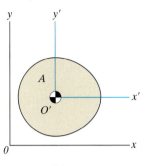

Problem 8.80

Design Experience

Problems 8.81–8.84 are related to Design Example 8.6.

8.81 Determine the moment of inertia of the beam cross section about the x axis. Compare your result with the moment of inertia of a solid square cross section of equal area and confirm the ratio shown in Fig. 8.14.

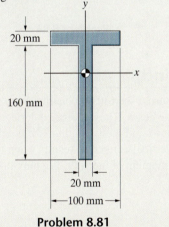

20 mm

160 mm

20 mm
100 mm

Problem 8.81

8.82 The area of the beam cross section is 5200 mm^2. Determine the moment of inertia of the beam cross section about the x axis. Compare your result with the moment of inertia of a solid square cross section of equal area and confirm the ratio shown in Fig. 8.14.

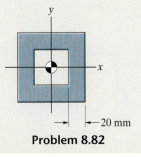

20 mm

Problem 8.82

8.83 (a) If I_x is expressed in m^4, R is in meters, and M is in N-m, what are the SI units of the modulus of elasticity E?

(b) A beam with the cross section shown is subjected to couples $M = 180$ N-m as shown in Fig. 8.13b. As a result, it bends into a circular arc with radius $R = 3$ m. What is the modulus of elasticity of the material?

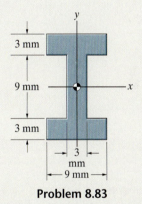

3 mm

9 mm

3 mm

3 mm
9 mm

Problem 8.83

8.84 Suppose that you want to design a beam made of material whose density is 8000 kg/m^3. The beam is to be 4 m in length and have a mass of 320 kg. Design a cross section for the beam so that $I_x = 3 \times 10^{-5}$ m^4.

8.85 The area in Fig. a is a C230×30 American Standard Channel beam cross section. Its cross sectional area is $A = 3790$ mm^2 and its moments of inertia about the x and y axes are $I_x = 25.3 \times 10^6$ mm^4 and $I_y = 1 \times 10^6$ mm^4. Suppose that two beams with C230×30 cross sections are riveted together to obtain a composite beam with the cross section shown in Fig. b. What are the moments of inertia about the x and y axes of the composite beam?

8.86 The area in Fig. a is an L152×102×12.7 Angle beam cross section. Its cross sectional area is $A = 3060$ mm^2 and its moments of inertia about the x and y axes are $I_x = 7.24 \times 10^6$ mm^4 and $I_y = 2.61 \times 10^6$ mm^4. Suppose that four beams with L152×102×12.7 cross sections are riveted together to obtain a composite beam with the cross section shown in Fig. b. What are the moments of inertia about the x and y axes of the composite beam?

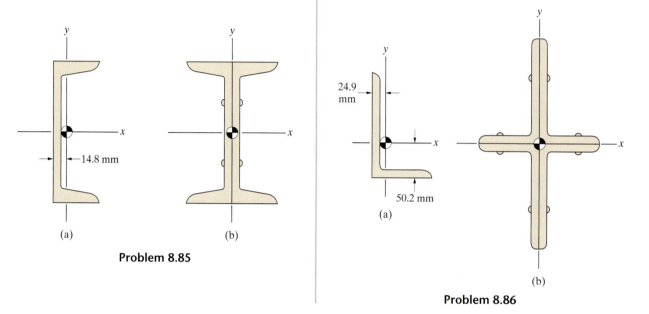

Problem 8.85

Problem 8.86

8.3 Rotated and Principal Axes

Suppose that Fig. 8.16a is the cross section of a cantilever beam. If you apply a vertical force to the end of the beam, a larger vertical deflection results if the cross section is oriented as shown in Fig. 8.16b than if it is oriented as shown in Fig. 8.16c. The minimum vertical deflection results when the beam's cross section is oriented so that the moment of inertia I_x is a maximum (Fig. 8.16d).

Figure 8.16
(a) A beam cross section.
(b)-(d) Applying a lateral load with different orientations of the cross section.

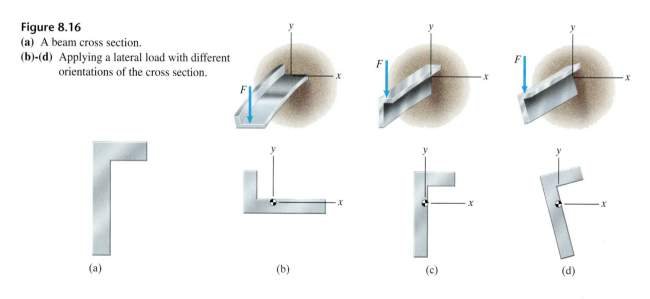

In many engineering applications you must determine moments of inertia of areas with various angular orientations relative to a coordinate system and also determine the orientation for which the value of a moment of inertia is a maximum or minimum. We discuss these procedures in this section.

Rotated Axes

Consider an area A, a coordinate system xy, and a second coordinate system $x'y'$ that is rotated through an angle θ relative to the xy coordinate system (Fig. 8.17a.) Suppose that we know the moments of inertia of A in terms of the xy coordinate system. Our objective is to determine the moments of inertia in terms of the $x'y'$ coordinate system.

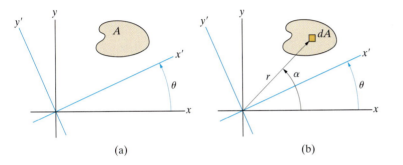

(a)   (b)

Figure 8.17
(a) The $x'y'$ coordinate system is rotated through an angle θ relative to the xy coordinate system.
(b) A differential element of area dA.

In terms of the radial distance r to a differential element of area dA and the angle α in Fig. 8.17b, the coordinates of dA in the xy coordinate system are

$$x = r\cos\alpha, \tag{8.14}$$
$$y = r\sin\alpha. \tag{8.15}$$

The coordinates of dA in the $x'y'$ coordinate system are

$$x' = r\cos(\alpha - \theta) = r(\cos\alpha\cos\theta + \sin\alpha\sin\theta), \tag{8.16}$$
$$y' = r\sin(\alpha - \theta) = r(\sin\alpha\cos\theta - \cos\alpha\sin\theta). \tag{8.17}$$

In Eqs. (8.16) and (8.17), we use identities for the cosine and sine of the difference of two angles (Appendix A). By substituting Eqs. (8.14) and (8.15) into Eqs. (8.16) and (8.17), we obtain equations relating the coordinates of dA in the two coordinate systems:

$$x' = x\cos\theta + y\sin\theta, \tag{8.18}$$
$$y' = -x\sin\theta + y\cos\theta. \tag{8.19}$$

We can use these expressions to derive relations between the moments of inertia of A in terms of the xy and $x'y'$ coordinate systems.

Moment of Inertia About the x' Axis

$$I_{x'} = \int_A (y')^2\, dA = \int_A (-x\sin\theta + y\cos\theta)^2\, dA$$
$$= \cos^2\theta \int_A y^2\, dA - 2\sin\theta\cos\theta \int_A xy\, dA + \sin^2\theta \int_A x^2\, dA.$$

From this equation we obtain

$$I_{x'} = I_x\cos^2\theta - 2I_{xy}\sin\theta\cos\theta + I_y\sin^2\theta. \tag{8.20}$$

Moment of Inertia About the y′ Axis

$$I_{y'} = \int_A (x')^2 \, dA = \int_A (x\cos\theta + y\sin\theta)^2 \, dA$$

$$= \sin^2\theta \int_A y^2 \, dA + 2\sin\theta\cos\theta \int_A xy \, dA + \cos^2\theta \int_A x^2 \, dA.$$

This equation gives us the result

$$I_{y'} = I_x \sin^2\theta + 2I_{xy}\sin\theta\cos\theta + I_y\cos^2\theta. \tag{8.21}$$

Product of Inertia In terms of the $x'y'$ coordinate system, the product of inertia of A is

$$I_{x'y'} = (I_x - I_y)\sin\theta\cos\theta + (\cos^2\theta - \sin^2\theta)I_{xy}. \tag{8.22}$$

Polar Moment of Inertia From Eqs. (8.20) and (8.21), the polar moment of inertia in terms of the $x'y'$ coordinate system is

$$J'_O = I_{x'} + I_{y'} = I_x + I_y = J_O.$$

Thus the value of the polar moment of inertia is unchanged by a rotation of the coordinate system.

Principal Axes

We have seen that the moments of inertia of A in terms of the $x'y'$ coordinate system depend on the angle θ in Fig. 8.17a. Consider the following question: For what values of θ is the moment of inertia $I_{x'}$ a maximum or minimum?

To answer this question, it is convenient to use the identities

$$\sin 2\theta = 2\sin\theta\cos\theta,$$

$$\cos 2\theta = \cos^2\theta - \sin^2\theta = 1 - 2\sin^2\theta = 2\cos^2\theta - 1.$$

With these expressions, we can write Eqs. (8.20)–(8.22) in the forms

$$I_{x'} = \frac{I_x + I_y}{2} + \frac{I_x - I_y}{2}\cos 2\theta - I_{xy}\sin 2\theta, \tag{8.23}$$

$$I_{y'} = \frac{I_x + I_y}{2} - \frac{I_x - I_y}{2}\cos 2\theta + I_{xy}\sin 2\theta, \tag{8.24}$$

$$I_{x'y'} = \frac{I_x - I_y}{2}\sin 2\theta + I_{xy}\cos 2\theta. \tag{8.25}$$

We will denote a value of θ at which $I_{x'}$ is a maximum or minimum by θ_p. To determine θ_p, we evaluate the derivative of Eq. (8.23) with respect to 2θ and equate it to zero, obtaining

$$\tan 2\theta_p = \frac{2I_{xy}}{I_y - I_x}. \tag{8.26}$$

If we set the derivative of Eq. (8.24) with respect to 2θ equal to zero to determine a value of θ for which $I_{y'}$ is a maximum or minimum, we again obtain Eq. (8.26). The second derivatives of $I_{x'}$ and $I_{y'}$ with respect to 2θ are opposite in sign; that is,

$$\frac{d^2 I_{x'}}{d(2\theta)^2} = -\frac{d^2 I_{y'}}{d(2\theta)^2},$$

which means that at an angle θ_p for which $I_{x'}$ is a maximum, $I_{y'}$ is a minimum; and at an angle θ_p for which $I_{x'}$ is a minimum, $I_{y'}$ is a maximum.

A rotated coordinate system $x'y'$ that is oriented so that $I_{x'}$ and $I_{y'}$ have maximum or minimum values is called a set of *principal axes* of the area A. The corresponding moments of inertia $I_{x'}$ and $I_{y'}$ are called the *principal moments of inertia*. In the next section we show that the product of inertia $I_{x'y'}$ corresponding to a set of principal axes equals zero.

Because the tangent is a periodic function, Eq. (8.26) does not yield a unique solution for the angle θ_p. We show, however, that it does determine the orientation of the principal axes within an arbitrary multiple of 90°. Observe in Fig. 8.18 that if $2\theta_0$ is a solution of Eq. (8.26), then $2\theta_0 + n(180°)$ is also a solution for any integer n. The resulting orientations of the $x'y'$ coordinate system are shown in Fig. 8.19.

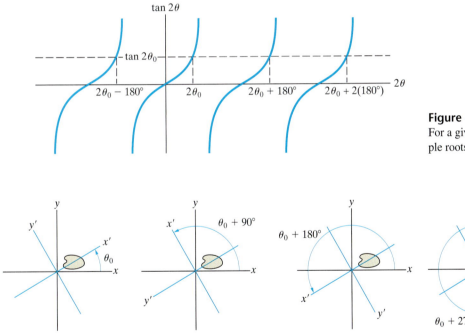

Figure 8.18
For a given value of $\tan 2\theta_0$, there are multiple roots $2\theta_0 + n(180°)$.

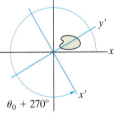

Figure 8.19
The orientation of the $x'y'$ coordinate system is determined within a multiple of 90°.

Determining principal axes and principal moments of inertia of an area involves three steps:

1. Determine I_x, I_y, and I_{xy}—You must determine the moments of inertia of the area in terms of the xy coordinate system.

2. Determine θ_p—Solve Eq. (8.26) to determine the orientation of the principal axes within an arbitrary multiple of 90°.

3. Calculate $I_{x'}$ and $I_{y'}$—Once you have chosen the orientation of the principal axes, you can use Eqs. (8.20) and (8.21) or Eqs. (8.23) and (8.24) to determine the principal moments of inertia.

Example 8.7 Determining Principal Axes and Moments of Inertia

Determine a set of principal axes and the corresponding principal moments of inertia for the triangular area in Fig. 8.20.

Strategy

We can obtain the moments of inertia of the triangular area from Appendix B. Then we can use Eq. (8.26) to determine the orientation of the principal axes and evaluate the principal moments of inertia with Eqs. (8.23) and (8.24).

Figure 8.20

Solution

Determine I_x, I_y, and I_{xy} The moments of inertia of the triangular area are

$$I_x = \frac{1}{12}(4 \text{ m})(3 \text{ m})^3 = 9 \text{ m}^4,$$

$$I_y = \frac{1}{4}(4 \text{ m})^3(3 \text{ m}) = 48 \text{ m}^4,$$

$$I_{xy} = \frac{1}{8}(4 \text{ m})^2(3 \text{ m})^2 = 18 \text{ m}^4.$$

Determine θ_p From Eq. (8.26),

$$\tan 2\theta_p = \frac{2I_{xy}}{I_y - I_x} = \frac{2(18)}{48 - 9} = 0.923,$$

and we obtain $\theta_p = 21.4°$. The principal axes corresponding to this value of θ_p are shown in Fig. a.

Calculate $I_{x'}$, and $I_{y'}$ Substituting $\theta_p = 21.4°$ into Eqs. (8.23) and (8.24), we obtain(with moments of inertia in m^4)

$$I_{x'} = \frac{I_x + I_y}{2} + \frac{I_x - I_y}{2}\cos 2\theta - I_{xy}\sin 2\theta$$

$$= \left(\frac{9 + 48}{2}\right) + \left(\frac{9 - 48}{2}\right)\cos[2(21.4°)] - (18)\sin[2(21.4°)]$$

$$= 1.96 \text{ m}^4,$$

$$I_{y'} = \frac{I_x + I_y}{2} - \frac{I_x - I_y}{2}\cos 2\theta + I_{xy}\sin 2\theta$$

$$= \left(\frac{9 + 48}{2}\right) - \left(\frac{9 - 48}{2}\right)\cos[2(21.4°)] + (18)\sin[2(21.4°)]$$

$$= 55.0 \text{ m}^4.$$

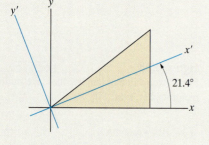

(a) The principal axes corresponding to $\theta = 21.4°$.

Critical Thinking

The product of inertia corresponding to a set of principal axes is zero. In this example, substituting $\theta_p = 21.4°$ into Eq. (8.25) confirms that $I_{x'y'} = 0$.

| Example 8.8 | Rotated and Principal Axes |

The moments of inertia of the area in Fig. 8.21 in terms of the xy coordinate system shown are $I_x = 22 \text{ ft}^4$, $I_y = 10 \text{ ft}^4$, and $I_{xy} = 6 \text{ ft}^4$. (a) Determine $I_{x'}$, $I_{y'}$, and $I_{x'y'}$ for $\theta = 30°$. (b) Determine a set of principal axes and the corresponding principal moments of inertia.

Strategy

(a) We can determine the moments of inertia in terms of the $x'y'$ coordinate system by substituting $\theta = 30°$ into Eqs. (8.23)–(8.25).
(b) The orientation of the principal axes is determined by solving Eq. (8.26) for θ_p. Once θ_p has been determined, the moments of inertia about the principal axes can be determined from Eqs. (8.23) and (8.24).

Figure 8.21

Solution

(a) Determine $I_{x'}$, $I_{y'}$, and $I_{x'y'}$ By setting $\theta = 30°$ in Eqs. (8.23)–(8.25), we obtain (with moments of inertia in ft^4)

$$I_{x'} = \frac{I_x + I_y}{2} + \frac{I_x - I_y}{2} \cos 2\theta - I_{xy} \sin 2\theta$$

$$= \left(\frac{22 + 10}{2}\right) + \left(\frac{22 - 10}{2}\right) \cos[2(30°)] - (6) \sin[2(30°)] = 13.8 \text{ ft}^4,$$

$$I_{y'} = \frac{I_x + I_y}{2} - \frac{I_x - I_y}{2} \cos 2\theta + I_{xy} \sin 2\theta$$

$$= \left(\frac{22 + 10}{2}\right) - \left(\frac{22 - 10}{2}\right) \cos[2(30°)] + (6) \sin[2(30°)] = 18.2 \text{ ft}^4,$$

$$I_{x'y'} = \frac{I_x - I_y}{2} \sin 2\theta + I_{xy} \cos 2\theta$$

$$= \left(\frac{22 - 10}{2}\right) \sin[2(30°)] + (6) \cos[2(30°)] = 8.2 \text{ ft}^4.$$

(b) Determine θ_p   We substitute the moments of inertia in terms of the xy coordinate system into Eq. (8.26), yielding

$$\tan 2\theta_p = \frac{2I_{xy}}{I_y - I_x} = \frac{2(6)}{10 - 22} = -1.$$

Thus, $\theta_p = -22.5°$. The principal axes corresponding to this value of θ_p are shown in Fig. a.

Calculate $I_{x'}$ and $I_{y'}$ We substitute $\theta_p = -22.5°$ into Eqs. (8.23) and (8.24), obtaining the principal moments of inertia:

$$I_{x'} = 24.5 \text{ ft}^4, \quad I_{y'} = 7.5 \text{ ft}^4.$$

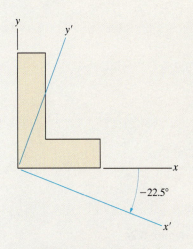

(a) The set of principal axes corresponding to $\theta_p = -22.5°$

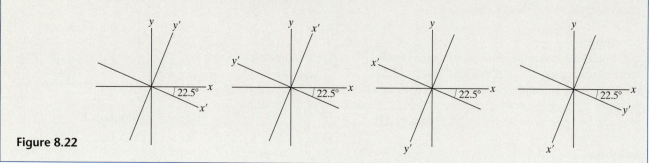

Critical Thinking

Remember that the orientation of the principal axes is only determined within an arbitrary multiple of 90°. In this example we chose to designate the axes in Fig. a as the positive x' and y' axes, but any of the four choices in Fig. 8.22 is equally valid.

Figure 8.22

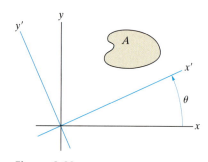

Figure 8.23
The xy coordinate system and the rotated $x'y'$ coordinate system.

Mohr's Circle

Given the moments of inertia of an area in terms of a particular coordinate system, we have presented equations that determine the moments of inertia in terms of a rotated coordinate system, the orientation of the principal axes, and the principal moments of inertia. We can also obtain this information by using a graphical method called *Mohr's circle*, which is very useful for visualizing the solutions of Eqs. (8.23)–(8.25).

Determining $I_{x'}$, $I_{y'}$, and $I_{x'y'}$ We first describe how to construct Mohr's circle and then explain why it works. Suppose we know the moments of inertia I_x, I_y, and I_{xy} of an area in terms of a coordinate system xy and we want to determine the moments of inertia for a rotated coordinate system $x'y'$ (Fig. 8.23). Constructing Mohr's circle involves three steps:

1. Establish a set of horizontal and vertical axes and plot two points: point 1 with coordinates (I_x, I_{xy}) and point 2 with coordinates $(I_y, -I_{xy})$ as shown in Fig. 8.24a.

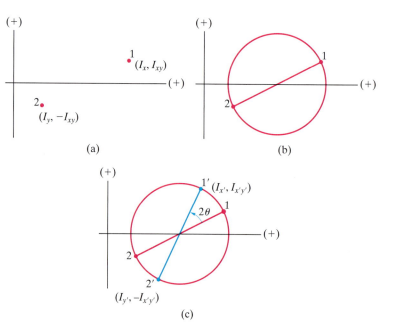

Figure 8.24
(a) Plotting the points 1 and 2.
(b) Drawing Mohr's circle. The center of the circle is the intersection of the line from 1 to 2 with the horizontal axis.
(c) Finding the points $1'$ and $2'$.

2. Draw a straight line connecting points 1 and 2. Using the intersection of the straight line with the horizontal axis as the center, draw a circle that passes through the two points (Fig. 8.24b).

3. Draw a straight line through the center of the circle at an angle 2θ measured counterclockwise from point 1. This line intersects the circle at point $1'$ with coordinates $(I_{x'}, I_{x'y'})$ and point $2'$ with coordinates $(I_{y'}, -I_{x'y'})$, as shown in Fig. 8.24c.

Thus, for a given angle θ, the coordinates of points $1'$ and $2'$ determine the moments of inertia in terms of the rotated coordinate system. Why does this graphical construction work? In Fig. 8.25, we show the points 1 and 2 and Mohr's circle. Notice that the horizontal coordinate of the center of the circle is $(I_x + I_y)/2$. The sine and cosine of the angle β are

$$\sin\beta = \frac{I_{xy}}{R}, \quad \cos\beta = \frac{I_x - I_y}{2R},$$

where R, the radius of the circle, is given by

$$R = \sqrt{\left(\frac{I_x - I_y}{2}\right)^2 + (I_{xy})^2}.$$

Figure 8.26 shows the construction of the points $1'$ and $2'$. The horizontal coordinate of point $1'$ is

$$\frac{I_x + I_y}{2} + R\cos(\beta + 2\theta)$$

$$= \frac{I_x + I_y}{2} + R(\cos\beta\cos 2\theta - \sin\beta\sin 2\theta)$$

$$= \frac{I_x + I_y}{2} + \frac{I_x - I_y}{2}\cos 2\theta - I_{xy}\sin 2\theta = I_{x'},$$

and the horizontal coordinate of point $2'$ is

$$\frac{I_x + I_y}{2} - R\cos(\beta + 2\theta)$$

$$= \frac{I_x + I_y}{2} - R(\cos\beta\cos 2\theta - \sin\beta\sin 2\theta)$$

$$= \frac{I_x + I_y}{2} - \frac{I_x - I_y}{2}\cos 2\theta + I_{xy}\sin 2\theta = I_{y'}.$$

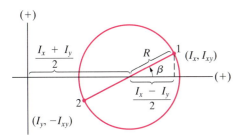

Figure 8.25
The points 1 and 2 and Mohr's circle.

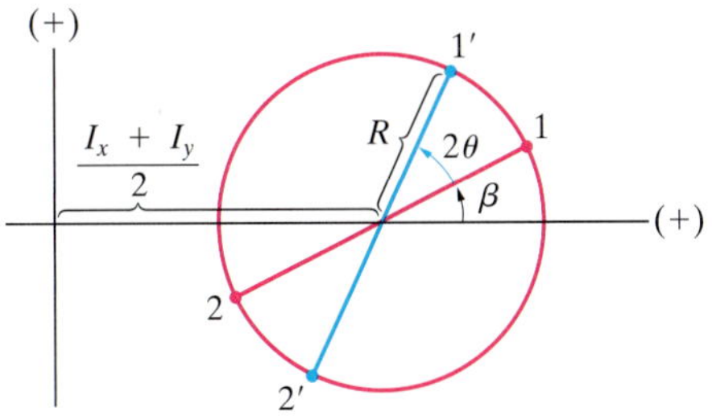

Figure 8.26
The points $1'$ and $2'$.

The vertical coordinate of point $1'$ is

$$R \sin(\beta + 2\theta) = R(\sin \beta \cos 2\theta + \cos \beta \sin 2\theta)$$

$$= I_{xy} \cos 2\theta + \frac{I_x - I_y}{2} \sin 2\theta = I_{x'y'},$$

and the vertical coordinate of point $2'$ is

$$-R \sin(\beta + 2\theta) = -I_{x'y'}.$$

We have shown that the coordinates of point $1'$ are $(I_{x'}, I_{x'y'})$ and the coordinates of point $2'$ are $(I_{y'}, -I_{x'y'})$.

Determining Principal Axes and Principal Moments of Inertia

Because the moments of inertia $I_{x'}$ and $I_{y'}$ are the horizontal coordinates of points $1'$ and $2'$ of Mohr's circle, their maximum and minimum values occur when points $1'$ and $2'$ coincide with the intersections of the circle with the horizontal axis (Fig. 8.27). (Which intersection you designate as $1'$ is arbitrary. In Fig. 8.27, we have designated the minimum moment of inertia as point $1'$.) You can determine the orientation of the principal axes by measuring the angle $2\theta_p$ from point 1 to point $1'$, and the coordinates of points $1'$ and $2'$ are the principal moments of inertia.

Notice that Mohr's circle demonstrates that the product of inertia $I_{x'y'}$ corresponding to a set of principal axes (the vertical coordinate of point $1'$ in Fig. 8.27) is always zero. Furthermore, we can use Fig. 8.25 to obtain an analytical expression for the horizontal coordinates of the points where the circle intersects the horizontal axis, which are the principal moments of inertia:

$$\text{Principal moments of inertia} = \frac{I_x + I_y}{2} \pm R$$

$$= \frac{I_x + I_y}{2} \pm \sqrt{\left(\frac{I_x - I_y}{2}\right)^2 + (I_{xy})^2}.$$

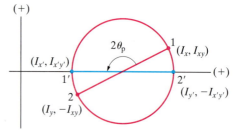

Figure 8.27
To determine the orientation of a set of principal axes, let points $1'$ and $2'$ be the points where the circle intersects the horizontal axis.

| Example 8.9 | Moments of Inertia by Mohr's Circle |

The moments of inertia of the area in Fig. 8.28 in terms of the xy coordinate system are $I_x = 22$ ft^4, $I_y = 10$ ft^4, and $I_{xy} = 6$ ft^4. Determine (a) the moments of inertia $I_{x'}$, $I_{y'}$, and $I_{x'y'}$ for $\theta = 30°$; (b) a set of principal axes and the corresponding principal moments of inertia.

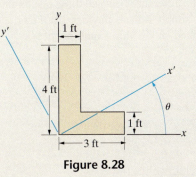

Figure 8.28

Strategy

By constructing Mohr's circle, we can determine the moments of inertia for a coordinate system oriented at $\theta = 30°$ and also determine the principal axes and principal moments of inertia.

Solution

(a) First we plot point 1 with coordinates $(I_x, I_{xy}) = (22, 6)$ ft^4 and point 2 with coordinates $(I_y, -I_{xy}) = (10, -6)$ ft^4 (Fig. a). Then we draw a straight line between points 1 and 2 and, using the intersection of the line with the horizontal axis as the center, draw a circle that passes through the points (Fig. b).

To determine the moments of inertia for $\theta = 30°$, we measure an angle $2\theta = 60°$ counterclockwise from point 1 (Fig. c). From the coordinates of points $1'$ and $2'$, we obtain

$$I_{x'} = 14 \text{ ft}^4, \quad I_{x'y'} = 8 \text{ ft}^4, \quad I_{y'} = 18 \text{ ft}^4.$$

(b) To determine the principal axes, we let the points $1'$ and $2'$ be the points where the circle intersects the horizontal axis (Fig. d). Measuring the angle from point 1 to point $1'$, we determine that $2\theta_p = 135°$. From the coordinates of points $1'$ and $2'$, we obtain the principal moments of inertia:

$$I_{x'} = 7.5 \text{ ft}^4, \quad I_{y'} = 24.5 \text{ ft}^4.$$

The principal axes are shown in Fig. e.

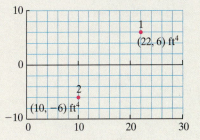

(a) Plot point 1 with coordinates (I_x, I_{xy}) and point 2 with coordinates $(I_y, -I_{xy})$.

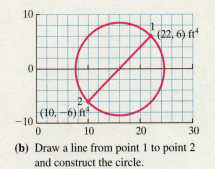

(b) Draw a line from point 1 to point 2 and construct the circle.

Critical Thinking

In Example 8.8 we solved this problem by using Eqs. (8.23)–(8.26). For $\theta = 30°$, we obtained $I_{x'} = 13.8$ ft^4, $I_{x'y'} = 8.2$ ft^4, and $I_{y'} = 18.2$ ft^4. The differences between these results and the ones we obtained using Mohr's circle are due to the errors inherent in measuring the answer graphically. By using Eq. (8.26) to determine the orientation of the principal axes, we obtained the principal axes shown in Fig. a of Example 8.8 and the principal moments of inertia $I_{x'} = 24.5$ ft^4 and $I_{y'} = 7.5$ ft^4. The difference between those results and the ones we obtained using Mohr's circle simply reflects the fact that the orientation of the principal axes can be determined only within a multiple of 90°.

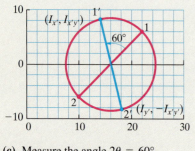

(c) Measure the angle $2\theta = 60°$ counterclockwise from point 1 to determine the points 1′ and 2′.

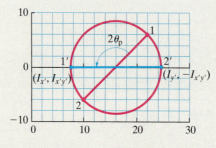

(d) Determine the principal axes by letting points 1′ and 2′ correspond to the points where the circle intersects the horizontal axis.

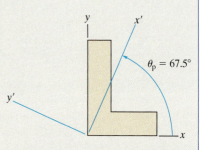

(e) The principal axes corresponding to $\theta_p = 67.5°$.

Problems

8.87 Determine $I_{x'}$, $I_{y'}$, and $I_{x'y'}$. (Do not use Mohr's circle.)

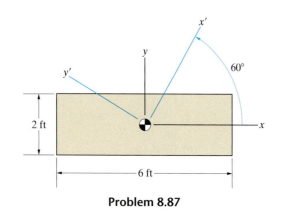

Problem 8.87

8.88 The area is an L152X102X12.7 Angle beam cross section. Its moments of inertia are $I_x = 7.24 \times 10^6$ mm^4, $I_y = 2.61 \times 10^6$ mm^4, and $I_{xy} = -2.53 \times 10^6$ mm^4. Determine the moments of inertia $I_{x'}$, $I_{y'}$, and $I_{x'y'}$ if $\theta = 60°$.

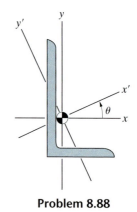

Problem 8.88

8.89 For the beam cross section in Problem 8.88, determine a set of principal axes and the corresponding principal moments of inertia. (Do not use Mohr's circle.)

8.90 Determine the moments of inertia $I_{x'}$, $I_{y'}$, and $I_{x'y'}$ if $\theta = 50°$. (Do not use Mohr's circle.)

8.91 Determine a set of principal axes and the corresponding principal moments of inertia. (Do not use Mohr's circle.)

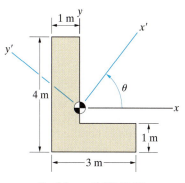

Problems 8.90/8.91

8.92* Determine a set of principal axes and the corresponding principal moments of inertia. (Do not use Mohr's circle.)

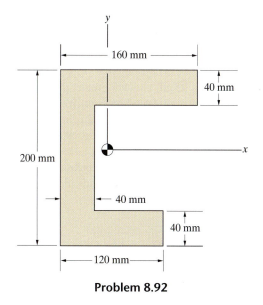

Problem 8.92

8.93 Solve Problem 8.87 by using Mohr's circle.

8.94 Solve Problem 8.88 by using Mohr's circle.

8.95 Solve Problem 8.89 by using Mohr's circle.

8.96 Solve Problem 8.90 by using Mohr's circle.

8.97 Solve Problem 8.91 by using Mohr's circle.

8.98* Solve Problem 8.92 by using Mohr's circle.

8.99 Derive Eq. (8.22) for the product of inertia by using the same procedure we used to derive Eqs. (8.20) and (8.21).

MASSES

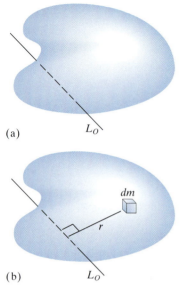

(a)

(b)

Figure 8.29
(a) An object and axis L_O.
(b) A differential element of mass dm.

The acceleration of an object that results from the forces acting on it depends on its mass. The angular acceleration, or rotational acceleration, that results from the forces and couples acting on an object depends on quantities called the mass moments of inertia of the object. In this section we discuss methods for determining mass moments of inertia of particular objects. We show that for special classes of objects, their mass moments of inertia can be expressed in terms of moments of inertia of areas, which explains how the names of those area integrals originated.

An object and a line or "axis" L_O are shown in Fig. 8.29a. The *moment of inertia* of the object about the axis L_O is defined by

$$I_O = \int_m r^2 \, dm, \tag{8.27}$$

where r is the perpendicular distance from the axis to the differential element of mass dm (Fig. 8.29b). Often L_O is an axis about which the object rotates, and the value of I_O is required to determine the angular acceleration, or the rate of change of the rate of rotation, caused by a given couple about L_O.

The dimensions of the moment of inertia of an object are $(\text{mass}) \times (\text{length})^2$. Notice that the definition implies that its value must be positive.

8.4 Simple Objects

The moments of inertia of complicated objects can be determined by summing the moments of inertia of their individual parts. We therefore begin by determining moments of inertia of some simple objects. Then in the next section we describe the parallel-axis theorem, which makes it possible to determine moments of inertia of objects composed of combinations of parts.

Slender Bars

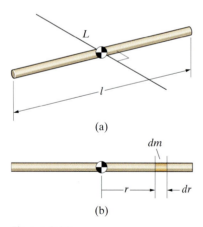

(a)

(b)

Figure 8.30
(a) A slender bar.
(b) A differential element of length dr.

Let us determine the moment of inertia of a straight, slender bar about a perpendicular axis L through the center of mass of the bar (Fig. 8.30a). "Slender" means that we assume that the bar's length is much greater than its width. Let the bar have length l, cross-sectional area A, and mass m. We assume that A is uniform along the length of the bar and that the material is homogeneous.

Consider a differential element of the bar of length dr at a distance r from the center of mass (Fig. 8.30b). The element's mass is equal to the product of its volume and the density: $dm = \rho A \, dr$. Substituting this expression into Eq. (8.27), we obtain the moment of inertia of the bar about a perpendicular axis through its center of mass:

$$I = \int_m r^2 \, dm = \int_{-l/2}^{l/2} \rho A r^2 \, dr = \frac{1}{12} \rho A l^3.$$

The mass of the bar equals the product of the mass density and the volume of the bar, $m = \rho A l$, so we can express the moment of inertia as

$$I = \frac{1}{12} m l^2. \tag{8.28}$$

We have neglected the lateral dimensions of the bar in obtaining this result. That is, we treated the differential element of mass dm as if it were concentrated on the axis of the bar. As a consequence, Eq. (8.28) is an approximation for the moment of inertia of a bar. Later in this section we determine the moments of inertia for a bar of finite lateral dimension and show that Eq. (8.28) is a good approximation when the width of the bar is small in comparison to its length.

Thin Plates

Consider a homogeneous flat plate that has mass m and uniform thickness T. We will leave the shape of the cross-sectional area of the plate unspecified. Let a cartesian coordinate system be oriented so that the plate lies in the x–y plane (Fig. 8.31a). Our objective is to determine the moments of inertia of the plate about the x, y, and z axes.

We can obtain a differential element of volume of the plate by projecting an element of area dA through the thickness T of the plate (Fig. 8.31b). The resulting volume is $T\,dA$. The mass of this element of volume is equal to the product of the density and the volume: $dm = \rho T\,dA$. Substituting this expression into Eq. (8.27), we obtain the moment of inertia of the plate about the z axis in the form

$$I_{z\text{ axis}} = \int_m r^2\,dm = \rho T \int_A r^2\,dA,$$

where r is the distance from the z axis to dA. Since the mass of the plate is $m = \rho T A$, where A is the cross-sectional area of the plate, $\rho T = m/A$. The integral on the right is the polar moment of inertia J_O of the cross-sectional area of the plate. We can therefore write the moment of inertia of the plate about the z axis as

$$I_{z\text{ axis}} = \frac{m}{A}J_O. \tag{8.29}$$

From Fig 8.31b, we see that the perpendicular distance from the x axis to the element of area dA is the y coordinate of dA. Therefore, the moment of inertia of the plate about the x axis is

$$I_{x\text{ axis}} = \int_m y^2\,dm = \rho T \int_A y^2\,dA = \frac{m}{A}I_x, \tag{8.30}$$

where I_x is the moment of inertia of the cross-sectional area of the plate about the x axis. The moment of inertia of the plate about the y axis is

$$I_{y\text{ axis}} = \int_m x^2\,dm = \rho T \int_A x^2\,dA = \frac{m}{A}I_y, \tag{8.31}$$

where I_y is the moment of inertia of the cross-sectional area of the plate about the y axis.

Because the sum of the area moments of inertia I_x and I_y is equal to the polar moment of inertia J_O, the mass moment of inertia of the thin plate about the z axis is equal to the sum of its moments of inertia about the x and y axes:

$$I_{z\text{ axis}} = I_{x\text{ axis}} + I_{y\text{ axis}}. \qquad \text{Thin plate} \tag{8.32}$$

We have expressed the moments of inertia of a thin homogeneous plate of uniform thickness in terms of the moments of inertia of the cross-sectional area of the plate. In fact, these results explain why the area integrals I_x, I_y, and J_O are called moments of inertia. The use of the same terminology and similar symbols for moments of inertia of areas and moments of inertia of objects can be confusing, but is entrenched in engineering practice. The type of moment of inertia being referred to can be determined either from the context or from the units: $(\text{length})^4$ for moments of inertia of areas and $(\text{mass}) \times (\text{length})^2$ for moments of inertia of masses.

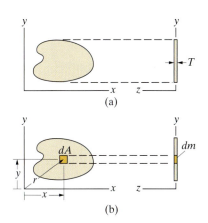

(a)

(b)

Figure 8.31
(a) A plate of arbitrary shape and uniform thickness T.
(b) An element of volume obtained by projecting an element of area dA through the plate.

Example 8.10   Moments of Inertia of a Slender Bar

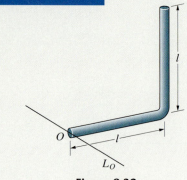

Figure 8.32

Two homogeneous slender bars, each of length l, mass m, and cross-sectional area A, are welded together to form the L-shaped object in Fig. 8.32. Determine the moment of inertia of the object about the axis L_O through point O. (The axis L_O is perpendicular to the two bars.)

Strategy
Using the same integration procedure we used for a single bar, we will determine the moment of inertia of each bar about L_O and sum the results.

Solution
Our first step is to introduce a coordinate system with the z axis along L_O and the x axis collinear with bar 1 (Fig. a). The mass of the differential element of bar 1 of length dx is $dm = \rho A\, dx$. The moment of inertia of bar 1 about L_O is

$$(I_O)_1 = \int_m r^2\, dm = \int_0^l \rho A x^2\, dx = \frac{1}{3}\rho A l^3.$$

In terms of the mass of the bar, $m = \rho A l$, we can write this result as

$$(I_O)_1 = \frac{1}{3}ml^2.$$

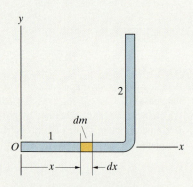

(a) Differential element of bar 1.

The mass of an element of bar 2 of length dy, shown in Fig. b, is $dm = \rho A\, dy$. From the figure we see that the perpendicular distance from L_O to the element is $r = \sqrt{l^2 + y^2}$. Therefore, the moment of inertia of bar 2 about L_O is

$$(I_O)_2 = \int_m r^2\, dm = \int_0^l \rho A(l^2 + y^2)\, dy = \frac{4}{3}\rho A l^3.$$

In terms of the mass of the bar, we obtain

$$(I_O)_2 = \frac{4}{3}ml^2.$$

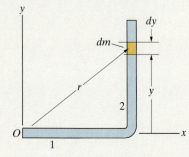

(b) Differential element of bar 2.

The moment of inertia of the L-shaped object about L_O is

$$I_O = (I_O)_1 + (I_O)_2 = \frac{1}{3}ml^2 + \frac{4}{3}ml^2 = \frac{5}{3}ml^2.$$

Critical Thinking
In this example we used integration to determine a moment of inertia of an object consisting of two straight bars. The same procedure could be applied to more complicated objects made of such bars, but it would obviously be cumbersome. Once we have used integration to determine a moment of inertia of a single bar, such as Eq. (8.28), it would be very convenient to use that result to determine moments of inertia of composite objects made of bars without having to resort to integration. We show how this can be done in the next section.

| Example 8.11 | **Moments of Inertia of a Triangular Plate** |

The thin homogeneous plate in Fig. 8.33 is of uniform thickness and mass m. Determine its moments of inertia about the x, y, and z axes.

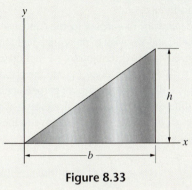

Figure 8.33

Strategy
The moments of inertia about the x and y axes are given by Eqs. (8.30) and (8.31) in terms of the moments of inertia of the cross-sectional area of the plate. We can determine the moment of inertia of the plate about the z axis from Eq. (8.32).

Solution
From Appendix B, the moments of inertia of the triangular area about the x and y axes are $I_x = \frac{1}{12}bh^3$ and $I_y = \frac{1}{4}hb^3$. Therefore, the moments of inertia of the plate about the x and y axes are

$$I_{x \text{ axis}} = \frac{m}{A}I_x = \left(\frac{m}{\frac{1}{2}bh}\right)\left(\frac{1}{12}bh^3\right) = \frac{1}{6}mh^2,$$

$$I_{y \text{ axis}} = \frac{m}{A}I_y = \left(\frac{m}{\frac{1}{2}bh}\right)\left(\frac{1}{4}hb^3\right) = \frac{1}{2}mb^2.$$

The moment of inertia about the z axis is

$$I_{z \text{ axis}} = I_{x \text{ axis}} + I_{y \text{ axis}} = m\left(\frac{1}{6}h^2 + \frac{1}{2}b^2\right).$$

Critical Thinking
As this example demonstrates, you can use the moments of inertia of areas tabulated in Appendix B to determine moments of inertia of thin homogeneous plates. For plates with more complicated shapes, you can use the methods for determining moments of inertia of composite areas illustrated in Examples 8.4–8.6.

8.5 Parallel-Axis Theorem

The parallel-axis theorem allows us to determine the moment of inertia of an object about any axis when the moment of inertia about a parallel axis through the center of mass is known. This theorem can be used to calculate the moment of inertia of a composite object about an axis given the moments of inertia of each of its parts about parallel axes.

Suppose that we know the moment of inertia I about an axis L through the center of mass of an object, and we wish to determine its moment of inertia I_O about a parallel axis L_O (Fig. 8.34a). To determine I_O, we introduce parallel coordinate systems xyz and $x'y'z'$ with the z axis along L_O and the z' axis along L, as shown in Fig. 8.34b. (In this figure the axes L_O and L are perpendicular to the page.) The origin O of the xyz coordinate system is contained in the $x'-y'$ plane. The terms d_x and d_y are the coordinates of the center of mass relative to the xyz coordinate system.

The moment of inertia of the object about L_O is

$$I_O = \int_m r^2 \, dm = \int_m (x^2 + y^2) \, dm, \tag{8.33}$$

where r is the perpendicular distance from L_O to the differential element of mass dm, and x, y are the coordinates of dm in the x–y plane. The coordinates of dm in the two coordinate systems are related by

$$x = x' + d_x, \quad y = y' + d_y.$$

By substituting these expressions into Eq. (8.33), we can write it as

$$I_O = \int_m [(x')^2 + (y')^2] \, dm + 2d_x \int_m x' \, dm + 2d_y \int_m y' \, dm$$

$$+ \int_m (d_x^2 + d_y^2) \, dm. \tag{8.34}$$

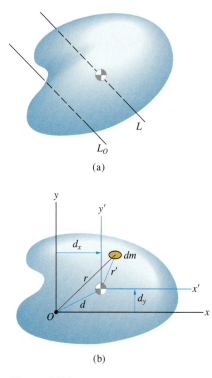

Figure 8.34
(a) An axis L through the center of mass of an object and a parallel axis L_O.
(b) The xyz and $x'y'z'$ coordinate systems.

Since $(x')^2 + (y')^2 = (r')^2$, where r' is the perpendicular distance from L to dm, the first integral on the right side of this equation is the moment of inertia I of the object about L. Recall that the x' and y' coordinates of the center of mass of the object relative to the $x'y'z'$ coordinate system are defined by

$$\bar{x}' = \frac{\int_m x' \, dm}{\int_m dm}, \quad \bar{y}' = \frac{\int_m y' \, dm}{\int_m dm}.$$

Because the center of mass of the object is at the origin of the $x'y'z'$ system, $\bar{x}' = 0$ and $\bar{y}' = 0$. Therefore the integrals in the second and third terms on the right side of Eq. (8.34) are equal to zero. From Fig. 8.34b, we see that $d_x^2 + d_y^2 = d^2$, where d is the perpendicular distance between the axes L and L_O. Therefore, we obtain

$$I_O = I + d^2 m. \tag{8.35}$$

This is the *parallel-axis theorem* for moments of inertia of objects. Equation (8.35) relates the moment of inertia I of an object about an axis *through the center of mass* to its moment of inertia I_O about any parallel axis, where d is the perpendicular distance between the two axes and m is the mass of the object.

Determining the moment of inertia of an object about a given axis L_O typically requires three steps:

1. Choose the parts—Try to divide the object into parts whose mass moments of inertia you know or can easily determine.

2. Determine the moments of inertia of the parts—You must first determine the moment of inertia of each part about the axis through its center of mass parallel to L_O. Then you can use the parallel-axis theorem to determine its moment of inertia about L_O.

3. Sum the results—Sum the moments of inertia of the parts (or subtract in the case of a hole or cutout) to obtain the moment of inertia of the composite object.

Example 8.12 | Moment of Inertia of a Composite Bar

Two homogeneous slender bars, each of length l and mass m, are welded together to form the L-shaped object in Fig. 8.35. Determine the moment of inertia of the object about the axis L_O through point O. (The axis L_O is perpendicular to the two bars.)

Strategy
The moment of inertia of a straight slender bar about a perpendicular axis through its center of mass is given by Eq. (8.28). We can use the parallel-axis theorem to determine the moments of inertia of the bars about the axis L_O and sum them to obtain the moment of inertia of the composite bar.

Solution
Choose the Parts   The parts are the two bars, which we call bar 1 and bar 2 (Fig. a).

Determine the Moments of Inertia of the Parts   The moment of inertia of each bar about a perpendicular axis through its center of mass is $I = \frac{1}{12}ml^2$. The distance from L_O to the parallel axis through the center of mass of bar 1 is $\frac{1}{2}l$ (Fig. a). Therefore, the moment of inertia of bar 1 about L_O is

$$(I_O)_1 = I + d^2m = \frac{1}{12}ml^2 + \left(\frac{1}{2}l\right)^2 m = \frac{1}{3}ml^2.$$

The distance from L_O to the parallel axis through the center of mass of bar 2 is $[l^2 + (\frac{1}{2}l)^2]^{1/2}$. The moment of inertia of bar 2 about L_O is

$$(I_O)_2 = I + d^2m = \frac{1}{12}ml^2 + \left[l^2 + \left(\frac{1}{2}l\right)^2\right]m = \frac{4}{3}ml^2.$$

Sum the Results   The moment of inertia of the L-shaped object about L_O is

$$I_O = (I_O)_1 + (I_O)_2 = \frac{1}{3}ml^2 + \frac{4}{3}ml^2 = \frac{5}{3}ml^2.$$

Critical Thinking
Compare this solution to Example 8.10, in which we used integration to determine the moment of inertia of this object about L_O. We obtained the result much more easily with the parallel-axis theorem, but of course we needed to know the moments of inertia of the bars about the axes through their centers of mass.

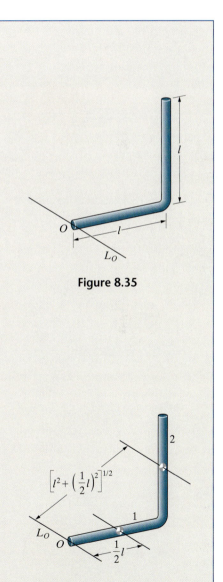

Figure 8.35

(a) The distances from L_O to parallel axes through the centers of mass of bars 1 and 2.

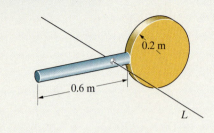

Figure 8.36

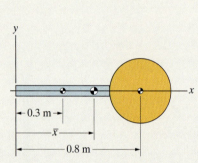

(a) The coordinate $\bar{x}$ of the center of mass of the object.

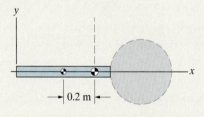

(b) Distance from L to the center of mass of the bar.

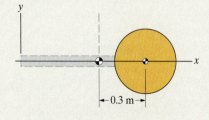

(c) Distance from L to the center of mass of the disk.

Example 8.13   Moment of Inertia of a Composite Object

The object in Fig. 8.36 consists of a slender, 3-kg bar welded to a thin, circular 2-kg disk. Determine its moment of inertia about the axis L through its center of mass. (The axis L is perpendicular to the bar and disk.)

Strategy
We must first locate the center of mass of the composite object and then apply the parallel-axis theorem to the parts separately and sum the results.

Solution
Choose the Parts   The parts are the bar and the disk. Introducing the coordinate system in Fig. a, the x coordinate of the center of mass of the composite object is

$$\bar{x} = \frac{\bar{x}_{bar} m_{bar} + \bar{x}_{disk} m_{disk}}{m_{bar} + m_{disk}} =$$

$$\frac{(0.3\ \text{m})(3\ \text{kg}) + (0.6\ \text{m} + 0.2\ \text{m})(2\ \text{kg})}{(3\ \text{kg}) + (2\ \text{kg})} = 0.5\ \text{m}.$$

Determine the Moments of Inertia of the Parts   The distance from the center of mass of the bar to the center of mass of the composite object is 0.2 m (Fig. b). Therefore, the moment of inertia of the bar about L is

$$I_{bar} = \frac{1}{12}(3\ \text{kg})(0.6\ \text{m})^2 + (3\ \text{kg})(0.2\ \text{m})^2 = 0.210\ \text{kg-m}^2.$$

The distance from the center of mass of the disk to the center of mass of the composite object is 0.3 m (Fig. c). The moment of inertia of the disk about L is

$$I_{disk} = \frac{1}{2}(2\ \text{kg})(0.2\ \text{m})^2 + (2\ \text{kg})(0.3\ \text{m})^2 = 0.220\ \text{kg-m}^2.$$

Sum the Results   The moment of inertia of the composite object about L is

$$I = I_{bar} + I_{disk} = 0.430\ \text{kg-m}^2.$$

Critical Thinking
This example demonstrates the most common procedure for determining moments of inertia of objects in engineering applications. Objects usually consist of assemblies of parts. The center of mass of each part and its moment of inertia about the axis through its center of mass must be determined. (It may be necessary to determine this information experimentally, or it is sometimes supplied by manufacturers of subassemblies.) Then the center of mass of the composite object is determined and the parallel-axis theorem is used to determine the moment of inertia of each part about the axis through the center of mass of the composite object. Finally, the individual moments of inertia are summed to obtain the moment of inertia of the composite object.

Example 8.14   Moments of Inertia of a Cylinder

The homogeneous cylinder in Fig. 8.37 has mass m, length l, and radius R. Determine its moments of inertia about the x, y, and z axes.

Strategy
We first determine the moments of inertia about the x, y, and z axes of an infinitesimal element of the cylinder consisting of a disk of thickness dz. We then integrate the results with respect to z to obtain the moments of inertia of the cylinder. We must apply the parallel-axis theorem to determine the moments of inertia of the disk about the x and y axes.

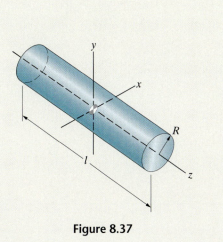

Figure 8.37

Solution
Consider an element of the cylinder of thickness dz at a distance z from the center of the cylinder (Fig. a). (You can imagine obtaining this element by "slicing" the cylinder perpendicular to its axis.) The mass of the element is equal to the product of the mass density and the volume of the element, $dm = \rho(\pi R^2 \, dz)$. We obtain the moments of inertia of the element by using the values for a thin circular plate given in Appendix C. The moment of inertia about the z axis is

$$dI_{z \text{ axis}} = \frac{1}{2} dm R^2 = \frac{1}{2}(\rho \pi R^2 \, dz) R^2.$$

By integrating this result with respect to z from $-l/2$ to $l/2$, we sum the mass moments of inertia of the infinitesimal disk elements that make up the cylinder. The result is the moment of inertia of the cylinder about the z axis:

$$I_{z \text{ axis}} = \int_{-l/2}^{l/2} \frac{1}{2} \rho \pi R^4 \, dz = \frac{1}{2} \rho \pi R^4 l.$$

We can write this result in terms of the mass of the cylinder, $m = \rho(\pi R^2 l)$, as

$$I_{z \text{ axis}} = \frac{1}{2} m R^2.$$

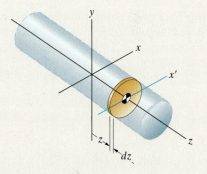

(a) A differential element of the cylinder in the form of a disk.

The moment of inertia of the disk element about the x' axis is

$$dI_{x' \text{ axis}} = \frac{1}{4} dm \, R^2 = \frac{1}{4}(\rho \pi R^2 \, dz) R^2.$$

We can use this result and the parallel-axis theorem to determine the moment of inertia of the element about the x axis:

$$dI_{x \text{ axis}} = dI_{x' \text{ axis}} + z^2 \, dm = \frac{1}{4}(\rho \pi R^2 \, dz) R^2 + z^2(\rho \pi R^2 \, dz).$$

Integrating this expression with respect to z from $-l/2$ to $l/2$, we obtain the moment of inertia of the cylinder about the x axis:

$$I_{x \text{ axis}} = \int_{-l/2}^{l/2} \left(\frac{1}{4} \rho \pi R^4 + \rho \pi R^2 z^2 \right) dz = \frac{1}{4} \rho \pi R^4 l + \frac{1}{12} \rho \pi R^2 l^3.$$

In terms of the mass of the cylinder,

$$I_{x \text{ axis}} = \frac{1}{4} m R^2 + \frac{1}{12} m l^2.$$

Due to the symmetry of the cylinder,

$$I_{y \text{ axis}} = I_{x \text{ axis}}.$$

Critical Thinking

When the cylinder is very long in comparison to its width, $l \gg R$, the first term in the equation for $I_{x \text{ axis}}$ can be neglected, and we obtain the moment of inertia of a slender bar about a perpendicular axis, Eq. (8.28). Conversely, when the radius of the cylinder is much greater than its length, $R \gg l$, the second term in the equation for $I_{x \text{ axis}}$ can be neglected, and we obtain the moment of inertia for a thin circular disk about an axis parallel to the disk. This indicates the sizes of the terms you neglect when you use the approximate expressions for the moments of inertia of a "slender" bar and a "thin" disk.

Problems

8.100 The axis L_O is perpendicular to both segments of the L-shaped slender bar. The mass of the bar is 6 kg and the material is homogeneous. Use integration to determine its moment of inertia about L_O.

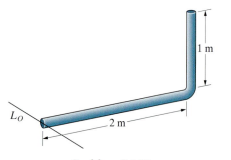

Problem 8.100

8.101 Two homogeneous slender bars, each of mass m and length l, are welded together to form the T-shaped object. Use integration to determine the moment of inertia of the object about the axis through point 0 that is perpendicular to the bars.

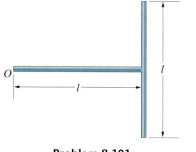

Problem 8.101

8.102 The slender bar lies in the x–y plane. Its mass is 6 kg and the material is homogeneous. Use integration to determine its moment of inertia about the z axis.

8.103 Use integration to determine the moment of inertia of the slender 6-kg bar about the y axis.

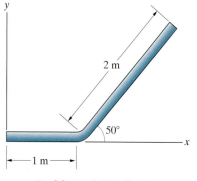

Problems 8.102/8.103

8.104 The homogeneous thin plate has mass $m = 12$ kg and dimensions $b = 1$ m and $h = 2$ m. Determine its moments of inertia about the x, y, and z axes.

Strategy: The moments of inertia of a thin plate of arbitrary shape are given by Eqs. (8.30)–(8.32) in terms of the moments of inertia of the cross-sectional area of the plate. You can obtain the moments of inertia of the triangular area from Appendix B.

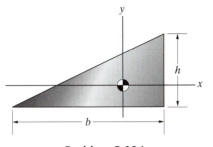

Problem 8.104

8.105 The homogeneous thin plate is of uniform thickness and mass m.

(a) Determine its moments of inertia about the x and z axes.

(b) Let $R_i = 0$ and compare your results with the values given in Appendix C for a thin circular plate.

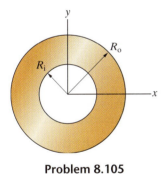

Problem 8.105

8.106 The homogeneous thin plate is of uniform thickness and weighs 20 lb. Determine its moment of inertia about the y axis.

8.107 Determine the moment of inertia of the plate about the x axis.

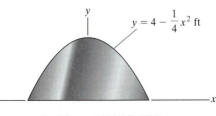

$$y = 4 - \frac{1}{4}x^2 \text{ ft}$$

Problems 8.106/8.107

8.108 The mass of the object is 10 kg. Its moment of inertia about L_1 is 10 kg-m². What is its moment of inertia about L_2? (The three axes lie in the same plane.)

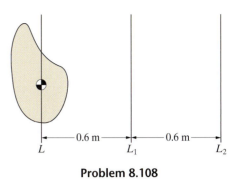

Problem 8.108

8.109 An engineer gathering data for the design of a maneuvering unit determines that the astronaut's center of mass is at $x = 1.01$ m, $y = 0.16$ m and that her moment of inertia about the z axis is 105.6 kg-m². Her mass is 81.6 kg. What is her moment of inertia about the z' axis through her center of mass?

Problem 8.109

8.110 Two homogeneous slender bars, each of mass m and length l, are welded together to form the T-shaped object. Use the parallel-axis theorem to determine the moment of inertia of the object about the axis through point O that is perpendicular to the bars.

8.111 Use the parallel-axis theorem to determine the moment of inertia of the T-shaped object about the axis through the center of mass of the object that is perpendicular to the two bars.

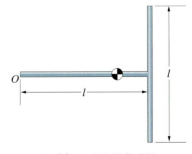

Problems 8.110/8.111

8.112 The mass of the homogeneous slender bar is 20 kg. Determine its moment of inertia about the z axis.

8.113 Determine the moment of inertia of the 20-kg bar about the z' axis through its center of mass.

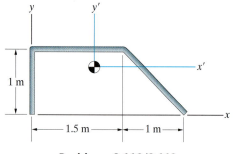

Problems 8.112/8.113

8.114 The homogeneous slender bar weighs 5 lb. Determine its moment of inertia about the z axis.

8.115 Determine the moment of inertia of the 5-lb bar about the z' axis through its center of mass.

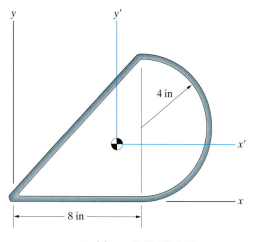

Problems 8.114/8.115

8.116 The rocket is used for atmospheric research. Its weight and its moment of inertia about the z axis through its center of mass (including its fuel) are 10 kip and 10,200 slug-ft^2, respectively. The rocket's fuel weighs 6000 lb, its center of mass is located at $x = -3$ ft, $y = 0$, $z = 0$, and the moment of inertia of the fuel about the axis through the fuel's center of mass parallel to z is 2200 slug-ft^2. When the fuel is exhausted, what is the rocket's moment of inertia about the axis through its new center of mass parallel to z?

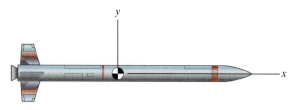

Problem 8.116

8.117 The mass of the homogeneous thin plate is 36 kg. Determine its moment of inertia about the x axis.

8.118 Determine the moment of inertia of the 36-kg plate about the z axis.

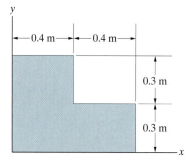

Problems 8.117/8.118

8.119 The homogeneous thin plate weighs 10 lb. Determine its moment of inertia about the x axis.

8.120 Determine the moment of inertia of the 10-lb plate about the y axis.

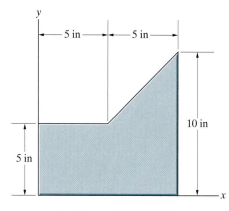

Problems 8.119/8.120

8.121 The thermal radiator (used to eliminate excess heat from a satellite) can be modeled as a homogeneous thin rectangular plate. Its mass is 5 slugs. Determine its moments of inertia about the x, y, and z axes.

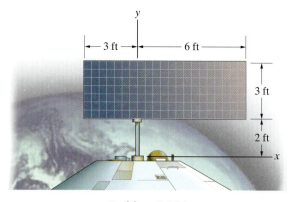

Problem 8.121

8.122 The mass of the homogeneous thin plate is 2 kg. Determine its moment of inertia about the axis L_O through point O that is perpendicular to the plate.

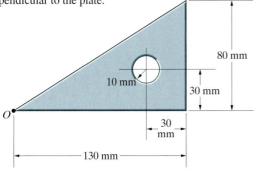

Problem 8.122

8.123 The homogeneous cone is of mass m. Determine its moment of inertia about the z axis, and compare your result with the value given in Appendix C.

Strategy: Use the same approach we used in Example 8.14 to obtain the moments of inertia of a homogeneous cylinder.

8.124 Determine the moments of inertia of the homogeneous cone of mass m about the x and y axes, and compare your results with the values given in Appendix C.

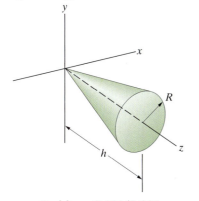

Problems 8.123/8.124

8.125 The homogeneous object has the shape of a truncated cone and consists of bronze with mass density $\rho = 8200 \text{ kg/m}^3$. Determine its moment of inertia about the z axis.

8.126 Determine the moment of inertia of the object described in Problem 8.125 about the x axis.

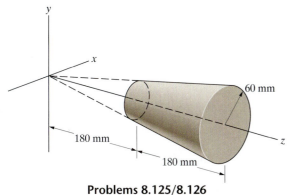

Problems 8.125/8.126

8.127 The homogeneous rectangular parallelepiped is of mass m. Determine its moments of inertia about the x, y, and z axes, and compare your results with the values given in Appendix C.

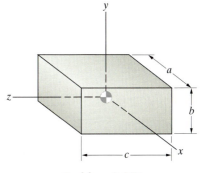

Problem 8.127

8.128 The L-shaped machine part is composed of two homogeneous bars. Bar 1 is tungsten alloy with density 14,000 kg/m³, and bar 2 is steel with density 7800 kg/m³. Determine its moment of inertia about the x axis.

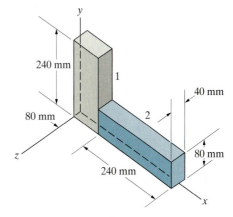

Problem 8.128

8.129 The sphere-capped cone consists of material with density 7800 kg/m^3. The radius $R = 80$ mm. Determine its moments of inertia about the x and y axes.

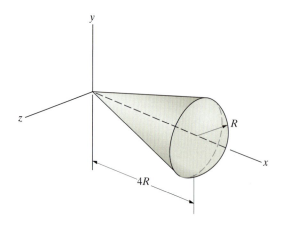

Problem 8.129

8.130 The circular cylinder is made of aluminum (Al) with density 2700 kg/m^3 and iron (Fe) with density 7860 kg/m^3. Determine its moments of inertia about the x' and y' axes.

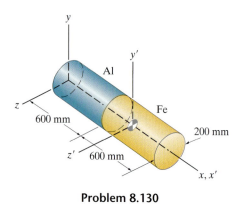

Problem 8.130

8.131 The homogeneous half-cylinder is of mass m. Determine its moment of inertia about the axis L through its center of mass.

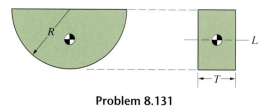

Problem 8.131

8.132 The homogeneous machine part is made of aluminum alloy with density $\rho = 2800$ kg/m^3. Determine its moment of inertia about the z axis.

8.133 Determine the moment of inertia of the machine part described in Problem 8.132 about the x axis.

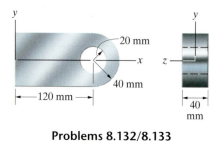

Problems 8.132/8.133

8.134 The object consists of steel of density $\rho = 7800$ kg/m^3. Determine its moment of inertia about the axis L_O.

8.135 Determine the moment of inertia of the object in Problem 8.134 about the axis through the center of mass of the object parallel to L_O.

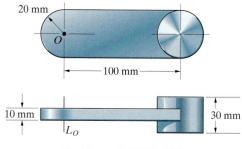

Problems 8.134/8.135

8.136 The thick plate consists of steel of density $\rho = 15$ slug/ft^3. Determine its moment of inertia about the z axis.

8.137 Determine the moment of inertia of the plate in Problem 8.136 about the x axis.

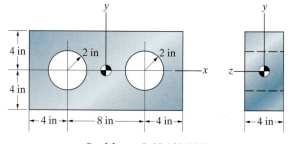

Problems 8.136/8.137

CHAPTER SUMMARY

Areas

Four moments of inertia of an area are defined (Fig. a):

1. The moment of inertia about the x axis:

$$I_x = \int_A y^2 \, dA. \tag{8.1}$$

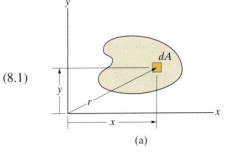

(a)

2. The moment of inertia about the y axis:

$$I_y = \int_A x^2 \, dA. \tag{8.3}$$

3. The product of inertia:

$$I_{xy} = \int_A xy \, dA. \tag{8.5}$$

4. The polar moment of inertia:

$$J_O = \int_A r^2 \, dA. \tag{8.6}$$

The *radii of gyration* about the x- and y-axes are defined by $k_x = \sqrt{I_x/A}$ and $k_y = \sqrt{I_y/A}$, respectively, and the radius of gyration about the origin O is defined by $k_O = \sqrt{J_O/A}$.

The polar moment of inertia is equal to the sum of the moments of inertia about the x and y axes: $J_O = I_x + I_y$. If an area is symmetric about either the x axis or the y axis, its product of inertia is zero.

Let $x'y'$ be a coordinate system with its origin at the centroid of an area A, and let xy be a parallel coordinate system. The moments of inertia of A in terms of the two systems are related by the *parallel-axis theorems* [Eqs. (8.10)–(8.13)], namely:

$$I_x = I_{x'} + d_y^2 A,$$

$$I_y = I_{y'} + d_x^2 A,$$

$$I_{xy} = I_{x'y'} + d_x d_y A,$$

$$J_O = J_O' + (d_x^2 + d_y^2)A = J_O' + d^2 A,$$

where d_x and d_y are the coordinates of the centroid of A in the xy coordinate system.

Masses

The moment of inertia of an object about an axis L_O is (Fig. b)

$$I_O = \int_m r^2 \, dm, \tag{8.27}$$

where r is the perpendicular distance from L_O to the differential element of mass dm.

Let L be an axis through the center of mass of an object, and let L_O be a parallel axis (Fig. c). The moment of inertia I_O about L_O is given in terms of the moment of inertia I about L by the parallel-axis theorem,

$$I_O = I + d^2 m, \tag{8.35}$$

where m is the mass of the object and d is the perpendicular distance between L and L_O.

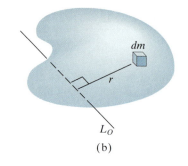

(b)

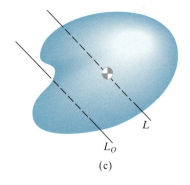

(c)

Review Problems

8.138 Determine I_y and k_y.

8.139 Determine I_x and k_x.

8.140 Determine J_O and k_O.

8.141 Determine I_{xy}.

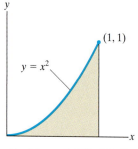

Problems 8.138–8.141

8.142 Determine I_y and k_y.

8.143 Determine I_x and k_x.

8.144 Determine I_{xy}.

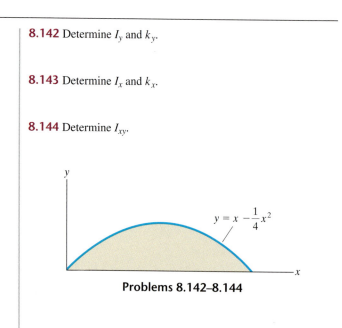

Problems 8.142–8.144

8.145 Determine $I_{y'}$ and $k_{y'}$.

8.146 Determine $I_{x'}$ and $k_{x'}$.

8.147 Determine $I_{x'y'}$.

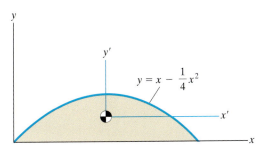
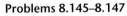

$$y = x - \frac{1}{4}x^2$$

Problems 8.145–8.147

8.148 Determine I_y and k_y.

8.149 Determine I_x and k_x.

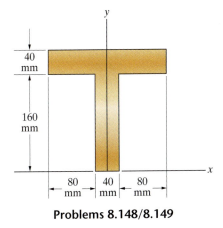

Problems 8.148/8.149

8.150 Determine I_x and k_x.

8.151 Determine J_O and k_O.

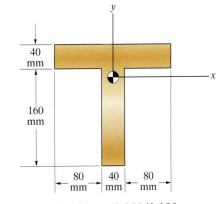

Problems 8.150/8.151

8.152 Determine I_y and k_y.

8.153 Determine J_O and k_O.

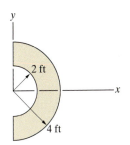

Problems 8.152/8.153

8.154 Determine I_x and k_x.

8.155 Determine I_y and k_y.

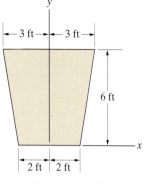
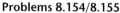

y

3 ft — 3 ft

6 ft

2 ft 2 ft

x

Problems 8.154/8.155

8.156 The moments of inertia of the area are $I_x = 36 \text{ m}^4$, $I_y = 145 \text{ m}^4$, and $I_{xy} = 44.25 \text{ m}^4$. Determine a set of principal axes and the principal moments of inertia.

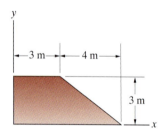

y

3 m — 4 m

3 m

x

Problem 8.156

8.157 The moment of inertia of the 31-oz bat about a perpendicular axis through point B is 0.093 slug-ft^2. What is the bat's moment of inertia about a perpendicular axis through point A? (Point A is the bat's "instantaneous center," or center of rotation, at the instant shown.)

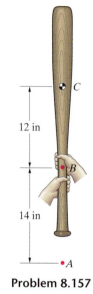

C

12 in

B

14 in

A

Problem 8.157

8.158 The mass of the thin homogeneous plate is 4 kg. Determine its moment of inertia about the y axis.

8.159 Determine the moment of inertia of the 4-kg plate about the z axis.

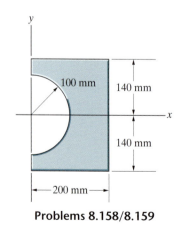

y

100 mm

140 mm

x

140 mm

200 mm

Problems 8.158/8.159

8.160 The homogeneous pyramid is of mass m. Determine its moment of inertia about the z axis.

8.161 Determine the moments of inertia of the homogeneous pyramid of mass m about the x and y axes.

Problems 8.160/8.161

8.162 The homogeneous object weighs 400 lb. Determine its moment of inertia about the x axis.

8.163 Determine the moments of inertia of the 400-lb object about the y and z axes.

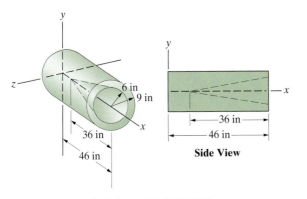

Problems 8.162/8.163

8.164 Determine the moment of inertia of the 14-kg flywheel about the axis L.

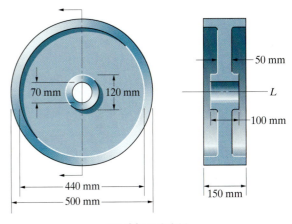

Problem 8.164

CHAPTER

9

Friction

Friction forces have many important effects, both desirable and undesirable, in engineering applications. The Coulomb theory of friction allows us to estimate the maximum friction forces that can be exerted by contacting surfaces and the friction forces exerted by sliding surfaces. This opens the path to the analysis of important new classes of supports and machines, including wedges (shims), threaded connections, bearings, and belts.

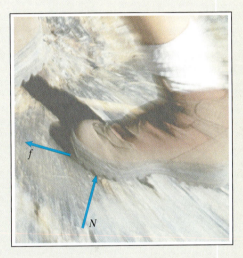

◄ Shoe soles are designed to support the friction forces necessary to prevent slipping. In this chapter we analyze friction forces between surfaces in contact.

9.1 Theory of Dry Friction

Suppose that a person climbs a ladder that leans against a smooth wall. Fig. 9.1 shows the free-body diagram of the person and ladder. If the person is stationary on the ladder, we can use the equilibrium equations to determine the friction force. But there is another question that we cannot answer using the equilibrium equations alone: Will the ladder remain in place, or will it slip on the floor? If a truck is parked on an incline, the total friction force exerted on its tires by the road prevents it from sliding down the incline (Fig. 9.1b). We can use the equilibrium equations to determine the total friction force. But here too there is another question that we cannot answer: What is the steepest incline on which the truck could be parked without slipping?

Friction force

(a)

Friction force

(b)

Figure 9.1
Objects supported by friction forces.

To answer these questions, we must examine the nature of friction forces in more detail. Place a book on a table and push it with a small horizontal force, as shown in Fig. 9.2a. If the force you exert is sufficiently small, the book does not move. The free-body diagram of the book is shown in Fig. 9.2b. The force W is the book's weight, and N is the total normal force exerted by the table on the surface of the book that is in contact with the table. The force F is the horizontal force you apply, and f is the total friction force exerted by the table. Because the book is in equilibrium, $f = F$.

Figure 9.2
(a) Exerting a horizontal force on a book.
(b) The free-body diagram of the book.

(a)

(b)

Now slowly increase the force you apply to the book. As long as the book remains in equilibrium, the friction force must increase correspondingly, since it equals the force you apply. When the force you apply becomes too large, the book moves. It slips on the table. After reaching some maximum value, the friction force can no longer maintain the book in equilibrium. Also, notice that the force you must apply to keep the book moving on the table is smaller than the force required to cause it to slip. (You are familiar with this phenomenon if you've ever pushed a piece of furniture across a floor.)

How does the table exert a friction force on the book? Why does the book slip? Why is less force required to slide the book across the table than is required to start it moving? If the surfaces of the table and the book are magnified sufficiently, they will appear rough (Fig. 9.3). Friction forces arise in part from the interactions of the roughnesses, or *asperities*, of the contacting surfaces. We can gain insight into this mechanism of friction by considering a simple two-dimensional model of the rough surfaces of the book and table.

Suppose that we idealize the asperities of the book and table as the mating two-dimensional "saw-tooth" profiles in Fig. 9.4a. As the horizontal force F increases, the book will remain stationary until the force is sufficiently large to cause the book to slide upward as shown in Fig. 9.4b. What horizontal force is necessary for this to occur? To find out, we must determine the value of F necessary for the book to be in equilibrium in the "slipped" position in Fig. 9.4b. The normal force C_i exerted on the ith saw-tooth asperity of the book is shown in Fig. 9.4c. (Notice that in this simple model we assume the contacting surfaces of the asperities to be smooth.) Denoting the sum of the normal forces exerted on the asperities of the book by the table by $C = \sum_i C_i$, we obtain the equilibrium equations

$$\Sigma F_x = F - C \sin \alpha = 0,$$

$$\Sigma F_y = C \cos \alpha - W = 0.$$

Eliminating C from these equations, we obtain the force necessary to cause the book to slip on the table:

$$F = (\tan \alpha)W.$$

We see that *the force necessary to cause the book to slip is proportional to the force pressing the saw-tooth surfaces together* (the book's weight). Think about stacking increasing numbers of books and applying a horizontal force to them. A progressively larger force is required to cause them to slip as the number of books increases. Also, in our two-dimensional thought experiment, *the angle α is a measure of the roughness of the saw-tooth surfaces.* As $\alpha \rightarrow 0$, the surfaces become smooth and the force necessary to cause the book to slip approaches zero. As α increases, the roughness increases and the force necessary to cause the book to slip increases.

In the sections that follow, we present a theory that incorporates the basic phenomena we have just described and that has been found useful for determining friction forces between dry surfaces. (Friction between lubricated surfaces is a hydrodynamic phenomenon and must be analyzed in the context of fluid mechanics.)

Coefficients of Friction

The theory of dry friction, or *Coulomb friction*, predicts the maximum friction forces that can be exerted by dry, contacting surfaces that are stationary relative to each other. It also predicts the friction forces exerted by the surfaces when they are in relative motion, or sliding. We first consider surfaces that are not in relative motion.

The Static Coefficient The magnitude of the *maximum* friction force that can be exerted between two plane, dry surfaces in contact that are not in motion relative to one another is

$$f = \mu_s N, \tag{9.1}$$

where N is the normal component of the contact force between the surfaces and μ_s is a constant called the *coefficient of static friction*. The value of μ_s is assumed

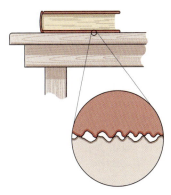

Figure 9.3
The roughnesses of the surfaces can be seen in a magnified view.

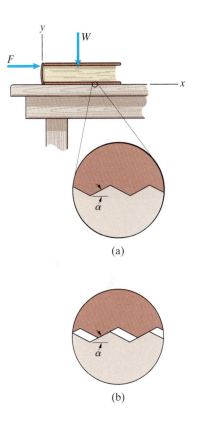

(a)

(b)

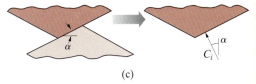

(c)

Figure 9.4
(a) Two-dimensional model of rough surfaces in contact.
(b) Slip of the book relative to the table.
(c) Normal force on one of the book's asperities.

TABLE 9.1 Typical values of the coefficient of static friction.

Materials	Coefficient of Static Friction μ_s
Metal on metal	0.15–0.20
Masonry on masonry	0.60–0.70
Wood on wood	0.25–0.50
Metal on masonry	0.30–0.70
Metal on wood	0.20–0.60
Rubber on concrete	0.50–0.90

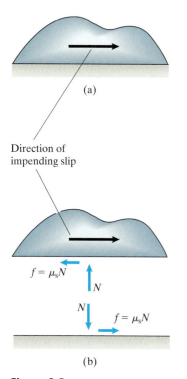

(a)

Direction of impending slip

$f = \mu_s N$

N

N

$f = \mu_s N$

(b)

Figure 9.5
(a) The upper surface is on the verge of slipping to the right.
(b) Directions of the friction forces.

to depend only on the materials of the contacting surfaces and the conditions (smoothness and degree of contamination by other materials) of the surfaces. Typical values of μ_s for various materials are shown in Table 9.1. The relatively large range of values for each pair of materials reflects the sensitivity of μ_s to the conditions of the surfaces. In engineering applications it is usually necessary to measure the value of μ_s for the actual surfaces used.

Let us return to the example of the book on the table (Fig. 9.2). If a *specified* horizontal force F is applied to the book, and the book remains in equilibrium, what friction force is exerted on the book by the table? We can see from the free-body diagram in Fig. 9.2b that $f = F$. Notice that we do not use Eq. (9.1) to answer this question. But suppose that we want to know the *largest* force F that can be applied to the book without causing it to slip. If we know the coefficient of static friction μ_s between the book and the table, Eq. (9.1) tells us the largest friction force that the table can exert on the book. Therefore, the largest force F that can be applied without causing the book to slip is $F = f = \mu_s N$. We also know from the free-body diagram in Fig. 9.2b that $N = W$, so the largest force that will not cause the book to slip is $F = \mu_s W$.

Equation (9.1) determines the magnitude of the maximum friction force but not its direction. The friction force is a maximum, and Eq. (9.1) is applicable, when two surfaces are on the verge of slipping relative to each other. We say that slip is *impending*, and the friction forces resist the impending motion. In Fig. 9.5a, suppose that the lower surface is fixed and slip of the upper surface toward the right is impending. The friction force on the upper surface resists its impending motion (Fig. 9.5b). The friction force on the lower surface is in the opposite direction.

The Kinetic Coefficient According to the theory of dry friction, the magnitude of the friction force between two plane dry contacting surfaces that are in motion (sliding) relative to each other is

$$f = \mu_k N, \tag{9.2}$$

where N is the normal force between the surfaces and μ_k is the *coefficient of kinetic friction*. The value of μ_k is assumed to depend only on the compositions of the surfaces and their conditions. For a given pair of surfaces, its value is generally smaller than that of μ_s.

Once you have caused the book in Fig. 9.2 to begin sliding on the table, the friction force $f = \mu_k N = \mu_k W$. Therefore, the force you must exert to keep the book in uniform motion is $F = f = \mu_k W$.

When two surfaces are sliding relative to each other, the friction forces resist the relative motion. In Fig. 9.6a, suppose that the lower surface is fixed and the upper surface is moving to the right. The friction force on the upper surface acts in the direction opposite to its motion (Fig. 9.6b). The friction force on the lower surface is in the opposite direction.

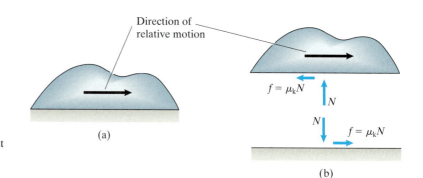

Direction of relative motion

$f = \mu_k N$

N

N

$f = \mu_k N$

(a)

(b)

Figure 9.6
(a) The upper surface is moving to the right relative to the lower surface.
(b) Directions of the friction forces.

Angles of Friction

We have expressed the reaction exerted on a surface due to its contact with another surface in terms of its components parallel and perpendicular to the surface, the friction force f and normal force N (Fig. 9.7a). In some situations it is more convenient to express the reaction in terms of its magnitude R and the *angle of friction* θ between the reaction and the normal to the surface (Fig. 9.7b). The forces f and N are related to R and θ by

$$f = R \sin\theta, \tag{9.3}$$

$$N = R \cos\theta. \tag{9.4}$$

The value of θ when slip is impending is called the *angle of static friction* θ_s, and its value when the surfaces are sliding relative to each other is called the *angle of kinetic friction* θ_k. By using Eqs. (9.1)–(9.4), we can express the angles of static and kinetic friction in terms of the coefficients of friction:

$$\tan\theta_s = \mu_s, \tag{9.5}$$

$$\tan\theta_k = \mu_k. \tag{9.6}$$

In summary, if slip is impending, the magnitude of the friction force is given by Eq. (9.1) and the angle of friction by Eq. (9.5). If surfaces are sliding relative to each other, the magnitude of the friction force is given by Eq. (9.2) and the angle of friction by Eq. (9.6). Otherwise, the friction force must be determined from the equilibrium equations. The sequence of decisions in evaluating the friction force and angle of friction is summarized in Fig. 9.8.

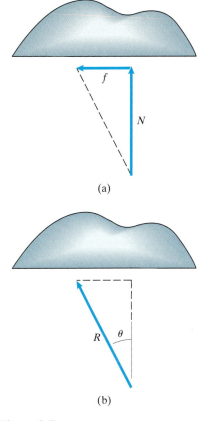

(a)

(b)

Figure 9.7
(a) The friction force f and the normal force N.
(b) The magnitude R and the angle of friction θ.

Study Questions

1. How is the coefficient of static friction defined?
2. How is the coefficient of kinetic friction defined?
3. If relative slip of two dry surfaces in contact is impending, what do you know about the friction forces they exert on each other?
4. If two dry surfaces in contact are sliding relative to each other, what do you know about the resulting friction forces?

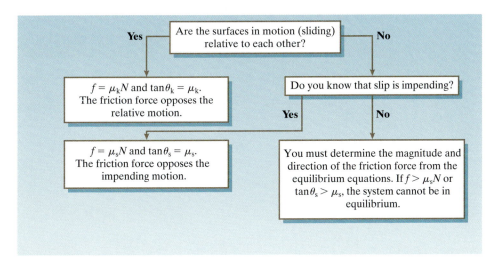

Figure 9.8
Evaluating the friction force.

Example 9.1 | Determining the Friction Force

The arrangement in Fig. 9.9 exerts a horizontal force on the stationary 180-lb crate. The coefficient of static friction between the crate and the ramp is $\mu_s = 0.4$.

(a) If the rope exerts a 90-lb force on the crate, what is the friction force exerted on the crate by the ramp?

(b) What is the largest force the rope can exert on the crate without causing it to slide up the ramp?

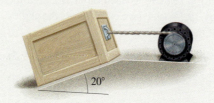

Figure 9.9

Strategy

(a) We can follow the logic in Fig. 9.8 to decide how to evaluate the friction force. The crate is not sliding on the ramp, and we don't know whether slip is impending, so we must determine the friction force by using the equilibrium equations.

(b) We want to determine the value of the force exerted by the rope that causes the crate to be on the verge of slipping up the ramp. When slip is impending, the magnitude of the friction force is $f = \mu_s N$ and the friction force opposes the impending slip. We can use the equilibrium equations to determine the force exerted by the rope.

Solution

(a) We draw the free-body diagram of the crate in Fig. a, showing the force T exerted by the rope, the weight W of the crate, and the normal force N and friction force f exerted by the ramp. We can choose the direction of f arbitrarily, and our solution will indicate the actual direction of the friction force. By aligning the coordinate system with the ramp as shown, we obtain the equilibrium equation

$$\Sigma F_x = f + T \cos 20° - W \sin 20° = 0.$$

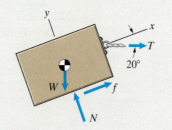

(a) Free-body diagram of the crate.

Solving for the friction force, we obtain

$$f = -T \cos 20° + W \sin 20°$$

$$= -(90 \text{ lb})\cos 20° + (180 \text{ lb})\sin 20°$$

$$= -23.0 \text{ lb}.$$

The minus sign indicates that the direction of the friction force on the crate is down the ramp.

(b) In this case the friction force is $f = \mu_s N$, and it opposes the impending slip. To simplify our solution for T, we align the coordinate system as shown in Fig. b, obtaining the equilibrium equations

$$\Sigma F_x = T - N \sin 20° - \mu_s N \cos 20° = 0,$$

$$\Sigma F_y = N \cos 20° - \mu_s N \sin 20° - W = 0.$$

Solving the second equilibrium equation for N, we obtain

$$N = \frac{W}{\cos 20° - \mu_s \sin 20°}$$

$$= \frac{180 \text{ lb}}{\cos 20° - 0.4 \sin 20°}$$

$$= 224 \text{ lb}.$$

Then, from the first equilibrium equation, the force T is

$$T = N(\sin 20° + \mu_s \cos 20°)$$

$$= (224 \text{ lb})(\sin 20° + 0.4 \cos 20°)$$

$$= 161 \text{ lb}.$$

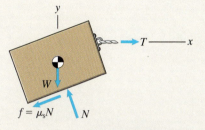

(b) The free-body diagram when slip up the ramp is impending.

Critical Thinking

When you use the equilibrium equations to determine a friction force, often you will not know its direction beforehand. We designed part (a) of this example to illustrate this. Depending on the value of the force T exerted on the crate by the rope, the friction force exerted on the crate by the ramp can point either up or down the ramp. In drawing the free-body diagram of the crate in Fig. a, we arbitrarily assumed that the friction force pointed up the ramp. The negative value we obtained from the equilibrium equations, $f = -23.0$ lb, tells us that the force is in the opposite direction, down the ramp. In contrast, when you use the equation $f = \mu_s N$, *the friction force must point in the correct direction on the free-body diagram*. In drawing the free-body diagram in Fig. b, how did we know the direction of the friction force? We wanted to determine the largest force T that would not cause the crate to slide up the ramp, so we assumed that the slip of the crate *up* the ramp was impending. This told us that the friction force, resisting the impending slip, pointed *down* the ramp.

Example 9.2   Determining Whether an Object Will Tip Over

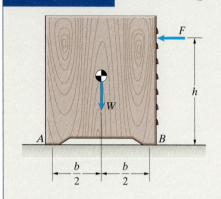

Figure 9.10

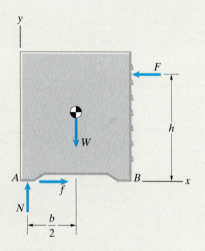

(a) The free-body diagram when the chest is on the verge of tipping over.

Suppose that we want to push the tool chest in Fig. 9.10 across the floor by applying the horizontal force F. If we apply the force at too great a height h, the chest will tip over before it slips. If the coefficient of static friction between the floor and the chest is μ_s, what is the largest value of h for which the chest will slip before it tips over?

Strategy

When the chest is on the verge of tipping over, it is in equilibrium with no reaction at B. We can use this condition to determine F in terms of h. Then, by determining the value of F that will cause the chest to slip, we will obtain the value of h that causes the chest to be on the verge of tipping over *and* on the verge of slipping.

Solution

We draw the free-body diagram of the chest when it is on the verge of tipping over in Fig. a. Summing moments about A, we obtain

$$\Sigma M_{\text{point } A} = Fh - W\left(\frac{1}{2}b\right) = 0.$$

Equilibrium also requires that $f = F$ and $N = W$.

When the chest is on the verge of slipping,

$$f = \mu_s N,$$

so

$$F = f = \mu_s N = \mu_s W.$$

Substituting this expression into the moment equation, we obtain

$$\mu_s Wh - W\left(\frac{1}{2}b\right) = 0.$$

Solving this equation for h, we find that when the chest is on the verge of slipping, it is also on the verge of tipping over if it is pushed at the height

$$h = \frac{b}{2\mu_s}.$$

If h is smaller than this value, the chest will begin sliding before it tips over.

Critical Thinking

Notice that the largest value of h for which the chest will slip before it tips over is independent of F. Whether the chest will tip over depends only on where the force is applied, not how large it is. What is the motivation for the solution in this example? The possibility of heavy objects falling over is an obvious safety hazard, and analyses of this kind can influence their design. Once they are in use, safety engineers can establish guidelines (for example, by marking a horizontal line on a vertical cabinet or machine above which it should not be pushed) to prevent tipping.

Example 9.3 Analyzing a Friction Brake

The motion of the disk in Fig. 9.11 is controlled by the friction force exerted at
C by the brake ABC. The hydraulic actuator BE exerts a horizontal force of
magnitude F on the brake at B. The coefficients of friction between the disk
and the brake are μ_s and μ_k. What couple M is necessary to rotate the disk at a
constant rate in the counterclockwise direction?

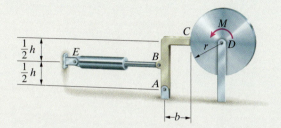

Figure 9.11

Strategy
We can use the free-body diagram of the disk to obtain a relation between M and
the reaction exerted on the disk by the brake, then use the free-body diagram of
the brake to determine the reaction in terms of F.

Solution
We draw the free-body diagram of the disk in Fig. a, representing the force exerted by the brake by a single force R. The force R opposes the counterclockwise rotation of the disk, and the friction angle is the angle of kinetic friction
$\theta_k = \arctan \mu_k$. Summing moments about D, we obtain

$$\Sigma M_{\text{point } D} = M - (R \sin \theta_k)r = 0.$$

(a) The free-body diagram of the disk.

Then, from the free-body diagram of the brake (Fig. b), we get

$$\Sigma M_{\text{point } A} = -F\left(\frac{1}{2}h\right) + (R \cos \theta_k)h - (R \sin \theta_k)b = 0.$$

We can solve these two equations for M and R. The solution for the couple M is

$$M = \frac{(1/2)hr\,F \sin \theta_k}{h \cos \theta_k - b \sin \theta_k} = \frac{(1/2)hr\,F\mu_k}{h - b\mu_k}.$$

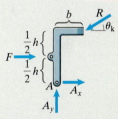

(b) The free-body diagram of the brake.

Critical Thinking
If the friction coefficient μ_k is sufficiently small, the denominator in our solution for the couple M, the term $h \cos \theta_k - b \sin \theta_k$, is positive. As μ_k increases, the denominator becomes smaller, because $\cos \theta_k$ decreases and $\sin \theta_k$
increases. As the denominator approaches zero, the couple required to rotate
the disk approaches infinity. To understand this result, notice that the denominator equals zero when $\tan \theta_k = h/b$, which means that the line of action of the
force R passes through point A (Fig. c). As μ_k increases and the line of action
of R approaches point A, the magnitude of R necessary to balance the moment
due to F about A approaches infinity. As a result, the *analytical prediction* for
M approaches infinity. Of course, at some value of M, the forces F and R would
exceed the values the brake could support.

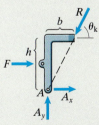

(c) The line of action of R passing through
point A.

Example 9.4   A Friction Problem in Three Dimensions

The 80-kg climber at A in Fig. 9.12 is supported on an icy slope by friends. The tensions in ropes AB and AC are 130 N and 220 N, respectively. The y axis is vertical, and the unit vector $\mathbf{e} = -0.182\mathbf{i} + 0.818\mathbf{j} + 0.545\mathbf{k}$ is perpendicular to the ground where the climber stands. What minimum coefficient of static friction between the climber's shoes and the ground is necessary to prevent him from slipping?

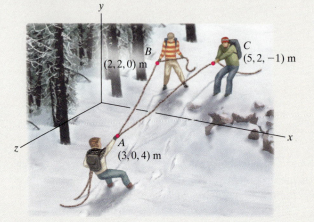

Figure 9.12

Strategy

We know the forces exerted on the climber by the two ropes and by his weight, so we can use equilibrium to determine the force $\mathbf{R}$ exerted on him by the ground. The components of $\mathbf{R}$ normal and parallel to the ground are the normal and friction forces exerted on him by the ground. By calculating them, we can obtain the minimum necessary coefficient of static friction.

Solution

We draw the free-body diagram of the climber in Fig. a, showing the forces $\mathbf{T}_{AB}$ and $\mathbf{T}_{AC}$ exerted by the ropes, the force $\mathbf{R}$ exerted by the ground, and his weight. The sum of the forces equals zero:

$$\mathbf{R} + \mathbf{T}_{AB} + \mathbf{T}_{AC} - mg\mathbf{j} = \mathbf{0}.$$

By expressing $\mathbf{T}_{AB}$ and $\mathbf{T}_{AC}$ in terms of their components, we can solve this equation for the components of $\mathbf{R}$. The force $\mathbf{T}_{AB}$ is

$$\mathbf{T}_{AB} = |\mathbf{T}_{AB}| \left[\frac{(2-3)\mathbf{i} + (2-0)\mathbf{j} + (0-4)\mathbf{k}}{\sqrt{(2-3)^2 + (2-0)^2 + (0-4)^2}} \right]$$

$$= (130 \text{ N})(-0.218\mathbf{i} + 0.436\mathbf{j} - 0.873\mathbf{k})$$

$$= -28.4\mathbf{i} + 56.7\mathbf{j} - 113.5\mathbf{k} \text{ (N)},$$

and the force $\mathbf{T}_{AC}$ is

$$\mathbf{T}_{AC} = |\mathbf{T}_{AC}| \left[\frac{(5-3)\mathbf{i} + (2-0)\mathbf{j} + (-1-4)\mathbf{k}}{\sqrt{(5-3)^2 + (2-0)^2 + (-1-4)^2}} \right]$$

$$= (220 \text{ N})(0.348\mathbf{i} + 0.348\mathbf{j} - 0.870\mathbf{k})$$

$$= 76.6\mathbf{i} + 76.6\mathbf{j} - 191.5\mathbf{k} \text{ (N)}.$$

Substituting these expressions into the equilibrium equation and solving for **R**, we obtain

$$\mathbf{R} = -48.2\mathbf{i} + 651.5\mathbf{j} + 305.0\mathbf{k} \text{ (N)}.$$

The normal force on the climber is the component of **R** perpendicular to the surface, which is the component of **R** parallel to the unit vector **e**:

$$\mathbf{N} = (\mathbf{e} \cdot \mathbf{R})\mathbf{e}$$

$$= [(-0.182)(-48.2 \text{ N}) + (0.818)(651.5 \text{ N}) + (0.545)(305.0 \text{ N})]\mathbf{e}$$

$$= -129\mathbf{i} + 579\mathbf{j} + 386\mathbf{k} \text{ (N)}.$$

The magnitude of the normal force is $N = |\mathbf{N}| = 708$ N. The friction force on the climber is the magnitude of the component of **R** parallel to the surface:

$$f = |\mathbf{R} - \mathbf{N}| = 135 \text{ N}.$$

The minimum coefficient of friction necessary to prevent the climber from slipping is therefore

$$\mu_s = \frac{f}{N} = \frac{135 \text{ N}}{708 \text{ N}} = 0.191.$$

Critical Thinking

Notice the role of the unit vector **e** in this example. By using equilibrium, we were able to determine the *total* force **R** exerted on the climber by the ground. This force consists of components normal and parallel to the ground (the normal and friction forces respectively). How could we determine the normal and friction forces in order to calculate the minimum coefficient of friction necessary to hold the climber in place? The unit vector **e** specified the orientation of the ground on which he stood and also allowed us to determine the normal and friction forces.

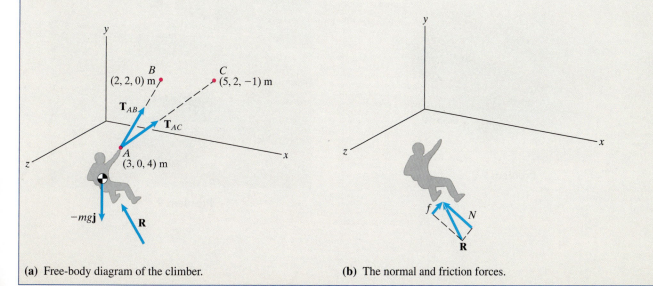

(a) Free-body diagram of the climber.   **(b)** The normal and friction forces.

Problems

9.1 The coefficients of static and kinetic friction between the 2-lb book and the table are $\mu_s = 0.30$ and $\mu_k = 0.28$. A person exerts a horizontal force on the book.

(a) If the person exerts a 0.4-lb force, what is the magnitude of the friction force the table exerts on the book?

(b) What is the largest force the person can exert without causing the book to slip?

(c) If the person pushes the book across the table at a constant speed, what is the magnitude of the friction force on the book?

Problem 9.1

9.2 The coefficients of static and kinetic friction between the 2-lb book and the table are $\mu_s = 0.30$ and $\mu_k = 0.28$. The table is tilted at 15° relative to the horizontal. A person exerts a force on the book that is *parallel to the table*.

(a) If the person exerts a 0.4-lb force, what is the magnitude of the friction force the table exerts on the book?

(b) What is the largest force the person can exert without causing the book to slip?

(c) If the person pushes the book up the table at a constant speed, what is the magnitude of the friction force on the book?

Problem 9.2

9.3 A student pushes a 200-lb box of books across the floor. The coefficient of kinetic friction between the carpet and the box is $\mu_k = 0.15$.

(a) If he exerts the force F at angle $\alpha = 25°$, what is the magnitude of the force he must exert to slide the box across the floor?

(b) If he bends his knees more and exerts the force F at angle $\alpha = 10°$, what is the magnitude of the force he must exert to slide the box?

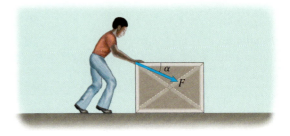

Problem 9.3

9.4 The 2975-lb car is parked on a 30° slope. The brakes are applied to both its front and rear wheels.

(a) What is the magnitude of the total friction force exerted on the car by the road?

(b) What minimum coefficient of static friction is necessary for the car to remain in equilibrium in this position?

   Strategy: Represent the sums of the friction forces and normal forces on the car's tires by a single friction force and a single normal force.

Problem 9.4

9.5 The truck's winch exerts a horizontal force on the 200-kg crate in an effort to pull it down the ramp. The coefficient of static friction between the crate and the ramp is $\mu_s = 0.6$.

(a) If the winch exerts a 200-N horizontal force on the crate, what is the magnitude of the friction force exerted on the crate by the ramp?

(b) What is the magnitude of the horizontal force the winch must exert on the crate to cause it to start moving down the ramp?

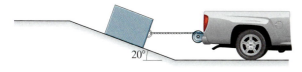

Problem 9.5

9.6 The device shown is designed to position pieces of luggage on a ramp. It exerts a force parallel to the ramp. The mass of the suitcase is 9 kg. The coefficients of friction between the suitcase and ramp are $\mu_s = 0.20$ and $\mu_k = 0.18$.

(a) Will the suitcase remain stationary on the ramp when the device exerts no force on it?

(b) What force must the device exert to start the suitcase moving up the ramp?

(c) What force must the device exert to move the suitcase up the ramp at a constant speed?

Problem 9.6

9.7 The coefficient of static friction between the 50-kg crate and the ramp is $\mu_s = 0.35$. The unstretched length of the spring is 800 mm, and the spring constant is $k = 660$ N/m. What is the minimum value of x at which the crate can remain stationary on the ramp?

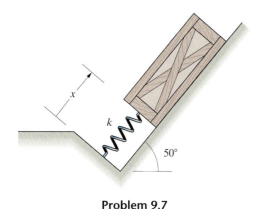

Problem 9.7

9.8 The coefficient of kinetic friction between the 40-kg crate and the slanting floor is $\mu_k = 0.3$. If the angle $\alpha = 20°$, what tension must the person exert on the rope to move the crate at constant speed?

9.9 In Problem 9.8, for what angle α is the tension necessary to move the crate at constant speed a minimum? What is the necessary tension?

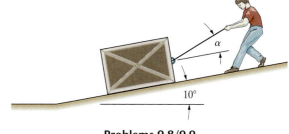

Problems 9.8/9.9

9.10 Box A weighs 100 lb and box B weighs 30 lb. The coefficients of friction between box A and the ramp are $\mu_s = 0.30$ and $\mu_k = 0.28$. What is the magnitude of the friction force exerted on box A by the ramp?

9.11 Box A weighs 100 lb, and the coefficients of friction between box A and the ramp are $\mu_s = 0.30$ and $\mu_k = 0.28$. For what range of weights of the box B will the system remain stationary?

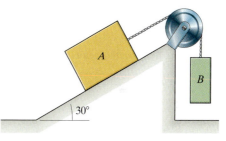

Problems 9.10/9.11

9.12 The mass of the box on the left is 30 kg, and the mass of the box on the right is 40 kg. The coefficient of static friction between each box and the inclined surface is $\mu_s = 0.2$. Determine the minimum angle α for which the boxes will remain stationary.

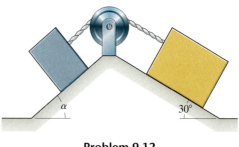

Problem 9.12

9.13 The coefficient of kinetic friction between the 100-kg box and the inclined surface is 0.35. Determine the tension T necessary to pull the box up the surface at a constant rate.

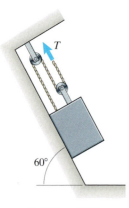

60°

Problem 9.13

9.14 The box is stationary on the inclined surface. The coefficient of static friction between the box and the surface is μ_s.

(a) If the mass of the box is 10 kg, $\alpha = 20°$, $\beta = 30°$, and $\mu_s = 0.24$, what force T is necessary to start the box sliding up the surface?

(b) Show that the force T necessary to start the box sliding up the surface is a minimum when $\tan \beta = \mu_s$.

9.15 In explaining observations of ship launchings at the port of Rochefort in 1779, Coulomb analyzed the system shown to determine the minimum force T necessary to hold the box stationary on the inclined surface. Show that the result is

$$T = \frac{(\sin \alpha - \mu_s \cos \alpha)mg}{\cos \beta - \mu_s \sin \beta}.$$

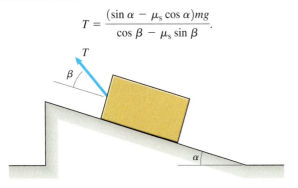

Problems 9.14/9.15

9.16 Two sheets of plywood A and B lie on the bed of the truck. They have the same weight W, and the coefficient of static friction between the two sheets of wood and between sheet B and the truck bed is μ_s.

(a) If you apply a horizontal force to sheet A and apply no force to sheet B, can you slide sheet A off the truck without causing sheet B to move? What force is necessary to cause sheet A to start moving?

(b) If you prevent sheet A from moving by exerting a horizontal force on it, what horizontal force on sheet B is necessary to start it moving?

Problem 9.16

9.17 The weights of the two boxes are $W_1 = 100$ lb and $W_2 = 50$ lb. The coefficients of kinetic friction between the left box and the inclined surface are $\mu_s = 0.12$ and $\mu_k = 0.10$. Determine the tension the man must exert on the rope to pull the boxes upward at a constant rate.

9.18 In Problem 9.17, for what range of tensions exerted on the rope by the man will the boxes remain stationary?

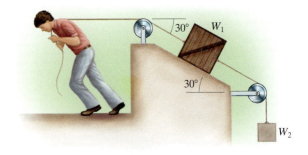

Problems 9.17/9.18

9.19 Each box weighs 10 lb. The coefficient of static friction between box A and box B is 0.24, and the coefficient of static friction between box B and the inclined surface is 0.3. What is the largest angle α for which box B will not slip?

Strategy: Draw individual free-body diagrams of the two boxes and write their equilibrium equations assuming that slip of box B is impending.

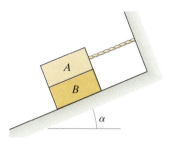

Problem 9.19

9.20 The masses of the boxes are $m_A = 15$ kg and $m_B = 60$ kg. The coefficient of static friction between boxes A and B and between box B and the inclined surface is 0.12. What is the largest force F for which the boxes will not slip?

9.21 In Problem 9.20, what is the smallest force F for which the boxes will not slip?

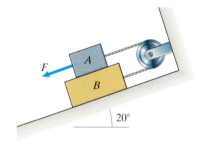

Problems 9.20/9.21

9.22 The weights of the boxes are $W_A = 65$ lb and $W_B = 130$ lb. The coefficient of static friction between boxes A and B and between box B and the floor is 0.12. What is the largest force F for which the boxes will not slip?

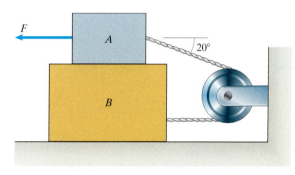

Problem 9.22

9.23 A sander consists of a rotating cylinder with sandpaper bonded to the outer surface. The normal force exerted on the workpiece A by the sander is 30 lb. The workpiece A weighs 50 lb. The coefficients of friction between the sander and the workpiece A are $\mu_s = 0.65$ and $\mu_k = 0.60$. The coefficients of friction between the workpiece A and the table are $\mu_s = 0.35$ and $\mu_k = 0.30$. Will the workpiece remain stationary while it is being sanded?

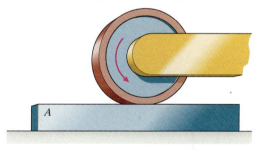

Problem 9.23

9.24 Suppose that you want the bar of length L to act as a simple brake that will allow the workpiece A to slide to the left but will not allow it to slide to the right no matter how large a horizontal force is applied to it. The weight of the bar is W, and the coefficient of static friction between it and the workpiece A is μ_s. You can neglect friction between the workpiece and the surface it rests on.

(a) What is the largest angle α for which the bar will prevent the workpiece from moving to the right?

(b) If α has the value determined in (a), what horizontal force is necessary to slide the workpiece A toward the left at a constant rate?

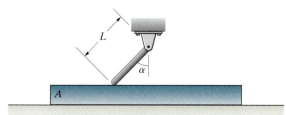

Problem 9.24

9.25 The mass of the bar is 4 kg. The coefficient of static friction between the bar and the floor is 0.3. Neglect friction between the bar and the wall.

(a) If $\alpha = 20°$, what is the magnitude of the friction force exerted on the bar by the floor?

(b) What is the maximum angle α for which the bar will not slip?

9.26 The coefficient of static friction between the bar and the floor and between the 4-kg bar and the wall is 0.3. What is the maximum angle α for which the bar will not slip?

1 m

α

Problems 9.25/9.26

9.27 The ladder and the person weigh 30 lb and 180 lb, respectively. The center of mass of the 12-ft ladder is at its midpoint. The angle $\alpha = 30°$. Assume that the wall exerts a negligible friction force on the ladder.

(a) If $x = 4$ ft, what is the magnitude of the friction force exerted on the ladder by the floor?

(b) What minimum coefficient of static friction between the ladder and the floor is necessary for the person to be able to climb to the top of the ladder without slipping?

9.28 The ladder and the person weigh 30 lb and 180 lb, respectively. The center of mass of the 12-ft ladder is at its midpoint. The coefficient of static friction between the ladder and the floor is $\mu_s = 0.5$. What is the largest value of the angle α for which the person could climb to the top of the ladder without it slipping?

9.29 The ladder and the person weigh 30 lb and 180 lb, respectively. The center of mass of the 12-ft ladder is at its midpoint. The coefficient of static friction between the ladder and the floor is 0.5 and the coefficient of friction between the ladder and the wall is 0.3. What is the largest value of the angle α for which the person could climb to the top of the ladder without it slipping? Compare your answer to the answer to Problem 9.28.

α

x

Problems 9.27–9.29

9.30 The disk weighs 50 lb and the bar weighs 25 lb. The coefficient of friction between the disk and the inclined surface are $\mu_s = 0.6$ and $\mu_k = 0.5$.

(a) What is the largest couple M that can be applied to the stationary disk without causing it to start rotating?

(b) What couple M is necessary to rotate the disk at a constant rate?

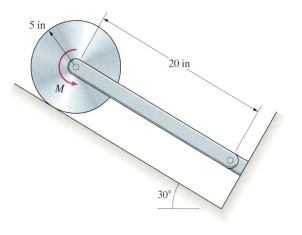

5 in

20 in

M

30°

Problem 9.30

9.31 The cylinder has weight W. The coefficient of static friction between the cylinder and the floor and between the cylinder and the wall is μ_s. What is the largest couple M that can be applied to the stationary cylinder without causing it to rotate?

9.32 Suppose that $\alpha = 30°$ and a couple $M = 0.5RW$ is required to turn the cylinder of weight W at a constant rate. What is the coefficient of kinetic friction?

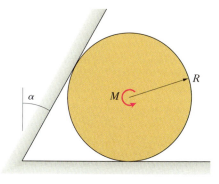

α

M

R

Problems 9.31/9.32

9.33 The disk of weight W and radius R is held in equilibrium on the circular surface by a couple M. The coefficient of static friction between the disk and the surface is μ_s. Show that the largest value M can have without causing the disk to slip is

$$M = \frac{\mu_s RW}{\sqrt{1 + \mu_s^2}}.$$

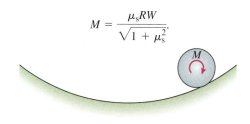

M

Problem 9.33

9.34 The coefficient of static friction between the blades of the shears and the object they are gripping is 0.36. What is the largest value of the angle α for which the object will not slip out? Neglect the object's weight.

Strategy: Draw the free-body diagram of the object and assume that slip is impending.

α

Problem 9.34

9.35 A stationary disk of 300-mm radius is attached to a pin support at D. The disk is held in place by the brake ABC in contact with the disk at C. The hydraulic actuator BE exerts a horizontal 400-N force on the brake at B. The coefficients of friction between the disk and the brake are $\mu_s = 0.6$ and $\mu_k = 0.5$. What couple must be applied to the stationary disk to cause it to slip in the counterclockwise direction?

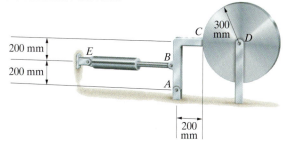

200 mm

200 mm

E

B

A

C

300 mm

D

200 mm

Problem 9.35

9.36 The figure shows a preliminary conceptual idea for a device to exert a braking force on a rope when the rope is pulled downward by the force T. The coefficient of kinetic friction between the rope and the two bars is $\mu_k = 0.28$. Determine the force T necessary to pull the rope downward at a constant rate if $F = 10$ lb and (a) $\alpha = 30°$; (b) $\alpha = 20°$.

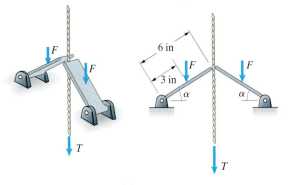

Problem 9.36

9.37 The mass of block B is 8 kg. The coefficient of static friction between the surfaces of the clamp and the block is $\mu_s = 0.2$. When the clamp is aligned as shown, what minimum force must the spring exert to prevent the block from slipping out?

9.38 By altering its dimensions, redesign the clamp in Problem 9.37 so that the minimum force the spring must exert to prevent the block from slipping out is 180 N. Draw a sketch of your new design.

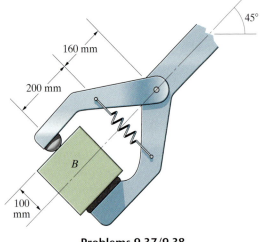

Problems 9.37/9.38

9.39 The horizontal bar is attached to a collar that slides on the smooth vertical bar. The collar at P slides on the smooth horizontal bar. The total mass of the horizontal bar and the two collars is 12 kg. The system is held in place by the pin in the circular slot. The pin contacts only the lower surface of the slot, and the coefficient of static friction between the pin and the slot is 0.8. If the system is in equilibrium and $y = 260$ mm, what is the magnitude of the friction force exerted on the pin by the slot?

9.40 In Problem 9.39, what is the minimum height y at which the system can be in equilibrium?

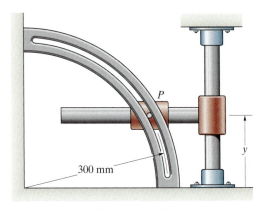

Problems 9.39/9.40

9.41 The rectangular 100-lb plate is supported by the pins A and B. If friction can be neglected at A and the coefficient of static friction between the pin at B and the slot is $\mu_s = 0.4$, what is the largest angle α for which the plate will not slip?

9.42 If you can neglect friction at B and the coefficient of static friction between the pin at A and the slot is $\mu_s = 0.4$, what is the largest angle α for which the 100-lb plate will not slip?

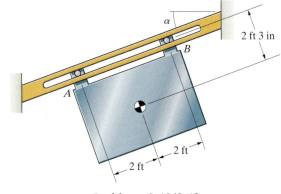

Problems 9.41/9.42

9.43 The airplane's weight is $W = 2400$ lb. Its brakes keep the rear wheels locked, and the coefficient of static friction between the wheels and the runway is $\mu_s = 0.6$. The front (nose) wheel can turn freely and so exerts a negligible friction force on the runway. Determine the largest horizontal thrust force T the plane's propeller can generate without causing the rear wheels to slip.

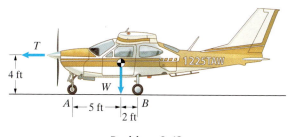

Problem 9.43

9.44 The mass of the refrigerator is 110 kg. It is supported at A and B. The distances $h = 1520$ mm and $b = 360$ mm. The coefficient of static friction at A and B is 0.2.

(a) What force F is necessary for impending slip?

(b) Will the refrigerator tip over before it slips?

9.45 In Problem 9.44, what is the maximum coefficient of static friction at A and B for which the refrigerator will slip before it tips over?

Problems 9.44/9.45

9.46 To obtain a preliminary evaluation of the stability of a turning car, imagine subjecting the stationary car to an increasing lateral force F at the height of its center of mass, and determine whether the car will slip (skid) laterally before it tips over. Show that this will be the case if $b/h > 2\mu_s$. (Notice the importance of the height of the center of mass relative to the width of the car. This reflects on recent discussions of the stability of sport utility vehicles and vans that have relatively high centers of mass.)

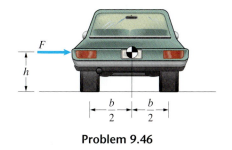

Problem 9.46

9.47 The man exerts a force P on the car at an angle $\alpha = 20°$. The 1760-kg car has front wheel drive. The driver spins the front wheels, and the coefficient of kinetic friction is $\mu_k = 0.02$. Snow behind the rear tires exerts a horizontal resisting force S. Getting the car to move requires overcoming a resisting force $S = 420$ N. What force P must the man exert?

9.48 In Problem 9.47, what value of the angle α minimizes the magnitude of the force P the man must exert to overcome the resisting force $S = 420$ N exerted on the rear tires by the snow? What force must he exert?

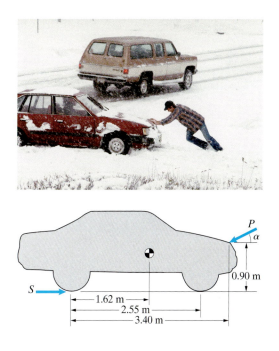

Problems 9.47/9.48

9.49 The coefficient of static friction between the 3000-lb car's tires and the road is $\mu_s = 0.5$. Determine the steepest grade (the largest value of the angle α) the car can drive up at constant speed if the car has (a) rear-wheel drive; (b) front-wheel drive; (c) four-wheel drive.

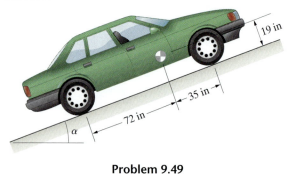

Problem 9.49

9.50 The stationary cabinet has weight W. Determine the force F that must be exerted to cause it to move if (a) the coefficient of static friction at A and at B is μ_s; (b) the coefficient of static friction at A is μ_{sA} and the coefficient of static friction at B is μ_{sB}.

Problem 9.50

9.51 The table weighs 50 lb and the coefficient of static friction between its legs and the inclined surface is 0.7.

(a) If you apply a force at A parallel to the inclined surface to push the table up the inclined surface, will the table tip over before it slips? If not, what force is required to start the table moving up the surface?

(b) If you apply a force at B parallel to the inclined surface to push the table down the inclined surface, will the table tip over before it slips? If not, what force is required to start the table moving down the surface?

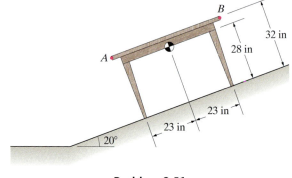

Problem 9.51

9.52 The coefficient of static friction between the right bar and the surface at A is $\mu_s = 0.6$. Neglect the weights of the bars. If $\alpha = 20°$, what is the magnitude of the friction force exerted at A?

9.53 The coefficient of static friction between the right bar and the surface at A is $\mu_s = 0.6$. Neglect the weights of the bars. What is the largest angle α at which the truss will remain stationary without slipping?

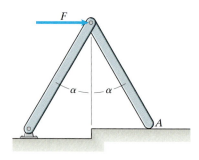

Problems 9.52/9.53

9.54 Each of the uniform 2-ft bars weighs 4 lb. Neglect the weight of the collar at P. The coefficient of static friction between the collar and the horizontal bar is $\mu_s = 0.6$. If the system is in equilibrium and the angle $\theta = 45°$, what is the magnitude of the friction force exerted on the collar by the horizontal bar?

9.55 In Problem 9.54, what is the minimum coefficient of static friction between the collar P and the horizontal bar necessary for the system to be in equilibrium when $\theta = 45°$?

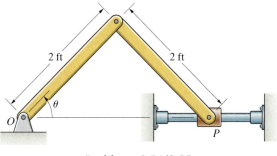

Problems 9.54/9.55

9.56 The weight of the box is 20 lb and the coefficient of static friction between the box and the floor is $\mu_s = 0.65$. Neglect the weights of the bars. What is the largest value of the force F that will not cause the box to slip?

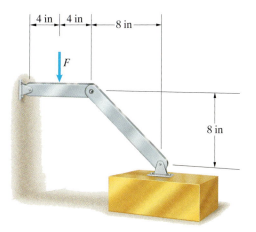

Problem 9.56

9.57 The mass of the suspended object is 6 kg. The structure is supported at B by the normal and friction forces exerted on the plate by the wall. Neglect the weights of the bars.

(a) What is the magnitude of the friction force exerted on the plate at B?

(b) What is the minimum coefficient of static friction at B necessary for the structure to remain in equilibrium?

9.58 Suppose that the lengths of the bars in Problem 9.57 are $L_{AB} = 1.2$ m and $L_{AC} = 1.0$ m and their masses are $m_{AB} = 3.6$ kg and $m_{AC} = 3.0$ kg.

(a) What is the magnitude of the friction force exerted on the plate at B?

(b) What is the minimum coefficient of static friction at B necessary for the structure to remain in equilibrium?

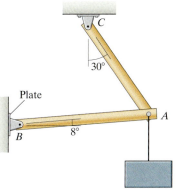

Problems 9.57/9.58

9.59 The frame is supported by the normal and friction forces exerted on the plates at A and G by the fixed surfaces. The coefficient of static friction at A is $\mu_s = 0.6$. Will the frame slip at A when it is subjected to the loads shown?

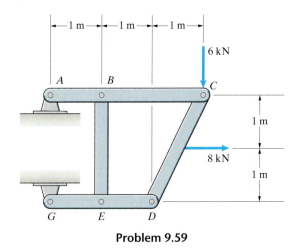

Problem 9.59

9.60 The frame is supported by the normal and friction forces exerted on the plate at A by the wall.

(a) What is the magnitude of the friction force exerted on the plate at A?

(b) What is the minimum coefficient of static friction at A necessary for the structure to remain in equilibrium?

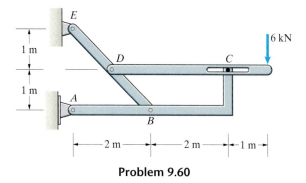

6 kN

E

D

C

A

B

2 m — 2 m — 1 m

Problem 9.60

9.61 The direction cosines of the crane's cable are $\cos \theta_x = 0.588$, $\cos \theta_y = 0.766$, $\cos \theta_z = 0.260$. The y axis is vertical. The stationary caisson to which the cable is attached weighs 2000 lb and rests on horizontal ground. If the coefficient of static friction between the caisson and the ground is $\mu_s = 0.4$, what tension in the cable is necessary to cause the caisson to slip?

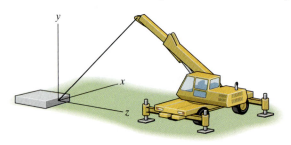

Problem 9.61

9.62* The 10-lb metal disk A is at the center of the inclined surface. The tension in the string AB is 5 lb. What minimum coefficient of static friction between the disk and the surface is necessary to keep the disk from slipping?

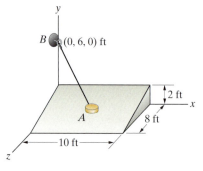

B (0, 6, 0) ft

2 ft

x

A

8 ft

10 ft

z

Problem 9.62

9.63* The 5-kg box is at rest on the sloping surface. The y axis points upward. The unit vector $0.557\mathbf{i} + 0.743\mathbf{j} + 0.371\mathbf{k}$ is perpendicular to the sloping surface. What is the magnitude of the friction force exerted on the box by the surface?

9.64* In Problem 9.63, what is the minimum coefficient of static friction necessary for the box to remain at rest on the sloping surface?

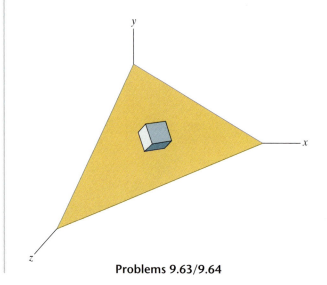

y

x

z

Problems 9.63/9.64

9.2 Applications

Effects of friction forces, such as wear, loss of energy, and generation of heat, are often undesirable. But many devices cannot function properly without friction forces and may actually be designed to create them. A car's brakes work by exerting friction forces on the rotating wheels, and its tires are designed to maximize the friction forces they exert on the road under various weather conditions. In this section we analyze several types of devices in which friction forces play important roles.

Wedges

A *wedge* is a bifacial tool with the faces set at a small acute angle (Figs. 9.13a and b). When a wedge is pushed forward, the faces exert large normal forces as a result of the small angle between them (Fig. 9.13c). In various forms, wedges are used in many engineering applications.

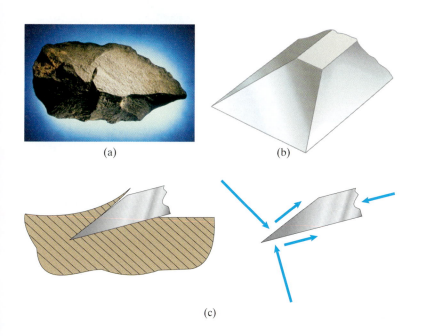

(a)   (b)

(c)

Figure 9.13
(a) An early wedge tool—a bifacial "hand axe" from Olduvai Gorge, East Africa.
(b) A modern chisel blade.
(c) The faces of a wedge can exert large lateral forces.

The large lateral force generated by a wedge can be used to lift a load (Fig. 9.14a). Let W_L be the weight of the load and W_W the weight of the wedge. To determine the force F necessary to start raising the load, we assume that slip of the load and wedge are impending (Fig. 9.14b). From the free-body diagram of the load, we obtain the equilibrium equations

$$\Sigma F_x = Q - N \sin \alpha - \mu_s N \cos \alpha = 0$$

and

$$\Sigma F_y = N \cos \alpha - \mu_s N \sin \alpha - \mu_s Q - W_L = 0.$$

From the free-body diagram of the wedge, we obtain the equations

$$\Sigma F_x = N \sin \alpha + \mu_s N \cos \alpha + \mu_s P - F = 0$$

and

$$\Sigma F_y = P - N \cos \alpha + \mu_s N \sin \alpha - W_W = 0.$$

These four equations determine the three normal forces Q, N, and P, and the force F. The solution for F is

$$F = \mu_s W_W + \left[\frac{(1 - \mu_s^2) \tan \alpha + 2\mu_s}{(1 - \mu_s^2) - 2\mu_s \tan \alpha} \right] W_L.$$

Suppose that $W_W = 0.2 W_L$ and $\alpha = 10°$. If $\mu_s = 0$, the force necessary to lift the load is only $0.176 W_L$. But if $\mu_s = 0.2$, the force becomes $0.680 W_L$, and if $\mu_s = 0.4$, it becomes $1.44 W_L$. From this standpoint, friction is undesirable. But if there were no friction, the wedge would not remain in place when the force F is removed.

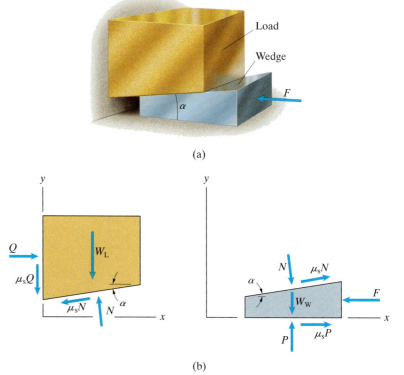

(a)

(b)

Figure 9.14
(a) Raising a load with a wedge.
(b) Free-body diagrams of the load and the wedge when slip is impending.

Example 9.5	Forces on a Wedge

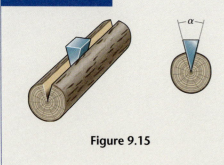

Figure 9.15

Splitting a log must have been among the first applications of the wedge (Fig. 9.15). Although it is a dynamic process—the wedge is hammered into the wood—you can get an idea of the forces involved from a static analysis. Suppose that $\alpha = 10°$ and the coefficients of friction between the surfaces of the wedge and the log are $\mu_s = 0.22$ and $\mu_k = 0.20$. Neglect the weight of the wedge.

(a) If the wedge is driven into the log at a constant rate by a vertical force F, what are the magnitudes of the normal forces exerted on the log by the wedge?
(b) Will the wedge remain in place in the log when the force is removed?

Strategy
(a) The friction forces resist the motion of the wedge into the log and are equal to $\mu_k N$, where N is the normal force the log exerts on the faces. We can use equilibrium to determine N in terms of F.
(b) By assuming that the wedge is on the verge of slipping out of the log, we can determine the minimum value of μ_s necessary for the wedge to stay in place.

Solution

(a) In Fig. a we draw the free-body diagram of the wedge as it is pushed into the log by a force F. The faces of the wedge are subjected to normal forces and friction forces by the log. The friction forces resist the motion of the wedge. From the equilibrium equation

$$2N\sin\left(\frac{\alpha}{2}\right) + 2\mu_k N\cos\left(\frac{\alpha}{2}\right) - F = 0,$$

we obtain the normal force N:

$$N = \frac{F}{2[\sin(\alpha/2) + \mu_k \cos(\alpha/2)]} = \frac{F}{2[\sin(10°/2) + (0.20)\cos(10°/2)]}$$

$$= 1.75F.$$

(b) In Fig. b we draw the free-body diagram when $F = 0$ and the wedge is on the verge of slipping out. From the equilibrium equation

$$2N\sin\left(\frac{\alpha}{2}\right) - 2\mu_s N\cos\left(\frac{\alpha}{2}\right) = 0,$$

we obtain the minimum coefficient of friction necessary for the wedge to remain in place:

$$\mu_s = \tan\left(\frac{\alpha}{2}\right) = \tan\left(\frac{10°}{2}\right) = 0.087.$$

Critical Thinking

We can also obtain this result by representing the reaction exerted on the wedge by the log as a single force (Fig. c). When the wedge is on the verge of slipping out, the friction angle is the angle of static friction θ_s. The sum of the forces in the vertical direction is zero only if

$$\theta_s = \arctan(\mu_s) = \frac{\alpha}{2} = 5°,$$

so $\mu_s = \tan 5° = 0.087$. Thus we conclude that the wedge will remain in place.

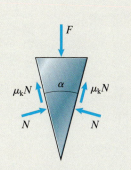

(a) Free-body diagram of the wedge with a vertical force F applied to it.

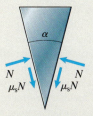

(b) Free-body diagram of the wedge when it is on the verge of slipping out.

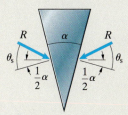

(c) Representing the reactions by a single force.

Threads

Threads are familiar from their use on wood screws, machine screws, and other machine elements. We show a shaft with square threads in Fig. 9.16a. The axial distance p from one thread to the next is called the *pitch* of the thread, and the angle α is its *slope*. We will consider only the case in which the shaft has a single continuous thread, so the relation between the pitch and slope is

$$\tan\alpha = \frac{p}{2\pi r}, \tag{9.7}$$

where r is the mean radius of the thread.

Suppose that the threaded shaft is enclosed in a fixed sleeve with a mating groove and is subjected to an axial load F (Fig. 9.16b). Applying a couple M in the direction shown will tend to cause the shaft to start rotating and moving in the axial direction opposite to F. Our objective is to determine the couple M necessary to cause the shaft to start rotating.

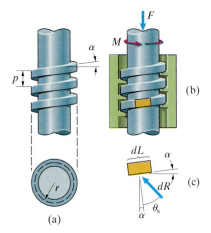

Figure 9.16
(a) A shaft with a square thread.
(b) The shaft within a sleeve with a mating groove and the direction of M that can cause the shaft to start moving in the axial direction opposite to F.
(c) A differential element of the thread when slip is impending.

We draw the free-body diagram of a differential element of the thread of length dL in Fig. 9.16c, representing the reaction exerted by the mating groove by the force dR. If the shaft is on the verge of rotating, dR resists the impending motion and the friction angle is the angle of static friction θ_s. The vertical component of the reaction on the element is $dR \cos(\theta_s + \alpha)$. To determine the total vertical force on the thread, we must integrate this expression over the length L of the thread. For equilibrium, the result must equal the axial force F acting on the shaft:

$$\cos(\theta_s + \alpha) \int_L dR = F. \qquad (9.8)$$

The moment about the center of the shaft due to the reaction on the element is $r\, dR \sin(\theta_s + \alpha)$. The total moment must equal the couple M exerted on the shaft:

$$r \sin(\theta_s + \alpha) \int_L dR = M.$$

Dividing this equation by Eq. (9.8), we obtain the couple M necessary for the shaft to be on the verge of rotating and moving in the axial direction opposite to F:

$$M = rF \tan(\theta_s + \alpha). \qquad (9.9)$$

Replacing the angle of static friction θ_s in this expression with the angle of kinetic friction θ_k gives the couple required to cause the shaft to rotate at a constant rate.

If the couple M is applied to the shaft in the opposite direction (Fig. 9.17a), the shaft tends to start rotating and moving in the axial direction of the load F. Figure 9.17b shows the reaction on a differential element of the thread of length dL when slip is impending. The direction of the reaction opposes the rotation of the shaft. In this case, the vertical component of the reaction on the element is $dR \cos(\theta_s - \alpha)$. Equilibrium requires that

$$\cos(\theta_s - \alpha) \int_L dR = F. \qquad (9.10)$$

The moment about the center of the shaft due to the reaction is $r\, dR \sin(\theta_s - \alpha)$, so

$$r \sin(\theta_s - \alpha) \int_L dR = M.$$

Dividing this equation by Eq. (9.10), we obtain the couple M necessary for the shaft to be on the verge of rotating and moving in the direction of the force F:

$$M = rF \tan(\theta_s - \alpha). \qquad (9.11)$$

Replacing θ_s with θ_k in this expression gives the couple necessary to rotate the shaft at a constant rate.

Notice in Eq. (9.11) that the couple required for impending motion is zero when $\theta_s = \alpha$. When the angle of static friction is less than this value, the shaft will rotate and move in the direction of the force F with no couple applied.

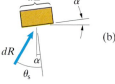

Figure 9.17
(a) The direction of M that can cause the shaft to move in the axial direction of F.
(b) A differential element of the thread when slip is impending.

Study Questions

1. How is the slope α of a thread defined?
2. If you know the pitch and mean radius of a thread, how do you determine its slope?
3. If a threaded shaft is subjected to an axial load, how do you determine the couple necessary to rotate the shaft at a constant rate and cause it to move in the direction opposite to the direction of the axial load?

Example 9.6   Rotating a Threaded Collar

The right end of bar AB in Fig. 9.18 is pinned to an unthreaded collar B that rests on a threaded collar C. The mean radius of the thread is $r = 1.6$ in and its pitch is $p = 0.2$ in. The coefficients of static and kinetic friction between the threads of the collar and the mating threads of the vertical shaft are $\mu_s = 0.25$ and $\mu_k = 0.22$. The 400-lb suspended object can be raised or lowered by rotating the collar C.

(a) When the system is in the position shown, what couple must be applied to the collar C to rotate it at a constant rate and cause the suspended object to move upward?
(b) Will the system remain in equilibrium in the position shown if no couple is applied to the collar C?

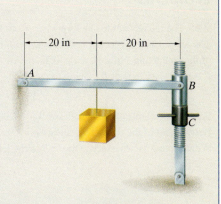

Figure 9.18

Strategy

(a) By drawing the free-body diagram of the bar and collar B, we can determine the axial force exerted on collar C. Then we can use Eq. (9.9), with θ_s replaced by θ_k, to determine the required couple.
(b) From Eq. (9.11), the collar C is on the verge of rotating and moving in the direction of the axial load when no couple is exerted on it if $\theta_s = \alpha$. If the angle of static friction θ_s is greater than or equal to the slope α, the system will remain in equilibrium with no couple applied.

Solution

(a) We draw the free-body diagram of the bar AB and the collar B in Fig. a, where F is the force exerted on the collar B by the collar C. From the equilibrium equation

$$\Sigma M_{\text{point } A} = (40 \text{ in})F - (20 \text{ in})(40 \text{ lb}) = 0,$$

we obtain $F = 200$ lb. This is the axial force exerted on collar C (Fig. b). Replacing θ_s by θ_k in Eq. (9.9), we obtain the couple necessary to rotate collar C at a constant rate:

$$M = rF \tan(\theta_k + \alpha).$$

The slope α is related to the pitch and mean radius of the thread by Eq. (9.7):

$$\tan\alpha = \frac{p}{2\pi r} = \frac{0.2 \text{ in}}{2\pi(1.6 \text{ in})} = 0.0199.$$

Therefore, $\alpha = \arctan(0.0199) = 1.14°$. The angle of kinetic friction is

$$\theta_k = \arctan(\mu_k) = \arctan(0.22) = 12.41°.$$

Using these values, we find that the required couple is

$$\begin{aligned} M &= rF \tan(\theta_k + \alpha) \\ &= (1.6 \text{ in})(200 \text{ lb}) \tan(12.41° + 1.14°) \\ &= 77.1 \text{ in-lb}. \end{aligned}$$

(b) The angle of static friction is

$$\theta_s = \arctan(\mu_s) = \arctan(0.25) = 14.04°.$$

We see that θ_s is greater than the slope α, so we conclude from Eq. (9.11) that the system will remain in equilibrium with no couple applied to the collar C.

(a) Free-body diagram of bar AB and the collar B.

(b) The threaded shaft and the collar C.

Critical Thinking

How would you determine the couple that must be applied to the collar C to cause the suspended object to move downward? In that case the collar C moves in the direction of the applied force F. Therefore Eq. (9.11), with θ_s replaced by θ_k, yields the necessary couple:

$$\begin{aligned} M &= rF \tan(\theta_k - \alpha) \\ &= (1.6 \text{ in})(200 \text{ lb}) \tan(12.41° - 1.14°) \\ &= 63.8 \text{ in-lb}. \end{aligned}$$

Problems

9.65 The masses of the blocks are $m_A = 40$ kg and $m_B = 10$ kg. The left surface of block A is in contact with the wall. Between all of the contacting surfaces, $\mu_s = 0.28$ and $\mu_k = 0.26$. What force F is necessary to start B moving to the left?

9.66 The weights of the blocks are $W_A = 100$ lb and $W_B = 25$ lb. Between all of the contacting surfaces, $\mu_s = 0.32$ and $\mu_k = 0.30$. What force F is necessary to move B to the left at a constant rate?

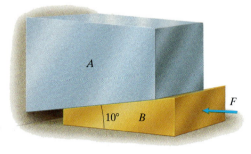

Problems 9.65/9.66

9.67 The wedge shown is being used to split the log. The wedge weighs 20 lb and the angle α equals 30°. The coefficient of kinetic friction between the faces of the wedge and the log is 0.28. If the normal force exerted by each face of the wedge must equal 150 lb to split the log, what vertical force F is necessary to drive the wedge into the log at a constant rate?

9.68 The coefficient of static friction between the faces of the wedge and the log in Problem 9.67 is 0.30. Will the wedge remain in place in the log when the vertical force F is removed?

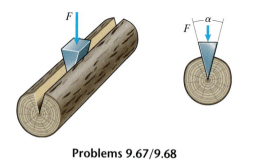

Problems 9.67/9.68

9.69 The masses of the blocks are $m_A = 50$ kg and $m_B = 42$ kg. Between all of the contacting surfaces, $\mu_s = 0.22$ and $\mu_k = 0.20$. What force F is necessary to start B moving to the left?

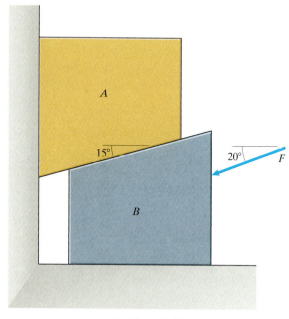

Problem 9.69

9.70 The stationary blocks A, B, and C each have a mass of 200 kg. Between all contacting surfaces, $\mu_s = 0.6$. What force F is necessary to start B moving downward?

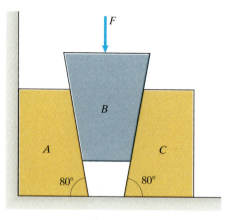

Problem 9.70

9.71 Small wedges called *shims* can be used to hold an object in place. The coefficient of kinetic friction between the contacting surfaces is 0.4. What force F is needed to push the shim downward until the horizontal force exerted on the object A is 200 N?

9.72 The coefficient of static friction between the contacting surfaces is 0.44. If the shims are in place and exert a 200-N horizontal force on the object A, what upward force must be exerted on the left shim to loosen it?

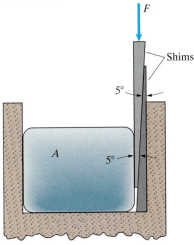

Problems 9.71/9.72

9.73 The crate A weighs 600 lb. Between all contacting surfaces, $\mu_s = 0.32$ and $\mu_k = 0.30$. Neglect the weights of the wedges. What force F is required to move A to the right at a constant rate?

9.74 Suppose that between all contacting surfaces, $\mu_s = 0.32$ and $\mu_k = 0.30$. Neglect the weights of the 5° wedges. If a force $F = 800$ N is required to move A to the right at a constant rate, what is the mass of A?

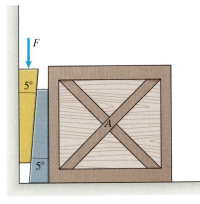

Problems 9.73/9.74

9.75 The box A has a mass of 80 kg, and the wedge B has a mass of 40 kg. Between all contacting surfaces, $\mu_s = 0.15$ and $\mu_k = 0.12$. What force F is required to raise A at a constant rate?

9.76 Suppose that A weighs 800 lb and B weighs 400 lb. The coefficients of friction between all of the contacting surfaces are $\mu_s = 0.15$ and $\mu_k = 0.12$. Will B remain in place if the force F is removed?

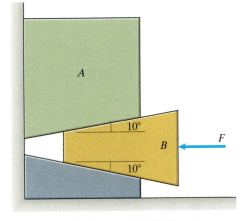

Problems 9.75/9.76

9.77 Between A and B, $\mu_s = 0.20$, and between B and C, $\mu_s = 0.18$. Between C and the wall, $\mu_s = 0.30$. The weights $W_B = 20$ lb and $W_C = 80$ lb. What force F is required to start C moving upward?

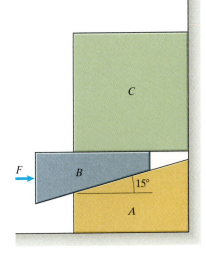

Problem 9.77

9.78 The masses of A, B, and C are 8 kg, 12 kg, and 80 kg, respectively. Between all contacting surfaces, $\mu_s = 0.4$. What force F is required to start C moving upward?

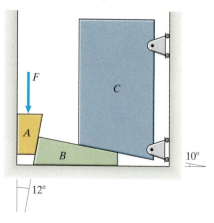

Problem 9.78

9.79 The vertical threaded shaft fits into a mating groove in the tube C. The pitch of the threaded shaft is $p = 0.1$ in, and the mean radius of the thread is $r = 0.5$ in. The coefficients of friction between the thread and the mating groove are $\mu_s = 0.15$ and $\mu_k = 0.10$. The weight $W = 200$ lb. Neglect the weight of the threaded shaft.

(a) Will the stationary threaded shaft support the weight if no couple is applied to the shaft?

(b) What couple must be applied to the threaded shaft to raise the weight at a constant rate?

9.80 The pitch of the threaded shaft is $p = 2$ mm and the mean radius of the thread is $r = 20$ mm. The coefficients of friction between the thread and the mating groove are $\mu_s = 0.22$ and $\mu_k = 0.20$. The weight $W = 500$ N. Neglect the weight of the threaded shaft. What couple must be applied to the threaded shaft to lower the weight at a constant rate?

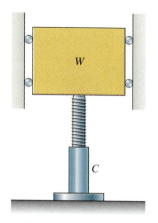

Problems 9.79/9.80

9.81 The position of the horizontal beam can be adjusted by turning the machine screw A. Neglect the weight of the beam. The pitch of the screw is $p = 1$ mm, and the mean radius of the thread is $r = 4$ mm. The coefficients of friction between the thread and the mating groove are $\mu_s = 0.20$ and $\mu_k = 0.18$. If the system is initially stationary, determine the couple that must be applied to the screw to cause the beam to start moving (a) upward; (b) downward.

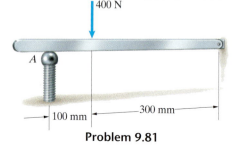

Problem 9.81

9.82 The pitch of the threaded shaft of the C clamp is $p = 0.05$ in, and the mean radius of the thread is $r = 0.15$ in. The coefficients of friction between the threaded shaft and the mating collar are $\mu_s = 0.18$ and $\mu_k = 0.16$.

(a) What maximum couple must be applied to the shaft to exert a 30-lb force on the clamped object?

(b) If a 30-lb force is exerted on the clamped object, what couple must be applied to the shaft to begin loosening the clamp?

Problem 9.82

9.83 The mass of block A is 60 kg. Neglect the weight of the 5° wedge. The coefficient of kinetic friction between the contacting surfaces of the block A, the wedge, the table, and the wall is $\mu_k = 0.4$. The pitch of the threaded shaft is 5 mm, the mean radius of the thread is 15 mm, and the coefficient of kinetic friction between the thread and the mating groove is 0.2. What couple must be exerted on the threaded shaft to raise the block A at a constant rate?

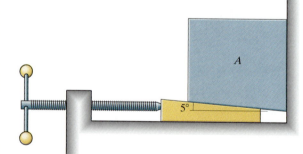

Problem 9.83

9.84 The vise exerts 80-lb forces on A. The threaded shafts are subjected only to axial loads by the jaws of the vise. The pitch of their threads is $p = 1/8$ in, the mean radius of the threads is $r = 1$ in, and the coefficient of static friction between the threads and the mating grooves is 0.2. Suppose that you want to loosen the vise by turning one of the shafts. Determine the couple you must apply (a) to shaft B; (b) to shaft C.

9.85 Suppose that you want to tighten the vise in Problem 9.84 by turning one of the shafts. Determine the couple you must apply
(a) to shaft B;
(b) to shaft C.

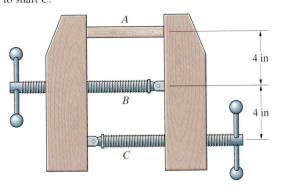

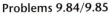

Problems 9.84/9.85

9.86 The threaded shaft has a ball and socket support at B. The 400-lb load A can be raised or lowered by rotating the threaded shaft, causing the threaded collar at C to move relative to the shaft. Neglect the weights of the members. The pitch of the shaft is $p = \frac{1}{4}$ in, the mean radius of the thread is $r = 1$ in, and the coefficient of static friction between the thread and the mating groove is 0.24. If the system is stationary in the position shown, what couple is necessary to start the shaft rotating to raise the load?

9.87 In Problem 9.86, if the system is stationary in the position shown, what couple is necessary to start the shaft rotating to lower the load?

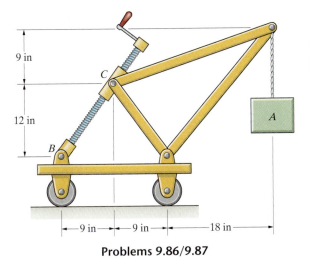

Problems 9.86/9.87

9.88 The car jack is operated by turning the threaded shaft at A. The threaded shaft fits into a mating collar at B. As the shaft turns, points A and B move closer together or farther apart, thereby raising or lowering the jack. The pitch of the threaded shaft is $p = 2.5$ mm, the mean radius of the thread is $r = 5.5$ mm, and the coefficient of kinetic friction between the threaded shaft and the mating collar is 0.15. What couple must be applied at A to rotate the shaft at a constant rate and raise the jack when it is in the position shown if the load $L = 6.5$ kN?

9.89 In Problem 9.88, what couple must be applied at A to rotate the shaft at a constant rate and lower the jack when it is in the position shown if the load $L = 6.5$ kN?

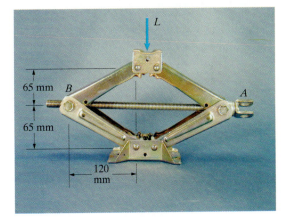

Problems 9.88/9.89

9.90 A *turnbuckle*, used to adjust the length or tension of a bar or cable, is threaded at both ends. Rotating it draws threaded segments of a bar or cable together or moves them apart. Suppose that the pitch of the threads is $p = 3$ mm, their mean radius is $r = 25$ mm, and the coefficient of static friction between the threads and the mating grooves is 0.24. If $T = 800$ N, what couple must be exerted on the turnbuckle to start tightening it?

9.91 In Problem 9.90, what couple must be exerted on the turnbuckle to start loosening it?

Problems 9.90/9.91

9.92 Member *BE* of the frame has a turnbuckle. (See Problem 9.90.) The threads have pitch $p = 1$ mm, their mean radius is $r = 6$ mm, and the coefficient of static friction between the threads and the mating grooves is 0.2. What couple must be exerted on the turnbuckle to start loosening it?

9.93 In Problem 9.92, what couple must be exerted on the turnbuckle to start tightening it?

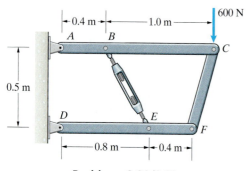

Problems 9.92/9.93

9.94 Members *CD* and *DG* of the truss have turnbuckles. (See Problem 9.90.) The pitch of the threads is $p = 4$ mm, their mean radius is $r = 10$ mm, and the coefficient of static friction between the threads and the mating grooves is 0.18. What couple must be exerted on the turnbuckle of member *CD* to start loosening it?

9.95 In Problem 9.94, what couple must be exerted on the turnbuckle of member *DG* to start loosening it?

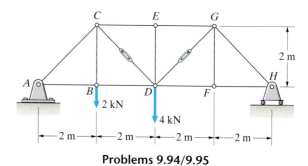

Problems 9.94/9.95

9.96* The load $W = 800$ N can be raised or lowered by rotating the threaded shaft. The distances are $b = 75$ mm and $h = 200$ mm. The pinned bars are each 300 mm in length. The pitch of the threaded shaft is $p = 5$ mm, the mean radius of the thread is $r = 15$ mm, and the coefficient of kinetic friction between the thread and the mating groove is 0.2. When the system is in the position shown, what couple must be exerted to turn the threaded shaft at a constant rate, raising the load?

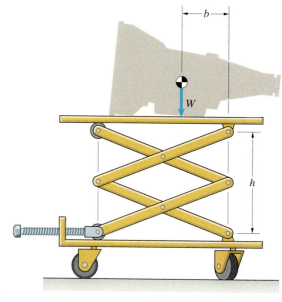

Problem 9.96

Journal Bearings

A *bearing* is a support. This term usually refers to supports designed to allow the supported object to move. For example, in Fig. 9.19a, a horizontal shaft is supported by two *journal bearings*, which allow the shaft to rotate. The shaft can then be used to support a load perpendicular to its axis, such as that subjected by a pulley (Fig. 9.19b).

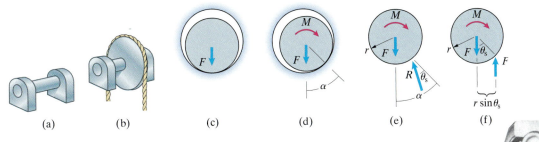

(a) (b) (c) (d) (e) (f)

Figure 9.19
(a) A shaft supported by journal bearings.
(b) A pulley supported by the shaft.
(c) The shaft and bearing when no couple is applied to the shaft.
(d) A couple causes the shaft to roll within the bearing.
(e) Free-body diagram of the shaft.
(f) The two forces on the shaft must be equal and opposite.

Here we analyze journal bearings consisting of brackets with holes through which the shaft passes. The radius of the shaft is slightly smaller than the radius of the holes in the bearings. Our objective is to determine the couple that must be applied to the shaft to cause it to rotate in the bearings. Let F be the total load supported by the shaft including the weight of the shaft itself. When no couple is exerted on the shaft, the force F presses it against the bearings as shown in Fig. 9.19c. When a couple M is exerted on the shaft, it rolls up the surfaces of the bearings (Fig. 9.19d). The term α is the angle from the original point of contact of the shaft to its point of contact when M is applied.

In Fig. 9.19e, we draw the free-body diagram of the shaft when M is sufficiently large that slip is impending. The force R is the total reaction exerted on the shaft by the two bearings. Since R and F are the only forces acting on the shaft, equilibrium requires that $\alpha = \theta_s$ and $R = F$ (Fig. 9.19f). The reaction exerted on the shaft by the bearings is displaced a distance $r \sin \theta_s$ from the vertical line through the center of the shaft. By summing moments about the center of the shaft, we obtain the couple M that causes the shaft to be on the verge of slipping:

$$M = rF \sin \theta_s. \tag{9.12}$$

This is the largest couple that can be exerted on the shaft without causing it to start rotating. Replacing θ_s in this expression by the angle of kinetic friction θ_k gives the couple necessary to rotate the shaft at a constant rate.

The simple type of journal bearing we have described is too primitive for most applications. The surfaces where the shaft and bearing are in contact would quickly become worn. Designers usually incorporate "ball" or "roller" bearings in journal bearings to minimize friction (Fig. 9.20).

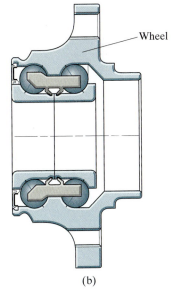

(a)

(b)

Figure 9.20
(a) A journal bearing with one row of balls.
(b) Journal bearing assembly of the wheel of a car. There are two rows of balls between the rotating wheel and the fixed inner cylinder.

Example 9.7 **Pulley Supported by Journal Bearings**

The weight of the suspended load in Fig. 9.21 is $W = 1000$ lb. The pulley P has a 6-in radius and is rigidly attached to a horizontal shaft that is supported by journal bearings. The radius of the horizontal shaft is 0.5 in, and the coefficient of kinetic friction between the shaft and the bearings is 0.2. The weights of the pulley and shaft are negligible. What tension must the winch A exert on the cable to raise the load at a constant rate?

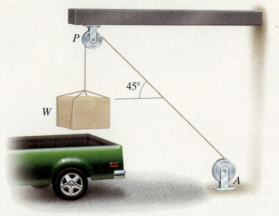

45°

W

P

A

Figure 9.21

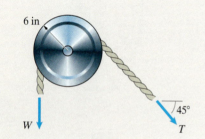

6 in

W

45°

T

(a) Free-body diagram of the pulley.

45°

W

T

F

(b) The total force F on the shaft.

Strategy

Equation (9.12) with θ_s replaced by θ_k relates the couple M required to turn the pulley at a constant rate to the total force F on the shaft. By expressing M and F in terms of the load and the tension exerted by the winch, we can obtain an equation for the required tension.

Solution

Let T be the tension exerted by the winch (Fig. a). By calculating the magnitude of the sum of the forces exerted by the tension and the load (Fig. b), we obtain an expression for the total force F on the shaft supporting the pulley:

$$F = \sqrt{(W + T\sin 45°)^2 + (T\cos 45°)^2}.$$

The (clockwise) couple exerted on the pulley by the tension and the load is

$$M = (6\text{ in})(T - W).$$

The radius of the shaft is $r = 0.5$ in and the angle of kinetic friction is $\theta_k = \arctan(0.2) = 11.3°$. We substitute our expressions for F and M into Eq. (9.12) to get

$$M = rF\sin\theta_k:$$

$$(6\text{ in})(T - W) = (0.5\text{ in})\sqrt{(W + T\sin 45°)^2 + (T\cos 45°)^2}\sin 11.3°.$$

Setting $W = 1000$ lb and solving for the tension, we obtain $T = 1031$ lb.

Critical Thinking

How can you determine the tension the winch A must exert on the cable to lower the load at a constant rate? The only change in the analysis is that the (counterclockwise) couple exerted on the pulley by the tension and the load is $M = (6\text{ in})(W - T)$. From the equation

$$M = rF\sin\theta_k:$$

$$(6\text{ in})(W - T) = (0.5\text{ in})\sqrt{(W + T\sin 45°)^2 + (T\cos 45°)^2}\sin 11.3°,$$

we obtain $T = 970$ lb.

Thrust Bearings and Clutches

A *thrust bearing* supports a rotating shaft that is subjected to an axial load. In the type shown in Figs. 9.22a and 9.22b, the conical end of the shaft is pressed against the mating conical cavity by an axial load F. Let us determine the couple M necessary to rotate the shaft.

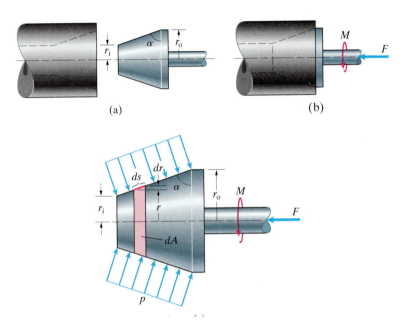

Figure 9.22
(a), **(b)** A thrust bearing supports a shaft subjected to an axial load.
(c) The differential element dA and the uniform pressure p exerted by the cavity.

The differential element of area dA in Fig. 9.22c is

$$dA = 2\pi r\, ds = 2\pi r\left(\frac{dr}{\cos \alpha}\right).$$

Integrating this expression from $r = r_i$ to $r = r_o$, we obtain the area of contact:

$$A = \frac{\pi(r_o^2 - r_i^2)}{\cos \alpha}.$$

If we assume that the mating surface exerts a uniform pressure p, the axial component of the total force due to p must equal F: $pA \cos \alpha = F$. Therefore, the pressure is

$$p = \frac{F}{A \cos \alpha} = \frac{F}{\pi(r_o^2 - r_i^2)}.$$

As the shaft rotates about its axis, the moment about the axis due to the friction force on the element dA is $r\mu_k(p\, dA)$. The total moment is

$$M = \int_A \mu_k rp\, dA = \int_{r_i}^{r_o} \mu_k r\left[\frac{F}{\pi(r_o^2 - r_i^2)}\right]\left(\frac{2\pi r\, dr}{\cos \alpha}\right).$$

Integrating, we obtain the couple M necessary to rotate the shaft at a constant rate:

$$M = \frac{2\mu_k F}{3 \cos \alpha}\left(\frac{r_o^3 - r_i^3}{r_o^2 - r_i^2}\right). \tag{9.13}$$

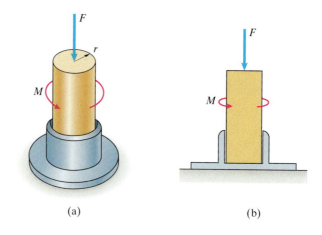

Figure 9.23
A thrust bearing that supports a flat-ended shaft.

(a)   (b)

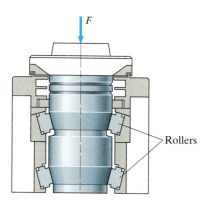

Figure 9.24
A thrust bearing with two rows of cylindrical rollers between the shaft and the fixed support.

Rollers

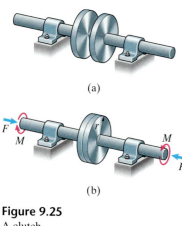

(a)

(b)

Figure 9.25
A clutch.
(a) Disengaged position.
(b) Engaged position.

A simpler thrust bearing is shown in Figs. 9.23a and 9.23b. The bracket supports the flat end of a shaft of radius r that is subjected to an axial load F. We can obtain the couple necessary to rotate the shaft at a constant rate from Eqs. (9.13) by setting $\alpha = 0$, $r_i = 0$, and $r_o = r$:

$$M = \frac{2}{3}\mu_k Fr. \tag{9.14}$$

Although they are good examples of the analysis of friction forces, the thrust bearings we have described would become worn too quickly to be used in most applications. The designer of the thrust bearing in Fig. 9.24 minimizes friction by incorporating "roller" bearings.

A *clutch* is a device used to connect and disconnect two coaxial rotating shafts. The type shown in Figs. 9.25a and 9.25b consists of disks of radius r attached to the ends of the shafts. When the disks are separated (Fig. 9.25a), the clutch is *disengaged*, and the shafts can rotate freely relative to each other. When the clutch is engaged by pressing the disks together with axial forces F (Fig. 9.25b), the shafts can support a couple M due to the friction forces between the disks. If the couple M becomes too large, the clutch slips.

The friction forces exerted on one face of the clutch by the other face are identical to the friction forces exerted on the flat-ended shaft by the bracket in Fig. 9.23. We can therefore determine the largest couple the clutch can support without slipping by replacing μ_k with μ_s in Eq. (9.14):

$$M = \frac{2}{3}\mu_s Fr. \tag{9.15}$$

Study Questions

1. What is a journal bearing?

2. If the shaft of a journal bearing is subjected to a lateral force F, how do you determine the couple M necessary to rotate the shaft at a constant rate?

3. When the axis of a clutch is subjected to an axial force F (Fig. 9.25b), how do you determine the largest couple M the clutch can support without slipping?

Example 9.8 Friction on a Disk Sander

The handheld sander in Fig. 9.26 has a rotating disk D of 4-in radius with sand-paper bonded to it. The total downward force exerted by the operator and the weight of the sander is 15 lb. The coefficient of kinetic friction between the sandpaper and the surface is $\mu_k = 0.6$. What couple (torque) M must the motor exert to turn the sander at a constant rate?

Strategy
As the disk D rotates, it is subjected to friction forces analogous to the friction forces exerted on the flat-ended shaft by the bracket in Fig. 9.23. We can determine the couple required to turn the disk D at a constant rate from Eq. (9.14).

Solution
The couple required to turn the disk at a constant rate is

$$M = \frac{2}{3}\mu_k rF = \frac{2}{3}(0.6)(4 \text{ in})(15 \text{ lb}) = 24 \text{ in-lb}.$$

Critical Thinking
Equations (9.13)–(9.15) were derived under the assumption that the normal force (and consequently the friction force) is uniformly distributed over the contacting surfaces. Evaluating and improving upon this assumption would require analysis of the deformations of the contacting surfaces in specific applications such as the disk sander in this example.

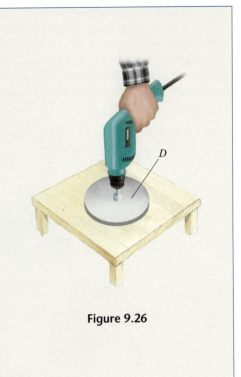

Figure 9.26

Problems

9.97 The horizontal shaft is supported by two journal bearings. The coefficient of kinetic friction between the shaft and the bearings is $\mu_k = 0.2$. The radius of the shaft is 20 mm, and its mass is 5 kg. Determine the couple M necessary to rotate the shaft at a constant rate.

    Strategy: You can obtain the couple necessary to rotate the shaft at a constant rate by replacing θ_s with θ_k in Eq. (9.12).

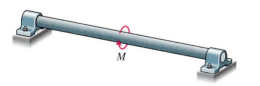

Problem 9.97

9.98 The radius of the pulley is 4 in. The pulley is rigidly attached to the horizontal shaft, which is supported by two journal bearings. The radius of the shaft is 1 in, and the combined weight of the pulley and shaft is 20 lb. The coefficients of friction between the shaft and the bearings are $\mu_s = 0.30$ and $\mu_k = 0.28$. Determine the largest weight W that can be suspended as shown without causing the stationary shaft to slip in the bearings.

9.99 In Problem 9.98, suppose that the weight $W = 4$ lb. What couple would have to be applied to the horizontal shaft to raise the weight at a constant rate?

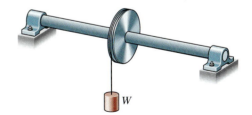

Problems 9.98/9.99

9.100 The pulley is mounted on a horizontal shaft supported by journal bearings. The coefficient of kinetic friction between the shaft and the bearings is $\mu_k = 0.3$. The radius of the shaft is 20 mm, and the radius of the pulley is 150 mm. The mass $m = 10$ kg. Neglect the masses of the pulley and shaft. What force T must be applied to the cable to move the mass upward at a constant rate?

9.101 In Problem 9.100, what force T must be applied to the cable to lower the mass at a constant rate?

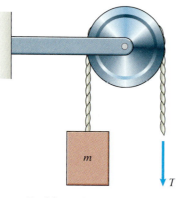

Problems 9.100/9.101

9.102 The pulley of 8-in radius is mounted on a shaft of 1-in radius. The shaft is supported by two journal bearings. The coefficient of static friction between the bearings and the shaft is $\mu_s = 0.15$. Neglect the weights of the pulley and shaft. The 50-lb block A rests on the floor. If sand is slowly added to the bucket B, what do the bucket and sand weigh when the shaft slips in the bearings?

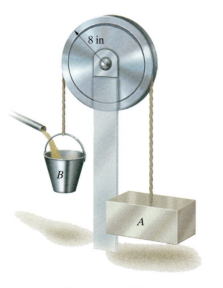

Problem 9.102

9.103 The pulley of 50-mm radius is mounted on a shaft of 10-mm radius. The shaft is supported by two journal bearings. The mass of the block A is 8 kg. Neglect the weights of the pulley and shaft. If a force $T = 84$ N is necessary to raise block A at a constant rate, what is the coefficient of kinetic friction between the shaft and the bearings?

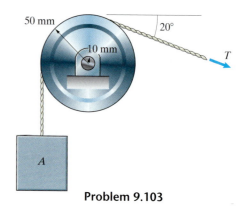

Problem 9.103

9.104 The mass of the suspended object is 4 kg. The pulley has a 100-mm radius and is rigidly attached to a horizontal shaft supported by journal bearings. The radius of the horizontal shaft is 10 mm and the coefficient of kinetic friction between the shaft and the bearings is 0.26. What tension must the person exert on the rope to raise the load at a constant rate?

9.105 In Problem 9.104, what tension must the person exert to lower the load at a constant rate?

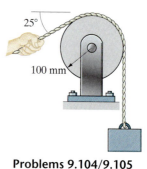

Problems 9.104/9.105

9.106 The radius of the pulley is 200 mm, and it is mounted on a shaft of 20-mm radius. The coefficient of static friction between the pulley and shaft is $\mu_s = 0.18$. If $F_A = 200$ N, what is the largest force F_B that can be applied without causing the pulley to turn? Neglect the weight of the pulley.

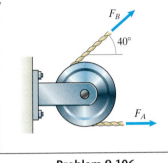

Problem 9.106

9.107 The masses of the boxes are $m_A = 15$ kg and $m_B = 60$ kg. The coefficient of static friction between boxes A and B and between box B and the inclined surface is 0.12. The pulley has a radius of 60 mm and is mounted on a shaft of 10-mm radius. The coefficient of static friction between the pulley and shaft is 0.16. What is the largest force F for which the boxes will not slip?

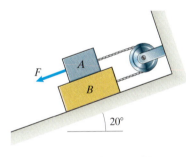

Problem 9.107

9.108 The two pulleys have a radius of 4 in and are mounted on shafts of 1-in radius supported by journal bearings. Neglect the weights of the pulleys and shafts. The tension in the spring is 40 lb. The coefficient of kinetic friction between the shafts and the bearings is $\mu_k = 0.3$. What couple M is required to turn the left pulley at a constant rate?

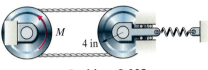

Problem 9.108

9.109 The weights of the boxes are $W_A = 65$ lb and $W_B = 130$ lb. The coefficient of static friction between boxes A and B and between box B and the floor is 0.12. The pulley has a radius of 4 in and is mounted on a shaft of 0.8-in radius. The coefficient of static friction between the pulley and shaft is 0.16. What is the largest force F for which the boxes will not slip?

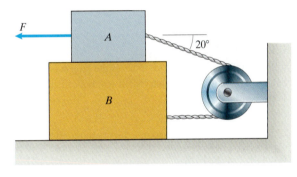

Problem 9.109

9.110 The coefficient of kinetic friction between the 100-kg box and the inclined surface is 0.35. Each pulley has a radius of 100 mm and is mounted on a shaft of 5-mm radius supported by journal bearings. The coefficient of kinetic friction between the shafts and the journal bearings is 0.18. Determine the tension T necessary to pull the box up the surface at a constant rate.

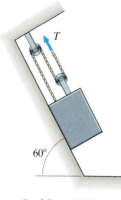

60°

Problem 9.110

9.111 The circular flat-ended shaft is pressed into the thrust bearing by an axial load of 100 N. Neglect the weight of the shaft. The coefficients of friction between the end of the shaft and the bearing are $\mu_s = 0.20$ and $\mu_k = 0.15$. What is the largest couple M that can be applied to the stationary shaft without causing it to rotate in the bearing?

9.112 In Problem 9.111, what couple M is required to rotate the shaft at a constant rate?

9.113 Suppose that the end of the shaft in Problem 9.111 is supported by a thrust bearing of the type shown in Fig. 9.22, where $r_o = 30$ mm, $r_i = 10$ mm, $\alpha = 30°$, and $\mu_k = 0.15$. What couple M is required to rotate the shaft at a constant rate?

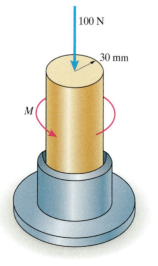

100 N

30 mm

M

Problems 9.111–9.113

9.114 The disk D is rigidly attached to the vertical shaft. The shaft has flat ends supported by thrust bearings. The disk and the shaft together have a mass of 220 kg and the diameter of the shaft is 50 mm. The vertical force exerted on the end of the shaft by the upper thrust bearing is 440 N. The coefficient of kinetic friction between the ends of the shaft and the bearings is 0.25. What couple M is required to rotate the shaft at a constant rate?

9.115 Suppose that the ends of the shaft in Problem 9.114 are supported by thrust bearings of the type shown in Fig. 9.22, where $r_o = 25$ mm, $r_i = 6$ mm, $\alpha = 45°$, and $\mu_k = 0.25$. What couple M is required to rotate the shaft at a constant rate?

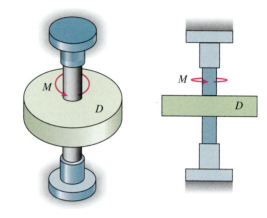

M

D

M

D

Problems 9.114/9.115

9.116 The shaft is supported by thrust bearings that subject it to an axial load of 800 N. The coefficients of kinetic friction between the shaft and the left and right bearings are 0.20 and 0.26, respectively. What couple is required to rotate the shaft at a constant rate?

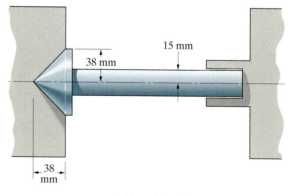

15 mm

38 mm

38 mm

Problem 9.116

9.117 A motor is used to rotate a paddle for mixing chemicals. The shaft of the motor is coupled to the paddle using a friction clutch of the type shown in Fig. 9.25. The radius of the disks of the clutch is 120 mm, and the coefficient of static friction between the disks is 0.6. If the motor transmits a maximum torque of 15 N-m to the paddle, what minimum normal force between the plates of the clutch is necessary to prevent slipping?

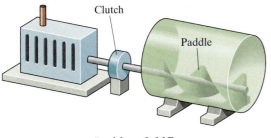

Problem 9.117

9.118 The thrust bearing is supported by contact of the collar C with a fixed plate. The area of contact is an annulus with an inside diameter $D_1 = 40$ mm and an outside diameter $D_2 = 120$ mm. The coefficient of kinetic friction between the collar and the plate is $\mu_k = 0.3$. The force $F = 400$ N. What couple M is required to rotate the shaft at a constant rate?

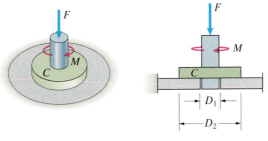

Problem 9.118

9.119 An experimental automobile brake design works by pressing the fixed red annular plate against the rotating wheel. If $\mu_k = 0.6$, what force F pressing the plate against the wheel is necessary to exert a couple of 200 N-m on the wheel?

9.120 An experimental automobile brake design works by pressing the fixed red annular plate against the rotating wheel. Suppose that $\mu_k = 0.65$ and the force pressing the plate against the wheel is $F = 2$ kN.

(a) What couple is exerted on the wheel?

(b) What percentage increase in the couple exerted on the wheel is obtained if the outer radius of the brake is increased from 90 mm to 100 mm?

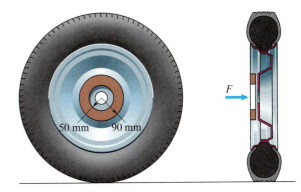

Problems 9.119/9.120

9.121 The coefficient of static friction between the plates of the car's clutch is 0.8. If the plates are pressed together with a force $F = 2.60$ kN, what is the maximum torque the clutch will support without slipping?

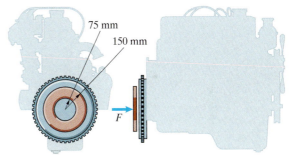

Problem 9.121

9.122* The "Morse taper" is used to support the workpiece on a machinist's lathe. The taper is driven into the spindle and is held in place by friction. If the spindle exerts a uniform pressure $p = 15$ psi on the taper and $\mu_s = 0.2$, what couple must be exerted about the axis of the taper to loosen it?

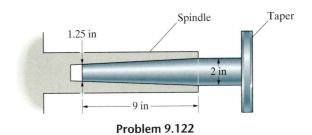

Problem 9.122

Figure 9.27
A rope wrapped around a post.

Belt Friction

If a rope is wrapped around a fixed post as shown in Fig. 9.27, a large force T_2 exerted on one end can be supported by a relatively small force T_1 applied to the other end. In this section we analyze this familiar phenomenon. It is referred to as *belt friction* because a similar approach can be used to analyze belts used in machines, such as the belts that drive alternators and other devices in a car.

Let's consider a rope wrapped through an angle β around a fixed cylinder (Fig. 9.28a). We will assume that the tension T_1 is known. Our objective is to determine the largest force T_2 that can be applied to the other end of the rope without causing the rope to slip.

We begin by drawing the free-body diagram of an element of the rope whose boundaries are at angles α and $\alpha + \Delta\alpha$ from the point where the rope comes into contact with the cylinder (Figs. 9.28b and 9.28c). The force T is the tension in the rope at the position defined by the angle α. We know that the tension in the rope varies with position, because it increases from T_1 at $\alpha = 0$ to T_2 at $\alpha = \beta$. We therefore write the tension in the rope at the position $\alpha + \Delta\alpha$ as $T + \Delta T$. The force ΔN is the normal force exerted on the element by the cylinder. Because we want to determine the largest value of T_2 that will not cause the rope to slip, we assume that the friction force is equal to its maximum possible value $\mu_s \Delta N$, where μ_s is the coefficient of static friction between the rope and the cylinder.

The equilibrium equations in the directions tangential to and normal to the centerline of the rope are

$$\Sigma F_{\text{tangential}} = \mu_s \Delta N + T \cos\left(\frac{\Delta\alpha}{2}\right) - (T + \Delta T)\cos\left(\frac{\Delta\alpha}{2}\right) = 0,$$

$$\Sigma F_{\text{normal}} = \Delta N - (T + \Delta T)\sin\left(\frac{\Delta\alpha}{2}\right) - T\sin\left(\frac{\Delta\alpha}{2}\right) = 0. \qquad (9.16)$$

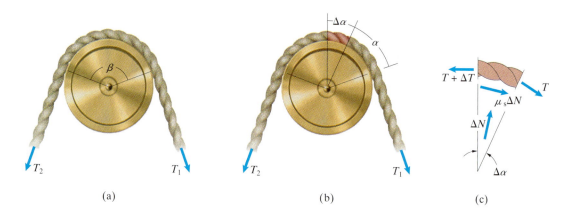

(a) (b) (c)

Figure 9.28
(a) A rope wrapped around a fixed cylinder.
(b) A differential element with boundaries at angles α and $\alpha + \Delta\alpha$.
(c) Free-body diagram of the element.

Eliminating ΔN, we can write the resulting equation as

$$\left[\cos\left(\frac{\Delta\alpha}{2}\right) - \mu_s \sin\left(\frac{\Delta\alpha}{2}\right)\right]\frac{\Delta T}{\Delta\alpha} - \mu_s T\frac{\sin(\Delta\alpha/2)}{\Delta\alpha/2} = 0.$$

Evaluating the limit of this equation as $\Delta\alpha \to 0$ and observing that

$$\frac{\sin(\Delta\alpha/2)}{\Delta\alpha/2} \to 1,$$

we obtain

$$\frac{dT}{d\alpha} - \mu_s T = 0.$$

This differential equation governs the variation of the tension in the rope. Separating variables yields

$$\frac{dT}{T} = \mu_s \, d\alpha,$$

We can now integrate to determine the tension T_2 in terms of the tension T_1 and the angle β:

$$\int_{T_1}^{T_2} \frac{dT}{T} = \int_0^\beta \mu_s \, d\alpha.$$

Thus, we obtain the largest force T_2 that can be applied without causing the rope to slip when the force on the other end is T_1:

$$T_2 = T_1 e^{\mu_s \beta}. \tag{9.17}$$

The angle β in this equation must be expressed in radians. Replacing μ_s by the coefficient of kinetic friction μ_k gives the force T_2 required to cause the rope to slide at a constant rate.

Equation (9.17) explains why a large force can be supported by a relatively small force when a rope is wrapped around a fixed support. The force required to cause the rope to slip increases exponentially as a function of the angle through which the rope is wrapped. Suppose that $\mu_s = 0.3$. When the rope is wrapped one complete turn around the post ($\beta = 2\pi$), the ratio $T_2/T_1 = 6.59$. When the rope is wrapped four complete turns around the post ($\beta = 8\pi$), the ratio $T_2/T_1 = 1880$.

Study Questions

1. What is the definition of the term β in Eq. (9.17)?

2. If a rope is wrapped through a given angle around a fixed post and one end is subjected to a given tension T_1, how can you determine the tension T_2 necessary to cause the rope to be on the verge of slipping in the direction of T_2? How can you determine the smallest value of T_2 that will prevent the rope from slipping in the direction of T_1?

Example 9.9   Rope Wrapped Around Two Cylinders

The 100-lb crate in Fig. 9.29 is suspended from a rope that passes over two fixed cylinders. The coefficient of static friction is 0.2 between the rope and the left cylinder and 0.4 between the rope and the right cylinder. What is the smallest force the woman can exert and support the crate?

Figure 9.29

Strategy

She exerts the smallest possible force when slip of the rope is impending on both cylinders. Because we know the weight of the crate, we can use Eq. (9.17) to determine the tension in the rope between the two cylinders and then use Eq. (9.17) again to determine the force she exerts.

Solution

The weight of the crate is $W = 100$ lb. Let T be the tension in the rope between the two cylinders (Fig. a). The rope is wrapped around the left cylinder through an angle $\beta = \pi/2$ rad. The tension T necessary to prevent the rope from slipping on the left cylinder is related to W by

$$W = Te^{\mu_s\beta} = Te^{(0.2)(\pi/2)}.$$

Solving for T, we obtain

$$T = We^{-(0.2)(\pi/2)} = (100 \text{ lb})e^{-(0.2)(\pi/2)} = 73.0 \text{ lb.}$$

The rope is also wrapped around the right cylinder through an angle $\beta = \pi/2$ rad. The force F the woman must exert to prevent the rope from slipping on the right cylinder is related to T by

$$T = Fe^{\mu_s\beta} = Fe^{(0.4)(\pi/2)}.$$

The solution for F is

$$F = Te^{-(0.4)(\pi/2)} = (73.0 \text{ lb}) \, e^{-(0.4)(\pi/2)} = 39.0 \text{ lb.}$$

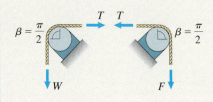

(a) The tensions in the rope.

Critical Thinking

How could you determine the force that would need to be exerted on the rope to cause the crate to begin moving upward? In this case, you would assume that slip of the rope is impending on both cylinders, but in the opposite direction to our analysis in this example. For the left cylinder, the tension T necessary for slip of the rope to be impending in the direction that would cause the crate to move upward is

$$T = We^{(0.2)(\pi/2)} = (100 \text{ lb})e^{(0.2)(\pi/2)} = 137 \text{ lb.}$$

For the right cylinder, the force F necessary for slip to be impending in the direction that would cause the crate to move upward is

$$F = Te^{(0.4)(\pi/2)} = (137 \text{ lb})e^{(0.4)(\pi/2)} = 257 \text{ lb.}$$

Although the young woman would be able to support the stationary crate, she would clearly need to call for help to raise it.

Design Example 9.10 | Belts and Pulleys

The pulleys in Fig. 9.30 turn at a constant rate. The large pulley is attached to a fixed support. The small pulley is supported by a smooth horizontal slot and is pulled to the right by the force $F = 200$ N. The coefficient of static friction between the pulleys and the belt is $\mu_s = 0.8$, the dimension $b = 500$ mm, and the radii of the pulleys are $R_A = 200$ mm and $R_B = 100$ mm. What are the largest values of the couples M_A and M_B for which the belt will not slip?

Figure 9.30

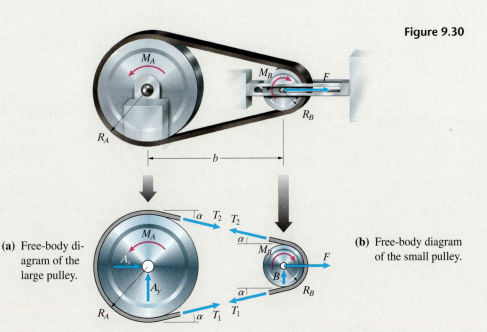

(a) Free-body diagram of the large pulley.

(b) Free-body diagram of the small pulley.

Strategy
By drawing free-body diagrams of the pulleys, we can use the equilibrium equations to relate the tensions in the belt to M_A and M_B and obtain a relation between the tensions in the belt and the force F. When slip is impending, the tensions are also related by Eq. (9.17). From these equations we can determine M_A and M_B.

Solution
From the free-body diagram of the large pulley (Fig. a), we obtain the equilibrium equation

$$M_A = R_A(T_2 - T_1), \qquad (1)$$

and from the free-body diagram of the small pulley (Fig. b), we obtain

$$F = (T_1 + T_2) \cos \alpha, \qquad (2)$$

$$M_B = R_B(T_2 - T_1). \qquad (3)$$

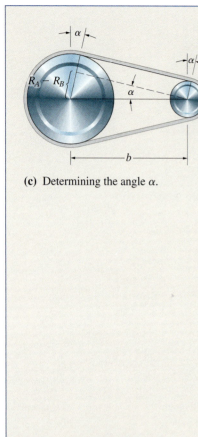

(c) Determining the angle α.

The belt is in contact with the small pulley through the angle $\pi - 2\alpha$ (Fig. c). From the dashed line parallel to the belt, we see that the angle α satisfies the relation

$$\sin \alpha = \frac{R_A - R_B}{b} = \frac{200 \text{ mm} - 100 \text{ mm}}{500 \text{ mm}} = 0.2.$$

Therefore, $\alpha = 11.5° = 0.201$ rad. If we assume that slip impends between the small pulley and the belt, Eq. (9.17) states that

$$T_2 = T_1 e^{\mu_s \beta} = T_1 e^{0.8(\pi - 2\alpha)} = 8.95 T_1.$$

We solve this equation together with Eq. (2) for the two tensions, obtaining $T_1 = 20.5$ N and $T_2 = 183.6$ N. Then from Eqs. (1) and (3), the couples are $M_A = 32.6$ N-m and $M_B = 16.3$ N-m.

If we assume that slip impends between the large pulley and the belt, we obtain $M_A = 36.3$ N-m and $M_B = 18.1$ N-m, so the belt slips on the small pulley at smaller values of the couples.

Design Issues

Belts and pulleys are used to transfer power in cars and many other types of machines, including printing presses, farming equipment, and industrial robots. Because two pulleys of different diameters connected by a belt are subjected to different torques and have different rates of rotation, they can be used as a mechanical "transformer" to alter torque or rotation rate.

In this example we assumed that the belt was flat, but "V-belts" that fit into matching grooves in the pulleys are often used in applications (Fig. 9.31a). This configuration keeps the belt in place on the pulleys and also decreases the tendency of the belt to slip. Suppose that a V-belt is wrapped through an angle β around a pulley (Fig. 9.31b). If the tension T_1 is known, what is the largest tension T_2 that can be applied to the other end of the belt without causing it to slip relative to the pulley?

In Fig. 9.31c, we draw the free-body diagram of an element of the belt whose boundaries are at angles α and $\alpha + \Delta\alpha$ from the point where the belt comes into contact with the pulley. (Compare this figure with Fig. 9.28c.)

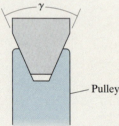

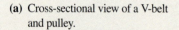

(a) Cross-sectional view of a V-belt and pulley.

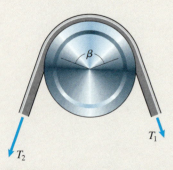

(b) V-belt wrapped around a pulley.

Figure 9.31

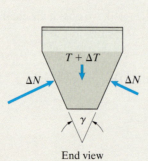

End view

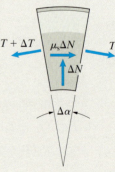

Side view

(c) Free-body diagram of an element of the belt.

The equilibrium equations in the directions tangential to and normal to the centerline of the belt are

$$\Sigma F_{\text{tangential}} = 2\mu_s\Delta N + T\cos\left(\frac{\Delta\alpha}{2}\right) - (T + \Delta T)\cos\left(\frac{\Delta\alpha}{2}\right) = 0,$$

(4)

$$\Sigma F_{\text{normal}} = 2\Delta N\sin\left(\frac{\gamma}{2}\right) - (T + \Delta T)\sin\left(\frac{\Delta\alpha}{2}\right) - T\sin\left(\frac{\Delta\alpha}{2}\right) = 0.$$

By the same steps leading from Eqs. (9.16) to Eq. (9.17), it can be shown that

$$T_2 = T_1 e^{\mu_s\beta/\sin(\gamma/2)}.$$

(5)

Thus, using a V-belt effectively increases the coefficient of friction between the belt and pulley by the factor $1/\sin(\gamma/2)$.

When it is essential that the belt not slip relative to the pulley, a belt with cogs and a matching pulley (Fig. 9.32a) or a chain and sprocket wheel (Fig. 9.32b) can be used. The chains and sprocket wheels in bicycles and motorcycles are examples.

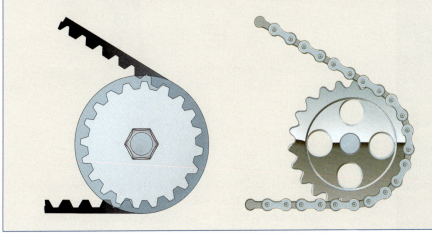

Figure 9.32
Designs that prevent slip of the belt relative to the pulley.

Problems

9.123 Suppose that you want to lift a 50-lb crate off the ground by using a rope looped over a tree limb as shown. The coefficient of static friction between the rope and the limb is 0.4, and the rope is wound 120° around the limb. What force must you exert to lift the crate?

Strategy: The tension necessary to cause impending slip of the rope on the limb is given by Eq. (9.17), with $T_1 = 50$ lb, $\mu_s = 0.4$, and $\beta = (\pi/180)(120)$ rad.

9.124 In Problem 9.123, once you have lifted the crate off the ground, what is the minimum force you must exert on the rope to keep it suspended?

Problems 9.123/124

9.125 *Winches* are used on sailboats to help support the forces exerted by the sails on the ropes (*sheets*) holding them in position. The winch shown is a post that will rotate in the clockwise direction (seen from above), but will not rotate in the counterclockwise direction. The sail exerts a tension $T_S = 800$ N on the sheet, which is wrapped two complete turns around the winch. The coefficient of static friction between the sheet and the winch is $\mu_s = 0.2$. What tension T_C must the crew member exert on the sheet to prevent it from slipping on the winch?

9.126 The coefficient of kinetic friction between the sheet and the winch in Problem 9.125 is $\mu_k = 0.16$. If the crew member wants to let the sheet slip at a constant rate, releasing the sail, what initial tension T_C must he exert on the sheet as it begins slipping?

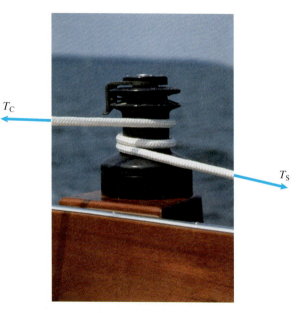

Problems 9.125/9.126

9.127 The box A weighs 20 lb. The rope is wrapped one and one-fourth turns around the fixed wooden post. The coefficients of friction between the rope and post are $\mu_s = 0.15$ and $\mu_k = 0.12$.
(a) What minimum force does the man need to exert to support the stationary box?
(b) What force would the man have to exert to raise the box at a constant rate?

Problem 9.127

9.128 The weight of block A is W. The disk is supported by a smooth bearing. The coefficient of kinetic friction between the disk and the belt is μ_k. What couple M is necessary to turn the disk at a constant rate?

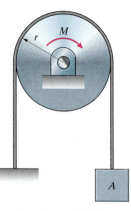

Problem 9.128

9.129 The couple required to turn the wheel of the exercise bicycle is adjusted by changing the weight W. The coefficient of kinetic friction between the wheel and the belt is μ_k. Assume the wheel turns clockwise.

(a) Show that the couple M required to turn the wheel is $M = WR\left(1 - e^{-3.4\mu_k}\right)$.

(b) If $W = 40$ lb and $\mu_k = 0.2$, what force will the scale S indicate when the bicycle is in use?

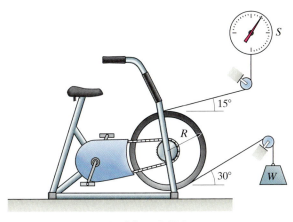

Problem 9.129

9.130 The box B weighs 50 lb. The coefficients of friction between the cable and the fixed round supports are $\mu_s = 0.4$ and $\mu_k = 0.3$.

(a) What is the minimum force F required to support the box?

(b) What force F is required to move the box upward at a constant rate?

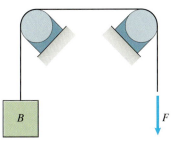

Problem 9.130

9.131 The coefficient of static friction between the 50-lb box and the inclined surface is 0.10. The coefficient of static friction between the rope and the fixed cylinder is 0.05. Determine the force the woman must exert on the rope to cause the box to start moving up the inclined surface.

9.132 In Problem 9.131, what is the minimum force the woman must exert on the rope to hold the box in equilibrium on the inclined surface?

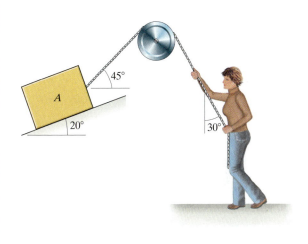

Problems 9.131/9.132

9.133* The mass of the block A is 14 kg. The coefficient of kinetic friction between the rope and the cylinder is 0.2. If the cylinder is rotated at a constant rate, first in the counterclockwise direction and then in the clockwise direction, the difference in the height of block A is 0.3 m. What is the spring constant k?

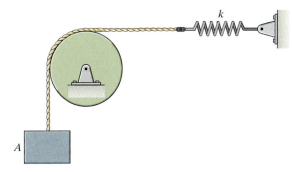

Problem 9.133

Design Experience

Problems 9.134–9.138 are related to Design Example 9.10.

9.134 If the force F in Example 9.10 is increased to 400 N, what are the largest values of the couples M_A and M_B for which the belt will not slip?

9.135 If the belt in Example 9.10 is a V-belt with angle $\gamma = 45°$, what are the largest values of the couples M_A and M_B for which the belt will not slip?

9.136 The spring exerts a 320-N force on the left pulley. The coefficient of static friction between the flat belt and the pulleys is $\mu_s = 0.5$. The right pulley cannot rotate. What is the largest couple M that can be exerted on the left pulley without causing the belt to slip?

9.137 Suppose that the belt in Problem 9.136 is a V-belt with angle $\gamma = 30°$. What is the largest couple M that can be exerted on the left pulley without causing the belt to slip?

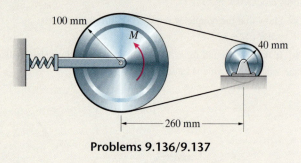

Problems 9.136/9.137

9.138 Beginning with Eqs. (4) in Design Example 9.10, derive Eq. (9.22):

$$T_2 = T_1 e^{\mu_s \beta / \sin(\gamma/2)}.$$

COMPUTATIONAL MECHANICS

The following example and problems are designed for the use of a programmable calculator or computer.

Computational Example 9.11

The mass of the block A in Fig. 9.33 is 20 kg, and the coefficient of static friction between the block and the floor is $\mu_s = 0.3$. The spring constant $k = 1$ kN/m, and the spring is unstretched. How far can the slider B be moved to the right without causing the block to slip.

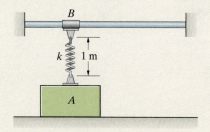

Figure 9.33

Strategy

We will draw the free-body diagram of block A assuming that the slider B is moved a distance x to the right and slip of block A is impending. Then by applying the equilibrium equations, we can obtain an equation for the distance x corresponding to impending slip

Solution

Suppose that moving the slider B a distance x to the right causes impending slip of the block (Fig. a). The resulting stretch of the spring is $\sqrt{1 + x^2} - 1$ m, so the magnitude of the force exerted on the block by the spring is

$$F_s = k\left(\sqrt{1 + x^2} - 1\right). \qquad (1)$$

From the free-body diagram of the block (Fig. b), we obtain the equilibrium equations

$$\Sigma F_x = \left(\frac{x}{\sqrt{1 + x^2}}\right)F_s - \mu_s N = 0,$$

$$\Sigma F_y = \left(\frac{1}{\sqrt{1 + x^2}}\right)F_s + N - mg = 0.$$

Substituting Eq. (1) into these two equations and then eliminating N, we can write the resulting equation in the form

$$h(x) = k(x + \mu_s)\left(\sqrt{1 + x^2} - 1\right) - \mu_s mg\sqrt{1 + x^2} = 0.$$

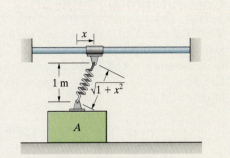

(a) Moving the slider to the right a distance x.

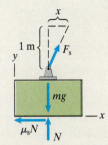

(b) Free-body diagram of the block when slip is impending.

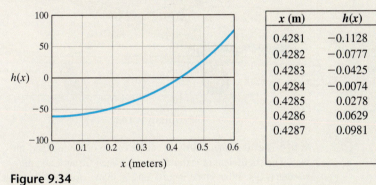

x (m)	$h(x)$
0.4281	−0.1128
0.4282	−0.0777
0.4283	−0.0425
0.4284	−0.0074
0.4285	0.0278
0.4286	0.0629
0.4287	0.0981

Figure 9.34
Graph of the function $h(x)$.

We must obtain the root of this function to determine the value of x corresponding to impending slip of the block. From the graph of $h(x)$ in Fig. 9.34, we estimate that $h(x) = 0$ at $x = 0.43$ m. By examining computed results near this value of x (see table), we see that $h(x) = 0$, and slip is impending, when x is approximately 0.4284 m.

Critical Thinking

Many software packages are available that allow you to obtain solutions to nonlinear algebraic equations such as the one we obtained in this example. Even when you have access to such software, it is a good idea to examine graphical results like those we have presented. One reason is that nonlinear equations sometimes have multiple roots, and you want to insure that you have obtained all of the solutions within the range of interest and have identified the one you want. In addition, you can often gain insight by examining the behavior of an equation over a range of its variables instead of obtaining just a single solution.

Computational Problems

9.139 The mass of the block A is 20 kg, and the coefficient of static friction between the block and the floor is $\mu_s = 0.3$. The spring constant $k = 1$ kN/m, and the spring is unstretched. How far can the slider B be moved to the right without causing the block to slip?

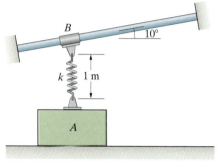

Problem 9.139

9.140 The slender circular ring of weight W is supported by normal and friction forces at A. If slip is impending when the vertical force $F = 0.4W$, what is the coefficient of static friction between the ring and the support?

9.141 Suppose that the vertical force on the ring is $F = KW$ and slip is impending. Draw a graph of K as a function of the coefficient of static friction between the ring and the support for $0 \le \mu_s \le 1$.

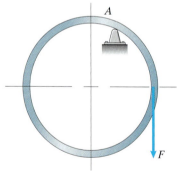

Problems 9.140/9.141

9.142 The mass of the 3-m bar is 20 kg, and the coefficient of static friction between the ends of the bar and the circular surface is $\mu_s = 0.3$. What is the largest value of the angle α for which the bar will not slip?

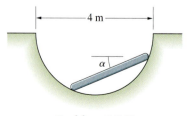

Problem 9.142

9.143 The load $W = 800$ N can be raised or lowered by rotating the threaded shaft. The distance $b = 75$ mm, and the pinned bars are each 300 mm in length. The pitch of the threaded shaft is $p = 5$ mm, the mean radius of the thread is $r = 15$ mm, and the coefficient of kinetic friction between the thread and the mating groove is 0.2. Draw a graph of the moment that must be exerted to turn the threaded shaft at a constant rate, raising the load, as a function of the height h from $h = 100$ mm to $h = 400$ mm.

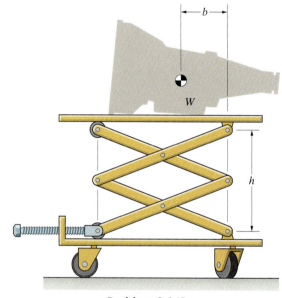

Problem 9.143

9.144 The 10-lb metal disk A is at the center of the inclined surface. The coefficient of static friction between the disk and the surface is 0.3. What is the largest tension in the string AB that will not cause the disk to slip?

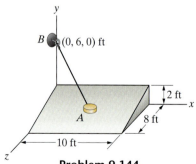

Problem 9.144

9.145 The direction cosines of the crane's cable are $\cos \theta_x = 0.588$, $\cos \theta_y = 0.766$, and $\cos \theta_z = 0.260$. The y axis is vertical. The stationary caisson to which the cable is attached weighs 2000 lb. The unit vector $\mathbf{e} = 0.260\mathbf{i} + 0.940\mathbf{j} - 0.221\mathbf{k}$ is perpendicular to the ground where the caisson rests. If the coefficient of static friction between the caisson and the ground is $\mu_s = 0.4$, what is the largest tension in the cable that will not cause the caisson to slip?

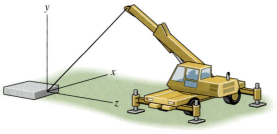

Problem 9.145

9.146 The thrust bearing is supported by contact of the collar C with a fixed plate. The area of contact is an annulus with inside diameter D_1 and outside diameter D_2. Suppose that because of thermal constraints, you want the area of contact to be 0.02 m². The coefficient of kinetic friction between the collar and the plate is $\mu_k = 0.3$. The force $F = 600$ N, and the couple M required to rotate the shaft at a constant rate is 10 N-m. What are the diameters D_1 and D_2?

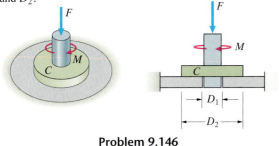

Problem 9.146

9.147 The block A weighs 30 lb, and the spring constant $k = 30$ lb/ft. If the cylinder is rotated at a constant rate, first in the counterclockwise direction and then in the clockwise direction, the difference in the height of block A is 2 ft. What is the coefficient of kinetic friction between the rope and the cylinder?

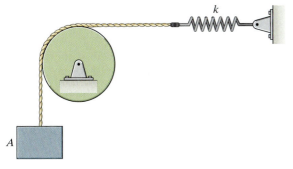

Problem 9.147

9.148 The coefficient of static friction between the 1-kg slider and the vertical bar is $\mu_s = 0.6$. The constant of the spring is $k = 20$ N/m, and its unstretched length is 1 m. Determine the range of values of y at which the slider will remain stationary on the bar.

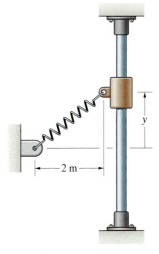

Problem 9.148

9.149 The axial force on the thrust bearing is $F = 200$ lb, and the dimension $b = 6$ in. The uniform pressure exerted by the mating surface is $p = 7$ psi, and the coefficient of kinetic friction is $\mu_k = 0.28$. If a couple $M = 360$ in-lb is required to turn the shaft, what are the dimensions D_o and D_i?

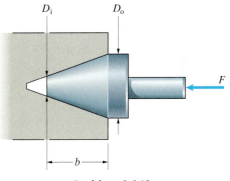

Problem 9.149

CHAPTER SUMMARY

Dry Friction

The forces resulting from the contact of two plane surfaces can be expressed in terms of the normal force N and friction force f (Fig. a) or the magnitude R and angle of friction θ (Fig. b).

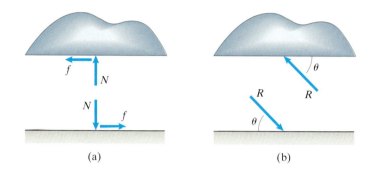

(a)   (b)

If slip is impending, the magnitude of the friction force is

$$f = \mu_s N, \tag{9.1}$$

and its direction opposes the impending slip. The angle of friction equals the angle of static friction $\theta_s = \arctan(\mu_s)$.

If the surfaces are sliding, the magnitude of the friction force is

$$f = \mu_k N, \tag{9.2}$$

and its direction opposes the relative motion. The angle of friction equals the angle of kinetic friction $\theta_k = \arctan(\mu_k)$.

Threads

The slope α of the thread (Fig. c) is related to its pitch p by

$$\tan \alpha = \frac{p}{2\pi r}. \tag{9.7}$$

The couple required for impending rotation and axial motion opposite to the direction of F is

$$M = rF \tan(\theta_s + \alpha), \tag{9.9}$$

and the couple required for impending rotation and axial motion of the shaft in the direction of F is

$$M = rF \tan(\theta_s - \alpha). \tag{9.11}$$

When $\theta_s < \alpha$, the shaft will rotate and move in the direction of the force F with no couple applied.

Journal Bearings

The couple required for impending slip of the circular shaft (Fig. d) is

$$M = rF \sin \theta_s, \tag{9.12}$$

where F is the total load on the shaft.

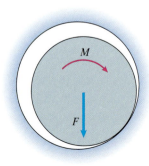

(c)

(d)

Thrust Bearings and Clutches

The couple required to rotate the shaft at a constant rate (Fig. e) is

$$M = \frac{2\mu_k F}{3 \cos \alpha}\left(\frac{r_o^3 - r_i^3}{r_o^2 - r_i^2}\right). \tag{9.13}$$

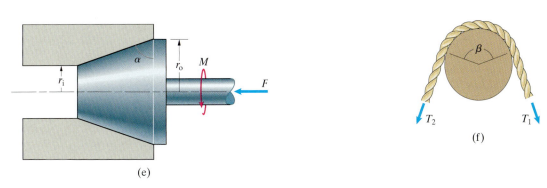

(e)

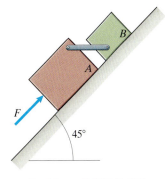

(f)

Belt Friction

The force T_2 required for impending slip in the direction of T_2 (Fig. f) is

$$T_2 = T_1 e^{\mu_s \beta}, \tag{9.17}$$

where β is in radians.

Review Problems

9.150 The weight of the box is $W = 30$ lb, and the force F is perpendicular to the inclined surface. The coefficient of static friction between the box and the inclined surface is $\mu_s = 0.2$.

(a) If $F = 30$ lb, what is the magnitude of the friction force exerted on the stationary box?

(b) If $F = 10$ lb, show that the box cannot remain at rest on the inclined surface.

9.151 In Problem 9.150, what is the smallest force F necessary to hold the box stationary on the inclined surface?

9.152 Blocks A and B are connected by a horizontal bar. The coefficient of static friction between the inclined surface and the 400-lb block A is 0.3. The coefficient of static friction between the surface and the 300-lb block B is 0.5. What is the smallest force F that will prevent the blocks from slipping down the surface?

9.153 What force F is necessary to cause the blocks in Problem 9.152 to start sliding up the plane?

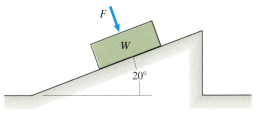

Problems 9.150/9.151

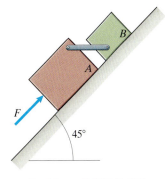

Problems 9.152/9.153

9.154 The masses of crates A and B are 25 kg and 30 kg, respectively. The coefficient of static friction between the contacting surfaces is $\mu_s = 0.34$. What is the largest value of α for which the crates will remain in equilibrium?

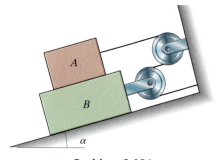

Problem 9.154

9.155 The side of a soil embankment has a 45° slope (Fig. a). If the coefficient of static friction of soil on soil is $\mu_s = 0.6$, will the embankment be stable or will it collapse? If it will collapse, what is the smallest slope that can be stable?

Strategy: Draw a free-body diagram by isolating part of the embankment as shown in Fig. b.

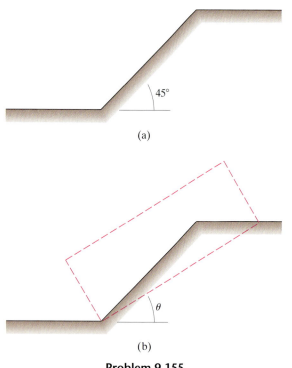

(a)

(b)

Problem 9.155

9.156 The mass of the van is 2250 kg, and the coefficient of static friction between its tires and the road is 0.6. If its front wheels are locked and its rear wheels can turn freely, what is the largest value of α for which it can remain in equilibrium?

9.157 In Problem 9.156, what is the largest value of α for which the van can remain in equilibrium if it points up the slope?

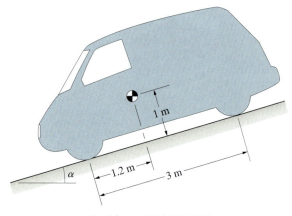

Problems 9.156/9.157

9.158 The shelf is designed so that it can be placed at any height on the vertical beam. The shelf is supported by friction between the two horizontal cylinders and the vertical beam. The combined weight of the shelf and camera is W. If the coefficient of static friction between the vertical beam and the horizontal cylinders is μ_s, what is the minimum distance b necessary for the shelf to stay in place?

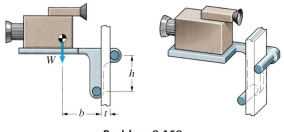

Problem 9.158

9.159 The 20-lb homogeneous object is supported at A and B. The distance $h = 4$ in, friction can be neglected at B, and the coefficient of static friction at A is 0.4. Determine the largest force F that can be exerted without causing the object to slip.

9.160 In Problem 9.159, suppose that the coefficient of static friction at B is 0.36. What is the largest value of h for which the object will slip before it tips over?

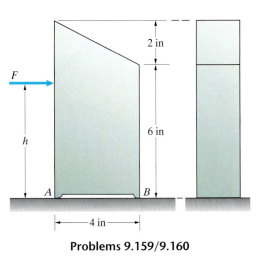

Problems 9.159/9.160

9.161 The 180-lb climber is supported in the "chimney" by the normal and friction forces exerted on his shoes and back. The static coefficients of friction between his shoes and the wall and between his back and the wall are 0.8 and 0.6, respectively. What is the minimum normal force his shoes must exert?

Problem 9.161

9.162 The sides of the 200-lb door fit loosely into grooves in the walls. Cables at A and B raise the door at a constant rate. The coefficient of kinetic friction between the door and the grooves is $\mu_k = 0.3$. What force must the cable at A exert to continue raising the door at a constant rate if the cable at B breaks?

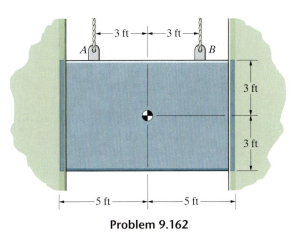

Problem 9.162

9.163 The coefficients of static friction between the tires of the 1000-kg tractor and the ground and between the 450-kg crate and the ground are 0.8 and 0.3, respectively. Starting from rest, what torque must the tractor's engine exert on the rear wheels to cause the crate to move? (The front wheels can turn freely.)

9.164 In Problem 9.163, what is the most massive crate the tractor can cause to move from rest if its engine can exert sufficient torque? What torque is necessary?

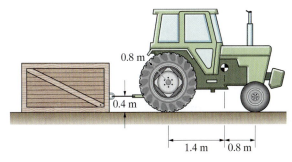

Problems 9.163/9.164

9.165 The mass of the vehicle is 900 kg, it has rear-wheel drive, and the coefficient of static friction between its tires and the surface is 0.65. The coefficient of static friction between the crate and the surface is 0.4. If the vehicle attempts to pull the crate up the incline, what is the largest value of the mass of the crate for which it will slip up the incline before the vehicle's tires slip?

Problem 9.165

9.166 Each 1-m bar has a mass of 4 kg. The coefficient of static friction between the bar and the surface at B is 0.2. If the system is in equilibrium, what is the magnitude of the friction force exerted on the bar at B?

9.167 Each 1-m bar has a mass of 4 kg. What is the minimum coefficient of static friction between the bar and the surface at B necessary for the system to be in equilibrium?

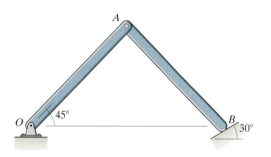

Problems 9.166/9.167

9.168 The collars A and B each have a mass of 2 kg. If friction between collar B and the bar can be neglected, what minimum coefficient of static friction between collar A and the bar is necessary for the collars to remain in equilibrium in the position shown?

9.169 If the coefficient of static friction has the same value μ_s between the 2-kg collars A and B and the bars, what minimum value of μ_s is necessary for the collars to remain in equilibrium in the position shown? (Assume that slip impends at A and B.)

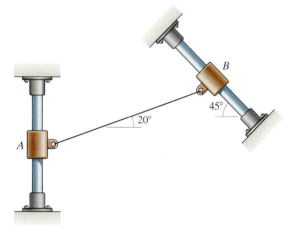

Problems 9.168/9.169

9.170 The clamp presses two pieces of wood together. The pitch of the threads is $p = 2$ mm, the mean radius of the thread is $r = 8$ mm, and the coefficient of kinetic friction between the thread and the mating groove is 0.24. What couple must be exerted on the threaded shaft to press the pieces of wood together with a force of 200 N?

9.171 In Problem 9.170, the coefficient of static friction between the thread and the mating groove is 0.28. After the threaded shaft is rotated sufficiently to press the pieces of wood together with a force of 200 N, what couple must be exerted on the shaft to loosen it?

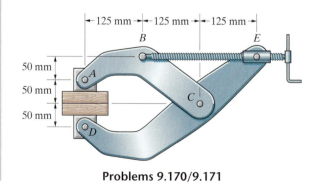

Problems 9.170/9.171

9.172 The axles of the tram are supported by journal bearing. The radius of the wheels is 75 mm, the radius of the axles is 15 mm, and the coefficient of kinetic friction between the axles and the bearings is $\mu_k = 0.14$. The mass of the tram and its load is 160 kg. If the weight of the tram and its load is evenly divided between the axles, what force P is necessary to push the tram at a constant speed?

Problem 9.172

9.173 The two pulleys have a radius of 6 in and are mounted on shafts of 1-in radius supported by journal bearings. Neglect the weights of the pulleys and shafts. The coefficient of kinetic friction between the shafts and the bearings is $\mu_k = 0.2$. If a force $T = 200$ lb is required to raise the man at a constant rate, what is his weight?

9.174 If the man in Problem 9.173 weighs 160 lb, what force T is necessary to lower him at a constant rate?

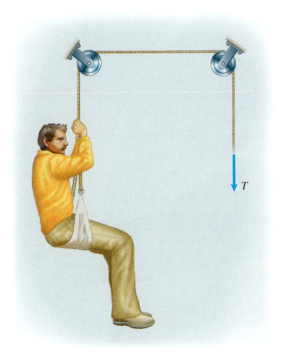

Problems 9.173/9.174

9.175 If the two cylinders are held fixed, what is the range of W for which the two weights will remain stationary?

9.176 If the system is initially stationary and the left cylinder is slowly rotated, determine the largest weight W that can be

(a) raised;
(b) lowered.

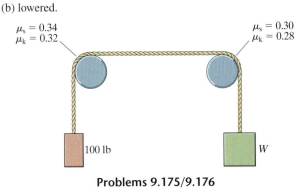

$\mu_s = 0.34$
$\mu_k = 0.32$

$\mu_s = 0.30$
$\mu_k = 0.28$

100 lb

W

Problems 9.175/9.176

Design Project 1

The wedge is used to split firewood by hammering it into a log as shown (see Example 9.5). Suppose that you want to design such a wedge to be marketed at hardware stores. Experiments indicate that the coefficient of static friction between the steel wedge and various types of wood varies from 0.2 to 0.4.

(a) Based on the given range of static friction coefficients, determine the maximum wedge angle α for which the wedge would remain in place in a log with no external force acting on it.

(b) Using the wedge angle determined in part (a), and assuming that the coefficient of kinetic friction is 0.9 times the coefficient of static friction, determine the range of vertical forces necessary to drive the wedge into a log at a constant rate.

(c) Write a brief report describing your analysis and recommending a wedge angle for the manufactured product. Consider whether a margin of safety in the chosen wedge angle might be appropriate.

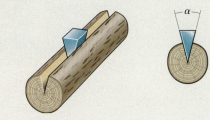

Design Project 2

Design and build a device to measure the coefficient of static friction μ_s between two materials. Use it to measure μ_s for several of the materials listed in Table 9.1 and compare your results with the values in the table. Discuss possible sources of error in your device and determine how closely your values agree when you perform repeated experiments with the same two materials.

CHAPTER

10

Internal Forces and Moments

We began our study of equilibrium by drawing free-body diagrams of individual objects to determine unknown forces and moments acting on them. In this chapter we carry this process one step further and draw free-body diagrams of parts of individual objects to determine internal forces and moments. In doing so, we arrive at the central concern of the structural design engineer: It is the forces within an object that determine whether it will support the external loads to which it is subjected.

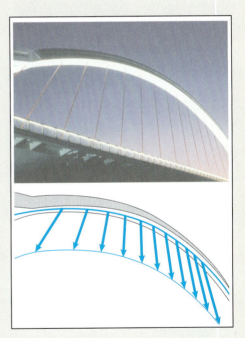

◀ The system of cables supporting the roadway subjects the bridge's arch to a distribution of internal forces and couples.

BEAMS

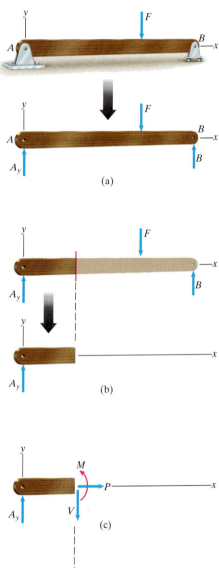

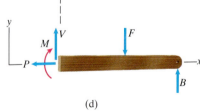

Figure 10.1
(a) A beam subjected to a load and reactions.
(b) Isolating a part of the beam.
(c), (d) The axial force, shear force, and bending moment.

10.1 Axial Force, Shear Force, and Bending Moment

To ensure that a structural member will not fail (break or collapse) due to the forces and moments acting on it, the design engineer must know not only the external loads and reactions acting on the member, but also the forces and moments acting *within* it.

Consider a beam subjected to an external load and reactions (Fig. 10.1a). How can we determine the forces and moments within the beam? In Fig. 10.1b, we "cut" the beam by a plane at an arbitrary cross section and isolate the part of the beam to the left of the plane. It is clear that the isolated part cannot be in equilibrium unless it is subjected to some system of forces and moments at the plane where it joins the other part of the beam. These are the internal forces and moments we seek.

In Chapter 4 we demonstrated that *any* system of forces and moments can be represented by an equivalent system consisting of a force and a couple. Since the system of external loads and reactions on the beam is two-dimensional, we can represent the internal forces and moments by an equivalent system consisting of two components of force and a couple (Fig. 10.1c). The component P parallel to the beam's axis is called the *axial force*. The component V normal to the beam's axis is called the *shear force*, and the couple M is called the *bending moment*. The axial force, shear force, and bending moment on the part of the beam to the right of the cutting plane are shown in Fig. 10.1d. Notice that they are equal in magnitude but opposite in direction to the internal forces and moment on the free-body diagram in Fig. 10.1c.

The directions of the axial force, shear force, and bending moment in Figs. 10.1c and 10.1d are the established definitions of the positive directions of these quantities. A positive axial force P subjects the beam to tension. A positive shear force V tends to rotate the axis of the beam clockwise (Fig. 10.2a). A positive bending moment M tends to cause upward curvature of the beam's axis (Fig. 10.2b). Notice that a positive bending moment subjects the upper part of the beam to compression, shortening the beam in the direction parallel to its axis, and subjects the lower part of the beam to tension, lengthening the beam in the direction parallel to its axis.

Determining the internal forces and moment at a particular cross section of a beam typically involves three steps:

1. Determine the external forces and moments—Draw the free-body diagram of the beam, and determine the reactions at its supports. If the beam is a member of a structure, you must analyze the structure.

2. Draw the free-body diagram of part of the beam—Cut the beam at the point at which you want to determine the internal forces and moment, and draw

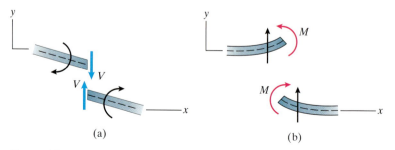

Figure 10.2
(a) Positive shear forces tend to rotate the axis of the beam clockwise.
(b) Positive bending moments tend to bend the axis of the beam upward.

the free-body diagram of one of the resulting parts. You can choose the part with the simplest free-body diagram. If your cut divides a distributed load, don't represent the distributed load by an equivalent force until after you have obtained your free-body diagram.

3. Apply the equilibrium equations—Use the equilibrium equations to determine P, V, and M.

Study Questions

1. What are the axial force, shear force, and bending moment?
2. How is the positive direction of the shear force defined?
3. How is the positive direction of the bending moment defined?

Example 10.1	**Determining the Internal Forces and Moment**

For the beam in Fig. 10.3, determine the internal forces and moment at C.

Strategy
After determining the reactions at the supports, we will cut the beam by a plane at point C and draw the free-body diagram of the part of the beam to the left of the plane. Then we can use the equilibrium equations to determine the internal forces and moments at C.

Solution

Determine the External Forces and Moments We begin by drawing the free-body diagram of the beam and determining the reactions at its supports; the results are shown in Fig. (a).

Draw the Free-Body Diagram of Part of the Beam We cut the beam at C (Fig. a) and draw the free-body diagram of the left part, including the internal forces and moment in their defined positive directions (Fig. b).

Apply the Equilibrium Equations From the equilibrium equations

$$\Sigma F_x = P_C = 0,$$

$$\Sigma F_y = \frac{1}{4}F - V_C = 0,$$

$$\Sigma M_{\text{point } C} = M_C - \left(\frac{1}{4}L\right)\left(\frac{1}{4}F\right) = 0,$$

we obtain $P_C = 0$, $V_C = \frac{1}{4}F$, and $M_C = \frac{1}{16}LF$.

Critical Thinking
We can check our results with the free-body diagram of the part of the beam to the right of point C (Fig. c). The equilibrium equations are

$$\Sigma F_x = -P_C = 0,$$

$$\Sigma F_y = V_C - F + \frac{3}{4}F = 0,$$

$$\Sigma M_{\text{point } C} = -M_C - \left(\frac{1}{2}L\right)F + \left(\frac{3}{4}L\right)\left(\frac{3}{4}F\right) = 0,$$

confirming that $P_C = 0$, $V_C = \frac{1}{4}F$, and $M_C = \frac{1}{16}LF$.

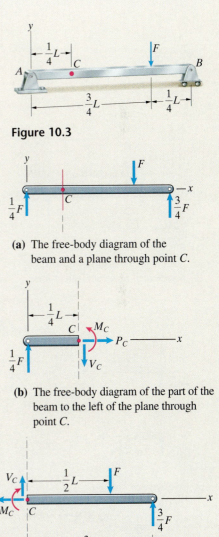

Figure 10.3

(a) The free-body diagram of the beam and a plane through point C.

(b) The free-body diagram of the part of the beam to the left of the plane through point C.

(c) The free-body diagram of the part of the beam to the right of the plane through point C.

Example 10.2	Determining the Internal Forces and Moment

For the beam in Fig. 10.4, determine the internal forces and moment at B.

Strategy

To determine the reactions at the supports, we will represent the triangular distributed load by an equivalent force. Then we will determine the internal forces and moment at B by cutting the beam by a plane at B and drawing the free-body diagram of the part of the beam to the left of the plane, *including the part of the distributed load to the left of the plane.*

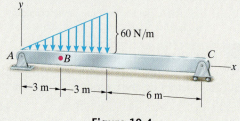

Figure 10.4

Solution

Determine the External Forces and Moments We draw the free-body diagram of the beam and represent the distributed load by an equivalent force in Fig. a. The equilibrium equations are

$$\Sigma F_x = A_x = 0,$$

$$\Sigma F_y = A_y + C - 180 \text{ N} = 0$$

$$\Sigma M_{\text{point }A} = (12 \text{ m})C - (4 \text{ m})(180 \text{ N}) = 0$$

Solving them, we obtain $A_x = 0$, $A_y = 120$ N, and $C = 60$ N.

Draw the Free-Body Diagram of Part of the Beam We cut the beam at B, obtaining the free-body diagram in Fig. b. Because point B is at the midpoint of the triangular distributed load, the value of the distributed load at B is 30 N/m. By representing the distributed load in Fig. b by an equivalent force, we obtain the free-body diagram in Fig. c. From the equilibrium equations

$$\Sigma F_x = P_B = 0,$$

$$\Sigma F_y = 120 \text{ N} - 45 \text{ N} - V_B = 0,$$

$$\Sigma_{\text{point }B} = M_B + (1 \text{ m})(45 \text{ N}) - (3 \text{ m})(120 \text{ N}) = 0,$$

we obtain $P_B = 0$, $V_B = 75$ N, and $M_B = 315$ N-m.

Critical Thinking

If you attempt to determine the internal forces and moment at B by cutting the free-body diagram in Fig. a at B, you do *not* obtain correct results. (You can confirm that the resulting free-body diagram of the part of the beam to the left of B gives $P_B = 0$, $V_B = 120$ N, and $M_B = 360$ N-m.) The reason is that you do not properly account for the effect of the distributed load on your free-body diagram. You must wait until *after* you have isolated part of the beam before representing distributed loads acting on that part by equivalent forces.

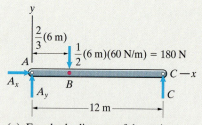

(a) Free-body diagram of the entire beam with the distributed load represented by an equivalent force.

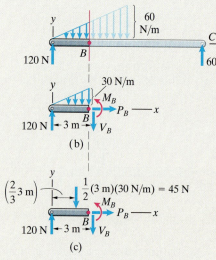

(b), (c) Free-body diagram of the part of the beam to the left of B.

Problems

10.1 Determine the reactions at the beam's fixed support. Then determine the internal forces and moment at A (a) by drawing the free-body diagram of the part of the beam to the left of A; (b) by drawing the free-body diagram of the part of the beam to the right of A.

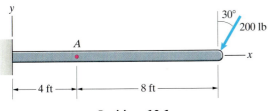

Problem 10.1

10.2 The magnitude of the triangular distributed load at $x = 1.2$ m is w_0. The shear force at A is $V_A = 100$ N.

(a) Determine w_0.
(b) What is the bending moment at A?

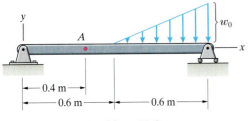

Problem 10.2

10.3 The C clamp exerts 30-lb forces on the clamped object. Determine the internal forces and moment in the clamp at A.

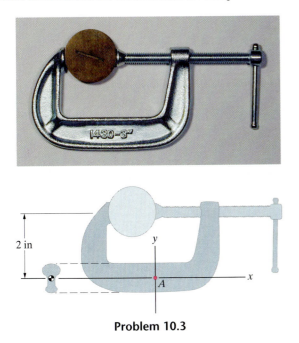

Problem 10.3

10.4 Determine the internal forces and moment at A.

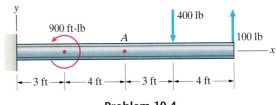

Problem 10.4

10.5 The pipe has a fixed support at the left end. Determine the internal forces and moment at A.

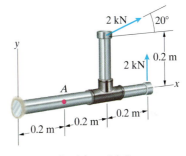

Problem 10.5

10.6 Determine the internal forces and moment at A for each loading.

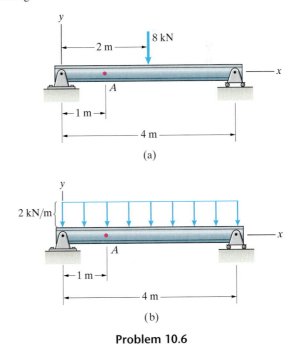

(a)

(b)

Problem 10.6

10.7 Model the ladder rung as a simply supported (pin support-ed) beam and assume that the 750-N load exerted by the person's shoe is uniformly distributed. Determine the internal forces and moment at A.

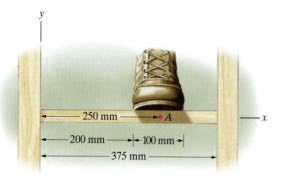

Problem 10.7

10.8 The shear force and bending moment at A are $V_A = 190$ N and $M_A = -130$ N-m. Determine the dimension b and the value of w_0.

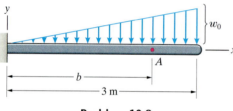

Problem 10.8

10.9 If $x = 3$ m, what are the internal forces and moment at A?

10.10 If $x = 8$ m, what are the internal forces and moment at A?

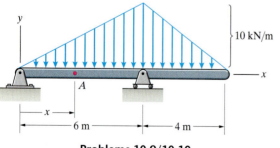

Problems 10.9/10.10

10.11 Determine the internal forces and moment at A for the loadings (a) and (b).

10.12 Determine the internal forces and moment at B for the loadings (a) and (b).

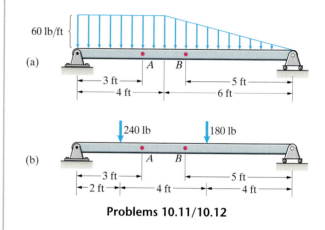

Problems 10.11/10.12

10.13 Determine the internal forces and moment at A.

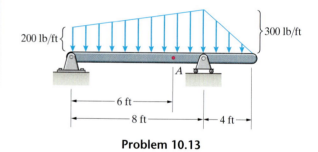

Problem 10.13

10.14 Determine the internal forces and moment at A.

10.15 Determine the internal forces and moment at B.

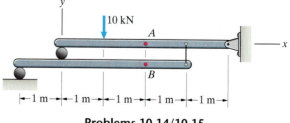

Problems 10.14/10.15

10.16 Determine the internal forces and moment at A.

10.17 Determine the internal forces and moment at B.

10.19 Determine the internal forces and moment at point A of the frame.

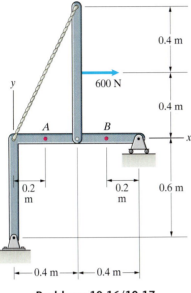

Problems 10.16/10.17

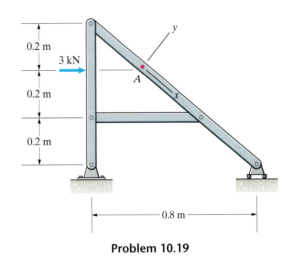

Problem 10.19

10.20 Determine the internal forces and moment at A.

10.21 Determine the internal forces and moment at B.

10.18 The tension in the rope is 10 kN. Determine the internal forces and moment at point A.

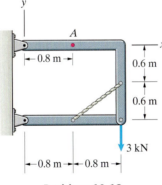

Problem 10.18

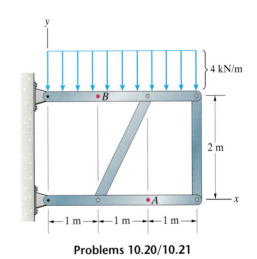

Problems 10.20/10.21

10.2 Shear Force and Bending Moment Diagrams

To design a beam, an engineer must know the internal forces and moments throughout its length. Of special concern are the maximum and minimum values of the shear force and bending moment and where they occur. In this section we show how the values of P, V, and M can be determined as functions of x and introduce shear force and bending moment diagrams.

Consider a simply supported beam loaded by a force (Fig. 10.5a). Instead of cutting the beam at a specific cross section to determine the internal forces

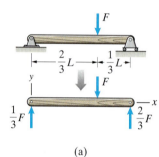

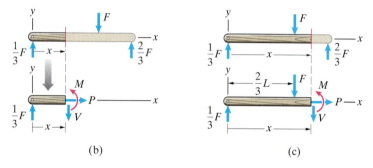

(a)   (b)   (c)

Figure 10.5

(a) A beam loaded by a force F and its free-body diagram.

(b) Cutting the beam at an arbitrary position x to the left of F.

(c) Cutting the beam at an arbitrary position x to the right of F.

and moment, we cut it at an arbitrary position x between the left end of the beam and the load F (Fig. 10.5b). Applying the equilibrium equations to this free-body diagram, we obtain

$$\left. \begin{array}{l} P = 0 \\[2mm] V = \dfrac{1}{3}F \\[2mm] M = \dfrac{1}{3}Fx \end{array} \right\} \quad 0 < x < \frac{2}{3}L.$$

To determine the internal forces and moment for values of x greater than $\frac{2}{3}L$, we obtain a free-body diagram by cutting the beam at an arbitrary position x between the load F and the right end of the beam (Fig. 10.5c). The results are

$$\left. \begin{array}{l} P = 0 \\[2mm] V = -\dfrac{2}{3}F \\[2mm] M = \dfrac{2}{3}F(L - x) \end{array} \right\} \quad \frac{2}{3}L < x < L.$$

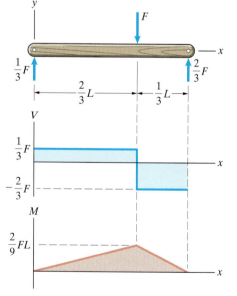

Figure 10.6

The shear force and bending moment diagrams indicating the maximum and minimum values of V and M.

The *shear force and bending moment diagrams* are simply the graphs of V and M, respectively, as functions of x (Fig. 10.6). They permit you to see the changes in the shear force and bending moment that occur along the beam's length as well as their maximum and minimum values. (By *maximum* we mean the least upper bound of the shear force or bending moment, and by *minimum* we mean the greatest lower bound.)

Thus we can determine the distributions of the internal forces and moment in a beam by considering a plane at an arbitrary distance x from the end of the beam and solving for P, V, and M as functions of x. Depending on the complexity of the loading, it may be necessary to draw several free-body diagrams to determine the distributions over the entire length of the beam. The resulting equations for V and M allow us to draw the shear force and bending moment diagrams.

Example 10.3   Shear Force and Bending Moment Diagrams

Determine the shear force and bending moment diagrams for the beam in Fig. 10.7.

Strategy
After determining the reactions at the supports, we can cut the beam at an arbitrary position between A and B to determine the internal forces and moment for $0 < x < 2$ m. Then by cutting the beam at an arbitrary position between B and C, we can determine the internal forces and moment for $2 < x < 4$ m.

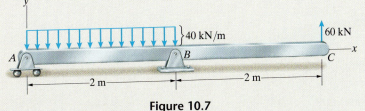

Figure 10.7

Solution
We begin by drawing the free-body diagram of the entire beam and representing the distributed force by an equivalent force (Fig. a). From the equilibrium equations

$$\Sigma F_x = B_x = 0,$$

$$\Sigma F_y = A + B_y - 80 \text{ kN} + 60 \text{ kN} = 0,$$

$$\Sigma M_{\text{point } A} = (2 \text{ m})B_y - (1 \text{ m})(80 \text{ kN}) + (4 \text{ m})(60 \text{ kN}) = 0,$$

we obtain the reactions $A = 100$ kN, $B_x = 0$, and $B_y = -80$ kN.

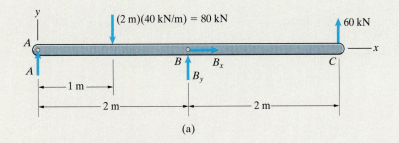

(a)

In Fig. b we obtain a free-body diagram by cutting the beam at an arbitrary position between A and B. From the equilibrium equations

$$\Sigma F_x = P = 0,$$

$$\Sigma F_y = 100 - 40x - V = 0,$$

$$\Sigma M_{\text{right end}} = M - 100x + (\tfrac{1}{2}x)(40x) = 0,$$

we obtain

$$\left.\begin{array}{l} P = 0 \\ V = 100 - 40x \text{ kN} \\ M = 100x - 20x^2 \text{ kN-m} \end{array}\right\} \quad 0 < x < 2 \text{ m}.$$

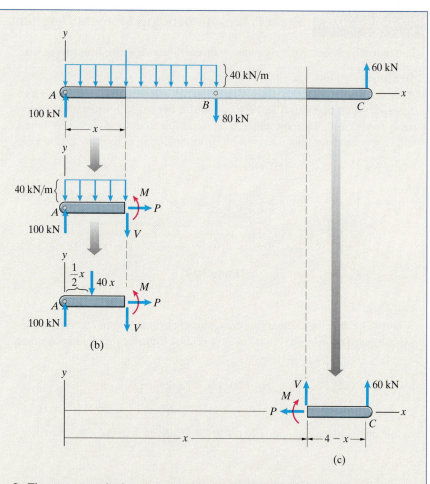

(b)

(c)

In Fig. c, we cut the beam at an arbitrary position between B and C and draw the free-body diagram of the part of the beam to the right of the cutting plane. From the equilibrium equations

$$\Sigma F_x = -P = 0,$$

$$\Sigma F_y = V + 60 = 0,$$

$$\Sigma M_{\text{left end}} = -M + 60(4 - x) = 0,$$

we obtain

$$\left.\begin{array}{l} P = 0 \\ V = -60 \text{ kN} \\ M = 60(4 - x) \text{ kN-m} \end{array}\right\} \quad 2 < x < 4 \text{ m.}$$

The shear force and bending moment diagrams, obtained by plotting the equations for V and M for the two ranges of x, are shown in Fig. 10.8.

Critical Thinking

When you obtain equations for the shear force and bending moment in a beam that apply to different parts of the beam, as we did in this example, there are two conditions you can often use to check your results. (We discuss the bases of these conditions in the next section.) The first one is that *the shear force diagram of a beam is continuous except at points where the beam is subjected to*

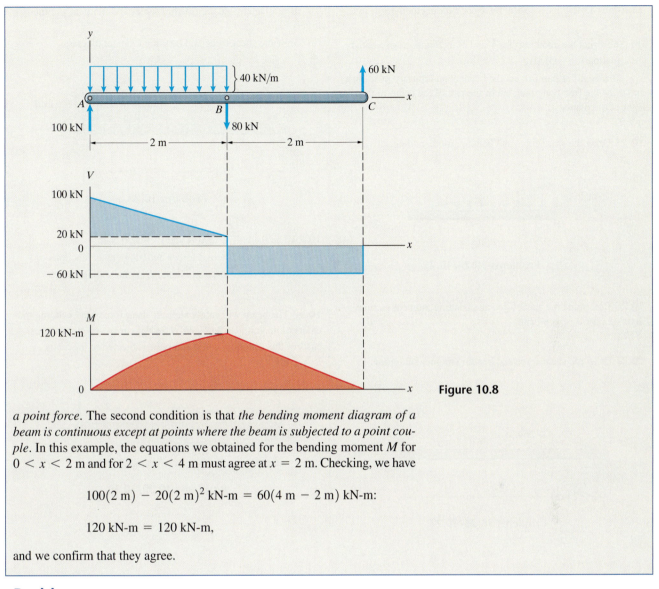

Figure 10.8

a point force. The second condition is that *the bending moment diagram of a beam is continuous except at points where the beam is subjected to a point couple*. In this example, the equations we obtained for the bending moment M for $0 < x < 2$ m and for $2 < x < 4$ m must agree at $x = 2$ m. Checking, we have

$$100(2 \text{ m}) - 20(2 \text{ m})^2 \text{ kN-m} = 60(4 \text{ m} - 2 \text{ m}) \text{ kN-m}:$$

$$120 \text{ kN-m} = 120 \text{ kN-m},$$

and we confirm that they agree.

Problems

10.22 Determine the shear force and bending moment as functions of x.

    Strategy: Cut the beam at an arbitrary position x and draw the free-body diagram of the part of the beam to the left of the plane.

10.23 (a) Determine the shear force and bending moment as functions of x.

(b) Draw the shear force and bending moment diagrams.

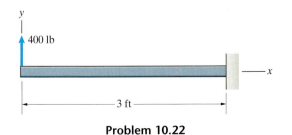

Problem 10.22

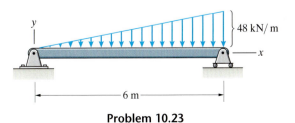

Problem 10.23

10.24 (a) Determine the shear force and bending moment as functions of x.

(b) Show that the equations for V and M as functions of x satisfy the equation $V = dM/dx$.

Strategy: For part (a), cut the beam at an arbitrary position x and draw the free-body diagram of the part of the beam to the right of the plane.

10.25 Draw the shear force and bending moment diagrams.

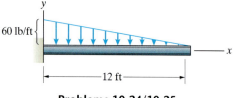

Problems 10.24/10.25

10.26 Determine the shear force and bending moment as functions of x for $0 < x < 2$ m.

10.27 Draw the shear force and bending moment diagrams.

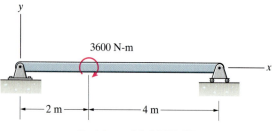

Problems 10.26/10.27

10.28 (a) Determine the internal forces and moment as functions of x.

(b) Draw the shear force and bending moment diagrams.

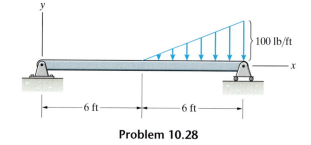

Problem 10.28

10.29 The loads $F = 200$ N and $C = 800$ N-m.

(a) Determine the internal forces and moment as functions of x.

(b) Draw the shear force and bending moment diagrams.

10.30 The beam will safely support shear forces and bending moments of magnitudes 2 kN and 6.5 kN-m, respectively. On the basis of this criterion, can it safely be subjected to the loads $F = 1$ kN, $C = 1.6$ kN-m?

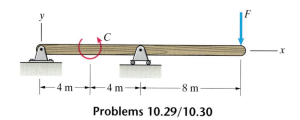

Problems 10.29/10.30

10.31 Model the ladder rung as a simply supported (pin-support-ed) beam and assume that the 750-N load exerted by the person's shoe is uniformly distributed. Draw the shear force and bending moment diagrams.

10.32 What is the maximum bending moment in the ladder rung in Problem 10.31 and where does it occur?

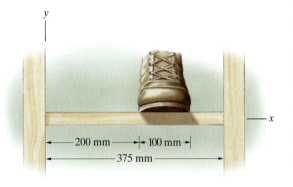

Problems 10.31/10.32

10.33 Assume that the surface the beam rests on exerts a uniformly distributed load. Draw the shear force and bending moment diagrams.

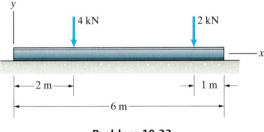

Problem 10.33

10.34 The homogeneous beams AB and CD weigh 600 lb and 500 lb, respectively. Draw the shear force and bending moment diagrams for beam AB.

10.35 The homogeneous beams AB and CD weigh 600 lb and 500 lb, respectively. Draw the shear force and bending moment diagrams for beam CD.

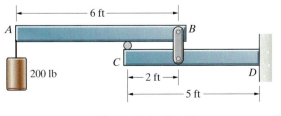

Problems 10.34/10.35

10.36 Determine the shear force as a function of x.

10.37 Draw the shear force and bending moment diagrams.

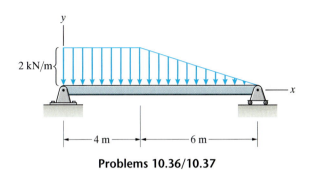

Problems 10.36/10.37

10.38 Draw the shear force and bending moment diagrams.

10.39* Determine the maximum bending moment and the value of x at which it occurs.

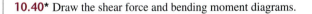
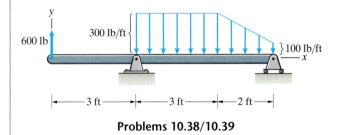

Problems 10.38/10.39

10.40* Draw the shear force and bending moment diagrams.

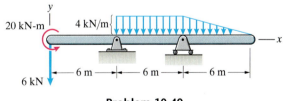

Problem 10.40

10.3 Relations Between Distributed Load, Shear Force, and Bending Moment

The shear force and bending moment in a beam subjected to a distributed load are governed by simple differential equations. In this section we derive these equations and show that they provide an interesting and enlightening way to obtain shear force and bending moment diagrams. These equations are also useful for determining deformations of beams.

Suppose that a portion of a beam is subjected to a distributed load w (Fig. 10.9a). In Fig. 10.9b we obtain a free-body diagram by cutting the beam at positions x and $x + \Delta x$. The terms ΔP, ΔV, and ΔM are the changes in the axial force, shear force, and bending moment, respectively, from x to $x + \Delta x$. The sum of the forces in the x direction is

$$\Sigma F_x = P + \Delta P - P = 0.$$

Dividing this equation by Δx and taking the limit as $\Delta x \rightarrow 0$, we obtain

$$\frac{dP}{dx} = 0,$$

which simply states that the axial force does not depend on x in a portion of a beam subjected only to a lateral distributed load. To sum the forces on the free-body diagram in the y direction, we must determine the force exerted by the distributed load. In Fig. 10.9b we introduce a coordinate $\hat{x}$ that measures distance from the left edge of the free-body diagram. In terms of this coordinate, the downward force exerted on the free-body diagram by the distributed load is

$$\int_0^{\Delta x} w(x + \hat{x}) \, d\hat{x},$$

where $w(x + \hat{x})$ denotes the value of w at $x + \hat{x}$. To evaluate this integral, we express $w(x + \hat{x})$ as a Taylor series in terms of $\hat{x}$:

$$w(x + \hat{x}) = w(x) + \frac{dw(x)}{dx}\hat{x} + \frac{1}{2}\frac{d^2w(x)}{dx^2}\hat{x}^2 + \cdots. \quad (10.1)$$

Substituting this equation into the integral expression for the downward force and integrating term by term, we obtain

$$\int_0^{\Delta x} w(x + \hat{x}) \, d\hat{x} = w(x)\Delta x + \frac{1}{2}\frac{dw(x)}{dx}(\Delta x)^2 + \cdots.$$

The sum of the forces on the free-body diagram in the y direction is therefore

$$\Sigma F_y = V - V - \Delta V - w(x)\Delta x - \frac{1}{2}\frac{dw(x)}{dx}(\Delta x)^2 + \cdots = 0.$$

Dividing by Δx and taking the limit as $\Delta x \rightarrow 0$, we obtain

$$\frac{dV}{dx} = -w, \quad (10.2)$$

where $w = w(x)$.

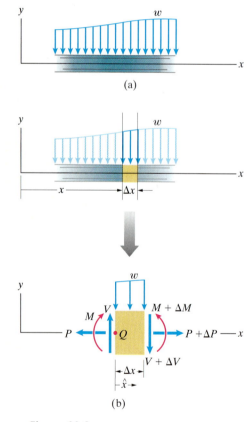

(a)

(b)

Figure 10.9
(a) A portion of a beam subjected to a distributed force w.
(b) Obtaining the free-body diagram of an element of the beam.

We now want to sum the moments about point Q on the free-body diagram in Fig. 10.9b. The clockwise moment about Q due to the distributed load is

$$\int_0^{\Delta x} \hat{x}\, w(x + \hat{x})\, d\hat{x}.$$

Substituting Eq. (10.1) and integrating term by term, the clockwise moment about Q is

$$\int_0^{\Delta x} \hat{x}\, w(x + \hat{x})\, d\hat{x} = \frac{1}{2} w(x)(\Delta x)^2 + \frac{1}{3}\frac{dw(x)}{dx}(\Delta x)^3 + \cdots .$$

The sum of the moments about Q is therefore

$$\Sigma M_{\text{point } Q} = M + \Delta M - M - (V + \Delta V)\, \Delta x$$

$$-\frac{1}{2} w(x)(\Delta x)^2 - \frac{1}{3}\frac{dw(x)}{dx}(\Delta x)^3 + \cdots = 0.$$

Dividing by Δx and taking the limit as $\Delta x \to 0$ gives

$$\frac{dM}{dx} = V. \qquad\qquad (10.3)$$

In principle, we can use Eqs. (10.2) and (10.3) to determine the shear force and bending moment diagrams for a beam. Equation (10.2) can be integrated to determine V as a function of x, then Eq. (10.3) can be integrated to determine M as a function of x. However, we derived these equations for a segment of beam subjected only to a distributed load. To apply them for a more general loading, we must account for the effects of any point forces and couples acting on the beam.

Let us determine what happens to the shear force and bending moment diagrams where a beam is subjected to a force F in the positive y direction (Fig. 10.10a). By cutting the beam just to the left and just to the right of the force, we obtain the free-body diagram in Fig. 10.10b, where the subscripts $-$ and $+$ denote values to the left and right of the force, respectively. Equilibrium requires that

$$V_+ - V_- = F,$$

$$M_+ - M_- = 0.$$

The shear force diagram undergoes a jump discontinuity of magnitude F (Fig. 10.10c), but the bending moment diagram is continuous (Fig. 10.10d). The jump in the shear force is positive if the force is in the positive y direction.

Now we consider what happens to the shear force and bending moment diagrams when a beam is subjected to a counterclockwise couple C (Fig. 10.11a). Cutting the beam just to the left and just to the right of the couple (Fig. 10.11b), we determine that

$$V_+ - V_- = 0,$$

$$M_+ - M_- = -C.$$

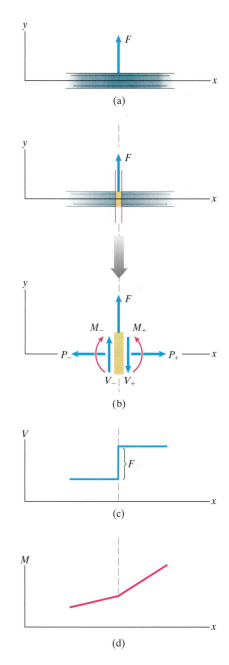

Figure 10.10
(a) A portion of a beam subjected to a distributed force F in the positive y direction.
(b) Obtaining a free-body diagram by cutting the beam to the left and right of F.
(c) The shear force diagram undergoes a positive jump of magnitude F.
(d) The bending moment diagram is continuous.

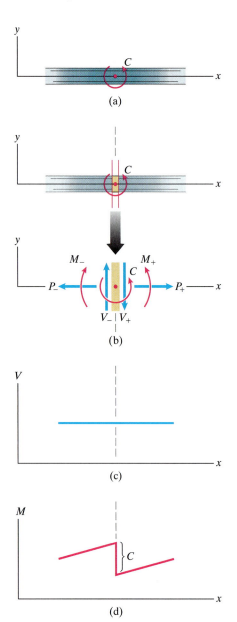

(a)

(b)

(c)

(d)

Figure 10.11
(a) A portion of a beam subjected to a counterclockwise couple C.
(b) Obtaining a free-body diagram by cutting the beam to the left and right of C.
(c) The shear force diagram is continuous.
(d) The bending moment diagram undergoes a *negative* jump of magnitude C.

The shear force diagram is continuous (Fig. 10.11c), but the bending moment diagram undergoes a jump discontinuity of magnitude C (Fig. 10.11d), where a beam is subjected to a couple. The jump in the bending moment is *negative* if the couple is in the counterclockwise direction.

We now have the results needed to construct shear force and bending moment diagrams.

Construction of the Shear Force Diagram

In a segment of a beam that is subjected only to a distributed load, we have shown that the shear force is related to the distributed load by

$$\frac{dV}{dx} = -w. \tag{10.4}$$

This equation states that the derivative, or slope, of the shear force with respect to x is equal to the negative of the distributed load. Notice that if there is no distributed load ($w = 0$) throughout the segment, the slope is zero and the shear force is constant. If w is a constant throughout the segment, the slope of the shear force is constant, which means that the shear force diagram for the segment is a straight line. Integrating Eq. (10.4) with respect to x from a position x_A to a position x_B, that is,

$$\int_{x_A}^{x_B} \frac{dV}{dx}\,dx = -\int_{x_A}^{x_B} w\,dx,$$

yields

$$V_B - V_A = -\int_{x_A}^{x_B} w\,dx.$$

The change in the shear force between two positions is equal to the negative of the area defined by the loading curve between those positions (Fig. 10.12):

$$V_B - V_A = -(\text{area defined by the distributed load from } x_A \text{ to } x_B). \tag{10.5}$$

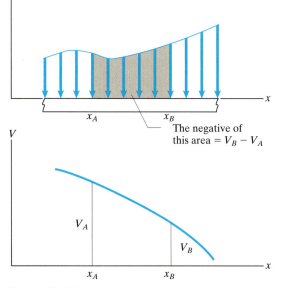

Figure 10.12
The change in the shear force is equal to the negative of the area defined by the loading curve.

Where a beam is subjected to a point force of magnitude F in the positive y direction, we have shown that the shear force diagram undergoes an increase of magnitude F. Where a beam is subjected to a couple, the shear force diagram is unchanged (continuous).

Let us demonstrate these results by determining the shear force diagram for the beam in Fig. 10.13. The beam is subjected to a downward force F that results in upward reactions at A and C. Notice that there is no distributed load. Our procedure is to begin at the left end of the beam and construct the diagram from left to right. Figure 10.14a shows the increase in the value of V due to the upward reaction at A. Because there is no distributed load, the value of V remains constant between A and B (Fig. 10.14b). At B, the value of V decreases due to the downward force (Fig. 10.14c). The value of V remains constant between B and C, which completes the shear force diagram (Fig. 10.14d). Compare Fig. 10.14d with the shear force diagram we obtained in Fig. 10.6 by drawing free-body diagrams and applying the equilibrium equations.

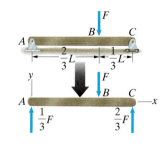

Figure 10.13
Beam loaded by a force F and its free-body diagram.

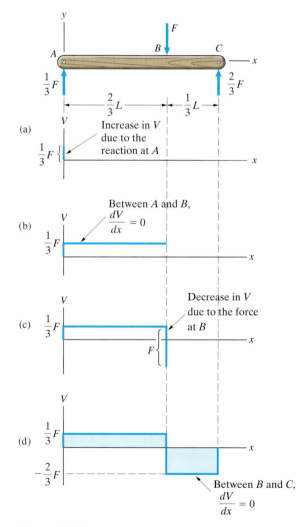

Figure 10.14
Constructing the shear force diagram for the beam in Fig. 10.13.

Construction of the Bending Moment Diagram

In a segment of a beam subjected only to a distributed load, the bending moment is related to the shear force by

$$\frac{dM}{dx} = V, \tag{10.6}$$

which states that the slope of the bending moment with respect to x is equal to the shear force. If V is constant throughout the segment, the bending moment diagram for the segment is a straight line. Integrating Eq. (10.6) with respect to x from a position x_A to a position x_B yields

$$M_B - M_A = \int_{x_A}^{x_B} V \, dx.$$

The change in the bending moment between two positions is equal to the area defined by the shear force diagram between those positions (Fig. 10.15):

$$M_B - M_A = \text{area defined by the shear force from } x_A \text{ to } x_B. \tag{10.7}$$

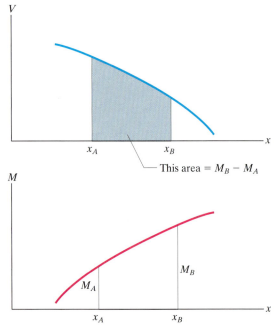

Figure 10.15
The change in the bending moment is equal to the area defined by the shear force diagram.

Where a beam is subjected to a counterclockwise couple of magnitude C, the bending moment diagram undergoes a decrease of magnitude C. Where a beam is subjected to a point force, the bending moment diagram is unchanged.

As an example, we will determine the bending moment diagram for the beam in Fig. 10.13. We begin with the shear force diagram we have already determined (Fig. 10.16a) and proceed to construct the bending moment diagram from left to right. The beam is not subjected to a couple at A, so $M_A = 0$. Between A and B, the slope of the bending moment is constant $(dM/dx = V = F/3)$, which tells us that the bending moment diagram between

A and B is a straight line (10.16b). The change in the bending moment from A to B is equal to the area defined by the shear force from A to B:

$$M_B - M_A = (\tfrac{2}{3}L)(\tfrac{1}{3}F) = \tfrac{2}{9}LF.$$

Therefore, $M_B = 2LF/9$. The slope of the bending moment is also constant between B and C $(dM/dx = V = -2F/3)$, so the bending moment diagram between B and C is a straight line. The change in the bending moment from B to C is equal to the area defined by the shear force from B to C, or

$$M_C - M_B = (\tfrac{1}{3}L)(-\tfrac{2}{3}F) = -\tfrac{2}{9}LF,$$

from which we obtain $M_C = M_B - 2LF/9 = 0$. (Notice that we did not actually need this calculation to conclude that $M_C = 0$, because the beam is not subjected to a couple at C.) The completed bending moment diagram is shown in Fig. 10.16c. Compare it with the bending moment diagram we obtained in Fig. 10.6 by drawing free-body diagrams and applying the equilibrium equations.

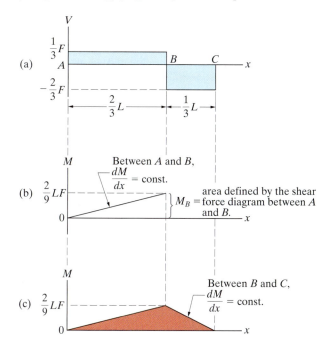

Figure 10.16
Constructing the bending moment diagram for the beam in Fig. 10.13.

Study Questions

1. If a segment of a beam is subjected only to a constant distributed load w, what does Eq. (10.4) tell you about the shear force diagram?

2. If a segment of a beam is not subject to any loads, what can you conclude from Eqs. (10.4) and (10.6) about the bending moment diagram?

3. When Eqs. (10.4)–(10.7) are used to determine the shear force and bending moment diagrams for a beam, how can you account for point forces and couples acting on the beam?

Example 10.4	Shear Force and Bending Moment Diagrams Using Eqs. (10.4)–(10.7)

Determine the shear force and bending moment diagrams for the beam in Fig. 10.17.

Strategy

We can begin with the free-body diagram of the beam and use Eqs. (10.4) and (10.5) to construct the shear force diagram. Then we can use the shear force diagram and Eqs. (10.6) and (10.7) to construct the bending moment diagram. In determining both the shear force and bending moment diagrams, we must account for the effects of point forces and couples acting on the beam.

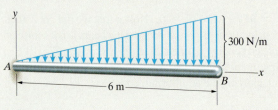

Figure 10.17

Solution

Shear force diagram The first step is to draw the free-body diagram of the beam and determine the reactions at the built-in support A. Using the results of this step, shown in Fig. a, we proceed to construct the shear force diagram from left to right. Figure b shows the increase in the value of V due to the upward force at A. Between A and B, the distributed load on the beam increases linearly from 0 to 300 N/m. Therefore, the slope of the shear force diagram decreases linearly from 0 to -300 N/m. At B, the shear force must be 0, because no force acts there. With this information, we can sketch the shear force diagram qualitatively (Fig. c).

We can also obtain an explicit equation for the shear force between A and B by integrating Eq. (10.4). The distributed load as a function of x is $w = (x/6)300 = 50x$ N/m. We write Eq. (10.4) as

$$dV = -w\,dx = -50x\,dx$$

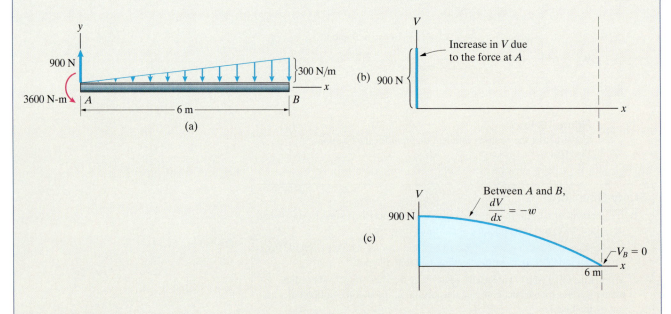

and integrate to determine V at an arbitrary position x:

$$\int_{V_A}^{V} dV = \int_{0}^{x} - 50x \, dx$$

$$V - V_A = -25x^2.$$

Due to the 900-N upward reaction at A, $V_A = 900$ N, so we obtain

$$V = 900 - 25x^2 \text{ N.} \qquad (1)$$

Bending moment diagram We construct the bending moment diagram from left to right. Figure d shows the initial decrease in the value of M due to the counterclockwise couple at A. Between A and B, the slope of the bending moment diagram is equal to the shear force V. We see from the shear force diagram (Fig. c) that at A, the slope of the bending moment diagram has a positive value (900 N). As x increases, the slope begins to decrease, and its rate of decrease grows until the value of the slope reaches zero at B. At B, we know that the value of the bending moment is zero, because no couple acts on the beam at B. Using this information, we can sketch the bending moment diagram qualitatively (Fig. e). Notice that its slope decreases from a positive value at A to zero at B, and the rate at which it decreases grows as x increases.

We can obtain an equation for the bending moment between A and B by integrating Eq. (10.6). The shear force as a function of x is given by Eq. (1). We write Eq. (10.6) as

$$dM = V \, dx = (900 - 25x^2) \, dx$$

and integrate:

$$\int_{M_A}^{M} dM = \int_{0}^{x} (900 - 25x^2) \, dx$$

$$M - M_A = 900x - \tfrac{25}{3}x^3.$$

As a result of the 3600 N-m counterclockwise couple at A, $M_A = -3600$ N-m, yielding the bending moment distribution

$$M = -3600 + 900x - \tfrac{25}{3}x^3 \text{ N-m.}$$

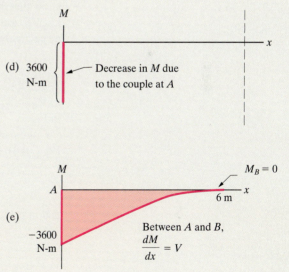

(d) 3600 N-m — Decrease in M due to the couple at A

(e)

$M_B = 0$

6 m

Between A and B, $\dfrac{dM}{dx} = V$

−3600 N-m

Critical Thinking

As demonstrated in this example, Eqs. (10.4)–(10.7) can be applied in two ways. They provide a basis for rapidly obtaining qualitative sketches of shear force and bending moment diagrams. In addition, explicit equations for the diagrams can be obtained by integrating Eqs. (10.4) and (10.6).

Example 10.5 **Shear Force and Bending Moment Diagrams Using Eqs. (10.4)–(10.7)**

Determine the shear force and bending moment diagrams for the beam in Fig. 10.18.

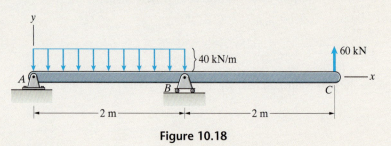

Figure 10.18

Strategy

Just as in Example 10.4, we can begin with the free-body diagram of the beam and use Eqs. (10.4) and (10.5) to construct the shear force diagram, then use the shear force diagram and Eqs. (10.6) and (10.7) to construct the bending moment diagram.

Solution

Shear force diagram   We determined the reactions at the supports of this beam in Example 10.3. Using the results shown in Fig. a, we proceed to construct the shear force diagram from left to right. Due to the 100-kN upward force at A, $V_A = 100$ kN (Fig. b). Between A and B, the distributed load on the beam is constant, which means that the slope of the shear force diagram is constant. Therefore, the shear force diagram between A and B is a straight line. We know that the shear force will be discontinuous at B due to the 80-kN reaction. Let V_B^- and V_B^+ denote the values of V to the left and right of the 80-kN force, respectively. From Eq. (10.5), the change in V from A to B is

$$V_B^- - V_A = -(2 \text{ m})(40 \text{ kN/m}) = -80 \text{ kN},$$

so $V_B^- = V_A - 80 = 20$ kN. The shear force diagram between A and B is shown in Fig. c.

We can also determine the shear force between A and B by integrating Eq. (10.4). We write Eq. (10.4) as

$$dV = -w \, dx = -40 \, dx$$

and integrate:

$$\int_{V_A}^{V} dV = \int_{0}^{x} -40 \, dx$$

$$V - V_A = -40x.$$

Due to the 100 kN upward reaction at A, $V_A = 100$ kN, yielding the shear force distribution

$$V = 100 - 40x \text{ kN.} \tag{1}$$

Notice that at $x = 2$ m, this equation gives $V_B^- = 20$ kN. The effect of the 80-kN downward force at B is shown in Fig. d. The value of V to the right

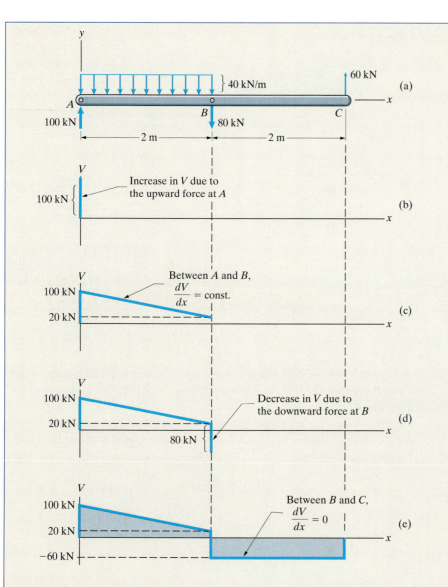

of the 80-kN force is

$$V_B^+ = V_B^- - 80 = -60 \text{ kN}.$$

Because there is no distributed load between B and C, the value of V remains constant between B and C, completing the shear force diagram (Fig. e).

Bending moment diagram   The beam is not subjected to a couple at A, so $M_A = 0$. Between A and B, the slope of the bending moment diagram equals the shear force. From the shear force diagram (Fig. f), we see that the slope is positive between A and B and decreases linearly from A to B. The change in the bending moment between A and B is equal to the area defined by the shear force diagram between A and B, namely,

$$M_B - M_A = (2 \text{ m})(20 \text{ kN}) + \tfrac{1}{2}(2 \text{ m})(80 \text{ kN}) = 120 \text{ kN-m},$$

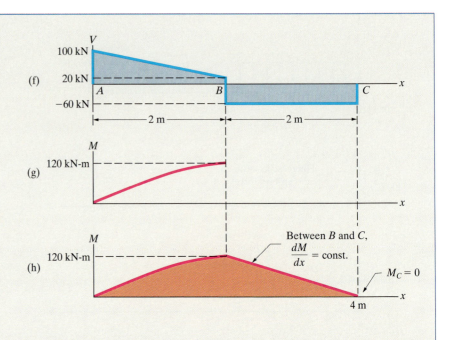

so $M_B = 120$ kN-m. With this information, we can sketch the diagram between A and B qualitatively (Fig. g). Observe that the slope is positive, but decreases from A to B.

We can obtain an equation for the bending moment between A and B by integrating Eq. (10.6). The shear force as a function of x is given by Eq. (1). We write Eq. (10.6) as

$$dM = V\,dx = (100 - 40x)\,dx$$

and integrate:

$$\int_{M_A}^{M} dM = \int_{0}^{x} (100 - 40x)\,dx$$

$$M - M_A = 100x - 20x^2 \text{ kN-m.}$$

We know that $M_A = 0$, so the bending moment distribution between A and B is

$$M = 100x - 20x^2 \text{ kN-m.}$$

Because no couple is applied to the beam at C, $M_C = 0$. The slope of the bending moment is constant between B and C ($dM/dx = V = -60$ kN), so the bending moment diagram between B and C is a straight line (Fig. h).

Critical Thinking

Compare this example with Example 10.3, in which we use free-body diagrams and the equilibrium equations to determine the shear force and bending moment diagrams for this beam and loading.

Problems

The following problems are to be solved using Eqs. (10.4)–(10.7).

10.41 Draw the shear force and bending moment diagrams.

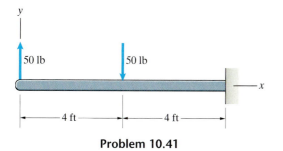

Problem 10.41

10.42 Draw the shear force and bending moment diagrams.

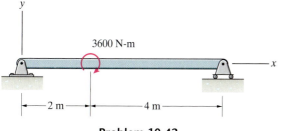

Problem 10.42

10.43 This arrangement is used to subject a segment of a beam to a uniform bending moment. Draw the shear force and bending moment diagrams.

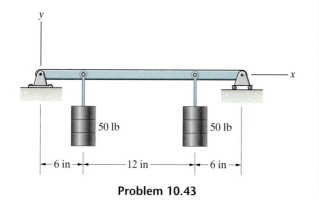

Problem 10.43

10.44 Draw the shear force and bending moment diagrams.

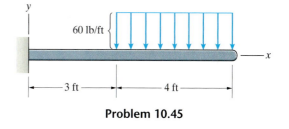

Problem 10.44

10.45 Draw the shear force and bending moment diagrams.

Problem 10.45

10.46 Draw the shear force and bending moment diagrams.

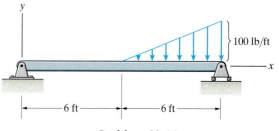

Problem 10.46

10.47* Draw the shear force and bending moment diagrams.

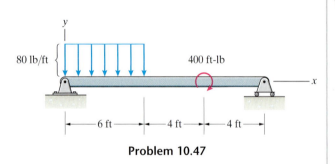

Problem 10.47

10.48* Draw the shear force and bending moment diagrams.

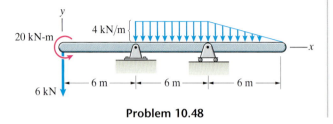

Problem 10.48

10.49 Draw the shear force and bending moment diagrams for the beam *AB*.

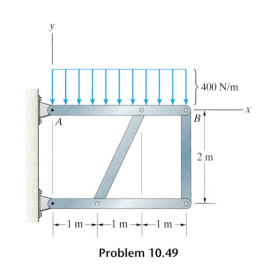

Problem 10.49

CABLES

Because of their unique combination of strength, lightness, and flexibility, ropes and cables are often used to support loads and transmit forces in structures, machines, and vehicles. The great suspension bridges are supported by enormous steel cables. Architectural engineers use cables to create aesthetic structures with open interior spaces (Fig. 10.19). In the following sections we determine the tensions in ropes and cables subjected to distributed and discrete loads.

Figure 10.19
The use of cables to suspend the roof of this sports stadium provides spectators with a view unencumbered by supporting columns.

10.4 Loads Distributed Uniformly Along Straight Lines

The main cable of a suspension bridge is the classic example of a cable subjected to a load uniformly distributed along a straight line (Fig. 10.20). The weight of the bridge is (approximately) uniformly distributed horizontally. The load, transmitted to the main cable by the large number of vertical cables, can be modeled as a distributed load. In this section we determine the shape and the variation in the tension of a cable loaded in this way.

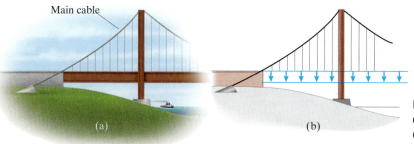

Main cable

(a)

(b)

Figure 10.20
(a) Main cable of a suspension bridge.
(b) The load is distributed horizontally.

Consider a suspended cable subjected to a load distributed uniformly along a horizontal line (Fig. 10.21a). We neglect the weight of the cable. The origin of the coordinate system is located at the cable's lowest point. Let the function $y(x)$ be the curve described by the cable in the x–y plane. Our objective is to determine the curve $y(x)$ and the tension in the cable.

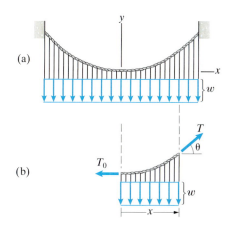

(a)

(b)

Figure 10.21
(a) A cable subjected to a load uniformly distributed along a horizontal line.
(b) Free-body diagram of the cable between $x = 0$ and an arbitrary position x.

Shape of the Cable

We obtain a free-body diagram by cutting the cable at its lowest point and at an arbitrary position x (Fig. 10.21b). The term T_0 is the tension in the cable at its lowest point, and T is the tension at x. The downward force exerted by the distributed load is wx. From this free-body diagram, we obtain the equilibrium equations

$$T \cos \theta = T_0,$$

$$T \sin \theta = wx. \qquad (10.8)$$

We eliminate the tension T by dividing the second equation by the first one, obtaining

$$\tan \theta = \frac{w}{T_0}x = ax,$$

where

$$a = \frac{w}{T_0}.$$

The slope of the cable at x is $dy/dx = \tan \theta$, so we obtain a differential equation governing the curve described by the cable:

$$\frac{dy}{dx} = ax. \tag{10.9}$$

We have chosen the coordinate system so that $y = 0$ at $x = 0$. Integrating Eq. (10.9), we get

$$\int_0^y dy = \int_0^x ax\, dx,$$

we find that the curve described by the cable is the parabola

$$y = \frac{1}{2}ax^2. \tag{10.10}$$

Tension of the Cable

To determine the distribution of the tension in the cable, we square both sides of Eqs. (10.8) and then sum them, obtaining

$$T = T_0\sqrt{1 + a^2x^2}. \tag{10.11}$$

The tension is a minimum at the lowest point of the cable and increases monotonically with distance from the lowest point.

Length of the Cable

In some applications it is useful to have an expression for the length of the cable in terms of x. We can write the relation $ds^2 = dx^2 + dy^2$, where ds is an element of length of the cable (Fig. 10.22), in the form

$$ds = \sqrt{1 + \left(\frac{dy}{dx}\right)^2}\, dx.$$

Substituting Eq. (10.9) into this expression and integrating, we obtain an equation for the length s of the cable in the horizontal interval from 0 to x:

$$s = \frac{1}{2}\left\{ x\sqrt{1 + a^2x^2} + \frac{1}{a}\ln\left[ax + \sqrt{1 + a^2x^2} \right] \right\}. \tag{10.12}$$

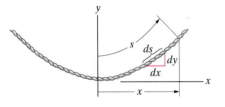

Figure 10.22
The length s of the cable in the horizontal interval from 0 to x.

Study Questions

1. If a cable is subjected to a load that is uniformly distributed along a straight line and its weight is negligible, what mathematical curve describes its shape?

2. Equation (10.10) describes the shape of a cable loaded as described in Question 1. Where must the origin of the x–y coordinate system be located?

Example 10.6	Cable with a Horizontally Distributed Load

The horizontal distance between the supporting towers of the Golden Gate Bridge in San Francisco, California, is 1280 m (Fig. 10.23). The tops of the towers are 160 m above the lowest point of the main supporting cables. Obtain the equation for the curve described by the cables.

Figure 10.23

Strategy

We know the coordinates of the cables' attachment points relative to their lowest points. By substituting the coordinates into Eq. (10.10), we can determine a. Once a is known, Eq. (10.10) describes the shapes of the cables.

Solution

The coordinates of the top of the right supporting tower relative to the lowest point of the support cables are $x_R = 640$ m, $y_R = 160$ m (Fig. a). By substituting these values into Eq. (10.10),

$$160 \text{ m} = \frac{1}{2}a(640 \text{ m})^2,$$

we obtain

$$a = 7.81 \times 10^{-4} \text{ m}^{-1}.$$

The curve described by the supporting cables is

$$y = \frac{1}{2}ax^2 = (3.91 \times 10^{-4})x^2.$$

Fig. a compares this parabola with a photograph of the supporting cables.

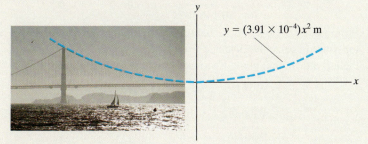

$$y = (3.91 \times 10^{-4})x^2 \text{ m}$$

(a) The theoretical curve superimposed on a photograph of the supporting cable.

Critical Thinking

Knowing the locations of the cable's highest and lowest points has allowed us to calculate the value of the coefficient a. This coefficient not only determines the equation describing the cable's shape, as we demonstrated in this example, but also is the ratio of the distributed load w acting on the cable to the tension T_0 in the cable at its lowest point. If the value of w was also known, the tension throughout the cable would be determined by Eq. (10.11). (See Example 10.7.)

Example 10.7 Maximum Tension in a Cable

The cable in Fig. 10.24 supports a distributed load of 100 lb/ft. What is the maximum tension in the cable?

Strategy

We are given the vertical coordinate of each attachment point, but we are told only the total horizontal span. However, the coordinates of each attachment point relative to a coordinate system with its origin at the lowest point of the cable must satisfy Eq. (10.10). This permits us to determine the horizontal coordinates of the attachment points. Once we know them, we can use Eq. (10.10) to determine $a = w/T_0$, which tells us the tension at the lowest point, and then use Eq. (10.11) to obtain the maximum tension.

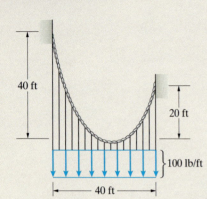

Figure 10.24

Solution

We introduce a coordinate system with its origin at the lowest point of the cable, denoting the coordinates of the left and right attachment points by (x_L, y_L) and (x_R, y_R), respectively (Fig. a). Equation (10.10) must be satisfied for both of these points:

$$y_L = 40 \text{ ft} = \frac{1}{2} a x_L^2,$$

$$y_R = 20 \text{ ft} = \frac{1}{2} a x_R^2. \tag{1}$$

We don't know a, but we can eliminate it by dividing the first equation by the second one, obtaining

$$\frac{x_L^2}{x_R^2} = 2.$$

We also know that

$$x_R - x_L = 40 \text{ ft}.$$

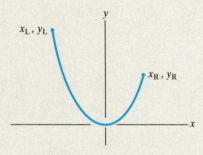

(a) A coordinate system with its origin at the lowest point and the coordinates of the left and right attachment points.

(The reason for the minus sign is that x_L is negative.) We therefore have two equations we can solve for x_L and x_R; the results are $x_L = -23.4$ ft and $x_R = 16.6$ ft.

We can now use either of Eqs. (1) to determine a. We obtain $a = 0.146 \text{ ft}^{-1}$, so the tension T_0 at the lowest point of the cable is

$$T_0 = \frac{w}{a} = \frac{100 \text{ lb/ft}}{0.146 \text{ ft}^{-1}} = 686 \text{ lb}.$$

From Eq. (10.11), we know that the maximum tension in the cable occurs at the maximum horizontal distance from its lowest point, which in this example is the left attachment point. The maximum tension is therefore

$$T_{max} = T_0 \sqrt{1 + a^2 x_L^2}$$
$$= (686 \text{ lb}) \sqrt{1 + (0.146 \text{ ft}^{-1})^2 (-23.4 \text{ ft})^2}$$
$$= 2440 \text{ lb}.$$

Critical Thinking

Why are calculations of this kind useful? To a structural engineer designing a cable suspension system, knowledge of the tension in the cable, and particularly the maximum tension, is crucial for specifying the properties of the cable to be used. This example demonstrates that knowing the gross dimensions of the suspended cable and the load it supports is sufficient to determine the tension throughout the cable.

10.5 Loads Distributed Uniformly Along Cables

A cable's own weight subjects it to a load that is distributed uniformly along its length. If a cable is subjected to equal, parallel forces spaced uniformly along its length, the load on the cable can often be modeled as a load distributed uniformly along its length. In this section we show how to determine both the cable's resulting shape and the variation in its tension.

Suppose that a cable is acted on by a distributed load that subjects each element ds of its length to a force $w\,ds$, where w is constant. In Fig. 10.25 we show the free-body diagram obtained by cutting the cable at its lowest point and at a point a distance s along its length. The terms T_0 and T are the tensions at the lowest point and at s, respectively. The distributed load exerts a downward force ws. The origin of the coordinate system is located at the lowest point of the cable. Let the function $y(x)$ be the curve described by the cable in the x–y plane. Our objective is to determine $y(x)$ and the tension T.

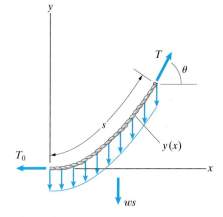

Figure 10.25
A cable subjected to a load distributed uniformly along its length.

Shape of the Cable

From the free-body diagram in Fig. 10.25, we obtain the equilibrium equations

$$T \sin \theta = ws, \tag{10.13}$$

$$T \cos \theta = T_0. \tag{10.14}$$

Dividing Eq. (10.13) by Eq. (10.14), we obtain

$$\tan \theta = \frac{w}{T_0}s = as, \tag{10.15}$$

where

$$a = \frac{w}{T_0}. \tag{10.16}$$

The slope of the cable $dy/dx = \tan \theta$, so Eq. (10.15) can be written

$$\frac{dy}{dx} = as.$$

The derivative of this equation with respect to x is

$$\frac{d}{dx}\left(\frac{dy}{dx}\right) = a\frac{ds}{dx}. \tag{10.17}$$

By using the relation

$$ds^2 = dx^2 + dy^2,$$

we can write the derivative of s with respect to x as

$$\frac{ds}{dx} = \sqrt{1 + \left(\frac{dy}{dx}\right)^2} = \sqrt{1 + \sigma^2}, \tag{10.18}$$

where

$$\sigma = \frac{dy}{dx} = \tan \theta.$$

is the slope. Now, with Eq. (10.18), we can write Eq. (10.17) as

$$ \frac{d\sigma}{\sqrt{1 + \sigma^2}} = a\, dx. $$

The slope $\sigma = 0$ at $x = 0$. Integrating this equation yields

$$ \int_0^\sigma \frac{d\sigma}{\sqrt{1 + \sigma^2}} = \int_0^x a\, dx, $$

and we obtain the slope as a function of x:

$$ \sigma = \frac{dy}{dx} = \frac{1}{2}(e^{ax} - e^{-ax}) = \sinh ax. \tag{10.19} $$

Then, integrating this equation with respect to x yields the curve described by the cable, which is called a *catenary*:

$$ y = \frac{1}{2a}(e^{ax} + e^{-ax} - 2) = \frac{1}{a}(\cosh ax - 1). \tag{10.20} $$

Tension of the Cable

Using Eq. (10.14) and the relation $dx = \cos\theta\, ds$, we obtain

$$ T = \frac{T_0}{\cos\theta} = T_0 \frac{ds}{dx}. $$

Substituting Eq. (10.18) into this expression and using Eq. (10.19) yields the tension in the cable as a function of x:

$$ T = T_0 \sqrt{1 + \frac{1}{4}(e^{ax} - e^{-ax})^2} = T_0 \cosh ax. \tag{10.21} $$

Length of the Cable

From Eq. (10.15), the length s of the cable from the origin to the point at which the angle between the cable and the x axis equals θ is

$$ s = \frac{1}{a}\tan\theta = \frac{\sigma}{a}. $$

Substituting Eq. (10.19) into this equation, we obtain an expression for the length s of the cable in the horizontal interval from its lowest point to x:

$$ s = \frac{1}{2a}(e^{ax} - e^{-ax}) = \frac{\sinh ax}{a}. \tag{10.22} $$

Example 10.8 Cable Loaded by Its Own Weight

The mass per unit length of the cable in Fig. 10.26 is 1 kg/m. The tension at its lowest point is 50 N. Determine the distance h and the maximum tension in the cable.

Strategy

The cable is subjected to a load $w = (9.81 \text{ m/s}^2)(1 \text{ kg/m}) = 9.81 \text{ N/m}$ distributed uniformly along its length. Since we know w and T_0, we can determine $a = w/T_0$. Then we can determine h from Eq. (10.20). Because the maximum tension occurs at the greatest distance from the lowest point of the cable, we can determine it by letting $x = 10$ m in Eq. (10.21).

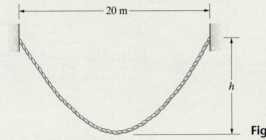

Figure 10.26

Solution

The parameter a is

$$a = \frac{w}{T_0} = \frac{9.81 \text{ N/m}}{50 \text{ N}} = 0.196 \text{ m}^{-1}$$

In terms of a coordinate system with its origin at the lowest point of the cable (Fig. a), the coordinates of the right attachment point are $x = 10$ m, $y = h$. From Eq. (10.20),

$$h = \frac{1}{a}(\cosh ax - 1)$$

$$= \frac{1}{0.196 \text{ m}^{-1}}\{\cosh[(0.196 \text{ m}^{-1})(10 \text{ m})] - 1\}$$

$$= 13.4 \text{ m}.$$

From Eq. (10.21), the maximum tension is

$$T_{max} = T_0\sqrt{1 + \sinh^2 ax}$$

$$= (50 \text{ N})\sqrt{1 + \sinh^2[(0.196 \text{ m}^{-1})(10 \text{ m})]}$$

$$= 181 \text{ N}.$$

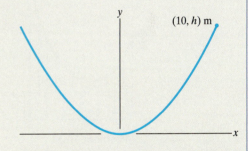

(a) A coordinate system with its origin at the lowest point of the cable.

Critical Thinking

In this example we specified the mass per unit length of the cable and the tension T_0 at the cable's lowest point and used them to determine the distance h in Fig. 10.26. In a real application, you would be much more likely to know the mass per unit length of the cable and the distance h and need to determine the tension T_0. How could you do that? If h is given, the x and y coordinates of the cable's highest points relative to the lowest point are known. Substituting the coordinates of one of these points into Eq. (10.20) yields an equation for the coefficient $a = w/T_0$, from which T_0 could be determined. However, notice that the transcendental Eq. (10.20) would have to be solved numerically for the value of a.

Problems

10.50 The cable supports a uniformly distributed load $w = 1$ kN/m.
(a) What is the maximum tension in the cable?
(b) What is the length of the cable?

    Strategy: You know the coordinates of the attachment points of the cable relative to its lowest point, so you can use Eq. (10.10) to determine the coefficient a and then use $a = w/T_0$ to determine the tension at the lowest point.

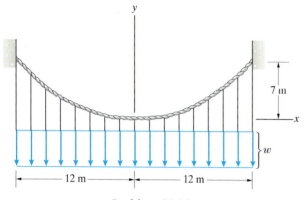

Problem 10.50

10.51 The cable supports a distributed load $w = 12,000$ lb/ft.
(a) What is the maximum tension in the cable?
(b) Determine the cable's length.

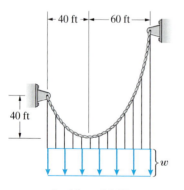

Problem 10.51

10.52 A cable is used to suspend a pipeline above a river. The towers supporting the cable are 36 m apart. The lowest point of the cable is 1.4 m below the tops of the towers. The mass of the suspended pipe is 2700 kg.
(a) What is the maximum tension in the cable?
(b) What is the suspending cable's length?

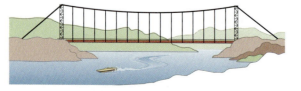

Problems 10.52/10.53

10.53 In Problem 10.52, let the lowest point of the cable be a distance h below the tops of the towers supporting the cable.
(a) If the cable will safely support a tension of 70 kN, what is the minimum safe value of h?
(b) If h has the value determined in part (a), what is the suspending cable's length?

10.54 The cable supports a uniformly distributed load $w = 750$ N/m. The lowest point of the cable is 0.18 m below the attachment points C and D. Determine the axial loads in the truss members AC and BC.

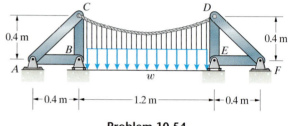

Problem 10.54

10.55 The cable supports a railway bridge between two tunnels. The distributed load is $w = 1$ MN/m, and $h = 40$ m.
(a) What is the maximum tension in the cable?
(b) What is the length of the cable?

10.56 The cable in Problem 10.55 will safely support a tension of 40 MN. What is the shortest cable that can be used, and what is the corresponding value of h?

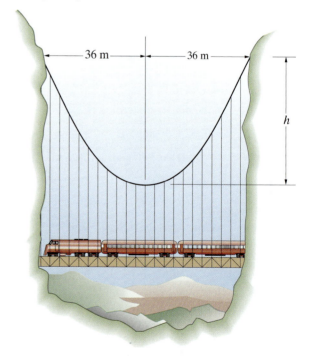

Problems 10.55/10.56

10.57 An oceanographic research ship tows an instrument package from a cable. Hydrodynamic drag subjects the cable to a uniformly distributed force $w = 2$ lb/ft. The tensions in the cable at 1 and 2 are 800 lb and 1300 lb, respectively. Determine the distance h.

10.58 Draw a graph of the shape of the cable in Problem 10.57.

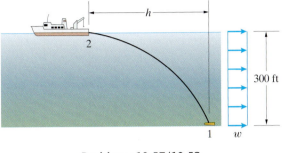

Problems 10.57/10.58

10.59 The mass of the rope per unit length is 0.10 kg/m. The tension at its lowest point is 4.6 N.

(a) What is the maximum tension in the rope?

(b) What is the rope's length?

　　Strategy: Use the given information to evaluate the coefficient $a = w/T_0$. Because the rope is loaded only by its own weight, the tension is given as a function of x by Eq. (10.21) and the length of the rope in the horizontal interval from its lowest point to x is given by Eq. (10.22).

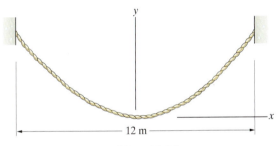

Problem 10.59

10.60 The stationary balloon's tether is horizontal at point O where it is attached to the truck. The mass per unit length of the tether is 0.45 kg/m. The tether exerts a 50-N horizontal force on the truck. The horizontal distance from point O to point A where the tether is attached to the balloon is 20 m. What is the height of point A relative to point O?

10.61 In Problem 10.60, determine the magnitudes of the horizontal and vertical components of the force exerted on the balloon at A by the tether.

Problems 10.60/10.61

10.62 The mass per unit length of lines AB and BC is 2 kg/m. The tension at the lowest point of cable AB is 1.8 kN. The two lines exert equal horizontal forces at B.

(a) Determine the sags h_1 and h_2.

(b) Determine the maximum tensions in the two lines.

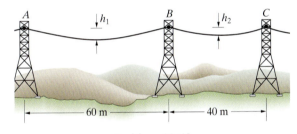

Problem 10.62

10.63 The rope is loaded by 2-kg masses suspended at 1-m intervals along its length. The mass of the rope itself is negligible. The tension in the rope at its lowest point is 100 N. Determine h and the maximum tension in the rope.

　　Strategy: Obtain an approximate answer by modeling the discrete loads on the rope as a load uniformly distributed along its length.

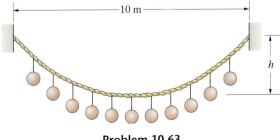

Problem 10.63

10.6 Discrete Loads

Our first applications of equilibrium in Chapter 3 involved determining the tensions in cables supporting suspended objects. In this section we consider the case of an arbitrary number N of objects suspended from a cable (Fig. 10.27a). We assume that the weight of the cable can be neglected in comparison to the suspended weights and that the cable is sufficiently flexible that we can approximate its shape by a series of straight segments.

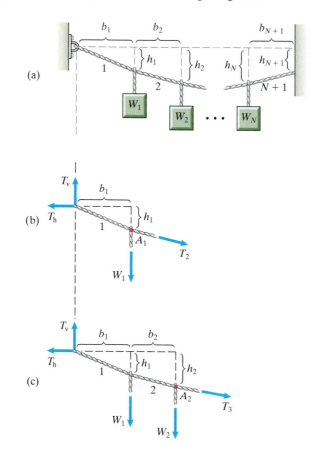

Figure 10.27
(a) N weights suspended from a cable.
(b) The first free-body diagram.
(c) The second free-body diagram.

Determining the Configuration and Tensions

Suppose that the horizontal distances $b_1, b_2, \ldots, b_{N+1}$ are known and that the vertical distance h_{N+1} specifying the position of the cable's right attachment point is known. We have two objectives: (1) to determine the configuration (shape) of the cable by solving for the vertical distances $h_1, h_2, \ldots, h_N$ specifying the positions of the attachment points of the weights and (2) to determine the tensions in the segments $1, 2, \ldots, N + 1$ of the cable.

We begin by drawing a free-body diagram, cutting the cable at its left attachment point and just to the right of the weight W_1 (Fig. 10.27b). We resolve the tension in the cable at the left attachment point into its horizontal and vertical components T_h and T_v. Summing moments about the attachment point A_1, we obtain the equation

$$\Sigma M_{\text{point } A_1} = h_1 T_h - b_1 T_v = 0.$$

Our next step is to obtain a free-body diagram by cutting the cable at its left attachment point and just to the right of the weight W_2 (Fig. 10.27c). Summing moments about A_2, we obtain

$$\Sigma M_{\text{point } A_2} = h_2 T_{\text{h}} - (b_1 + b_2)T_{\text{v}} + b_2 W_1 = 0.$$

Proceeding in this way, cutting the cable just to the right of each of the N weights, we obtain N equations. We can also draw a free-body diagram by cutting the cable at its left and right attachment points and sum moments about the right attachment point. In this way, we obtain $N + 1$ equations in terms of $N + 2$ unknowns: the two components of the tension T_{h} and T_{v} and the vertical positions of the attachment points $h_1, h_2 \ldots, h_N$. If the vertical position of just one attachment point is also specified, we can solve the system of equations for the vertical positions of the other attachment points, determining the configuration of the cable.

Once we know the configuration of the cable and the force T_{h}, we can determine the tension in any segment by cutting the cable at the left attachment point and within the segment and summing forces in the horizontal direction.

Comments on Continuous and Discrete Models

By comparing cables subjected to distributed and discrete loads, we can make some observations about how continuous and discrete systems are modeled in engineering. Consider a cable subjected to a horizontally distributed load w (Fig. 10.28a). The total force exerted on it is wL. Since the cable passes through the point $x = L/2$, $y = L/2$, we find from Eq. (10.10) that $a = 4/L$, so the equation for the curve described by the cable is $y = (2/L)x^2$.

In Fig. 10.28b, we compare the shape of the cable with the distributed load to that of a cable of negligible weight subjected to three discrete loads $W = wL/3$ with equal horizontal spacing. (We chose the dimensions of the cable with discrete loads so that the heights of the two cables would be equal at their midpoints.) In Fig. 10.28c, we compare the shape of the cable with the distributed load to that of a cable subjected to five discrete loads $W = wL/5$ with equal horizontal spacing. In Figs. 10.29a and 10.29b, we compare the tension in the cable subjected to the distributed load to those in the cables subjected to three and five discrete loads.

The shape and the tension in the cable with a distributed load are approximated by the shapes and tensions in the cables with discrete loads. Although the approximation of the tension is less impressive than the approximation of the shape, it is clear that the former can be improved by increasing the number of discrete loads.

This approach, approximating a continuous distribution by a discrete model, is very important in engineering. It is the starting point of the finite difference and finite element methods. The opposite approach, modeling discrete systems by continuous models, is also widely used, for example when the forces exerted on a bridge by traffic are modeled as a distributed load.

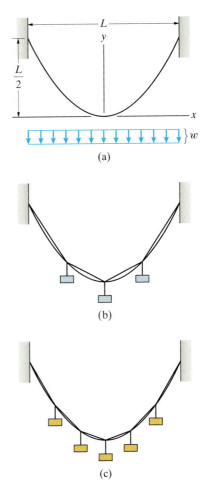

Figure 10.28
(a) Cable subjected to a continuous load.
(b) Cable with three discrete loads.
(c) Cable with five discrete loads.

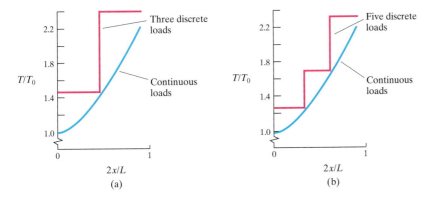

Figure 10.29
(a) The tension in a cable with a continuous load compared to the cable with three discrete loads.
(b) The tension in a cable with a continuous load compared to the cable with five discrete loads.

Example 10.9 | Cable Subjected to Discrete Loads

Two masses $m_1 = 10$ kg and $m_2 = 20$ kg are suspended from the cable in Fig. 10.30.

    (a) Determine the vertical distance h_2.
    (b) Determine the tension in cable segment 2.

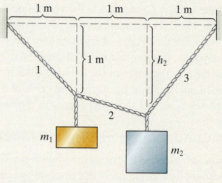

Figure 10.30

Strategy

We will obtain three free-body diagrams by cutting the cable at the left attachment point and (1) just to the right of the mass m_1; (2) just to the right of the mass m_2; and (3) at the right attachment point. By writing a moment equation for each free-body diagram, we will obtain three equations in terms of the two components of the tension at the left attachment point and the unknown vertical distance h_2. Once the geometry of the cable is determined, we can use equilibrium to determine the tension in segment 2.

Solution

(a) We begin by cutting the cable at the left attachment point and just to the right of the mass m_1, and resolve the tension at the left attachment point into horizontal and vertical components (Fig. a). Summing moments about A_1 yields

$$\Sigma M_{\text{point } A_1} = (1 \text{ m})T_h - (1 \text{ m})T_v = 0.$$

We then cut the cable just to the right of the mass m_2 (Fig. b) and sum moments about A_2:

$$\Sigma M_{\text{point } A_2} = h_2 T_h - (2 \text{ m})T_v + (1 \text{ m})m_1 g = 0.$$

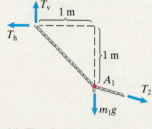

(a) First free-body diagram

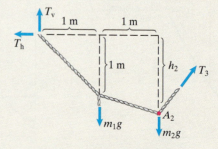

(b) Second free-body diagram.

The last step is to cut the cable at the right attachment point (Fig. c) and sum moments about A_3:

$$\Sigma M_{\text{point } A_3} = -(3 \text{ m})T_v + (2 \text{ m})m_1g + (1 \text{ m})m_2g = 0.$$

We have three equations in terms of the unknowns T_h, T_v, and h_2. Solving them yields $T_h = T_v = 131$ N and $h_2 = 1.25$ m.

(b) To determine the tension in segment 2, we use the free-body diagram in Fig. a. The angle between the force T_2 and the horizontal is

$$\arctan [(h_2 - 1)/1] = 14.0°.$$

Summing forces in the horizontal direction gives

$$T_2 \cos 14.0° - T_h = 0.$$

Solving, we obtain

$$T_2 = \frac{T_h}{\cos 14.0°} = 135 \text{ N}.$$

Critical Thinking

The systematic solution procedure we applied to this cable system with three segments resulted in three equations. In addition to the two components of the tension at the left attachment point, we were able to determine the unknown vertical distance h_2. That is, there was one "excess" equation with which to determine h_2. Instead of the height h_2, some other parameter of the system could have been left unspecified, such as one of the masses or the horizontal position of one of the masses. In a cable system with N segments, $N - 2$ parameters can be left unspecified.

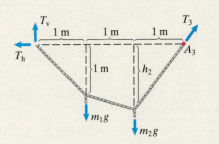

(c) Free-body diagram of the entire cable.

Problems

10.64 In Example 10.9, what are the tensions in cable segments 1 and 3?

10.65 Each lamp weighs 12 lb.

(a) What is the length of the wire $ABCD$ needed to suspend the lamps as shown?

(b) What is the maximum tension in the wire?

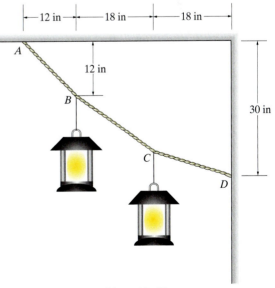

Problem 10.65

10.66 Two weights, $W_1 = W_2 = 50$ lb, are suspended from a cable. The vertical distance $h_1 = 4$ ft.

(a) Determine the vertical distance h_2.
(b) What is the maximum tension in the cable?

10.67 The weights are $W_1 = 50$ lb and $W_2 = 100$ lb, and the vertical distance $h_1 = 4$ ft.

(a) Determine the vertical distance h_2.
(b) What is the maximum tension in the cable?

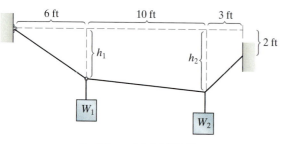

Problems 10.66/10.67

10.68 Three identical masses $m = 10$ kg are suspended from the cable. Determine the vertical distances h_1 and h_3 and draw a sketch of the configuration of the cable.

10.69 In Problem 10.68, what are the tensions in cable segments 1 and 2?

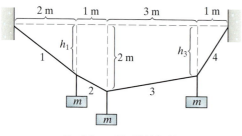

Problem 10.68/10.69

10.70 Three masses are suspended from the cable, where $m = 30$ kg, and the vertical distance $h_1 = 400$ mm. Determine the vertical distances h_2 and h_3.

10.71 In Problem 10.70, what is the maximum tension in the cable, and where does it occur?

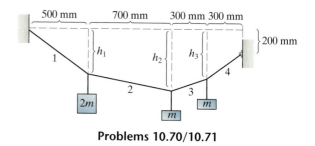

Problems 10.70/10.71

10.72 Each suspended object has the same weight W. Determine the vertical distances h_2 and h_3.

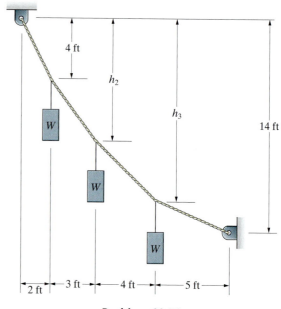

Problem 10.72

COMPUTATIONAL MECHANICS

The following example and problems are designed to be solved using a programmable calculator or computer.

Computational Example 10.10

As the first step in constructing a suspended pedestrian bridge, a cable is suspended across the span from attachment points of equal height (Fig. 10.31). The cable weighs 5 lb/ft and is 42 ft long. Determine the maximum tension in the cable and the vertical distance from the attachment points to the cable's lowest point.

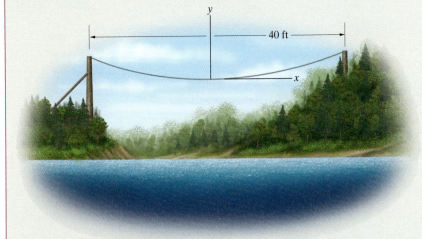

Figure 10.31

Strategy

Equation (10.22) gives the length s of the cable as a function of the horizontal distance x from the cable's lowest point and the parameter $a = w/T_0$. The term w is the weight per unit length, and T_0 is the tension in the cable at its lowest point. We know that the half-span of the cable is 20 ft, so we can draw a graph of s as a function of a and estimate the value of a for which $s = 21$ ft. Then we can determine the maximum tension from Eq. (10.21) and the vertical distance to the cable's lowest point from Eq. (10.20).

Solution

Setting $x = 20$ ft in Eq. (10.22),

$$s = \frac{\sinh[(20 \text{ ft})a]}{a},$$

we compute s as a function of a (Fig. 10.32). The length $s = 21$ ft when the parameter a is approximately 0.027 ft^{-1}. By examining the computed results near $a = 0.027$ ft^{-1},

a (ft^{-1})	s (ft)
0.0269	20.9789
0.0270	20.9863
0.0271	20.9937
0.0272	21.0012
0.0273	21.0086
0.0274	21.0162
0.0275	21.0237

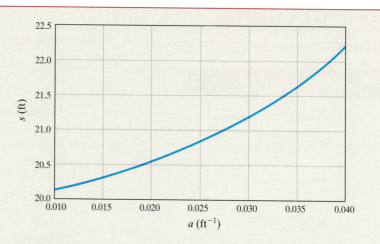

Figure 10.32
Graph of the length s as a function of the parameter a.

we see that s is approximately 21 ft when $a = 0.0272$ ft^{-1}. Therefore, the tension at the cable's lowest point is

$$T_0 - \frac{w}{a} = \frac{5 \text{ lb/ft}}{0.0272 \text{ ft}^{-1}} = 184 \text{ lb},$$

and the maximum tension is

$$T_{\max} = T_0 \cosh ax$$

$$= (184 \text{ lb}) \cosh[(0.0272 \text{ ft}^{-1})(20 \text{ ft})]$$

$$= 212 \text{ lb}.$$

From Eq. (10.20), the vertical distance from the cable's lowest point to the attachment points is

$$y_{\max} = \frac{1}{a}(\cosh ax - 1)$$

$$= \frac{1}{0.0272 \text{ ft}^{-1}}\{\cosh[(0.0272 \text{ ft}^{-1})(20 \text{ ft})] - 1\}$$

$$= 5.58 \text{ ft}.$$

Critical Thinking

From this example you can see how computational results can be used in the design of a suspended cable system. By using a as a parameter, we could determine the maximum tension and the vertical distance to the cable's lowest point for cables with a range of lengths. By calculations of this kind, the design engineer can choose the properties the cable must have to satisfy the criteria of a particular application.

Computational Problems

10.73 The beam's length is $L = 10$ m and the distributed load is

$$w = 20x\left(1 - \frac{x^3}{L^3}\right) \text{N/m}.$$

What is the maximum bending moment in the beam, and where does it occur?

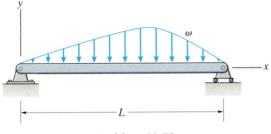

Problem 10.73

10.74 The rope weighs 1 N/m and is 16 m in length.

(a) What is the maximum tension?

(b) What is the vertical distance from the attachment points to the lowest point of the rope?

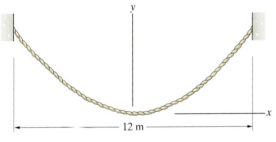

Problem 10.74

10.75 A chain weighs 20 lb and is 20 ft long. It is suspended from two points of equal height that are 10 ft apart.

(a) Determine the maximum tension in the chain.

(b) Draw a sketch of the shape of the chain.

10.76 An engineer wants to suspend high-voltage power lines between poles 200 m apart. Each line has a mass of 2 kg/m.

(a) If the engineer wants to subject the lines to a tension no greater than 10 kN, what should be the maximum allowable sag between poles? That is, what is the largest allowable vertical distance between the attachment points and the lowest point of the line?

(b) What is the length of each line?

10.77 The mass per unit length of lines AB and BC is 2 kg/m. The length of line AB is 62 m. The two lines exert equal horizontal forces at B.

(a) Determine the sags h_1 and h_2.

(b) Determine the maximum tensions in the two lines.

10.78 The mass per unit length of the lines AB and BC is 2 kg/m. The sag $h_1 = 4.5$ m, but the length of line AB is unknown. The two lines exert equal horizontal forces at B.

(a) Determine the sag h_2.

(b) Determine the maximum tensions in the two lines.

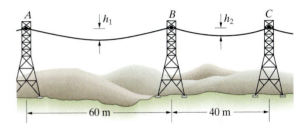

Problems 10.77/10.78

10.79 Two 30-ft cables A and B are suspended from points of equal height that are 20 ft apart. Cable A is subjected to a 200-lb load uniformly distributed horizontally. Cable B is subjected to a 200-lb load distributed uniformly along its length. What are the maximum tensions in the two cables?

10.80 Draw a graph of the two cables in Problem 10.79, comparing their shapes.

10.81 The masses $m_1 = 10$ kg and $m_2 = 20$ kg. The total length of the three segments of rope is 5 m.

(a) What are h_1 and h_2?

(b) What is the maximum tension in the rope?

   Strategy: If you choose a value of h_1, you can determine h_2 and then L. By drawing a graph of L as a function of h_1, you can determine the value of h_1 that corresponds to $L = 5$ m.

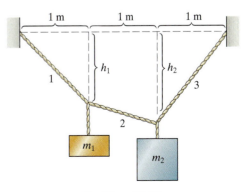

Problem 10.81

LIQUIDS AND GASES

Wind forces on buildings and aerodynamic forces on cars and airplanes are examples of forces that are distributed over areas. The downward force exerted on the bed of a dump truck by a load of gravel is distributed over the area of the bed. The upward force that supports a building is distributed over the area of its foundation. Loads distributed over the roofs of buildings by accumulated snow can be hazardous. Many forces of concern in engineering are distributed over areas. In this section we analyze the most familiar example, the force exerted by the pressure of a gas or liquid.

10.7 Pressure and the Center of Pressure

A surface immersed in a gas or liquid is subjected to forces exerted by molecular impacts. If the gas or liquid is stationary, the load can be described by a function p, the *pressure*, defined such that the normal force exerted on a differential element dA of the surface is $p\, dA$ (Figs. 10.33a and 10.33b). (Notice the parallel between the pressure and a load w distributed along a line, which is defined such that the force on a differential element dx of the line is $w\, dx$.)

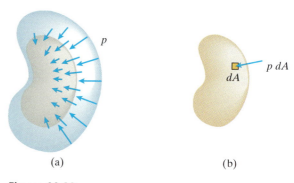

(a)   (b)

Figure 10.33
(a) The pressure on an area.
(b) The force on an element dA is $p\, dA$.

The dimensions of p are (force)/(area). In U.S. Customary units, pressure can be expressed in pounds per square foot or pounds per square inch (psi). In SI units, pressure can be expressed in newtons per square meter, which are called pascals (Pa).

In some applications, it is convenient to use the *gage pressure*

$$p_g = p - p_{\text{atm}}, \qquad (10.23)$$

where p_{atm} is the pressure of the atmosphere. Atmospheric pressure varies with location and climatic conditions. Its value at sea level is approximately 1×10^5 Pa in SI units and 14.7 psi or 2120 lb/ft^2 in U.S. Customary units.

If the distributed force due to pressure on a surface is represented by an equivalent force, the point at which the line of action of the force intersects the surface is called the *center of pressure*. Consider a plane area A subjected to a pressure p and introduce a coordinate system such that the area lies in the x–y

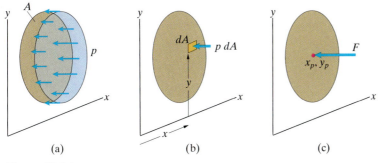

Figure 10.34
(a) A plane area subjected to pressure.
(b) The force on a differential element dA.
(c) The total force acting at the center of pressure.

plane (Fig. 10.34a). The normal force on each differential element of area dA is $p\,dA$ (Fig. 10.34b), so the total normal force on A is

$$F = \int_A p\,dA. \tag{10.24}$$

Now we will determine the coordinates (x_p, y_p) of the center of pressure (Fig. 10.34c). Equating the moment of F about the origin to the total moment due to the pressure about the origin gives

$$(x_p\mathbf{i} + y_p\mathbf{j}) \times (-F\mathbf{k}) = \int_A (x\mathbf{i} + y\mathbf{j}) \times (-p\,dA\mathbf{k}),$$

and using Eq. (10.24), we obtain

$$x_p = \frac{\displaystyle\int_A xp\,dA}{\displaystyle\int_A p\,dA}, \quad y_p = \frac{\displaystyle\int_A yp\,dA}{\displaystyle\int_A p\,dA}. \tag{10.25}$$

These equations determine the position of the center of pressure when the pressure p is known. If the pressure p is uniform, the total normal force is $F = pA$ and Eqs. (10.25) indicate that the center of pressure is the centroid of A.

In Chapter 7 it was shown that if we calculate the "area" defined by a load distributed along a line and place the resulting force at its centroid, the force is equivalent to the distributed load. A similar result holds for a pressure distributed on a plane area. The term $p\,dA$ in Eq. (10.24) is equal to a differential element dV of the "volume" between the surface defined by the pressure distribution and the area A (Fig. 10.35a). The total force exerted by the pressure is therefore equal to this "volume":

$$F = \int_V dV = V.$$

Substituting $p\,dA = dV$ into Eqs. (10.25), we obtain

$$x_p = \frac{\displaystyle\int_V x\,dV}{\displaystyle\int_V dV}, \quad y_p = \frac{\displaystyle\int_V y\,dV}{\displaystyle\int_V dV}.$$

The center of pressure coincides with the x and y coordinates of the centroid of the "volume" (Fig. 10.35b).

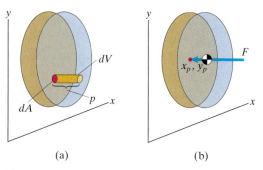

(a)   (b)

Figure 10.35
(a) The differential element $dV = p\,dA$.
(b) The line of action of F passes through the centroid of V.

Study Questions

1. What is the definition of the pressure p?
2. What is the gage pressure?
3. What is the center of pressure? How can the "volume" defined by the pressure distribution be used to determine the location of the center of pressure?

10.8 Pressure in a Stationary Liquid

Designers of pressure vessels and piping, ships, dams, and other submerged structures must be concerned with forces and moments exerted by water pressure. The pressure in a liquid at rest increases with depth, which you can confirm by descending to the bottom of a swimming pool and noting the effect of the pressure on your ears. If we restrict ourselves to changes in depth for which changes in the density of the liquid can be neglected, we can determine the dependence of the pressure on depth by using a simple free-body diagram.

Introducing a coordinate system with its origin at the surface of the liquid and the positive x axis downward (Fig. 10.36a), we draw a free-body diagram of a cylinder of liquid that extends from the surface to a depth x (Fig. 10.36b). The top of the cylinder is subjected to the pressure at the surface, which we call p_0. The sides and bottom of the cylinder are subjected to pressure by the

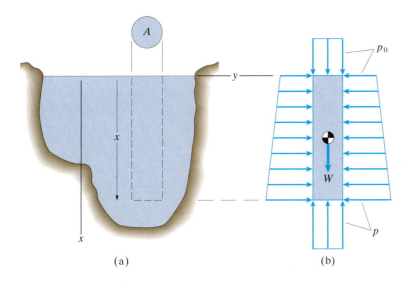

Figure 10.36
(a) A cylindrical volume that extends to a depth x in a body of stationary liquid.
(b) Free-body diagram of the cylinder.

surrounding liquid, which increases from p_0 at the surface to a value p at the depth x. The volume of the cylinder is Ax, where A is its cross-sectional area. Therefore, its weight is $W = \gamma Ax$, where γ is the weight density of the liquid. (Recall that the weight and mass densities are related by $\gamma = \rho g$.) Since the liquid is stationary, the cylinder is in equilibrium. From the equilibrium equation

$$\Sigma F_x = p_0 A - pA + \gamma Ax = 0,$$

we obtain a simple expression for the pressure p of the liquid at depth x:

$$p = p_0 + \gamma x. \tag{10.26}$$

Thus, the pressure increases linearly with depth, and the derivation we have used illustrates why: The pressure at a given depth literally holds up the liquid above that depth. If the surface of the liquid is open to the atmosphere, $p_0 = p_{atm}$, and we can write Eq. (10.26) in terms of the gage pressure $p_g = p - p_{atm}$ as

$$p_g = \gamma x. \tag{10.27}$$

In SI units, the density of water at sea level conditions is $\rho = 1000 \text{ kg/m}^3$, so its weight density is approximately $\gamma = \rho g = 9.81 \text{ kN/m}^3$. In U.S. Customary units, the weight density of water is approximately 62.4 lb/ft^3.

We have seen that the force and moment due to the pressure on a submerged plane area can be determined in two ways:

1. Integration—Integrate Eq. (10.26) or Eq. (10.27).

2. Volume analogy—Determine the total force by calculating the "volume" between the surface defined by the pressure distribution and the area A. The center of pressure coincides with the x and y coordinates of the centroid of the volume.

Example 10.11 **Pressure Force and Center of Pressure**

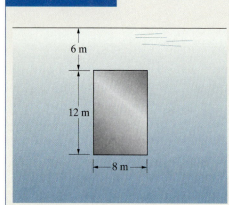

Figure 10.37

An engineer making preliminary design studies for a canal lock needs to determine the total pressure force on a submerged rectangular plate (Fig. 10.37) and the location of the center of pressure. The top of the plate is 6 m below the surface. Atmospheric pressure is $p_{atm} = 1 \times 10^5$ Pa, and the weight density of the water is $\gamma = 9.81$ kN/m^3.

Strategy
We will determine the pressure force on a differential element of area of the plate in the form of a horizontal strip and integrate to determine the total force and moment exerted by the pressure.

Solution
In terms of a coordinate system with its origin at the surface and the positive x axis downward (Fig. a), the pressure of the water is $p = p_{atm} + \gamma x$. The horizontal strip $dA = (8 \text{ m}) \, dx$. Therefore, the total force exerted on the face of the plate by the pressure is

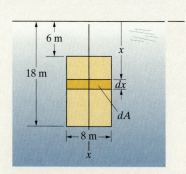

$$F = \int_A p \, dA = \int_6^{18} (p_{atm} + \gamma x)(8 \text{ m}) \, dx$$

$$= p_{atm}(8 \text{ m}) \int_6^{18} dx + \gamma(8 \text{ m}) \int_6^{18} x \, dx$$

$$= \left(1 \times 10^5 \frac{\text{N}}{\text{m}^2}\right)(8 \text{ m})(12 \text{ m}) + \left(9810 \frac{\text{N}}{\text{m}^3}\right)(8 \text{ m})(144 \text{ m}^2)$$

$$= 20.9 \times 10^6 \text{ N}.$$

(a) An element of area in the form of a horizontal strip.

The moment about the y axis due to the pressure on the plate is

$$M = \int_A xp \, dA = \int_6^{18} x(p_{atm} + \gamma x)(8 \text{ m}) \, dx$$

$$= p_{atm}(8 \text{ m}) \int_6^{18} x \, dx + \gamma(8 \text{ m}) \int_6^{18} x^2 \, dx$$

$$= 262 \times 10^6 \text{ N-m}.$$

The force F acting at the center of pressure (Fig. b) exerts a moment about the y axis equal to M:

$$x_p F = M.$$

Therefore, the location of the center of pressure is

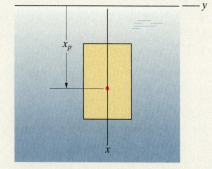

(b) The center of pressure.

$$x_p = \frac{M}{F} = \frac{262 \text{ MN-m}}{20.9 \text{ MN}} = 12.5 \text{ m}.$$

Critical Thinking
Notice that the center of pressure does not coincide with the centroid of the area. The center of pressure of a plane area generally coincides with the centroid of the area only when the pressure is uniformly distributed. In this example, the pressure increases with depth, and as a result, the center of pressure is below the centroid.

Example 10.12   Gate Loaded by a Pressure Distribution

The gate AB in Fig. 10.38 has water of 2-ft depth on the right side. The width of the gate (the dimension into the page) is 3 ft, and its weight is 100 lb. The weight density of the water is $\gamma = 62.4 \ \text{lb/ft}^3$. Determine the reactions on the gate at the supports A and B.

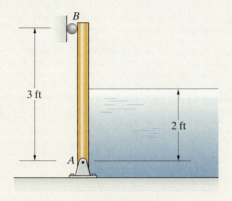

Figure 10.38

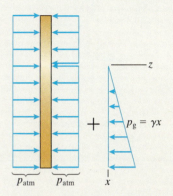

(a) The pressures acting on the faces of the gate.

Strategy

The left face of the gate and the right face above the level of the water are exposed to atmospheric pressure. From Eqs. (10.23) and (10.26), the pressure in the water is the sum of atmospheric pressure and the gage pressure $p_g = \gamma x$, where x is measured downward from the surface of the water. The effects of atmospheric pressure cancel (Fig. a), so we need to consider only the forces and moments exerted on the gate by the gage pressure. We will determine them by integrating and also by calculating the "volume" of the pressure distribution.

Solution

Integration   The face of the gate is shown in Fig. b. In terms of the differential element of area dA, the force exerted on the gate by the gage pressure is

$$F = \int_A p_g \, dA = \int_0^2 (\gamma x)(3 \ \text{ft}) \, dx = 374 \ \text{lb},$$

and the moment about the y axis is

$$M = \int_A x p_g \, dA = \int_0^2 x(\gamma x)(3 \ \text{ft}) \, dx = 499 \ \text{ft-lb}.$$

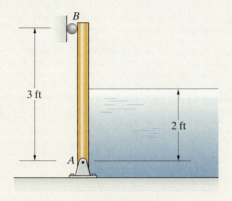

(b) The face of the gate and the differential element dA.

The position of the center of pressure is

$$x_p = \frac{M}{F} = \frac{499 \ \text{ft-lb}}{374 \ \text{lb}} = 1.33 \ \text{ft}.$$

Volume Analogy   The gage pressure at the bottom of the gate is $p_g = (2 \ \text{ft})\gamma$ (Fig. c), so the "volume" of the pressure distribution is

$$F = \frac{1}{2}[2 \ \text{ft}][(2 \ \text{ft})(62.4 \ \text{lb/ft}^3)][3 \ \text{ft}] = 374 \ \text{lb}.$$

The x coordinate of the centroid of the triangular distribution, which is the center of pressure, is $\frac{2}{3}(2) = 1.33 \ \text{ft}$.

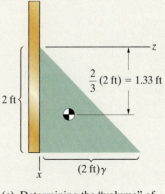

(c) Determining the "volume" of the pressure distribution and its centroid.

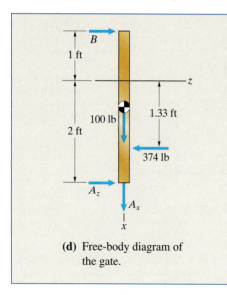

(d) Free-body diagram of the gate.

Determining the Reactions   We draw the free-body diagram of the gate in Fig. d. From the equilibrium equations

$$\Sigma F_x = A_x + 100 \text{ lb} = 0,$$

$$\Sigma F_z = A_z + B - 374 \text{ lb} = 0$$

$$\Sigma M_{y \text{ axis}} = (1 \text{ ft})B - (2 \text{ ft})A_z + (1.33 \text{ ft})(374 \text{ lb}) = 0,$$

we obtain $A_x = -100$ lb, $A_z = 291.2$ lb, and $B = 83.2$ lb.

Critical Thinking
The motivation for defining the gage pressure is demonstrated by this example. In many applications it is the difference between the pressure and atmospheric pressure, not the pressure itself, that is significant.

Example 10.13   Determination of a Pressure Force

The container in Fig. 10.39 is filled with a liquid with weight density γ. Determine the force exerted by the pressure of the liquid on the cylindrical wall AB.

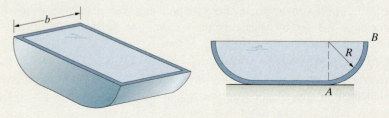

Figure 10.39

(a) The pressure of the liquid on the wall AB.

Strategy
The pressure of the liquid on the cylindrical wall varies with depth (Fig. a). It is the force exerted by this pressure distribution we want to determine. We could determine it by integrating over the cylindrical surface, but we can avoid that by drawing a free-body diagram of the quarter-cylinder of liquid to the right of A.

Solution
We draw the free-body diagram of the quarter-cylinder of liquid in Fig. b. The pressure distribution on the cylindrical surface of the liquid is the same one that acts on the cylindrical wall. If we denote the force exerted on the liquid by this pressure distribution by $\mathbf{F}_p$, the force exerted by the liquid on the cylindrical wall is $-\mathbf{F}_p$.

The other forces parallel to the x–y plane that act on the quarter-cylinder of liquid are its weight, atmospheric pressure at the upper surface, and the pressure distribution of the liquid on the left side. The volume of liquid is $(\frac{1}{4}\pi R^2)b$, so the force exerted on the free-body diagram by the weight of the liquid is $\frac{1}{4}\gamma\pi R^2 b\mathbf{i}$. The force exerted on the upper surface by atmospheric pressure is $Rbp_{atm}\mathbf{i}$.

We can integrate to determine the force exerted by the pressure on the left side of the free-body diagram. Its magnitude is

$$\int_A p\, dA = \int_0^R (p_{atm} + \gamma x)b\, dx = Rb\left(p_{atm} + \frac{1}{2}\gamma R\right).$$

From the equilibrium equation

$$\Sigma\mathbf{F} = \frac{1}{4}\gamma\pi R^2 b\mathbf{i} + Rbp_{atm}\mathbf{i} + Rb\left(p_{atm} + \frac{1}{2}\gamma R\right)\mathbf{j} + \mathbf{F}_p = \mathbf{0},$$

we obtain the force exerted on the wall AB by the pressure of the liquid:

$$-\mathbf{F}_p = Rb\left(p_{atm} + \frac{\pi}{4}\gamma R\right)\mathbf{i} + Rb\left(p_{atm} + \frac{1}{2}\gamma R\right)\mathbf{j}.$$

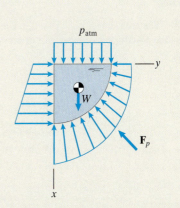

(b) Free-body diagram of the liquid to the right of A.

Critical Thinking
The need to integrate over a curved surface to calculate a pressure force can often be avoided by choosing a suitable free-body diagram as we have done in this example. See Problem 10.101.

Problems

10.82 A deep submersible research vehicle operates at a depth of 1000 m. The average mass density of the water is $\rho = 1025$ kg/m³. Atmospheric pressure is $p_{atm} = 1 \times 10^5$ Pa. Determine the pressure on the vehicle's surface (a) in pascals (Pa); (b) in pounds per square inch (psi).

10.83 An engineer planning a water system for a new community estimates that at maximum expected usage, the pressure drop between the central system and the farthest planned fire hydrant will be 25 psi. Firefighting personnel indicate that a gage pressure of 40 psi at the fire hydrant is required. The weight density of the water is $\gamma = 62.4$ lb/ft³. How tall would a water tower at the central system have to be to provide the needed pressure?

10.84 A cube of material is suspended below the surface of a liquid of weight density γ. By calculating the forces exerted on the faces of the cube by pressure, show that their sum is an upward force of magnitude γb^3.

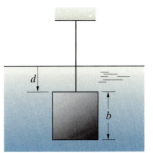

Problem 10.84

10.85 The area shown is subjected to a *uniform* pressure $p_{atm} = 1 \times 10^5$ Pa.
(a) What is the total force exerted on the area by the pressure?
(b) What is the moment about the y-axis due to the pressure on the area?

10.86 The area shown is subjected to a *uniform* pressure. Determine the coordinates of the center of pressure.

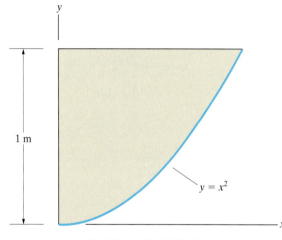

$y = x^2$

1 m

Problems 10.85/10.86

10.87 The area shown is subjected to a *uniform* pressure $p_{atm} = 14.7$ psi.
(a) What is the total force exerted on the area by the pressure?
(b) What is the moment about the y axis due to the pressure on the area?

10.88 The area shown is subjected to a *uniform* pressure. Determine the coordinates of the center of pressure.

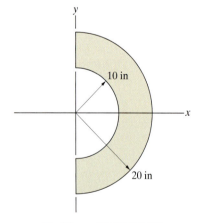

10 in

20 in

Problems 10.87/10.88

10.89 The top of the rectangular plate is 2 m below the sur¬face of a lake. Atmospheric pressure is $p_{atm} = 1 \times 10^5$ Pa and the mass density of the water is $\rho = 1000$ kg/m^3.
(a) What is the maximum pressure exerted on the plate by the water?
(b) Determine the force exerted on a face of the plate by the pressure of the water.

10.90 In Problem 10.89, how far below the top of the plate is the center of pressure located?

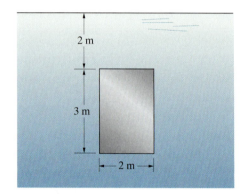

2 m

3 m

2 m

Problems 10.89/10.90

10.91 The width of the dam (the dimension into the page) is 100 m. The mass density of the water is $\rho = 1000$ kg/m^3. Determine the force exerted on the dam by the gage pressure of the water
(a) by integration;
(b) by calculating the "volume" of the pressure distribution.

10.92 In Problem 10.91, how far down from the surface of the water is the center of pressure due to the gage pressure of the water on the dam?

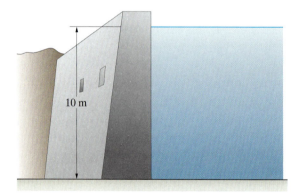

10 m

Problems 10.91/10.92

10.93 The width of the gate (the dimension into the page) is 3 m. Atmospheric pressure is $p_{atm} = 1 \times 10^5$ Pa and the mass density of the water is $\rho = 1000$ kg/m^3. Determine the horizontal force and couple exerted on the gate by its built-in support A.

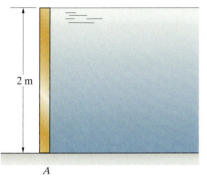

Problem 10.93

10.94 The homogeneous gate weighs 100 lb, and its width (the dimension into the page) is 3 ft. The weight density of the water is $\gamma = 62.4$ lb/ft^3, and atmospheric pressure is $p_{atm} = 2120$ lb/ft^2. Determine the reactions at A and B.

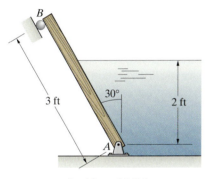

Problem 10.94

10.95 The width of the gate (the dimension into the page) is 2 m and there is water of depth $d = 1$ m on the right side. Atmospheric pressure is $p_{atm} = 1 \times 10^5$ Pa and the mass density of the water is $\rho = 1000$ kg/m^3. Determine the horizontal forces exerted on the gate at A and B.

10.96 The gate in Problem 10.95 is designed to rotate and release the water when the depth d exceeds a certain value. What is that depth?

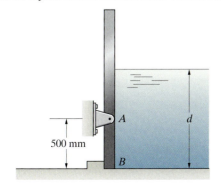

Problems 10.95/10.96

10.97* The dam has water of depth 4 ft on one side. The width of the dam (the dimension into the page) is 8 ft. The weight density of the water is $\gamma = 62.4$ lb/ft^3, and atmospheric pressure $p_{atm} = 2120$ lb/ft^2. If you neglect the weight of the dam, what are the reactions at A and B?

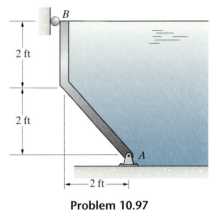

Problem 10.97

10.98* The dam has water of depth 4 ft on one side. The width of the dam (the dimension into the page) is 8 ft. The weight density of the water is $\gamma = 62.4$ lb/ft^3, and atmospheric pressure is $p_{atm} = 2120$ lb/ft^2. If you neglect the weight of the dam, what are the reactions at A and B?

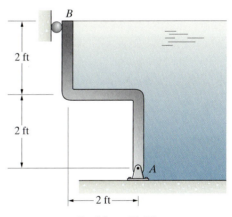

Problem 10.98

10.99 Consider a plane, vertical area A below the surface of a liquid. Let p_0 be the pressure at the surface.

(a) Show that the force exerted by pressure on the area is $F = \bar{p}A$, where $\bar{p} = p_0 + \gamma \bar{x}$ is the pressure of the liquid at the centroid of the area.

(b) Show that the x coordinate of the center of pressure is

$$x_p = \bar{x} + \frac{\gamma I_{y'}}{\bar{p}A},$$

where $I_{y'}$ is the moment of inertia of the area about the y' axis through its centroid.

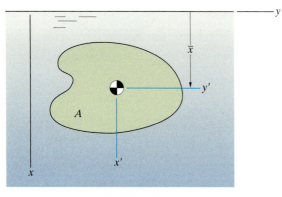

Problem 10.99

10.100 A circular plate of 1-m radius is below the surface of a stationary pool of water. Atmospheric pressure is $p_{atm} = 1 \times 10^5$ Pa, and the mass density of the water is $\rho = 1000$ kg/m^3. Determine (a) the force exerted on a face of the plate by the pressure of the water; (b) the x coordinate of the center of pressure. (See Problem 10.99.)

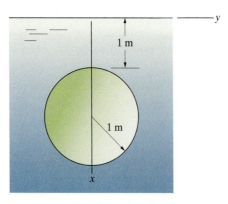

Problem 10.100

10.101* The tank consists of a cylinder with hemispherical ends. It is filled with water ($\rho = 1000$ kg/m^3). The pressure of the water at the top of the tank is 140 kPa. Determine the magnitude of the force exerted by the pressure of the water on each hemispherical end of the tank.

Strategy: Draw a free-body diagram of the water to the right of the dashed line in the figure. See Example 10.13.

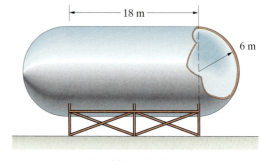

Problem 10.101

10.102 An object of volume V and weight W is suspended below the surface of a stationary liquid of weight density γ (Fig. a). Show that the tension in the cord is $W - V\gamma$. In other words, show that the pressure distribution on the surface of the object exerts an upward force equal to the product of the object's volume and the weight density of the water. This result is due to Archimedes (287–212 B.C.)

Strategy: Draw the free-body diagram of a volume of liquid that has the same shape and position as the object (Fig. b).

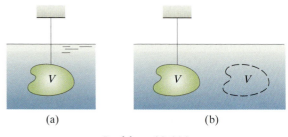

(a)   (b)

Problem 10.102

CHAPTER SUMMARY

Beams

The internal forces and moment in a beam are expressed as the *axial force P, shear force V*, and *bending moment M*. Their positive directions are defined in Fig. a.

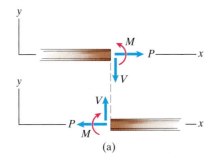

(a)

By cutting a beam at an arbitrary position x, the axial load P, shear force V, and bending moment M can be determined as functions of x. Depending on the loading and supports of the beam, it may be necessary to draw several free-body diagrams to determine the distributions for the entire beam. The graphs of V and M as functions of x are the *shear force and bending moment diagrams*.

The distributed load, shear force, and bending moment in a portion of a beam subjected only to a distributed load satisfy the relations

$$\frac{dV}{dx} = -w, \tag{10.4}$$

$$\frac{dM}{dx} = V. \tag{10.6}$$

For segments of a beam that are unloaded or are subjected to a distributed load, these equations can be integrated to determine V and M as functions of x. To obtain the complete shear force and bending moment diagrams, forces and couples must also be accounted for.

Cables

Loads Distributed Uniformly Along a Straight Line   If a suspended cable is subjected to a horizontally distributed load w (Fig. b), the curve described by the cable is the parabola

$$y = \frac{1}{2}ax^2, \tag{10.10}$$

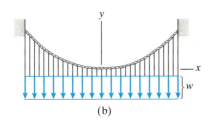

(b)

where $a = w/T_0$ and T_0 is the tension in the cable at $x = 0$. The tension in the cable at a position x is

$$T = T_0\sqrt{1 + a^2x^2}, \tag{10.11}$$

and the length of the cable in the horizontal interval from 0 to x is

$$s = \frac{1}{2}\left\{ x\sqrt{1 + a^2x^2} + \frac{1}{a}\ln\left[ax + \sqrt{1 + a^2x^2} \right] \right\}. \tag{10.12}$$

Loads Distributed Uniformly Along a Cable If a suspended cable is subjected to a load w distributed along its length, the curve described by the cable is the catenary

$$y = \frac{1}{2a}(e^{ax} + e^{-ax} - 2) = \frac{1}{a}(\cosh ax - 1), \tag{10.20}$$

where $a = w/T_0$ and T_0 is the tension in the cable at $x = 0$. The tension in the cable at a position x is

$$T = T_0\sqrt{1 + \frac{1}{4}(e^{ax} - e^{-ax})^2} = T_0 \cosh ax, \tag{10.21}$$

and the length of the cable in the horizontal interval from 0 to x is

$$s = \frac{1}{2a}(e^{ax} - e^{-ax}) = \frac{\sinh ax}{a}. \tag{10.22}$$

Discrete Loads If N known weights are suspended from a cable and the positions of the attachment points of the cable, the horizontal positions of the attachment points of the weights, and the vertical position of the attachment point of one of the weights are known, the configuration of the cable and the tension in each of its segments can be determined.

Liquids and Gases

The pressure p on a surface is defined so that the normal force exerted on an element dA of the surface is $p\,dA$. The total normal force exerted by pressure on a plane area A is

$$F = \int_A p\,dA. \tag{10.24}$$

The *center of pressure* is the point on A at which F must be placed to be equivalent to the pressure on A. The coordinates of the center of pressure are

$$x_p = \frac{\int_A xp\,dA}{\int_A p\,dA}, \qquad y_p = \frac{\int_A yp\,dA}{\int_A p\,dA}. \tag{10.25}$$

The pressure in a stationary liquid is

$$p = p_0 + \gamma x, \tag{10.26}$$

where p_0 is the pressure at the surface, γ is the weight density of the liquid, and x is the depth. If the surface of the liquid is open to the atmosphere, $p_0 = p_{atm}$, the atmospheric pressure.

Review Problems

10.103 Determine the internal forces and moment at B (a) if $x = 250$ mm; (b) if $x = 750$ mm.

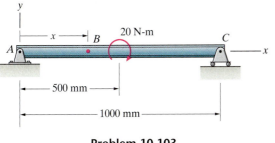

Problem 10.103

10.104 Determine the internal forces and moment (a) at B; (b) at C.

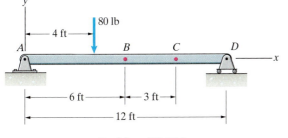

Problem 10.104

10.105 (a) Determine the maximum bending moment in the beam and the value of x where it occurs.

(b) Show that the equations for V and M as functions of x satisfy the equation $V = dM/dx$.

10.106 Draw the shear force and bending moment diagrams for the beam in Problem 10.105.

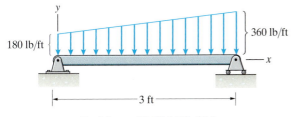

Problems 10.105/10.106

10.107 Determine the shear force and bending moment diagrams for the beam.

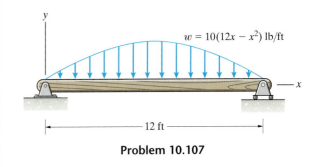

Problem 10.107

10.108 Determine V and M as functions of x for the beam ABC.

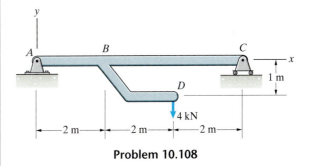

Problem 10.108

10.109 Draw the shear force and bending moment diagrams for beam ABC.

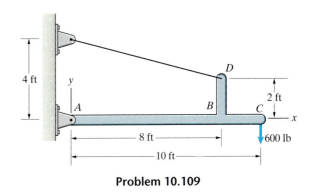

Problem 10.109

10.110 Determine the internal forces and moments at *A*.

10.111 Draw the shear force and bending moment diagrams of beam *BC*.

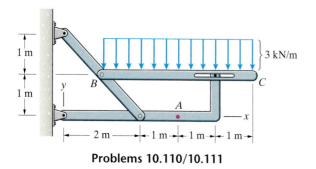

Problems 10.110/10.111

10.112 Determine the internal forces and moment at *B*
(a) if $x = 250$ mm;
(b) if $x = 750$ mm.

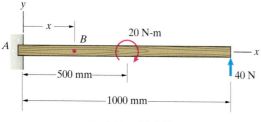

Problem 10.112

10.113 Draw the shear force and bending moment diagrams.

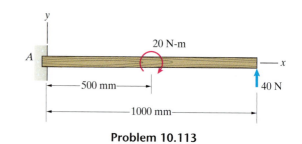

Problem 10.113

10.114 The homogeneous beam weighs 1000 lb. What are the internal forces and bending moment at its midpoint?

10.115 The homogeneous beam weighs 1000 lb. Draw the shear force and bending moment diagrams.

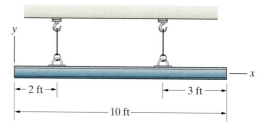

Problems 10.114/10.115

10.116 At *A* the main cable of the suspension bridge is horizontal and its tension is 1×10^8 lb.
(a) Determine the distributed load acting on the cable.
(b) What is the tension at *B*?

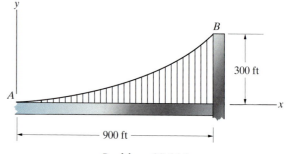

Problem 10.116

10.117 The power line has a mass of 1.4 kg/m. If the line will safely support a tension of 5 kN, determine whether it will safely support an ice accumulation of .4 kg/m.

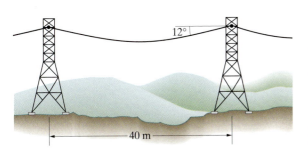

Problem 10.117

10.118 The water depth at the center of the elliptical window is 20 ft. Determine the magnitude of the net force exerted on the window by the pressure of the seawater ($\gamma = 64$ lb/ft^3) and the atmospheric pressure of the air on the opposite side. (See Problem 10.99.)

10.119 The water depth at the center of the elliptical window is 20 ft. Determine the magnitude of the net moment exerted on the window about the horizontal axis L by the pressure of the seawater ($\gamma = 64$ lb/ft^3) and the atmospheric pressure of the air on the opposite side. (See Problem 10.99.)

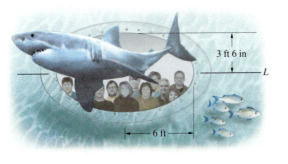

3 ft 6 in

L

6 ft

Problems 10.118/10.119

10.120* The gate has water of 2-m depth on one side. The width of the gate (the dimension into the page) is 4 m, and its mass is 160 kg. The mass density of the water is $\rho = 1000$ kg/m^3, and atmospheric pressure is $p_{atm} = 1 \times 10^5$ Pa. Determine the reactions on the gate at A and B. (The support at B exerts only a horizontal reaction on the gate.)

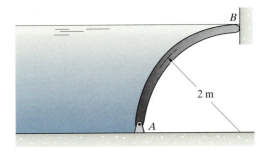

B

2 m

A

Problem 10.120

10.121 A spherical tank of 400-mm inner radius is full of water ($\rho = 1000$ kg/m^3). The pressure of the water at the top of the tank is 4×10^5 Pa.

(a) What is the pressure of the water at the bottom of the tank?

(b) What is the total force exerted on the inner surface of the tank by the pressure of the water?

Strategy: For (b), draw a free-body diagram of the sphere of water in the tank.

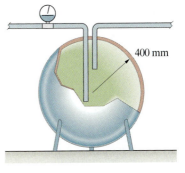

400 mm

Problem 10.121

Virtual Work and Potential Energy

When a spring is stretched, the work performed is stored in the spring as potential energy. Raising an extensible platform increases its gravitational potential energy. In this chapter we define work and potential energy and introduce a general and powerful result called the principle of virtual work.

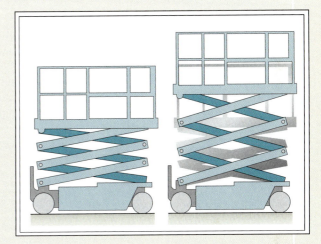

◀ The force necessary to hold the extensible platform in equilibrium can be determined by subjecting the platform to a hypothetical motion and applying the concept of virtual work.

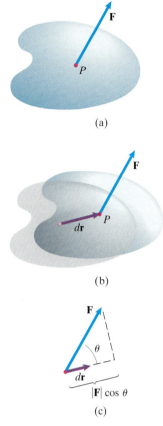

(a)

(b)

(c)

Figure 11.1
(a) A force $\mathbf{F}$ acting on an object.
(b) A displacement $d\mathbf{r}$ of P.
(c) The work $dU = (|\mathbf{F}| \cos \theta)|d\mathbf{r}|$.

11.1 Virtual Work

The principle of virtual work is a statement about work done by forces and couples when an object or structure is subjected to various hypothetical motions. Before we can introduce this principle, we must define work.

Work

Consider a force acting on an object at a point P (Fig. 11.1a). Suppose that the object undergoes an infinitesimal motion, so that P has a differential displacement $d\mathbf{r}$ (Fig. 11.1b). The *work* dU done by $\mathbf{F}$ as a result of the displacement $d\mathbf{r}$ is defined to be

$$dU = \mathbf{F} \cdot d\mathbf{r}. \tag{11.1}$$

From the definition of the dot product, $dU = (|\mathbf{F}| \cos \theta)|d\mathbf{r}|$, where θ is the angle between $\mathbf{F}$ and $d\mathbf{r}$ (Fig. 11.1c). The work is equal to the product of the component of $\mathbf{F}$ in the direction of $d\mathbf{r}$ and the magnitude of $d\mathbf{r}$. Notice that if the component of $\mathbf{F}$ parallel to $d\mathbf{r}$ points in the direction opposite to $d\mathbf{r}$, the work is negative. Also notice that if $\mathbf{F}$ is perpendicular to $d\mathbf{r}$, the work is zero. The dimensions of work are (force) $\times$ (length).

Now consider a couple acting on an object (Fig. 11.2a). The moment due to the couple is $M = Fh$ in the counterclockwise direction. If the object rotates through an infinitesimal counterclockwise angle $d\alpha$ (Fig. 11.2b), the points of application of the forces are displaced through differential distances $\frac{1}{2}h\,d\alpha$. Consequently, the total work done is $dU = F(\frac{1}{2}h\,d\alpha) + F(\frac{1}{2}h\,d\alpha) = M\,d\alpha$.

We see that when an object acted on by a couple M is rotated through an angle $d\alpha$ in the same direction as the couple (Fig. 11.2c), the resulting work is

$$dU = M\,d\alpha. \tag{11.2}$$

If the direction of the couple is opposite to the direction of $d\alpha$, the work is negative.

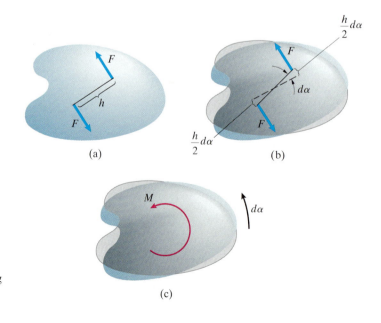

Figure 11.2
(a) A couple acting on an object.
(b) An infinitesimal rotation of the object.
(c) An object acted on by a couple M rotating through an angle $d\alpha$.

Principle of Virtual Work

Now that we have defined the work done by forces and couples, we can introduce the principle of virtual work. Before stating it, we first discuss an example to give you context for understanding the principle.

The homogeneous bar in Fig. 11.3a is supported by the wall and by the pin support at A and is loaded by a couple M. The free-body diagram of the bar is shown in Fig. 11.3b. The equilibrium equations are

$$\Sigma F_x = A_x - N = 0, \tag{11.3}$$

$$\Sigma F_y = A_y - W = 0, \tag{11.4}$$

$$\Sigma M_{\text{point } A} = NL \sin \alpha - W\frac{1}{2}L \cos \alpha - M = 0. \tag{11.5}$$

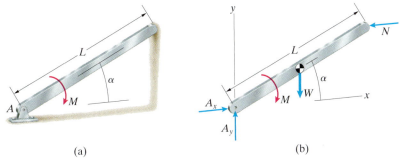

(a) (b)

Figure 11.3
(a) A bar subjected to a couple M.
(b) Free-body diagram of the bar.

We can solve these three equations for the reactions A_x, A_y, and N. However, we have a different objective.

Consider the following question: If the bar is acted on by the forces and couple in Fig. 11.3b and we subject it to a hypothetical infinitesimal translation in the x direction, as shown in Fig. 11.4, what work is done? The hypothetical displacement δx is called a *virtual displacement* of the bar, and the resulting work δU is called the *virtual work*. The pin support and the wall prevent the bar from actually moving in the x direction: the virtual displacement is a theoretical artifice. Our objective is to calculate the resulting virtual work:

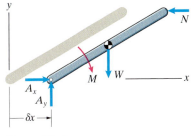

Figure 11.4
A virtual displacement δx.

$$\delta U = A_x \delta_x + (-N)\delta_x = (A_x - N)\delta_x. \tag{11.6}$$

The forces A_y and W do no work because they are perpendicular to the displacements of their points of application. The couple M also does no work, because the bar does not rotate. Comparing this equation with Eq. (11.3), we find that the virtual work equals zero.

Next, we give the bar a virtual translation in the y direction (Fig. 11.5). The resulting virtual work is

$$\delta U = A_y \delta y + (-W)\delta y = (A_y - W)\delta y. \tag{11.7}$$

From Eq. (11.4), the virtual work again equals zero.

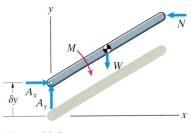

Figure 11.5
A virtual displacement δy.

Finally, we give the bar a virtual rotation while holding point A fixed (Fig. 11.6a). The forces A_x and A_y do no work because their point of application does not move. The work done by the couple M is $-M\,\delta\alpha$, because its direction is opposite to that of the rotation. The displacements of the points of application of the forces N and W are shown in Fig. 11.6b, and the components of the forces in the direction of the displacements are shown in Fig. 11.6c. The work done by N is $(N\sin\alpha)(L\,\delta\alpha)$, and the work done by W is $(-W\cos\alpha)(\frac{1}{2}L\,\delta\alpha)$. The total work is

$$\delta U = (N\sin\alpha)(L\delta\alpha) + (-W\cos\alpha)\left(\frac{1}{2}L\delta\alpha\right) - M\,\delta\alpha$$

$$= \left(NL\sin\alpha - W\frac{1}{2}L\cos\alpha - M\right)\delta\alpha. \qquad (11.8)$$

From Eq. (11.5), the virtual work resulting from the virtual rotation is also zero.

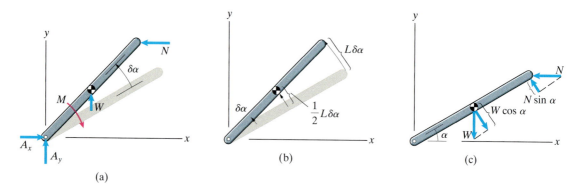

(a) (b) (c)

Figure 11.6
(a) A virtual rotation $\delta\alpha$.
(b) Displacements of the points of application of N and W.
(c) Components of N and W in the direction of the displacements.

We have shown that for three virtual motions of the bar, the virtual work is zero. These results are examples of a form of the principle of virtual work:

If an object is in equilibrium, the virtual work done by the external forces and couples acting on it is zero for any virtual translation or rotation:

$$\delta U = 0. \qquad (11.9)$$

As our example illustrates, this principle can be used to derive the equilibrium equations for an object. Subjecting the bar to virtual translations δx and δy and a virtual rotation $\delta\alpha$ results in Eqs. (11.6)–(11.8). Because the virtual work must be zero in each case, we obtain Eqs. (11.3)–(11.5). But there is no advantage to this approach compared to simply drawing the free-body diagram of the object and writing the equations of equilibrium in the usual way. The advantages of the principle of virtual work become evident when we consider structures.

Application to Structures

The principle of virtual work stated in the preceding section applies to each member of a structure. By subjecting certain types of structures in equilibrium to virtual motions and calculating the total virtual work, we can determine unknown reactions at their supports as well as internal forces in their members. The procedure involves finding virtual motions that result in virtual work being done both by known loads and by unknown forces and couples.

Suppose that we want to determine the axial load in member *BD* of the truss in Fig. 11.7a. The other members of the truss are subjected to the 4-kN load and the forces exerted on them by member *BD* (Fig. 11.7b). If we give the structure a virtual rotation $\delta\alpha$ as shown in Fig. 11.7c, virtual work is done by the force T_{BD} acting at *B* and by the 4-kN load at *C*. Furthermore, the virtual work done by these two forces is the total virtual work done on the members of the structure, because the virtual work done by the internal forces they exert on each other cancels out. For example, consider joint *C* (Fig. 11.7d). The force T_{BC} is the axial load in member *BC*. The virtual work done at *C* on member *BC* is $T_{BC}(1.4 \text{ m}) \delta\alpha$, and the work done at *C* on member *CD* is $(4 \text{ kN} - T_{BC})(1.4 \text{ m}) \delta\alpha$. When we add up the virtual work done on the members to obtain the total virtual work on the structure, the virtual work due to the internal force T_{BC} cancels out. (If the members exerted an internal *couple* on each other at *C*—for example, as a result of friction in the pin support—the virtual work would not cancel out.) Therefore, we can ignore internal forces in calculating the total virtual work on the structure:

$$\delta U = (T_{BD} \cos\theta)(1.4 \text{ m}) \delta\alpha + (4 \text{ kN})(1.4 \text{ m}) \delta\alpha = 0.$$

The angle $\theta = \arctan(1.4/1) = 54.5°$. Solving this equation, we obtain $T_{BD} = -6.88$ kN.

We have seen that using virtual work to determine reactions on members of structures involves two steps:

1. **Choose a virtual motion**—Identify a virtual motion that results in virtual work being done by known loads and by an unknown force or couple you want to determine.

2. **Determine the virtual work**—Calculate the total virtual work resulting from the virtual motion to obtain an equation for the unknown force or couple.

Study Questions

1. What work is done by a force **F** when its point of application undergoes a displacement $d\mathbf{r}$?

2. What work is done by a couple *M* when the object on which it acts rotates through an angle $d\alpha$ in the same direction as the couple?

3. What does the principle of virtual work say about the work done when an object in equilibrium is subjected to a virtual translation or rotation?

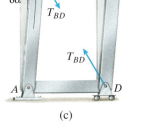

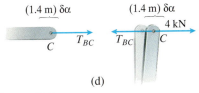

Figure 11.7
(a) A truss with a 4-kN load.
(b) Forces exerted by member *BD*.
(c) A virtual motion of the structure.
(d) Calculating the virtual work on members *BC* and *CD* at the joint *C*.

Example 11.1	Applying Virtual Work to a Structure

For the structure in Fig. 11.8, use the principle of virtual work to determine the horizontal reaction at C.

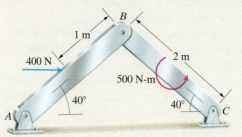

Figure 11.8

Strategy

Notice that even though the structure is fixed at A and C, it can be subjected to *hypothetical* virtual motions. We need to choose a virtual motion for which the horizontal reaction at C and the external force and couple acting on the structure do work. By calculating the resulting virtual work we can determine the horizontal reaction at C.

Solution

Choose a Virtual Motion   We draw the free-body diagram of the structure in Fig. a. Our objective is to determine C_x. If we hold point A fixed and subject bar AB to a virtual rotation $\delta\alpha$ while requiring point C to move horizontally (Fig. b), virtual work is done only by the external loads on the structure and by C_x. The reactions A_x and A_y do no work because A does not move, and the reaction C_y does no work because it is perpendicular to the virtual displacement of point C.

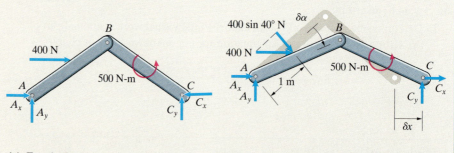

(a) Free-body diagram of the structure.

(b) A virtual displacement in which A remains fixed and C moves horizontally.

Determine the Virtual Work   The virtual work done by the 400-N force is $(400 \sin 40° \text{ N})(1 \text{ m}) \, \delta\alpha$. The bar BC undergoes a virtual rotation $\delta\alpha$ in the counterclockwise direction, so the work done by the couple is $(500 \text{ N-m}) \, \delta\alpha$. In terms of the virtual displacement δx of point C, the work done by the reaction C_x is $C_x\delta x$. The total virtual work is

$$\delta U = (400 \sin 40° \text{ N})(1 \text{ m}) \, \delta\alpha + (500 \text{ N-m})\delta\alpha + C_x\delta x = 0.$$

To obtain C_x from this equation we must determine the relationship between $\delta\alpha$ and δx. From the geometry of the structure (Fig. c), the relationship between the angle α and the distance x from A to C, in m, is

$$x = 2(2 \cos \alpha).$$

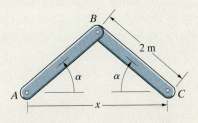

(c) The geometry for determining the relation between $\delta\alpha$ and δx.

The derivative of this equation with respect to α is

$$\frac{dx}{d\alpha} = -4 \sin \alpha.$$

Therefore, an infinitesimal change in x is related to an infinitesimal change in α by

$$dx = -4 \sin \alpha \, d\alpha.$$

Because the virtual rotation $\delta\alpha$ in Fig. b is a decrease in α, we conclude that δx is related to $\delta\alpha$ by

$$\delta x = 4 \sin 40° \, \delta\alpha.$$

Substituting this expression into our equation for the virtual work gives

$$\delta U = [(400 \sin 40° \text{ N-m}) + (500 \text{ N-m}) + C_x(4 \sin 40° \text{ m})] \, \delta\alpha = 0.$$

Solving, we obtain $C_x = -294$ N.

Critical Thinking

Notice that we ignored the internal forces the members exert on each other at B. The virtual work done by these internal forces cancels out. To obtain the solution, we needed to determine the relationship between the virtual displacements δx and $\delta\alpha$. Determining the geometrical relationships between virtual displacements is often the most challenging aspect of applying the principle of virtual work.

| **Example 11.2** | **Applying Virtual Work to a Machine** |

The extensible platform in Fig. 11.9 is raised and lowered by the hydraulic cylinder BC. The total weight of the platform and men is W. The weights of the beams supporting the platform can be neglected. What axial force must the hydraulic cylinder exert to hold the platform in equilibrium in the position shown?

Strategy

We can use a virtual motion that coincides with the actual motion of the platform and beams when the length of the hydraulic cylinder changes. By calculating the virtual work done by the hydraulic cylinder and by the weight of the men and platform, we can determine the force exerted by the hydraulic cylinder.

Solution

Choose a Virtual Motion

We draw the free-body diagram of the platform and beams in Fig. a. Our objective is to determine the force F exerted by the hydraulic cylinder. If we hold point A fixed and subject point C to a horizontal virtual displacement δx, the only external forces that do virtual work are F and the weight W. (The reaction due to the roller support at C is perpendicular to the virtual displacement.)

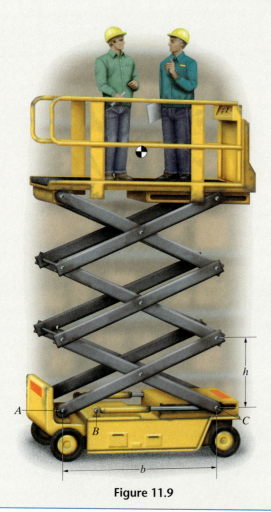

Figure 11.9

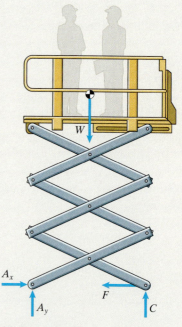

(a) Free-body diagram of the platform and supporting beams.

Determine the Virtual Work The virtual work done by the force F as point C undergoes a virtual displacement δx to the right (Fig. b) is $-F\,\delta x$. To determine the virtual work done by the weight W, we must determine the vertical displacement of point D in Fig. b when point C moves to the right a distance δx. The dimensions b and h are related by

$$b^2 + h^2 = L^2,$$

where L is the length of the beam AD. Taking the derivative of this equation with respect to b, we obtain

$$2b + 2h\frac{dh}{db} = 0,$$

which we can solve for dh in terms of db:

$$dh = -\frac{b}{h}db.$$

Thus, when b increases an amount δx, the dimension h *decreases* an amount $(b/h)\,\delta x$. Because there are three pairs of beams, the platform moves downward a distance $(3b/h)\,\delta x$, and the virtual work done by the weight is $(3b/h)\,W\,\delta x$. The total virtual work is

$$\delta U = \left[-F + \left(\frac{3b}{h}\right)W\right]\delta x = 0,$$

and we obtain $F = (3b/h)W$.

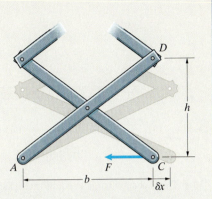

(b) A virtual displacement in which A remains fixed and C moves horizontally.

Critical Thinking

We designed this example to demonstrate how advantageous the method of virtual work can be for certain types of problems. You can see that it would be very tedious to draw the free-body diagrams of the individual members of the frame supporting the platform and solve the equilibrium equations to determine the force exerted by the hydraulic cylinder. In contrast, it was relatively simple to determine the virtual work done by the external forces acting on the frame.

Problems

The following problems are to be solved using the principle of virtual work.

11.1 Determine the reactions at A.

Strategy: Subject the beam to three virtual motions: **1.** a horizontal displacement δx; **2.** a vertical displacement δy; and **3.** a rotation $\delta\theta$ about A.

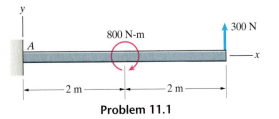

Problem 11.1

11.2 (a) Determine the virtual work done by the 2-kN force and the 2.4 kN-m couple when the beam is rotated through a counterclockwise angle $\delta\theta$ about point A.

(b) Use the result of (a) to determine the reaction at B.

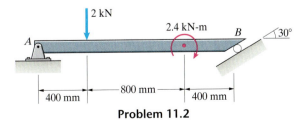

Problem 11.2

11.3 Determine the tension in the cable.

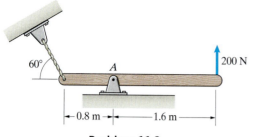

Problem 11.3

11.4 The L-shaped bar is in equilibrium. Determine F.

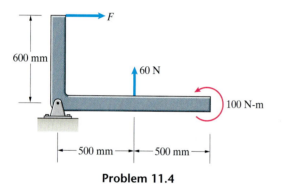

Problem 11.4

11.5 The dimension $L = 4$ ft and $w_0 = 300$ lb/ft. Determine the reactions at A and B.

Strategy: To determine the virtual work done by the distributed load, represent it by an equivalent force.

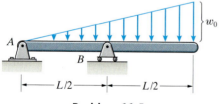

Problem 11.5

11.6 Determine the reactions at A and B.

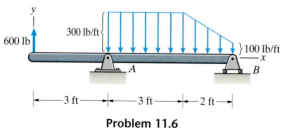

Problem 11.6

11.7 The mechanism is in equilibrium. Determine the force R in terms of F.

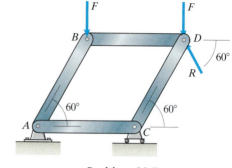

Problem 11.7

11.8 Determine the reaction at the roller support.

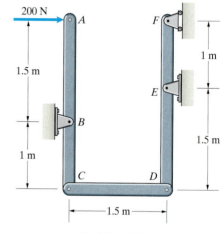

Problem 11.8

11.9 Determine the couple M necessary for the mechanism to be in equilibrium.

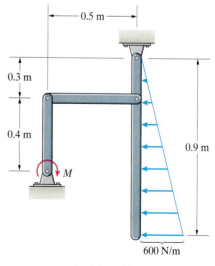

Problem 11.9

11.10 The system is in equilibrium. The total mass of the suspended load and assembly A is 120 kg.

(a) By using equilibrium, determine the force F.

(b) Using the result of (a) and the principle of virtual work, determine the distance the suspended load rises if the cable is pulled upward 300 mm at B.

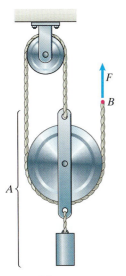

Problem 11.10

11.11 Determine the force P necessary for the mechanism to be in equilibrium.

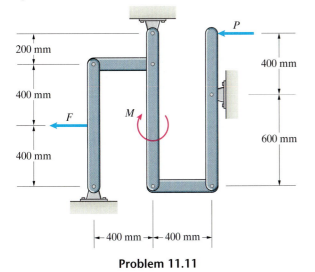

Problem 11.11

11.12 * Show that δx is related to $\delta \alpha$ by

$$\delta x = (L_1 \tan \beta)\, \delta \alpha.$$

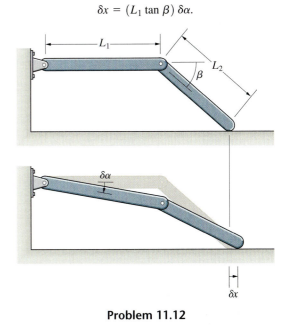

Problem 11.12

11.13 The horizontal surface is smooth. Determine the horizontal force F necessary for the system to be in equilibrium.

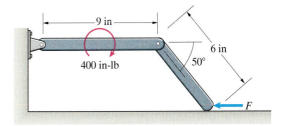

Problem 11.13

11.14 * Show that δx is related to $\delta \alpha$ by

$$\delta x = \frac{L_1 x \sin \alpha}{x - L_1 \cos \alpha} \delta \alpha.$$

Strategy: Write the law of cosines in terms of α and take the derivative of the resulting equation with respect to α. (See Example 11.2.)

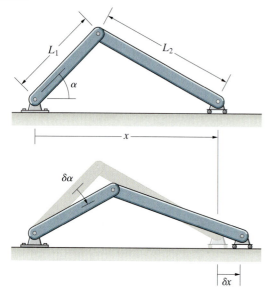

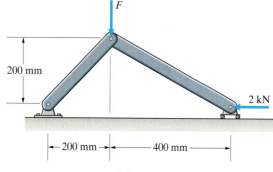

Problem 11.14

11.15 The linkage is in equilibrium. What is the force F?

Problem 11.15

11.16 The linkage is in equilibrium. What is the force F?

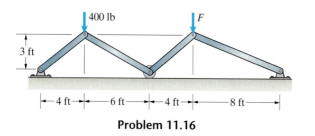

Problem 11.16

11.17 Bar AC is connected to bar BD by a pin that fits in the smooth vertical slot. The masses of the bars are negligible. If $M_A = 30$ N-m, what couple M_B is necessary for the system to be in equilibrium?

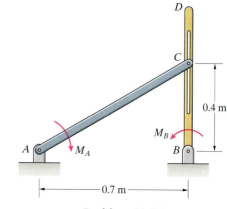

Problem 11.17

11.18 The angle $\alpha = 20°$, and the force exerted on the stationary piston by pressure is 4 kN toward the left. What couple M is necessary to keep the system in equilibrium?

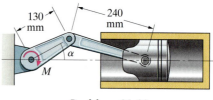

Problem 11.18

11.19 The structure is subjected to a 400-N load and is held in place by a horizontal cable. Determine the tension in the cable.

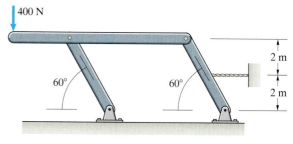

Problem 11.19

11.20 If the load on the car jack is $L = 6.5$ kN, what is the tension in the threaded shaft between A and B?

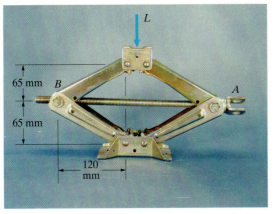

Problem 11.20

11.21 What are the reactions at A and B?

Strategy: Use the equilibrium equations to determine the horizontal components of the reactions, and use the principle of virtual work to determine the vertical components.

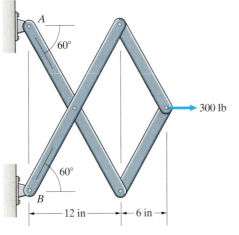

Problem 11.21

11.22 This device raises a load W by extending the hydraulic actuator DE. The bars AD and BC are each 2 m long, and the distances $b = 1.4$ m and $h = 0.8$ m. If $W = 4$ kN, what force must the actuator exert to hold the load in equilibrium?

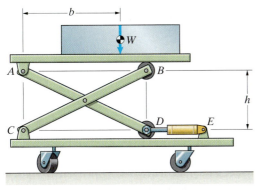

Problem 11.22

11.23 Determine the force P necessary for the mechanism to be in equilibrium.

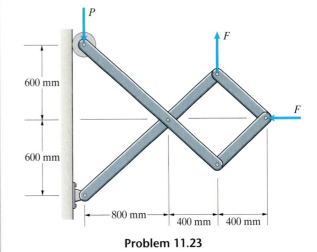

Problem 11.23

11.24 The collar A slides on the smooth vertical bar. The masses are $m_A = 20$ kg and $m_B = 10$ kg.

(a) If the collar A is given an upward virtual displacement δy, what is the resulting downward displacement of the mass B?

(b) Use virtual work to determine the tension in the spring.

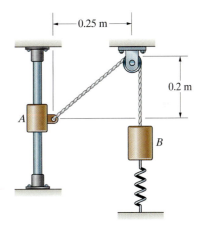

Problem 11.24

11.2 Potential Energy

The work of a force $\mathbf{F}$ due to a differential displacement of its point of application is

$$dU = \mathbf{F} \cdot d\mathbf{r}.$$

If a function of position V exists such that for any $d\mathbf{r}$,

$$dU = \mathbf{F} \cdot d\mathbf{r} = -dV, \qquad (11.10)$$

the function V is called the *potential energy* associated with the force $\mathbf{F}$, and $\mathbf{F}$ is said to be *conservative*. (The negative sign in this equation is in keeping with the interpretation of V as "potential" energy. Positive work results from a decrease in V.) If the forces that do work on a system are conservative, we will show that the total potential energy of the system can be used to determine its equilibrium positions.

Examples of Conservative Forces

Weights of objects and the forces exerted by linear springs are conservative. In the following sections, we derive the potential energies associated with these forces.

Weight In terms of a coordinate system with its y axis upward, the force exerted by the weight of an object is $\mathbf{F} = -W\mathbf{j}$ (Fig. 11.10a). If we give the object an arbitrary displacement $d\mathbf{r} = dx\mathbf{i} + dy\mathbf{j} + dz\mathbf{k}$ (Fig. 11.10b), the work done by its weight is

$$dU = \mathbf{F} \cdot d\mathbf{r} = (-W\mathbf{j}) \cdot (dx\mathbf{i} + dy\mathbf{j} + dz\mathbf{k}) = -W\,dy.$$

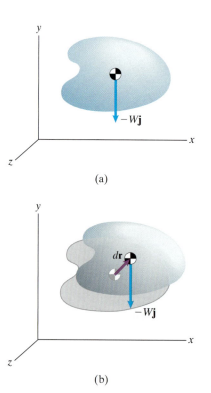

(a)

(b)

Figure 11.10
(a) Force exerted by the weight of an object.
(b) A differential displacement.

We seek a potential energy V such that

$$dU = -W\,dy = -dV \tag{11.11}$$

or

$$\frac{dV}{dy} = W.$$

If we neglect the variation in the weight with height and integrate, we obtain

$$V = Wy + C.$$

The constant C is arbitrary. Since this function satisfies Eq. (11.11) for any value of C, we will let $C = 0$. The position of the origin of the coordinate system can also be chosen arbitrarily. Thus, the potential energy associated with the weight of an object is

$$V = Wy, \tag{11.12}$$

where y is the height of the object above some chosen reference level, or *datum*.

Springs Consider a linear spring connecting an object to a fixed support (Fig. 11.11a). In terms of the stretch $S = r - r_0$, where r is the length of the spring and r_0 is its unstretched length, the force exerted on the object is kS (Fig. 11.11b). If the point at which the spring is attached to the object undergoes a differential displacement $d\mathbf{r}$ (Fig. 11.11c), the work done by the force on the object is

$$dU = -kS\,dS,$$

where dS is the increase in the stretch of the spring resulting from the displacement (Fig. 11.11d). We seek a potential energy V such that

$$dU = -kS\,dS = -dV \tag{11.13}$$

or

$$\frac{dV}{dS} = kS.$$

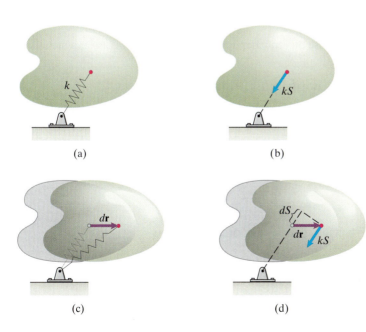

(a)

(b)

(c)

(d)

Figure 11.11
(a) A spring connected to an object.
(b) The force exerted on the object.
(c) A differential displacement of the object.
(d) The work done by the force is
$$dU = -kS\,dS.$$

Integrating this equation and letting the integration constant be zero, we obtain the potential energy associated with the force exerted by a linear spring:

$$V = \frac{1}{2}kS^2. \tag{11.14}$$

Notice that V is positive if the spring is either stretched (S is positive) or compressed (S is negative). Potential energy (the potential to do work) is stored in a spring by either stretching or compressing it.

Principle of Virtual Work for Conservative Forces

Because the work done by a conservative force is expressed in terms of its potential energy through Eq. (11.10), we can give an alternative statement of the principle of virtual work when an object is subjected to conservative forces:

> *Let an object be in equilibrium. If the forces and couples that do work on the object as the result of a virtual translation or rotation are conservative, the change in the total potential energy is zero:*

$$\delta V = 0. \tag{11.15}$$

We emphasize that it is not necessary that all of the forces and couples acting on the object be conservative for this result to hold; it is necessary only that the forces and couples that do work be conservative. This principle also applies to a system of interconnected objects if the external forces that do work are conservative and the internal forces at the connections between objects either do no work or are conservative. Such a system is called a *conservative system*.

If the position of a system can be specified by a single coordinate q, the system is said to have one *degree of freedom*. The total potential energy of a conservative, one-degree-of-freedom system can be expressed in terms of q, and we can write Eq. (11.15) as

$$\delta V = \frac{dV}{dq}\delta q = 0.$$

Thus, when the object or system is in equilibrium, the derivative of its total potential energy with respect to q is zero:

$$\frac{dV}{dq} = 0. \tag{11.16}$$

We can use this equation to determine the values of q at which the system is in equilibrium.

Stability of Equilibrium

Suppose that a homogeneous bar of weight W and length L is suspended from a pin support at one end. In terms of the angle α shown in Fig. 11.12a, the height of the center of mass relative to the pinned end is $-\frac{1}{2}L\cos\alpha$. Choosing the level of the pin support as the datum, we can therefore express the potential energy associated with the weight of the bar as

$$V = -\frac{1}{2}WL\cos\alpha.$$

When the bar is in equilibrium,

$$\frac{dV}{d\alpha} = \frac{1}{2}WL\sin\alpha = 0.$$

This condition is satisfied when $\alpha = 0$ (Fig. 11.12b) and also when $\alpha = 180°$ (Fig. 11.12c).

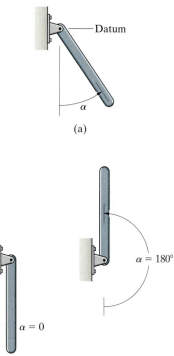

Figure 11.12
(a) A bar suspended from one end.
(b) The equilibrium position $\alpha = 0$.
(c) The equilibrium position $\alpha = 180°$.

There is a fundamental difference between the two equilibrium positions of the bar. In the position shown in Fig. 11.12b, if we displace the bar slightly from its equilibrium position and release it, the bar will remain near the equilibrium position. We say that this equilibrium position is *stable*. When the bar is in the position shown in Fig. 11.12c, if we displace it slightly and release it, the bar will move away from the equilibrium position. This equilibrium position is *unstable*.

The graph of the bar's potential energy V as a function of α is shown in Fig. 11.13a. The potential energy is a minimum at the stable equilibrium position $\alpha = 0$ and a maximum at the unstable equilibrium position $\alpha = 180°$. The derivative of V (Fig. 11.13b) equals zero at both equilibrium positions. The second derivative of V (Fig. 11.13c) is positive at the stable equilibrium position $\alpha = 0$ and negative at the unstable equilibrium position $\alpha = 180°$.

If a conservative, one-degree-of-freedom system is in equilibrium and the second derivative of V evaluated at the equilibrium position is positive, the equilibrium position is stable. If the second derivative of V is negative, it is unstable (Fig. 11.14).

$$\frac{dV}{dq} = 0, \qquad \frac{d^2V}{dq^2} > 0: \qquad \text{Stable equilibrium}$$

$$\frac{dV}{dq} = 0, \qquad \frac{d^2V}{dq^2} < 0: \qquad \text{Unstable equilibrium}$$

Proving these results requires analyzing the motion of the system near an equilibrium position.

Using potential energy to analyze the equilibrium of one-degree-of-freedom systems typically involves three steps:

1. Determine the potential energy—Express the total potential energy in terms of a single coordinate that specifies the position of the system.

2. Find the equilibrium positions—By calculating the first derivative of the potential energy, determine the equilibrium position or positions.

3. Examine the stability—Use the sign of the second derivative of the potential energy to determine whether the equilibrium positions are stable.

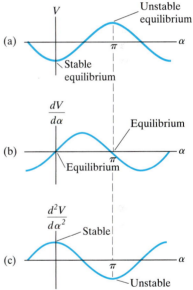

Figure 11.13
Graphs of V, $dV/d\alpha$, and $d^2V/d\alpha^2$.

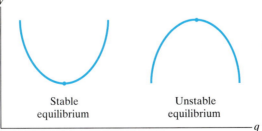

Figure 11.14
Graphs of the potential energy V as a function of the coordinate q that exhibit stable and unstable equilibrium positions.

Study Questions

1. What is the definition of the potential energy of a conservative force?

2. If an object in equilibrium is subjected only to conservative forces, what do you know about the total potential energy when the object undergoes a virtual translation or rotation?

3. What does it mean when an equilibrium position of an object is said to be stable or unstable? How can you distinguish whether an equilibrium position of a conservative, one-degree-of-freedom system is stable or unstable?

Example 11.3 Stability of a Conservative System

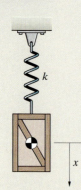

Figure 11.15

In Fig. 11.15 a crate of weight W is suspended from the ceiling by a wire modeled as a linear spring with constant k. The coordinate x measures the position of the centered mass of the crate relative to its position when the wire is unstretched. Find the equilibrium position of the crate, and determine whether it is stable or unstable.

Strategy

The forces acting on the crate—its weight and the force exerted by the spring—are conservative. Therefore, the system is conservative, and we can use the potential energy to determine both the equilibrium position and whether the equilibrium position is stable.

Solution

Determine the Potential Energy   We can use $x = 0$ as the datum for the potential energy associated with the weight. Because the coordinate x is positive downward, the potential energy is $-Wx$. The stretch of the spring equals x, so the potential energy associated with the force of the spring is $\frac{1}{2}kx^2$. The total potential energy is

$$V = \frac{1}{2}kx^2 - Wx.$$

Find the Equilibrium Positions   When the crate is in equilibrium,

$$\frac{dV}{dx} = kx - W = 0.$$

The equilibrium position is $x = W/k$.

Examine the Stability   The second derivative of the potential energy is

$$\frac{d^2V}{dx^2} = k.$$

The equilibrium position is stable.

Critical Thinking

Notice that the equation

$$\frac{dV}{dx} = kx - W = 0$$

is the equilibrium equation (the sum of the forces in the vertical direction) for the crate.

   Although we have confirmed that the equilibrium position of the crate is stable, you could have predicted this based on your intuitive understanding of the behavior of this system. If you give the crate a small vertical displacement from its equilibrium position, you know that it will oscillate, or vibrate, but remain near the equilibrium position. In other cases, intuition is not adequate to predict whether systems will be stable or unstable. (See Problems 11.31–11.40.)

Example 11.4 Stability of an Equilibrium Position

The homogeneous hemisphere in Fig. 11.16 is at rest on the plane surface. Show that it is in equilibrium in the position shown. Is the equilibrium position stable?

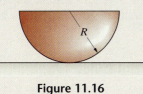

Figure 11.16

Strategy
To determine whether the hemisphere is in equilibrium and whether its equilibrium is stable, we must introduce a coordinate that specifies its orientation and express its potential energy in terms of that coordinate. We can use as the coordinate the angle of rotation of the hemisphere relative to the position shown.

Solution
Determine the Potential Energy Suppose that the hemisphere is rotated through an angle α relative to its original position (Fig. a). Then, from the datum shown, the potential energy associated with the weight W of the hemisphere is

$$V = -\frac{3}{8}RW\cos\alpha.$$

— Datum

$\frac{3}{8}R$

(a) The hemisphere rotated through an angle α.

Find the Equilibrium Positions When the hemisphere is in equilibrium,

$$\frac{dV}{d\alpha} = \frac{3}{8}RW\sin\alpha = 0,$$

which confirms that $\alpha = 0$ is an equilibrium position.

Examine the Stability The second derivative of the potential energy is

$$\frac{d^2V}{d\alpha^2} = \frac{3}{8}RW\cos\alpha.$$

This expression is positive at $\alpha = 0$, so the equilibrium position is stable.

Critical Thinking
Notice that we ignored the normal force exerted on the hemisphere by the plane surface. This force does no work and so does not affect the potential energy.

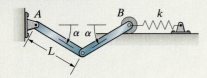

Example 11.5   Stability of an Equilibrium Position

Figure 11.17

The pinned bars in Fig. 11.17 are held in place by the linear spring. Each bar has weight W and length L. The spring is unstretched when $\alpha = 0$, and the bars are in equilibrium when $\alpha = 60°$. Determine the spring constant k, and determine whether the equilibrium position is stable or unstable.

Strategy
The only forces that do work on the bars are their weights and the force exerted by the spring. By expressing the total potential energy in terms of α and using Eq. (11.16), we will obtain an equation we can solve for the spring constant k.

Solution
Determine the Potential Energy   If we use the datum shown in Fig. a, the potential energy associated with the weights of the two bars is

$$W\left(-\frac{1}{2}L \sin \alpha\right) + W\left(-\frac{1}{2}L \sin \alpha\right) = -WL \sin \alpha.$$

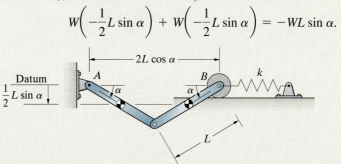

(a) Determining the total potential energy.

The spring is unstretched when $\alpha = 0$ and the distance between points A and B is $2L \cos \alpha$ (Fig. a), so the stretch of the spring is $2L - 2L \cos \alpha$. Therefore, the potential energy associated with the spring is $\frac{1}{2}k(2L - 2L \cos \alpha)^2$, and the total potential energy is

$$V = -WL \sin \alpha + 2kL^2(1 - \cos \alpha)^2.$$

When the system is in equilibrium,

$$\frac{dV}{d\alpha} = -WL \cos \alpha + 4kL^2 (\sin \alpha)(1 - \cos \alpha) = 0.$$

Because the system is in equilibrium when $\alpha = 60°$, we can solve this equation for the spring constant in terms of W and L:

$$k = \frac{W \cos \alpha}{4L (\sin \alpha)(1 - \cos \alpha)} = \frac{W \cos 60°}{4L (\sin 60°)(1 - \cos 60°)} = \frac{0.289W}{L}$$

Examine the Stability   The second derivative of the potential energy is

$$\frac{d^2V}{d\alpha^2} = WL \sin \alpha + 4kL^2 (\cos \alpha - \cos^2 \alpha + \sin^2 \alpha)$$

$$= WL \sin 60° + 4kL^2 (\cos 60° - \cos^2 60° + \sin^2 60°)$$

$$= 0.866WL + 4kL^2.$$

This is a positive number, so the equilibrium position is stable.

Critical Thinking
How do you know when you can apply the principle of virtual work for conservative forces to a system? The system must be conservative, which means that the forces and couples that do work when the system undergoes a virtual motion are conservative. Conservative forces are forces for which a potential energy exists. In this example, work is done by the weights of the bars and the force exerted by the spring, which are conservative forces.

Problems

11.25 The potential energy of a conservative system is given by $V = 2x^3 + 3x^2 - 12x$.

(a) For what values of x is the system in equilibrium?

(b) Determine whether the equilibrium positions you found in (a) are stable or unstable.

11.26 The potential energy of a conservative system is given by $V = 2q^3 - 21q^2 + 72q$.

(a) For what values of q is the system in equilibrium?

(b) Determine whether the equilibrium positions you found in (a) are stable or unstable.

11.27 The mass $m = 2$ kg and the spring constant $k = 100$ N/m. The spring is unstretched when $x = 0$.

(a) Determine the value of x for which the mass is in equilibrium.

(b) Is the equilibrium position stable or unstable?

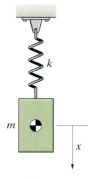

Problem 11.27

11.28 The *nonlinear* spring exerts a force $-kx + \varepsilon x^3$ on the mass, where k and ε are constants. Determine the potential energy V associated with the force exerted on the mass by the spring.

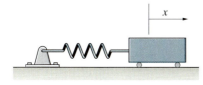

Problem 11.28

11.29 The 1-kg mass is suspended from the nonlinear spring described in Problem 11.28. The constants $k = 10$ and $\varepsilon = 1$, where x is in meters.

(a) Show that the mass is in equilibrium when $x = 1.12$ m and when $x = 2.45$ m.

(b) Determine whether the equilibrium positions are stable or unstable.

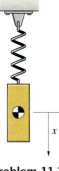

Problem 11.29

11.30 The two straight segments of the bar are each of weight W and length L. Determine whether the equilibrium position shown is stable if (a) $0 < \alpha_0 < 90°$; (b) $90° < \alpha_0 < 180°$.

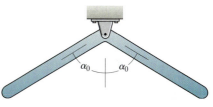

Problem 11.30

11.31 The homogeneous composite object consists of a hemisphere and a cylinder. It is at rest on the plane surface. Show that this equilibrium position is stable only if $L < R/\sqrt{2}$.

Problem 11.31

11.32 The homogeneous composite object consists of a half-cylinder and a triangular prism. It is at rest on the plane surface. Show that this equilibrium position is stable only if $h < \sqrt{2}\,R$.

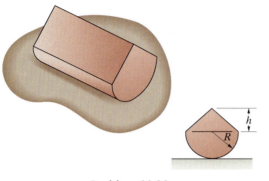

Problem 11.32

11.33 The homogeneous bar has weight W, and the spring is unstretched when the bar is vertical $(\alpha = 0)$.

(a) Use potential energy to show that the bar is in equilibrium when $\alpha = 0$.

(b) Show that the equilibrium position $\alpha = 0$ is stable only if $2kL > W$.

11.34 Suppose that the bar in Problem 11.33 is in equilibrium when $\alpha = 20°$.

(a) Show that the spring constant $k = 0.490\ W/L$.

(b) Determine whether the equilibrium position is stable.

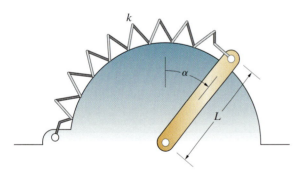

Problems 11.33/11.34

11.35 The bar AB has mass m and length L. The spring is unstretched when the bar is vertical $(\alpha = 0)$. The light collar C slides on the smooth vertical bar so that the spring remains horizontal. Show that the equilibrium position $\alpha = 0$ is stable only if $2kL > mg$.

11.36 The bar AB in Problem 11.35 has mass $m = 4$ kg, length 2 m, and the spring constant is $k = 12$ N/m.

(a) Determine the value of α in the range $0 < \alpha < 90°$ for which the bar is in equilibrium.

(b) Is the equilibrium position determined in part (a) stable?

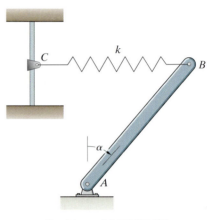

Problems 11.35/11.36

11.37 The bar AB has weight W and length L. The spring is unstretched when the bar is vertical $(\alpha = 0)$. The light collar C slides on the smooth horizontal bar so that the spring remains vertical. Show that the equilibrium position $\alpha = 0$ is unstable.

11.38 The bar AB described in Problem 11.37 has a mass of 2 kg, and the spring constant is $k = 80$ N/m.

(a) Determine the value of α in the range $0 < \alpha < 90°$ for which the bar is in equilibrium.

(b) Is the equilibrium position determined in (a) stable?

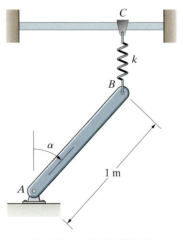

Problems 11.37/11.38

11.39 Each homogeneous bar is of mass m and length L. The spring is unstretched when $\alpha = 0$. If $mg = kL$, determine the value of α in the range $0 < \alpha < 90°$ for which the system is in equilibrium.

11.40 Determine whether the equilibrium position found in Problem 11.39 is stable or unstable.

Problems 11.39/11.40

11.41 The pinned bars are held in place by the linear spring. Each bar has weight W and length L. The spring is unstretched when $\alpha = 90°$. Determine the value of α in the range $0 < \alpha < 90°$ for which the system is in equilibrium.

11.42 Determine whether the equilibrium position found in Problem 11.41 is stable or unstable.

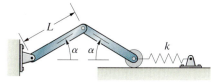

Problems 11.41/11.42

11.43 The bar weighs 15 lb. The spring is unstretched when $\alpha = 0$. The bar is in equilibrium when $\alpha = 30°$. Determine the spring constant k.

11.44 Determine whether the equilibrium positions of the bar in Problem 11.43 are stable or unstable.

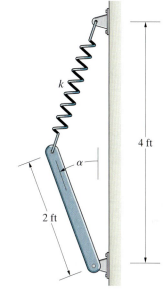

Problems 11.43/11.44

11.45 The spring is unstretched when $\alpha = 90°$. The mass $m = bk/2g$.

(a) Determine the value of α in the range $0 < \alpha < 90°$ for which the system is in equilibrium.

(b) Is the equilibrium position obtained in part (a) stable?

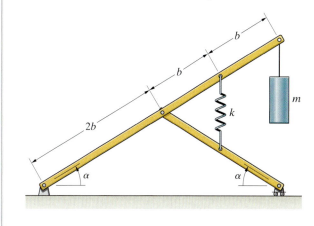

Problem 11.45

COMPUTATIONAL MECHANICS

The following example and problems are designed for the use of a programmable calculator or computer.

Computational Example 11.6

The two bars in Fig. 11.18 are held in place by the linear spring. Each bar has weight W and length L. The spring is unstretched when $\alpha = 0$. If $W = kL$, what is the value of α for which the bars are in equilibrium? Is the equilibrium position stable?

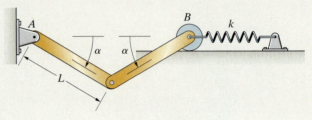

Figure 11.18

Strategy

By obtaining a graph of the derivative of the total potential energy as a function of α, we can estimate the value of α corresponding to equilibrium and determine whether the equilibrium position is stable.

Solution

We derived the total potential energy of the system and determined its derivative with respect to α in Example 11.5, obtaining

$$\frac{dV}{d\alpha} = -WL \cos \alpha + 4kL^2 (\sin \alpha)(1 - \cos \alpha).$$

Substituting $W = kL$, we obtain

$$\frac{dV}{d\alpha} = kL^2[-\cos \alpha + 4 (\sin \alpha)(1 - \cos \alpha)].$$

From the graph of this function (Fig. 11.19), we estimate that the system is in equilibrium when $\alpha = 43°$.

The slope of $dV/d\alpha$, which is the second derivative of V, is positive at $\alpha = 43°$. The equilibrium position is therefore stable.

Figure 11.19
Graph of the derivative of V.

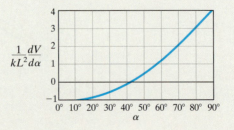

Critical Thinking

Part of our solution, the derivation of the potential energy and the determination of its derivative, was analytical. Part of it, the estimation of the value of α corresponding to equilibrium and the determination of whether that value was stable, was computational. As this example demonstrates, analytical solutions can often be augmented by numerical results to solve a given problem. Engineers must be familiar with both approaches.

Computational Problems

11.46 The 1-kg mass is suspended from a *nonlinear* spring that exerts a force $-10x + x^3$, where x is in meters.

(a) Draw a graph of the total potential energy of the system as a function of x from $x = 0$ to $x = 4$ m.

(b) Use your graph to estimate the equilibrium positions of the mass.

(c) Determine whether the equilibrium positions you obtained in (b) are stable or unstable.

Problem 11.46

11.47 Suppose that the homogeneous bar in Problem 11.33 weighs 20 lb and has length $L = 2$ ft, and that $k = 4$ lb/ft.

(a) Determine the value of α in the range $0 < \alpha < 90°$ for which the bar is in equilibrium.

(b) Is the equilibrium position found in (a) stable?

11.48 The bar in Problem 11.43 weighs 15 lb, and the spring is unstretched when $\alpha = 0$. The spring constant is $k = 6$ lb/ft.

(a) Determine the value of α in the range $0 < \alpha < 90°$ for which the bar is in equilibrium.

(b) Is the equilibrium position found in (a) stable?

11.49 The homogeneous bar has length L and mass $4m$.

(a) Determine the value of α in the range $0 < \alpha < 90°$ for which the bar is in equilibrium.

(b) Is the equilibrium position found in (a) stable?

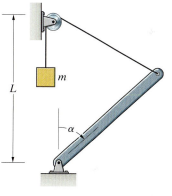

Problem 11.49

11.50 The 2-m long, 10-kg homogeneous bar is pinned at A and at its midpoint B to light collars that slide on a smooth bar. The spring attached at A is unstretched when $\alpha = 0$, and its constant is $k = 1.2$ kN/m.

(a) Determine the value of α when the bar is in equilibrium.

(b) Determine whether the equilibrium position found in (a) is stable.

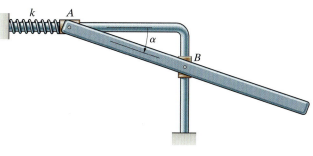

Problem 11.50

CHAPTER SUMMARY

Work

The *work* done by a force $\mathbf{F}$ as a result of a displacement $d\mathbf{r}$ of its point of application is defined by

$$dU = \mathbf{F} \cdot d\mathbf{r}. \tag{11.1}$$

The work done by a counterclockwise couple M due to a counterclockwise rotation $d\alpha$ is

$$dU = M\,d\alpha. \tag{11.2}$$

Principle of Virtual Work

If an object is in equilibrium, the *virtual work* done by the external forces and couples acting on it is zero for any *virtual translation* or *rotation*:

$$\delta U = 0. \tag{11.9}$$

Potential Energy

If a function of position V exists such that for any displacement $d\mathbf{r}$, the work done by a force $\mathbf{F}$ is

$$dU = \mathbf{F} \cdot d\mathbf{r} = -dV, \tag{11.10}$$

then V is called the *potential energy* associated with the force and $\mathbf{F}$ is said to be *conservative*.

The potential energy associated with the weight W of an object is

$$V = Wy, \tag{11.12}$$

where y is the height of the center of mass above some reference level, or *datum*.

The potential energy associated with the force exerted by a linear spring is

$$V = \frac{1}{2}kS^2, \tag{11.14}$$

where k is the spring constant and S is the stretch of the spring.

Principle of Virtual Work for Conservative Forces

An object or a system of interconnected objects is *conservative* if the external forces and couples that do work are conservative and internal forces at the connections between objects either do no work or are conservative. The change in the total potential energy resulting from any virtual motion of a conservative object or system is zero:

$$\delta V = 0. \tag{11.15}$$

If the position of an object or a system can be specified by a single coordinate q, it is said to have one *degree of freedom*. When a conservative, one-degree-of-freedom object or system is in equilibrium,

$$\frac{dV}{dq} = 0. \tag{11.16}$$

If the second derivative of V is positive, the equilibrium position is stable, and if the second derivative of V is negative, the equilibrium position is unstable.

Review Problems

11.51 (a) Determine the couple exerted on the beam at A.
(b) Determine the vertical force exerted on the beam at A.

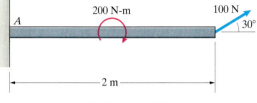

Problem 11.51

11.52 The structure is subjected to a 20 kN-m couple. Determine the horizontal reaction at C.

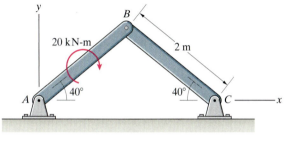

Problem 11.52

11.53 The "rack and pinion" mechanism is used to exert a vertical force on a sample at A for a stamping operation. If a force $F = 30$ lb is exerted on the handle, use the principle of virtual work to determine the force exerted on the sample.

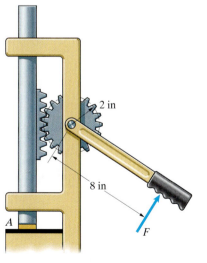

Problem 11.53

11.54 If you were assigned to calculate the force exerted on the bolt by the pliers when the grips are subjected to forces F as shown in Fig. a, you could carefully measure the dimensions, draw free-body diagrams, and use the equilibrium equations. But another approach would be to measure the change in the distance between the jaws when the distance between the handles is changed by a small amount. If your measurements indicate that the distance d in Fig. b decreases by 1 mm when D is decreased 8 mm, what is the approximate value of the force exerted on the bolt by each jaw when the forces F are applied?

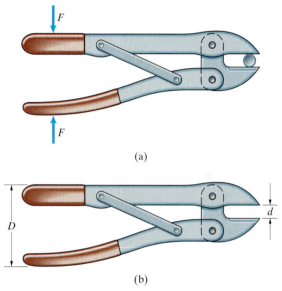

Problem 11.54

11.55 The system is in equilibrium. The total weight of the suspended load and assembly A is 300 lb.
(a) By using equilibrium, determine the force F.
(b) Using the result of (a) and the principle of virtual work, determine the distance the suspended load rises if the cable is pulled downward 1 ft at B.

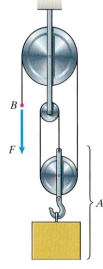

Problem 11.55

11.56 The system is in equilibrium.

(a) By drawing free-body diagrams and using equilibrium equations, determine the couple M.

(b) Using the result of (a) and the principle of virtual work, determine the angle through which pulley B rotates if pulley A rotates through an angle α.

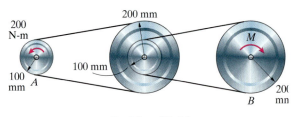

Problem 11.56

11.57 The mechanism is in equilibrium. Neglect friction between the horizontal bar and the collar. Determine M in terms of F, α, and L.

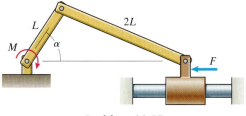

Problem 11.57

11.58 In an injection casting machine, a couple M applied to arm AB exerts a force on the injection piston at C. Given that the horizontal component of the force exerted at C is 4 kN, use the principle of virtual work to determine M.

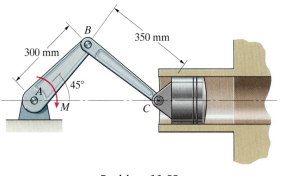

Problem 11.58

11.59 Show that if bar AB is subjected to a clockwise virtual rotation $\delta\alpha$, bar CD undergoes a counterclockwise virtual rotation $(b/a)\,\delta\alpha$.

11.60 The system is in equilibrium, $a = 800$ mm, and $b = 400$ mm. Use the principle of virtual work to determine the force F.

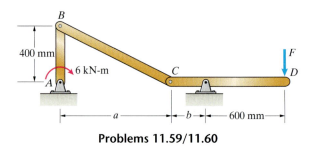

Problems 11.59/11.60

11.61 Show that if bar AB is subjected to a clockwise virtual rotation $\delta\alpha$, bar CD undergoes a clockwise virtual rotation $[ad/(ac + bc - bd)]\,\delta\alpha$.

11.62 The system is in equilibrium, $a = 300$ mm, $b = 350$ mm, $c = 350$ mm, and $d = 200$ mm. Use the principle of virtual work to determine the couple M.

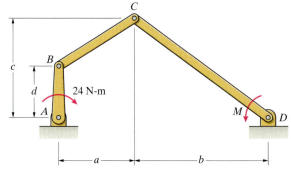

Problems 11.61/11.62

11.63 The mass of the bar is 10 kg, and it is 1 m in length. Neglect the masses of the two collars. The spring is unstretched when the bar is vertical ($\alpha = 0$), and the spring constant is $k = 100$ N/m. Determine the values of α at which the bar is in equilibrium.

11.64 Determine whether the equilibrium positions of the bar in Problem 11.63 are stable or unstable.

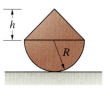

Problems 11.63/11.64

11.65 The spring is unstretched when $\alpha = 90°$. Determine the value of α in the range $0 < \alpha < 90°$ for which the system is in equilibrium.

11.66 Determine whether the equilibrium position found in Problem 11.65 is stable or unstable.

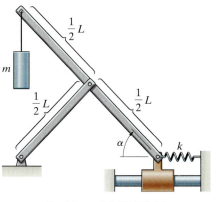

Problems 11.65/11.66

11.67 The hydraulic cylinder C exerts a horizontal force at A, raising the weight W. Determine the magnitude of the force the hydraulic cylinder must exert to support the weight in terms of W and α.

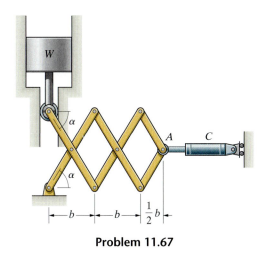

Problem 11.67

11.68 The homogeneous composite object consists of a hemisphere and a cone. It is at rest on the plane surface. Show that this equilibrium position is stable only if $h < \sqrt{3}R$.

Problem 11.68

APPENDIX

A

Review of Mathematics

A.1 Algebra

Quadratic Equations

The solutions of the quadratic equation

$$ax^2 + bx + c = 0$$

are

$$x = \frac{-b \pm \sqrt{b^2 - 4ac}}{2a}.$$

Natural Logarithms

The natural logarithm of a positive real number x is denoted by $\ln x$. It is defined to be the number such that

$$e^{\ln x} = x,$$

where $e = 2.7182\ldots$ is the base of natural logarithms.

Logarithms have the following properties:

$$\ln(xy) = \ln x + \ln y,$$
$$\ln(x/y) = \ln x - \ln y,$$
$$\ln y^x = x \ln y.$$

A.2 Trigonometry

The trigonometric functions for a right triangle are

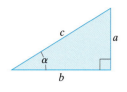

$$\sin \alpha = \frac{1}{\csc \alpha} = \frac{a}{c}, \quad \cos \alpha = \frac{1}{\sec \alpha} = \frac{b}{c}, \quad \tan \alpha = \frac{1}{\cot \alpha} = \frac{a}{b}.$$

The sine and cosine satisfy the relation

$$\sin^2 \alpha + \cos^2 \alpha = 1,$$

and the sine and cosine of the sum and difference of two angles satisfy

$$\sin(\alpha + \beta) = \sin \alpha \cos \beta + \cos \alpha \sin \beta,$$

$$\sin(\alpha - \beta) = \sin \alpha \cos \beta - \cos \alpha \sin \beta,$$

$$\cos(\alpha + \beta) = \cos \alpha \cos \beta - \sin \alpha \sin \beta,$$

$$\cos(\alpha - \beta) = \cos \alpha \cos \beta + \sin \alpha \sin \beta.$$

The **law of cosines** for an arbitrary triangle is

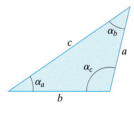

$$c^2 = a^2 + b^2 - 2ab \cos \alpha_c,$$

and the **law of sines** is

$$\frac{\sin \alpha_a}{a} = \frac{\sin \alpha_b}{b} = \frac{\sin \alpha_c}{c}.$$

A.3 Derivatives

$$\frac{d}{dx} x^n = n x^{n-1} \qquad\qquad \frac{d}{dx} \sin x = \cos x \qquad\qquad \frac{d}{dx} \sinh x = \cosh x$$

$$\frac{d}{dx} e^x = e^x \qquad\qquad \frac{d}{dx} \cos x = -\sin x \qquad\qquad \frac{d}{dx} \cosh x = \sinh x$$

$$\frac{d}{dx} \ln x = \frac{1}{x} \qquad\qquad \frac{d}{dx} \tan x = \frac{1}{\cos^2 x} \qquad\qquad \frac{d}{dx} \tanh x = \frac{1}{\cosh^2 x}$$

A.4 Integrals

$$\int x^n \, dx = \frac{x^{n+1}}{n+1} \quad (n \neq -1)$$

$$\int x^{-1} \, dx = \ln x$$

$$\int (a + bx)^{1/2} \, dx = \frac{2}{3b}(a + bx)^{3/2}$$

$$\int x(a + bx)^{1/2} \, dx = -\frac{2(2a - 3bx)(a + bx)^{3/2}}{15b^2}$$

$$\int (1 + a^2x^2)^{1/2} \, dx = \frac{1}{2}\left\{ x(1 + a^2x^2)^{1/2} \right.$$

$$\left. + \frac{1}{a}\ln\left[x + \left(\frac{1}{a^2} + x^2\right)^{1/2} \right] \right\}$$

$$\int x(1 + a^2x^2)^{1/2} \, dx = \frac{a}{3}\left(\frac{1}{a^2} + x^2\right)^{3/2}$$

$$\int x^2(1 + a^2x^2)^{1/2} \, dx = \frac{1}{4}ax\left(\frac{1}{a^2} + x^2\right)^{3/2}$$

$$- \frac{1}{8a^2}x(1 + a^2x^2)^{1/2} - \frac{1}{8a^3}\ln\left[x + \left(\frac{1}{a^2} + x^2\right)^{1/2} \right]$$

$$\int (1 - a^2x^2)^{1/2} \, dx = \frac{1}{2}\left[x(1 - a^2x^2)^{1/2} + \frac{1}{a}\arcsin ax \right]$$

$$\int x(1 - a^2x^2)^{1/2} \, dx = -\frac{a}{3}\left(\frac{1}{a^2} - x^2\right)^{3/2}$$

$$\int x^2(a^2 - x^2)^{1/2} \, dx = -\frac{1}{4}x(a^2 - x^2)^{3/2}$$

$$+ \frac{1}{8}a^2\left[x(a^2 - x^2)^{1/2} + a^2 \arcsin\frac{x}{a} \right]$$

$$\int \frac{dx}{(1 + a^2x^2)^{1/2}} = \frac{1}{a}\ln\left[x + \left(\frac{1}{a^2} + x^2\right)^{1/2} \right]$$

$$\int \frac{dx}{(1 - a^2x^2)^{1/2}} = \frac{1}{a}\arcsin ax \quad \text{or} \quad -\frac{1}{a}\arccos ax$$

$$\int \sin x \, dx = -\cos x$$

$$\int \cos x \, dx = \sin x$$

$$\int \sin^2 x \, dx = -\frac{1}{2}\sin x \cos x + \frac{1}{2}x$$

$$\int \cos^2 x \, dx = \frac{1}{2}\sin x \cos x + \frac{1}{2}x$$

$$\int \sin^3 x \, dx = -\frac{1}{3}\cos x(\sin^2 x + 2)$$

$$\int \cos^3 x \, dx = \frac{1}{3}\sin x(\cos^2 x + 2)$$

$$\int \cos^4 x \, dx = \frac{3}{8}x + \frac{1}{4}\sin 2x + \frac{1}{32}\sin 4x$$

$$\int \sin^n x \cos x \, dx = \frac{(\sin x)^{n+1}}{n+1} \quad (n \neq -1)$$

$$\int \sinh x \, dx = \cosh x$$

$$\int \cosh x \, dx = \sinh x$$

$$\int \tanh x \, dx = \ln \cosh x$$

$$\int e^{ax} \, dx = \frac{e^{ax}}{a}$$

$$\int xe^{ax} \, dx = \frac{e^{ax}}{a^2}(ax - 1)$$

A.5 Taylor Series

The Taylor series of a function $f(x)$ is

$$f(a + x) = f(a) + f'(a)x + \frac{1}{2!}f''(a)x^2 + \frac{1}{3!}f'''(a)x^3 + \cdots,$$

where the primes indicate derivatives.

Some useful Taylor series are

$$e^x = 1 + x + \frac{x^2}{2!} + \frac{x^3}{3!} + \cdots,$$

$$\sin(a + x) = \sin a + (\cos a)x - \frac{1}{2}(\sin a)x^2 - \frac{1}{6}(\cos a)x^3 + \cdots,$$

$$\cos(a + x) = \cos a - (\sin a)x - \frac{1}{2}(\cos a)x^2 + \frac{1}{6}(\sin a)x^3 + \cdots,$$

$$\tan(a + x) = \tan a + \left(\frac{1}{\cos^2 a}\right)x + \left(\frac{\sin a}{\cos^3 a}\right)x^2$$

$$+ \left(\frac{\sin^2 a}{\cos^4 a} + \frac{1}{3\cos^3 a}\right)x^3 + \cdots.$$

APPENDIX

B

Properties of Areas and Lines

B.1 Areas

The coordinates of the centroid of the area A are

$$\bar{x} = \frac{\displaystyle\int_A x \, dA}{\displaystyle\int_A dA}, \qquad \bar{y} = \frac{\displaystyle\int_A y \, dA}{\displaystyle\int_A dA}.$$

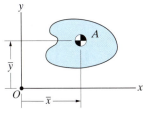

The moment of inertia about the x axis I_x, the moment of inertia about the y axis I_y, and the product of inertia I_{xy} are

$$I_x = \int_A y^2 \, dA, \qquad I_y = \int_A x^2 \, dA, \qquad I_{xy} = \int_A xy \, dA.$$

The polar moment of inertia about O is

$$J_O = \int_A r^2 \, dA = \int_A (x^2 + y^2) \, dA = I_x + I_y.$$

Area $= bh$

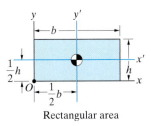

$$I_x = \frac{1}{3}bh^3, \qquad I_y = \frac{1}{3}hb^3, \qquad I_{xy} = \frac{1}{4}b^2h^2$$

$$I_{x'} = \frac{1}{12}bh^3, \qquad I_{y'} = \frac{1}{12}hb^3, \qquad I_{x'y'} = 0$$

Rectangular area

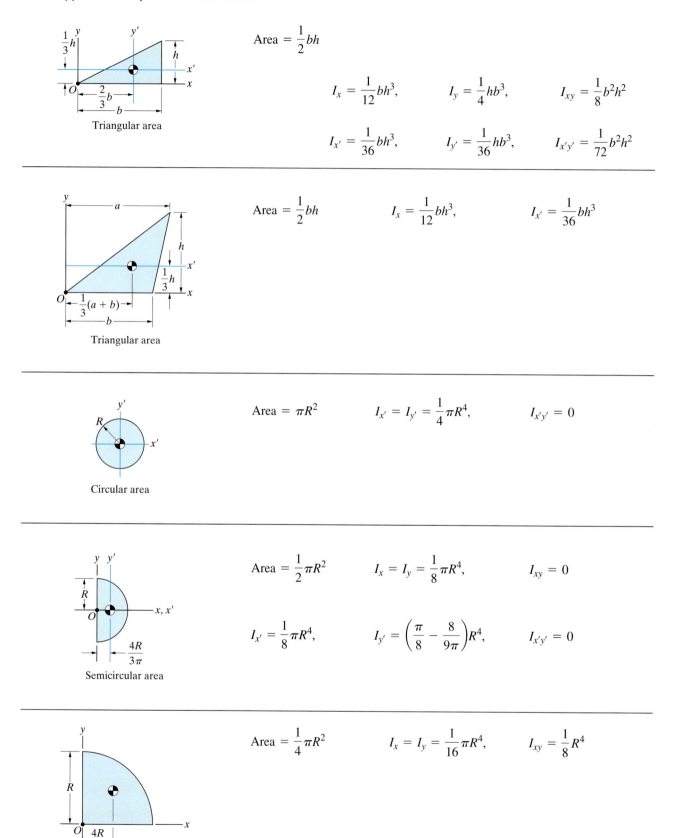

$$\text{Area} = \frac{1}{2}bh$$

$$I_x = \frac{1}{12}bh^3, \qquad I_y = \frac{1}{4}hb^3, \qquad I_{xy} = \frac{1}{8}b^2h^2$$

$$I_{x'} = \frac{1}{36}bh^3, \qquad I_{y'} = \frac{1}{36}hb^3, \qquad I_{x'y'} = \frac{1}{72}b^2h^2$$

Triangular area

$$\text{Area} = \frac{1}{2}bh \qquad I_x = \frac{1}{12}bh^3, \qquad I_{x'} = \frac{1}{36}bh^3$$

Triangular area

$$\text{Area} = \pi R^2 \qquad I_{x'} = I_{y'} = \frac{1}{4}\pi R^4, \qquad I_{x'y'} = 0$$

Circular area

$$\text{Area} = \frac{1}{2}\pi R^2 \qquad I_x = I_y = \frac{1}{8}\pi R^4, \qquad I_{xy} = 0$$

$$I_{x'} = \frac{1}{8}\pi R^4, \qquad I_{y'} = \left(\frac{\pi}{8} - \frac{8}{9\pi}\right)R^4, \qquad I_{x'y'} = 0$$

Semicircular area

$$\text{Area} = \frac{1}{4}\pi R^2 \qquad I_x = I_y = \frac{1}{16}\pi R^4, \qquad I_{xy} = \frac{1}{8}R^4$$

Quarter-circular area

Area $= \alpha R^2$

$$I_x = \frac{1}{4}R^4\left(\alpha - \frac{1}{2}\sin 2\alpha\right), \qquad I_y = \frac{1}{4}R^4\left(\alpha + \frac{1}{2}\sin 2\alpha\right),$$

$$I_{xy} = 0$$

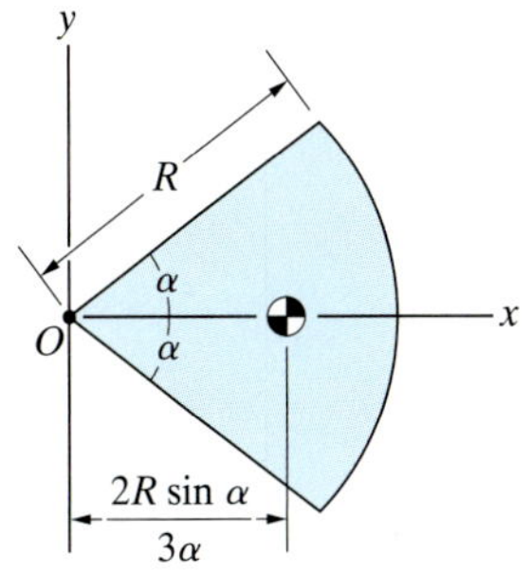

Circular sector

Area $= \frac{1}{4}\pi ab$

$$I_x = \frac{1}{16}\pi ab^3, \qquad I_y = \frac{1}{16}\pi a^3 b, \qquad I_{xy} = \frac{1}{8}a^2 b^2$$

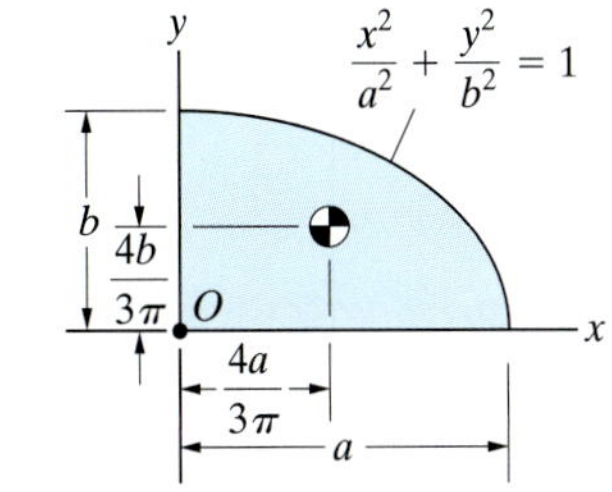

Quarter-elliptical area

Area $= \dfrac{cb^{n+1}}{n+1}$

$$I_x = \frac{c^3 b^{3n+1}}{9n+3}, \qquad I_y = \frac{cb^{n+3}}{n+3}, \qquad I_{xy} = \frac{c^2 b^{2n+2}}{4n+4}$$

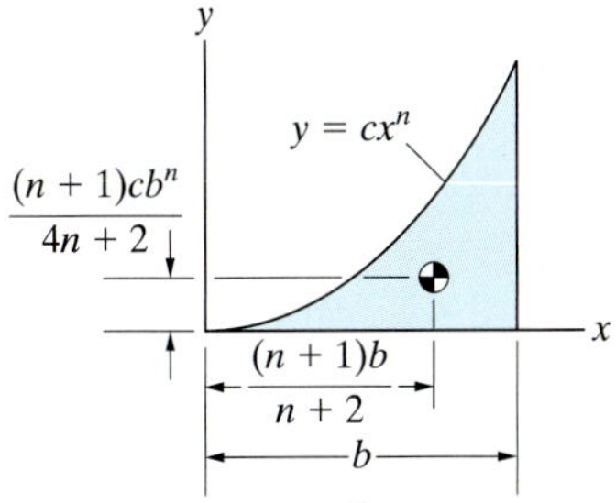

Spandrel

B.2 Lines

The coordinates of the centroid of the line L are

$$\bar{x} = \frac{\displaystyle\int_L x\,dL}{\displaystyle\int_L dL}, \qquad \bar{y} = \frac{\displaystyle\int_L y\,dL}{\displaystyle\int_L dL}, \qquad \bar{z} = \frac{\displaystyle\int_L z\,dL}{\displaystyle\int_L dL}.$$

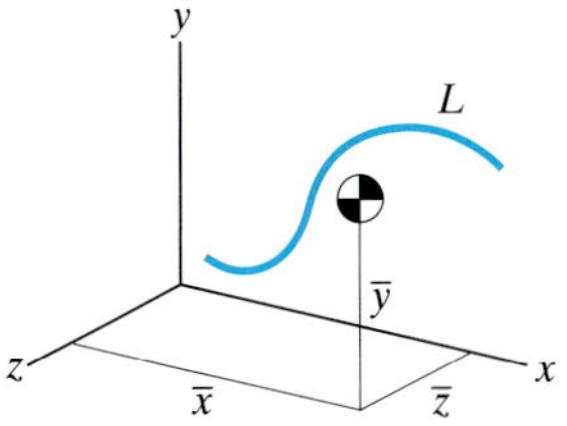

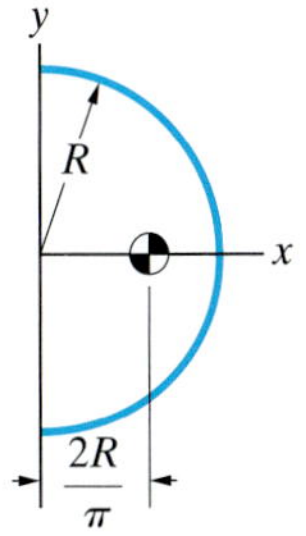

Semicircular arc

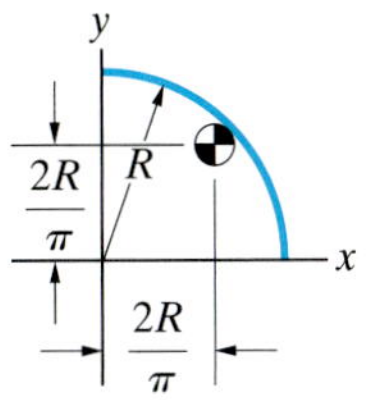

Quarter-circular arc

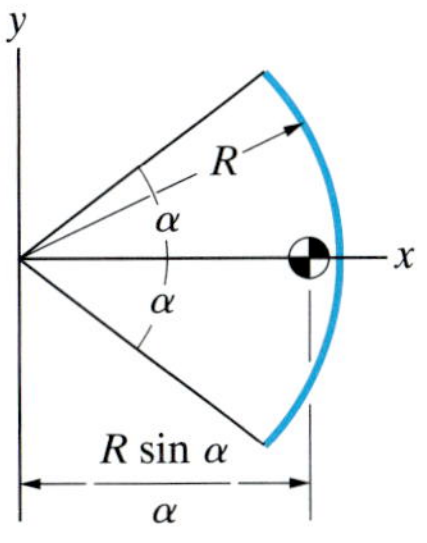

Circular arc

APPENDIX

C

Properties of Volumes and Homogeneous Objects

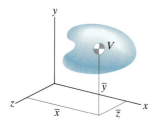

The coordinates of the centroid of the volume V are

$$\bar{x} = \frac{\int_V x \, dV}{\int_V dV}, \qquad \bar{y} = \frac{\int_V y \, dV}{\int_V dV}, \qquad \bar{z} = \frac{\int_V z \, dV}{\int_V dV}.$$

The center of mass of a homogeneous object coincides with the centroid of its volume.

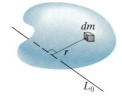

The moment of inertia of the object about the axis L_0 is

$$I_0 = \int_m r^2 \, dm.$$

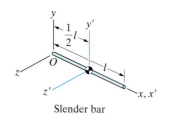

Slender bar

$$I_{x \text{ axis}} = 0, \qquad I_{y \text{ axis}} = I_{z \text{ axis}} = \frac{1}{3} m l^2$$

$$I_{x' \text{ axis}} = 0, \qquad I_{y' \text{ axis}} = I_{z' \text{ axis}} = \frac{1}{12} m l^2$$

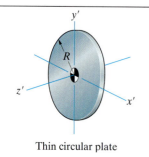

Thin circular plate

$$I_{x' \text{ axis}} = I_{y' \text{ axis}} = \frac{1}{4} m R^2, \qquad I_{z' \text{ axis}} = \frac{1}{2} m R^2$$

$$I_{x \text{ axis}} = \frac{1}{3}mh^2, \qquad I_{y \text{ axis}} = \frac{1}{3}mb^2, \qquad I_{z \text{ axis}} = \frac{1}{3}m(b^2 + h^2)$$

$$I_{x' \text{ axis}} = \frac{1}{12}mh^2, \qquad I_{y' \text{ axis}} = \frac{1}{12}mb^2, \qquad I_{z' \text{ axis}} = \frac{1}{12}m(b^2 + h^2)$$

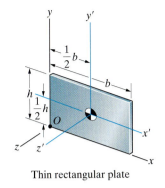

Thin rectangular plate

$$I_{x \text{ axis}} = \frac{m}{A}I_x, \qquad I_{y \text{ axis}} = \frac{m}{A}I_y, \qquad I_{z \text{ axis}} = I_{x \text{ axis}} + I_{y \text{ axis}}$$

The terms I_x and I_y are the moments of inertia of the plate's cross-sectional area A about the x and y axes.

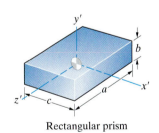

Thin plate

Volume $= abc$

$$I_{x' \text{ axis}} = \frac{1}{12}m(a^2 + b^2), \qquad I_{y' \text{ axis}} = \frac{1}{12}m(a^2 + c^2),$$

$$I_{z' \text{ axis}} = \frac{1}{12}m(b^2 + c^2)$$

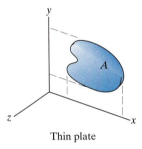

Rectangular prism

Volume $= \pi R^2 l$

$$I_{x \text{ axis}} = I_{y \text{ axis}} = m\left(\frac{1}{3}l^2 + \frac{1}{4}R^2\right), \qquad I_{z \text{ axis}} = \frac{1}{2}mR^2$$

$$I_{x' \text{ axis}} = I_{y' \text{ axis}} = m\left(\frac{1}{12}l^2 + \frac{1}{4}R^2\right), \qquad I_{z' \text{ axis}} = \frac{1}{2}mR^2$$

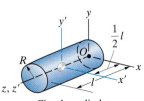

Circular cylinder

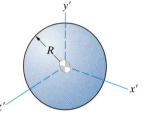

Circular cone

$$\text{Volume} = \frac{1}{3}\pi R^2 h$$

$$I_{x \text{ axis}} = I_{y \text{ axis}} = m\left(\frac{3}{5}h^2 + \frac{3}{20}R^2\right), \qquad I_{z \text{ axis}} = \frac{3}{10}mR^2$$

$$I_{x' \text{ axis}} = I_{y' \text{ axis}} = m\left(\frac{3}{80}h^2 + \frac{3}{20}R^2\right), \qquad I_{z' \text{ axis}} = \frac{3}{10}mR^2$$

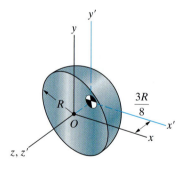

Sphere

$$\text{Volume} = \frac{4}{3}\pi R^3$$

$$I_{x' \text{ axis}} = I_{y' \text{ axis}} = I_{z' \text{ axis}} = \frac{2}{5}mR^2$$

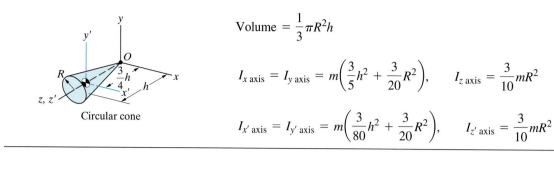

Hemisphere

$$\text{Volume} = \frac{2}{3}\pi R^3$$

$$I_{x \text{ axis}} = I_{y \text{ axis}} = I_{z \text{ axis}} = \frac{2}{5}mR^2$$

$$I_{x' \text{ axis}} = I_{y' \text{ axis}} = \frac{83}{320}mR^2, \quad I_{z' \text{ axis}} = \frac{2}{5}mR^2$$

Answers to Even-Numbered Problems

Chapter 1

1.2 (a) $e = 2.7183$; (b) $e^2 = 7.3891$; (c) $e^2 = 7.3892$

1.4 7.32 m wide, 2.44 m high.

1.6 The 1-in wrench fits the 25-mm nut.

1.8 343 mi/h.

1.10 (a) 5000 m/s; (b) 3.11 mi/s.

1.12 $g = 32.2$ ft/s^2.

1.14 0.310 m^2.

1.16 2.07×10^6 Pa.

1.18 $G = 3.44 \times 10^{-8}$ lb-ft^2/slug^2.

1.20 (a) The SI units of T are kg-m^2/s^2;
  (b) $T = 73.8$ slug-ft^2/s^2.

1.22 (a) $4.448W$; (b) $4.448W/9.81 = 0.453W$.

1.24 (a) 491 N; (b) 81.0 N.

1.26 163 lb.

1.28 32.1 km.

1.30 345,000 km.

Chapter 2

2.2 $\alpha = 55°$.

2.4 $|\mathbf{F}_A + \mathbf{F}_B + \mathbf{F}_C| = 83$ N.

2.6 $\theta = 35.2°$.

2.8 $|\mathbf{F}_B| = 86.6$ N, $|\mathbf{F}_C| = 50$ N.

2.10 $|\mathbf{L}| = 3170$ N, $|\mathbf{D}| = 1480$ N.

2.12 $|\mathbf{F}_{BA}| = 802$ N.

2.14 $|\mathbf{r}| = 390$ m, $\alpha = 21.2°$.

2.18 $F_y = -102$ MN.

2.20 $|\mathbf{F}| = 447$ kip.

2.22 $V_x = 16$, $V_y = 12$ or $V_x = -16$, $V_y = -12$.

2.24 $\mathbf{F} = 56.4\mathbf{i} + 20.5\mathbf{j}$ (lb).

2.26 $\mathbf{r}_{AD} = -1.8\mathbf{i} - 0.3\mathbf{j}$ (m), $|\mathbf{r}_{AD}| = 1.825$ m.

2.28 $\mathbf{r}_{AB} - \mathbf{r}_{BC} = \mathbf{i} - 1.73\mathbf{j}$ (m).

2.30 (a) $\mathbf{r}_{AB} = 48\mathbf{i} + 15\mathbf{j}$ (in);
  (b) $\mathbf{r}_{BC} = -53\mathbf{i} + 5\mathbf{j}$ (in);
  (c) $|\mathbf{r}_{AB} + \mathbf{r}_{BC}| = 20.6$ in.

2.32 (a) $\mathbf{r}_{AB} = 52.0\mathbf{i} + 30\mathbf{j}$ (mm);
  (b) $\mathbf{r}_{AB} = -42.4\mathbf{i} - 42.4\mathbf{j}$ (mm).

2.34 $x_B = 785$ m, $y_B = 907$ m or $x_B = 255$ m,
  $y_B = 1173$ m.

2.36 $\mathbf{e}_{CA} = 0.458\mathbf{i} - 0.889\mathbf{j}$.

2.38 $\mathbf{e} = 0.806\mathbf{i} + 0.593\mathbf{j}$.

2.40 $\mathbf{F} = -937\mathbf{i} + 750\mathbf{j}$ (N).

2.42 $\mathbf{e}_{EM} = 0.609\mathbf{i} - 0.793\mathbf{j}$.

2.44 $|\mathbf{F}_{BA}| = 802$ N.

2.46 $|\mathbf{F}_A| = 1720$ lb, $\alpha = 33.3°$.

2.48 $57.9° \le \alpha \le 90°$.

2.50 $|\mathbf{F}_A| = 10$ kN, $|\mathbf{F}_D| = 8.66$ kN.

2.52 $|\mathbf{L}| = 216.1$ lb, $|\mathbf{D}| = 78.7$ lb.

2.54 $|\mathbf{F}_A| = 68.2$ kN.

2.56 $|\mathbf{F}_{AC}| = 2.11$ kN, $|\mathbf{F}_{AD}| = 2.76$ kN.

2.58 $x = 75 - 0.880s$, $y = 12 + 0.476s$.

2.60 $\mathbf{r} = (0.814s - 6)\mathbf{i} + (0.581s + 1)\mathbf{j}$ (m).

2.62 $e_z = \dfrac{2}{3}$ or $e_z = -\dfrac{2}{3}$.

2.64 $U_x = 3.61$, $U_y = -7.22$,
  $U_z = -28.89$ or $U_x = -3.61$,
  $U_y = 7.22$, $U_z = 28.89$.

2.66 (a) $|\mathbf{U}| = 7$, $|\mathbf{V}| = 13$;
  (b) $|3\mathbf{U} + 2\mathbf{V}| = 27.5$.

2.68 (a) $\cos\theta_x = 0.333$, $\cos\theta_y = -0.667$,
  $\cos\theta_z = -0.667$;
  (b) $\mathbf{e} = 0.333\mathbf{i} - 0.667\mathbf{j} - 0.667\mathbf{k}$.

2.70 $\mathbf{F} = -0.5\mathbf{i} + 0.2\mathbf{j} + 0.843\mathbf{k}$.

2.72 $\mathbf{r}_{BD} = -\mathbf{i} + 3\mathbf{j} - 2\mathbf{k}$ (m), $|\mathbf{r}_{BD}| = 3.74$ m.

2.74 $\mathbf{e}_{CD} = -0.535\mathbf{i} + 0.802\mathbf{j} + 0.267\mathbf{k}$.

2.76 $\mathbf{F} = -21.4\mathbf{i} + 32.1\mathbf{j} + 10.7\mathbf{k}$ (kN).

2.78 (a) $|\mathbf{r}_{AB}| = 16.2$ m;
  (b) $\cos\theta_x = 0.615$, $\cos\theta_y = -0.492$,
  $\cos\theta_z = -0.615$.

2.80 $\mathbf{r}_{AR}$: $\cos\theta_x = 0.667$, $\cos\theta_y = 0.667$,
  $\cos\theta_z = 0.333$. $\mathbf{r}_{BR}$: $\cos\theta_x = -0.242$,
  $\cos\theta_y = 0.970$, $\cos\theta_z = 0$.

2.82 $h = 8848$ m (29,030 ft).

2.84 $\mathbf{r} = 2\mathbf{i} + 1.73\mathbf{j} + \mathbf{k}$ (m).

2.86 $\mathbf{r}_{OP} = R_E(0.612\mathbf{i} + 0.707\mathbf{j} + 0.354\mathbf{k})$.

2.88 (a) $\mathbf{e}_{BC} = -0.286\mathbf{i} - 0.857\mathbf{j} + 0.429\mathbf{k}$;
  (b) $\mathbf{F} = -2.29\mathbf{i} - 6.86\mathbf{j} + 3.43\mathbf{k}$ (kN).

2.90 $\cos\theta_x = -0.703$, $\cos\theta_y = 0.592$, $\cos\theta_z = 0.394$.

2.92 259 lb.

2.94 $|\mathbf{F}_{AC}| = 1116$ N, $|\mathbf{F}_{AD}| = 910$ N.

2.96 $\mathbf{T} = -15.4\mathbf{i} + 27.0\mathbf{j} + 7.7\mathbf{k}$ (lb).

2.98 $\mathbf{T} = -41.1\mathbf{i} + 28.8\mathbf{j} + 32.8\mathbf{k}$ (N).

2.100 $\mathbf{U} \cdot \mathbf{V} = -300$.

2.102 Either $|\mathbf{V}| = 0$ or $\mathbf{V}$ is perpendicular to $\mathbf{U}$.

2.104 $U_x = 2.857$, $V_y = 0.857$, $W_z = -3.143$.

2.108 $\theta = 62.3°$.

2.110 $\theta = 53.5°$.

2.112 Parallel component is $12\mathbf{i} - 4\mathbf{j} + 6\mathbf{k}$ (kN), normal
  component is $9\mathbf{i} + 18\mathbf{j} - 6\mathbf{k}$ (kN).

2.114 (a) 42.5°; (b) $-423\mathbf{j} + 604\mathbf{k}$ (lb).

2.116 $\mathbf{F}_p = 5.54\mathbf{j} + 3.69\mathbf{k}$ (N),
  $\mathbf{F}_n = 10\mathbf{i} + 6.46\mathbf{j} - 9.69\mathbf{k}$ (N).

2.118 $\mathbf{T}_n = -37.1\mathbf{i} + 31.6\mathbf{j} + 8.2\mathbf{k}$ (N).

2.120 $\mathbf{F}_p = -0.1231\mathbf{i} + 0.0304\mathbf{j} - 0.1216\mathbf{k}$ (lb).

2.122 $\mathbf{v}_p = -1.30\mathbf{i} - 1.68\mathbf{j} - 3.36\mathbf{k}$ (m/s).

2.124 (a) $\mathbf{U} \times \mathbf{V} = 132\mathbf{i} + 72\mathbf{j} + 96\mathbf{k}$;
  (b) $(\mathbf{U} \times \mathbf{V}) \cdot \mathbf{U} = 0$ and $(\mathbf{U} \times \mathbf{V}) \cdot \mathbf{V} = 0$.

2.126 $\mathbf{r} \times \mathbf{F} = -80\mathbf{i} + 120\mathbf{j} - 40\mathbf{k}$ (N-m).

2.128 Either $|\mathbf{V}| = 0$ or $\mathbf{V}$ is parallel to $\mathbf{U}$.

2.130 (a), (c) $\mathbf{U} \times \mathbf{V} = -51.8\mathbf{k}$; (b), (d) $\mathbf{V} \times \mathbf{U} = 51.8\mathbf{k}$.

2.134 (a) $\mathbf{r}_{OA} \times \mathbf{r}_{OB} = -4\mathbf{i} + 36\mathbf{j} + 32\mathbf{k}$ (m^2);
(b) $-0.083\mathbf{i} + 0.745\mathbf{j} + 0.662\mathbf{k}$
or $0.083\mathbf{i} - 0.745\mathbf{j} - 0.662\mathbf{k}$.

2.136 $\mathbf{r}_{AB} \times \mathbf{F} = -2400\mathbf{i} + 9600\mathbf{j} + 7200\mathbf{k}$ (ft-lb).

2.138 $\mathbf{r}_{CA} \times \mathbf{T} = -4.72\mathbf{i} - 3.48\mathbf{j} - 7.96\mathbf{k}$ (N-m).

2.140 $x_B = 2.81$ m, $y_B = 6.75$ m, $z_B = 3.75$ m.

2.144 1.8×10^6 mm^2.

2.146 $U_y = -2$.

2.148 $|\mathbf{A}| = 1110$ lb, $\alpha = 29.7°$.

2.150 $|\mathbf{E}| = 313$ lb, $|\mathbf{F}| = 140$ lb.

2.152 $\mathbf{e}_{AB} = 0.625\mathbf{i} - 0.469\mathbf{j} - 0.625\mathbf{k}$.

2.154 $\mathbf{F}_p = 8.78\mathbf{i} - 6.59\mathbf{j} - 8.78\mathbf{k}$ (lb).

2.156 $\mathbf{r}_{BA} \times \mathbf{F} = -70\mathbf{i} + 40\mathbf{j} - 100\mathbf{k}$ (ft-lb).

2.158 (a), (b) $686\mathbf{i} - 486\mathbf{j} - 514\mathbf{k}$ (ft-lb).

2.160 $\mathbf{F}_A = 18.2\mathbf{i} + 19.9\mathbf{j} + 15.3\mathbf{k}$ (N),
$\mathbf{F}_B = -7.76\mathbf{i} + 26.9\mathbf{j} + 13.4\mathbf{k}$ (N).

2.162 $\mathbf{F}_p = 1.29\mathbf{i} - 3.86\mathbf{j} + 2.57\mathbf{k}$ (kN),
$\mathbf{F}_n = -1.29\mathbf{i} - 2.14\mathbf{j} - 2.57\mathbf{k}$ (kN).

2.164 $\mathbf{r}_{AG} \times \mathbf{W} = -16.4\mathbf{i} - 82.4\mathbf{k}$ (N-m).

2.166 $\mathbf{r}_{BC} \times \mathbf{T} = -12.0\mathbf{i} - 138.4\mathbf{j} - 117.4\mathbf{k}$ (N-m).

Chapter 3

3.2 (a) $W = 19.6$ N; (b) $F_1 = 14.4$ N, $F_2 = 17.6$ N.

3.4 $T_{AB} = T_{AC} = 1.53$ kN.

3.6 $T = 763$ N, $M = 875$ N.

3.8 $k = 1960$ N/m, $m_A = 4$ kg, $m_B = 6$ kg.

3.10 (a) $|N_{crane}| = 197$ kN, $|f_{crane}| = 0.707$ kN;
(b) $|N_{caisson}| = 3.22$ kN, $|f_{caisson}| = 0.707$ kN.

3.12 (a) $|N| = 11.06$ kN, $|f| = 4.03$ kN;
(b) $\alpha = 31.0°$.

3.14 (a) 254 lb; (b) 41.8°.

3.16 5.91 kN.

3.18 (a) 128 N; (b) 98.1 N.

3.20 $T_{left} = 299$ lb, $T_{right} = 300$ lb.

3.22 188 N.

3.24 (a) 66.1 lb; (b) 12.2 lb.

3.26 $T_{AB} = 2.75$ kN, $T_{BC} = 2.06$ kN.

3.28 Upper cable tension is $0.828W$, lower cable tension is $0.132W$.

3.30 $T_{AB} = 1.21N$, $T_{AD} = 2.76$ N.

3.32 $m = 12.2$ kg.

3.34 $T_{AC} = 54.8$ N, $T_{AB} = 138.5$ N.

3.36 $h = b$.

3.38 $T_{AB} = 688$ lb.

3.40 $T_{AB} = 64.0$ kN, $T_{BC} = 61.0$ kN.

3.44 $\alpha = 79.7°$, $T_{AB} = 120$ N,
$T_{BC} = 21.4$ N, $T_{CD} = 62.6$ N.

3.46 $W_1 = 133$ lb.

3.48 (b) Left surface: 36.6 lb; right surface: 25.9 lb.

3.50 $k = 1420$ N/m.

3.52 Normal force = 13.29 kN, friction force = 4.19 kN.

3.54 $T = m_A g/7 - (4/7)mg$.

3.56 $m_2 = 12.5$ kg.

3.58 (a) $T = W/2$; (b) $T = W/4$; (c) $T = W/8$.

3.60 $L = 131.1$ kN, $D = 36.0$ kN.

3.62 (a) $\gamma = -14.0°$; (b) 4 km.

3.64 $T_{AB} = 780$ N, $T_{AC} = 1976$ N, $T_{AD} = 2568$ N.

3.66 $T_{AC} = 20.6$ lb, $T_{AD} = 21.4$ lb, $T_{AE} = 11.7$ lb.

3.68 Two at B, three at C, and three at D.

3.70 $T_{AB} = 10,270$ lb, $T_{AC} = 4380$ lb, $T_{AD} = 11,010$ lb.

3.72 $D = 1176$ N, $T_{OA} = 6774$ N.

3.74 $T_{BC} = 1.61$ kN, $T_{BD} = 1.01$ kN.

3.76 $T_{EF} = T_{EG} = 738$ kN.

3.78 (a) The tension = 2.70 kN;
(b) The force exerted by the bar = $1.31\mathbf{i} - 1.31\mathbf{k}$ (kN).

3.80 $T_{AB} = 357$ N.

3.82 $F = 36.6$ N.

3.84 18.0 ft.

3.86 (a)

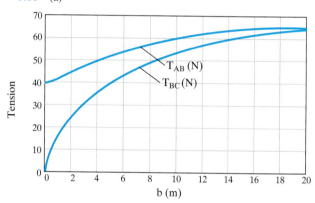

(b) $b \le 10.0$ m.

3.88 0.470 m.

3.90 $h = 1.66$ m.

3.92 $s = 0.305$ m.

3.94 $s = 2.65$ m.

3.96 $W = 25.0$ lb.

3.98 (a) 83.9 lb; (b) 230.5 lb.

3.100 $T = mg/26$.

3.102 $F = 162.0$ N.

3.104 $T_{AB} = 420$ N, $T_{AC} = 533$ N, $|\mathbf{F}_S| = 969$ N.

3.106 $T = mgL/(R + h)$.

3.108 $T_{AB} = 1.54$ lb, $T_{AC} = 1.85$ lb.

3.110 Normal force = 12.15 kN, friction force = 4.03 kN.

Chapter 4

4.2 (a) -124 N-m; (b) 165 N-m.

4.4 $F = 36.2$ N.

4.6 $F = 4.86$ kN.

4.8 $L = 2.4$ m.

4.10 $15.8° \le \alpha \le 37.3°$.

4.12 229 ft-lb.

4.14 $M_S = 611$ in-lb.

4.16 $M_P = 298$ N-m.

4.18 $G = 1400$ lb.

4.20 $F_1 = -30$ kN, $F_2 = 50$ kN.

4.22 (a) $A = 56.6$ lb, $B = 24.4$ lb, $C = 12.2$ lb;
(b) Zero.

4.24 $T = 1.2$ kN.

4.26 $M = 2.39$ kN-m.

4.28 (a) $A_x = 18.1$ kN, $A_y = -29.8$ kN,
$B = -20.4$ kN;
(b) Zero.

4.30 (a) $A_x = 300$ lb, $A_y = 240$ lb, $B = 280$ lb;
(b) Zero.

4.32 $\alpha = 67.2°$, $M_P = 39.5$ N-m.

4.34 -22.3 ft-lb.

4.36 $M = -2340$ N-m.

4.38 $T_{AB} = T_{AC} = 223$ kN.

4.40 617 N-m.

4.42 $M_A = -3.00$ kN-m, $M_D = 7.50$ kN-m.

4.44 796 N.

4.46 (a), (b) $480\mathbf{k}$ (N-m).

4.48 (a) $800\mathbf{k}$ (kN-m);
(b) $-400\mathbf{k}$ (kN-m).

4.50 $\mathbf{F} = 20\mathbf{i} + 40\mathbf{j}$ (N).

4.52 $\mathbf{M}_O = -5600\mathbf{k}$ (ft-lb).

4.54 (a), (b) 1270 N-m.

4.56 $|\mathbf{M}_O| = 4.15$ N-m.

4.58 $\mathbf{F} = 40\mathbf{i} + 40\mathbf{j} + 70\mathbf{k}$ (N) or
$\mathbf{F} = -40\mathbf{i} - 40\mathbf{j} - 70\mathbf{k}$ (N).

4.60 58.0 kN.

4.62 (a) $|\mathbf{F}| = 1586$ N;
(b) $|\mathbf{F}| = 1584$ N.

4.64 $-16.4\mathbf{i} - 111.9\mathbf{k}$ (N-m).

4.66 $\mathbf{F} = 4\mathbf{i} - 4\mathbf{j} + 2\mathbf{k}$ (kN) or
$\mathbf{F} = 4\mathbf{i} - 3.38\mathbf{j} + 2.92\mathbf{k}$ (kN).

4.68 $\mathbf{M}_D = 1.25\mathbf{i} + 1.25\mathbf{j} - 6.25\mathbf{k}$ (kN-m).

4.70 $T_{AC} = 2.23$ kN, $T_{AD} = 2.43$ kN.

4.72 $T_{AB} = 1.60$ kN, $T_{AC} = 1.17$ kN.

4.74 $T_{BC} = 886$ N, $T_{BD} = 555$ N.

4.76 $\mathbf{M} = 482\mathbf{k}$ (kN-m).

4.78 (a) $\mathbf{M}_{x\,\text{axis}} = 80\mathbf{i}$ (N-m);
(b) $\mathbf{M}_{y\,\text{axis}} = -140\mathbf{j}$ (N-m);
(c) $\mathbf{M}_{z\,\text{axis}} = \mathbf{0}$.

4.80 (a) Zero; (b) $2.7\,\mathbf{k}$ (kN-m).

4.82 (a) $\mathbf{M}_{x\,\text{axis}} = -16\mathbf{i}$ (kN-m);
(b) $\mathbf{M}_{z\,\text{axis}} = 15\mathbf{k}$ (kN-m).

4.84 $\mathbf{F} = 80\mathbf{i} + 80\mathbf{j} + 40\mathbf{k}$ (lb).

4.86 $-16.4\mathbf{i}$ (N-m).

4.88 (a), (b) $\mathbf{M}_{AB} = -76.1\mathbf{i} - 95.1\mathbf{j}$ (N-m).

4.90 $\mathbf{M}_{AO} = 119.1\mathbf{j} + 79.4\mathbf{k}$ (N-m).

4.92 $\mathbf{M}_{AB} = 77.1\mathbf{j} - 211.9\mathbf{k}$ (ft-lb).

4.94 $\mathbf{M}_{y\,\text{axis}} = 215\mathbf{j}$ (N-m).

4.96 $\mathbf{M}_{x\,\text{axis}} = 44\mathbf{i}$ (N-m).

4.98 $-338\mathbf{j}$ (ft-lb).

4.100 $|\mathbf{F}| = 13$ lb.

4.102 $\mathbf{M}_{\text{axis}} = -478\mathbf{i} - 174\mathbf{k}$ (N-m).

4.104 1 N-m.

4.106 $124\,\mathbf{k}$ (ft-lb).

4.108 (a), (b) 3.2 N-m clockwise, or $-3.2\mathbf{k}$ (N-m).

4.110 $\alpha = 30.9°$ or $\alpha = 71.8°$.

4.112 (a), (b) $-400\mathbf{k}$ (N-m).

4.114 40 ft-lb clockwise, or $-40\mathbf{k}$ (ft-lb).

4.116 2200 ft-lb clockwise.

4.118 (a) $C = 26$ kN-m; (b) Zero.

4.120 (a) $\mathbf{M} = -14\mathbf{i} - 10\mathbf{j} - 8\mathbf{k}$ (kN-m); (b) $D = 6.32$ m.

4.122 356 ft-lb.

4.124 $|\mathbf{M}| = 6.13$ kN-m.

4.126 $M_{Cy} = 7$ kN-m, $M_{Cz} = -2$ kN-m.

4.128 Yes.

4.130 Systems 1, 2, and 4 are equivalent.

4.134 $F = 265$ N.

4.136 $F = 70$ lb, $M = 130$ in-lb.

4.138 (a) $\mathbf{F} = -10\mathbf{j}$ (lb), $M = -10$ ft-lb;
(b) $D = 1$ ft.

4.140 $\mathbf{F} = 200\mathbf{i} + 180\mathbf{j}$ (N), $d = 0.317$ m.

4.142 (a) $A_x = 12$ kip, $A_y = 10$ kip, $B = -10$ kip;
(b) $\mathbf{F} = -12\mathbf{i}$ (kip), intersects at $y = 5$ ft;
(c) They are both zero.

4.144 $\mathbf{F} = 104\mathbf{j}$ (kN), $M = 13.2$ kN-m counterclockwise.

4.146 $\mathbf{F} = 100\mathbf{j}$ (lb), $\mathbf{M} = \mathbf{0}$.

4.148 (a) $\mathbf{F} = 920\mathbf{i} - 390\mathbf{j}$ (N), $M = -419$ N-m;
(b) intersects at $y = 456$ mm.

4.150 $\mathbf{F} = 800\mathbf{j}$ (lb), intersects at $x = 7.5$ in.

4.152 (a) $-360\mathbf{k}$ (in-lb);
(b) $-36\mathbf{j}$ (in-lb);
(c) $\mathbf{F} = 10\mathbf{i} - 30\mathbf{j} + 3\mathbf{k}$ (lb),
$\mathbf{M} = -36\mathbf{j} - 360\mathbf{k}$ (in-lb).

4.154 (a) $\mathbf{F} = 600\mathbf{i}$ (lb), $\mathbf{M} = 1400\mathbf{j} - 1800\mathbf{k}$ (ft-lb);
(b) $\mathbf{F} = 600\mathbf{i}$ (lb), intersects at $y = 3$ ft, $z = 2.33$ ft.

4.156 $\mathbf{F} = 100\mathbf{j} + 80\mathbf{k}$ (N), $\mathbf{M} = 240\mathbf{j} - 300\mathbf{k}$ (N-m).

4.158 (a) $\mathbf{F} = \mathbf{0}$, $\mathbf{M} = rA\mathbf{i}$;
(b) $\mathbf{F}' = \mathbf{0}$, $\mathbf{M}' = rA\mathbf{i}$.

4.160 (a) $\mathbf{F} = \mathbf{0}$, $\mathbf{M} = 4.60\mathbf{i} + 1.86\mathbf{j} - 3.46\mathbf{k}$ (kN-m);
(b) 6.05 kN-m.

4.162 $\mathbf{F} = -20\mathbf{i} + 20\mathbf{j} + 10\mathbf{k}$ (lb),
$\mathbf{M} = 50\mathbf{i} + 250\mathbf{j} + 100\mathbf{k}$ (in-lb).

4.164 (a) $\mathbf{F} = 28\mathbf{k}$ (kip), $\mathbf{M} = 96\mathbf{i} - 192\mathbf{j}$ (ft-kip);
(b) $x = 6.86$ ft, $y = 3.43$ ft.

4.166 $\mathbf{F} = 100\mathbf{i} + 20\mathbf{j} - 20\mathbf{k}$ (N),
$\mathbf{M} = -143\mathbf{i} + 406\mathbf{j} - 280\mathbf{k}$ (N-m).

4.168 $\mathbf{M}_p = \mathbf{0}$, line of action intersects at $y = 0$, $z = 2$ ft.

4.170 $x = 2.41$ m, $y = 3.80$ m.

4.172 $\mathbf{F} = 40.8\mathbf{i} + 40.8\mathbf{j} + 81.6\mathbf{k}$ (N),
$\mathbf{M} = -179.6\mathbf{i} + 391.9\mathbf{j} - 32.7\mathbf{k}$ (N-m).

4.174 (a) $320\mathbf{i}$ (in-lb);
(b) $\mathbf{F} = -20\mathbf{k}$ (lb), $\mathbf{M} = 320\mathbf{i} + 660\mathbf{j}$ (in-lb);
(c) $\mathbf{M}_t = \mathbf{0}$, $x = 33$ in, $y = -16$ in.

4.176 $k = 124$ lb/ft.

4.178 $M_A = 13{,}200$ ft-lb at $\alpha = 48.2°$.

4.180 $d = 13.0$ ft, moment is $265\,\mathbf{k}$ (ft-lb).

4.182 $T_{AB} = 155$ N, $T_{CD} = 445$ N.

4.184 (a) 160 N-m;
(b) $160\mathbf{k}$ (N-m).

4.186 $|\mathbf{M}_P| = 244$ N-m.

4.188 (a) -76.2 N-m;
(b) -66.3 N-m.

4.190 $|\mathbf{F}| = 224$ lb, $|\mathbf{M}| = 1600$ ft-lb.

4.192 671 lb.

4.194 $-228.1\mathbf{i} - 68.4\mathbf{k}$ (N-m).

4.196 $\mathbf{M}_{x\,\text{axis}} = -153\mathbf{i}$ (ft-lb).

4.198 $\mathbf{M}_{CD} = -173\mathbf{i} + 1038\mathbf{k}$ (ft-lb).

4.200 (a) $T_{AB} = T_{CD} = 173$ lb;
(b) $\mathbf{F} = 300\mathbf{j}$ (lb), intersects at $x = 4$ ft.

4.202 $\mathbf{F} = -20\mathbf{i} + 70\mathbf{j}$ (N), $M = 22$ N-m.

609

4.204 $\mathbf{F}' = -100\mathbf{i} + 40\mathbf{j} + 30\mathbf{k}$ (lb),
$\mathbf{M} = -80\mathbf{i} + 200\mathbf{k}$ (in-lb).

4.206 $\mathbf{F} = 1166\mathbf{i} + 566\mathbf{j}$ (N), $y = 13.9$ m.

4.208 $\mathbf{F} = 190\mathbf{j}$ (N), $\mathbf{M} = -98\mathbf{i} + 184\mathbf{k}$ (N-m).

4.210 $\mathbf{F} = -0.364\mathbf{i} + 4.908\mathbf{j} + 1.090\mathbf{k}$ (kN),
$\mathbf{M} = -0.131\mathbf{i} - 0.044\mathbf{j} + 1.112\mathbf{k}$ (kN-m).

Chapter 5

5.2 (b) $A_x = -1$ kN, $A_y = -1.73$ kN,
$M_A = 12.9$ kN-m clockwise.

5.4 (b) $A_y = 334$ lb, $B_x = 300$ lb, $B_y = 186$ lb.

5.6 (b) $A_x = 0$, $A_y = -1.85$ kN, $B_y = 2.74$ kN.

5.8 (b) $A_x = 0$, $A_y = -5$ kN, $B_y = 15$ kN.

5.10 (b) $A = 100$ lb, $B = 200$ lb.

5.12 (b) $A_x = 502$ N, $A_y = 870$ N.

5.14 (b) $A_x = 4$ kN, $A_y = -2.8$ kN, $B_y = 2.8$ kN.

5.16 (a) 293.3 N; (b) 99.1 N.

5.18 $A_x = 0$, $A_y = 144$ kN, $M_A = 288$ kN-m.

5.20 $k = 3380$ N/m, $B_x = -188.0$ N, $B_y = 98.7$ N.

5.22 5.93 kN.

5.24 $R = 12.5$ lb, $B_x = 11.3$ lb, $B_y = 15.3$ lb.

5.26 (a) $A = 53.8$ lb, $B = 46.2$ lb; (b) $F = 21.2$ lb.

5.28 $A_x = 9211$ N, $B_x = 0$, $B_y = 789$ N.

5.30 $T = 4.71$ lb.

5.34 $T_{AE} = 33.0$ lb, $D_x = -31.0$ lb, $D_y = 30.7$ lb.

5.36 $A_x = -1.83$ kN, $A_y = 2.10$ kN, $B_y = 2.46$ kN.

5.38 $A_x = -200$ lb, $A_y = -100$ lb, $M_A = 1600$ ft-lb.

5.40 $0.354W$.

5.42 $A_x = 3.46$ kN, $A_y = -2$ kN,
$B_x = -3.46$ kN, $B_y = 2$ kN.

5.44 (a) $\mathbf{F} = 30\mathbf{i} - 11.96\mathbf{j}$ (lb) at $x = 33.3$ in;
(b) $A_x = -30$ lb, $A_y = 7.28$ lb, $B_y = 4.68$ lb.

5.46 $A_x = -1.57$ kN, $A_y = 1.57$ kN, $E_x = 1.57$ kN.

5.48 $A_x = 0$, $A_y = 200$ lb, $M_A = 900$ ft-lb.

5.50 $A_x = 57.7$ lb, $A_y = -13.3$ lb, $B = 15.3$ lb.

5.52 $W = 15$ kN.

5.54 (b) $C_x = 500$ N, $C_y = -200$ N.

5.56 $T_{BC} = 5.45$ lb, $A_x = 5.03$ lb, $A_y = 7.90$ lb.

5.58 20.3 kN.

5.60 $W_2 = 2484$ lb, $A_x = -2034$ lb, $A_y = 2425$ lb.

5.62 $W = 46.2$ N, $A_x = 22.3$ N, $A_y = 61.7$ N.

5.64 $F = 44.5$ lb, $A_x = 25.3$ lb, $A_y = -1.9$ lb.

5.66 $W = 132$ lb.

5.68

5.76 (1) and (2) are improperly supported. For (3), reactions
are $A = F/2$, $B = F/2$, $C = F$.

5.78 (b) $A_x = -6.53$ kN, $A_y = -3.27$ kN,
$A_z = 3.27$ kN, $M_{Ax} = 0$, $M_{Ay} = -6.53$ kN-m,
$M_{Az} = -6.53$ kN-m.

5.80 $T_{BC} = 20.3$ kN.

5.82 $C_x = -349$ lb, $C_y = 698$ lb,
$C_z = 175$ lb, $M_{Cx} = -3490$ ft-lb,
$M_{Cy} = -2440$ ft-lb, $M_{Cz} = 2790$ ft-lb.

5.84 (a) $-17.8\mathbf{i} - 62.8\mathbf{k}$ (N-m);
(b) $A_x = 0$, $A_y = 360$ N, $A_z = 0$,
$M_{Ax} = 17.8$ N-m, $M_{Ay} = 0$, $M_{Az} = 62.8$ N-m.

5.86 $O_x = \pm900$ N, $O_y = \pm900$ N,
$O_z = 0$, $M_{Ox} = \pm135$ N-m,
$M_{Oy} = \pm135$ N-m, $M_{Oz} = \pm288$ N-m.

5.88 $|\mathbf{F}| = 10.9$ kN.

5.90 $T_{AB} = 553$ lb, $T_{AC} = 289$ lb,
$O_x = 632$ lb, $O_y = 574$ lb, $O_z = 0$.

5.92 $x = 0.1$ m, $z = 0.133$ m.

5.94 $T_{BD} = 50.2$ lb, $A_x = -34.4$ lb,
$A_y = 17.5$ lb, $A_z = -24.1$ lb,
$M_{Ax} = 0$, $M_{Ay} = 192.5$ in-lb.

5.96 $\mathbf{F} = 4\mathbf{j}$ (kN) at $x = 0$, $z = 0.15$ m.

5.98 (b) $A_x = -0.74$ kN, $A_y = 1$ kN, $A_z = -0.64$ kN,
$B_x = 0.74$ kN, $B_z = 0.64$ kN.

5.100 $F_y = 34.5$ lb.

5.102 $T_{BD} = 1.47$ kN, $T_{BE} = 1.87$ kN,
$A_x = 0$, $A_y = 4.24$ kN, $A_z = 0$.

5.104 $F = 22.5$ kN.

5.106 Tension is 60 N, $B_x = -10$ N, $B_y = 90$ N,
$B_z = 10$ N, $M_{By} = 1$ N-m, $M_{Bz} = -3$ N-m.

5.108 Tension is 60 N, $B_x = -10$ N, $B_y = 75$ N,
$B_z = 15$ N, $C_y = 15$ N, $C_z = -5$ N.

5.110 $A_x = -2.86$ kip, $A_y = 17.86$ kip, $A_z = -8.10$ kip,
$B_y = 3.57$ kip, $B_z = 12.38$ kip.

5.112 $E_x = 0.67$ kN, $E_y = -1.33$ kN, $E_z = 2.67$ kN,
$F_x = 4.67$ kN, $F_y = 6.67$ kN.

5.114 $|\mathbf{A}| = 8.54$ kN, $|\mathbf{B}| = 10.75$ kN.

5.116 $A_x = 3.62$ kN, $A_y = 5.89$ kN, $A_z = 5.43$ kN,
$C_x = 8.15$ kN, $C_y = 0$, $C_z = 0.453$ kN.

5.118 $T_{AB} = 488$ lb, $T_{CD} = 373$ lb, reaction is
$31\mathbf{i} + 823\mathbf{j} - 87\mathbf{k}$ (lb).

5.120 $A_x = -76.7$ N, $A_y = 97.0$ N, $A_z = -54.3$ N,
$M_{Ax} = -2.67$ N-m, $M_{Ay} = 6.39$ N-m,
$M_{Az} = 2.13$ N-m.

5.122 $\alpha = 52.4°$.

5.124 Tension is 33.3 lb; magnitude of reaction is 44.1 lb.

5.126 $\alpha = 10.9°$, $F_A = 1.96$ kN, $F_B = 2.27$ kN.

5.128 (a) No, because of the 3 kN-m couple; (b) magnitude at
A is 7.88 kN; magnitude at B is 6.66 kN; (c) no.

5.130 (b) $A_x = -8$ kN, $A_y = 2$ kN, $C_x = 8$ kN.

5.134 $\alpha = 24.1°$, $A_x = 5.42$ lb, $A_y = 7.43$ lb.

5.136 $h = 298$ mm, $B_x = -439$ N, $B_y = 270$ N.

5.138 $\alpha = 30.8°$.

5.140 (b) $T_A = 7.79$ lb, $T_B = 10.28$ lb; (c) 6.61 lb.

5.142 (a) There are four unknown reactions and three equilibrium equations; (b) $A_x = -50$ lb, $B_x = 50$ lb.

5.144 (b) Force on nail = 55 lb, normal force = 50.77 lb,
friction force = 9.06 lb.

5.146 $k = 13,500$ N/m.

5.148 $A_y = 727$ lb, $H_x = 225$ lb, $H_y = 113$ lb.

5.150 $\alpha = 0$ and $\alpha = 59.4°$.

5.152 $A_x = -32.0$ kN, $A_y = -61.7$ kN.

5.154 The force is 800 N upward; its line of action passes through the midpoint of the plate.

5.156 $m = 67.2$ kg.

5.158 $\alpha = 90°$, $T_{BC} = W/2$, $A = W/2$.

Chapter 6

6.2 AB: 915 N (C); AC: 600 N (C); BC: 521 N (T).

6.4 AB: 0.577F (T); AC: 0.289F (C); BC: 0.577F (C); BD: 0.577F (T); CD: 1.155F (C).

6.6 (a) Tension: 2.43 kN in AB and BD.
Compression: 2.88 kN in CD.
(b) Tension: 1.74 kN in BD.
Compression: 1.60 kN in CD.

6.8 Tension, 31.9 kip in AC, CE, EG, and GH. Compression, 42.5 kip in BD and DF.

6.10 BC: 0; BD: 14.7 kN (C); BE: 5 kN (T).

6.12 (a) Tension: 5540 lb in BD. Compression: 7910 lb in CE.
(b) Tension: 2770 lb in BD. Compression: 3760 lb in CE.

6.14 $F = 8.33$ kN.

6.16 DE: 3.66 kN (C); DF: 1.45 kN (C); DG: 3.36 kN (T).

6.18 AB: 10.56 kN (T); AC: 17.58 kN (C); BC: 6.76 kN (T); BD: 1.81 kN (T); CD: 16.23 kN (C).

6.20 AB: 375 lb (C); AC: 625 lb (T); BC: 300 lb (T).

6.22 BC: 90.1 kN (T); CD: 90.1 kN (C); CE: 300 kN (T).

6.24 BC: 1200 kN (C); BI: 300 kN (T); BJ: 636 kN (T).

6.26 AB: 2520 lb (C); BC: 2160 lb (C); CD: 1680 lb (C).

6.34 (a), (b) 141 kN (C).

6.36 AB: 1.33F (C); BC: 1.33F (C); CE: 1.33F (T).

6.38 BD: 95.6 kip (C); BE: 41.1 kip (T); CE: 58.4 kip (T).

6.40 DF: 69.1 kip (C); DG: 29.4 kip (C); EG: 95.6 kip (T).

6.42 96.2 kN (T).

6.44 BD: 3.33 kN (C); CD: 1.18 kN (T); CE: 1.66 kN (T).

6.46 CE: 35.0 kN (C); DE: 11.7 kN (T); DG: 25.0 kN (T).

6.48 2.50 kN (C).

6.50 CE: 680 kN (T); CF: 374 kN (C); DF: 375 kN (C).

6.52 (a) 1160 lb (C).

6.54 IL: 16 kN (C); KM: 24 kN (T).

6.58 AD: 4.72 kN (C); BD: 4.16 kN CD (C); CD: 4.85 kN (C).

6.60 AB, AC, AD: 0.408F (C).

6.62 AB: 379 lb (C); AC: 665 lb (C); AD: 160 lb (C).

6.64 BC: 32.7 kN (T); BD: 45.2 kN (T); BE: 112.1 kN (C).

6.66 $P_3 = -315$ kN.

6.68 5.59 kN (C) in each member.

6.70 $A_x = 100$ N, $A_y = 100$ N.

6.72 $A_x = 57.2$ lb, $A_y = 42.8$ lb, $M_A = 257$ ft-lb, $B_x = -57.2$ lb, $B_y = -42.8$ lb.

6.74 $F = 50$ kN.

6.76 $A_x = 0$, $A_y = -400$ N, $C_x = -600$ N, $C_y = -300$ N, $D_x = 0$, $D_y = 1000$ N.

6.78 $D_x = -1475$ N, $D_y = -516$ N, $E_x = 0$, $E_y = -516$ N, $M_E = 619$ N-m.

6.80 $A_x = -2.35$ kN, $A_y = 2.35$ kN, $B_x = 0$, $B_y = -4.71$ kN, $C_x = 2.35$ kN, $C_y = 2.35$ kN.

6.82 Tension = 62.5 lb, $F_x = -75$ lb, $F_y = 25$ lb.

6.84 $B_x = -400$ lb, $B_y = -300$ lb, $C_x = 400$ lb, $C_y = 200$ lb, $D_x = 0$, $D_y = 100$ lb.

6.86 $A_x = -150$ lb, $A_y = 120$ lb, $B_x = 180$ lb, $B_y = -30$ lb, $D_x = -30$ lb, $D_y = -90$ lb.

6.88 $A_x = -310$ lb, $A_y = -35$ lb, $B_x = 80$ lb, $B_y = -80$ lb, $C_x = 310$ lb, $C_y = 195$ lb, $D_x = -80$ lb, $D_y = -80$ lb.

6.90 $A_x = 170$ lb, $A_y = 129$ lb, $B_x = -170$ lb, $B_y = -209$ lb.

6.94 $A_x = -22$ lb, $A_y = 15$ lb, $C_x = -14$ lb, $C_y = 3$ lb.

6.96 300 lb (C).

6.98 B: 73.5 N; C: 88.8 N.

6.100 539 N.

6.102 $A_x = 2$ kN, $A_y = -1.52$ kN, $B_x = -2$ kN, $B_y = 1.52$ kN.

6.104 Axial force is 4 kN compression, reaction at A is 4.31 kN.

6.106 100 N.

6.108 At B: 1750 N. DE: 1320 N (C).

6.110 742 lb.

6.112 $A_x = -8$ kN, $A_y = 2$ kN, axial force = 8 kN.

6.114 $K_x = 847$ N, $K_y = 363$ N.

6.116 (a)

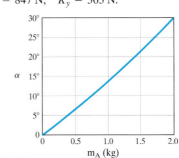

(b) 13.9°.

6.118 (a)

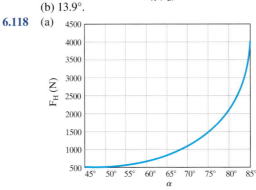

(b) $\alpha = 79.5°$.

6.120 $h = 1.15$ ft.

6.122 3.54 m.

6.124

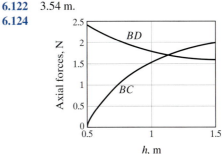

6.126 (a) $B_y = 82.9$ N, $C_x = 40$ N, $C_y = -22.9$ N;
(b) AB: 82.9 N (C); BC: zero; AC: 46.1 N (T).

6.128 $T_{AB} = 7.14$ kN (C), $T_{AC} = 5.71$ kN (T), $T_{BC} = 10$ kN (T).

6.130 BC: 120 kN (C); BG: 42.4 kN (T); FG: 90 kN (T).

6.132 AB: 125 lb (C); AC: zero; BC: 188 lb (T); BD: 225 lb (C); CD: 125 lb (C); CE: 225 lb (T).

6.134 $T_{BD} = 13.3$ kN (T), $T_{CD} = 11.7$ kN (T), $T_{CE} = 28.3$ kN (C).

6.136 AC: 480 N (T); CD: 240 N (C); CF: 300 N (T).

6.138 Tension: member AC, 480 lb (T); Compression: member BD, 633 lb (C).

6.140 CD: 11.42 kN (C); CJ: 4.17 kN (C); IJ: 12.00 kN (T).

6.142 AB: 7.20 kN (C); AC: 4.56 kN (C).

6.144 $A_x = -1.57$ kN, $A_y = 1.18$ kN, $B_x = 0$, $B_y = -2.35$ kN, $C_x = 1.57$ kN, $C_y = 1.18$ kN.

6.146 The force on the bolt is 972 N. The force at A is 576 N.

6.148 973 N.

6.150 $A_x = -52.33$ kN, $A_y = -43.09$ kN, $E_x = 0.81$ kN, $E_y = -14.86$ kN.

Chapter 7

7.2 $\bar{y} = 3.22$.

7.4 $\bar{y} = 0.987$.

7.8 $\bar{x} = 0.711$ ft, $y = 0.584$ ft.

7.10 $\bar{x} = 0$, $y = 1.6$ ft.

7.12 $\bar{x} = 8, \bar{y} = 3.6$.

7.14 $\bar{x} = 0.533$.

7.16 $\bar{x} = \bar{y} = 9/20$.

7.18 $\bar{y} = -7.6$.

7.20 $\bar{y} = 2.53$.

7.22 $a = 0.656$, $b = 6.56 \times 10^{-5}$ m^{-2}.

7.24 $\bar{x} = \bar{y} = 4R/3\pi$.

7.26 $\bar{x} = 3.31$.

7.28 $\bar{x} = 0$, $\bar{y} = 47.5$ mm.

7.30 $\bar{x} = 9.90$ in, $\bar{y} = 0$.

7.32 $\bar{x} = 0.0497$ m, $\bar{y} = 0$.

7.34 $\bar{x} = 9$ in, $\bar{y} = 13.5$ in.

7.36 $\bar{x} = 3.67$ mm, $\bar{y} = 21.52$ mm.

7.38 $b = 39.6$ mm, $h = 18.2$ mm.

7.40 $\bar{x} = 9.64$ m, $\bar{y} = 4.60$ m.

7.44 $\bar{x} = 6.47$ ft, $\bar{y} = 10.60$ ft.

7.46 (a) 360 N; (b) 360 N at $x = 5$ m; (c) $A_x = 0$, $A_y = 60$ N, $B_y = 300$ N.

7.48 $A_x = 0, A_y = 450$ N, $M_A = 2100$ N-m counterclockwise.

7.50 $A_x = 0$, $A_y = 10$ kN, $M_A = -31.3$ kN-m.

7.52 $A_x = 0$, $A_y = 4.17$ kN, $B_y = 8.83$ kN.

7.54 $A_y = 3267$ lb, $B_x = -800$ lb, $B_y = -1267$ lb.

7.56 BD: 21.3 kN (C); CD: 3.77 kN (C); CE: 24 kN (T).

7.58 $A_x = -18$ kN, $A_y = 20$ kN, $B_x = 0$, $B_y = -4$ kN, $C_x = 18$ kN, $C_y = -16$ kN.

7.60 $V = 275$ m^3, height $= 2.33$ m.

7.62 $V = 4.16$ m^3, $\bar{x} = 1.41$ m.

7.64 $\bar{x} = 0.675R$, $\bar{y} = 0$, $\bar{z} = 0$.

7.66 $\bar{x} = h[(2R/3) + a/4]/(R + a/3)$, $\bar{y} = 0$, $\bar{z} = 0$.

7.68 $\bar{x} = 3.24$.

7.70 $\bar{x} = R(\sin \alpha)/\alpha$, $\bar{y} = R(1 - \cos \alpha)/\alpha$

7.72 $\bar{x} = 485$ mm, $\bar{y} = 233$ mm, $\bar{z} = 376$ mm.

7.74 $\bar{x} = -128$ mm, $\bar{y} = \bar{z} = 0$.

7.76 $\bar{x} = 0$, $\bar{y} = 43.7$ mm, $\bar{z} = 38.2$ mm.

7.78 $\bar{x} = 229.5$ mm, $\bar{y} = \bar{z} = 0$.

7.80 $\bar{x} = 23.65$ mm, $\bar{y} = 36.63$ mm, $\bar{z} = 3.52$ mm.

7.82 $\bar{x} = 6$ m, $\bar{y} = 1.83$ m.

7.84 $\bar{x} = 65.9$ mm, $\bar{y} = 21.7$ mm, $\bar{z} = 68.0$ mm.

7.86 $A = \frac{3}{4}\pi R \sqrt{h^2 + R^2}$.

7.88 $V = \pi/5$.

7.90 $\bar{y} = 0.410$.

7.92 $A = 138$ ft^2.

7.94 $V = 0.0377$ m^3.

7.96 $V = 2.48 \times 10^6$ mm^3.

7.98 Volume $= 0.0266$ m^3.

7.100 $A_x = 0$, $A_y = 294$ N, $B_y = 196$ N.

7.102 $A_x = 0$, $A_y = 316$ N, $B = 469$ N.

7.104 $\bar{x} = 6.59$ in, $\bar{y} = 2.17$ in, $\bar{z} = 6.80$ in.

7.106 $A_x = 0$, $A_y = 3.16$ kN, $M_A = 1.94$ kN-m.

7.108 $\bar{x} = 121$ mm, $\bar{y} = 0$, $\bar{z} = 0$.

7.110 $\bar{x}_3 = 82$ mm, $\bar{y}_3 = 122$ mm, $\bar{z}_3 = 16$ mm.

7.112 No, the center of mass of the airplane with the engine and propeller removed is 0.74 ft to the left of point B.

7.114 Mass $= 408$ kg, $\bar{x} = 2.5$ m, $\bar{y} = -1.5$ m.

7.116 $\bar{x} = 20.10$ in, $\bar{y} = 8.03$ in, $\bar{z} = 15.35$ in.

7.118 $\bar{x} = 3/8$, $\bar{y} = 3/5$.

7.120 $\bar{x} = 87.3$ mm, $\bar{y} = 55.3$ mm.

7.122 917 N (T).

7.124 $A_x = 7$ kN, $A_y = -6$ kN, $D_x = 4$ kN, $D_y = 0$.

7.126 $\bar{x} = 1.87$ m.

7.128 $A = 682$ in^2.

7.130 $\bar{x} = 110$ mm.

7.132 $\bar{x} = 1.70$ m.

7.134 $\bar{x} = 25.24$ mm, $\bar{y} = 8.02$ mm, $\bar{z} = 27.99$ mm.

7.136 (a) $\bar{x} = 1.511$ m; (b) $\bar{x} = 1.611$ m.

7.138 $A = 80.7$ kN, $B = 171.6$ kN.

Chapter 8

8.2 $I_x = 0.0288$ m^4, $k_x = 0.346$ m.

8.4 (a) $I_y = 12.8 \times 10^5$ mm^4; (b) $I_{y'} = 3.2 \times 10^5$ mm^4.

8.6 $I_y = 0.175$ m^4, $k_y = 0.624$ m.

8.8 $I_{xy} = 0.0638$ m^4.

8.10 $I_x = 34.9$ in^4.

8.12 $I_{xy} = 17.3$ in^4.

8.14 $I_x = 1330$, $k_x = 4.30$.

8.16 $I_{xy} = 2070$.

8.18 $I_x = 953$, $k_x = 6.68$.

8.20 (a) $I_x = \frac{1}{8}\pi R^4$, $k_x = \frac{1}{2}R$.

8.22 $I_y = 49.09$ m^4, $k_y = 2.50$ m.

8.24 $I_y = 522$, $k_y = 2.07$.

8.28 $I_y = 10.00$ m^4, $k_y = 1.29$ m.

8.30 $I_y = 0.0125 \text{ m}^4, \quad k_y = 0.177 \text{ m}.$

8.32 $I_y = 0.0125 \text{ m}^4, \quad k_y = 0.177 \text{ m}.$

8.34 $I_y = 3.6 \times 10^5 \text{ mm}^4, \quad J_O = 1 \times 10^6 \text{ mm}^4.$

8.36 $I_x = 2.65 \times 10^8 \text{ mm}^4, \quad k_x = 129 \text{ mm}.$

8.38 $I_x = 7.79 \times 10^7 \text{ mm}^4, \quad k_x = 69.8 \text{ mm}.$

8.40 $I_{xy} = 1.08 \times 10^7 \text{ mm}^4.$

8.42 $J_O = 363 \text{ ft}^4, \quad k_O = 4.92 \text{ ft}.$

8.44 $I_x = 10.7 \text{ ft}^4, \quad k_x = 0.843 \text{ ft}.$

8.46 $I_{xy} = 7.1 \text{ ft}^4.$

8.48 $J_O = 5.63 \times 10^7 \text{ mm}^4, \quad k_O = 82.1 \text{ mm}.$

8.50 $I_x = 1.08 \times 10^7 \text{ mm}^4, \quad k_x = 36.0 \text{ mm}.$

8.52 $J_O = 1.58 \times 10^7 \text{ mm}^4, \quad k_O = 43.5 \text{ mm}.$

8.54 $J_O = 2.35 \times 10^5 \text{ in}^4, \quad k_O = 15.1 \text{ in}.$

8.56 $I_x = 49.7 \text{ m}^4, \quad k_x = 2.29 \text{ m}.$

8.58 $I_y = 0.596 \text{ m}^4, \quad k_y = 0.807 \text{ m}.$

8.60 $I_{xy} = 0.219 \text{ m}^4.$

8.62 $I_x = 0.0525 \text{ m}^4, \quad k_x = 0.240 \text{ m}.$

8.64 $I_y = 4.34 \times 10^4 \text{ in}^4, \quad k_y = 10.5 \text{ in}.$

8.66 $I_{xy} = 4.83 \times 10^4 \text{ in}^4.$

8.68 $J_O = 4.01 \times 10^4 \text{ in}^4, \quad k_O = 14.6 \text{ in}.$

8.70 $I_x = 8.89 \times 10^3 \text{ in}^4, \quad k_x = 7.18 \text{ in}.$

8.72 $I_y = 3.52 \times 10^3 \text{ in}^4, \quad k_y = 4.52 \text{ in}.$

8.74 $I_{xy} = 995 \text{ in}^4.$

8.76 $J_O = 5.80 \times 10^6 \text{ mm}^4, \quad k_O = 37.5 \text{ mm}.$

8.78 $I_x = 1470 \text{ in}^4, \quad I_y = 3120 \text{ in}^4.$

8.80 $I_x = 4020 \text{ in}^4, \quad I_y = 6980 \text{ in}^4, \text{ or}$
$I_x = 6820 \text{ in}^4, \quad I_y = 4180 \text{ in}^4.$

8.82 $I_x = 4.01 \times 10^6 \text{ mm}^4.$

8.86 $I_x = 59.8 \times 10^6 \text{ mm}^4, \quad I_y = 18.0 \times 10^6 \text{ mm}^4.$

8.88 $I_{x'} = 5.96 \times 10^6 \text{ mm}^4, \quad I_{y'} = 3.89 \times 10^6 \text{ mm}^4,$
$I_{x'y'} = 3.27 \times 10^6 \text{ mm}^4.$

8.90 $I_{x'} = 8.81 \text{ m}^4, \quad I_{y'} = 3.69 \text{ m}^4, \quad I_{x'y'} = 2.74 \text{ m}^4.$

8.92 $\theta_p = -12.1°,$ principal moments of inertia are
$80.2 \times 10^{-6} \text{ m}^4$ and $27.7 \times 10^{-6} \text{ m}^4.$

8.94 $I_{x'} = 5.96 \times 10^6 \text{ mm}^4, \quad I_{y'} = 3.89 \times 10^6 \text{ mm}^4,$
$I_{x'y'} = 3.27 \times 10^6 \text{ mm}^4.$

8.96 $I_{x'} = 8.81 \text{ m}^4, \quad I_{y'} = 3.69 \text{ m}^4, \quad I_{x'y'} = 2.74 \text{ m}^4.$

8.98 $\theta_p = -12.1°,$ principal moments of inertia are
$80.2 \times 10^{-6} \text{ m}^4$ and $27.7 \times 10^{-6} \text{ m}^4.$

8.100 $I_O = 14 \text{ kg-m}^2.$

8.102 $I_{z \text{ axis}} = 15.1 \text{ kg-m}^2.$

8.104 $I_{x \text{ axis}} = 2.667 \text{ kg-m}^2, \quad I_{y \text{ axis}} = 0.667 \text{ kg-m}^2,$
$I_{z \text{ axis}} = 3.333 \text{ kg-m}^2.$

8.106 $I_{y \text{ axis}} = 1.99 \text{ slug-ft}^2.$

8.108 $20.8 \text{ kg-m}^2.$

8.110 $I_O = \frac{17}{12} ml^2.$

8.112 $I_{z \text{ axis}} = 47.0 \text{ kg-m}^2.$

8.114 $I_{z \text{ axis}} = 0.0803 \text{ slug-ft}^2.$

8.116 $3810 \text{ slug-ft}^2.$

8.118 $I_{z \text{ axis}} = 9.00 \text{ kg-m}^2.$

8.120 $I_{y \text{ axis}} = 0.0881 \text{ slug-ft}^2.$

8.122 $I_O = 0.0188 \text{ kg-m}^2.$

8.124 $I_{x \text{ axis}} = I_{y \text{ axis}} = m\left(\frac{3}{20}R^2 + \frac{3}{5}h^2\right).$

8.126 $I_{x \text{ axis}} = 0.844 \text{ kg-m}^2.$

8.128 $I_{x \text{ axis}} = 0.221 \text{ kg-m}^2.$

8.130 $I_{x'} = 0.995 \text{ kg-m}^2, \quad I_{y'} = 20.1 \text{ kg-m}^2.$

8.132 $I_{z \text{ axis}} = 0.00911 \text{ kg-m}^2.$

8.134 $I_O = 0.00367 \text{ kg-m}^2.$

8.136 $I_{z \text{ axis}} = 0.714 \text{ slug-ft}^2.$

8.138 $I_y = \frac{1}{5}, \quad k_y = \sqrt{\frac{3}{5}}.$

8.140 $J_O = \frac{26}{105}, \quad k_O = \sqrt{\frac{26}{35}}.$

8.142 $I_y = 12.8, \quad k_y = 2.19.$

8.144 $I_{xy} = 2.13.$

8.146 $I_{x'} = 0.183, \quad k_{x'} = 0.262$

8.148 $I_y = 2.75 \times 10^7 \text{ mm}^4, \quad k_y = 43.7 \text{ mm}.$

8.150 $I_x = 5.03 \times 10^7 \text{ mm}^4, \quad k_x = 59.1 \text{ mm}.$

8.152 $I_y = 94.2 \text{ ft}^4, \quad k_y = 2.24 \text{ ft}.$

8.154 $I_x = 396 \text{ ft}^4, \quad k_x = 3.63 \text{ ft}.$

8.156 $\theta_p = 19.5°, \quad 20.3 \text{ m}^4, \quad 161 \text{ m}^4.$

8.158 $I_{y \text{ axis}} = 0.0702 \text{ kg-m}^2.$

8.160 $I_{z \text{ axis}} = \frac{1}{10} mw^2.$

8.162 $I_{x \text{ axis}} = 3.83 \text{ slug-ft}^2.$

8.164 $0.537 \text{ kg-m}^2.$

Chapter 9

9.2 (a) 0.118 lb; (b) 1.10 lb; (c) 0.541 lb.

9.4 (a) 1490 lb; (b) $\mu_s = 0.577.$

9.6 (a) No; (b) 46.8 N; (c) 45.1 N.

9.8 177 N.

9.10 20 lb.

9.12 $\alpha = 14.0°.$

9.14 (a) $T = 56.5 \text{ N}.$

9.16 (a) Yes. The force is $\mu_s W$; (b) $3\mu_s W.$

9.18 $89.6 \le T \le 110.4 \text{ lb}.$

9.20 $F = 267 \text{ N}.$

9.22 $F = 41.1 \text{ lb}.$

9.24 (a) $\alpha = \arctan(\mu_s)$; (b) $\mu_k W/4.$

9.26 $\alpha = 33.4°.$

9.28 $\alpha = 28.3°.$

9.30 (a) $M = 162 \text{ in-lb}$; (b) $M = 135 \text{ in-lb}.$

9.32 $\mu_k = 0.35.$

9.34 $\alpha = 39.6°.$

9.36 (a) $T = 9.42 \text{ lb}$; (b) $T = 33.3 \text{ lb}.$

9.40 $y = 234 \text{ mm}.$

9.42 $\alpha = 9.27°.$

9.44 (a) $F = 216 \text{ N}$; (b) No.

9.48 $\alpha = 1.54°, \quad P = 202 \text{ N}.$

9.50 (a) $F = \mu_s W$;
(b) $F = (W/2)(\mu_{sA} + \mu_{sB})/[1 + (h/b)$
$(\mu_{sA} - \mu_{sB})].$

9.52 $F/2.$

9.54 $f = 2 \text{ lb}.$

9.56 $F = 74.3 \text{ lb}.$

9.58 (a) $f = 24.5 \text{ N}$; (b) $\mu_s = 0.503.$

9.60 (a) $f = 8 \text{ kN}$; (b) $\mu_s = 0.533.$

9.62 $\mu_s = 0.432.$

9.64 $\mu_s = 0.901$.

9.66 $F = 102$ lb.

9.68 Yes. It is necessary that $\mu_s \geq 0.268$.

9.70 $F = 2.30$ kN.

9.72 $F = 156$ N.

9.74 343 kg.

9.76 No. The minimum value of μ_s required is 0.176.

9.78 $F = 1160$ N.

9.80 1.84 N-m.

9.82 (a) 0.967 in-lb; (b) 0.566 in-lb.

9.84 (a) 2.39 ft-lb; (b) 1.20 ft-lb.

9.86 11.8 ft-lb.

9.88 14.8 N-m.

9.90 10.4 N-m.

9.92 4.18 N-m.

9.94 4.88 N-m.

9.96 17.4 N-m.

9.98 $W = 1.55$ lb.

9.100 106 N.

9.102 51.9 lb.

9.104 $T = 40.9$ N.

9.106 $F_B = 207$ N.

9.108 $M = 1.92$ ft-lb.

9.110 $T = 346$ N.

9.112 $M = 0.3$ N-m.

9.114 $M = 12.7$ N-m.

9.116 $M = 7.81$ N-m.

9.118 $M = 5.20$ N-m.

9.120 (a) $M = 93.5$ N-m; (b) 8.17 percent.

9.122 9.51 ft-lb.

9.124 21.6 lb.

9.126 $T_C = 107$ N.

9.128 $M = rW(e^{\pi \mu_k} - 1)$.

9.130 (a) 14.2 lb; (b) 128.3 lb.

9.132 13.1 lb.

9.134 $M_A = 65.2$ N-m, $M_B = 32.6$ N-m.

9.136 $M = 19.2$ N-m.

9.138 $T_2 = T_1 e^{[\mu_s \beta / \sin(\gamma/2)]}$.

9.140 $\mu_s = 0.298$.

9.142 $\alpha = 37.8°$.

9.144 $T = 3.84$ lb.

9.146 $D_1 = 29.2$ mm, $D_2 = 162.2$ mm.

9.148 $-1.963 \leq y \leq 0.225$ m.

9.150 (a) $f = 10.3$ lb.

9.152 $F = 290$ lb.

9.154 $\alpha = 65.7°$.

9.156 $\alpha = 24.2°$.

9.158 $b = (h/\mu_s - t)/2$.

9.160 $h = 5.82$ in.

9.162 286 lb.

9.164 1130 kg, torque $= 2.67$ kN-m.

9.166 $f = 2.63$ N.

9.168 $\mu_s = 0.272$.

9.170 $M = 1.13$ N-m.

9.172 $P = 43.5$ N.

9.174 146 lb.

9.176 (a) $W = 106$ lb; (b) $W = 273$ lb.

Chapter 10

10.2 (a) $w_0 = 2$ kN/m; (b) $M_A = 40$ N-m.

10.4 $P_A = 0$, $V_A = 300$ lb, $M_A = -500$ ft-lb.

10.6 (a) $P_A = 0$, $V_A = 4$ kN, $M_A = 4$ kN-m;
(b) $P_A = 0$, $V_A = 2$ kN, $M_A = 3$ kN-m.

10.8 $b = 1.74$ m, $w_0 = 191$ N/m.

10.10 $P_A = 0$, $V_A = 5.00$ kN, $M_A = -3.33$ kN-m.

10.12 (a) $P_B = 0$, $V_B = -31$ lb, $M_B = 572$ ft-lb;
(b) $P_B = 0$, $V_B = 24$ lb, $M_B = 600$ ft-lb.

10.14 $P_A = 0$, $V_A = -2$ kN, $M_A = 6$ kN-m.

10.16 $P_A = 300$ N, $V_A = -150$ N, $M_A = 330$ N-m.

10.18 $P_A = 4$ kN, $V_A = 6$ kN, $M_A = 4.8$ kN-m.

10.20 $P_A = 0$, $V_A = -6$ kN, $M_A = 6$ kN-m.

10.22 $V = 400$ lb, $M = 400x$ ft-lb.

10.24 (a) $V = (5/2)(12 - x)^2$ lb,
$M = -(5/6)(12 - x)^3$ ft-lb.

10.26 $V = -600$ N, $M = -600x$ N-m.

10.28 (a) $0 < x < 6$ ft, $P = 0$, $V = 50$ lb,
$M = 50x$ ft-lb; $6 < x < 12$ ft,
$P = 0$, $V = 50 - (25/3)(x - 6)^2$ lb,
$M = 50x - (25/9)(x - 6)^3$ ft-lb;
(b)

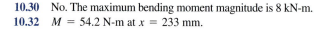

10.30 No. The maximum bending moment magnitude is 8 kN-m.

10.32 $M = 54.2$ N-m at $x = 233$ mm.

10.34

10.36 $0 < x < 4$ m: $V = -2x + 8.8$ kN.
$4 < x < 10$ m: $V = -5.2 + \frac{1}{6}(10 - x)^2$ kN.

10.38

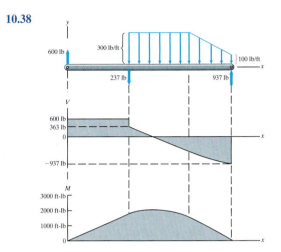

10.40

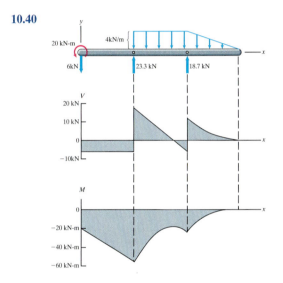

10.42

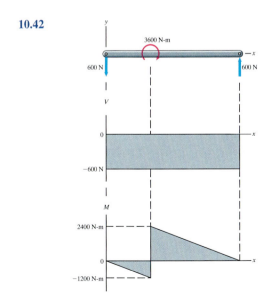

10.44

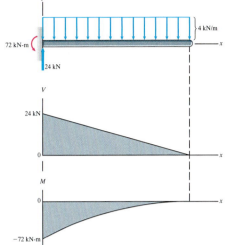

10.46

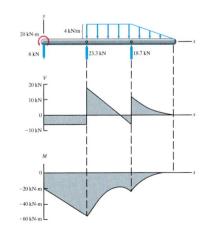

10.48

10.50 (a) 15.8 kN; (b) 28.7 m.

10.52 (a) $T_{max} = 86.2$ kN;   (b) 36.14 m.

10.54 *AC*: 1061 N (T),   *BC*: 1200 N (C).

10.56 Length = 108.3 m, $h = 37.2$ m.

10.58

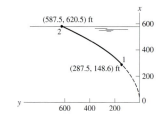

10.60 22.8 m.

10.62 (a) $h_1 = 4.95$ m, $h_2 = 2.19$ m;
(b) $T_{AB} = 1.90$ kN, $T_{BC} = 1.84$ kN.

10.64 $T_1 = 185$ N, $T_3 = 209$ N.

10.66 (a) $h_2 = 4$ ft; (b) 90.1 lb.

10.68 $h_1 = 1.739$ m, $h_3 = 0.957$ m.

10.70 $h_2 = 464$ mm, $h_3 = 385$ mm.

10.72 $h_2 = 8.38$ ft, $h_3 = 12.08$ ft.

10.74 (a) 9.15 N; (b) 4.71 m.

10.76 (a) 10.0 m; (b) 201 m.

10.78 (a) $h_2 = 1.99$ m; (b) $T_{AB} = 2.06$ kN, $T_{BC} = 2.01$ kN.

10.80

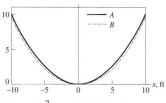

10.82 (a) 1.016×10^7 Pa. (b) 1470 psi.

10.86 $x_p = 3/8$ m, $y_p = 3/5$ m.

10.88 $x_p = 9.90$ in, $y_p = 0$.

10.90 1.55 m.

10.92 6.67 m.

10.94 A: 257 lb to the right, 248 lb upward; B: 136 lb.

10.96 $d = 1.5$ m.

10.98 $A_x = 2160$ lb, $A_y = 2000$ lb, $B_x = 1830$ lb.

10.100 (a) 376 kN; (b) $x_p = 2.02$ m.

10.104 (a) $P_B = 0$, $V_B = -26.7$ lb, $M_B = 160$ ft-lb;
(b) $P_C = 0$, $V_C = -26.7$ lb, $M_C = 80$ ft-lb.

10.106

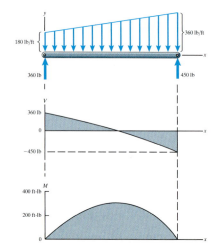

10.108 $0 < x < 2$ m, $P = 0$, $V = 1.33$ kN,
$M = 1.33x$ kN-m;
$2 < x < 6$ m, $P = 0$, $V = -2.67$ kN,
$M = 2.67(6 - x)$ kN-m.

10.110 $P_A = 0$, $V_A = 8$ kN, $M_A = -8$ kN-m.

10.112 (a) $P_B = 0$, $V_B = -40$ N, $M_B = 10$ N-m;
(b) $P_B = 0$, $V_B = -40$ N, $M_B = 10$ N-m.

10.114 $P = 0$, $V = -100$ lb, $M = -50$ ft-lb.

10.116 (a) $w = 74{,}100$ lb/ft; (b) 1.20×10^8 lb.

10.118 84.4 kip.

10.120 A: 44.2 kN to the left, 35.3 kN upward; B: 34.3 kN.

Chapter 11

11.2 (a) Work $= -3.20 \, \delta\theta$ kN-m; (b) $B = 2.31$ kN.

11.4 $F = 217$ N.

11.6 $A_x = 0$, $A_y = -237$ lb, $B_y = 937$ lb.

11.8 $F = 450$ N.

11.10 (a) $F = 392$ N; (b) 100 mm.

11.16 $F = 360$ lb.

11.18 $M = 270$ N-m.

11.20 12 kN.

11.22 9.17 kN.

11.24 (a) $0.625 \, \delta y$; (b) 216 N.

11.26 (a) $q = 3$, $q = 4$;
(b) $q = 3$ is unstable and $q = 4$ is stable.

11.28 $V = \frac{1}{2}kx^2 - \frac{1}{4}\varepsilon x^4$.

11.30 (a) Stable; (b) Unstable.

11.34 (b) It is stable.

11.36 (a) $\alpha = 35.2°$; (b) No.

11.38 (a) $\alpha = 28.7°$; (b) Yes.

11.40 Stable.

11.42 Unstable.

11.44 $\alpha = 0$ is unstable and $\alpha = 30°$ is stable.

11.46 (b) $x = 1.12$ m and $x = 2.45$ m;
(c) $x = 1.12$ m is stable and $x = 2.45$ m is unstable.

11.48 (a) $\alpha = 43.9°$; (b) Yes.

11.50 (a) $\alpha = 30.5°$; (b) Yes.

11.52 $C_x = -7.78$ kN.

11.54 $8F$.

11.56 (a) $M = 800$ N-m; (b) $\alpha/4$.

11.58 $M = 1.50$ kN-m.

11.60 $F = 5$ kN.

11.62 $M = 63$ N-m.

11.64 $\alpha = 0$ is unstable and $\alpha = 59.4°$ is stable.

11.66 Unstable.

11.68 $\alpha = 30°$.

Index

A

Acceleration, 6
    due to gravity, 14
Addition, of vectors, 21–22
Algebra, 597
Analysis of forces, 87–89
Angle:
    determining components in terms of, 31
    in terms of radians, 9
    of kinetic friction, 451
    of static friction, 451
    unit conversions, 9
Angular units, 9
Arch, 285
Area analogy, 352
Areas, 441
    centroids of, 337–338
        with a cutout, 347
        by integration, 339
    properties of, 601–603
Asperities, 449
Atmospheric pressure, 550
Axial force, 278, 329, 508–509, 561

B

Ball and socket supports, 243
Base units:
    International System, 8
    U.S. Customary, 8
Beams, 508–532, 561
    cantilever, 409, 416
    defined, 409
    design, 409, 415
    I-beam, 409
    simply supported, 409
Bearings, 245, 447. *See also* Journal bearings
    defined, 479
    journal, 479, 480, 500
    thrust, 481–482, 500
Belt friction, 488–489, 500
Belts, 447
Belts and pulleys, 491–493
Bending moment, 508–509, 561
    relations between shear force, distributed loads and, 520–525
Bending moment diagram, 514–517
    construction of, 524–530
Body forces, 82, 120

Box beam, 410
Bridges:
    arch, 285
    concrete, 285
    design, 284–285, 292
    structures, examples of, 278
    trusses, 279
Buckling, 410
Built-in supports, 214–215, 245, 247 *See also* Fixed supports

C

Cables, 532–546, 561–562
    discrete loads, 562
    with a horizontally distributed load, 535
    loaded by their own weight, 539
    loads distributed uniformly along, 537–538, 562
        cable length, 538
        cable shape, 537–538
        cable tension, 538
    loads distributed uniformly along straight lines, 533–534, 561–562
        cable length, 534
        cable shape, 533–534
        cable tension, 534–535
    maximum tension in, 536
    subjected to discrete loads, 544–545
Cantilever beams, 409, 416
Car jack, 477, 579
Car tires, reactions on, 221
Cartesian components, 28–34
    position vectors, 76
        in terms of components, 44–45
    in three dimensions, 42–53
        determining, 48–49, 51–52
    in two dimensions, 28–29
Cartesian coordinates, 28–29
Catenary, 538
Center of mass, 273, 374–387, 388–389
    of a composite object, 380–385
    coordinates of, 389
    cylinder with nonuniform density, 378
    defined, 335, 374–375
    of objects, 375–376
    representing the weight of an L-shaped bar, 377
    of vehicles, 383–385, 387
Center of pressure, 550–552
    and pressure force, 554

Centroids, 335–374, 388
of areas, 337–338
with a cutout, 347
by integration, 339
of composite areas, 344–347
of a cone by integration, 359
defined, 335
distributed loads, 351–355
of lines, 358–361
composite, 363–364, 366
by integration, 360
Pappus-Guldinus theorems, 369–372
of a semicircular line, by integration, 360–361
of volumes, 358–361
composite, 364
containing a cutout, 364–365
Circular area, moment of inertia, 399
Clutches, 481–482, 501
defined, 482
disengaged, 482
engaged, 482
Coefficient of kinetic friction, 450
Coefficient of static friction, 449–450
typical values of, 449
Components, 45
Cartesian, 28–34
determining, 32–33
parallel to a line, 45
perpendicular to a plane, 72
scalar, 28, 76
vector, 28, 45, 72
Composite, 363–364
areas, 344–347
lines, 363–364
objects, 380–385, 433
volumes, 364
Composite areas:
centroids of, 344–347
determining, 345
defined, 344
Composite bar, moment of inertia, 433
Compression, 278
Concrete bridges, 285
Concurrent forces:
represented by a force, 183
system of, 204–205
Concurrent system of forces, 82, 120
Conservative forces, 582, 586
Conservative system, 582, 586
stability of, 584
Contact forces, 83–86, 120
Coordinate systems, 42–43, 76
Coplanar forces, 82
Coplanar system of forces, 120
Cosines:
direction, 43–44, 46–47, 76
law of, 598

Coulomb theory of friction, 447–449
Couples, 168–173, 204
sum of the moments due to, 173
Cross products, 66–69, 77
calculating, 70–71
definition, 66–67
evaluating a 3×3 determinant, 68–69
in terms of components, 67–68
units of, 67
Cylinder, moment of inertia, 435–436

D

Dams, 558-559
Datum, 592
Degree of freedom, 582, 592
Degree of redundancy, 238, 271
Degrees, angles expressed in, 9
Density, 388
Derivatives, 598
Derived units, 8
Design, and engineering, 5
Dimensionally homogeneous equation, defined, 11
Direction cosines, 43–44, 46–47, 76
Direction, moments, 141–143
Discrete loads, 542–545, 562
cable subjected to, 544–545
configuration/tensions, determining, 542–543
continuous and discrete models, 543–544
Disk sander, friction on, 483
Displacement, 21
Distributed loads, 351–355, 561
area analogy, 352
beam subjected to, 354
beam with a triangular, 353
describing, 351
force/moment, determining, 351–352
relations between shear force, bending moment and, 520–525
Dot products, 59–66, 77
application, 108–109
definition, 59
in terms of components, 59–60
using to determine an angle, 61
vector components parallel and normal to a line, 60–62
normal component, 61
parallel component, 60–61
Dynamics, 4

E

Einstein's special theory of relativity, 8
Engineering, and mechanics, 4
Engineering applications, mechanics, 5

Equilibrium, 87, 121, 127
  applying in three dimensions, 106–107
  computational mechanics, 267–268
  objects in, 211–275
  stability of, 582–583
  statically indeterminate objects, 271
  structures in, 277–333
  three-dimensional applications, 243–247, 271
  three-force members, 262–263, 271
  two-dimensional applications, 270–271
  two-force members, 260–262, 271
  using to determine forces on an object, 90–91
Equilibrium equations, 87, 212
Equilibrium position, stability of, 585–586
Equivalent systems, 178–182, 204
  conditions for equivalence, 178
  demonstration of equivalence, 178–179
  determining equivalence of systems, 180–182
  representing systems by, 183–190, 204
External forces, 82, 120

F

Feet per second (ft/s), 9
Feet per second squared (ft/s^2), 6
First Pappus-Guldinus theorem, 369–370, 388
Fixed (built-in) supports, 214–215, 245, 247
  reactions at, 215, 247
  three-dimensional applications, 245
Flight, steady, 93–94, 105
Flight path angle, 93
Flywheel, 445
Forces, 7, 81–125
  body, 82
  concurrent, represented by a force, 183
  concurrent system of, 82, 120
  contact, 83–86, 120
  coplanar, 82
  determining the moment of, 130
  equilibrium, 87, 121
  external, 82, 120
  free-body diagrams, 87–89, 120, 121
  friction, 83, 120
  gravitational, 82–83, 120
  internal, 82, 120
  line of action, 82, 120
  moment of a system of, 131
  normal, 83, 120
  parallel, represented by a force, 184
  representing by a force and couple, 183, 184–185
  ropes/cables, 84–85
  springs, 121
  surface, 82
  systems of, 82
  tensile, 290

  terminology, 82
  types of, 82–86
  unit conversion, 9
  as a vector quantity, 20
Frames, 277, 302–315, 329
  analyzing, 308–309
  determining forces on members of, 310–311
  entire structure, analyzing, 303
  forces and couples on the members of, 307
  loads applied at joints, 304–307
  members, analyzing, 303–304
  two-force members in, 304
Free-body diagrams, 87–89, 120, 121, 127, 302–303
  choosing, 91
  freed objects, 87–88
  steps in drawing, 88
Friction, 447–505
  angles of, 451–452
  applications, 469–473
    belts and pulleys, 491–493
    clutches, 481–482, 500
    journal bearings, 479–480, 500
    threads, 471–473, 500
    thrust bearings, 481–482
    wedges (shims), 447, 469–471
  belt friction, 488–489, 500
  coefficients of, 449–450
    kinetic coefficient, 450
    static coefficient, 449–450
  Coulomb friction, theory of, 449–457
  determining tip-over potential, 454
  on a disk sander, 483
  dry friction, see Coulomb friction.
  problem in three dimensions, 456–457
Friction brake, analyzing, 455
Friction force, 83, 120, 447–505

G

Gage pressure, 550
giga-, 8
Gravitation, Newtonian, 14
Gravitational forces, 82–83, 120

H

Hinge axis, 244
Hinge supports, 244
  reactions at, 250–251
Hinges, properly aligned, reactions at, 252–253
Homogeneous objects, 388–389
  properties of, 604–606
Hour (h), 6
Howe truss, 288, 291, 295, 296, 331
Human factors, design for, 222–223, 237

I

I-beam, 409
Impending slip, 450
Improper supports, 227, 237, 240–241, 271
Inertial reference frame, 87
Integrals, 598
Internal forces, 82, 120
Internal forces and moment, 507–565
    determining, 509–510
    shear force and bending moment diagrams, 514–517
International System of units (SI units), 6, 8, 9
    prefixes used in, 8
    for pressure, 550
Isolated objects, free-body diagrams, 87–88

J

Joints, 278
    method of, 280–282, 329
        applying, 282–283
    truss, 281
Journal bearings, 500
    defined, 479
    pulley supported by, 480

K

kilo-, 8
Kilogram (kg), 7
Kilopound (kip), 9
Kinetic coefficient of friction, 450
Kinetic friction:
    angle of, 451
    coefficient of, 450

L

Law of cosines, 598
Law of sines, 598
Laws of motion, 6–7
Length, unit conversions, 9
Line of action, 82, 120
    determining perpendicular distance to, 146–147
Linear spring, 86
Lines:
    centroids of, 358–361
        composite, 363–364, 366
        by integration, 360
    moment of a force about, 203
    properties of, 603
Liquids/gases, 550–557, 562
    center of pressure, 550–552
    pressure, 550–552
    stationary liquid, pressure in, 552–553
Loading curve, 351–352
Loads, 213
Logarithms, natural, 597

M

Machines, 277, 302–315, 329
    analyzing, 314–315
    forces and couples on the members of, 307
Magnitude, 20,
    moment about a line, 154
    moment about a point, 128, 141
    vectors, 20, 28, 43
Masonry, and bridges, 285
Mass, 7, 428–440, 442
    center of, 273, 374–387, 388–389
    as a scalar quantity, 20
    moments of inertia, 428
    unit conversion, 9
    in U.S. Customary units, 8
Mathematics review, 597–598
    algebra, 597
    derivatives, 598
    integrals, 598
    natural logarithms, 597
    quadratic equations, 597
    Taylor series, 600
    trigonometry, 598
Meaningful digits, 5
Mechanical advantage, 226
Mechanics:
    chronology of developments in mechanics, up to Newton's
        Principia, 7
    and engineering, 4
    engineering applications, 5
    fundamental concepts, 5–8
        numbers, 5–6
        space, 6
        time, 6
    learning, 4–5
    Newtonian gravitation, 14
    Newton's laws, 6–8
    principles of, 4
    problem solving, 4–5
    as science, 4
    units:
        angular units, 9
        conversion of, 9–11
mega-, 8
Members, 277
Meters per second (m/s), 6
Meters per second squared (m/s^2), 6
Method of joints, 280–282, 329
    applying, 282–283
Method of sections, 292–294, 329
Mile (mi), 9
Mile per hour (mi/h), 9
Minute (min), 6
Mixed triple products, 69, 77
Mohr's circle, 422–426
    determining principal axes and principal moments of inertia, 424
    moments of inertia by, 425–426

Moment of a couple, 168. *See also* Couples
    determining, 170–171
Moment of a force:
    about a line, 154–157
    about a point, 128–129, 141–144
        two-dimensional description, 128
        vector description, 141
Moment vector, 141–148
    applying, 147–148
    two-dimensional description and, 145
Moments, 127–209
    choosing the point about which to evaluate, 222
    direction of, 128, 141
    due to couples, 172
    of a force about a line, 154–157
    of a force about a point, 128–129, 141–144
    magnitude of, 141
    relation to two-dimensional description, 143–144
    rotating machines, 160–161
    two-dimensional description, 128–132
    Varignon's theorem, 144
    vector description, 141
Moments of inertia, 395–445, 601–606
    areas, 366–427, 601–603
        about the x axis, 396, 441
        about the y axis, 396, 441
        circular area, 399
        composite area, 404–405
        Mohr's circle, 422–426
        parallel-axis, 418–422
        polar moment of inertia, 418–422
        principal, 419
        product of inertia, 396
        radius of gyration, 396–397
        rotated axes, 416–418
        table, 601–603
        triangular area, 397
        of a triangular plate, 431
    masses, 428–445, 604–606
        composite objects, 432–434
        cylinder, 435–436
        definition, 428
        parallel-axis theorem, 432–436
        slender bars, 428–429
        table, 604–606
        thin plates, 429
        triangular plate, 431

N

nano-, 8
Natural logarithms, 597
Newton, Isaac, 6–7
Newtonian reference frame, 87
Newtonian gravitation, 14
Newton-meters (N-m), 128
Newton's laws, 6–8

Newtons per meter (N/m), 351
Newtons per square meter (N/m^2), 550. *See also* Pascals
Newton's second law, 7, 8
Newton's third law, 83
Normal force, 83, 120
Numbers, 5–6
    significant digits, 5–6
    use of, in book, 6

O

Objects in equilibrium, 211–275
One-degree-of-freedom systems, using potential energy to
    analyze the equilibrium of, 583

P

Pappus–Guldinus theorems, 369–374, 388
Parallel-axis theorems,
    areas, 402–410
    masses, 432–436
Parallel forces:
    represented by a force, 184
    representing by a single force, 186–187
Parallelogram rule, vector addition, 22, 76
Pascals (Pa), 10, 550
Philosophiae Naturalis Principia Mathematica (Newton), 6
Pin supports, 213, 215
    reactions at, 213–219
Piston, 322, 578, 594
Pitch, threads, 471
Point, moment of a force about a point, 128–129, 141–144
Polar moment of inertia, 396–397, 404–405, 418, 441
Position vectors, 20
    in terms of components,
        in two demensions, 29
        in three demensions, 44
Potential energy, 567, 580–586, 592
    conservative forces:
        springs, 581–582
        weight, 580–581
    defined, 580
    stability of equilibrium, 582–583
    virtual work, principle of, for conservative forces, 582
Pound (lb), 8
Pounds per foot (lb/ft), 351
Pounds per square foot (lb/ft^2), 550
Pounds per square inch $(lb/in^2$ or psi), 550
Pratt truss, 286, 288, 290, 295
Prefixes, used in SI units, 8
Pressure. *See also* Center of pressure
    converting units of, 10–11
    in a stationary liquid, 552–553
Pressure force, 550–557
    and center of pressure, 550–551
    determination of, 556–557

Principal axes of areas, 418–422
  defined, 419
    determining, 420
  and moments of inertia, 421–422
Principal moments of inertia of areas, 419–420, 424
Principle of transmissibility, 191
Problem solving, 4–5
Products of inertia of areas, 396, 418, 441
Products of vectors, 59–75
  cross products, 66–69, 77
  dot products, 59–66, 77
  mixed triple products, 69, 77
Proper supports, 240–241
Properly aligned hinges, reactions at, 252–253
Pulleys, 84–85, 92, 120, 491–493
Pythagorean theorem, 28

Q

Quadratic equations, 597
Quantum mechanics, 8

R

radians (rad), 9
Radius of gyration, 396–397
Reactions:
  exerted by supports in two dimensions, 212–215
  exerted by supports in three-dimensional, 243–247
Redundant supports, 237, 238–239, 241, 271
Right triangle, trigonometric functions for, 598
Right-hand rule, 66–67, 77
Right-handed coordinate system, 42, 76
Roller supports, 213–215, 243, 246
  reactions at, 215, 246
  three-dimensional applications, 243
Roof structures, examples of, 278
Roof trusses, 279
Ropes/cables, 84–85, 120
Rotated axes, 417–418
  moment of inertia about the x' axis, 417–418
  moment of inertia about the y' axis, 418
  and moments of inertia, 421–422
Rotating machines, 160–161
Rough surfaces, 83, 120
Rounding off numbers, 5–6
Russell's traction, 121

S

Safety, and engineering, 5
Scalar components, 28, 76
Scalar equilibrium equations, 89, 105, 217, 248
Scalars:
  defined, 20
  product of a vector and a scalar, 22–23

Second Pappus-Guldinus theorem, 370–372, 388
Second (s), 6
Sections:
  appropriate, choosing, 294–295
  method of, 292–294, 329
    applying, 293
Shear force, 508–509, 561
  relations between distributed loads, bending moment and, 520–525
Shear force and bending moment diagrams, 514–517, 526–530, 561
Shear force diagram, construction of, 522–523
Shims, 469–471
SI system of units, *See* International System of units (SI units).
Significant digits, 5–6
Simply supported beam, 409
Sines, law of, 598
Slender bar, 428–429, 436
  moment of inertia, 428
Slip, impending, 450
Slope, threads, 471
Smooth surfaces, 83, 120
Space, 6
Space trusses, 298–300, 329
Span, bridges, 285
Spring constant, 121
Springs, 85–86, 121, 581–582
  linear, 86
  spring constant, 86
Stability of equilibrium, 582–583
Stable equilibrium position, 583
Static friction, 83–84, 446–505
  angle of, 451
  coefficient of, 449–450
    typical values of, 449
Statically indeterminate objects, 237–242, 271
  improper supports, 237, 240–241
  recognizing, 239
  redundant supports, 237, 238–239, 241
Statically indeterminate problems, 107, 237–241
Statics, 4
Steady translation, 87
Strategies, for problem solving, 4
Structures, 277
Structures in equilibrium, 277–333
Subtraction, of vectors, 23
Support conventions, 270
Supports, 212–215, 246–247
  ball and socket, 243, 259
  bearings, 245
  conventions, 213
  fixed (built-in), 214–215, 245
  hinge, 244, 252–253
    reactions at, 220
  improper, 237, 240–241, 271
  loads, 213
  pin, 213, 215
    reactions at, 218–219

reactions, 212–215, 243–247
redundant, 237–239, 241, 271
roller, 213–214, 215, 243
reactions at, 218–219
tables, 215, 246–247
three-dimensional applications, 243–247
two-dimensional applications, 212–215
Surfaces, 120
Suspended cables, 20, 532–546
loaded uniformly along a straight line:
length, 534
shape, 534
tension, 534
loaded uniformly along a length:
length, 538
shape, 537
tension, 538
subjected to discrete loads, 542–546
Systems of forces and moments, 82, 127–209
defined, 178
equivalent systems, 178–183
conditions for equivalence, 178
demonstration of equivalence, 178–179
determining equivalence of systems, 180–182
representing systems by, 183–190
force, representing by a force and couple, 184–185
parallel forces:
represented by a force, 184
representing by a single force, 186–187
representing a system by a force and couple, 186
representing a system by a simpler equivalent system, 185
representing by a wrench, 188–190

T

Taylor series, 600
Tensile force, 290
Tension, 84, 278
in a suspended cable, 532–546
in a truss member, 278–294
Thin plates, moments of inertia of, 429
Threads, 471–473, 500
Three-dimensional applications, 243–247, 271
fixed supports, 245
reactions, 248–249
scalar equilibrium equations, 248
supports, 243–247
ball and socket, 243
bearings, 245
hinge, 244
roller, 243
Three-dimensional coordinate systems:
drawing, 42
right-handed, 42
vector components in terms of, 42–53
Three-force members, 262–263, 271
example, 264

Thrust bearings, 481–482, 499, 501
Time, 6
unit conversions, 9
Translation:
steady, 87
virtual, 592
Transmissibility, principle of, 191
Triangle rule, for vector addition, 21, 76
Triangular area, moment of inertia, 397–398
Triangular plate, moment of inertia, 431
Trigonometry, 598
Trusses, 231, 277, 278–279
compression, 278
defined, 378
determining the largest force supported by, 284
Howe, 288, 291, 295, 296, 331
method of joints, 280–282
method of sections, 292–294
Pratt, 286, 288, 290, 295
space trusses, 298–300, 329
tension, 278, 329
Warren, 278, 280, 290, 292, 296
Turnbuckle, 478
Two-dimensional applications, 270–271
equilibrium equations, 212, 217–218
free-body diagrams, 212, 216
proper supports, 240–241
supports, 212–215
fixed (built-in), 214–215
improper, 237, 240–241
pin, 213, 215
redundant, 237, 238–239, 241
roller, 213–214, 215
Two-force members, 260–265, 271
examples, 263–264
in frames, 304

U

Unit vectors, 23–24, 76
Units:
angular units, 9
conversion of, 9–11
International System of units, 6, 8
U.S. Customary units, 6, 8–9
Universal gravitational constant, 14
Unstable equilibrium position, 583
U.S. Customary units, 6, 8–9
for pressure, 550

V

Varignon's theorem, 144
Vector products, 59–72
Vectors, 19–71
addition of, 21–22, 24–25
parallelogram rule, 22, 76

Vectors, addition of, *continued*
    in terms of components, 30
    triangle rule, 21, 76
      addition of, 21, 76
    Cartesian components, 28–34
    components, 28
      parallel to a given line, 45
      perpendicular to a plane, 72
    defined, 20
    direction cosines, 43–44, 76
    direction specified by two points, 50–51
    dot products, 59–66
      vector components parallel and normal to a line, 60–62
    graphical representation of, 20
    magnitude, in terms of components, 28, 43
    manipulating in terms of components, 28–29
    position vectors, in terms of components, 44–45
    product of a scalar and a vector, 22–23
    products of, 59–75
      cross products, 66–69, 77
      dot products, 59–66, 77
      mixed triple products, 69, 77
    projection onto a line, 60–61
    rules for manipulating, 20–25
    subtraction of, 23
    unit vectors, 23–24, 76
Velocity, 6
Virtual displacement, 569
Virtual translation or rotation, 592

Virtual work, 568–575
    applying to a machine, 574–575
    applying to structures, 571–575
    defined, 569
    principle of, 569–570, 592
      for conservative forces, 582
Vise, 477
Volume analogy, and force/moment due to pressure, 553
Volumes:
    centroids of, 358–361
      composite, 364
      containing a cutout, 364–365
    defined, 358
    properties of, 604–606

W

Warren truss, 278, 280, 290, 292, 296
Wedges (shims), 447, 469–471, 505
    forces on, 470–471
Weight, 377, 580–581
    determining, 15
Weight density, 388
Weighted average position. *See* Centroids
Winch, 494
Work, 592, *See also* Virtual work
Wrench, 204
    defined, 189
    representing a force and couple by, 190

Properties of Areas and Lines

Areas

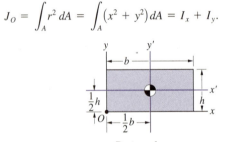

The coordinates of the centroid of the area A are

$$\bar{x} = \frac{\int_A x\,dA}{\int_A dA}, \qquad \bar{y} = \frac{\int_A y\,dA}{\int_A dA}.$$

The moment of inertia about the x axis I_x, the moment of inertia about the y axis I_y, and the product of inertia I_{xy} are

$$I_x = \int_A y^2\,dA, \qquad I_y = \int_A x^2\,dA, \qquad I_{xy} = \int_A xy\,dA.$$

The polar moment of inertia about O is

$$J_O = \int_A r^2\,dA = \int_A (x^2 + y^2)\,dA = I_x + I_y.$$

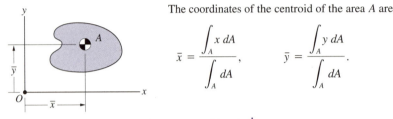

Triangular area

$$\text{Area} = \frac{1}{2}\,bh \qquad I_x = \frac{1}{12}\,bh^3, \qquad I_{x'} = \frac{1}{36}\,bh^3$$

Rectangular area

$$\text{Area} = bh$$

$$I_x = \frac{1}{3}\,bh^3, \qquad I_y = \frac{1}{3}\,hb^3, \qquad I_{xy} = \frac{1}{4}\,b^2h^2$$

$$I_{x'} = \frac{1}{12}\,bh^3, \qquad I_{y'} = \frac{1}{12}\,hb^3, \qquad I_{x'y'} = 0$$

Circular area

$$\text{Area} = \pi R^2 \qquad I_{x'} = I_{y'} = \frac{1}{4}\,\pi R^4, \qquad I_{x'y'} = 0$$

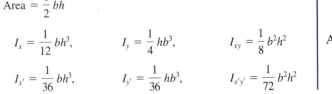

Triangular area

$$\text{Area} = \frac{1}{2}\,bh$$

$$I_x = \frac{1}{12}\,bh^3, \qquad I_y = \frac{1}{4}\,hb^3, \qquad I_{xy} = \frac{1}{8}\,b^2h^2$$

$$I_{x'} = \frac{1}{36}\,bh^3, \qquad I_{y'} = \frac{1}{36}\,hb^3, \qquad I_{x'y'} = \frac{1}{72}\,b^2h^2$$

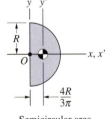

Semicircular area

$$\text{Area} = \frac{1}{2}\,\pi R^2 \qquad I_x = I_y = \frac{1}{8}\,\pi R^4, \qquad I_{xy} = 0$$

$$I_{x'} = \frac{1}{8}\,\pi R^4, \qquad I_{y'} = \left(\frac{\pi}{8} - \frac{8}{9\pi}\right)R^4, \qquad I_{x'y'} = 0$$